Methods in Enzymology

Volume XLVI

AFFINITY LABELING

METHODS IN ENZYMOLOGY

EDITORS-IN-CHIEF

Sidney P. Colowick   Nathan O. Kaplan

Methods in Enzymology

Volume XLVI

Affinity Labeling

EDITED BY

William B. Jakoby

NATIONAL INSTITUTES OF HEALTH
BETHESDA, MARYLAND

Meir Wilchek

THE WEIZMANN INSTITUTE OF SCIENCE
REHOVOT, ISRAEL

1977

ACADEMIC PRESS   New York   San Francisco   London

A Subsidiary of Harcourt Brace Jovanovich, Publishers

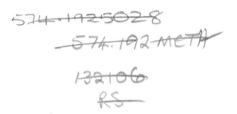

ACADEMIC PRESS, INC.
111 Fifth Avenue, New York, New York 10003

United Kingdom Edition published by
ACADEMIC PRESS, INC. (LONDON) LTD.
24/28 Oval Road, London NW1

Library of Congress Cataloging in Publication Data

Main entry under title:

Affinity labeling.

    (Methods in enzymology ; 46)
    Includes bibliographical references.
    1.   Proteins–Affinity labeling.   I.   Jakoby,
William B., Date   II.   Wilchek, Meir.
III.   Series.   [DNLM: 1.   Chemistry, Physical.
W1 ME9615K v. 46 / QD505.5 A256]
QP601.M49   vol. 46   [QP551]   574.1$'$925$'$08s
                 [574.1$'$925$'$028]   77-5422
ISBN 0–12–181946–9

Table of Contents

Section I. General Methodology

Section II. Specific Procedures for Enzymes, Antibodies, and Other Proteins

A. Proteolytic Systems

B. Nucleotide and Nucleic Acid Systems

Contributors to Volume XLVI

Article numbers are in parentheses following the names of contributors.
Affiliations listed are current.

JUDITH M. ANDREWS (9), *Center for Blood Research, Boston, Massachusetts*

ROSS S. ANTONOFF (34), *Department of Biochemistry and Biophysics, University of Pennsylvania School of Medicine, Philadelphia, Pennsylvania*

VIC ARMSTRONG (36), *Abteilung Chemie, Max-Planck-Institut für Experimentelle Medizin, Göttingen, West Germany*

DAPHNE ATLAS (66, 69), *Department of Biological Chemistry, The Hebrew University of Jerusalem, Jerusalem, Israel*

ANDREA BARTA (81), *Institut für Biochemie, Universität Wien, Vienna, Austria*

F. H. BATZOLD (51), *Department of Pharmacology and Experimental Therapeutics, The Albany Medical College of Union University, Albany, New York*

EDWARD A. BAYER (72), *Department of Biophysics, The Weizmann Institute of Science, Rehovot, Israel*

HAGAN BAYLEY (8), *Department of Chemistry, Harvard University, Cambridge, Massachusetts*

WILLIAM F. BENISEK (52), *Department of Biological Chemistry, School of Medicine, University of California, Davis, California*

ANN M. BENSON (51), *Department of Pharmacology and Experimental Therapeutics, The Johns Hopkins University School of Medicine, Baltimore, Maryland*

DAVID H. BING (9), *Center for Blood Research, Boston, Massachusetts*

WALTER BIRCHMEIER (4), *Department of Biology, University of California, La Jolla, California*

LUDWIG BISPINK (74), *Abteilung Molekulare Genetik, Max-Planck-Institut für Experimentelle Medizin, Göttingen, West Germany*

ELENA S. BOCHKAREVA (78), *Institute of Protein Research, Academy of Sciences of the USSR, Poustchino, Moscow Region, USSR*

RONALD T. BORCHARDT (64), *Department of Biochemistry, McCollum Laboratories, University of Kansas, Lawrence, Kansas*

BRUCE R. BRANCHINI (18, 61), *Department of Chemistry, University of Wisconsin, Parkside Kenosha, Wisconsin*

JOHNNY BRANDT (65), *Institute of Biochemistry, University of Uppsala, Uppsala, Sweden*

GERALD P. BUDZIK (14), *Department of Biology and Chemistry, Massachusetts Institute of Technology, Cambridge, Massachusetts*

CHARLES R. CANTOR (15), *Department of Biological Sciences, Columbia University, New York, New York*

JOHN J. CEBRA (54), *Department of Biology, The Johns Hopkins University, Baltimore, Maryland*

SHEUE-YANN CHENG (48), *Clinical Endocrinology Branch, National Institute of Arthritis, Metabolism, and Digestive Diseases, National Institutes of Health, Bethesda, Maryland*

GIANNI CHINALI (82), *Cattedra di Chimica, Universita Degli Studi di Napoli, Naples, Italy*

PHILIPP CHRISTEN (4, 5), *Biochemisches Institut der Universität, Zürich, Switzerland*

ROBERTA F. COLMAN (23), *Chemistry Department, University of Delaware, Newark, Delaware*

BARRY S. COOPERMAN (85), *Department of Chemistry, University of Pennsylvania, Philadelphia, Pennsylvania*

MICHAEL CORY (9), *Stanford Research Institute, Menlo Park, California*

DOUGLAS F. COVEY (51), *Department of Pharmacology and Experimental Therapeutics, The Johns Hopkins University School of Medicine, Baltimore, Maryland*

BRIAN M. COX (70), *Addiction Research Foundation, Palo Alto, California*

PEDRO CUATRECASAS (38), *Burroughs Wellcome Co., Research Triangle Park, North Carolina*

ANTOINE DANCHIN (31), *Institut Pasteur, Paris, France*

FRANK DAVIDOFF (63), *University of Connecticut Health Center, Farmington, Connecticut*

FRITZ ECKSTEIN (32, 36), *Abteilung Chemie, Max-Planck-Institut für Experimentelle Medizin, Göttingen, West Germany*

LEWIS L. ENGEL (6), *Laboratory of Human Reproduction and Reproductive Biology, and the Department of Biological Chemistry, Harvard Medical School, Boston, Massachusetts*

YUVAL ESHDAT (44), *Department of Biophysics, The Weizmann Institute of Science, Rehovot, Israel*

HUGO FASOLD (26), *Mehrzweckgebäude Chemie, Institut für Biochemie der Universität Frankfurt am Main, Frankfurt am Main-Niederrad, West Germany*

J. J. FERGUSON, JR. (34), *Department of Biochemistry and Biophysics, University of Pennsylvania School of Medicine, Philadelphia, Pennsylvania*

A. GIARTOSIO (49), *Instituto di Chimica Biologica, Facolta di Farmacia, Universita di Roma, Rome, Italy*

ALEXANDER S. GIRSHOVICH (77, 78), *Institute of Protein Research, Academy of Sciences of the USSR, Poustchino, Moscow Region, USSR*

DAVID GIVOL (11, 53, 55), *Department of Immunochemistry, The Weizmann Institute of Science, Rehovot, Israel*

G. I. GLOVER (71), *Department of Chemistry, Texas A & M University, College Station, Texas*

P. V. GOPALAKRISHNAN (58), *152 North Mansfield Boulevard, Cherry Hill, New Jersey*

ERNEST V. GROMAN (6), *Laboratory of Human Reproduction and Reproductive Biology, and the Department of Biological Chemistry, Harvard Medical School, Boston, Massachusetts*

RICHARD JOHN GUILLORY (25), *Department of Biochemistry and Biophysics, John A. Burns School of Medicine, University of Hawaii, Honolulu, Hawaii*

B. FRANK GUPTON (17), *School of Chemistry, Georgia Institute of Technology, Atlanta, Georgia*

BOYD E. HALEY (35), *Department of Chemistry and Biochemistry, University of Wyoming, Laramie, Wyoming*

ALEXANDER HAMPTON (27, 28, 29), *The Institute for Cancer Research, The Fox Chase Cancer Center, Philadelphia, Pennsylvania*

ROBERT E. HANDSCHUMACHER (47), *Department of Pharmacology, Yale University School of Medicine, New Haven, Connecticut*

PETER J. HARPER (29), *Union Carbide Co., Sydney, Australia*

FRED C. HARTMAN (10, 42), *Biology Division, Oak Ridge National Laboratory, Oak Ridge, Tennessee*

GEORGE D. HEGEMAN (62), *Department of Microbiology, Indiana University, Bloomington, Indiana*

STEPHEN S. HIXSON (82), *Department of Chemistry, University of Massachusetts, Amherst, Massachusetts*

SUSAN H. HIXSON (82), *Department of Chemistry, Mount Holyoke College, South Hadley, Massachusetts*

JOHN HOBBS (32), *Abteilung Chemie, Max-Planck-Institut für Experimentelle Medizin, Göttingen, West Germany*

PAUL HOWGATE (29), *The Institute for Cancer Research, The Fox Chase Cancer Center, Philadelphia, Pennsylvania*

FRANZ W. HULLA (26), *Mehrzweckgebäude Chemie, Institut für Biochemie der Universität Frankfurt am Main, Frankfurt am Main-Niederrad, West Germany*

TADASHI INAGAMI (22), *Department of Biochemistry, Vanderbilt University School of Medicine, Nashville, Tennessee*

REINHARD JECK (24), *Gustav-Emden Zentrum der Biologischen Chemie, Klinikum der Johann Wolfgang Goethe Universität, Frankfurt am Main, West Germany*

STELLA JYHLIH JENG (25), *Department of Biochemistry and Biophysics, John A. Burns School of Medicine, University of Hawaii, Honolulu, Hawaii*

ARTHUR E. JOHNSON (15), *Department of Chemistry, Columbia University, New York, New York*

ARTHUR KARLIN (68), *Department of Neurology, College of Physicians and Surgeons, Columbia University, New York, New York*

FRED KARUSH (58), *Department of Microbiology, University of Pennsylvania School of Medicine, Philadelphia, Pennsylvania*

B. KEIL (21), *Service de Chimie des Proteines, Institut Pasteur, Paris, France*

GEORGE L. KENYON (62), *Department of Pharmaceutical Chemistry, University of California, San Francisco, California*

A. P. KIMBALL (37), *Department of Biophysical Sciences, University of Houston, Houston, Texas*

JEREMY R. KNOWLES (8), *Department of Chemistry, Harvard University, Cambridge, Massachusetts*

WILLIAM H. KONIGSBERG (57), *Department of Molecular Biochemistry and Biophysics, Yale University School of Medicine, New Haven, Connecticut*

ERNST KUECHLER (81), *Institut für Biochemie, Universität Wien, Vienna, Austria*

JACK KYTE (59), *Department of Chemistry, University of California at San Diego, La Jolla, California*

ROBERT G. LANGDON (13), *Department of Biochemistry, University of Virginia, Charlottesville, Virginia*

ERICH LANKA (80), *Max-Planck-Institut für Molekulare Genetik, Berlin-Dahlem, West Germany*

G. LEGLER (40), *Institute of Biochemistry, University of Cologne, Cologne, West Germany*

ANTONIO LUCACCHINI (33), *Instituto di Chimica Biologica, Universita di Pisa, Pisa, Italy*

J. A. MAASSEN (76), *Laboratory for Physiological Chemistry, Leiden, The Netherlands*

HEINRICH MATTHAEI (74), *Abteilung Molekulare Genetik, Max-Planck-Institut für Experimentelle Medizin, Göttingen, West Germany*

ALTON MEISTER (46), *Department of Biochemistry, Cornell University Medical College, New York, New York*

W. MÖLLER (76), *Laboratory for Physiological Chemistry, Leiden, The Netherlands*

J. ROBERT MUELLER (50), *Department of Obstetrics and Gynecology, Washington University School of Medicine, St. Louis, Missouri*

I. LUCILE NORTON (42), *Biology Division, Oak Ridge National Laboratory, Oak Ridge, Tennessee*

TOM OBRIG (34), *Department of Biochemistry and Biophysics, University of Pennsylvania School of Medicine, Philadelphia, Pennsylvania*

EDWARD L. O'CONNELL (41), *The Institute for Cancer Research, The Fox Chase Cancer Center, Philadelphia, Pennsylvania*

JAMES OFENGAND (82), *Department of Biochemistry, Roche Institute of Molecular Biology, Nutley, New Jersey*

HIROSHI OGAWARA (60), *Department of Biochemistry, Meiji College of Pharmacy Graduate School, Nozawa, Stagaya-ku, Tokyo, Japan*

FRANZ ORTANDERL (26), *Mehrzweckgebäude Chemie, Institut für Biochemie der Universität Frankfurt am Main, Frankfurt am Main-Niederrad, West Germany*

YURI A. OVCHINNIKOV (77, 78), *Institute of Protein Research, Academy of Sciences of the USSR, Poustchino, Moscow Region, USSR*

PRANAB K. PAL (23), *Chemistry Department, University of Delaware, Newark, Delaware*

ALLEN T. PHILLIPS (7), *Department of*

Biochemistry and Biophysics, Pennsylvania State University, University Park, Pennsylvania

LAWRENCE M. PINKUS (45), Laboratory of Metabolism and Endocrinology, National Institute of Arthritis, Metabolism, and Digestive Diseases, National Institutes of Health, Bethesda, Maryland

PAUL H. PLOTZ (56), Arthritis and Rheumatism Branch, National Institute of Arthritis, Metabolism, and Digestive Diseases, National Institutes of Health, Bethesda, Maryland

IEVA R. POLITZER (18), Department of Chemistry, University of New Orleans, New Orleans, Louisiana

OLAF PONGS (79, 80), MRC Laboratory of Molecular Biology, Cambridge, England

JAMES C. POWERS (16, 17), School of Chemistry, Georgia Institute of Technology, Atlanta, Georgia

VALERY A. POZDNYAKOV (77), Institute of Protein Research, Academy of Sciences of the USSR, Poustchino, Moscow Region, USSR

MICHAEL RACK (26), Mehrzweckgebäude Chemie, Institut für Biochemie der Universität Frankfurt am Main, Frankfurt am Main-Niederrad, West Germany

ROBERT R. RANDO (3, 12), Department of Pharmacology, Harvard Medical School, Boston, Massachusetts

PARLANE REID (63), University of Connecticut Health Center, Farmington, Connecticut

ERWIN REINWALD (79), Max-Planck-Institut für Molekulare Genetik, Berlin-Dahlem, West Germany

NOEL M. RELYEA (46), Memorial Sloan Kettering Cancer Center, New York, New York

MANUEL J. RICARDO (54), Department of Microbiology, Center for Health Sciences, University of Tennessee, Memphis, Tennessee

FRANK F. RICHARDS (57), Department of Internal Medicine, Yale University

School of Medicine, New Haven, Connecticut

F. RIVA (49), Instituto di Chimica Biologica, Facolta di Farmacia, Universita di Roma, Rome, Italy

C. H. ROBINSON (51), Department of Pharmacology and Experimental Therapeutics, The Johns Hopkins University School of Medicine, Baltimore, Maryland

GIOVANNI RONCA (33), Instituto di Chimica Biologica, Universita di Pisa, Pisa, Italy

IRWIN A. ROSE (41), The Institute for Cancer Research, The Fox Chase Cancer Center, Philadelphia, Pennsylvania

CARLO ALFONSO ROSSI (33), Instituto di Chimica Biologica, Universita di Bologna, Bologna, Italy

DAVID F. ROSWELL (18), Department of Chemistry, Loyola College, Baltimore, Maryland

ARNOLD RUOHO (59), Department of Pharmacology, University of Wisconsin, Madison, Wisconsin

DANIEL V. SANTI (30), Department of Biochemistry and Biophysics, and Department of Pharmaceutical Chemistry, University of California, San Francisco, California

TAKUMA SASAKI (29), National Cancer Research Institute, Tokyo, Japan

PAUL R. SCHIMMEL (14), Department of Biology, Massachusetts Institute of Technology, Cambridge, Massachusetts

HENRI SCHMITT (66), Departments of Neurobiology, Membrane Research, and Biophysics, The Weizmann Institute of Science, Rehovot, Israel

RICHARD M. SCHULTZ (6), Laboratory of Human Reproduction and Reproductive Biology, and the Department of Biological Chemistry, Harvard Medical School, Boston, Massachusetts

IRA SCHWARTZ (82), Department of Biochemistry, University of Massachusetts, Amherst, Massachusetts

NATHAN SHARON (44), Department of Biophysics, The Weizmann Institute of Science, Rehovot, Israel

LEWIS A. SLOTIN (27), *Division of Biological Sciences, National Research Council of Canada, Ottawa, Ontario, Canada*

M. SOKOLOVSKY (20, 67), *Department of Biochemistry, The George S. Wise Center of Life Sciences, Tel-Aviv University, Tel Aviv, Israel*

N. SONENBERG (83, 84), *Department of Biochemistry, The Weizmann Institute of Science, Rehovot, Israel*

HANS STERNBACH (36), *Abteilung Chemie, Max-Planck-Institut für Experimentelle Medizin, Göttingen, West Germany*

YASUNOBU SUKETA (22), *Shizuoka College of Pharmacy, Oshi-ka, Shizuoka, Japan*

PAUL TALALAY (51), *Department of Pharmacology and Experimental Therapeutics, The Johns Hopkins University School of Medicine, Baltimore, Maryland*

SURESH S. TATE (46), *Department of Biochemistry, Cornell University Medical College, New York, New York*

DHIREN R. THAKKER (64), *Department of Biochemistry, McCollum Laboratories, University of Kansas, Lawrence, Kansas*

LARS THELANDER (32), *Department of Biochemistry, Karolinska Institute, Stockholm, Sweden*

E. W. THOMAS (39), *Department of Chemistry and Applied Chemistry, University Salford, Salford, England*

ROBERT C. THOMPSON (19), *Department of Chemistry, Temple University, Philadelphia, Pennsylvania*

C. TURANO (49), *Instituto di Chimica Biologica, Facolta di Farmacia, Universita di Roma, Rome, Italy*

JAMES C. WARREN (50), *Department of Obstetrics and Gynecology, Washington University School of Medicine, St. Louis, Missouri*

YUSUKE WATAYA (30), *Department of Biochemistry and Biophysics, and Department of Pharmaceutical Chemistry, University of California, San Francisco, California*

MOSHE M. WERBER (31), *Polymer Department, The Weizmann Institute of Science, Rehovot, Israel*

EMIL H. WHITE (18, 61), *Department of Chemistry, The Johns Hopkins University, Baltimore, Maryland*

WILLIAM E. WHITE, JR. (75), *Laboratory of Molecular Biology, University of Alabama Medical Center, Birmingham, Alabama*

MEIR WILCHEK (11, 38, 53, 55, 72, 83, 84), *Department of Biophysics, The Weizmann Institute of Science, Rehovot, Israel*

CHRISTOPH WOENCKHAUS (24), *Gustav-Embden Zentrum der Biologischen Chemie, Klinikum der Johann Wolfgang Goethe Universität, Frankfurt am Main, West Germany*

FINN WOLD (1), *Department of Biochemistry, University of Minnesota, St. Paul, Minnesota*

RICHARD WOLFENDEN (2), *Department of Biochemistry, University of North Carolina, Chapel Hill, North Carolina*

JAMES L. WYATT (23), *Cancer Research Institute, University of California School of Medicine, San Francisco, California*

JOSEPH YARIV (43), *Department of Biophysics, The Weizmann Institute of Science, Rehovot, Israel*

K. LEMONE YIELDING (75), *Laboratory of Molecular Biology, University of Alabama Medical Center, Birmingham, Alabama*

ADA ZAMIR (73, 83, 84), *Biochemistry Department, The Weizmann Institute of Science, Rehovot, Israel*

U. J. ZIMMERMAN (58), *Department of Microbiology, University of Pennsylvania School of Medicine, Philadelphia, Pennsylvania*

NAVA ZISAPEL (67), *Department of Biochemistry, The George S. Wise Center for Life Sciences, Tel-Aviv University, Tel Aviv, Israel*

Preface

Few investigators entertaining the use of affinity labels will be interested in working with the enzymes, antibodies, receptors, or other macromolecular structures that are reported on here. Indeed, if the work has been done well—and this volume contains a number of elegant examples of affinity labeling—there is little value in repeating identical procedures. What then is the rationale for presenting this material in the context of the Methods in Enzymology series?

The answer requires acknowledgment that the state of the art is such that the detailed instructions for designing effective affinity labels cannot be given with predictive success. Although the first section of this volume, that on general methodology, attempts to present both a critical basis for design as well as a number of synthetic methods of wide applicability, these serve only as guidelines at best. The specific instances that are analyzed in the subsequent articles represent both the triumphs and shortcomings of the technique of affinity labeling in its various guises, including those in which the labeling has little to do with the specific interactions that are basic to the method.

The answer, then, is that our aim in collecting this large number of illustrative methods is to offer to the investigator contemplating the design of a specific affinity label a background to the type of problems and complexities that have been encountered by others. Our expectation is that at least some of the difficulties may be avoided and others may be correctly interpreted. Since some of the affinity reagents are ligands for proteins other than those recorded here, the fortunate investigator may find in this volume the very compound suitable for the purpose. To aid in the search for such compounds lists of ligands and of macromolecules that are discussed in this volume have been compiled.

It would be unfair to blame the contributors to Volume XLVI for failure to include the specifics of the actual labeling experiments. Our instructions to the authors were to stress the design aspects and synthesis of the reagents, including those instances that were not successful, rather than the details of their use; the latter aspect is readily available in the primary journals.

The investigator may also wish to consult Volume XXXIV of this series in which a number of synthetic methods and concepts are discussed from the standpoint of affinity methods applicable to the separation of proteins.

William B. Jakoby
Meir Wilchek

METHODS IN ENZYMOLOGY

EDITED BY

Sidney P. Colowick and Nathan O. Kaplan

VANDERBILT UNIVERSITY
SCHOOL OF MEDICINE
NASHVILLE, TENNESSEE

DEPARTMENT OF CHEMISTRY
UNIVERSITY OF CALIFORNIA
AT SAN DIEGO
LA JOLLA, CALIFORNIA

METHODS IN ENZYMOLOGY

EDITORS-IN-CHIEF

Sidney P. Colowick   Nathan O. Kaplan

List of Substances for Which Affinity Analogs Are Presented

Article numbers follow each entry

List of Macromolecules Subjected to Affinity Labeling

Article numbers follow each entry

A

Acetylcholine receptor, 68
Acrosin, 16
Actin, 26
Adenosine deaminase, 33
Adenylate kinase, 27
Adenylosuccinate AMP lyase, 29
ADP-binding protein, 23
β-Adrenergic receptor, 69
Albumin, 65
Alcohol dehydrogenase, 24, 25
Aldolase, 5, 10
Amidotransferases, 45
Amino acid-tRNA synthetase, 82
γ-Aminobutyrate-ketoglutanate amino-
    transferase, 3, 12
Aminotransferases, 3, 49
    (see also individual enzymes)
AMP aminohydrolase, 29
Anthranilate synthetase, 45
Antibodies to benzenearsonate, 54
Antibodies to dinitrophenol, 53, 55, 57
Antibodies to lactase, 58
Antibodies to p-nitrophenol, 56
L-Asparaginase, 47
Aspartate aminotransferase, 4
ATPase, 26, 31, 59

B

Biotin transport system, 72

C

C-1 esterase, 9
Cacoonase, 16
Carbamylphosphate synthetase, 45
Carbonic anhydrase, 11
Carboxypeptidase, 16, 20
Catachol-O-methyltransferase, 64
Cathepsin, 16
Chymotrypsin, 11, 16–18
Clostripain, 16, 21
Coenzyme Q reductase, 25
Complement (component C-1), 9
Cyclic nucleotide receptors, 34, 35
Cystathionase, 3, 49

D

Decarboxylases, 3, 49
Dehydrogenases, 23
    (see also individual enzymes)
DNA, 75
DNA-dependent RNA polymerase, 36, 37

E

Elastase, 23
Elongation factors, 77, 83
Enolase, 41
Estradiol dehydrogenase, 6, 50

F

Flavine-linked enzymes, 49
Formylglycinamide ribonucleotide amido-
    transferase, 45

G

β-Galactosidase, 39, 43
D-Glucose transport, 13
Glucosidases, 40
Glutamate decarboxylase, 49
Glutamate dehydrogenase, 23
Glutamate synthase, 45
Glyceraldehyde-3-phosphate dehydro-
    genase, 41
GMP kinase, 28
GTPase, 77

I

IMP kinase, 28
Initiation factor 3, 85

K

Kallikrein, 16
2-Keto-3-deoxy-6-phosphogluconate aldo-
    lase, 10
3-Ketosteroid isomerase, 51, 52
Kinases, 23
    (see also individual enzymes)

Section I

General Methodology

[1] Affinity Labeling—An Overview

By FINN WOLD

The purpose of this chapter is to present a descriptive overview of the current status of affinity labeling; the evolution of the concepts and techniques, the most general picture of the principles involved, the terminology and the criteria commonly applied, and the current and prospective applications of affinity labeling techniques. The discussion is kept general and is based on hypothetical examples; all the other chapters of this volume focus on real systems and give the proper illustrations and examples covering this rapidly expanding field.

Background

It is important to establish from the start that affinity labeling by definition applies to all molecules possessing a site that binds another molecule with a certain degree of specificity and affinity. In biology the proteins are the obvious ligand-binding molecules, and in practice the affinity labeling techniques have involved primarily proteins. Whether we consider enzymes, immune proteins, transport proteins, receptor proteins, or contractile proteins, the key step in their biological function is the specific recognition and binding of a ligand (substrate, antigen, hapten, transported ligand, hormone, regulatory substance, etc.). A small, and well-defined, part of the total protein surface is presumed to be involved in this selective, high affinity interaction with the specific ligand, and this part is referred to as the binding site (or active site in the case of enzymes). The complete understanding of how proteins carry out their biological function requires a detailed understanding of how this binding site is constituted, and this then has become the key problem in the area of protein structure–function analysis.

The early studies of the active sites of enzymes established some of the fundamental principles of affinity labeling. The use of competitive inhibitors illustrated the importance of substrate analogs in exploring the topology of the active site and at the time impressed upon us the concept of proximity effects, illustrating the high effective concentration in the binding site of a given functional group in the ligand when the ligand is bound at the site. An interesting early illustration of this principle can be found in the studies by London et al.[1] on acid phosphatase. This

---

[1] M. London, R. McHugh, and P. B. Hudson, Arch. Biochem. Biophys. **73,** 72 (1958).

enzyme is competitively inhibited by compounds like glycolate ($K_I = 60$ mM), D-glycerate ($K_I = 0.1$ mM), and D-malate ($K_I = 1.6$ mM) when glycerol phosphate is the substrate. Assuming a binding site made up of alternating OH-binding groups and COO^--(PO_3^-)-binding groups in the proper geometrical array, London *et al.* calculated the dissociation constant for the substrate analogs by taking into account the concentration effect. As illustrated with glycolate, the K_I should be $K_I = K_{(OH)} \cdot K_{(COO^-)}$, where $K_{COO^-} \approx K_{OH} \approx 1$. However, when the weak interaction of COO^- with its binding group is established, the concentration of OH groups is no longer that of the bulk solution. The OH group of the bound glycolate is restricted to a half-sphere of a volume $\frac{2}{3}\pi r^3$, where r^3 is 2.9 Å, the distance between the OH and COO^- in glycolate, and its "concentration" in this half-sphere is 32 M. Thus $K_I = K_{COO^-} \cdot K_{OH}/32 = 31$ mM. Applying the same concentration corrections to glycerate and malate, calculated K_I were found to be 0.11 mM and 1.1 mM, respectively, in excellent agreement with the experimentally determined values.

The simple, reversible competitive inhibitors have yielded much useful information about the topology of binding sites, but it became clear very early that in order to actually identify the specific amino acid residues making up the binding site, stable, covalent derivatives had to be introduced.

Thus the general area of protein modification was rapidly developing along with the applications of competitive inhibitors. A large number of chemical reagents specific for different amino acid residues were established and applied to the exploration of the active sites of proteins. The general approach was often a hit or miss proposition: Try different reagents with known amino acid residue specificity, and if the protein is inactivated by one of them, the type of reagent will indicate which residue is crucial for that protein. For this kind of conclusion to be valid, it was essential that the inactivation be associated with the chemical modification of single residues. Unfortunately, it was usually found that the inactivation was accompanied by reaction of several residues, and elaborate and ingenious variations had to be developed to make the results subject to meaningful interpretation. One of the most direct and easily interpreted of these variations was the *differential labeling* technique (this volume [7]). If the ligand protects the binding site residues from chemical reaction with a given reagent, then the chemical modification can first be carried out with the ligand present to modify all the peripheral residues. When the ligand is subsequently removed and a second round of modification is carried out, ideally with the same reagent containing a radioactive marker, only the residue(s) originally protected by the ligand should react. All activity should be lost with the incorporation of a single mole of radioactive reagent, and the radioactive marker should distinguish the critical residue. These kinds of approaches yielded, and still yield, significant information about protein binding sites and at the same time helped to build up a large fund of knowledge and understanding of chemical reactivity of amino acid residues in proteins.

It was inevitable that the substrate (ligand) analog (competitive inhibitor) concepts and the chemical modification techniques eventually would be combined under the general heading of affinity labeling. Let us put the chemically reactive group right on a substratelike compound and bring the covalent modification reagent right to the active site by using the very function we wish to elucidate, namely the protein's specificity of binding. It is perhaps not surprising that the idea occurred to different workers in widely different fields, including pharmacy, chemistry, enzymology, and immunochemistry. The initial elegant demonstrations of the feasibility and utility of the general approach of "affinity labeling" or "active-site-directed reagents"[2-7] certainly gave a tremendous impact to the study of protein structure–function relationships and has contributed and will continue to contribute significantly both to our fundamental understanding of how proteins carry out their biological functions and to the applied areas of drug design and chemotherapy.

Principles and Types of Reagents

Considering affinity labeling in its broadest sense, we must include all reactions through which an actual or potentially reactive compound is bound selectively to a protein binding site in such a manner that a covalent bond can be formed. The definition of affinity labeling thus must contain two distinct steps: selective binding and covalent bond formation.

Meloche[8] has presented a concise analysis of this two-step process. The treatment emphasizes the main experimental criteria for true affinity labeling: rate saturation effect on the rate of inactivation of the enzyme by the affinity labeling reagent, protection against activation by substrate or competitive inhibitors, and the stoichiometric incorporation of one reagent molecule per binding site.

The general reactions for a protein, P, which binds ligand (substrate),

[2] B. R. Baker, W. W. Lee, E. Tong, and L. O. Ross, *J. Am. Chem. Soc.* **83**, 3713 (1961).
[3] G. Schoellmann and E. Shaw, *Biochem. Biophys. Res. Commun.* **7**, 36 (1962).
[4] W. B. Lawson and H. J. Schramm, *J. Am. Chem. Soc.* **84**, 2017 (1962).
[5] L. Wofsy, H. Metzger, and S. J. Singer, *Biochemistry* **1**, 1031 (1962).
[6] A. Singh, E. R. Thornton, and F. H. Westheimer, *J. Biol. Chem.* **237**, 3006 (1962).
[7] G. Gundlach and F. Turba, *Biochem. Z.* **335**, 573 (1962).
[8] H. P. Meloche, *Biochemistry* **6**, 2273 (1967).

S, and is inactivated by a ligand-analog, R, through irreversible affinity labeling are

$$P + S \underset{}{\overset{K_s}{\rightleftarrows}} P \cdot S \tag{1}$$

$$P + R \underset{k_{-1}}{\overset{k_1 \quad k_2}{\rightleftarrows}} P \cdot R \rightarrow P\text{–}R \tag{2}$$

The steady-state rate equation for the inactivation of P through the formation P–R is

$$v_{inact} = \frac{V_{inact}}{1 + K_{inact}(1 + [S]/K_S)/[R]} \tag{3}$$

where K_{inact}, in analogy with a Michaelis constant, is defined as $(k_2 + k_{-1})k_1$; $v_{inact} = k_2 [P \cdot R]$; and V_{inact} is the inactivation velocity at infinitely high concentration of R. This is completely analogous to the Michaelis–Menten equation for competitive inhibition and shows that a plot of V_{inact} against [R] is the rectangular hyperbola (showing rate saturation at high concentration of R) expected for a reaction proceeding through the $P \cdot R$ intermediate and also that S and R compete for the same site.

In the experimental assessment of these criteria, it is more common to determine the half-time (τ) of the inactivation of the protein as a function of either [R] or [S], and to use the linear form of the rate equation in either of the two following (or other) forms[8]:

$$\tau = \frac{T}{[R]}\left(K_{inact} + \frac{K_{inact}[S]}{K_S}\right) + T \tag{4}$$

$$\tau = \frac{[S]T}{[R]}\left(\frac{K_{inact}}{K_S}\right) + \left(T + T\frac{K_{inact}}{[R]}\right) \tag{5}$$

Here the half-time of inactivation, τ, is $\tau \propto (1/v_{inact})$ ($\tau = \ln 2/k_2$) and the minimum half-time of inactivation at infinite concentration of R, T, is $T \propto (1/V_{inact})$. Thus, for a true affinity labeling process, a plot of τ against 1/[R] [Eq. (4)] should give straight lines whose slope is determined by K_{inact} and the concentration of competitive inhibitor(s) and whose common intercept is T. Similarly, a plot of τ against [S] [Eq. (5)] should give straight lines whose slopes are proportional to 1/[R] and whose intercepts reflect the concentration of R.

This type of kinetic analysis, together with careful analyses of reagent incorporation, represents the most common way of assessing the basic criteria for true affinity labeling. It will become evident in subsequent chapters in this volume that there are variations, exceptions, and experimental restrictions in the strict application of these criteria, but basically they do apply and should always be assessed.

Reagent Types

The reactions in Fig. 1 illustrate currently known types of affinity labeling processes and will serve as a basis for a brief consideration and classification of different types of reagents.

Reaction 1 is the normal enzyme-catalyzed conversion of substrate, S, to product, P, through a *covalent E–S intermediate*. In the normal reaction, the intermediate is transient and would serve no useful purpose as a means of introducing a permanent label into the active site of the

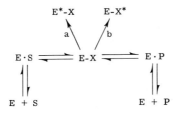

Reaction 1

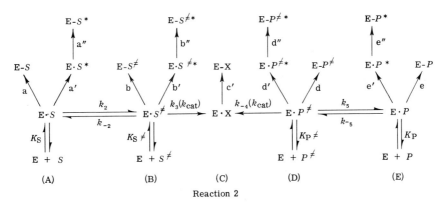

Reaction 2

FIG. 1. Reaction schemes illustrating different means by which affinity labels can be generated and incorporated into proteins. S and P symbolize normal substrates and products. *S* and *P* and the other letters in italics symbolize substrate (product) analogs. A dot (E·S) signifies a noncovalent complex; a bond (E–S) signifies a covalent complex. Reactions (D) and (E) are completely analogous to (B) and (A) and are really redundant; they were included to emphasize the fact that, for enzymes, the design of affinity labeling reagents of any type can be approached with either the substrates or the products as models.

enzyme. However, it is often possible to "freeze" or fix the intermediate, either by an inactivation of the enzyme (1a) or by a chemical stabilization of the complex itself (1b). A good example of the first approach is the stabilization of the acyl enzyme intermediate in the serine esterases by lowering the pH of the reaction mixture to a point at which the rate of the deacylation step approaches zero. Under these conditions, the intermediate ester is stable; the protein can be degraded, and the covalently labeled peptide can be isolated. A good illustration of reaction 1b is the conversion of unstable Schiff-base intermediates to stable secondary amines by reduction with $NaBH_4$. It is probably proper to add here that, in a large number of cases in which covalent intermediates have been suspected or predicted, attempts to demonstrate their existence have been unsuccessful. Since negative results are meaningless in such cases, it is impossible to establish whether covalent E–S intermediates are rare or whether the methods used to convert them from transient intermediates to stable end products are inadequate. As an approach to introducing affinity labels into enzymes, this arrest of the actual enzyme–substrate intermediate should in principle be the most direct and most easily interpretable; but as in all studies of this kind, caution must be exercised. It has been well established, for example, that although the aldolase reaction goes through a Schiff-base intermediate which can be stabilized by $NaBH_4$ reduction, the carbonyls of both substrate and products will slowly form Schiff bases with any primary amino group in the protein; under the wrong kinds of conditions, virtually all amino groups can be labeled, thereby completely obscuring the true affinity-determined active-site label.

Reaction 2 in Fig. 1 is meant to survey all possible real and hypothetical variations on the theme: affinity labeling with substrate (ligand) analogs. S (and P) in the scheme represents analogs that are sufficiently similar to the real substrate (ligand) to be recognized and bound as such by the protein in question. The analog has further been designed to contain one or several chemically reactive groups (either preformed or activated *in situ*) that can cause covalent bond formation with amino acid residues in the protein binding site. In other words, the analogs in reaction 2 are designed as affinity labels.

Reaction 2A (2E)—The K_s Reagents. These are the typical affinity labels designed as direct analogs of known substrates or ligands, and the majority of the reagents presently known fit in this category. They would normally be expected to bind to the protein with the same or somewhat lower affinity than that of the real substrate (ligand). The K_s reagents do not depend on subsequent activations or transformations involving enzyme catalysis, and both catalytic and noncatalytic protein sites can be explored with these reagents. The reagents are relatively easy to design as simple analogs of known ligands, but the K_s reagents are probably also the least specific affinity labeling reagents. This is due in part to the relatively low affinity of these compounds for the active site, and also in part to the fact that the reactive function most often is pre-formed and can react at low affinity sites as well. In addition, since virtu-ally all natural substrates are produced as products by the preceding enzyme in the metabolic chain, any simple substrate analog would be expected to interact with at least two enzymes. On the other hand, it is also clear that each enzyme has at least two model compounds, the sub-strate and the product, on the basis of which the K_s reagents can be fashioned.

Reaction 2B (2D)—The $K_s^{\ddagger}$ Reagents. These are the transition state analogs (this volume [2]), reagents designed to match reasonable inter-mediates in the activation process through which the substrate's conver-sion to product is catalyzed. There is a double bonus involved in work with the transition state analogs: first, the intellectual satisfaction of having been able to predict and reproduce the proper intermediate in the catalytic process and the information about the reaction path that this entails; and, second, the production of a very highly specific reagent, more likely to react uniquely with a single enzyme than any other kind of reagent. Characteristically, the transition state analogs interact with the enzyme with an affinity several orders of magnitude greater than that of the regular substrate. It is probably not useful to consider $K_s^{\ddagger}$ reagents for noncatalytic proteins. In principle, any ligand must have a conforma-tion which gives optimal fit and which can be better derived from an analog than from the ligand itself. In that sense, noncatalytic proteins would have $K_s^{\ddagger}$-like analogs, but in the absolute sense they would not be true transition state intermediates. Unfortunately, there are no known in-stances of a real $K_s^{\ddagger}$ reagent. Although a number of reversible transition state analogs are known, only a couple of examples of such analogs capable of irreversible interaction have been described. However, in terms of specificity and effective reaction with enzymes, this represents one of the most attractive types of reagents for future exploration.

Reaction 2C—The k_{cat} Reagents (Suicide Reagents). These reagents are in most instances very similar to those considered under reaction 1. They are produced through the action of the enzyme itself, and therefore must be considered as pseudo-substrates. In contrast to the situation in reaction 1, which requires an obligatory covalent enzyme–substrate inter-mediate, the k_{cat} reagents require only an enzyme-catalyzed activation by which a chemically reactive form of the analog is produced in situ. The subsequent reaction of this activated analog, with a properly interposed reactive residue in the active site, leads to the covalent affinity-labeled derivative. Since the specificity of enzymes is derived from the k_{cat} step much more than from the K_S step, it is clear that the k_{cat} reagents are much more specific than are the K_S reagents. A given reagent may bind to a number of enzymes, but only the enzyme capable of catalyzing the activation step will produce the reactive form of the analog and be covalently modified. The affinity of an enzyme for a k_{cat} reagent will probably be of about the same order of magnitude as (or less than) the affinity of that enzyme for its substrate. The k_{cat} reagents at the first glance may seem to represent a rather small and somewhat limited group of affinity labels. In fact, many such reagents have been well studied (this volume [3]).

The Covalent Bond Formation

It is well beyond the scope of this chapter to review the many types of functional groups that have been used or could be used in affinity labeling reagents, and how they react with the different amino acid side chains in proteins. However, it may be worth emphasizing that, with the introduction of reagents with such functional groups as carbenes and nitrenes, i.e., reagents that react almost indiscriminately with so many types of side chains, the old prejudices that some residues (Val, Ile, Leu) are inert in modification reactions can be discarded. Reaction 2 in Fig. 1 shows two distinct steps in the covalent bond formation, and it may be useful to elaborate briefly on the difference they signify. The unprimed letters a, b, d, and e denote reactions in which the reagents were synthesized and used with the chemically reactive group present. Once the reagent binds, it can react directly to form the covalent bond, provided that a suitable amino acid residue is properly positioned for a reaction to occur. The proximity effect of placing the functional group of the analog in direct juxtaposition to its reaction partner is very significant here. It means that the effective reagent concentration is much higher near the active site residue than near any of the other residues of the same kind in the protein. Even for sluggish reagents and for relatively low affinity sites, the reaction at the active site residue will consequently be favored.

The double-primed letters in reaction 2 (a″, b″, c″, d″, and e″) denote covalent bond formation after an activation process (indicated by the corresponding single-primed letters). This activation process may be the enzyme-catalyzed step in C (k_3 or $k_{-4} = k_{cat}$) or it may be a physical or chemical activation step in A, B, D, and E. In principle, many such activation processes (pH, temperature, dielectric changes) should apply here, but most of them have the disadvantage of drastically perturbing the protein–ligand complex and probably have only limited application. However, one very important and successful activation process has emerged that yields highly reactive products without any perturbation of the protein–ligand complex. This is activation by irradiation with visible or near-ultraviolet light, leading to an overall reaction that is appropriately referred to as *photoaffinity labeling* (this volume [8]). The products of the photolysis step (carbenes from diazoacyl derivatives, nitrenes from aryl azides, and aromatic ketone triplet-state reagents) are, as already pointed out, very reactive and nonselective in terms of amino acid residues.

Interpretations, Variations, and Special Considerations

As most fields of research become established, their practitioners develop refinements of concepts and language that may not convey the intended nuances of experimental and conceptual detail to the outsider. It may, therefore, be useful to attempt to briefly consider some of the fine points of affinity labeling.

Specific Inactivation in the Absence of Affinity Labeling—Uniquely Reactive Residues in the Active Site. This is undoubtedly the most common reason for highly specific reagent incorporation leading to inactivation. The classical example is the diisopropylphosphorofluoridate inactivation of the serine esterases, in which the active-site serine residue, which appears to be the strongest nucleophile in the entire esterase molecule, is selectively phosphorylated to give an inactive derivative with a single reagent molecule incorporated. There is probably no problem in identifying these cases for what they are. If the reaction is slow enough to permit a kinetic analysis, it can be shown that the inactivation is bimolecular at all concentrations of reagent (no rate saturation) and in addition, that the uniquely reactive residue should be subject to derivatization with several different reagents, none of which bears any structural relation to the protein's true ligand.

Syncatalytic Inactivation. Syncatalytic inactivation (this volume [4]) represents a special case of a uniquely reactive single residue in the active site of an enzyme. In this case the high reactivity is induced by substrates; the inactivation reaction is synchronous with catalysis, i.e., syncatalytic. Rather than decreasing the rate of inactivation, substrates in this case enhance it.

Catalytic Competence of the E-S Derivative (this volume[6]). It has recently been demonstrated that a substrate analog, covalently bound in the active site of any enzyme, can be catalytically transformed to a product analog, still covalently bound, by the addition of the proper coenzyme (cosubstrate) to the inactive E-S derivative. Clearly this represents a most satisfactory and compelling criterion for a case of true affinity labeling. Unfortunately, it is unlikely that this situation will be found to be a very general one. The substrate analog will have to be located in the proper orientation to permit the approach of the cosubstrate, and the components of the catalytic apparatus of the enzyme cannot have been involved in the covalent modification reaction.

Exoaffinity and Endoaffinity Labels (this volume [9]). These are reagents forming covalent bonds with residues "outside" and "inside" the binding site, respectively. This concept represents an important point in the interpretation of affinity labeling data, and deserves careful attention in formulating conclusions regarding the active-site components.

The most rational approach to the design of reagents to react with residues within or outside the active site would appear to involve the length of the "arm" holding the reactive group. Thus, the typical endoaffinity label would presumably have a short arm with the reactive group close to the affinity group; the exoaffinity label should have a long arm permitting covalent attachment away from the affinity site. In either case the derivative should satisfy all the criteria of an affinity label-inactivated system, but any interpretation in terms of relation of the modified residue to the active site will obviously be more tenuous the longer the arm. Since the exoaffinity labels are less likely to modify components of the catalytic apparatus than are the endoaffinity labels, it seems reasonable to predict that the criterion of catalytic competence (above) might be favored by the exoaffinity labels.

Polyfunctional Affinity Labels. It is impossible for an enthusiast for protein cross-linking reagents not to include a brief consideration of affinity labeling with such reagents. Any affinity label reagent, by definition, is at least bifunctional: one function consists of the affinity group, and the other of a chemically reactive group. If reagents can be produced that, in addition to the affinity group, contain two or more reactive groups, active-site residues can be cross-linked and the knowledge of the reagent dimensions would subsequently allow establishment of inter-residue distances in the active site. Polyfunctional reagents have indeed been made,[9,10] and there is no obvious reason why they should not be readily prepared and serve as useful reagents.

Affinity-Sensitized Photooxidation. This is a rather specialized ramification of affinity labeling, which deserves attention primarily as an illustration of secondary reaction possibilities, once an affinity label has

[9] J. J. Bechet, A. Dupaix, and J. Yon, *Eur. J. Biochem.* **35,** 527 (1973).
[10] S. S. Husain and G. Lowe, *Biochem. J.* **117,** 341 (1970).

been covalently attached. If the affinity label contains a proper chromophore, either as a natural constituent or an an added appendage, the chromophore may act as a sensitizer for photooxidation, i.e., the generation of singlet oxygen affected by energy transfer from the photoactivated chromophore. Residues such as tryptophan, tyrosine, cysteine, methionine, and histidine are readily modified by the singlet oxygen and, if present in the active site, would be subject to a very highly specific modification reaction. Thus, by very simple means it should be possible to broaden the information available from the affinity labeling of the primary reaction site to include the nature of neighboring groups susceptible to the secondary modification.

Information from and Use of Affinity Labeling

In Vitro Information. The most significant contributions of affinity labeling techniques to date have been in the elucidation of active site structures and in unraveling mechanistic details in the catalytic processes. The incorporation of the affinity label provides the necessary handle to pull out the proper peptide sequences from the labeled proteins and, with due care, to draw conclusions about the amino acid sequences in or around the active sites. By the use of differently designed reagents with the reactive group in various geometrical relationships to the affinity group, it has been possible to involve a variety of residues in the covalent modification and in this way to build up an inventory of the residues in different peptide sequences that presumably constitute the active site proper. Because the method focuses directly on those parts of the molecule that are of most immediate interest, it has been possible to accumulate relevant sequence data on families of proteins and do comparative studies on active-site peptides in cases in which complete sequences would take years to produce. With the more specific $K_s^{\ddagger}$ reagents and k_{cat} reagents, it has been, or will be, possible to test hypothetical intermediates and catalytically produced analogs arrested in the modification reactions, and thus to glean from the data mechanistic detail with a degree of confidence that is unique in the study of enzyme action. As a general single approach for obtaining information about protein structure and function, affinity labeling can stand on its own in terms of the conclusions drawn. However, there is no doubt that a good part of the excitement with this field derives from the fact that the techniques have developed in parallel with other very powerful physical and chemical methods. The combination of affinity labeling with powerful electromagnetic spectroscopic analyses, and ultimately with complete structure determination through X-ray diffraction analysis, has led to both static and dynamic models of biologically active proteins at a resolution few would have dared to hope for a decade ago.

In Vivo Applications. There is no doubt that a good deal of the ideas and work that have gone into developing affinity labeling reagents and techniques has been inspired by the tremendous potential of these reagents in drug design and in chemotherapy. Whereas the usual type of K_s reagent may not offer very unique properties of specificity, it certainly must represent a more permanent, and thus presumably more powerful, version of the usual competitive inhibitors currently used in chemotherapy. The k_{cat} reagents, however, appear to offer some truly unique and impressive possibilities, based on the fact that they are essentially inert as chemical reagents until they are specifically activated by the enzyme that is to be inactivated. Thus, if an enzyme can be identified as being unique to an infective agent or a malignant cell, a k_{cat} reagent designed for that enzyme should be a specific antibiotic against the infective agent or malignant cell in question, and should have no effect on host cells or tissues.

Epilogue

A vaguely remembered Norwegian fairy tale relates a story about a magic goose. Its magic is that every stranger who touches it or anyone associated with it becomes permanently attached. The hero with his goose collects quite a following: A pretty maiden pats the goose and becomes attached; a fellow pinches the girl and his hand remains permanently bound at the site of pinching; a lady expresses her disgust over that spectacle with a wooden spoon to the fellow's head and thus joins the nascent chain; the lady's husband manages to get in a swift kick of disapproval as the train goes by—and ends up hopping along with one foot attached to his better half's posterior. The train becomes longer, but memory fails as to the details. However, the delightful mental picture still remains, to represent a happy model of affinity labeling: a series of separate dynamic and transient actions, initiated through specific and definite affinity interactions (the girl's hand to the goose, the boy's hand . . . and so on) and then fixed at the point of optimal spatial and temporal association by the covalent, permanent bond (or Scandinavian magic). Once fixed, the interactions are there for everyone to study and dissect, and a good deal of new and unique information about each reaction step and each reactant has been made available. It works extremely well in science, but it is perhaps just as well that magic geese have become extinct. The hero, by the way, succeeded in making a very sober princess laugh with his affinity-labeled entourage, and as a consequence he won the princess and half the kingdom and lived happily ever after.

[2] Transition State Analogs as Potential Affinity Labeling Regents

By RICHARD WOLFENDEN

The principle of transition state affinity requires the development during enzyme catalysis of very powerful forces of attraction between the enzyme and the altered form of the substrate that is present in the transition state.[1,2] This principle has served as a basis for the synthesis of some very strong reversible enzyme inhibitors,[3-6] and may also in principle be useful in the design of affinity labeling reagents. The latter possibility remains to be explored, but seems already to be partly substantiated by a few serendipitous examples of inhibitors of glycosidases and proteases.

Principle of Transition State Affinity

Enzymes and other catalysts, in bringing about an increase in the rates of reactions, are considered, in the language of transition state theory, to lower their free energies of activation. The equilibrium constant $K_n^{\ddagger}$, for the activation of a substrate to its rate-determining transition state $S^{\ddagger}$, is accordingly more favorable in the presence of the catalyst than in its absence (Scheme 1).

$$S \xrightarrow{\;k_n\;} P \qquad S \underset{\longleftarrow}{\overset{K_N^{\ddagger}}{\longrightarrow}} S^{\ddagger} \xrightarrow{\;10^{13}\,s^{-1}\;} P$$

$$ES \xrightarrow{\;k_{es}\;} E + P \qquad ES \underset{\longleftarrow}{\overset{K_{ES}^{\ddagger}}{\longrightarrow}} ES^{\ddagger} \xrightarrow{\;10^{13}\,s^{-1}\;} E + P$$

$$\frac{K_{ES}^{\ddagger}}{K_S^{\ddagger}} = \frac{k_{es}}{k_n} > 10^7$$

SCHEME 1

[1] W. P. Jencks, *in* "Current Aspects of Biochemical Energetics" (N. O. Kaplan and E. P. Kennedy, eds.), p. 273. Academic Press, New York, 1966.
[2] R. Wolfenden, *Nature (London)* **223**, 704 (1969).
[3] R. Wolfenden, *Acc. Chem. Res.* **5**, 10 (1972).
[4] G. E. Lienhard, *Science* **180**, 149 (1973).
[5] R. N. Lindquist, *in* "Drug Design" (E. J. Ariens, ed.), Vol. 5, p. 24. Academic Press, New York, 1975.
[6] R. Wolfenden, *Annu. Rev. Biophys.* **5**, 271 (1976).

If this proposal is examined with the aid of the thermodynamic cycle shown in Scheme 2, it is found to require the development of very great powers of attraction, which achieve a maximum during the catalytic process and subside with the generation of products.[7] The inequality between $K_{ES}^{\ddagger}$ and $K_S^{\ddagger}$, equilibrium constants for activation that lie in the same ratio as the rate constants for the catalyzed and uncatalyzed reactions, is found to be matched by an inequality between K_S (the dissociation constant of ES, the Michaelis complex) and K_{TX} (a dissociation constant for the hypothetical process of separating the enzyme from the altered substrate in the transition state). The ratio $K_{ES}^{\ddagger}:K_S^{\ddagger}$ exceeds 10^7 in all known cases, and may sometimes attain values as large as 10^{15}–10^{20}. If the K_S of the substrate is 10^{-4} M, the dissociation constant of the altered substrate from the enzyme in the transition state should thus be no greater than 10^{-11} M, and may attain values as low as 10^{-24} M.[8] A stable analog of such an activated form of the substrate would be expected to be an extremely powerful inhibitor.

In the simple situation described by Scheme 2, analogs may be designed to resemble $S^{\ddagger}$ directly. Many cases are more complicated. If the enzyme reaction involves several substrates reacting simultaneously at

---

[7] If the enzymic and nonenzymic reactions under comparison differ in the detailed structure of the altered substrate in the transition state, then Scheme 2 seems to provide only a lower limiting value for the affinity that must be developed at some stage in order to explain the catalysis observed.[3,6] If the rate of the enzymic reaction is determined, for example, by a conformation change or by product release, then chemical processes in substrate transformation (and hence the affinity developed between E and $S^{\ddagger}$) must presumably be even greater at some point than the rate ratio suggests. This should be true whenever the the rate-limiting transition state for substrate transformation is reached at a different point on the reaction coordinate in the enzymic and nonenzymic reactions. If there is a radical difference in mechanism when the two reactions are compared, then the rate observed for the nonenzymic reaction is presumably larger than that of any hypothetical reaction that might resemble the reaction as it occurs on the enzyme, since the nonenzymic reaction must follow the pathway with the lowest available free energy of activation. Again transition state affinity (as it might be expressed in an ideal analog, perfectly resembling the substrate part of the transition state for the enzymatic reaction) will have been underestimated. There would of course be no reason to expect the enzyme to show appreciable affinity for analogs resembling the transition state for the nonenzymic reaction, in such a case.

[8] It has long been understood that the kinetic constants k_{cat} and K_M, though sometimes equivalent in meaning to that assigned to them by Michaelis and Menten, may be complex in their physical significance. Nonproductive binding, for example, tends to reduce k_{cat} and K_M below the values that would be observed if only productive binding occurred. The bimolecular rate constant for reaction between free enzyme and substrate, k_{cat}/K_M, appears to be relatively free of ambiguities of this kind, and is therefore used in Scheme 2.

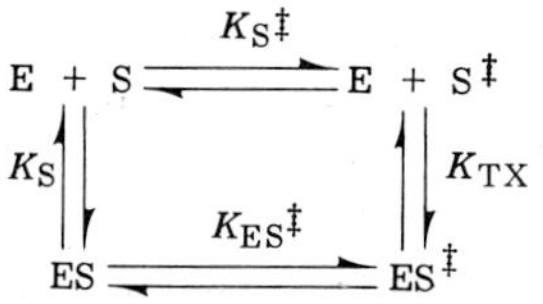

$$E + S \xrightleftharpoons[]{K_S^{\ddagger}} E + S^{\ddagger} \qquad K_{TX} = \frac{K_S^{\ddagger}}{K_{ES}^{\ddagger}} \quad K_S < 10^{-7}\, K_S$$

$$K_S \updownarrow \qquad\qquad \updownarrow K_{TX}$$

$$ES \xrightleftharpoons[]{K_{ES}^{\ddagger}} ES^{\ddagger} \qquad\qquad = \frac{k_n\, K_m}{k_{cat}}$$

SCHEME 2

the active site, a "multisubstrate analog" may be designed to resemble this combination. In such reactions, activation may be achieved by concentration, orientation, and chemical activation of the substrates, and analogs may in principle be designed to incorporate features corresponding to all these forms of activation. In other cases, the enzyme reaction may proceed through covalently bound intermediates, in which the enzyme may form an integral part of ES‡. Analogs may then be designed to be transformed, through a chemical reaction with the enzyme, into similar complexes. Where covalent bonds are formed with the enzyme, or where several substrates are simultaneously present in the transition state, schemes describing the affinity of an ideal transition state analog are more complex than Scheme 2. Theoretical problems arising in the quantitative analysis of these cases are considered elsewhere.[1-6,9,10] In multisubstrate reactions, very substantial activation can be achieved by purely "physical" means, and these may be difficult or impossible to distinguish from effects due to chemical or electronic activation.[9] Likewise, in reactions involving covalent intermediates, there is no entirely satisfying way of calculating the rate of a nonenzymic reaction that would be suitable for comparison with the reaction on the enzyme, since the enzyme is an integral part of the transition state. These theoretical difficulties are of considerable mechanistic interest, and analogs of these special kinds can be used to explore and exploit the various possible sources of catalytic affinity[3-6].

Methods of Examining Potential Transition State Analogs

When an inhibitor is found to be very much more tightly bound than a substrate, it is important to inquire whether its high affinity is based on a resemblance to activated intermediates in catalysis, or to factors which may be irrelevant to the mechanism (e.g., reaction of enzyme sulfhydryl groups with organic mercurial inhibitors). Among other criteria, the structure of the inhibitor should possess features that would be expected to distinguish activated intermediates from the substrates themselves.

[9] W. P. Jencks, *Adv. Enzymol.* **43,** 220 (1975).
[10] R. Wolfenden, *in* "Transition States of Biochemical Processes" (R. L. Schowen and R. D. Gandour, eds.), in press. Plenum, New York, 1977.

It may be difficult in some cases to decide whether tight binding of an inhibitor, instead of reflecting a structural resemblance to activated intermediates in catalysis (including all the possible forms of activation mentioned above), may be based on the hydrophobic character of the inhibitor, or on a diminution in space-filling requirements as compared with those of the substrate. Such effects need not be dismissed as trivial, since they may contribute to catalysis itself. It is apparent that one of the important questions that must be asked about enzyme catalysis is how such a large difference in affinity can exist between the complexes ES and ES‡. The design of transition state analogs is a relativistic procedure that may permit an independent evaluation of structural characteristics of altered forms of the substrate that serve to distinguish these species from the substrate itself, contributing to catalytic affinity.

Regardless of detailed interpretation, Scheme 2 suggests that K_{TX} should vary as does $k_n K_{\mathrm{M}}/k_{\mathrm{cat}}$, with changing experimental conditions. If it is found that there is an experimental variable that affects k_{cat} only, without appreciably affecting K_{M} or k_n, then this variation should also be reflected in the observed K_{I} of a transition state analog. This provides one means of testing potential transition state analogs without resort to mechanistic assumptions. Variables may include enzyme origin, substrate structure, temperature, pressure, and solvent composition (e.g., pH, ionic strength, dielectric constant).

Triosephosphate isomerase, for example, catalyzes the interconversion of glyceraldehyde 3-phosphate and dihydroxyacetone phosphate with k_{cat} values that vary with pH, whereas K_{M} is pH independent in the neutral range. This behavior appears consistent with mechanisms in which the enzyme combines with either monoanionic or dianionic forms of the substrate with similar affinity, but only the latter forms a productive complex, ES^{2-}, which goes on to form products. Unlike an ordinary substrate analog, a transition-state analog for this reaction would be expected to be bound tightly only as a dianionic species, and this appears to be the case for the inhibitor 2-phosphoglycolic acid.[11]

$$E + S^{\ominus} \rightleftharpoons ES^{\ominus} \quad \text{(nonproductive)}$$
$$E + S^{2-} \rightleftharpoons ES^{2-} \rightleftharpoons [ES^{\ddagger\,2-}] \rightarrow E + P \quad \text{(productive)}$$
$$E + A^{\ddagger\,2-} \rightleftharpoons EA^{\ddagger\,2-} \quad \text{(binding of analog)}$$

From studies of this kind, involving equilibrium measurements alone, it is possible to check for consistency of behavior with that expected for an ideal analog, but not to prove the validity of the hypothesis. Nor do

[11] F. Hartman, G. M. LaMuraglia, Y. Tomozawa, and R. Wolfenden, *Biochemistry* **14**, 5274 (1975).

such studies make clear the state of the bound analog. This important information can in principle be obtained from the results of spectroscopic studies of enzyme–inhibitor complexes. In the above example, the dianionic form of the bound inhibitor could involve any of the species shown below. Nuclear magnetic resonance (NMR) studies of bound ligands suggest that bound 2-phosphoglycolate is present as the trianion, the enzyme itself taking up a proton as the transition state analog is bound.[11a]

$$
\text{EH}^{\oplus}
\begin{bmatrix}
\text{COO}^{\ominus} \\
| \\
\text{CH}_2 \\
| \\
\text{OP}^{\,2\ominus}
\end{bmatrix}
\qquad
\text{E}
\begin{bmatrix}
\text{COO}^{\ominus} \\
| \\
\text{CH}_2 \\
| \\
\text{OP}^{\ominus}
\end{bmatrix}
\qquad
\text{E}
\begin{array}{c}
\text{COOH} \\
| \\
\text{CH}_2 \\
| \\
\text{OP}^{\,2\ominus}
\end{array}
$$

$$(1) \qquad\qquad\qquad (2) \qquad\qquad\qquad (3)$$

Boronic acid derivatives of proteases have recently been shown to combine with active-site nucleophiles in the manner shown below, both by X-ray and infrared spectroscopy.[12,13] The pH dependence of binding is also consistent with this view, according to which these analogs are considered to resemble tetrahedral intermediates in the formation and breakdown of covalent acyl–enzyme intermediates in double displacement.[14]

$$
\begin{array}{c}
\text{OH} \\
| \\
\text{R}-\text{B}: \\
| \\
\text{OH}
\end{array}
+ \text{EOH} \rightleftharpoons
\begin{array}{c}
\text{OH} \\
|^{\ominus} \\
\text{R}-\text{B}-\text{OE} \\
| \\
\text{OH}
\end{array}
+ \text{H}^+
$$

Subtle variations in substrate structure, at points rather distantly removed from bonds being formed or broken, are sometimes found to affect k_{cat} strongly without much affecting K_M. Adenosine deaminase, for example, exhibits a high degree of specificity for the ribose substituent of the substrate, such that k_{cat} for adenosine exceeds that for adenine by a factor of about 10^5, whereas K_M values for these substrates differ by only a small factor. Model reactions suggest that the nonenzymic reactivities of adenosine and adenine should be very similar. It is thus of interest that methanol photoadducts of purine, thought to resemble tetrahedral intermediates in substrates hydrolysis, exhibit very large differences in K_I, the presence of ribose serving to enhance affinity by a factor

[11a] I. D. Campbell, P. A. Kiener, S. G. Waley, and R. Wolfenden, unpublished results.

[12] G. P. Hess, D. Seybert, A. Lewis, J. Spoonhower, and R. Cookingham, *Science* **189**, 384 (1975).

[13] D. A. Matthews, R. A. Alden, J. J. Birktoft, S. T. Freer, and J. Kraut, *J. Biol. Chem.* **250**, 7120 (1975).

[14] K. A. Koehler and G. E. Lienhard, *Biochemistry* **10**, 2477 (1971).

of about 10^4.[15] Another example of this behavior is elastase, where k_{cat} for substrates, and K_I for aldehyde inhibitors that may form adducts resembling tetrahedral intermediates in substrate hydrolysis, vary markedly but in parallel as the structure of the substrate (or inhibitor) is varied at points distant from the scissile bond.[16] Cases of this kind are especially interesting in that they seem to imply the occurrence of conformation changes in the enzyme and/or the substrate during the catalytic process. In principle, such changes can be examined directly with transition-state analogs, using exact structural methods to compare atomic coordinates in complexes corresponding to various stages in the catalytic process.

Possible Transition State Analogs and Multisubstrate Analogs

Some examples of potential transition state analogs of various types are shown in the table, arranged according to categories of reaction. By way of introduction, it may be useful to consider two inhibitors of dehydrogenases that seem to have received less attention than they deserve. Lactate dehydrogenase has long been known to be inhibited strongly by oxalate.[17] The strong affinity of the enzyme for this inhibitor may be due to its resemblance to an alcoholate intermediate, such as might be generated by hydride transfer from NADH to pyruvate. The state of charge of the bound inhibitor does not appear to have been determined, and it may have a bearing on the puzzling observation that oxamate is also very effective as an inhibitor, though differing somewhat in its detailed mode of inhibition.[17]

$$
\begin{array}{cccc}
\underset{\text{Pyruvate}}{\overset{\displaystyle \mathrm{COO(H)}}{\underset{\displaystyle \mathrm{CH_3}}{\mid\; \mathrm{C}{=}\mathrm{O} \;\mid}}}
&
\xrightleftharpoons{(\mathrm{H}^-)}
&
\left[\underset{\text{Lactate}\atop\text{dehydrogenase}}{\overset{\displaystyle \mathrm{COO(H)}}{\underset{\displaystyle \mathrm{CH_3}}{\mid\; \mathrm{HC}{-}\mathrm{O}^\ominus \;\mid}}}\right]
&
\xrightleftharpoons{(\mathrm{H}^+)}
&
\underset{\text{Lactate}}{\overset{\displaystyle \mathrm{COO(H)}}{\underset{\displaystyle \mathrm{CH_3}}{\mid\; \mathrm{HC}{-}\mathrm{OH} \;\mid}}}
&
\underset{\text{Oxalate}}{\overset{\displaystyle \mathrm{COO(H)}}{\mathrm{C}{-}\mathrm{O}^\ominus}}
\end{array}
$$

A more complex example, triosephosphate dehydrogenase, is strongly inhibited by a tetrose bisphosphate that contaminates preparations of glycolaldehyde phosphate, of which it is a condensation product.[18,19] The

---

[15] D. F. Wentworth and R. Wolfenden, unpublished results.

[16] R. C. Thompson, *Biochemistry* **13**, 5495 (1974).

[17] W. B. Novoa, A. D. Winer, A. J. Glaid, and G. W. Schwert, *J. Biol. Chem.* **234**, 1143 (1959).

[18] E. Racker, V. Klybas, and M. Schramm, *J. Biol. Chem.* **234**, 2510 (1958).

[19] A. L. Fluharty and C. E. Ballou, *J. Biol. Chem.* **234**, 2517 (1958).

SOME POTENTIAL TRANSITION STATE AND MULTISUBSTRATE ANALOGS

Enzyme	Inhibitor	Possible intermediate	References
Alcohol dehydrogenase			a
Lactate dehydrogenase			a
Malate dehydrogenase			a
Glutamate dehydrogenase			a
Lactate dehydrogenase	 Oxalate	See text	b
Triosephosphate dehydrogenase	D-Threose 2,4-bisphosphate	See text	c, d
Transcarboxylase	Oxalate	 Pyruvate enolate	e
Pyruvate kinase			f
Aspartate transcarbamylase	 PALA		g
Pyridoxamine-pyruvate transaminase	 Pyridoxylalanine		h
Adenylate kinase	5′ A—OPOPOPOPOPO—5′ A	A—OPOP—-OPO--PO—A	i
Creatine kinase	A—OPOPO--NO₂⁻--NH₂R	[A—OPOPO--PO₃⁻--NH₂R]	j
Arginine kinase			k

Some Potential Transition State and Multisubstrate Analogs (*Continued*)

Enzyme	Inhibitor	Possible intermediate	References
Acetylcholin-esterase	Boronic acids	See test	*l*
Chymotrypsin			*m*
Subtilisin			*n*

Enzyme	Inhibitor	Possible intermediate	References
Papain			*o*
Elastase	R—C—E (with H top, OH bottom) Aldehydes (hemiacetals)	[R—C—E with X top, OH bottom]	*p*
Asparaginase			*q*
Amidase			*r*

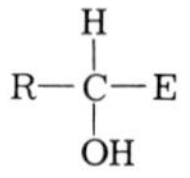

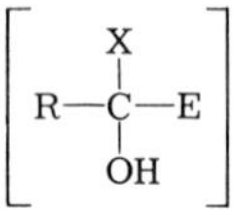

Glycosidases

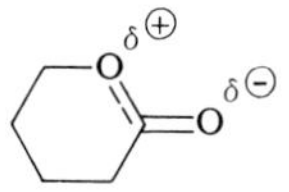

Glycono-lactone

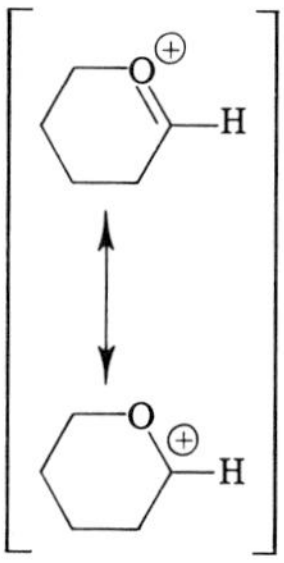

1-Amino-glycoside

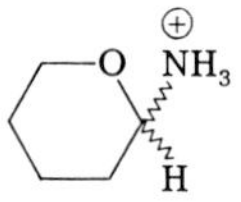

s

t

Purine deaminases

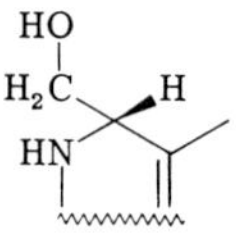

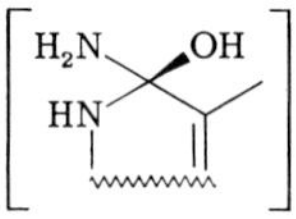

u

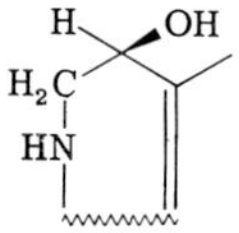

v

Pyrimidine deaminases

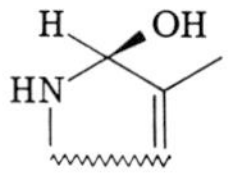

w

Carboxy-peptidase A

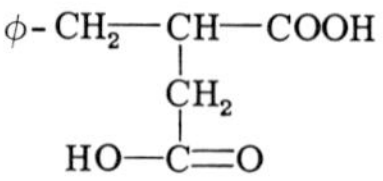

ϕ-CH$_2$—CH—COOH
CH$_2$
HO—C=O

Benzyl succinate

[ϕ-CH$_2$—CH—COOH
NH$_2$
HO—C=O
R]

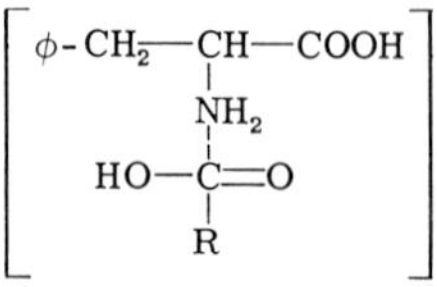

x

Some Potential Transition State and Multisubstrate Analogs (*Continued*)

Enzyme	Inhibitor	Possible intermediate	References
Oxaloacetate decarboxylase	Oxalate	See transcarboxylase	*y*
Ribulose bis-phosphate carboxylase	CH_2OP $C(OH)COOH$ $HCOH$ $HCOH$ CH_2OP	$\begin{bmatrix} CH_2OP \\ C(OH)COOH \\ C=O \\ HCOH \\ CH_2OP \end{bmatrix}$	*z*
Aldolase	$\overset{NHOH}{\underset{\underset{OP}{\overset{\|}{CH_2}}}{\overset{\|}{C}-O^{\ominus}}}$ 2-Phosphoglycolo-hydroxamate	$\begin{bmatrix} HC-OH \\ C-O^{\ominus} \\ CH_2 \\ OP \end{bmatrix}$	*aa*
Enolase	$\overset{COOH}{\underset{\underset{\delta^{\oplus} NH}{\overset{\|}{CH}}}{\delta^{\ominus} \overset{\|}{C}-OP}}$	$\begin{bmatrix} COOH \\ \overset{\ominus}{C}-OP \\ CH_2OH \end{bmatrix}$	*bb*
Proline racemase	(pyrrole-2-carboxylate) $COO^{\ominus}$	$\begin{bmatrix} \text{(pyrrolidine-2-carboxylate)} COO^{\ominus} \end{bmatrix}$	*cc*
Steroid Δ-isomerase	(naphthol structure)	$\begin{bmatrix} \text{(decalone structure)} \end{bmatrix}$	*dd*
Triosephosphate isomerase	$\overset{O}{\underset{R}{\overset{\|}{C}-O^{\ominus}}}$	$\begin{bmatrix} HC-OH \\ C-O^{\ominus} \\ R \end{bmatrix}$	*ee*
Glucose-6-phosphate isomerase	$\overset{NH-OH}{\underset{R}{\overset{\|}{C}-O^{\ominus}}}$		*ff*

SOME POTENTIAL TRANSITION STATE AND MULTISUBSTRATE ANALOGS (*Continued*)

Enzyme	Inhibitor	Possible intermediate	References
Amino acid activating enzymes	$AMP-O-CH_2-CHR-NH_3^{\oplus}$	$\left[AMP-O-\overset{\overset{\textstyle O}{\|}}{C}-CHR-NH_3^{\oplus} \right]$	*gg*
Glutamine synthetase	$O{=}\overset{\overset{\textstyle CH_3}{\|}}{\underset{\underset{\textstyle NH_2-\overset{\textstyle H}{\underset{}{C}}-COOH}{\overset{\textstyle (CH_2)_2}{\|}}}{S}}{-}\overset{(H^+)}{N}-P^{\textcircled{\scriptsize 2}\ominus}$	$\left[\overset{\overset{\textstyle NH_3^{\oplus}}{\|}}{HO-\underset{\underset{\textstyle NH_2-\overset{\textstyle H}{\underset{}{C}}-COOH}{\overset{\textstyle (CH_3)_2}{\|}}}{C}-OP^{\textcircled{\scriptsize 2}\ominus}} \right]$	*hh*

[a] J. Everse, E. C. Zoll, L. Kahan, and N. O. Kaplan, *Bioorg. Chem.* **1**, 207 (1971).

[b] W. B. Noroa, A. D. Winer, A. J. Glaid, and G. W. Schwert, *J. Biol. Chem.* **234**, 1143 (1959).

[c] E. Racker, V. Klybas, and M. Schramm, *J. Biol. Chem.* **234**, 2510 (1958).

[d] A. L. Fluharty and C. E. Ballou, *J. Biol. Chem.* **234**, 2517 (1958).

[e] D. Northrop and H. G. Wood, *J. Biol. Chem.* **244**, 5810 (1969).

[f] G. H. Reed and S. D. Morgan, *Biochemistry* **13**, 3537 (1974).

[g] K. D. Collins and G. R. Stark, *J. Biol. Chem.* **246**, 6599 (1971).

[h] W. B. Dempsey and E. E. Snell, *Biochemistry* **2**, 1414 (1973).

[i] G. E. Lienhard and I. I. Secemski, *J. Biol. Chem.* **248**, 1121 (1973).

[j] E. J. Milner-White and D. C. Watts, *Biochem. J.* **122**, 727 (1971).

[k] D. H. Buttlaire and M. Cohn, *J. Biol. Chem.* **249**, 5733 (1974).

[l] K. A. Koehler and G. P. Hess, *Biochemistry* **13**, 5345 (1974).

[m] K. A. Koehler and G. E. Lienhard, *Biochemistry* **10**, 2477 (1971).

[n] R. N. Lindquist and C. Terry, *Arch. Biochem. Biophys.* **160**, 135 (1974).

[o] J. O. Westerik and R. Wolfenden, *J. Biol. Chem.* **247**, 8195 (1971).

[p] R. C. Thompson, *Biochemistry* **12**, 47 (1973).

[q] J. O. Westerik and R. Wolfenden, *J. Biol. Chem.* **249**, 6351 (1974).

[r] J. D. Findlater and B. A. Orsi, *FEBS Lett.* **35**, 109 (1973)

[s] J. Conchie, A. J. Hay, I. Strachan, and G. A. Levvy, *Biochem. J.* **102**, 929 (1967).

[t] H. L. Lai and B. Axelrod, *Biochem. Biophys. Res. Commun.* **54**, 463 (1973).

[u] B. E. Evans and R. Wolfenden, *J. Am. Chem. Soc.* **92**, 4751 (1970).

[v] P. W. K. Woo, H. W. Dion, S. M. Lange, L. F. Dahl, and L. J. Durham, *J. Heterocycl. Chem.* **11**, 641 (1974).

[w] R. M. Cohen and R. Wolfenden, *J. Biol. Chem.* **246**, 7561 (1971).

[x] L. D. Byers and R. Wolfenden, *J. Biol. Chem.* **247**, 606 (1972).

[y] A. Schmitt, I. Bottke, and G. Siebert, *Hoppe-Seyler's Z. Physiol. Chem.* **347**, 18 (1966).

[z] M. Wishnick, M. D. Lane, and M. C. Scrutton, *J. Biol. Chem.* **245**, 4939 (1970).

[aa] K. D. Collins, *J. Biol. Chem.* **249**, 136 (1974).

[bb] T. G. Spring and F. Wold, *Biochemistry* **10**, 4655 (1971).

[cc] G. J. Cardinale and R. H. Abeles, *Biochemistry* **7**, 397 (1968).

[dd] S. Wang, F. S. Kawahara, and P. Talalay, *J. Biol. Chem.* **238**, 576 (1973).

[ee] R. Wolfenden, *Biochemistry* **9**, 3404 (1970).

[ff] J. M. Chirgwin and E. A. Noltmann, *Fed. Proc.* **32**, 667 (1973).

[gg] D. Cassio, F. LeMoine, J. P. Waller, E. Sandrin, and R. A. Boissonas, *Biochemistry* **6**, 827 (1967).

[hh] W. B. Rowe, R. A. Ronzio, and A. Meister, *Biochemistry* **8**, 2674 (1969).

active-site sulfhydryl group of this enzyme, thought to serve as a point of attachment for the developing carboxylic acid in the normal reaction catalyzed by this enzyme, is also a possible site of attachment of the inhibitor, as shown below. It can be shown with models that a thiohemiacetal of this kind might occupy a space almost coextensive with that which would be occupied by a tetrahedral intermediate such as might be generated during phosphoryl attack on a hypothetical 3-phosphoglyceryl-enzyme intermediate.

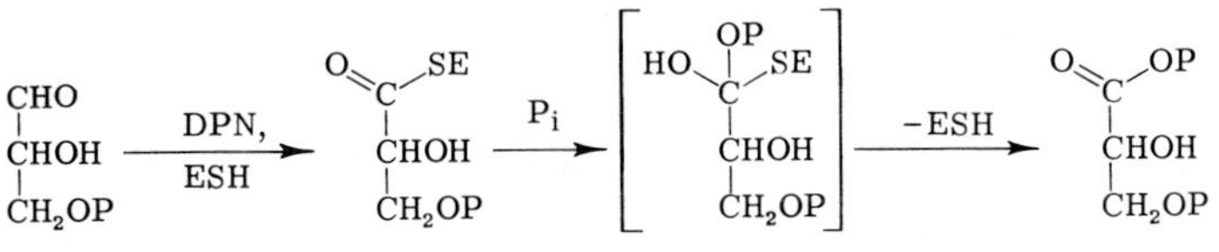

$$
\begin{array}{ccccc}
\underset{\substack{\text{CHO} \\ | \\ \text{CHOH} \\ | \\ \text{CH}_2\text{OP}}}{}
& \xrightarrow[\text{ESH}]{\text{DPN,}}
& \underset{\substack{\text{O}=\text{C}-\text{SE} \\ | \\ \text{CHOH} \\ | \\ \text{CH}_2\text{OP}}}{}
& \xrightarrow{P_i}
& \left[\ \underset{\substack{\text{HO}-\overset{\text{OP}}{\text{C}}-\text{SE} \\ | \\ \text{CHOH} \\ | \\ \text{CH}_2\text{OP}}}{}\ \right]
& \xrightarrow{-\text{ESH}}
& \underset{\substack{\text{O}=\text{C}-\text{OP} \\ | \\ \text{CHOH} \\ | \\ \text{CH}_2\text{OP}}}{}
\end{array}
$$

Glyceraldehyde 3-phosphate   Triosephosphate dehydrogenase   1,3-Diphosphoglyceric acid

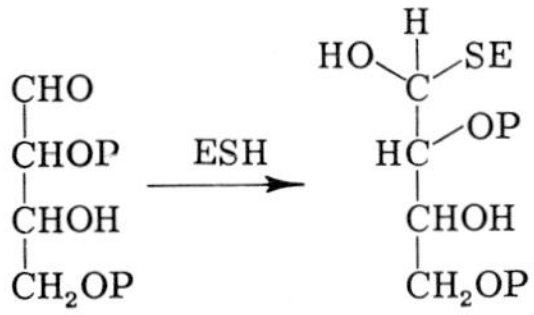

$$
\underset{\substack{\text{CHO} \\ | \\ \text{CHOP} \\ | \\ \text{CHOH} \\ | \\ \text{CH}_2\text{OP}}}{}
\xrightarrow{\text{ESH}}
\underset{\substack{\text{HO}-\overset{\text{H}}{\underset{}{\text{C}}}-\text{SE} \\ | \\ \text{HC}-\text{OP} \\ | \\ \text{CHOH} \\ | \\ \text{CH}_2\text{OP}}}{}
$$

Tetrose diphosphate   Adduct

In both these examples, K_I values for the inhibitors are found to be some thousandfold lower than K_M values for the corresponding substrates. There are many examples of comparable magnitude, and some in which the relative affinity of the inhibitor appears to be considerably greater. Among the more impressive examples are 3,4,5,6-tetrahydrouridine 5'-phosphate, bound more tightly than CMP to CMP deaminase by a factor of about 10^4 [20]; trimethylammonium methyleneboronic acid, bound 10^4 times more tightly than acetylcholine by acetylcholinesterase[21]; coformycin, bound some 10^5 times more tightly than adenosine by adenosine deaminase[22]; and 2-phosphoglycolohydroxamic acid, bound more than 10^4 times more tightly than substrates by yeast aldolase.[23] Very large binding ratios are also exhibited by multisubstrate analog inhibitors of adenylate kinase (Ap$_5$A),[24] glutamine synthetase (methionine sulfoxi-

[20] F. Maley and G. F. Maley, Arch. Biochem. Biophys. 144, 723 (1971).
[21] K. A. Koehler and G. P. Hess, Biochemistry 13, 5345 (1974).
[22] S. Cha, R. P. Agarwal, and R. E. Parks, Jr., Biochem. Pharmacol., in press.
[23] K. D. Collins, J. Biol. Chem. 249, 136 (1974).
[24] G. E. Lienhard and I. I. Secemski, J. Biol. Chem. 248, 1121 (1973).

mine phosphate),[25] and amino acid activating enzymes (aminoalkyl esters of AMP).[26]

Although binding ratios as large as 10^5–10^6 are found in some cases, it should be emphasized that true K_S values are not often known for productive ES complexes, and may be higher or lower than the observed K_M values on which the apparent binding ratios are based. In addition, it seems probable that calculations according to expressions such as Scheme 2 provide only an upper limiting value for K_{TX}, in the likely event that the enzymic and nonenzymic reactions differ in the detailed structure of the altered substrate in the transition state.[5]

Affinity Labeling with Transition State Analogs

The possibility of combining the features of a transition state analog with those of an affinity labeling agent remains to be explored in detail, but it appears to have been realized unexpectedly in two cases.

Bacterial β-galactosidase is rather strongly inhibited by D-galactal.[27]

D-Galactal

This inhibition was ascribed originally to a possible resemblance between this inhibitor and a possible carbonium ion intermediate in substrate transformation. Although inhibition was reversible and competitive as expected, it was later shown to be measurably slow in its onset, and both the rate of onset and the rate of release of inhibitor could be determined by measurement of residual catalytic activity using ordinary steady-state methods. Rates of onset and release were in reasonable agreement with the measured K_I value of galactal, but the released inhibitor was found to consist entirely of 2-deoxygalactose.[28] This could be explained if the enzyme reacted covalently with the inhibitor to yield a 2-deoxygalactosyl-enzyme derivative, which is hydrolyzed slowly with recovery of activity.

[25] W. B. Rowe, R. A. Ronzio, and A. Meister, *Biochemistry* **8**, 2674 (1969).

[26] D. Cassio, F. LeMoine, J. P. Waller, E. Sandrin, and R. A. Boissonas, *Biochemistry* **6**, 827 (1967).

[27] Y. C. Lee, *Biochem. Biophys. Res. Commun.* **35**, 161 (1969).

[28] D. F. Wentworth and R. Wolfenden, *Biochemistry* **13**, 4715 (1975).

Inhibition:

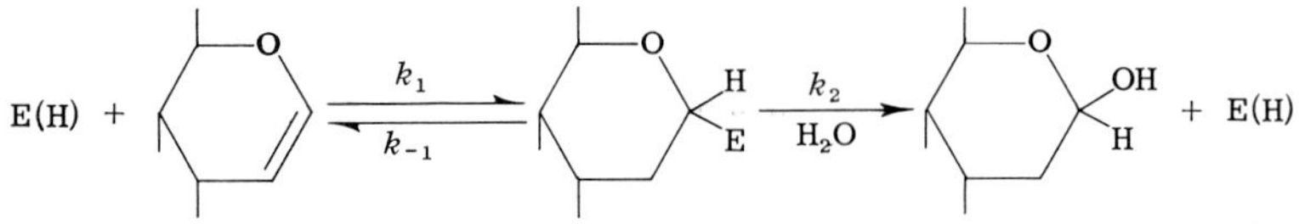

Normal reaction:

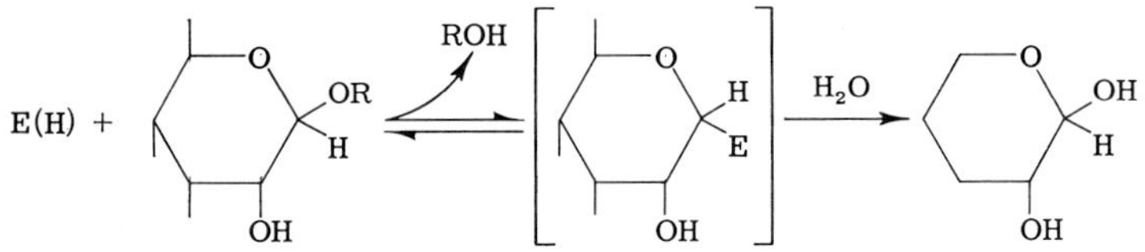

Consistent with this hypothesis, rates of onset and release from inhibition were retarded in D_2O, but to quite different extents, K_I itself showing a pronounced solvent isotope effect.

The observed K_I of galactal is substantially lower than K_M values for substrates of β-galactosidase. It is of particular interest that K_I can be accounted for entirely in terms of the rate constants for galactal addition and hydrolysis of the presumed glycosylated enzyme. In the equation below, k_2 is therefore presumably greater than k_{-1}, and this implies that the reversible binding of galactal, as described by the dissociation constant (k_{-1}/k_1), is even more favorable than the observed K_I suggests. This appears reasonable for a complex resembling a high-energy intermediate (galactosylgalactosidase) in substrate hydrolysis. Like aldehyde and boronic acid inhibitors of proteolytic enzymes, galactal can be viewed as a compound which *becomes* a transition state analog when it reacts covalently with the enzyme. Unlike these inhibitors, its release is quite slow, and it may thus be regarded as an affinity labeling agent as well.

$$K_T = \frac{k_{-1} + k_2}{k_1} \qquad k_2 > k_{-1}$$

An interesting example of an affinity labeling agent, apparently sharing some characteristics of a transition state analog, is provided by halomethylketones in their reaction with subtilisin BPN.[29] These agents

[29] T. L. Poulos, R. A. Alden, S. T. Freer, J. J. Birktoft, and J. Kraut, *Biochemistry* **251**, 1097 (1976).

are found to yield, not the simple alkylated enzymes expected, but hemi-ketals in which the ketone has undergone further addition of the side chain of Ser-221:

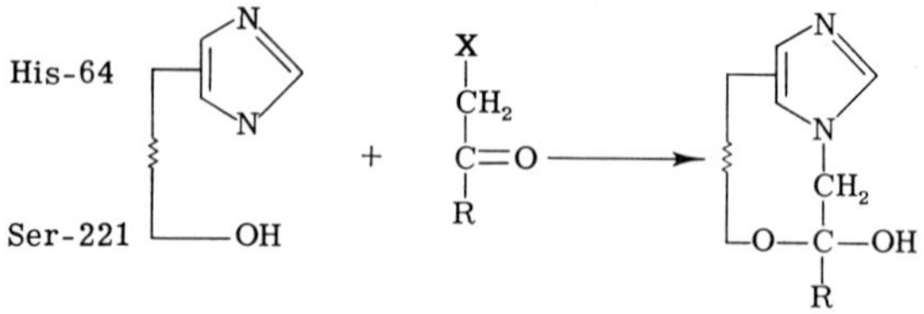

The covalent bond resulting from alkylation presumably imposes a geometry appropriate for the addition observed, resulting in a complex somewhat resembling tetrahedral intermediates in acylation and deacylation of the enzyme. The actual sequences of events in inhibition is ambiguous, since the halogen atom might tend to promote prior addition of serine at the carbonyl group of the ketone. In this connection it is of interest that several halomethylketones have been found to serve as reversible but very potent inhibitors of leucine aminopeptidase.[30]

These examples, encountered accidentally, suggest that deliberate efforts to synthesize affinity labels, with features resembling activated intermediates in substrate transformation, may prove fruitful. In addition to the extra affinity offered by transition state analogs, inhibition may in principle be more specific than in the case of substrate analogs, since reactions are more distinctively characterized by their transition states than by their reactants and products.

[30] C. Kettner, G. I. Glover, and J. M. Prescott, *Arch. Biochem. Biophys.* **165,** 739 (1974).

[3] Mechanism-Based Irreversible Enzyme Inhibitors

By Robert R. Rando

There are two extreme forms of mechanism for specific irreversible enzyme inhibitors. One comprises the classical affinity labeling agents, which are substrate analogs containing chemically reactive functional groups.[1] The specific binding of these inhibitors to the active-site regions of the enzyme ensures a greater probability of chemical reaction with a residue within this region as opposed to nonspecific chemical reactions.

[1] E. Shaw, *in* "The Enzymes" (P. D. Boyer, ed.), 3rd ed., Vol. 1, Chap. 2. Academic Press, New York, 1970.

Thus the specificity, and hence effectiveness, of these molecules is determined solely by their binding affinity for the enzyme that is to be inhibited. A second type of irreversible enzyme inhibitor includes those whose specificity is determined not only by binding affinity, but also by its effectiveness to serve as a substrate for the target enzyme. The substrate itself is chemically unreactive, but the product of the enzymic conversion is a highly reactive molecule. This product immediately reacts with an active-site moiety, resulting in the irreversible inhibition of the enzyme.[2] We have termed inhibitors of this type k_{cat} inhibitors because they require catalytic conversion by the target enzyme. The overall scheme for this mode of inhibition is that of Eq. (1).

$$E + S \underset{k_2}{\overset{k_1}{\rightleftharpoons}} ES \overset{k_{cat}}{\longrightarrow} E \cdot I \overset{k_{inh}}{\longrightarrow} E - I \tag{1}$$

whereas for affinity labeling it is that of Eq. (2).

$$E + I \underset{k_2}{\overset{k_1}{\rightleftharpoons}} E \cdot I \overset{k_{inh}}{\longrightarrow} E - I \tag{2}$$

Inhibitors of the k_{cat} (mechanism-based) type offer many advantages over affinity labeling agents. Primarily, they are more specific because they are chemically unreactive to foreign biomolecules. This means that they can be used with crude *in vitro* and *in vivo* systems without fear of competing with possibly predominant nonspecific reactions. Furthermore, since the chemically reactive inhibitor is generated at the active site, exceedingly reactive functionalities can be used that would not be possible with an affinity labeling agent. On a more pragmatic level, these reagents are, in general, more stable than affinity labeling agents because they are relatively chemically inert prior to enzymic activation.

These considerations suggest that the k_{cat} inhibitors will offer advantages over affinity labeling agents mainly when crude systems are used. They, in fact, offer no great advantage in carrying out structural work on pure enzymes, e.g., in determining the sequence of the active-site region. The choice of an affinity labeling agent or a k_{cat} inhibitor in these systems will depend primarily on which reagent happens to be a good inhibitor of closely related enzymes. For example, acetylenes are known to be powerful k_{cat} inhibitors of flavin-linked enzymes and, therefore, would be a likely choice.[3] Similarly, an affinity labeling chloroketone would be used to label the active site of a serine protease.

The successful design of mechanism-based irreversible enzyme inhi-

[2] R. R. Rando, *Science* **185**, 320 (1974).
[3] C. T. Walsh, A. Schonbrunn, O. Lockridge, V. Massey, and R. H. Abeles, *J. Biol. Chem.* **247**, 6004 (1972).

bition is dependent on the following conditions: (1) the molecule first and foremost must be chemically unreactive with Lewis bases; (2) it must be a substrate for the target enzyme; and (3) it must be converted into a chemically reactive product that must immediately react with an active-site residue of the enzyme without diffusing into solution. Condition (1) is obvious and requires no comment. Condition (2) is an important boundary condition because it means that the less the chemical modification compared to the natural substrates, the more likely it is that the inhibitor will work. The structural restrictions here depend on the specificity of the enzyme. The more specific the enzyme, the more stringent the structural requirements for the inhibitor. In addition, the less structurally complicated the inhibitor, the less likely that it will be specific, since many enzymes will use it as substrate. As a general rule, molecules designed to inhibit an enzyme that uses a metabolically ubiquitous substrate are not likely to be very specific. For example, it is hard to see how we would design a glutamate or glucose analog that is specific for a single enzyme. Condition (3) is also crucial to the design of these inhibitors and involves several important points: (a) the simple formation of a highly reactive molecule at the active site does not ensure that a chemical reaction will occur with an active-site residue; (b) a chemically compatible residue must be within bonding distance of the reactive moiety; and (c) the breakdown rate of the enzyme inhibitor complex must not be so rapid that a chemical reaction with an active-site residue is precluded. In fact, most successfully designed k_{cat} inhibitors make use of covalent catalysis, so that the reactive grouping is held by covalent bonds to the active site; this ensures that there will be sufficient time available for a chemical reaction to occur.

It is important to have experimental criteria by which to judge whether an inhibitor requires catalytic conversion and whether direct inactivation of the enzyme occurs. The following criteria may be used to decide these points: (1) The kinetics of inactivation should be first order in inhibitor concentration. (2) The pH vs. rate profile for inhibition may be similar to that for substrate turnover. (3) If C—H bonds are cleared, a deuterium isotope effect should be manifest during the inhibition. (4) Trapping agents, such as mercaptans, should not decrease the rate of inactivation by inhibitor. The rate[4] would decrease if the reactive molecule first diffused into solution and only later inactivated the enzyme by an affinity labeling or nonspecific mode of inactivation. (5) If a second aliquot of fresh enzyme is added to a solution containing inactivated enzyme and excess inhibitor, the rate of inactivation of the second

[4] G. Schoellmann and E. Shaw, *Biochem. Biophys. Res. Commun.* **7**, 36 (1962).

enzyme "pulse" should be the same as for the first. If its rate of inactivation is faster, it would be expected that an inhibitory metabolite is being formed in solution. (6) If stable, the product of the enzyme–inhibitor conversion should be an affinity labeling agent of the enzyme. (7) The stoichiometry of inhibitor to active subunits should be one-to-one in the inactivated enzyme. (8) Inactivation of the enzyme by the inhibitor should be competitively blocked by substrate. Clearly, conditions (6)–(8) will be true also of affinity labeling agents.

Specific Reagents

In this section, all the specific reagents thus far utilized will be considered, with the exception of acetylenes. This latter class of inhibitors will be treated subsequently in this volume [12].

β, γ-Unsaturated Molecules

Inhibitors of the structural type shown in Eq. (3) can be converted into highly reactive Michael acceptors [shown on the right-hand side of Eq. (3)]

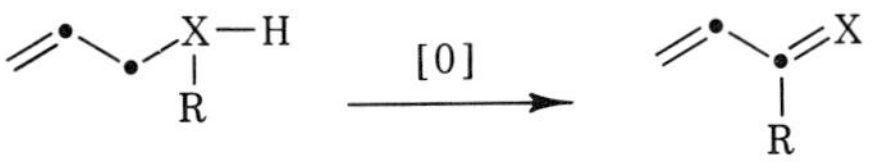

$$\overset{X-H}{\underset{R}{|}} \quad \xrightarrow{[O]} \quad \overset{X}{\underset{R}{|}} \qquad \text{where } X = 0, N, \text{ or } S \quad (3)$$

by a wide variety of enzymes. The reactants are relatively inert whereas the products of these reactions are highly reactive alkylating agents capable of undergoing facile additions of nucleophiles at the terminus of the double bond.

Inhibitors of this type have been used in the irreversible inhibition of pyridoxal-linked aspartate aminotransferase,[5] γ-cystathionase,[6] and tryptophan synthetase.[7] These inhibitors are β, γ-unsaturated amino acids. Aspartate aminotransferase is inhibited by molecules 1 and 2,

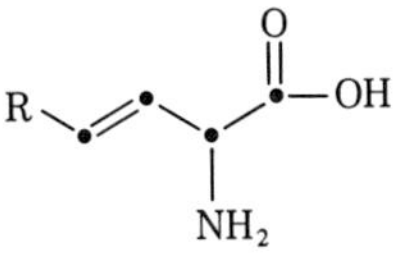

(1) R = H-vinylglycine
(2) R = OCH$_3$-2-amino-4-methoxy-trans-3-butenoic acid
(3) R = HOCH$_2$—CH—CH$_2$—O-rhizobitoxine
                      |
                     NH$_2$

[5] R. R. Rando, Nature (London) 250, 586 (1974); R. R. Rando, Biochemistry 13, 3859 (1974); R. R. Rando, N. Relyea, and L. Chang, J. Biol. Chem. 251, 3306 (1976).
[6] I. Giovanelli, L. D. Owens, and S. H. Mudd, Biochim. Biophys. Acta 227, 671 (1971).
[7] E. W. Miles, Biochem. Biophys. Res. Commun. 66, 94 (1975).

tryptophan synthetase by *2*, and γ-cystathionase by *3*. The inhibition of aspartate aminotransferase by *1* and *2* involves the sequence of steps shown in Eq. (4).

(4)

Since both inhibitors must be converted into the active form *4* by the enzyme, neither one inhibits the holoenzyme in the pyridoxamine form or the apoenzyme.[5] The reactive enzyme product itself must be an affinity labeling agent of the enzyme. As expected, β, γ-unsaturated α-ketoacids are irreversible inhibitors of the enzyme in the pyridoxamine form.[5] Finally, the pH rate profile for the inhibitor is the same as that for substrate turnover[5]; this need not be the case, but does obtain in this example.

Gabaculine, 1,3-cyclohexadienyl-5-aminocarboxylic acid, is a naturally occurring β, γ-unsaturated amino acid isolated from *Streptomyces toyocaenis*.[8]

Gabaculine	γ-Aminobutyric acid

This molecule specifically and irreversibly inhibits γ-aminobutyrate-α-ketoglutarate transaminase.[9] The K_I for this molecule is 2.8 μM, some three orders of magnitude lower than the K_m for the natural substrate, γ-aminobutyrate. The $t_{1/2}$ for inactivation of the bacterial enzyme is 9 min at 0.3 μM gabaculine. As expected, gabaculine is not an irreversible inhibitor of the enzyme in the pyridoxamine form, and the mechanism of the irreversible inhibition by gabaculine is completely in accord with the hypothesis that enzymic conversion before inhibition. The possible mechanisms proposed for the inhibition are those in Eq. (5).[9]

The reactive, transaminated product (5) can either engage in alkylation reaction with an active-site residue or can aromatize. Both processes would yield an inactivated enzyme.

β, γ-Unsaturated amino acids are not inhibitors of flavin-linked enzymes but are good substrates. These inhibitors appear to be excellent choices as inhibitors of pyridoxal-linked enzyme. A synthesis of the parent β, γ-unsaturated amino acid, vinyl glycine, is included in this article under the heading "Synthesis."

Other unsaturated substrate analogs that have been tried as enzyme inhibitors include allyl amine and allyl alcohol. Allylamine is a pseudo-irreversible inhibitor of flavin-linked monoamine oxidase[10]; i.e., in the presence of allylamine, the enzyme shows a time-dependent inactivation that cannot be reversed by dialysis. When radiolabeled allylamine is used, radioactivity is incorporated at the same rate as the enzyme is inhibited. However, inhibition is relieved and radioactivity is removed from the enzyme upon incubation with the substrate, benzylamine.

Allyl alcohol is oxidized by yeast alcohol dehydrogenase to afford the highly reactive molecule acrolein. During this process, the enzyme

[8] H. Mishima, H. Kurihara, K. Kobayashi, S. Miyazawa, and A. Terahara, *Tetrahedron Lett.* **7**, 537 (1976).
[9] R. R. Rando and F. W. Bangeter, *J. Am. Chem. Soc.* **98**, 6762 (1976).
[10] R. R. Rando, manuscript in preparation.

$$(5)$$

becomes progressively and irreversibly inactivated.[11] However, inhibition can be completely prevented by adding an exogenous mercaptan, such as β-mercaptoethanol. We have here, then, an interesting example, in which the highly reactive molecule, once formed, diffuses into solution more rapidly than it reacts covalently with an active-site residue. Thus the acrolein concentration increases to inhibit the enzyme, but only as an affinity label. The addition of exogenous trapping agent is a good test for this mode of inhibition. However, it will be appreciated that between the extremes of this mode of inhibition and that of direct stoichiometric inactivation all shades of possibilities can occur. For example, it is pos-

[11] R. R. Rando, *Biochem. Pharmacol.* **23**, 2328 (1974).

sible that during every 1000 turnovers one enzyme molecule will be inactivated.

Diazoesters and Ketones

The prototypical inhibitors in this series are azaserine, 5-diazo-4-oxo-L-norvaline (DONV), and 6-diazo-5-ketonorleucine (DON). All these inhibitors contain the diazo group as their active moiety.

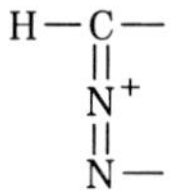

$$
\begin{array}{c}
\text{H}-\text{C}- \\
\quad \| \\
\text{N}^+ \\
\quad \| \\
\text{N}-
\end{array}
\qquad \text{Diazo group}
$$

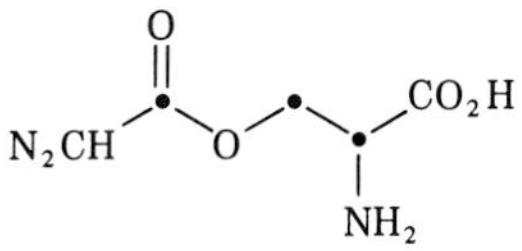

Azaserine

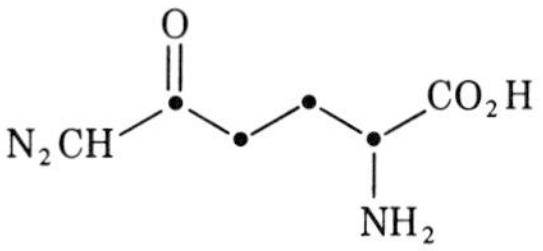

DON

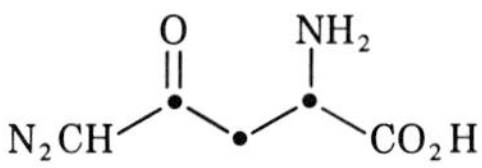

DONV

This grouping is quite stable to nucleophilic attack, but on enzymic protonation, it is converted to the reactive diazonium ion, which reacts readily with nucleophiles.[12] Therefore, in instances in which acid cataly-

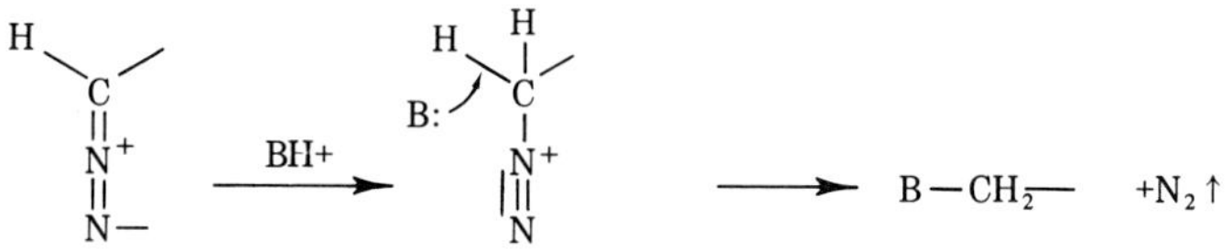

$$
\begin{array}{c}
\text{H} \\
\quad\diagdown \\
\text{C} \\
\quad\| \\
\text{N}^+ \\
\quad\| \\
\text{N}-
\end{array}
\xrightarrow{\text{BH}+}
\begin{array}{c}
\text{H}\quad\text{H} \\
\diagdown\ | \\
\text{B:}\diagup\text{C} \\
\quad | \\
\quad \text{N}^+ \\
\quad\| \\
\quad \text{N}
\end{array}
\longrightarrow
\text{B}-\text{CH}_2- \quad +\text{N}_2\uparrow
\qquad (6)
$$

sis is taking place, a diazo reagent of this type would be a good choice as an inhibitor. The aspariginases from *Escherichia coli* and pig serum are irreversibly inactivated by DONV, an asparagine analog.[13] DNO, a glutamine analog, is an irreversible inhibitor of *E. coli* glutaminase.[13]

---

[12] T. C. French, I. B. Dawid, R. A. Day, and J. M. Buchanan, *J. Biol. Chem.* **238**, 217 (1963); I. B. Dawid, T. C. French, and J. M. Buchanan, *ibid.* **238**, 2186 (1963). See also this volume [45].

[13] R. Handschumacher, C. Bates, P. Chang, A. Andrews, and G. Fisher, *Proc. Am. Assoc. Cancer Res.* **8**, 25 (1967). See also this volume [47].

DONV does not affect this enzyme. This is as expected, since DONV should not be a substrate for the enzyme. Both azaserine and DON irreversibly inhibit formylglycinamide ribotide amidotransferase.[12] Again, enzymic protonation precedes inactivation. In general, acid proteases seem to be quite sensitive to inhibition by diazo compounds, although cupric ion may be required for a facile reaction.[14] Recently, β-galactosidase has been reported to be irreversibly inhibited by the diazo sugar 2,6-anhydro-1-diazo-1-deoxy-D-glycero-L-*manno*-heptitol.[15] Protonation of the diazo group to yield the highly reactive diazonium ion is thought to mediate the inhibition.

These specific reagents have not been sufficiently exploited and could be used fruitfully with many different kinds of enzymes in which acid catalysis occurs. A synthesis of p-nitrophenyl diazoacetate is included in the section "Synthesis." This molecule is a good starting point for the synthesis of a variety of diazo compounds. The compound is a good acylating agent and can transfer the diazo grouping of a wide variety of nucleophilic functional groups[16] resulting in the synthesis of a large number of diazo inhibitors.

Allyl Chloride and Sulfates

Allyl chlorides and sulfates that have been often used include the following:

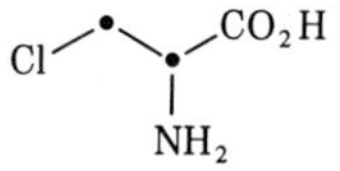

β-Chloroalanine 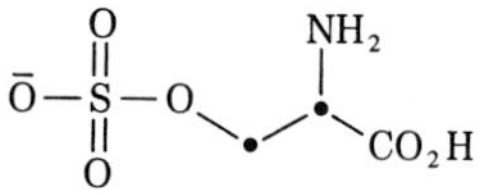   O-Serinesulfate

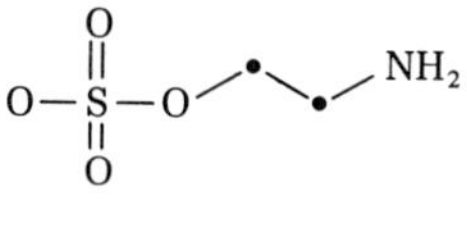

Aminoethanesulfate

---

[14] B. F. Erlanger, S. M. Vratsanos, M. Wasserman, and A. G. Cooper, *Biochem. Biophys. Res. Commun.* **23**, 243 (1966); T. G. Rajogopalan, W. H. Stein, and S. Moore, *J. Biol. Chem.* **241**, 4295 (1966).

[15] M. Brockhaus and J. Lehmann, *FEBS Lett.* **62**, 154 (1976).

[16] H. Chaimovich, R. J. Vaughan, and F. H. Westheimer, *J. Am. Chem. Soc.* **90**, 4088 (1968); J. Shafer, P. Baronowsky, R. Lawsen, F. Finn, and F. H. Westheimer, *J. Biol. Chem.* **241**, 421 (1966).

These compounds are thought to function by undergoing elimination reaction with certain enzymes to give rise to reactive Michael acceptors.

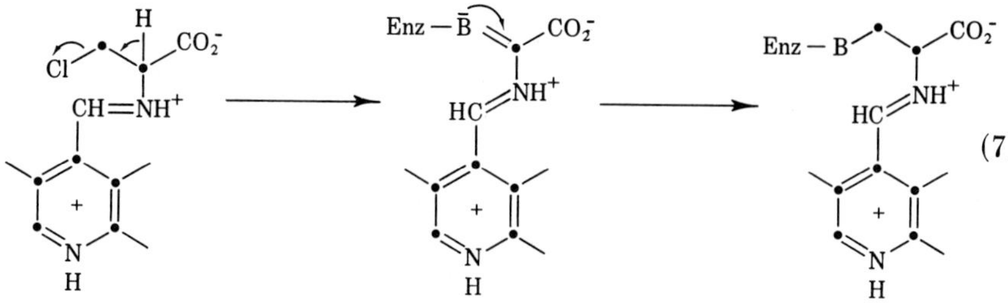

$$(7)$$

β-Chloroalanine has been found to be an irreversible inhibitor of the pyridoxal phosphate-linked β-aspartate decarboxylase,[17] aspartate amino-transferase,[18] and alanine racemase.[19] The mechanism of inhibition is shown above by Eq. (7) (the sulfate reacts in the same manner): amino-ethane sulfonate irreversibly inhibits pyridoxal phosphate-linked GABA transaminase[20] and L-serine-O-sulfate irreversibly inhibits aspartate aminotransferase.[21]

The inhibition by these compounds exemplifies interesting facets of the general mechanism by which these inhibitors act. As mentioned many enzyme turnovers may occur before the enzyme becomes inactivated. This need not be disastrous, because the product diffusing away may be chemically inert. For example, the dissociation of the aminoacrylate intermediate in Eq. (7) gives rise to pyruvate and the pyridoxal form of the enzyme. Although it might be considered that β-chloroglutamic acid would be a better inhibitor of aspartate aminotransferase than would β-chloroalanine, this is not the case since the glutamic acid derivative, although an excellent substrate for the enzyme, is not an inhibitor of it.[22] Similarly, the chloroamino acids are not inhibitors of the flavin-linked amino acid oxidases, but are substrates of them.[23]

Vinyl Chlorides

Two compounds of this type have been reported to be irreversible enzyme inhibitors. cis-3-Chloroallylamine is an irreversible inhibitor of

[17] E. W. Miles and A. Meister, Biochemistry 6, 1735 (1967).
[18] Y. Morino and M. Okamato, Biochem. Biophys. Res. Commun. 50, 1061 (1973).
[19] J. M. Manning, N. E. Merrifield, W. M. Jones, and E. C. Gotschlich, Proc. Natl. Acad. Sci. U.S.A. 71, 417 (1974).
[20] L. J. Fowler and R. A. John, Biochem. J. 130, 569 (1972).
[21] R. A. John and P. Fasella, Biochemistry 8, 4477 (1969).
[22] P. Fasella, in "Pyridoxal Catalysis: Enzymes and Model Systems" (E. Snell, ed.), p. 1. Wiley, New York, 1968.
[23] C. T. Walsh, A. Schonbrunn, and R. H. Abeles, J. Biol. Chem. 246, 6855 (1971).

cis-3-Chloroallylamine   2-Chloroallylamine

flavin-linked monoamine oxidase and 2-chloroallylamine is an irreversible inhibitor and nonflavin-linked monoamine oxidase.[24,25] The former reagent is thought to function by first isomerizing to the reactive allylic chloride [Eq. (8)], and the latter by undergoing an elimination reaction to provide the reactive allene derivative [Eq. (9)].

$$\text{Cl} \longrightarrow \text{H} \sim\!\sim\!\sim \text{Enz} \quad (8)$$

$$= \bullet = \bullet \longrightarrow \text{NH} \sim\!\sim\!\sim \text{Enz} \quad (9)$$

In both cases, the starting materials, being vinyl chlorides, are not susceptible to nucleophilic attack. These strategies of elimination or isomerization should be applicable to a large number of enzymes capable of abstracting an α–C–H bond.

Other Reagents

There are several reagents whose uses may be more restricted than those of inhibitors mentioned above but that nevertheless deserve discussion. Flavin-linked monoamine oxidase has been a fruitful enzyme for the development of new inhibitors. Besides being inhibited by acetylenic amines and olefinic amines, the enzyme is also inactivated by hydrazides and cyclopropyl amines,[26] e.g., the antidepressant drug tranylcypromine. Both of the latter reagents are first turned over by the enzyme before inhibition ensues, but the mechanisms of inhibition remain obscure. Recently, N-nitroso compounds have been introduced as irreversible inhibitors of proteolytic enzymes.[27]

[24] R. R. Rando, J. Am. Chem. Soc. **95**, 4438 (1973).
[25] R. C. Hevey, J. Babson, A. L. Maycock, and R. Abeles, J. Am. Chem. Soc. **95**, 6125 (1973).
[26] B. T. Ho, J. Pharm. Sci. **61**, 821 (1972).
[27] E. H. White, D. F. Roswell, I. R. Politzer, and B. R. Branchini, J. Am. Chem. Soc. **97**, 2290 (1975). See also this volume [18].

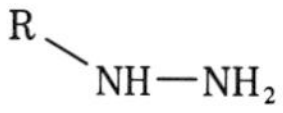

R
$\diagdown$
NH—NH$_2$

ϕ

NH$_2$

Alkyl hydrazide    Tranylcypromine

It has been shown that N-nitrosolactams can inhibit chymotrypsin irreversibly, possibly by a mechanism that involves the formation of a carbonium ion [Eq. (10)].

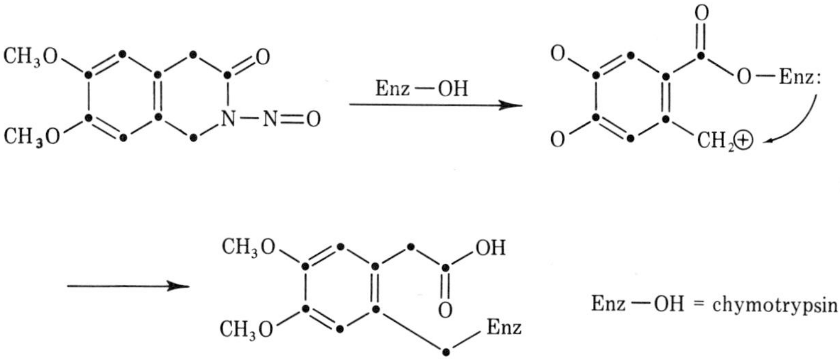

CH$_3$O

CH$_3$O
N—N=O

$\xrightarrow{\text{Enz—OH}}$

O

O
O—Enz:

CH$_2\oplus$

$$(10)$$

$\xrightarrow{\hspace{2cm}}$

CH$_3$O

CH$_3$O
O

OH

Enz

Enz—OH = chymotrypsin

Interestingly, the open-chain analog of the above inhibitor is a substrate for the enzyme but not an inhibitor; this underlines the importance of holding the inhibitor at the active site to ensure enough time for chemical reaction.

Synthesis

Vinyl Glycine[5]

The synthesis of vinyl glycine involves the sequence shown in Eq. (11).

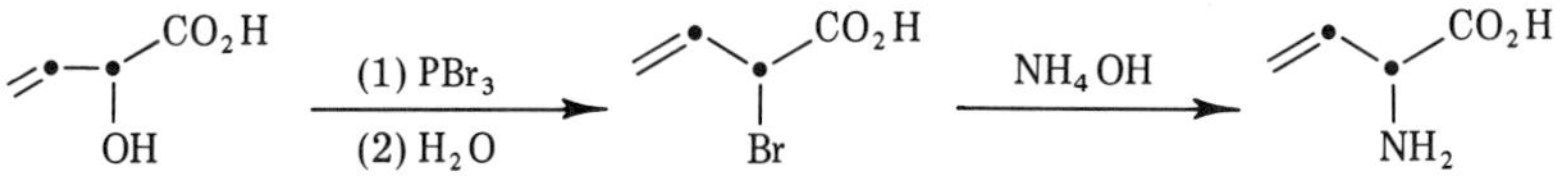

CO$_2$H

OH
$\xrightarrow[\text{(2) H}_2\text{O}]{\text{(1) PBr}_3}$
CO$_2$H

Br
$\xrightarrow{\text{NH}_4\text{OH}}$
CO$_2$H

NH$_2$
$$(11)$$

2-Hydroxy-3-butenoic acid, 3 mmoles (306 mg), dissolved in 1 ml of dry ether is added slowly with stirring to a solution of 1.67 mmoles (450 mg) of phosphorus tribromide in 2 ml of dry ether and 20 μl of pyridine in a 10-ml dry single-necked flask fitted with a dropping funnel. After 20 min at room temperature, 5 ml of water are added and the mixture is stirred for an additional hour. The ether layer is separated, dried over anhydrous sodium sulfate, and evaporated to dryness. The 2-bromoacid is not purified but is directly added to 9 ml of cold (5°), concentrated ammonium hydroxide. The solution is allowed to warm up to room temperature and allowed to remain there with stirring for 24 hr. The ammonia is removed with a rotary evaporator, and the remaining aqueous solution is lyophilized to dryness. The resulting white solid is desalted on an Amberlite IR-120 HCP resin (H⁺ form). After thorough washing with distilled water, the vinyl glycine may be eluted with 2 M NH₄OH and taken to dryness. The vinyl glycine, recrystallized from water–ethanol mixtures to afford exquisite white needles, melts at 215° with decomposition (yield 44% based on consumed 2-hydroxy-3-butenoic acid). Vinyl glycine has an R_f value (descending) of 0.21 on Whatman No. 1 paper with butanol–water–acetic acid (18:5:2). The amino acid shows color changes of yellow to gray to purple with ninhydrin. This color change is characteristic of simple β, γ-unsaturated amino acids. Vinyl glycine can be quantitatively reduced with hydrogen to afford DL-2-aminobutanoic acid.

p-Nitrophenyl Diazoacetate[16]

A solution of 48.4 mg of *p*-nitrophenyl chloroformate is added dropwise to 10 ml of a cold, stirred solution of diazomethane in ether (at least 4 mmoles of diazomethane, or a 16-fold excess, are used). The solution is allowed to stand overnight at 4°. Excess diazomethane is removed in a stream of nitrogen, and the ether is evaporated under reduced pressure. The yellow oil that remains is dissolved in 2 ml of benzene and chromatographed on a column (1 × 5 cm) containing 4 g of Woelm Grade IV alumina. The first fraction eluted by benzene weighs 48 mg. This fraction is recrystallized from 1 ml of chloroform plus 10 ml of hot hexane. The melting point is usually about 89°–90°, although 92°–94° has been obtained.

Phenyldiazoacetate, 2-chlorethyldiazoacetate, phenylthiol diazoacetate, and *p*-nitrophenylthiol diazoacetate have been synthesized by this procedure.

Conclusions

It is now clear that mechanism-based irreversible enzyme inhibitors are more than an enzymological curiosity and can be useful as general selective irreversible enzyme inhibitors. This is not to say that there are not many pitfalls in the successful design of these inhibitors. However, they are likely to be more specific than affinity labeling agents, so that they will fill a gap in studies requiring inhibition of a given enzyme *in vivo* or in crude extracts. As general reagents, these inhibitors have been most successfully applied to the inhibition of enzymes that function by covalent catalysis, especially pyridoxal phosphate and flavin-linked enzymes. This is as expected, since in each of these categories the reactive intermediate is allowed sufficient time to engage in chemical reactions with new neighboring residues.

[4] Syncatalytic Enzyme Modification: Characteristic Features and Differentiation from Affinity Labeling

By WALTER BIRCHMEIER *and* PHILIPP CHRISTEN

The term "syncatalytic modification" has been proposed for modifications of enzymes by group-specific reagents that are facilitated when the enzyme is in the "working" state, i.e., is undergoing substrate turnover. By definition, the increase in reactivity of particular protein side-chain groups underlying syncatalytic modification is linked to the occurrence and the course of the covalent phase of catalysis. Mechanistically, the enhanced susceptibility toward modification is thought to be due to either increased exposure to the solvent or other alterations in the microenvironment consequent to catalysis-linked adaptations of the enzyme conformation.[1]

In this chapter two aspects of syncatalytic modification are discussed: (1) its application as an approach to the dynamic conformational behavior of enzymes during catalysis; (2) the possible interference of syncatalytic modification with affinity labeling of enzymes and the differentiation of the two types of enzyme labeling.

General Experimental Procedure for Syncatalytic Modification

Using group-specific reagents, the surface of the enzyme is probed for side-chain groups that acquire increased reactivity when the enzyme in the presence of the substrate is passing through the covalent phase of catalysis. Although changes in protein conformation, syncatalytic ones included, may manifest themselves either as an increase or a decrease in the reactivity of a particular side chain (see Citri[2]), modification by definition confines itself to the study of reactivity enhancement. This event is easier to interpret since it does not require differentiation from direct steric hindrance of modification by enzyme-bound substrate. In principle,

[1] P. Christen, *Experientia* **26**, 337 (1970).
[2] N. Citri, *Adv. Enzymol.* **37**, 397 (1973).

any of the group-specific reagents suitable for protein modification may be employed. When oxidants are used the possibility of the occurrence of paracatalytic enzyme modification should be kept in mind; differentiation of syncatalytic from paracatalytic modification is noted elsewhere in this volume.[3]

The reactivity of enzyme–substrate intermediates toward side-chain modifying reagents is preferably examined with an equilibrium mixture of substrate and product, thus preventing a shift in the relative concentrations of the different enzyme–substrate intermediates during an experiment. Substrate analogs that pass beyond the adsorption complex, but only through a limited segment of the total enzymic pathway, may prove a valuable means for ascribing alterations in reactivity to individual steps. In order to maximize their effects on the reactivity of enzyme side chains, substrates, products, and substrate analogs should always be used at saturating concentrations. One of the main problems attending the design and the interpretation of syncatalytic modification experiments is the differentiation of syncatalytic increase in reactivity, i.e., the increase accompanying the actual catalytic steps, from ligand-induced increase in reactivity, i.e., the increase concomitant with the binding of substrate or of a nonfunctional substrate analog. The productive enzyme–substrate adsorption complex cannot, for trivial reasons, be examined in isolated form, i.e., without the presence of the ensuing enzyme–substrate intermediates. It thus has to be substituted by complexes formed with nonfunctional substrate analogs whose mode of binding approximates that of the substrate, but which do not form any intermediate beyond the adsorption complex and resist all covalent alteration. Competitive inhibitors or, in the case of enzymes catalyzing two-substrate reactions according to a double-displacement mechanism, the respective unproductive substrate, may serve this role, as exemplified by aspartate aminotransferases.

Syncatalytic Modification of Aspartate Aminotransferases

The specific features of syncatalytic modification are illustrated by studies on cytosolic aspartate aminotransferase from pig heart.[4-6] In this enzyme the reactivity of cysteinyl residue 390 toward sulfhydryl reagents changes within a range of two orders of magnitude during the course of catalysis (see the table). In the free pyridoxal form of the

[3] This volume [5].
[4] W. Birchmeier and P. Christen, *J. Biol. Chem.* **248**, 1751 (1973).
[5] W. Birchmeier, P. E. Zaoralek, and P. Christen, *Biochemistry* **12**, 2874 (1973).
[6] K. J. Wilson, W. Birchmeier, and P. Christen, *Eur. J. Biochem.* **41**, 471 (1974).

SYNCATALYTIC CYSTEINYL MODIFICATIONS OF ASPARTATE AMINOTRANSFERASES: RATE OF
MODIFICATION IN DIFFERENT TRANSAMINATION INTERMEDIATES

Conditions for modification[a]	Transamination intermediates formed[b] PLP-enzyme ⇌ Adsorption complex ⇌ Aldimine intermediate ⇌ Semiquinoide intermediate ⇌ Ketimine intermediate ⇌ Adsorption complex ⇌ PMP-enzyme	Relative rate of modification		
		Cytosolic enzyme from pig[c]	Mitochondrial enzyme from pig[d]	Mitochondrial enzyme from chicken[d]
PLP-enzyme without substrate		1.0	1.0	1.0
PLP-enzyme + 2-keto-glutarate	(unproductive)	2.5	1.0	1.6
PLP-enzyme + 2-methyl-aspartate		10.8	11.8	10.5
PLP-enzyme + erythro-3-hydroxyaspartate		5.8	13.2	13.3
Enzyme + glutamate + 2-ketoglutarate		83.3	3.0	4.6
PMP-enzyme + glutamate	(unproductive)	6.7	1.8	2.3
PMP-enzyme without substrate		6.7	1.5	1.9

[a] PLP pyridoxal-5′-P; PMP pyridoxamine-5′-P. Substrate and substrate analogs were used at saturating concentrations. Length of rule indicates reactions occurring.

[b] For a review of enzymic transamination, see A. Braunstein, in "The Enzymes" (P. D. Boyer, ed.), 3rd ed., Vol. 9, p. 379. Academic Press, New York, 1973.

[c] Modifying reagent: N-ethylmaleimide [W. Birchmeier and P. Christen, J. Biol. Chem. 248, 1751 (1973)].

[d] Modifying reagent: 5,5′-dithiobis-(2-nitrobenzoate) [H. Gehring and P. Christen, Biochem. Biophys. Res. Commun. 63, 441 (1975)].

enzyme the rate of modification of Cys-390 is at a minimum. The rate increases as the enzyme passes through the consecutive enzyme–substrate intermediates and decreases again when the free pyridoxamine form is reached. The rate of Cys-390 modification is fastest when, in the presence of a substrate pair (glutamate plus 2-ketoglutarate or aspartate plus oxaloacetate), covalent enzyme–substrate intermediates are formed. Since the first two of the three principal covalent intermediates of enzymic transamination failed to show high reactivity when formed separately with suitable substrate analogs (2-methylaspartate and *erythro*-3-hydroxyaspartate), the maximum reactivity of Cys-390 appears to be reached in the third covalent intermediate, the ketimine.

Recently, these studies were extended to aspartate aminotransferases of other origin, revealing the existence of syncatalytically sensitive amino acid residues in these enzymes as well[7,8] (see the table). In cytosolic aspartate aminotransferase from pig heart, the modification of Cys-390 with bulky or charged substituents markedly reduces enzymic activity. Substitution with the small and uncharged cyano group has, however, virtually no effect on activity, indicating clearly that Cys-390 is not functional.[4] In the mitochondrial isoenzymes examined, the modification of the syncatalytically responsive cysteinyl residue does not impair the enzymic activity at all, even when bulky substituents are present.[7] Therefore, the syncatalytically modified mitochondrial enzymes allowed an important additional experiment. With a reversible sulfhydryl reagent, the rate of release of the substituent from the modified enzyme was found to be altered syncatalytically in a manner quite analogous to that of the rate of the primary syncatalytic modification.[9] Correspondence of the syncatalytic alterations in rate of both the modification and the demodification supports the interpretation of the reactivity changes as due to conformational adaptations of the enzyme. This interpretation has gained additional affirmation from independent experimental approaches. Thus, spectroscopic observation of the coenzyme chromophore,[10] differential denaturation with urea,[11] and recent hydrogendeuterium exchange experiments[12] have confirmed the occurrence of substantial syncatalytic conformational adaptations in aspartate aminotransferases.

[7] H. Gehring and P. Christen, *Biochem. Biophys. Res. Commun.* **63**, 441 (1975).
[8] V. M. Kochkina and Y. M. Torchinsky, *Biochem. Biophys. Res. Commun.* **63**, 392 (1975).
[9] H. Gehring and P. Christen, *Experientia* **32**, 768 (1976).
[10] A. L. Bocharov, G. A. Kogan, and M. Y. Karpeisky, *Mol. Biol.* **8**, 443 (1974).
[11] V. I. Ivanov, A. L. Bocharov, M. V. Volkenstein, M. Y. Karpeisky, S. Mora, E. I. Okina, and L. V. Yudina, *Eur. J. Biochem.* **40**, 519 (1973).
[12] K. Pfister, J. H. R. Kägi, and P. Christen, *Experientia* **32**, 773 (1976).

Interference of Syncatalytic Modification with Affinity Labeling and Differentiation of the Two Types of Enzyme Labeling

Although differing in mechanism, syncatalytic modification and affinity labeling have one feature in common: the simultaneous interaction of the enzyme with a substrate moiety and with a reagent moiety. Syncatalytic modification requires two separate types of molecules to interact with the enzyme: a substrate to convert it into more reactive enzyme–substrate intermediates and a modifying reagent to gauge side-chain reactivity. An affinity labeling reagent, on the other hand, characteristically combines the two functions, that of substrate and that of reagent, in a single molecule. Therefore, it is possible that an experimental procedure designed to achieve affinity labeling of an enzyme might inadvertently result in its syncatalytic modification. The ambiguity arises from the possibility that some of the molecules of the affinity label might act in the role of substrate while others might function as a nonactive site-directed reagent modifying the enzyme syncatalytically. An example of such ambiguous behavior of a presumed affinity label has indeed been found in the reaction of aspartate aminotransferase and bromopyruvate.[13]

The inactivation of aspartate aminotransferase by this substrate analog had previously been interpreted[14] to result from affinity labeling (Scheme 1). According to this model, bromopyruvate molecules bound

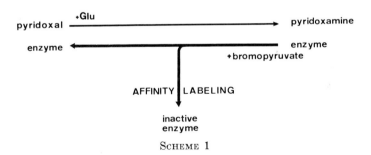

SCHEME 1

to the active site of the pyridoxamine form of the enzyme would either, as a substrate, convert it to the pyridoxal form or, as an affinity label, alkylate a neighboring functional side chain. For progressive affinity labeling to occur, the presence of an amino acid substrate, e.g., glutamate, would be required in order to convert nonalkylated enzyme molecules back to the starting form, the pyridoxamine enzyme. Indeed, the enzyme was

[13] W. Birchmeier and P. Christen, *J. Biol. Chem.* **249**, 6311 (1974).
[14] M. Okamoto and Y. Morino, *J. Biol. Chem.* **248**, 82 (1973).

found to be inactivated by bromopyruvate only in the presence of amino acid cosubstrates, such as glutamate.[13,14] However, in a reexamination of this reaction with the cytosolic isoenzyme of aspartate aminotransferase from pig,[13] the presence of a second keto acid substrate in addition to bromopyruvate (e.g., 2-ketoglutarate in concentrations of up to 50 times K_m) failed to inhibit inactivation by bromopyruvate, the latter being used in a concentration of 3 times K_m. Apparently, attachment of bromopyruvate to the substrate binding site is not a prerequisite for inactivation to occur. This conclusion was confirmed by a comparison of the kinetic properties of bromopyruvate transamination and enzyme inactivation by bromopyruvate. Transamination followed saturation kinetics whereas the rate of inactivation increased linearly with increasing concentration of bromopyruvate up to 7 times K_m. Furthermore, inactivation of the enzyme was shown to be accompanied by selective modification of the syncatalytically modifiable Cys-390.

These data unequivocally excluded the action of bromopyruvate as an affinity label in the presence of a dicarboxylic amino acid cosubstrate. They are, however, fully compatible with syncatalytic modification (Scheme 2). According to this alternative model nonidentical bromo-

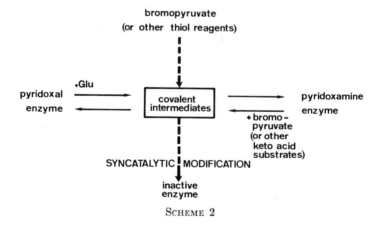

SCHEME 2

pyruvate molecules serve as a substrate, on the one hand, and as an alkylating agent on the other. In its function as substrate, bromopyruvate together with a dicarboxylic amino acid forms the covalent enzyme–substrate intermediates. In these intermediates Cys-390 acquires its markedly enhanced reactivity and, consequently, is alkylated syncatalytically by bromopyruvate molecules not attached to the active site.

However, if reaction conditions are altered, bromopyruvate may act as an affinity label of aspartate aminotransferase. Recent experiments of

Morino and collaborators[15] have indicated that a residue other than Cys-390 is affinity labeled when the normal dicarboxylic amino acid cosubstrate is replaced by alanine in the presence of formate.

Similarly ambiguous behavior of an affinity label might also arise if its mere binding to the enzyme were to induce an increase in reactivity of enzyme side chains. To our knowledge no such case has been reported. However, the possibility that both ligand-induced and syncatalytic modifications might interfere with affinity labeling emphasizes the importance of proper control experiments. Testing for inhibition of the rate of labeling by the normal, nonmodifying substrate of the enzyme is the most important control. Hitherto, this control has been applied in order to estimate the reactivity of nonactive-site residues toward the modifying moiety of the affinity label. However, the available data indicate that values of reactivity toward a modifying agent, affinity labels included, as determined in the presence of a particular ligand, e.g., the substrate, might critically differ from those pertaining to another ligand, such as a substrate analog. The other criteria that were applied to the reaction of aspartate aminotransferase and bromopyruvate, i.e., kinetic characteristics of the modification and the nature of the modified amino acid residue, might prove to be indispensable for unequivocal interpretation of the experimental findings.

Concluding Remarks

Much of the present knowledge on conformational adaptations of enzymes has been obtained from experiments with nonfunctional ligands.[2] Mapping of the enzyme surface for syncatalytic changes in reactivity of side-chain groups, as exemplified with aspartate aminotransferases, would seem to provide a supplementary approach to the dynamic behavior of enzymes during the catalytic bond-making and bond-breaking.

Similar to affinity labeling, syncatalytic modification is based on specific interactions of the enzyme with the substrate or a substrate analog, respectively. However, the mechanisms leading to a selective chemical modification of a specific enzyme side-chain group are different. In affinity labeling it is ensured by the high concentration of the reagent at the active site, in syncatalytic modification by a catalysis-synchronous increase in reactivity of the group concerned. The necessity to differentiate between affinity labeling, on the one hand, and ligand-induced or syncatalytic modification on the other, exists in any experiment employing a

[15] A. M. Osman, T. Yamano, and Y. Morino, *Biochem. Biophys. Res. Commun.* **70,** 153 (1976).

substrate analog with a protein-modifying group. This task may not be an easy one, and it is conceivable that the two types of modification may coexist. The recent work of Morino and collaborators[15] (see above) provides a very informative illustration of a substrate analog that acts as a syncatalytic modifier or as an affinity label, depending only on seemingly slight alterations of reaction conditions.

[5] Paracatalytic Enzyme Modification by Oxidation of Enzyme–Substrate Carbanion Intermediates

By PHILIPP CHRISTEN

The catalytic effect of enzymes is based in part on their capacity to endow the substrates with highly increased chemical reactivity. This superreactivity of enzyme-activated substrates manifests itself in the increased rate of the chemical transformation of the specific substrate into the specific product; it may, however, also comprise enhanced reactivity toward extrinsic reagents that are not constituents of the normal enzyme–substrate system. Enzyme–substrate intermediates may thus react with extrinsic reagents and branch off the normal catalytic pathway. The term "paracatalytic enzyme reaction" is suggested to describe these reactions that, except in the first stages of their pathway, do not correspond with the normal catalytic specificity of the enzyme.

Paracatalytic enzyme reactions on their part may entail "paracatalytic enzyme modifications."[1] This new mode of catalysis-linked modification is due to intramolecular reactions of unusual metastable intermediates formed during the paracatalytic reactions. It appears to afford an approach to specific active-site-directed labeling of certain enzymes. The present article describes the principal features of paracatalytic enzyme modification accompanying the oxidation of enzyme–substrate carbanion intermediates, discusses its differentiation from other substrate-dependent and catalysis-linked modifications, and concludes with an account of detailed modification procedures for several enzymes.

Paracatalytic Oxidation of Enzyme-Substrate Carbanion Intermediates

Enzyme–substrate carbanions such as formed consequent to the cleavage of C–C or C–H bonds are susceptible to oxidation by electron

[1] P. Christen, M. Cogoli-Greuter, M. J. Healy, and D. Lubini, *Eur. J. Biochem.* **63**, 223 (1976).

acceptors, e.g., oxidation-reduction indicators of relatively high redox potential,[2] tetranitromethane,[3,4] or H_2O_2.[5] Thus, in the presence of a suitable extrinsic electron acceptor, the carbanion intermediate is partitioned between the normal enzymic reaction pathway and an additional oxidative pathway leading to a new product.

Scheme 1 shows, as an example, the reactions of fructose-1,6-bisphos-

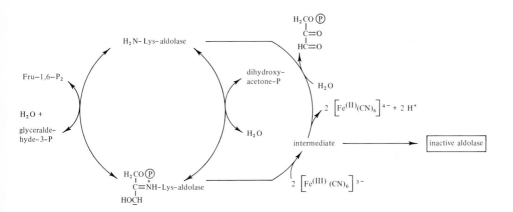

SCHEME 1

phate aldolase from rabbit muscle in the presence of substrate and hexacyanoferrate(III) (a recent review of the reaction mechanism of aldolase is available[6]). The carbanion intermediate formed by cleavage of the C3–C4 bond of fructose-1,6-bisphosphate or, in the reverse reaction, by deprotonation at C3 of dihydroxyacetone phosphate, is readily oxidized by hexacyanoferrate(III).[2] In the presence of 0.5 mM hexacyanoferrate(III) 4% of the aldolase-dihydroxyacetone phosphate carbanion intermediate formed with fructose-1,6-bisphosphate as substrate is oxidized and bypasses the aldol cleavage–condensation cycle. Dihydroxyacetone phosphate is thereby oxidized to the paracatalytic product, the corresponding ketoaldehyde, i.e., hydroxypyruvaldehyde phosphate. The oxidation occurs stoichiometrically: 1 mole of dihydroxyacetone phosphate is consumed per mole of hydroxypyruvaldehyde phosphate pro-

[2] M. J. Healy and P. Christen, *Biochemistry* **12**, 35 (1973).
[3] P. Christen and J. F. Riordan, *Biochemistry* **7**, 1531 (1968).
[4] J. F. Riordan and P. Christen, *Biochemistry* **8**, 2381 (1969).
[5] P. Christen, T. K. Anderson, and M. J. Healy, *Experientia* **30**, 603 (1974).
[6] B. L. Horecker, O. Tsolas, and C. Y. Lai, *in* "The Enzymes" (P. D. Boyer, ed.), 3rd ed., Vol. 7, p. 213. Academic Press, New York, 1972.

duced, and 2 moles of hexacyanoferrate(III) are reduced to hexacyano-ferrate(II).[2,7]

Paracatalytic Enzyme Modification

Concomitant to the oxidation of the carbanion intermediate, aldolase is progressively inactivated. The inactivation is strictly linked to the paracatalytic oxidation; with hexacyanoferrate(III) alone or with substrate alone, only an insignificant decrease in enzymic activity ensues. Likewise, the products of the reaction, hydroxypyruvaldehyde phosphate or hexacyanoferrate(II), fail to induce rapid inactivation. The kinetic features of the inactivation[1] indicate that it is due to a dead-end side reaction of the oxidative bypass (Scheme 1). With hexacyanoferrate(III) as oxidant, one out of every 130 aldolase molecules passing through the oxidative cycle is inactivated. The rate of inactivation is independent of enzyme concentration. Apparently, an unusually reactive substrate derivative of as yet unidentified nature is formed during the oxidation of the substrate carbanion and reacts *in situ* without being released from the active site with a neighboring group of the enzyme; paracatalytic modification is thus thought to label a side chain at or near the active site.[1]

Similar enzyme inactivation accompanying the oxidation of an enzyme–substrate carbanion intermediate also occurs with class II fructose-1,6-bisphosphate aldolase from yeast, transaldolase, transketolase, and pyruvate decarboxylase from yeast (see the table).

The properties of the inactivated enzymes are consistent with covalent modifications. Enzymic activity is not restored by gel filtration or extensive dialysis. In the case of muscle aldolase, radioactivity has been found to be incorporated when uniformly labeled [^{14}C]fructose-1,6-bisphosphate is used as substrate and hexacyanoferrate(III) as oxidant; the amount of radioactive material incorporated approximately corresponds to one 3-carbon fragment per subunit.[8]

Characteristically, paracatalytic substrate oxidation and inactivation occur only in the presence of the specific substrate of the enzyme. However, both processes are unspecific with respect to the oxidant. In fact, hexacyanoferrate(III), 2,6-dichlorophenolindophenol, porphyrindin, tetranitromethane, and H_2O_2 have been successfully employed with various enzymes (see the table). The oxidants differ considerably with respect to the ratio of the rate of oxidation of the carbanion intermediate to the rate of enzyme inactivation, e.g., the inactivation of transaldolase in the

[7] M. J. Healy and P. Christen, *J. Am. Chem. Soc.* **94**, 7911 (1972).
[8] D. Lubini and P. Christen, in preparation.

TABLE

PARACATALYTIC MODIFICATION OF VARIOUS ENZYMES. CONDITIONS AND EFFECT ON ENZYMIC ACTIVITY[a]

Enzyme	Substrate	Oxidant	Buffer	Reaction time (min)	Enzymic activity % of original	
					With substrate	Without substrate
Fructose-1,6-P_2 aldolase from rabbit muscle,[b] 0.1 unit/ml	Fructose-1,6 P_2, 2 mM	Hexacyanoferrate(III), 0.5 mM[d]	Tris-Cl, 100 mM, pH 7.6	60	7	80
Fructose-1,6-P_2 aldolase from yeast,[b] 4 units/ml	Fructose-1,6 P_2, 1.5 mM	Hexacyanoferrate(III), 1.25 mM	Glycylglycine, 50 mM/potassium acetate, 125 mM, pH 7.5	50	20	65
Transaldolase,[b] 1.1 units/ml	Fructose-6 P, 3 mM	Hexacyanoferrate(III), 0.5 mM[e]	Triethanolamine-Cl, 100 mM/EDTA, 10 mM, pH 7.6	30	12	90
Transketolase,[b] 0.25 unit/ml	Fructose-6 P, 3 mM	Hexacyanoferrate(III), 0.5 mM[d]	Potassium phosphate, 10 mM/KCl, 100 mM/MgSO$_4$, 2 mM/thiamine-PP, 0.1 mM, pH 7.4	30	15	48
Pyruvate decarboxylase from yeast,[c] 0.3 unit/ml	Pyruvate, 30 mM	Hexacyanoferrate(III), 0.5 mM[f]	Sodium citrate, 200 mM, pH 6.0	60	25	68

[a] All experiments were performed at 25°.

[b] P. Christen, M. Cogoli-Greuter, M. J. Healy, and D. Lubini, *Eur. J. Biochem.* **63**, 223 (1976).

[c] M. Cogoli, U. Hausner, and P. Christen, *Experientia* **32**, 766 (1976).

[d] Paracatalytic inactivation was also observed with H_2O_2 as oxidant.

[e] Paracatalytic inactivation was also observed with H_2O_2, porphyrindin, or 2,6-dichlorophenolindophenol.

[f] Paracatalytic inactivation was also observed with tetranitromethane, porphyrindin, or 2,6-dichlorophenolindophenol.

presence of fructose 6-phosphate occurs much faster when tetranitromethane instead of hexacyanoferrate(III) is used as an oxidant, allowing
the enzyme to perform only a few catalytic cycles.[1,9]

Differentiation from Other Types of Substrate-Dependent and Catalysis-Linked Enzyme Modifications

The facilitating effect of substrate is not an exclusive feature of paracatalytic enzyme modification. Although the substrate generally protects
the enzyme toward inactivation by chemical modification, numerous cases
have been observed in which the binding of the substrate or a nonfunctional substrate analog enhances the susceptibility of particular protein
side-chain groups toward a modifying reagent.[10] Such ligand-induced
increase in reactivity of particular groups is generally thought to reflect
conformational adaptations of the protein to the binding of the ligand.
Moreover, recent studies with aspartate aminotransferases have demonstrated that the reactivity of particular groups of these enzymes does not
increase on formation of the enzyme–substrate adsorption complex but
increases markedly during the subsequent covalent phase of catalysis
allowing the selective syncatalytic modification of these groups.[11,12]

The crucial criterion for verifying a paracatalytic modification, i.e.,
for its differentiation from ligand-induced or syncatalytic modification,
is its strict linkage to the underlying paracatalytic oxidation of the substrate. Paracatalytic modification may be diagnosed if appropriate control experiments, under conditions precluding paracatalytic oxidation,
fail to result in enzyme modification. Thus, substrate analogs that do not
form an oxidation-susceptible intermediate must not induce the modification of the enzyme in the presence of the oxidant. The most stringent
controls are those with substrate analogs that interact with the enzyme
in a manner similar to that of the substrate yet without forming the
oxidizable intermediate. In the case of fructose-1,6-bisphosphate aldolase,
dihydroxyacetone sulfate is such an analog of dihydroxyacetone phosphate. It forms a Schiff base with the active-site lysyl residue but is not
deprotonated.[13] Hence, it does not form the oxidizable intermediate[14]
and, in fact, fails to induce modification of aldolase.[1] Since the oxidizing

[9] K. Kobashi and K. Brand, *Arch. Biochem. Biophys.* **148**, 169 (1972).

[10] N. Citri, *Adv. Enzymol.* **37**, 397 (1973).

[11] W. Birchmeier, K. J. Wilson, and P. Christen, *J. Biol. Chem.* **248**, 1751 (1973).

[12] This volume [4].

[13] E. Grazi, C. Sivieri-Pecorari, R. Gagliano, and G. Trombetta, *Biochemistry* **12**, 2583 (1973).

[14] E. Grazi, *Biochem. Biophys. Res. Commun.* **56**, 106 (1974).

agents used for paracatalytic oxidation react preferentially with sulfhydryl groups of proteins,[15,16] the reactions of nonoxidizing sulfhydryl reagents with the enzyme are additional important control experiments. Again, the nonoxidizing reagents must fail to modify the enzyme in a catalysis-promoted manner.

Experimental Procedure for Paracatalytic Enzyme Modification

All paracatalytic enzyme modifications with oxidizing reagents examined to date (see the table) have been performed in the range of the pH optimum of the enzyme. The rate of paracatalytic modification is independent of enzyme concentration which, however, should not exceed a certain limit in order to ensure a sufficient concentration of substrate for the duration of paracatalytic modification. This consideration is particularly applicable if the product of the enzymic reaction does not form the oxidizable intermediate with the enzyme, i.e., if the reaction is not reversible. Similar precautions apply to the choice of concentration of the substrate. Furthermore, in order to achieve the maximum rate of paracatalytic enzyme modification, the substrate should be present at saturating concentration. If there is reason for a high enzyme concentration, and, consequently, the concentration of substrate and/or oxidant cannot be sufficiently high to last for complete modification, substrate and oxidant may be added repeatedly or the enzyme may be confined to a dialysis bag in a larger reservoir of substrate and oxidant solution.

At present hexacyanoferrate(III) is the most widely applicable oxidant (see the table). Oxidants are generally used at a concentration of 0.5 mM, thus avoiding excessive oxidation of sulfhydryl groups and other side chains of the enzyme while still allowing the paracatalytic modification to proceed at an acceptably rapid rate. In some instances, the inactivation of the enzyme may be faster in the absence of substrate than in the presence of oxidant plus substrate. In such instances, direct oxidation of essential amino acids, cysteinyl residues in particular, apparently precedes paracatalytic modification.

Preferably, the paracatalytic modification is started by the addition of the enzyme to a solution of substrate and oxidant. The oxidation of the carbanion intermediate may then be followed by measuring photometrically the reduction of the redox indicator to its leuco form.[2] The progress of paracatalytic modification is most conveniently estimated by determining the concomitant decrease in enzymic activity. The reaction

[15] E. S. G. Barron, *Adv. Enzymol* **11**, 201 (1951).
[16] F. W. Putnam, *in* "The Proteins" (H. Neurath and K. Bailey, eds.), Vol. IB, p. 807. Academic Press, New York, 1953.

may be stopped by gel filtration, dialysis, or precipitation with trichloroacetic acid.

Concluding Remarks

Paracatalytic enzyme modification is a new type of catalysis-linked and, hence, substrate-dependent enzyme modification. In all instances in which the substrate promotes inactivation of an enzyme by a chemical reagent, particularly by an oxidant, paracatalytic modification should be considered to be the underlying mechanism. In contrast to ligand-induced and syncatalytic modifications, paracatalytic modification involves a *direct* chemical interaction between enzyme-activated substrate and extrinsic reagent. In this respect, it is similar to the chemical trapping of covalent enzyme–substrate intermediates, e.g., the reduction of enzyme-substrate Schiff bases by sodium borohydride in class I fructose-1,6-bisphosphate aldolases[6,17] or in acetoacetate decarboxylase.[18]

Potentially, paracatalytic enzyme modification might afford a highly specific method for the active-site-directed chemical modification of certain enzymes. In contrast to other types of active-site-directed modifications, such as affinity labeling and modification by enzyme-activated inhibitors,[19] paracatalytic modification involves a binary combination of reagents, the *normal* substrate and the trapping reagent as a second component. Its specificity derives from both the binding and the reaction specificity of the enzyme for its *normal* substrate.

[17] C. Y. Lai, P. Hoffee, and B. L. Horecker, this series, Vol. 11 [77], 1967.
[18] I. Fridovich and F. H. Westheimer, *J. Am. Chem. Soc.* **84**, 3208 (1962).
[19] See this volume [3].

[6] Catalytic Competence: A Direct Criterion for Affinity Labeling[1]

By ERNEST V. GROMAN, RICHARD M. SCHULTZ, and LEWIS L. ENGEL

Affinity labels or active-site-directed irreversible enzyme inhibitors are designed by combining in the same molecule a moiety that resembles the substrate for an enzyme and a reactive group that will attack some amino acid residue in the protein with the formation of a covalent bond. The

[1] This is Publication No. 1514 of the Cancer Commission of Harvard University. Supported by Grant CA01393 from the National Cancer Institute, United States Public Health Service.

complementarity between the substratelike moiety of the affinity label and the binding site of the enzyme promotes, but does not necessarily determine, the binding of the compound at or near the site normally occupied by substrate. The amino acid residue attacked may be one directly involved in catalysis. It may also be a residue in the neighborhood of the active site, so that when the covalent link with the affinity label is formed, the substratelike moiety either rests in the substrate binding site or is sufficiently close so that substrate is blocked from access to its binding site and the enzyme is thus inactivated. The success of affinity labeling depends upon the fortuitous presence of a reactive amino acid residue at such a distance from the substrate binding site that the covalently bound substratelike moiety can rest either in the active site or very close to it.

Several criteria have been established for the definition of affinity labels. The first is that the compound be either a substrate for the enzyme or closely resemble a substrate. This is a necessary but not a sufficient condition for concluding that the label is at the substrate binding site.

A second criterion is that one molecule of label should be bound for each active site of the enzyme. If the labeling reagent contains a sensitive linkage joining the reactive moiety with the substratelike moiety, stoichiometry must be established for both parts of the molecule. Control experiments must be performed to establish that the substratelike portion of the label is required for enzyme inactivation to occur.

A third criterion is that the addition of the real substrate to the reaction linking the affinity label covalently to the protein should diminish the rate of inactivation, which, in the presence of excess labeling reagent, is usually pseudo-first order.

Affinity labeling normally takes place in two steps; the first is the reversible formation of a Michaelis–Menten complex, and the second, the irreversible reaction linking the labeling reagent covalently to the protein. Therefore, a plot of the rate constant for inactivation against affinity label concentration will reach a plateau. Demonstration of saturation kinetics constitutes the fourth criterion.

Satisfaction of these four criteria provides indirect evidence that the affinity label is located in the substrate binding site. In no case, however, do they, singly or in combination, constitute direct proof. For instance, it is possible that the enzyme may have a second, catalytically inactive site, that binds substrate and forms a covalent link with the affinity label leading to inactivation, while the catalytically active site does not have an amino acid residue correctly positioned for covalent linkage. In the case of a steroid alcohol dehydrogenase, a possible example is the binding of a steroid in the hydrophobic region normally occupied by the adenine portion of the pyridine nucleotide.

To eliminate the ambiguities inherent in the four criteria listed, we have proposed a fifth criterion for affinity labeling, catalytic competence, which is defined as the ability of the affinity labeled enzyme to catalyze its normal reaction on its covalently bound ligand. Catalytic competence therefore constitutes a direct method of demonstrating that labeling is occurring in the binding site with proper geometry. When the reaction catalyzed results in the generation or destruction of a chiral center, it is important to establish that the stereochemical course of the reaction is the same as that in the normal reaction.

We have studied the reaction of the potential affinity label, 3-bromo-acetoxyestrone, with human placental estradiol dehydrogenase and have found that it satisfies the first four criteria for an affinity label. Further-more, the 17-keto group of this compound offers an opportunity to deter-mine whether, after reaction with the enzyme, the covalently bound steroid is suitably positioned to serve as a substrate in the dehydrogenase reaction. We provide evidence that such is indeed the case.

Synthesis of the Affinity Reagent
Bromo[1-^{14}C]acetoxy[2,4,6,7-H^3]estrone

Estrone (280 mg) is dissolved with 1 mCi of [2,4,6,7-^{3}H]estrone in 30 ml of methylene chloride and cooled to 0°. To this solution are added in sequence 280 mg of [1-^{14}C]bromoacetic acid dissolved in 10 ml of methylene chloride and 514 mg of dicyclohexylcarbodiimide dissolved in 10 ml of methylene chloride containing 0.1 ml of pyridine. This solution is stirred for 1 hr at 0° and 1 hr at room temperature, filtered, and the solvent evaporated. The residue is dissolved in 5 ml of acetone, filtered, and applied to two 1000-μm preparative silica-gel thin-layer plates. The plates are developed with cyclohexane/ethyl acetate (3:1). The most rapidly migrating band of each plate is removed, and the silica gel is extracted with 100 ml of chloroform. The product is crystallized from cyclohexane/ethyl acetate to yield 155 mg of pure bromo[1-^{14}C]acetoxy [2,4,6,7-^{3}H]estrone (40% yield). The specific activity of the product was 3.44×10^6 dpm of ^{3}H/μmole and 9.8×10^4 dpm of ^{14}C per micromole (m.p., 163; m/e, 390/392). This compound is hydrolyzed about 50% upon storage for 24 hr in 100 mM potassium phosphate (pH 6.5), containing 5 mM EDTA and 20% glycerol at 25°.

Enzymic Conversion of Covalently Bound Estrone to
Covalently Bound Estradiol

We have reported that bromoacetoxyestrone is a substrate for estra-diol dehydrogenase. Bromoacetoxyestrone inactivates this enzyme with pseudo-first-order kinetics and the addition of estradiol or of estradiol 3-methyl ether to bromoacetoxyestrone decreased the rate of inactivation. Inactivation of estradiol dehydrogenase by bromoacetoxyestrone is stoichiometric; 2 moles of bromoacetoxyestrone are bound covalently to 68 kg of enzyme.[2] Finally, a plot of the rate constant for inactivation against affinity label concentration exhibits saturation kinetics (unpublished results). Since 3-bromoacetoxyestrone serves as a substrate for estradiol dehydrogenase, it might be expected that, if the label is indeed at the active site, upon addition of reduced pyridine nucleotide, it would be reduced to covalently bound estradiol. Addition of TPNH or DPNH to covalently labeled enzyme results in the production of covalently bound estradiol in about 50% yield. Furthermore, the stereochemistry of the reduction of covalently bound estrone is the same as that of the native enzyme acting upon free estrone. The 4-*pro-S*-hydrogen of DPNH is transferred to estrone, becoming the 17α-hydrogen of estradiol-17β.[2]

In these experiments, 100% inactivation was not achieved; native enzyme was, therefore, always present. Consequently, it was necessary to determine whether the reduction of estrone was catalyzed by residual native enzyme acting upon enzyme-bound steroid [Eq. (1)] or by modified enzyme acting upon its own covalently bound steroid [Eq. (2)]. These alternatives may be distinguished by exploiting the transhydrogenase function of this enzyme and the fact that diethylstilbestrol inhibits estradiol dehydrogenase and its transhydrogenase function.

$$\text{Enz} + \text{Enz-AcE1} + \text{DPNH} + \text{H}^+ \rightarrow \text{DPN}^+ + \text{Enz} + \text{Enz-AcE2} \qquad (1)$$
$$\text{Enz-AcE1} + \text{DPNH} + \text{H}^+ \rightarrow \text{DPN}^+ + \text{Enz-AcE2} \qquad (2)$$

In the transhydrogenase system, all components for the assay are present except estradiol dehydrogenase and steroid. Addition of affinity labeled estradiol dehydrogenase to the assay mixture simultaneously introduces the potential catalytic agents, enzyme (either native or modified) and steroid (estrone covalently linked to estradiol dehydrogenase).

Diethylstilbestrol greatly inhibits the transhydrogenase function of estradiol dehydrogenase (see the table). Addition of diethylstilbestrol to the complete reaction mixture results in a diminished rate of transhydrogenation (experiment 5); the residual rate observed is attributed to the action of modified estradiol dehydrogenase acting upon its own bound steroid to catalyze the transhydrogenation reaction. The difference in rates is due to residual native estradiol dehydrogenase utilizing protein-bound steroid as substrate. That native estradiol dehydrogenase could catalyze the reduction of estrone covalently bound to the enzyme was shown in experiments 8–10. Treatment of the enzyme bearing estrone with *p*-mer-

---

[2] E. V. Groman, R. M. Schultz, and L. L. Engel, *J. Biol Chem.* **250**, 5450 (1975).

EFFECTS OF DIETHYLSTILBESTROL (DES) AND p-MERCURIBENZOATE ON THE RATE OF THE TRANSHYDROGENASE REACTION[a,b]

Expt. no.	Enzyme preparation assayed	Additions to assay	Velocity (nmole/min)
1	Native	None	0
2	Native	0.1 μM E1	0.48
3	Native	0.1 μM E1 10 μM DES	0.08
4	Enz-Ac E1	None	0.28
5	Enz-Ac E1	10 μM DES	0.26
6	Enz-Ac E1	Native enzyme	0.52
7	Enz-Ac E1	Native enzyme, 10 μM DES	0.35
8	Enz-Ac E1 treated with p-mercuribenzoate	None	0
9	Enz-Ac E1 treated with p-mercuribenzoate	Native enzyme	0.35
10	Enz-Ac E1 treated with p-mercuribenzoate	Native enzyme, 10 μM DES	0

[a] From E. V. Groman, R. M. Schultz, and L. L. Engel, *J. Biol. Chem.* **250**, 5450 (1975).

[b] The transhydrogenase assay was performed by the method of Jarabak.[c] In each experiment, estradiol dehydrogenase,[c,d] equivalent to 4 μg of protein, was added. Enzyme-acetoxyestrone (E-Ac E1) was prepared from estradiol dehydrogenase (0.1 mg/ml) by inactivation (88%) with 3-bromoacetoxyestrone as described in the text. In this case, 2-mercaptoethanol was not added. The sample was then dialyzed against two 1-liter changes of 100 mM potassium phosphate (pH 6.5) containing 20% glycerol at 4° to remove excess steroid. Control enzyme was treated identically, but with the omission of bromoacetoxyestrone. Preparation of p-mercuribenzoate-treated enzyme–acetoxyestrone (experiments 8–10): A portion of enzyme–acetoxyestrone was treated with p-mercuribenzoate (0.1 mM) for 30 min at room temperature. This treatment resulted in the total loss of enzymic activity. The sample was then dialyzed (4°) against 1000 ml of 100 mM potassium phosphate (pH 6.5) containing 20% glycerol.

[c] J. Jarabak, this series, Vol. 15, p. 746 (1969).

[d] D. J. W. Burns, L. L. Engel, and J. L. Bethune, *Biochemistry* **11**, 2699 (1972).

curibenzoate abolished the transhydrogenase reaction (experiment 8). Addition of native enzyme reestablished transhydrogenation (experiment 9) and addition of diethylstilbestrol to this reaction mixture (experiment 10) inhibited transhydrogenation completely. The covalently bound steroid is thus susceptible to oxidoreduction catalyzed either by the enzyme molecule to which it is bound or by native enzyme.

These data show that the steroid molecule in the labeled estradiol dehydrogenase is capable of assuming the geometric relations to enzyme and cofactor that are required for the catalytic event to occur. The affinity label is therefore catalytically competent.

[7] Differential Labeling: A General Technique for Selective Modification of Binding Sites

By ALLEN T. PHILLIPS

Definition and Applications

Given knowledge of the principal steric and contact requirements for binding of a ligand to a protein, specific affinity reagents (reactive ligand analogs) can be designed that will utilize their inherent binding preference to produce covalent alteration of residues associated with the ligand binding site. In the best cases, reactions at unintended sites are absent or minimal because an efficient binding of the analog to its preferred site permits the use of small reagent excesses; subsequent identification of an affected amino acid residue and its sequence position can be attempted in a straightforward manner.

There are, however, instances in which affinity-type reagents are somewhat less selective as a result of weaker binding to the desired site and the possibility for interactions with regions of the protein molecule outside of the desired site of attachment. More commonly, it is not practical to construct suitable reactive analogs, either because of the formidable chemistry involved, or because there are insufficient data available to allow intelligent design. It is for these latter situations that the technique of differential labeling has primarily been used. Differential labeling is intended to be of assistance in enhancing the specificity of group modification, whether employed in connection with an affinity labeling reagent or with reagents of lower selectivity, by utilizing the natural binding properties of a protein to reduce interference from other reactive areas. Differential labeling is achieved through the use of a nonreactive, tightly bound ligand (substrate, coenzyme, or competitive inhibitor) which masks its own binding site against modification and thereby permits reactions to occur only at other reactive sites. Removal of the ligand, followed by retreatment with the same reactive agent in a radioactive form, thus labels only the site originally protected by the ligand.

The armamentarium of the protein chemists contains many group-selective reagents,[1] from which one capable of modifying some residue located at the desired site can often be chosen. Such common reagents

---

[1] A list of 18 references describing a broad range of group-specific reagents and their uses in protein modification is given in the review by D. S. Sigman and G. Mooser, *Annu. Rev. Biochem.* **44**, 889 (1975).

as fluorodinitrobenzene, substituted maleimides, haloacids, acyl imidazoles, imidoesters, and diazonium salts are readily obtained in isotopic form and can attack a range of functional groups with varying degrees of selectivity for the type of group modified. An obvious limitation to their direct use is their potential nonspecificity with respect to the location of residues attacked, but this is circumvented in the differential labeling procedure.

The intent of this chapter is to describe the general approach to differential labeling, explore its limitations, and provide examples of the type of results that can be expected with this technique.

A General Experimental Procedure for Differential Labeling

The differential labeling method, as originally described by Cohen and Warringa[2] and by Koshland et al.,[3] consists of allowing the protein to react with a covalent modification agent in two stages (Fig. 1). Stage 1, hereafter referred to as the prelabeling step, involves incubation of the protein with a large excess of a site-specific ligand or ligand analog, preferably unreactive, as in the case of a competitive reversible inhibitor or a coenzyme; the ligand must have a high affinity for its site under the incubation conditions. It is necessary to establish in preliminary studies

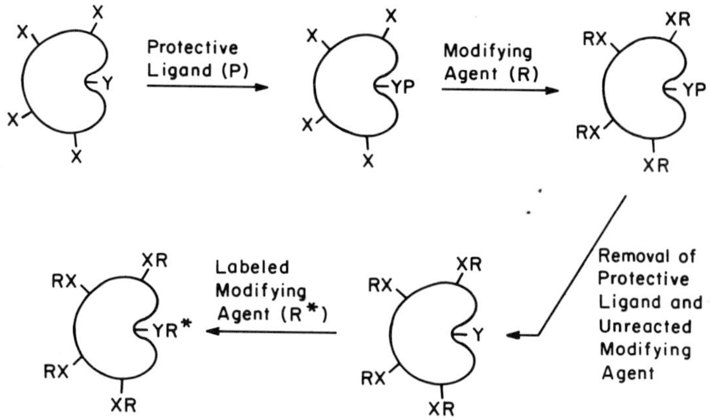

FIG. 1. Schematic illustration of the differential labeling procedure. The ligand-binding residue to be ultimately identified is indicated as Y, and other reactive sites are indicated as X. The actual chemical nature of X and Y groups may be identical or not, so long as they are susceptible to attack by modifying agent R.

[2] J. A. Cohen and M. G. P. J. Warringa, Biochim. Biophys. Acta 11, 52 (1953).
[3] M. E. Koshland, F. Englberger, and D. E. Koshland, Jr., Proc. Natl. Acad. Sci. U.S.A. 45, 1470 (1959).

that the ligand chosen can, in fact, protect the site of interest from inactivation by the reactive agent to be employed. The criteria for adequate protection will be described in the following section. This combination of protein plus protective ligand is then treated with the desired modifying reagent at a concentration equal to or higher (usually the latter) than will be required in the specific labeling step to follow. The complete mixture is incubated for a period of time such that all readily reacted sites are attacked, but not so long that protection of the ligand-binding site is overcome. The time should not be much less than that to be used in the second stage, particularly if concentrations of reactive agent are to be equal in the two stages. Some method for monitoring the degree of protection of the site of interest is extremely helpful during this prelabeling step. This is uncomplicated in the case of an active site but can be more challenging for regulatory or other noncatalytic sites.

To terminate the prelabeling step, several methods can be employed depending on the nature of the modifying reagent used. One simple approach is to add a large excess of an innocuous compound which consumes any remaining modification agent. The reaction mixture can then be exhaustively dialyzed, thereby removing both the added ligand and the defunct modifying agent. In the absence of an appropriate compound for reaction with the modifying agent, dialysis can be employed directly, but initial dialysis fluids should contain the protective ligand to guard against premature ligand removal before modifying agent has been eliminated. Further dialysis then can release the bound ligand.

Stage 2, the specific labeling step, is accomplished by incubation of nonspecifically labeled protein with radioactive modifying reagent at a concentration equal to or less than that used for prelabeling. Aside from this possible change in concentration, all other conditions should be identical to those employed in the initial stage, particularly with respect to pH, temperature, and concentrations of all accessory materials (e.g., buffer, cofactors, etc.), except that the protective ligand is not present. It is advisable when practical to follow the progress of the inactivation by periodic removal of small portions for assay. The reaction is terminated at the desired extent of inactivation by gel filtration, which has the added advantage of allowing an estimate to be made of the amount of isotope incorporated. Assuming that some knowledge exists of the number of specific binding sites, this information is invaluable for assessing the specificity of labeling in the second stage.

Control incubations should be carried through these same steps. The most useful control consists of an identical mixture in which the protective ligand or ligand analog is omitted from the prelabeling step, thereby allowing attack of all sites by nonradioactive modifying agent. The time required for complete loss of specific binding activity in this unprotected control incubation can be used as a crude guide for the expected course of the reaction during stage 2 reaction between radioactive reagent and prelabeled protein, but keeping in mind that rates may be influenced by the extent of prior modification and reagent excess.

Further operations on the control (unprotected) sample are those used for the protected sample. In the specific labeling step which follows, any isotope incorporated by the control sample represents either an exchange process or failure to achieve complete modification of all sites in the pelabeling step. Appreciable incorporation of label in the control sample relative to the normal experimental sample suggests the need to reevaluate the operational design or to choose a different modifying agent.

Other Variations of the Differential Labeling Principle

In recent years, the term differential labeling has also been applied to a simpler technique wherein a radioactive reagent is incubated in parallel with two samples of protein, only one of which contains a protective ligand. A comparison of label uptake in the unprotected and protected samples thereby permits conclusions to be drawn concerning the number of sites protected. This can also lead, upon subtractive amino analysis of the modified protein, to identification of the nature of the group(s) protected. However, location of the protected group is generally not possible since that residue is unlabeled, in contrast to the previously described form of differential labeling.

A still different procedure was used by Pressman and Roholt[4] to identify peptides containing hapten binding sites in antibody preparations directed against a p-azobenzoate hapten. This method consisted of iodination of one sample of antibody with ^{125}I, while a second sample containing p-nitrobenzoate as a protector group was iodinated with ^{131}I. Upon termination of iodination, the samples were mixed and digested with pepsin and the peptides were separated. The ratio of ^{125}I to ^{131}I was similar for most peptides, as predicted for groups outside of the hapten site, but several showed an elevated ratio, suggesting that these contained residues involved in the binding of p-nitrobenzoate. A similar approach was taken by Hart and Titus[5] for the identification of the component of rabbit brain microsomal ATPase involved in the K$^+$-dependent dephosphorylation of umbelliferone phosphate. In this instance, labeling was performed with N-ethyl-[^{14}C]maleimide in the absence of ligand and

[4] D. Pressman and O. Roholt, *Proc. Natl. Acad. Sci. U.S.A.* **47**, 1606 (1961).
[5] W. M. Hart, Jr. and E. O. Titus, *J. Biol. Chem.* **248**, 1365 (1973).

with N-ethyl-[³H]maleimide in the presence of ATP and Na⁺ as protective ligands. The labeled samples were mixed and analyzed by sodium dodecyl sulfate (SDS) gel electrophoresis to reveal which of several subunits contained a depressed ³H:¹⁴C ratio.

Principles of Differential Labeling

A description of the quantitative protection obtainable in differential labeling was provided by Singer[6] and is briefly restated here for illustrative purposes.

The rate of modification of a target residue Y by a reagent R which is in large excess over Y is

$$-(dY/dt) = k[Y][R] \tag{1}$$

where k is the second-order rate constant, [Y] is the concentration of target sites present at time t, and [R] is the essentially invariant initial concentration of modifying reagent. Since $[Y]/[Y]_{total} = 1 - ([Y]_{modified}/[Y]_{total})$, Eq. (1) can be rewritten as

$$\dot{x} = k(1 - x)[R] \tag{2}$$

where x equals the fraction of Y that is modified and $(1 - x)$ is the fraction that is unmodified.

The rate of modification in the presence of a protective ligand, P, is the sum of the reactions of R with Y at sites that are free and sites that are protected:

$$-dY_p/dt = k[Y]_{free}[R] + fk[YP][R] \tag{3}$$

The factor f is introduced as a measure of the extent of obstruction of Y by P against the attacking reagent and ranges from 0 (complete obstruction) to 1 (no obstruction).

Taking into account that $[YP] = [Y]_{free}K_p[P]$, where K_p is the equilibrium association constant for P with Y, and that $([Y]_{free} + [YP])/[Y]_{total} = 1 - x$, Eq. (3) can be rewritten in a form similar to (2):

$$\dot{x}_p = \frac{k[1 - x][R]}{1 + K_p[P]} (1 + fK_p[P]) \tag{4}$$

Although Eqs. (2) and (4) give the rates predicted for modification of a single site in the absence and in the presence of a protective ligand, the quantity of interest is the ratio of these rates, which will thus indicate the effectiveness of P in masking Y against attack.

$$\dot{x}/\dot{x}_p = (1 + K_p[P])/(1 + fK_p[P]) \tag{5}$$

[6] S. J. Singer, *Adv. Protein Chem.* **22**, 1 (1967).

Obviously, Eq. 5 is strictly valid only if the fraction of unmodified enzyme is identical in the two cases, but this is approximately correct in the initial stages of reaction.

In any differential labeling experiment, one strives for maximization of this ratio of rates through the combination of near zero values of f, and high values of the $K_p[P]$ product. Further inspection of Eq. 5 reveals that for reasonable values of $K_p[P]$, i.e., 10^2 or greater, the ratio is approximately given by the reciprocal of f, the obstruction parameter.

Considerations for Success in Differential Labeling

Few investigations have been concerned with a systematic comparison of f values for various modifying agents and protective ligands, because intuitively one seeks the combination of protective ligand and modifying agent which yields the most dramatic protection. However, it is likely that some investigations have been aborted for failure to realize the direct relationship between this parameter and the successful completion of a differential labeling procedure.

The following data, collected in the author's laboratory, will illustrate some variations noted in f as a function of the modifying agent used.

In a study of differential labeling of the allosteric site for 5'-ADP binding in the L-threonine dehydratase of *Clostridium tetanomorphum*, six reagents were compared for their ability to produce a loss in ADP binding in the absence or in the presence of ADP as a protective ligand.

The table illustrates that a rather broad range of f values can be

COMPARISON OF PROTECTION OFFERED BY ADP AGAINST VARIOUS MODIFYING AGENTS ACTING ON THREONINE DEHYDRATASE[a]

Reagent	Conc. (mM)	$\dot{x}$	$\dot{x}_p$	f
Iodine	2.0	12.5	0.1	0.008
Diazosulfanilic acid	0.12	1.1	0.6	0.54
Fluorodinitrobenzene	2.5	2.2	0.07	0.032
Glyoxal	5.0	0.5	0.03	0.06
N-Acetylimidazole	20.0	2.9	0.5	0.17
Tetranitromethane	1.0	0.8	0.07	0.09

[a] Rates are initial values (arbitrary units) for the change in ADP binding ability, measured separately as an activator constant. [See A. Vanquickenborne, J. D. Vidra, and A. T. Phillips, *J. Biol. Chem.* **244**, 4808 (1969).] In unmodified enzyme plus ADP, $K_p[P]$ was calculated to be 2.5×10^3. All modification reactions were performed at 0°, with pH varied between 6.0 and 9.0 to obtain the lowest f value for each reagent.

found for which there is no striking correlation with reagent size. In any event, based on f values obtained, the choice of iodine or possibly fluorodinitrobenzene as modifying agent would be logical, using ADP as a protector of its own binding site in the prelabeling step. Companion studies in which other nucleotides were tested as protectors of the ADP site have not been conducted in detail. It should be emphasized that these values are extremely sensitive to reaction conditions (pH, temperature, protective ligand) and thus cannot be extrapolated to other conditions, especially to other enzymes. They simply illustrate the importance of a careful examination of modifying agents and conditions for their use as a prelude to a differential labeling experiment.

The obvious question to raise at this point concerns how large an f value can be tolerated and still result in a satisfactory differential labeling of the desired site. Perhaps not surprising, there is no simple answer. For example, in the instance where there are but two groups on a protein capable of reacting with a modifying agent, should one of these groups be situated at the binding site to be protected, its obstruction need not be complete provided the reagent efficiently and completely attacks the exposed site remaining. Occasional penetration of the hemisphere of protection can be tolerated, since this merely reduces the amount of specific site to be labeled in the second phase of differential labeling. On the other hand, when many groups must be modified during the prelabeling phase, a much greater degree of protection is required in order to attain essentially complete modification of all unprotected groups.

Having begged the question in theory, one can attempt a more practical solution. First, a major criterion for successful labeling of the intended site is an appropriate stoichiometry for incorporation of radioactivity per quantity of site inactivated. This requires prior knowledge of the number of sites present, but the stoichiometry is readily assessed at the termination of the specific labeling step. In the experiments cited in the table, the stoichiometry values obtained for $^{131}I_2$ and [^{3}H]fluorodinitrobenzene, each employed separately in the specific labeling step that followed removal of ADP and nonradioactive inhibitor, were within 10% of the expected values based on the moles of ADP sites present. Thus it would appear in this instance that f values below 0.05 were satisfactory for modification of groups at the ADP binding site by these reagents.

However, a second consideration introduces greater complexity. It is conceivable that protection of a site will change as a function of secondary modifications of the protein. In other words, f may be time-dependent, and this is most pronounced when reactions are occurring rapidly at numerous other sites. Conformational changes resulting from these ancillary modifications can cause a decrease in specific ligand binding (K_p is reduced), thereby resulting in a breakdown in the protection of the desired site. Fortunately, this effect can often be detected in preliminary experiments conducted over the same time span as that to be used in the actual experiment. The problem is least troublesome when the reagent employed for labeling exhibits a very high reactivity toward the desired site. It is possible, perhaps even probable, that prelabeling does not result in the modification of all potentially reactive sites, but only those which are readily exposed and which react at reasonable rates. Then if the specific labeling step is performed quickly and at a somewhat lower reagent excess, reaction is limited to highly reactive groups that were protected in the first stage. This avoids the necessity of modifying every conceivable reaction site prior to specific-site labeling and maintains a more nearly native conformation throughout the prelabeling procedure. An estimate of the extent of secondary modifications occurring during prelabeling can be acquired through the use of radioactive modifying agent at this stage, with protective ligand present. Alternatively, a double label procedure can be used, as discussed below.

Evidence for major variations in the reactivity of target groups is widespread. Powers and Riordan[7] found that only 3 out of 25 arginyl groups in ovine brain glutamine synthetase were modified by [^{14}C]phenylglyoxal, concomitant with a 95% inactivation of catalytic activity. In the presence of ATP and $MnCl_2$, only 5% inactivation was observed over the same time period, with 2 arginyl residues undergoing reaction. Thus, 3 arginines are quite reactive toward phenylglyoxal, of which one is involved in ATP binding, and the remaining 22 arginyl residues do not exhibit reactivity under the conditions employed. Gregory[8] reported that iodoacetamide at a 200 molar excess inactivated bovine heart mitochondrial malate dehydrogenase 70% in 2 hr, whereas only 5% inactivation occurred if a 10 molar excess of NADH was also present. Analysis of the modified enzyme revealed that one carboxymethylhistidine was formed during the inactivation process and 12 histidyl residues remained unaffected. Iodoacetamide treatment in the presence of NADH produced no carboxymethylhistidine, and in neither case was there evidence of reaction with cysteinyl residues although the protein contains approximately 14 half-cystine groups.

Advantages of Dual Isotopes in Differential Labeling

From the preceding discussion, it is clear that a major problem faced in differential labeling is the need to assess the thoroughness of modifi-

[7] S. G. Powers and J. F. Riordan, *Proc. Natl. Acad. Sci. U.S.A.* **72**, 2616 (1975).
[8] E. M. Gregory, *J. Biol. Chem.* **250**, 5470 (1975).

cation achieved during prelabeling in the presence of protective ligand. This can be directly obtained in situations where the modification agent is available in different radioisotopic forms (e.g., ^{3}H- and ^{14}C-labeled fluorodinitrobenzene, $^{125}I_2$ and $^{131}I_2$) through the use of a labeled agent in the prelabeling stage as well as in the specific labeling phase. By this means, it is possible to monitor uptake of modifying agent and thereby terminate protective prelabeling at the point where nonspecific reactions have virtually ceased. This not only ensures a greater degree of selectivity during the specific labeling phase (conducted with a different isotopic form), but also provides data on the extent of total modification and the degree of protection of noncritical residues which may be partially shielded by the protector ligand.

Scope of the Use of Differential Labeling

There appear to be three basic situations in which differential labeling has been successfully employed. The first is in conjunction with reagents that are essentially site-specific, but of such high reactivity that unintended sites may also be modified to some degree. Two examples of this type are studies involving bromopyruvate[9,10] as an alkylating agent for cysteinyl residues located at the active site of several pyridoxal phosphate-dependent enzymes, and borohydride[11] as a reducing agent for the dehydroalanyl residue required for catalysis by phenylalanine ammonialyase. Such uses may not be completely dependent upon a protective pretreatment with nonradioactive modifying agent for satisfactory results, but this added step can provide confirmation of specificity and may reduce background (nonintentional) labeling to a very low level.

A less frequent use of differential labeling, but one with considerable potential, relies on its ability to select, out of complex mixtures, a particular protein or subunit involved in the binding of a specific ligand. In this instance the usual intent is not to determine the nature of groups at the ligand binding site, but merely to identify and separate the differentially labeled portion from other material. The experiments of Fox and Kennedy,[12] in which the membrane protein for transport of β-galactosides in *Escherichia coli* was differentially labeled by the combination of a galactoside protector and N-ethylmaleimide in a complex milieu of other proteins, stand as a clear example of the value of this approach.

[9] M. L. Fonda, *J. Biol. Chem.* **251**, 229 (1976).
[10] M. Okamoto and Y. Morino, *J. Biol. Chem.* **248**, 82 (1973).
[11] K. R. Hanson and E. A. Havir, *Arch. Biochem. Biophys.* **141**, 1 (1970).
[12] C. F. Fox and E. P. Kennedy, *Proc. Natl. Acad. Sci. U.S.A.* **54**, 891 (1965).

The final, and most widely utilized, situation is the labeling of active sites (or other binding sites) with moderately group-specific reagents where the intent is to identify only that fraction of reactive groups which is associated with the active site and ultimately to localize its position in the amino acid sequence. Several early examples of this application have been summarized by Singer.[6] A more recent study by Gleisner and Blakley[13] provides a penetrating look into the advantages, as well as the problems, of the technique.

Dihydrofolate reductase from *Streptococcus faecium* was subjected to differential labeling with iodo[14C]acetic acid after a pretreatment with unlabeled reagent in the presence of aminopterin. The availability of sequence data permitted establishing in preliminary work that iodoacetic acid modified portions of four methionine residues, located at positions 28, 36, 50, and 163, when the enzyme was inactivated (after 24 hr) to the extent of 90% in the absence of protective ligand (aminopterin). Three other methionine residues did not react at all, nor were any other residues modified. An analysis of the carboxymethylated enzyme after prelabeling with nonradioactive iodoacetic acid in the presence of aminopterin for 24 hours showed no loss in activity and no modification of residue 28, but reactions at the other sites were also incomplete. The extent of reaction was estimated to be 9% at position 50, 45% at position 36, and 18% at position 163. In retrospect, it would seem that the efficient protection observed at position 28 could have justified a longer prelabeling time or a higher concentration of iodoacetic acid. When aminopterin and unlabeled iodoacetate were replaced by iodo[14C]acetate and incubation was continued for an additional 24 hr, a 90% inactivation of catalytic activity occurred and residue 28 was modified to the extent of 93%. At the same time, methionines 36 and 163 underwent no additional carboxymethylation whereas residue 50 increased to 37% modified.

The conclusion from this work seems clear: carboxymethylation of methionine 28 is responsible for the loss in catalytic activity, presumably owing to interference with the binding of dihydrofolate. It is possible, however, that the partial protection noted on residue 50 could indicate its participation in the binding site as well, but the extent of its modification did not correlate well with loss in catalytic activity and it is equally probable that this residue was rendered partially inaccessible to iodoacetic acid owing to a conformational change associated with aminopterin binding. Distinction between these two explanations might be possible if more complete prelabeling could be conducted without appreciable loss in catalytic activity.

[13] J. M. Gleisner and R. L. Blakley, *Eur. J. Biochem.* **55**, 141 (1975).

Conclusions

Differential labeling offers a reasonable alternative to the use of affinity labeling techniques when proper affinity reagents are unavailable or inadequate. As with any complex technique, there are numerous problems that can be encountered: incomplete prelabeling; breakdown of protection during prelabeling; inactivation unrelated to inhibitor action, as in dialysis procedures; and excessive incorporation of labeled inhibitor during the supposedly specific-labeling phase. Most of these problems, however, are amenable to experimental solution if correctly diagnosed, either by variations in reaction conditions, or selection of a better combination of reactive agent and protective ligand. The technique is clearly of value as an adjunct to affinity labeling, and its potential for generating selective modifications of all types of binding sites using relatively nonspecific agents should not be underestimated.

[8] Photoaffinity Labeling

By HAGAN BAYLEY and JEREMY R. KNOWLES

The method of affinity labeling is based upon the fact that the binding of most biological ligands to their specific receptor sites involves a number of favorable interactions that together make up the ligand–receptor recognition process. The total binding free energy is usually high enough to allow some latitude in modifying the structure of the natural ligand without excessive sacrifice either of the selectivity or the strength of the overall binding interaction. This latitude often allows the incorporation of a chemically labile group into the ligand with the retention of most of the structural features that are required for recognition by the receptor macromolecule. The modified ligand can then become an affinity label, and (provided that a functionality of appropriate chemical reactivity is present at the ligand binding site) may result in the selective modification of the binding site, the selectivity deriving from the higher local concentration of reagent obtaining there. As much of this volume testifies, this approach has proved immensely valuable in the selective labeling of receptor sites in a large variety of biological systems. Yet, in addition to the possibility that a labile group of appropriate reactivity cannot be incorporated into the ligand molecule without excessive disturbance of the recognition process, there are two limitations to the affinity labeling approach. First, the range of chemical reactivity of groups that can be incorporated into the ligand is limited by the fact that these groups must not react so rapidly with water that they are destroyed hydrolytically before the ligand that carries them can reach the binding site. This is not always as stringent a requirement as it appears at first sight, and even, for instance, oxonium salts having half-lives in water of seconds have been used successfully in affinity labeling work.[1] But since the vast majority of chemical affinity labels are electrophilic species, their success depends upon the successful competition between a nucleophile at the binding site and the 55 M water in which the labeling reaction is necessarily carried out. Second, it is becoming clear that some biological problems require a reagent whose reactivity remains masked until the experimenter chooses to activate it. For example, how could one label the inside components of a cell or a vesicle, *without* modifying the outside? Ideally, a masked reagent would be incorporated in some appropriate way into the cytoplasm, and the outside of the cell would be washed clear of reagent before activation of the reagent remaining inside. With classical chemical reagents, it is obvious that all reactive groups on the outside would have to have been completely labeled with reagent before any attempt to incorporate radioactive reagent into the cytoplasm was made. This requirement stresses the need for a method that allows the experimenter to unleash the reagent at a particular time and place.

Both of the two limitations of classical chemical affinity labeling discussed above can in principle be circumvented by the use of a photo-generated reagent.[2] This method requires the modification of the natural ligand by the incorporation of a chemically inert but photochemically labile group. At a time determined by the experimenter, the system is irradiated and the photolabile group is converted to a species of very high chemical reactivity. Ideally, a very reactive reagent is generated only at the ligand binding site of interest, and reacts indiscriminately with whatever chemical groups it finds there. As will be seen from what follows, this ideal has not often been realized, but that does not diminish the conceptual attraction of the method, nor the practical rewards when the approach succeeds.

Areas of Usefulness of Photogenerated Reagents

There is a wide range of problems for which the approach is appropriate, including the following:

1. The location of particular macromolecules within a biological assembly or organelle. Which of the 54 or so proteins of the bacterial

[1] S. M. Parsons, L. Jao, F. W. Dahlquist, C. L. Borders, T. Groff, J. Raes, and M. A. Raftery, *Biochemistry* **8**, 700 (1969); A. K. Paterson and J. R. Knowles, *Eur. J. Biochem.* **31**, 510 (1972).

[2] For earlier reviews, see: J. R. Knowles, *Acc. Chem. Res.* **5**, 155 (1972); D. Creed, *Photochem. Photobiol.* **19**, 459 (1974); B. S. Cooperman, *in* "Proceedings of the International Conference on Proteins and Other Adducts to DNA" (K. C. Smith, ed.). Plenum, New York, in press.

ribosome are close to the peptidyl-tRNA binding site? Which membrane proteins actually span the erythrocyte membrane? Which of the subunits of RNA polymerase interact directly with the template?

2. The identification of a receptor macromolecule. With what membrane proteins does cyclic-AMP interact? What is the corticosteroid receptor? Can the site at which uncouplers of oxidative phosphorylation act be identified?

3. The location of a ligand binding site within a receptor macromolecule. What parts of an antibody molecule contribute to the specific binding of hapten? What areas of the lac repressor are involved in binding to the operator DNA?

4. Investigations of structure:function relationships. Can an allosteric enzyme be switched permanently to one state? Why does ethidium bromide induce petite mutants in *Saccharomyces?* While it is not the purpose of this article to discuss the scientific problems for which the method of photoaffinity labeling is most useful, it will be apparent from the above list that the nature of the problem will affect the choice of reagent. Thus, while nearly all the problems reduce to one of converting a reversible noncovalent binding interaction into a stable covalent attachment that allows the labeled receptor to be identified and characterized, some studies may require a reagent that will react rapidly with water as well as with receptor (thereby using the solvent as a scavenger for unbound reagent), and other studies will need a reagent that reacts preferentially with any or all macromolecules within reach rather than with solvent (thus providing more complete information about the macromolecular distribution around a particular ligand environment). The following section therefore summarizes both the formal characteristics of the available reagents, and the biological systems with which these reagents have been used.

Photogenerated Species and Their Precursors

Since the photoaffinity labeling method depends upon the generation of a highly reactive species by irradiation of a chemically stable precursor, the criteria for an effective reagent are as follows:

1. The precursor should be readily synthesized, chemically stable (which for these purposes often means a half-life in aqueous solution of hours or longer), and susceptible to smooth photolysis at wavelengths long enough (or for times short enough) not to cause photooxidative or other irrelevant photochemical damage to the system;

2. The photochemically derived species should be highly reactive, of extremely short half-life, and not susceptible to intramolecular rearrangement to a much less reactive compound.

Aside from the generation of electronically excited states, irradiation can give rise to two general classes of species, each of which is produced by the homolytic cleavage of chemical bonds. The absorption of ultraviolet or visible radiation may lead to fragmentation either at a single bond, resulting in the formation of two free radicals or a diradical, or at a double bond to carbon or nitrogen, which results in a carbene or a nitrene. All these approaches have been and are being used, even though the choice of reagent for particular systems has undoubtedly been based more on synthetic accessibility than on the desirability or effectiveness of the different reactive species. The problems of synthesis are discussed in the section entitled "Synthetic Approaches," and the rest of this section is devoted to the consideration of the advantages and disadvantages of the more promising reagents, without regard to their synthetic accessibility.

Carbenes and Carbene Precursors

Carbenes[3] react very rapidly with a variety of chemical functions: by coordination to nucleophilic centers (to give carbanions), by addition to multiple bonds (including those of aromatic systems), by insertion into single bonds (including carbon-hydrogen bonds), and by hydrogen abstraction (to give two free radicals that may then couple). However, if there is a hydrogen atom on the carbon atom adjacent to the carbene center, hydrogen migration readily occurs, and results in unreactive olefin:

$$R—CH_2—\overset{..}{C}H \rightarrow R—CH{=}CH_2$$

For this reason, the adjacent atom must bear no hydrogen, as in the following examples:

$$R—\overset{\overset{\displaystyle O}{\|}}{C}—\overset{..}{C}H \qquad \langle\!\!\bigcirc\!\!\rangle\!\!-\overset{..}{C}H \qquad R—CF_2—\overset{..}{C}H$$

In addition to this limitation, the α-ketocarbene species, although attractive for a number of other reasons, may undergo the intramolecular Wolff rearrangement,[4] resulting in a ketene:

$$R—\overset{\overset{\displaystyle O}{\|}}{C}—\overset{..}{C}H \xrightarrow[\text{rearrangement}]{\text{Wolff}} R—CH{=}C{=}O$$

[3] "Carbenes" (M. Jones, Jr. and R. A. Moss, eds.). Wiley (Interscience), New York, Vol I, 1973; Vol II, 1975.
[4] This rearrangement may occur, on photolysis, without the intermediacy of the carbene: H. D. Roth and M. L. Manion, *J. Am. Chem. Soc.* **98**, 3392 (1976).

Migratory aptitudes differ, with α-diazo oxygen esters and α-diazo-amides suffering 20–60% rearrangement[5] and α-diazo thiol esters undergoing essentially 100% sulfur migration.[6] The resulting ketene derivative is quite reactive and has been generated at the ligand binding site, but the very high indiscriminate reactivity of the initially formed carbene has now been reduced to that of mere acylating power that will act only upon nucleophiles.

A number of convenient precursor species for the photochemical generation of carbenes are shown in Table I.[3,5-10] Diazocarbonyl compounds are not, unfortunately, very stable, particularly at low pH, and although these derivatives have been used on a large number of systems (indeed, the very first example of photolabeling involved the use of a diazoacetyl compound by Westheimer and his group[11]), their utility is rather limited both by chemical instability and their susceptibility to the Wolff rearrangement.

Electron-withdrawing groups stabilize the parent diazo compounds, and diazomalonyl[7] derivatives are better candidates than the simpler diazoacetyl compounds, both for precursor stability and because the derived carbene is less susceptible to the Wolff rearrangement. However, the electrophilic nature of α-ketocarbenes, which cause a preference for O-H bonds, e.g., of water, even in the presence of a neighboring C-H bond, is exacerbated by such substitution. Substitution with fluorine[12] has some attraction, in that it both stabilizes the diazo compound remarkably (ethyl diazotrifluoropropionate is stable for 24 hr in 1 M HCl[8]), and sharply decreases the tendency to Wolff rearrangement.[8] Aryldiazomethanes, while satisfying the criterion of having no hydrogen on the adjacent carbon, are chemically much too reactive (and have been used in the dark as protein modification reagents[13]). Aryldiazirines, however, are kinetically more stable than the isomeric diazoalkanes and may prove to be a useful source of arylcarbenes.[10]

[5] H. Chaimovich, R. J. Vaughan, and F. H. Westheimer, *J. Am. Chem. Soc.* **90,** 4088 (1968).

[6] S. S. Hixson and S. H. Hixson, *J. Org. Chem.* **37,** 1279 (1972).

[7] C. S. Hexter and F. H. Westheimer, *J. Biol. Chem.* **246,** 3934 (1971).

[7a] R. J. Vaughan and F. H. Westheimer, *Anal. Biochem.* **29,** 305 (1969).

[8] V. Chowdhry, R. Vaughan, and F. H. Westheimer, *Proc. Natl. Acad. Sci. U.S.A.* **73,** 1406 (1976).

[9] J. A. Goldstein, C. McKenna, and F. H. Westheimer, *J. Am. Chem. Soc.* in press.

[10] R. A. G. Smith and J. R. Knowles, *J. Am. Chem. Soc.* **95,** 5072 (1975); *J. Chem. Soc., Perkin Trans. 2,* p. 686 (1975).

[11] A. Singh, E. R. Thornton, and F. H. Westheimer, *J. Biol. Chem.* **237,** PC 3006 (1962).

[12] C. G. Krespan and W. J. Middleton, *Fluorine Chem. Rev.* **5,** 57 (1971).

[13] G. R. Delpierre and J. S. Fruton, *Proc. Natl. Acad. Sci. U.S.A.* **56,** 1817 (1966).

TABLE I
Useful Carbene Precursors

Type[a]	Formula	Stability in the dark in neutral solution	Susceptibility to intramolecular rearrangement after photolysis
α-Diazoketones[3]	$R'—\overset{\overset{\displaystyle O}{\|}}{C}—CR=\overset{+}{N}=\overset{-}{N}$	Borderline	High
α-Diazoacetyl-[3,5,6]	$—\overset{\overset{\displaystyle O}{\|}}{C}—CH=\overset{+}{N}=\overset{-}{N}$	Borderline	High
α-Diazomalonyl-[3,7]	$—\overset{\overset{\displaystyle O}{\|}}{C}—\underset{\underset{\displaystyle COOR'}{\|}}{C}=\overset{+}{N}=\overset{-}{N}$	Fair	Reasonable
Trifluoromethyldiazoacetyl-[8]	$—\overset{\overset{\displaystyle O}{\|}}{C}—\underset{\underset{\displaystyle CF_3}{\|}}{C}=\overset{+}{N}=\overset{-}{N}$	Good	Rather low
α-Diazobenzylphosphonate[9]	$Ar—\underset{\underset{\displaystyle PO_3^{2-}}{\|}}{C}=\overset{+}{N}=\overset{-}{N}$	Poor	—
Aryldiazomethane[3]	$Ar—CH=\overset{+}{N}=\overset{-}{N}$	Poor	Low
Aryldiazirine[3,10]	$Ar—HC{\overset{\displaystyle N}{\underset{\displaystyle N}{\|\|}}}$	Good	Low[b]

[a] Superscript numbers refer to text footnotes.

[b] Rearrangement to the linear diazo compound can also occur, but this photolyzes in turn to the carbene.

Finally, the possibility of a Dimroth rearrangement in some N-acylated diazo compounds can reduce the effectiveness of these reagents.[14] A pH-dependent equilibrium is established:

$$\overset{-}{N}=\overset{+}{N}=C{\overset{\displaystyle COOEt}{\underset{\displaystyle \underset{\underset{\displaystyle R}{\|}}{NH}}{\|}{C=O}}} \rightleftharpoons \overset{N—C}{\underset{\underset{\displaystyle R}{\|}}{N\diagdown\diagup_{N}}}{\overset{\displaystyle COOEt}{\underset{\displaystyle C—O^-}{}}}$$

Because the triazole form predominates above pH 6.5–7 and does not photolyze smoothly,[14] such reagents should be irradiated at low pH[15] if other features of the system permit.

[14] D. J. Brunswick and B. S. Cooperman, *Biochemistry* 12, 4074 (1973).

[15] C. E. Guthrow, H. Rasmussen, D. J. Brunswick, and B. S. Cooperman, *Proc. Natl. Acad. Sci. U.S.A.* 70, 3344 (1973).

TABLE II
EXAMPLES OF PHOTOLABILE CARBENE PRECURSOR ANALOGS OF NATURAL LIGANDS

Derivative	Designed to label	Comments	References[a]
Diazoacetyl derivatives			
α-Chymotrypsin (on Ser-195)	Active site	80 % deacylation on photolysis; isolation of some labeled Ser, His, and Tyr	11, 16, 17
Subtilisin (on Ser-221)	Active site	19 % label incorporated and 23 % activity loss after irradiation	18
Glyceraldehyde phosphate dehydrogenase (on Cys)	Active site	Only products of Wolff rearrangement (but note development of diazotrifluoropropionyl chloride[8])	5, 6
Glucosamine	Lysozyme, α-lactalbumin binding sites	Analog is a galactose acceptor in lactose synthetase, but much nonspecific labeling seen	19
β-Hydroxymethyl-NAD+ analog	Coenzyme binding site in yeast alcohol dehydrogenase	≤9 % specific labeling, discussion of less successful analogs	20
4-Aminophenyl-α D-mannoside and 1-aminoglucose	Carbohydrate binding sites, e.g., concanavalin A	Synthesis only	21
Estrone and estradiol	Estrogen binding proteins in rat uterus	Some noncovalent binding affinity retained; photoattachment observed	22–24
Choline	Acetylcholine receptors	Synthesis, depolarization of neuromuscular junction	25, 26
Various steroids	Steroid receptors	Synthesis	27
Diazomalonyl derivatives			
α-Chymotrypsin (on Ser-195) and trypsin	Active sites	Isolation of some S-carboxymethyl-Cys; labeling of Ala in trypsin	7, 7a, 28
Cymarin	Ouabain site of Na+, K+-ATPase	Labeled the larger subunit	29
Puromycin	*Escherichia coli* ribosomes	Both protein and nucleic acid labeled	30
Phe-tRNA[Phe] (on α-N)	*E. coli* ribosomes	Labels 23 S RNA; protein labeling <5 %	31, 32
Cyclic AMP (cAMP) (on N-6)	Erythrocyte ghosts	Labels one protein; cAMP itself protects, low incorporation	14, 15

TABLE II (*Continued*)

Derivative	Designed to label	Comments	References[a]
cAMP (on *O-2'*)	Effector site of phosphofructokinase	Specific attachment to cAMP binding site; loss of cAMP effector function	14, 33, 34
Fatty acids and lipids	Membrane components	Synthesis	35
Various steroids *Diazoketone derivatives*	Steroid receptors	Synthesis	27
N-2,4-Dinitrophenylalanine	Anti-dinitrophenyl antibody	Heavy chain labeled	36
N-2,4-Dinitrophenylalanine	Dinitrophenyl binding site of mouse myeloma protein MOPC 460	Labels light chain Lys-54	37–40
Analog of menadione	Menadione binding site of mouse myeloma protein MOPC 460	67 % of sites labeled after repeated treatment; menadione protects	41
N-Tosylphenylalanine	Active site of α-chymotrypsin	Some increase in activity loss on photolysis with large molar excess of reagent; suggest loss of His	42
Progesterone and corticosterone (C_{21} diazoketone analog)	Corticosteroid receptor	Attachment to plasma proteins, but site specificity not yet shown	43
Estrone and estradiol derivatives	Estrogen receptor	Some noncovalent binding and photoattachment observed	22–24
Various steroids	Steroid receptors	Synthesis	27
N-2,4-dinitrophenyl-ϵ-aminocaproate	Purified anti-dinitrophenyl antibody	Photoattachment observed; partial localization of label	44, 45
N-2,4-dinitrophenyl-*p*-aminobenzoate	Purified anti-dinitrophenyl antibody	Labeling of His and Gly in heavy chain shown by mass spectroscopy	45

[a] Numbers refer to text footnotes.

α-Diazocarbonyl compounds show weak absorption maxima above 350 nm, and photolysis is normally smooth at these wavelengths. The low extinction coefficient (compared with that of the band around 250 nm) can lead to the need for long irradiation times, but these are usually acceptable for such long wavelengths.

We have listed in Table II most of the photolabile ligands that have been made to date (Spring 1976) which produce carbenes on irradiation.

The comments mostly reflect the authors' own claims[5-7a,11,14-45] rather than critical evaluations of their work.

[16] J. Shafer, P. Baronowsky, R. Laursen, F. Finn, and F. H. Westheimer, *J. Biol. Chem.* **241**, 421 (1966).

[17] C. S. Hexter and F. H. Westheimer, *J. Biol. Chem.* **246**, 3928 (1971).

[18] Y. Stefanovsky and F. H. Westheimer, *Proc. Natl. Acad. Sci. U.S.A.* **70**, 1132 (1973).

[19] A. E. Burkhardt, S. O. Russo, C. G. Rinehardt, and G. M. Loudon, *Biochemistry* **14**, 5465 (1975).

[20] D. J. Browne, S. S. Hixson, and F. H. Westheimer, *J. Biol. Chem.* **246**, 4477 (1971).

[21] E. W. Thomas, *Carbohydr. Res.* **31**, 101 (1973). See also this volume [39].

[22] J. A. Katzenellenbogen, H. J. Johnson, Jr., and H. N. Myers, *Biochemistry* **12**, 4085 (1973).

[23] J. A. Katzenellenbogen, H. N. Myers, and H. J. Johnson, Jr., *J. Org. Chem.* **38**, 3525 (1973).

[24] J. A. Katzenellenbogen, H. J. Johnson, Jr., K. E. Carlson, and H. N. Myers, *Biochemistry* **13**, 2986 (1974).

[25] J. Frank and R. Schwyzer, *Experientia* **26**, 1207 (1970).

[26] P. G. Waser, A. Hofmann, and W. Hopff, *Experientia* **26**, 1342 (1970).

[27] H. E. Smith, J. R. Neergaard, E. P. Burrows, R. G. Hardison, and R. G. Smith, this series, Vol. 36 [34].

[28] R. J. Vaughan and F. H. Westheimer, *J. Am. Chem. Soc.* **91**, 217 (1969).

[29] A. Ruoho and J. Kyte, *Proc. Natl. Acad. Sci. U.S.A.* **71**, 2352 (1974). See also this volume [59].

[30] B. S. Cooperman, E. N. Jaynes, Jr., D. J. Brunswick, and M. A. Luddy, *Proc. Natl. Acad. Sci. U.S.A.* **72**, 2974 (1975).

[31] L. Bispink and J. H. Matthaei, *FEBS Lett.* **37**, 291 (1973). See also this volume [74].

[32] J. H. Matthaei, A. Hagenberg, L. Bispink, and A. Gassen, *Int. Congr. Biochem., 9th.,* 3d 23, p. 144. Stockholm, 1973.

[33] D. J. Brunswick and B. S. Cooperman, *Proc. Natl. Acad. Sci. U.S.A.* **68**, 1801 (1971).

[34] B. S. Cooperman and D. J. Brunswick, *Biochemistry* **12**, 4079 (1973).

[35] P. Chakrabarti and H. G. Khorana, *Biochemistry* **14**, 5021 (1975).

[36] C. A. Converse and F. F. Richards, *Biochemistry* **8**, 4431 (1969).

[37] M. Yoshioka, J. Lifter, C.-L. Hew, C. A. Converse, M. Y. K. Armstrong, W. H. Konigsberg, and F. F. Richards, *Biochemistry* **12**, 4679 (1973). See also this volume [57].

[38] C.-L. Hew, J. Lifter, M. Yoshioka, F. F. Richards, and W. H. Konigsberg, *Biochemistry* **12**, 4685 (1973).

[39] J. Lifter, C.-L. Hew, M. Yoshioka, F. F. Richards, and W. H. Konigsberg, *Biochemistry* **13**, 3567 (1974).

[40] F. F. Richards, J. Lifter, C.-L. Hew, M. Yoshioka, and W. H. Konigsberg, *Biochemistry* **13**, 3572 (1974).

[41] R. W. Rosenstein and F. F. Richards, *J. Immunol.* **108**, 1467 (1972).

[42] H. Nakayama and Y. Kanaoka, *FEBS Lett.* **37**, 200 (1973).

[43] M. E. Wolff, D. Feldman, P. Catsoulacos, J. W. Funder, C. Hancock, Y. Amano, and I. S. Edelman, *Biochemistry* **14**, 1750 (1975).

[44] L. E. Cannon, D. K. Woodard, M. E. Woehler, and R. E. Lovins, *Immunology* **26**, 1183 (1974).

[45] J. G. Lindemann, D. K. Woodard, M. E. Woehler, G. E. Chism, and R. E. Lovins, *Immunochemistry* **12**, 849 (1975).

Nitrenes and Nitrene Precursors

The types of reaction open to a nitrene[46] are broadly similar to those undergone by carbenes, but the reactivity of nitrenes is considerably lower. Nitrenes are less indiscriminate than carbenes in their reactions with primary, secondary, and tertiary carbon–hydrogen bonds, for instance, and nitrenes are also somewhat electrophilic, preferring an O-H over a C-H bond. By far the most frequently used nitrene source is an aryl azide (see Table IV). The derived aryl nitrenes are much less reactive than α-keto, α-sulfonyl, or α-phosphoryl nitrenes, but the use of these acyl nitrenes is ruled out by the high chemical reactivity of the precursor species: acyl azides, sulfonyl azides, and phosphoryl azides.

The same principles concerning intramolecular rearrangement as have been discussed above for carbenes are relevant to nitrenes. The adjacent carbon should not bear a hydrogen atom, i.e., alkyl nitrenes can yield imines, and if an acyl azide were to be used, the derived α-keto nitrene would undergo the Curtius (or Schmidt) rearrangement to the isocyanate, which is analogous to the Wolff rearrangement of α-keto carbenes mentioned in the previous section. Two further problems have to be considered for aryl nitrenes which are not a concern if one is generating an aryl carbene. First, aryl nitrenes may undergo facile ring expansion (possibly via the azirine) and be trapped by a nucleophile from either the receptor or the solution to give the substituted azepine:

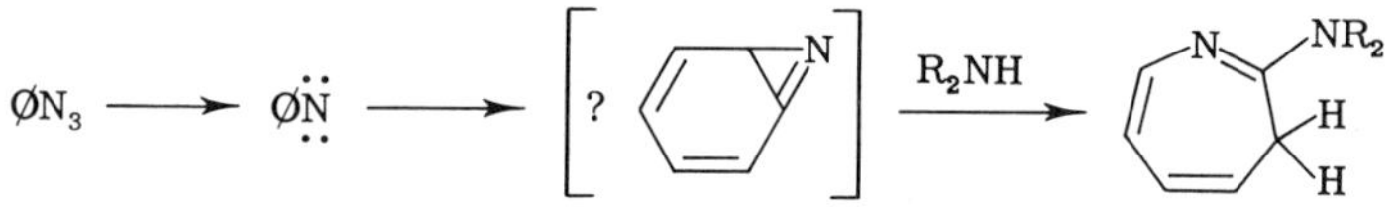

The importance of this rearrangement pathway for nitrenes generated in a biological receptor site is not known, however, and in model systems an aryl nitrene will insert into a carbon–hydrogen bond if there is one in the near neighborhood. Thus irradiation of 2-azidobiphenyl results, via both the singlet and the triplet nitrene, in the formation of carbazole[47]:

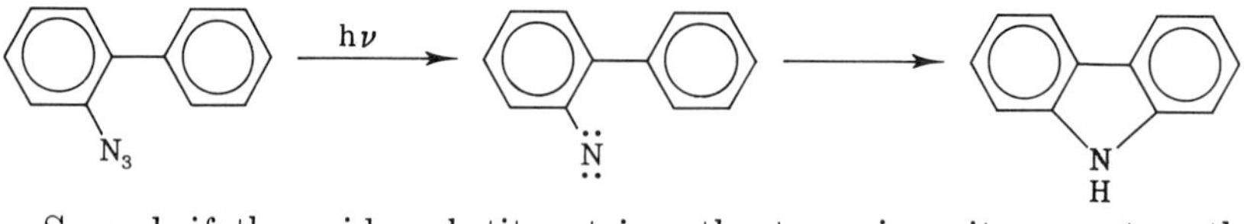

Second, if the azide substituent is ortho to a ring nitrogen atom the precursor aryl azide will normally be a mixture of the azide and its

[46] "Nitrenes" (W. Lwowski, ed.). Wiley (Interscience), New York, 1970.
[47] P. A. S. Smith and B. B. Brown, J. Am. Chem. Soc. 73, 2435 (1951).

isomer, the tetrazole:

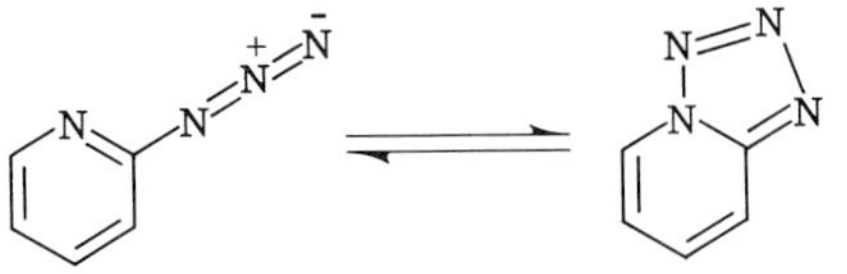

The tetrazole is very much less sensitive to photolysis than the azide, and for a number of such systems photochemical generation of the heteroaryl nitrene has failed.[48] This problem is important in nucleotide and nucleic acid chemistry, where azidopurines and azidopyrimidines are, at first sight, attractive candidates for photogenerated reagents. The position of the azide-tetrazole isomerization equilibrium is solvent dependent, and it is not possible to specify under what conditions (or in what kinds of receptor site) useful levels of photogenerated aryl nitrene can be produced. In general, therefore, it may be safer to avoid such systems if reasonable alternatives exist.

Finally, it has been nicely demonstrated by McRobbie et al.[49] that when alternative pathways are possible, singlet nitrenes react preferentially by insertion into oxygen–hydrogen or nitrogen–hydrogen bonds, and show electrophilic character. If intersystem crossing to the less reactive triplet nitrene occurs, then insertion into carbon–hydrogen bonds becomes more probable. If the aryl ring carrying the nitrene is substituted with electron-withdrawing groups, the reactivity and electrophilicity of the nitrene are increased. That the reactivity of aryl nitrenes is increased by the presence of electron-withdrawing substituents had already been suggested by the shorter lifetimes of nitrophenyl nitrenes compared with phenyl nitrenes in soft polystyrene matrices.[50] A shorter-lived and more reactive species will in general be preferred for photolabeling purposes, and the use of nitro-aryl precursors also pushes the λ_{max} to longer, more convenient, wavelengths. However, the increased electrophilicity of the derived nitrene is not necessarily desirable, since that will both result in a more facile reaction with solvent water compared with reaction with receptor carbon–hydrogen bonds and lead to selectivity

[48] J. A. Hyatt and J. S. Swenton, J. Heterocycl. Chem. 9, 409 (1972); J. Org. Chem. 37, 3216 (1972); T. Sasaki, K. Kanematsu, and M. Murata, Tetrahedron 27, 5121 (1971). See also: D. W. Allen, D. J. Buckland, and I. W. Nowell, J. Chem. Soc. Perkin Trans. 2, 1610 (1976); and M Rull and J. Vilarrasa, Tetrahedron Lett. p. 4175 (1976).

[49] I. M. McRobbie, O. Meth-Cohn, and H. Suschitzky, Tetrahedron Lett. p. 925 (1976); p. 929 (1976).

[50] A. Reiser, F. W. Willets, G. C. Terry, V. Williams, and R. Morley, Trans. Faraday Soc. 64, 3265 (1968); A. Reiser and L. Leyshon, J. Am. Chem. Soc. 92, 7487 (1970).

TABLE III
Useful Nitrene Precursors[46]

Type	Stability in the dark in neutral solution	Susceptibility to intramolecular rearrangement after photolysis	Wavelength for photolysis (nm)
Alkyl azides	Reasonable	High	$<300^a$
Aryl azides	Excellent	Usually low	≤ 300
Nitroaryl azides	Excellent	Usually low	>300

^{a} Even at these shorter wavelengths, photolysis can be inefficient.[51]

among the groups within the receptor site. Finally, it should be pointed out that substituents ortho to the azide group must be avoided when possible. For instance, the nitrene produced by irradiation of 2-azido-3,5-dinitrobiphenyl inserts into the nitro group to produce 3-nitro-5-phenylbenzofuroxan in 85% yield.[47]

The nitrene precursor species that have been used in or proposed for photolabeling studies are listed in Table III.[46,51] Table IV summarizes most of the actual nitrene precursors that have been made to date.[22-24,35,37-40,44,51-117]

[51] P. E. Nielsen, V. Leick, and O. Buchardt, *Acta Chem. Scand.* **B29**, 662 (1975).

[52] B. E. Haley, *Biochemistry* **14**, 3852 (1975). See also this volume [35].

[53] A. H. Pomerantz, S. A. Rudolph, B. E. Haley, and P. Greengard, *Biochemistry* **14**, 3858 (1975).

[54] B. E. Haley and J. F. Hoffman, *Proc. Natl. Acad. Sci. U.S.A.* **71**, 3367 (1974).

[55] G. Schäfer, E. Schrader, G. Rowohl-Quisthoudt, S. Penades, and M. Rimpler, *FEBS Lett.* **64**, 185 (1976).

[56] R. Koberstein, L. Cobianchi, and H. Sund, *FEBS Lett.* **64**, 176 (1976).

[57] S. S. Hixson and S. H. Hixson, *Photochem. Photobiol.* **18**, 135 (1973).

[58] W. G. Hanstein, *Fed. Proc. Fed. Am. Soc. Exp. Biol.* **32**, 515 (1973).

[59] Y. Hatefi and W. G. Hanstein, *Membr. Proteins Transp. Phosphorylation, Proc. Int. Symp.* (G. F. Azzone, M. E. Klingenberg, E. Qualiariello, and N. Siliprandi, eds.), p. 187. North-Holland, Amsterdam, 1974.

[60] W. G. Hanstein and Y. Hatefi, *J. Biol. Chem.* **249**, 1356 (1974).

[61] W. G. Hanstein, *Trends Biochem. Sci.* **1**, 65 (1976).

[62] N. J. Leonard, J. C. Greenfield, R. Y. Schmitz, and F. Skoog, *Plant Physiol.* **55**, 1057 (1975).

[63] R. Schwyzer and M. Calviezel, *Helv. Chim. Acta* **54**, 1395 (1971).

[64] E. Escher, R. Jost, H. Zuber, and R. Schwyzer, *Isr. J. Chem.* **12**, 129 (1974).

[65] E. Escher and R. Schwyzer, *FEBS Lett.* **46**, 347 (1974); *Helv. Chim. Acta* **58**, 1465 (1975).

[66] F. Fahrenholz and G. Schimmack, *Hoppe-Seyler's Z. Physiol. Chem.* **356**, 469 (1975).

[67] W. Fischli, M. Caviezel, A. Eberle, E. Escher, and R. Schwyzer, *Helv. Chim. Acta* **59**, 878 (1976).

[68] Wieland, A. von Dungen, and C. Birr, *Justus Liebigs Ann. Chem.* **752**, 109 (1971).

[69] G. R. Greenburg, P. Chakrabarti, and H. G. Khorana, *Proc. Natl. Acad. Sci. U.S.A.* **73**, 86 (1976).

[70] S. C. Hixon, W. E. White, Jr., and K. L. Yielding, *J. Mol. Biol.* **92**, 319 (1975). See also this volume [75].

[71] S. C. Hixon, W. E. White, Jr., and K. L. Yielding, *Biochem. Biophys. Res. Commun.* **66**, 31 (1975).

[72] R. de Nobrega Bastos, *J. Biol. Chem.* **250**, 7739 (1975).

[73] U. Das Gupta and J. S. Rieske, *Biochem. Biophys. Res. Commun.* **54**, 1247 (1973).

[74] F. Seela and F. Cramer, *Hoppe-Seyler's Z. Physiol. Chem.* **356**, 1185 (1975).

[75] A. J. Bridges and J. R. Knowles, *Biochem. J.* **143**, 663 (1974).

[76] A. S. Girshovich, E. S. Bochkareva, U. M. Kramarov, and Y. A. Ovchinnikov, *FEBS Lett.* **45**, 213 (1974).

[77] N. Hsiung, S. A. Reines, and C. R. Cantor, *J. Mol. Biol.* **88**, 841 (1974).

[78] N. Hsiung and C. R. Cantor, *Nucleic Acid Res.* **1**, 1753 (1974).

[79] I. Schwartz and J. Ofengand, *Proc. Natl. Acad. Sci. U.S.A.* **71**, 3951 (1974).

[80] J. Ofengand and I. Schwartz, *in* "Lipmann Symposium: Biosynthesis and Regulation in Molecular Biology," p. 456. de Gruyter, Berlin, Germany, 1974. See also this volume [83].

[81] I. Schwartz, E. Gordon, and J. Ofengand, *Biochemistry* **14**, 2907 (1975).

[82] V. G. Budker, D. G. Knorre, V. V. Kravchenko, O. I. Lavrik, G. A. Nevinsky, and N. M. Teplova, *FEBS Lett.* **49**, 159 (1974).

[83] J. A. Maasen and W. Möller, *Biochem. Biophys. Res. Commun.* **64**, 1175 (1975).

[84] J. A. Maasen and W. Möller, *Proc. Natl. Acad. Sci. U.S.A.* **71**, 1277 (1974). See also this volume [76].

[85] R. Vince, D. Weiss, and S. Pestka, *Antimicrob. Agents Chemother.* **9**, 131 (1976).

[86] M. Beppu, T. Terao, and T. Osawa, *J. Biochem. (Tokyo)* **78**, 1013 (1975).

[86a] A. R. Fraser, J. J. Hemperly, J. L. Wang, and G. M. Edelman, *Proc. Natl. Acad. Sci. U.S.A.* **73**, 790 (1976).

[87] R. E. Galardy, L. C. Craig, J. D. Jamieson, and M. P. Printz, *J. Biol. Chem.* **249**, 3510 (1974).

[88] D. Levy, *Biochim. Biophys. Acta* **322**, 329 (1973).

[89] B. Winter and A. Goldstein, *Mol. Pharmacol.* **6**, 601 (1972). See also this volume [70].

[90] H. Kiefer, J. Lindstrom, E. S. Lennox, and S. J. Singer, *Proc. Nat. Acad. Sci. U.S.A* **67**, 1688 (1970).

[91] S. J. Singer, *in* "Molecular Properties of Drug Receptors" (G. Wolstenholme, ed.), p. 229. CIBA, London, 1970.

[92] S. J. Singer, A. Ruoho, H. Kiefer, J. Lindstrom, and E. S. Lennox, *in* "Biological Council Symposium on Drug Receptors" (H. P. Rang, ed.), p. 183. Macmillan, New York, 1973.

[93] A. E. Ruoho, H. Kiefer, P. Roeder, and S. J. Singer, *Proc. Natl. Acad. Sci. U.S.A* **70**, 2567 (1973).

[94] F. Hucho, L. Bergman, J. M. Dubois, E. Rojas, and Kiefer, *Nature (London)* **260**, 802 (1976).

[95] G. Rudnick, H. R. Kaback, and R. Weil, *J. Biol. Chem.* **250**, 6847 (1975).

[96] G. Rudnick, H. R. Kaback, and R. Weil, *J. Biol. Chem.* **250**, 1371 (1975).

Free Radicals and Triplet States

The rationale behind the generation of two free radicals or a diradical at a receptor site is that atom abstraction from or radical addition to an appropriate group on the receptor will lead to a new radical that may then couple with the "other half" of the initially generated pair, for instance:

Other kinds of photoinduced reactions may be responsible for attachment of (some of) the reagent to the receptor site, but the chemistry is currently too ill-defined for a detailed discussion of the possibilities to be fruitful.

The members of this class of reagents (see Table V) that are best

[97] M. B. Perry and L. L. W. Heung, *Can. J. Biochem.* **50**, 510 (1972).

[98] E. Saman, M. Claeyssens, H. Kersters-Hilderson, and C. K. DeBruyne, *Carbohydr. Res.* **30**, 207 (1973).

[99] D. P. Witt and R. C. Woodworth, *in* "Proteins of Iron Storage and Transport in Biochemistry and Medicine" (R. R. Crighton, ed.), p. 133. North-Holland Publ., Amsterdam, 1975.

[100] D. F. Wilson, Y. Mukai, M. Erecińska, and J. M. Vanderkooi, *Arch. Biochem. Biophys.* **171**, 104 (1975).

[101] M. Erecińska, J. M. Vanderkooi, and D. F. Wilson, *Arch. Biochem Biophys.* **171**, 108 (1975).

[102] S. J. Jeng and R. J. Guillory, *J. Supramol. Struct.* **3**, 448 (1975).

[103] E. Saman, M. Claeyssens, H. Kersters-Hilderson, and C. DeBruyne, *FEBS Lett.* **63**, 211 (1976).

[104] G. W. J. Fleet, R. R. Porter, and J. R. Knowles, *Nature (London)* **224**, 511 (1969).

[105] G. W. J. Fleet, J. R. Knowles, and R. R. Porter, *Biochem. J.* **128**, 499 (1972).

[106] R. A. G. Smith and J. R. Knowles, *Biochem. J.* **141**, 51 (1974).

[107] C. E. Fisher and E. M. Press, *Biochem. J.* **139**, 135 (1974).

[108] A. Klip and C. Gitler, *Biochem. Biophys. Res. Commun.* **60**, 1155 (1974).

[109] J. V. Staros and F. M. Richards, *Biochemistry* **13**, 2720 (1974).

[110] J. V. Staros, F. M. Richards, and B. E. Haley, *J. Biol. Chem.* **250**, 8174 (1975).

[111] J. V. Staros, B. E. Haley, and F. M. Richards, *J. Biol. Chem.* **249**, 5004 (1974).

[112] A. Rothstein, Z. I. Cabantchik, and P. Knauf, *Fed. Proc., Fed. Am. Soc. Exp. Biol.* **35**, 3 (1976).

[113] P. A. Knauf, W. Brener, L. Davidson, and A. Rothstein, *Biophys. J.* **16**, 107a (1976).

[114] S. H. Hixson and S. S. Hixson, *Biochemistry* **14**, 114 (1975).

[115] W. E. White, Jr. and K. L. Yielding, *Biochem. Biophys. Res. Commun.* **52**, 1129 (1973).

[116] T. Tobin, T. Akera, T. M. Brady, and H. R. Taneja, *Eur. J. Pharmacol.* **35**, 69 (1976).

[117] R. Breslow, A. Feiring, and F. Herman, *J. Am. Chem. Soc.* **96**, 5938 (1974).

TABLE IV
EXAMPLES OF PHOTOLABILE NITRENE PRECURSOR ANALOGS OF NATURAL LIGANDS

Compound	Designed to label	Comments	References[a]
Analogs where an azide group (—N₃) has replaced an atom or group in the natural ligand			
8-Azido-cAMP	Erythrocyte membranes	Two proteins labeled; cAMP protects More efficient than the diazomalonyl derivative	52
8-Azido-cAMP	Protein kinase from bovine brain	Labels the component that can also be phosphorylated by ATP; cAMP protects	53
8-Azido-ATP	Erythrocyte membranes	Three proteins labeled; both Mg^{2+} and Na^+, K^+-ATPases inactivated; ATP protects	54
8-Azido-ADP	Adenine nucleotide transport system in mitochondria	Photoinduced inhibition of nucleotide exchange	55
8-Azido-ADP	Glutamate dehydrogenase	Photoattachment to the ADP effector site; ADP protects	56
β-Azido-NAD^+ analog	Cofactor binding site in yeast alcohol dehydrogenase	7% labeling of enzyme; 4% labeling in presence of excess NAD^+	57
2-Azido-4-nitrophenol	Oxidative phosphorylation uncoupling site in mitochondria	More effective uncoupler than dinitrophenol, attachment sites on inner mitochondrial membrane: subunit 1 of F_1-ATPase (MW 56,000) and a protein of MW 31,000	58–61
2,4-Dinitrophenyl azide	Dinitrophenyl binding site in mouse myeloma protein MOPC 460	Labels heavy-chain Tyr-33 and Tyr-88	37–40
2,4-Dinitrophenyl azide	Purified anti-dinitrophenyl antibodies	Identification of a labeled Phe in one case, and a labeled Ala in another, by mass spectrometry	44

TABLE IV (*Continued*)

Compound	Designed to label	Comments	References[a]
4-Azido-2-chloro-phenoxyacetic acid	Plant auxin binding site	Shows auxin activity	62
Azidophenylalanines and peptides containing them	Peptide receptors, etc.	Synthesis, labeling of α-chymotrypsin	63–67
Antamanide with 4-azido-Phe-6	Antitoxin site	Synthesis	68
Various azido alkyl and aryl fatty acids and lipids	Membrane components	Biosynthetic incorporation into phospholipids. Some cross-linking on irradiation of vesicles	35, 69
Monoazido ethidium bromide	Yeast	Increase in petite mutants of yeast after irradiation; attachment to DNA	70, 71
Diazido ethidium bromide	Yeast mitochondria	Labeling of 1 protein, probably subunit 9 of membrane ATPase	72
Azido estrones, estradiols, and hexestrols	Estrogen receptor of rat uterus	Noncovalent binding retained; some photoattachment shown	22–24
Azido-deformamido antimycin A	Complex III of mitochondrial respiratory chain	Binds to complex III; suggest 1 labeled protein	73
4-Denitro-4-azido-chloramphenicol	Ribosomes	Synthesis	51, 74
Azidoacetylchloramphenicol	Ribosomes	Synthesis and some photochemistry	51
4-Azidocinnamoyl-α-chymotrypsin (on Ser-195)	Specificity locus of α-chymotrypsin	60% labeling, of which 80% is in regions of the sequence that constitute the specificity locus	75

Analogs where an aryl azide group, ⬡—N_3, *has been affixed to the natural ligand*

N-(4-Azido-2-nitro-benzoyl)-Phe-tRNA[Phe]	E. coli ribosomes	With 70 S ribosomes + poly(U), labels RNA from 50 S subunit; with 30 S subunit + poly(U) labels proteins S_3, S_7, and S_{14}	76

TABLE IV (*Continued*)

Compound	Designed to label	Comments	References[a]
N-(4-Azido-2-nitro-phenoxy-4'-phenyl-acetyl)-Phe-tRNA^Phe	E. coli ribosomes	Labels L_{11} and L_{18} from 50 S subunit	77
N-(4-Azido-2-nitro-phenyl)glycyl-Phe-tRNA^Phe	E. coli ribosomes	Labels L_{11} and L_{18} from 50 S subunit	78
4-Azidophenacyl-Val-tRNA_1^Val (on the 4-thio U)	E. coli ribosomes	Labels 16 S RNA of 30 S subunit when at P site; also labels proteins when at A site	79–81
4-Azidoacetanilido-tRNA (on the 4-thio U)	Phenylalanyl tRNA synthetase	Attachment to enzyme, inactivation	82
γ-(4-Azidoanilide) of ATP	Phenylalanyl tRNA synthetase	Attachment (of 2.5 moles per mole of enzyme) and enzyme inactivation	82
β-(4-Azidophenyl ester) of GDP	Ribosomes	Compound inhibits GDP binding and labels 4 or 5 protein components; comparison of *Escherichia coli* and *Bacillus stearothermophilus*	83, 84
N-(4-Azido-2-nitro-phenyl) erythromy-cylamine	E. coli ribosomes (erythromycin binding site)	Analogs bind nearly as well as erythromycin itself	85
4-Azidophenyl, α-D-manno-pyranoside	Concanavalin A	Produced a monovalent Con A dimer	86, 86a
N-(4-Azidobenzoyl)-pentagastrin, and N-(5-azido-2-nitro-benzoyl)penta-gastrin	Gastrin receptor site	Evaluated by photo-attachment to BSA	87
N-(4-Azido-2-nitro-phenyl) insulins (on Phe-Bl, Gly-A1 and Lys-B29)	Insulin receptor	Hormonal activity retained after iodination	88
N-4-Azidophenyl-ethylnorlevorphanol	Opiate receptor site in mouse brain	Some protection by levorphanol	89
(4-Azido-2-nitro-benzyl)triethyl ammonium	Acetylcholine binding sites, e.g., in frog sartorius muscle	Inactivation of cholinesterase; much nonspecific labeling; use of scavenger	90–94

TABLE IV (*Continued*)

Compound	Designed to label	Comments	References[a]
N-4-Azido-2-nitro-phenylaminoethyl-1-thio-β-D-galacto-pyranoside	Lactose transport system in *E. coli*	Analog binds but is not transported; irradiation inactivates in presence of D-lactate; lactose protects	95
4-Azido-2-nitrophenyl-1-thio-β-D-galacto-pyranoside	Lactose transport system in *E. coli*	Analog is transported, competitive with lactose, irradiation inactivates in presence of D-lactate	96
Various azidoaryl sugar derivatives	—	Synthesis	97, 98
4-Azido-2-nitrophenyl conalbumin	Reticulocyte membranes	Photoattachment	99
3-Azido-4,6-dinitro-phenylcyto-chrome *c*	Mitochondrial membranes	Covalent binding to cytochrome *c* oxidase	100, 101
N-(4-Azido-2-nitro-phenyl) ω-amino alkyl esters of ATP	Myosin ATPase	Photoattachment	102
4-Azidophenyl-β-D-xylopyranoside	Xylosidase	Inactivates in the dark after irradiation	98, 103
Miscellaneous			
N-ϵ-(4-Azido-2-nitro-phenyl)lysine	Binding site of anti-4-azido-2-nitrophenyl antibody	Heavy chain labeled on Cys-92 and Ala-93	104, 105
N-ϵ-(4-Azido-2-nitro-phenyl)lysine and N-ϵ-(5-azido-2-nitrophenyl)lysine	Binding site of anti-azidonitrophenyl antibody	Overlap of subpopulations that bind the two reagents, but *not* between sites that are photo-labeled	106
N-ϵ-(4-Azido-2-nitro-phenyl)lysine	Binding site of anti-dinitrophenyl antibody	Heavy chain labeled between 29 and 34 and 95 and 114 (i.e., two of the hypervariable regions)	107
1-Azidonaphthalene and 4-iodophenyl azide	Liposomes and sarcoplasmic reticulum membranes	Half of label on membrane proteins, half on fatty acid chains	108
N-(4-Azido-2-nitro-phenyl)taurine	Erythrocytes (external surface)	Labels 8 proteins in intact cells	109
N-(4-Azido-2-nitro-phenyl)taurine	Erythrocytes (internal surface)	After allowing transport of reagent inside at 37°, labeling of cytoplasmic membrane proteins	110

TABLE IV (*Continued*)

Compound	Designed to label	Comments	References[a]
N-(4-Azido-2-nitrophenyl)taurine	Resealed erythrocyte ghosts	Labeling pattern not identical to intact cells	111
N-(4-Azido-2-nitrophenyl)taurine	Anion channel of erythrocyte membrane	Reagent binds to anion channel and is photoattached to it Cl^- partially protects	112, 113
4-Azido-α-bromoacetophenone	Glyceraldehyde phosphate dehydrogenase	Photoattachment after alkylation not shown	114
4-Azidobenzenesulfonamide	Carbonic anhydrase	Nonspecific labeling despite K_d of 1 μM	114
Azidoisophthalic acids	Glutamate dehydrogenase	Labels the enzyme; label becomes fluorescent on attachment	115
3-Azidoacetylstrophanthidin	Na^+,K^+-ATPase	15% irreversible inhibition of the enzyme on photolysis in the presence of the ligand	116
Phosphorylazide diesters	—	Synthesis and photochemistry	117

[a] Numbers refer to text footnotes.

understood are the α,β-unsaturated ketones. Excitation in the n-π* band produces, via the excited singlet, a diradical triplet state that is known to be an efficient hydrogen abstractor.[118] The abstraction is selective, and carbon–hydrogen bonds are normally attacked in preference to the stronger oxygen–hydrogen bonds of the solvent water. Thus even in aqueous solution, benzophenone will abstract the α-hydrogen of *N*-acetylglycine methyl ester, the two radicals then coupling in fair yield to give the α-benzhydryl derivative[87,119]:

$$\text{Ø—C(=O)—Ø} + CH_3CONHCH_2COOCH_3 \xrightarrow{h\nu} CH_3CONHCH(C(OH)Ø_2)COOCH_3$$

This preference for carbon–hydrogen bonds can be a real advantage, since both carbenes and nitrenes may react fruitlessly with the solvent, and result in extremely low labeling levels. (Reaction with solvent does

[118] N. J. Turro, "Molecular Photochemistry." Benjamin, New York, 1965.
[119] R. E. Galardy, L. C. Craig, and M. P. Printz, *Nature* (*London*) **242**, 127 (1973).

reduce the nonspecific labeling of receptor, and a balance between the selectivity of labeling and the proportion of sites labeled must be struck.)

Excitation of an α,β-unsaturated carbonyl function has been used with natural ligands that happen to contain one and with ligands to which such a function has been attached (see Table V). Thioketones have absorption maxima at much longer wavelengths than ketones, and while simple thiocarbonyl compounds rapidly trimerize in aqueous solution, derivatives of 4-thiouridine are much more stable and have been success-

fully used to label (poly)nucleotide receptor loci.[120–123]

Bromoaryl and iodoaryl compounds are sensitive to ultraviolet radiation and cleave first to give halogen atoms and aryl radicals.[118] Subsequent reactions of these radicals may lead to useful photoattachment. In one case, that of 5-iodo-2′-deoxyuridine, it has been shown that ^{125}I does not become attached to the receptor, and labeling derives from the attachment of the aryl group.[124]

Finally, various natural aromatic ligands have been irradiated in the presence of their receptors, and useful labeling has been achieved. Even though the chemistry involved is still largely unelucidated and the irradiation conditions are often harsh, cross-linking between parts of nucleic acids, or between nucleic acids and proteins, is proving helpful in delineating the regions of both the ligand and the receptor that are in contact. Further, the classical method of dye-sensitized photooxidation[125] has been shown in some cases to lead to covalent attachment of the dye to the protein when the dye is present in high concentration.[126]

[120] F. Sawada, *J. Biochem. (Tokyo)* **65**, 767 (1969).

[121] F. Sawada and N. Kanbayashi, *J. Biochem. (Tokyo)* **74**, 459 (1973).

[122] A. M. Frischauf and K. H. Scheit, *Biochem. Biophys. Res. Commun* **53**, 1227 (1973).

[123] I. Fiser, K. H. Scheit, G. Stöffler, and E. Kuechler, *Biochem. Biophys. Res. Commun.* **60**, 1112 (1974).

[124] R. Cysyk and W. H. Prusoff, *J. Biol. Chem.* **247**, 2522 (1972).

[125] G. Jori, *Photochem. Photobiol.* **21**, 463 (1975); G. Laustriat and G. Hasselmann, *Photochem. Photobiol.* **22**, 295 (1975).

[126] J. Brandt, M. Fredriksson, and L.-O. Andersson, *Biochemistry* **13**, 4758 (1974). See also this volume [65].

TABLE V
USE OF PHOTOCHEMICALLY GENERATED FREE RADICALS OR TRIPLET
STATES FROM α,β-UNSATURATED CARBONYL COMPOUNDS

Compound	Designed to label	Comments	References[a]
4-Acetylbenzoyl- or 4-benzoylbenzoyl-pentagastrin	Gastrin receptor in acinar cells	Photoattachment to BSA as a model receptor	87
3-(4-Benzoylphenyl) propionyl-Phe-tRNA[Phe]	E. coli ribosomes	Labels 23 S in presence of poly(U); localization within RNA	127
19-Nortestosterone acetate	Active site of Δ^5-3-ketosteroid isomerase	Photoattachment and enzyme inactivation	128
Quercetin (3,3',4,5,7-pentahydroxyflavone)	Glucose transport in erythrocytes	Photoinactivation; D-glucose protects	129
Phenacyl- and naphthacyl-α-chymotrypsin (on Met-192)	α-Chymotrypsin	Photodeacylation and reactivation, but some activity loss	130
Strophanthidin 3,5-bis-4-benzoylbenzoate	Ouabain binding site of brain Na^+,K^+-ATPase	Loss of enzyme activity on photolysis but much nonspecific attachment	131
6-Oxoestradiol	Estrogen receptor of rat uterus	Photoattachment	24
From thiocarbonyl compounds			
4-ThioUTP, or poly[d(4-thio T)]	RNA polymerase from E. coli	Photoinactivation and labeling of β and β' subunits	122
Poly[(4-thio U)]	E. coli ribosomes	Photoattachment to S1 of 30 S subunit in presence of tRNA[Phe]	123
4-Thio UMP	Ribonuclease	Inactivates, loss of amino acids	120, 121
From haloaryl compounds			
Iodohexestrols	Estrogen binding protein	Affinity for site decreases with increasing iodination; no photoattachment	24, 132
5-Iodo-2'-deoxy-uridine	Thymidine kinase	Enhanced activity loss on photolysis; thymidine protects	124
5-Bromo-2'-deoxyuridine-substituted *lac* operator	Complementary sites on *lac* repressor	Suggest intermolecular cross-linking	133
From various aromatic chromophores			
Poly(U)	E. coli ribosome	Labeling of S1 of 30 S subunit, and to 16 S RNA	134
Ribosomal RNA	E. coli ribosomes	Cross-linking of 50 S subunit proteins and ribosomal RNA	135

TABLE V (*Continued*)

Compound	Designed to label	Comments	References[a]
Chloramphenicol	50 S subunits of *E. coli* ribosomes	Photoattachment, but label distributed among most of the proteins; erythromycin protects against inactivation but not labeling	136
Puromycin	*E. coli* ribosomes	Labels protein L23	30
DNA	DNA polymerase	Intermolecular cross-linking observed	137
Poly[d(A-T)$_n$·d(A-T)$_n$]	RNA polymerase	Intermolecular cross-linking observed	138
Phage fd DNA	Gene 5 protein	Intermolecular cross-linking observed	139
tRNA^{Val}	tRNA^{Val}	Intramolecular cross-linking observed	140
tRNA$^{Tyr}_{1\ and\ 2}$	Tyrosyl-tRNA synthetase	Cross-linking involves 3 regions: dihydro U loop, anticodon loop, and "extra" loop	141
tRNA^{Ille}	Isoleucyl-tRNA synthetase and valyl tRNA synthetase	Cross-linking involves dihydro U stem and loop, and other regions	142
Pyrimidine nucleotides	Ribonuclease	Photosensitized intermolecular cross-linking observed, but not with denatured enzyme	143
Various dyes	Bovine serum albumin	Some photoattachment	126
cAMP	Receptors in testis and adrenal cortex	Some photoattachment	144
Peptides containing 4-nitrophenylalanine	α-Chymotrypsin	Labeling of the enzyme	65
Pyridoxal phosphate	Aspartate transcarbamylase, glutamate dehydrogenase	Photoattachment	145
Triarylethylene derivative	Estrogen receptor of rat uterus	More effective than diazo or azido derivatives	24
n-Butyryl-cAMP (on *N*-6)	Phosphofructokinase	Some photoattachment (compare *N*-6-diazomalonyl)	15

[a] Numbers refer to text footnotes.

As a general rule, if the natural ligand is aromatic, it seems advisable first to irradiate the natural ligand:receptor complex before embarking on the synthesis of a photolabile analog. It has been found in a few cases, e.g., with puromycin[30] or with N^6-butyryl-cAMP,[15] that the chromophore lacking a carbene or nitrene precursor group was nearly as effective in photolabeling as the photolabile ligand analog.

Photolabeling studies involving the generation of free radicals or triplet states are summarized in Table V.[15,24,30,67,87,127-145]

Synthetic Approaches

General Considerations

Structural Requirements

Most of the discussion in this section will center on the synthesis or modification of small molecules as photolabile ligands for macromolecular receptors. The derivatization of proteins and other large molecules will also receive mention.

Close examination of the structural requirements for ligand binding

[127] A. Barta, E. Kuechler, C. Branlant, J. Sriwidada, A. Krol, and J. P. Ebel, *FEBS Lett.* **56**, 170 (1975). See also this volume [81].
[128] R. J. Martyr and W. F. Benisek, *Biochemistry* **12**, 2172 (1973). See also this volume [52].
[129] R. A. Farley, K. D. Collins, and W. H. Konigsberg, *Biophys. J.* **16**, 169a (1976).
[130] G. I. Glover, P. S. Mariano, T. J. Wilkinson, R. A. Hildreth, and T. W. Lowe, *Arch. Biochem. Biophys.* **162**, 73 (1974). See also this volume [71].
[131] T. Tobin, T. Akera, T. M. Brody, and H. R. Taneja, *Res. Commun. Chem. Pathol. Pharmacol.* **10**, 605 (1975).
[132] J. A. Katzenellenbogen and H. M. Hsiung, *Biochemistry* **14**, 1736 (1975).
[133] S.-Y. Lin and A. D. Riggs, *Proc. Natl. Acad. Sci. U.S.A.* **71**, 947 (1974).
[134] M. Schenkman, D. C. Ward, and P. B. Moore, *Biochim. Biophys. Acta* **353**, 503 (1974).
[135] L. Gorelic, *Biochim. Biophys. Acta* **390**, 209 (1975).
[136] N. Sonenberg, A. Zamin, and M. Wilchek, *Biochem. Biophys. Res. Commun.* **59**, 693 (1974).
[137] A. Markovitz, *Biochim. Biophys. Acta* **281**, 522 (1972).
[138] G. F. Strniste and D. A. Smith, *Biochemistry* **13**, 485 (1974).
[139] E. Anderson, Y. Nakashiraa, and W. Konigsberg, *Nucl. Acid Res.* **2**, 361 (1975).
[140] S. A. Kumar, M. Krauskopf, and J. Ofengand, *J. Biochem. (Tokyo)* **74**, 341 (1973).
[141] H. J. Schoemaker and P. R. Schimmel, *J. Mol. Biol.* **84**, 503 (1974).
[142] G. P. Budzik, S. S. M. Lam, H. J. P. Schoemaker, and P. R. Schimmel, *J. Biol. Chem.* **250**, 443 (1975). See also this volume [14].
[143] J. Sperling and A. Havron, *Biochemistry* **15**, 1489 (1976).
[144] R. Antonoff and J. J. Ferguson, Jr., *J. Biol. Chem.* **249**, 3319 (1974). See also this volume [34].
[145] P. Greenwell, S. L. Jewett, and G. R. Stark, *J. Biol. Chem.* **248**, 5994 (1973); F. Hucho, U. Markau, and H. Sund, *Eur. J. Biochem.* **32**, 69 (1973).

should precede reagent design. The tighter the binding of the photolabile ligand to the receptor, the more successful labeling will be.[146] The extent and specificity of labeling[147] are also a function of the lifetime of the photogenerated intermediate, of its rate of dissociation from the receptor, and its relative reactivity towards different functional groups.[146,147] If the lifetime is long and the exchange rate is high, the ligand may become attached to parts of the receptor remote from the ligand binding site. The situation then becomes analogous to normal affinity labeling, and has been termed pseudo-photo-affinity labeling.[93]

The modification of a natural ligand to give a photolabile analog that retains good binding and functional properties may involve a part of the ligand that is unimportant in recognition and binding. This means that the modified part of the molecule may be exposed to the bulk solvent, to the lipid phase of a membrane, or to a neighboring macromolecule, and that little labeling of the target macromolecule will occur on photolysis. In practice then, it is advisable to synthesize a range of nonradioactive reagents to evaluate photoinactivation (for caveats, see below), before investing in a radioactive synthesis. The success of more than one reagent in the eventual labeling experiments must make the results of the work more certain and, in the case of the labeling of purified macromolecules, can in principle help to map the ligand binding site.

One great advantage of photolabile reagents over chemical affinity reagents should always be exploited: binding or activity assays can be carried out with the modified ligand before photoactivation, and such tests will often confirm the site-specificity of the ligand analog.

Absorption Characteristics

Normally, the ideal photolabile reagent will possess an absorption maximum well clear of the ultraviolet or visible absorption of the receptor system. However, many biological preparations are reasonably stable to irradiation at 254 nm, at least for short times, and photolysis of diazo compounds and aryl azides is often complete well before even the most

---

[146] If the chemical reactions of the photoactivated intermediate were completely nonspecific, the thermodynamic dissociation constant, K_d, would be the primary consideration (assuming that the reactive intermediate has the same K_d). Since there may be highly reactive entities outside the site, such as other macromolecules or water, the rate of dissociation and the chemical half-life of the intermediate are also important. Further, noncovalently bound reaction products can prevent stoichiometric labeling by their occupancy of the ligand binding site.

[147] Specificity may not always be required (e.g., see references cited in footnotes 109–111).

sensitive receptor components have been significantly photolysed.[34] However, in some cases irradiation in the visible is required, and a chromophoric label must be constructed accordingly. For preparations opaque to ultraviolet light, e.g., whole cells, irradiation at visible wavelengths may be helpful, although we have found that compounds buried in membranes may be rapidy photolysed in ultraviolet light.

Other Considerations

Radiolabeled molecules should be synthesized so that the labeled atom is located as close to the photolabile group as possible, and should not be separated from it by any linkages that might be cleaved during subsequent manipulations.

It is essential to have a reagent of high purity, both chemical and radiochemical, since small quantities of impurities with structures similar to that of the reagent under study may have tight binding constants to the target locus[22] or unsuspected binding properties to other loci and thereby prevent efficient labeling, or give misleading binding patterns.

In certain instances, e.g., if a photoprecursor is to be used as an antigen,[104,105,107] very stable photolabile ligands are required, and for such purposes, compounds as reactive as the diazoketones must be avoided.

Limitations of space preclude an exhaustive treatment of synthetic methods, and we therefore summarize only the most useful methods for directly introducing stable diazo and azido groups. Many of the reactions by which diazo groups are introduced have been described in detail for steroid analogs.[27]

Diazo Compounds

Diazoacetyl Compounds

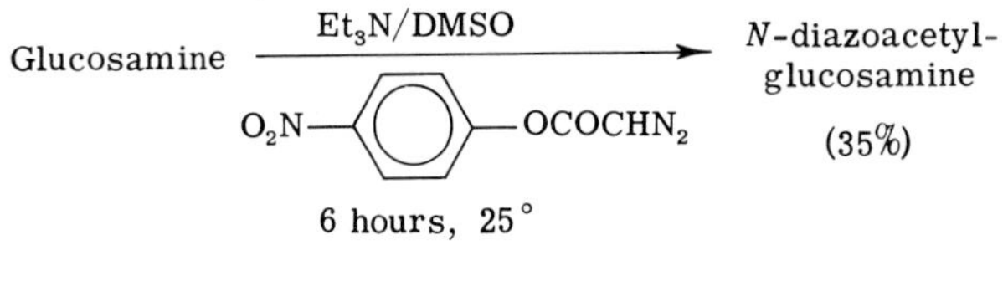

Although the diazoacetyl functionality has the disadvantages discussed earlier, it is one of the less bulky photolabile groups that can be attached to existing functionalities. (The introduction of a simple azido or diazo substituent is even better in steric terms.) The diazoacetyl group is not normally introduced by direct acylation, although in some cases this has been successful,[19] or even necessary.[11,16-18] The nitrophenyl ester can be synthesized with $^{14}COCl_2$ by the improved method of Shafer et al.,[16] but it is not a particularly reactive ester.[25]

Direct diazoacetylation can also be achieved with the tosylhydrazone of OHC-COCl in the presence of Et_3N,[23] but a more commonly used method has been the attachment of glycine, available radioactively labeled, to an amine or hydroxyl group, followed by diazotization with nitrous acid. Two examples are shown below.[20,21]

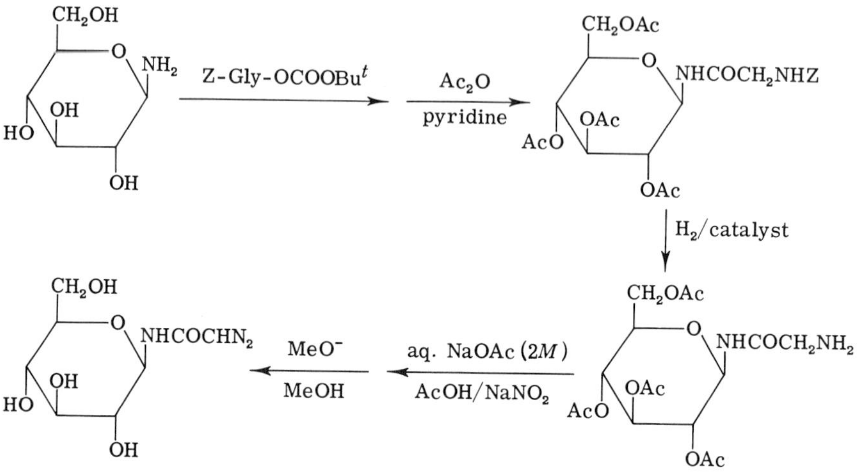

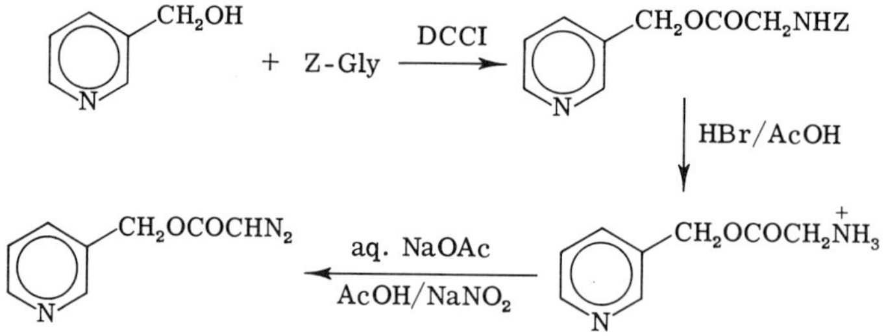

The product diazoacetyl ester is unstable in aqueous acid and must be quickly extracted or otherwise removed from the reaction mixture. The use of the N-carboxyanhydride of glycine is a useful alternative for introducing the glycyl residue.[27]

Franich et al.[148] have found that aromatic amines cannot be converted to the diazoacetyl compound via the 4-nitrophenyl ester or via the car-

[148] R. A. Franich, G. Lowe, and J. Parker, J. Chem. Soc. Perkin Trans. 1, p. 2034 (1972).

bamoyl chloride with diazomethane. However, the following route is useful for certain N-substituted anilines, even though simple anilines give

$$H_3C\!-\!NH\text{-}(C_6H_5) \xrightarrow{\text{diketene}} H_3C\,{\diagdown}_N\!(C_6H_5)\,COCH_2COCH_3$$

$$\xrightarrow[\text{CH}_3\text{CN/Et}_3\text{N}]{\text{CH}_3\text{SO}_3\text{N}_3} H_3C\,{\diagdown}_N\!(C_6H_5)\,COCN_2COCH_3$$

$$\xleftarrow[\text{MeOH}]{\text{MeO}^-} H_3C\,{\diagdown}_N\!(C_6H_5)\,COCHN_2$$

only the triazole rearrangement product.[148]

The trifluoromethyldiazoacetyl group of Chowdhry et al.[8] is likely to be a superior reagent and can be introduced with 3,3,3-trifluoro-2-diazo-propionyl chloride by acylation.

Diazomalonyl Compounds

The more bulky, but more dependable, diazomalonyl group is usually introduced by the reaction of a hydroxy or amino compound with ethyl 2-diazomalonyl chloride.[7,7a,14,27,29,30,35] The treatment of cAMP with this reagent illustrates a number of points.[14]

cAMP $\xrightarrow[\text{10 min, 25°}]{\substack{\text{4-fold excess} \\ \text{reagent,}}}$ O-2′ derivative (20–50%)

$\xrightarrow[\text{overnight, 4°}]{\substack{\text{11-fold excess} \\ \text{reagent,}}}$ O-2′, N-6 disubstituted derivative (20–35%) $\xrightarrow[\text{4 min, 0°}]{1\,N\text{ NaOH}}$ N-6 derivative (quantitative)

The reagent can be prepared with $^{14}COCl_2$, which puts the labeled atom close to the diazo group. Normally the labeled diazomalonyl chloride is generated and used *in situ*,[29,35] although this can lead to 2-chloromalonyl by-products.[29]

The *N*-hydroxysuccinimide ester of ethyl diazomalonate[31] is one of a number of succinimide esters that have been used selectively to acylate amino groups, e.g., the amino acid of charged tRNAs (see also Table VI):

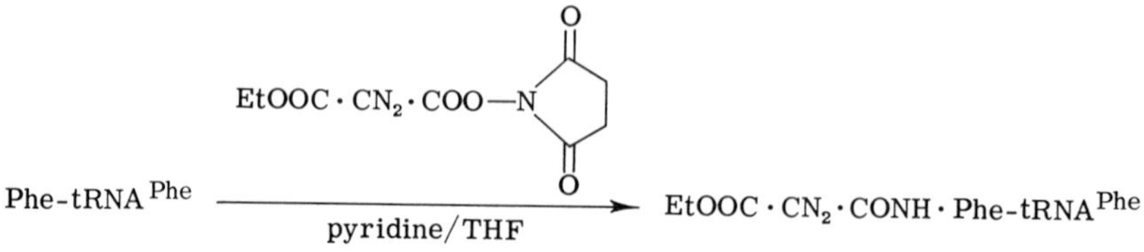
EtOOC · CN2 · COO—N
Phe-tRNA Phe
pyridine/THF
EtOOC · CN2 · CONH · Phe-tRNA Phe

The nitrophenyl ester has also been used to acylate trypsin[7,7a,28] and chymotrypsin.[7]

Diazoketones

Carboxyl groups are simply converted to α-diazoketones as follows[27,36,41,43,44]:

$$m\text{—COOH} \rightarrow m\text{—COCl} \xrightarrow{CH_2N_2} m\text{—COCHN}_2$$

A useful way of introducing radioactivity into a labeling reagent is to link the photolabile group to the ligand by a small radioactive molecule. For diazoketones this is exemplified as follows[41]:

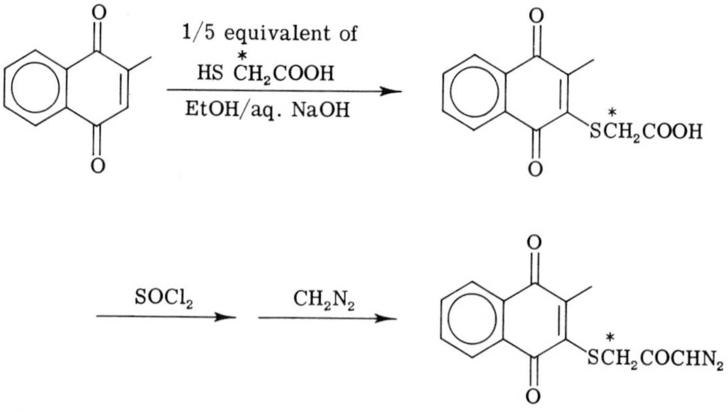
1/5 equivalent of
HS CH2COOH
EtOH/aq. NaOH
SCH2COOH
SOCl2
CH2N2
SCH2COCHN2

Diazo groups have been introduced α to keto groups in a number of steroids[23,27] by the route outlined below.[23] The undesirable presence of hydrogen atoms α to the diazo group should be considered in such molecules.

The hydrolytic lability of ester linkages may make the creation of a diazoketone preferable even in a situation in which a diazo ester could be made more easily.[23]

Estimation of Diazo Compounds

Diazo compounds have characteristic absorption spectra with a strong band in the ultraviolet (λ_{max} 240–280 nm) and a weak band at longer wavelength ($\lambda_{max} \sim 350$ nm, $\epsilon \leq 50$). The intense band can often be used to estimate the number of groups that are incorporated. Both the ultraviolet absorption and the characteristic infrared band, at approximately 2100 cm^{-1}, are useful in following photolyses.

Stability

Most diazo compounds are hydrolyzed or attacked by nucleophiles, e.g., Cl$^-$, in acidic solution. For example, α-diazoamides have half-lives of minutes at pH 5 and seconds at pH 3.[149] Most diazo compounds are

[149] G. Wybrandt, unpublished work.

light sensitive, and synthesis is best carried out in dim incandescent light, rather than in bright daylight or under fluorescent lighting. Some are thermally unstable even at room temperature. Small molecules carrying the diazo group are sometimes explosive and may be sensitive to shock, heat, sharp surfaces, and, occasionally, light. It is not advisable to prepare such reagents in gram quantities.[150]

Unless evidence of high stability has been obtained, these reagents should be stored in dry inert solvents in a freezer. Thin films of radioactive compounds are more susceptible to photolytic decomposition than are large amounts of solid, but it is bad practice to store such materials in solid form.

Aryl Azides

Unlike diazocarbonyl compounds, most aromatic azides are reasonably stable and can be carried through multistep syntheses that do not involve excessive heating, or strong oxidizing or reducing conditions.

Synthesis Directly from Ar-NH₂

The most convenient method is that which has been popularized by P. A. S. Smith, involving diazotization followed by treatment with azide ion.[151]

$$ArNH_3^+ \xrightarrow{HNO_2} ArN_2^+ \xrightarrow{HN_3} ArN_3$$

The aromatic amine is dissolved or suspended in excess hydrochloric or sulfuric acid (which in the case of sensitive compounds can be quite dilute, say 0.1 N)[152] and treated with slightly more than one equivalent of aqueous sodium nitrite at a temperature lower than 0°.[153] When diazotization is complete, slightly more than one equivalent of aqueous sodium azide is added slowly, and the mixture is stirred at room temperature. The azido compound precipitates or can be extracted into an appropriate solvent.

---

[150] Further information on the chemistry of the diazo group can be obtained from "Houben-Weyl: Methoden der Organischen Chemie." Thieme, Stuttgart; Band IV/5a, p. 1158 (1975); Band X/4, p. 473 (1965).

[151] See for instance, *Org. Synth. Coll.* IV, 75 (1963).

[152] But even N-t-butoxycarbonyl-p-aminophenylalanine can be converted to the aryl azide by this method in 2 N HCl without loss of the protecting group.[63]

[153] In certain cases, for instance in the preparation of 4-azido-2-nitrofluorobenzene,[105] lower temperatures (−20°) are required. In general −5° is preferred for fast diazotization.

The reaction has been used with success on many molecules of biological interest,[21,23,51,57,62-68,71-73,75,115] e.g.[51]:

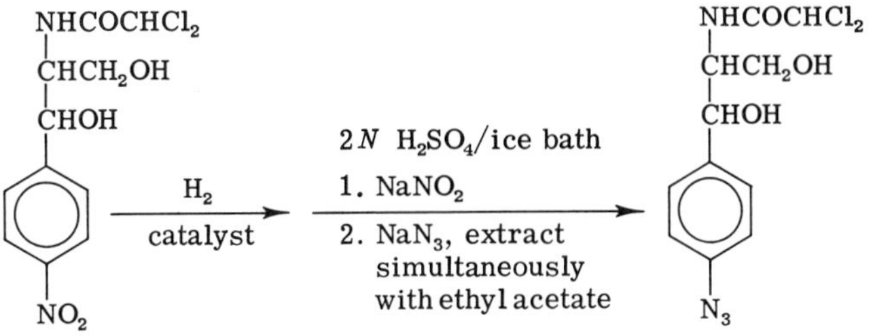

NHCOCHCl$_2$
CHCH$_2$OH
CHOH

$\overset{\text{H}_2}{\underset{\text{catalyst}}{\longrightarrow}}$

2N H$_2$SO$_4$/ice bath
1. NaNO$_2$
2. NaN$_3$, extract
   simultaneously
   with ethyl acetate

NHCOCHCl$_2$
CHCH$_2$OH
CHOH

NO$_2$ (left ring)   N$_3$ (right ring)

Chloramphenicol

If necessary, the procedure can be carried out in organic solvents[21,73,115] with an organic nitrite[21] and an azide with an organic counterion. In many procedures, methods of detecting and destroying excess nitrite and detecting the endpoint of the reaction with azide are employed, but these are often unnecessary.

Synthesis by Displacement of Halide Ion

The most useful reagents that have been prepared by this method are the azido-nucleotide phosphates,[52,54,56] e.g.[54]:

8-Bromo-AMP $\xrightarrow[\substack{\text{tri-}n\text{-octylammonium}\\ \text{azide}}]{\text{DMF}}$ 8-azido-AMP $\rightarrow$ 8-azido-ATP
                                         (80%)   (30%)

The method is of general utility; for instance it has been used to prepare various azidoquinones[154] and to prepare dinitrophenyl azide.[37]

Reagents for the Introduction of Arylazide Groups

Since many biological molecules do not contain aryl groups or do not contain groups that can be derivatized as described above, we have listed in Table VI some useful reagents for attaching photolabile groups to such molecules.

Estimation of Aryl Azides

The aryl azides have strong ultraviolet absorptions often with characteristic shoulders on the long wavelength side of the peak, e.g., 4-azido-benzoic acid esters, $\lambda_{max} \sim 276$ nm, and 4-azido-N-methyl-2-nitroaniline,[105] λ_{max} 258 nm (ϵ 26,750) and 458 nm (ϵ 4640). These bands usually disappear or are considerably diminished on photolysis.

[154] H. Bayley, unpublished work (1975).

TABLE VI

USEFUL INTERMEDIATES IN THE SYNTHESIS OF AZIDOARYL REAGENTS

Reagent	Special use	References[a]	Comments
(benzene ring with N$_3$, O$_2$N, F substituents)	Conjugation to protein amino groups	88, 99, 105, 107	Organic or partly organic solvent used Protein can be iodinated after derivatization
	Derivatization of amines, e.g., amino acids	85, 88, 105, 109, 110	DMSO/Et$_3$N (e.g., 88) is a superior medium for many such reactions[b]
	Derivatization of sugars (on O, S, N)	95, 97	
	Derivatization of lipids on head group or on fatty acid side chain	35	
	Other uses	36, 77, 78	
(benzene ring with N$_3$, O$_2$N, F substituents)	Comparative studies with 4-azido isomer	106	Similar methods to those described for the 4-azido isomer may be used in chemical synthesis
(benzene ring with NO$_2$, N$_3$, F substituents)	Conjugation with cytochrome c	100, 101	1000 times more reactive with amine than the mononitro compounds (may form furoxan on photolysis)

Structure	Application	References	Comments
ϕ—CH$_2$—C(NH$_2$)(H)—COOH	Peptide synthesis	63–65, 68 67	Various useful intermediates described. Tritiated amino acid of high specific activity prepared
N$_3$—〈ring, NO$_2$〉—CH$_2$—C(NH$_2$)(H)—COOH	Peptide synthesis	65, 66	Absorbs visible light (N.B., p-nitro-Phe itself, also appears to be useful as a photolabel)
HO—〈ring, NO$_2$〉—NHCOOCH$_2\phi$	Derivatization of sugars	21, 97	Protecting group subsequently removed and amine converted to azide
HN—(CH$_2$)$_n$—COOH ; 〈ring, NO$_2$〉—N$_3$	Coupling to ATP	102	These compounds can be better prepared by coupling the appropriate α-amino-acid to 4-azido-2-nitro-fluorobenzene in dry DMSO with triethylamine[b]

TABLE VI (Continued)

Reagent	Special use	References[a]	Comments
N_3—C$_6$H$_4$—O—P(=O)(OH)$_2$ (4-azidophenyl phosphate)	Coupling to the terminal phosphate groups of nucleotides	84	—
N_3—C$_6$H$_4$—COCH$_2$Br	Selective reaction with 4-thio-U residue on certain tRNA molecules, and of reactive protein nucleophiles	79, 114	Preparation of material with ^{14}C in carbonyl group described
N_3—C$_6$H$_4$—NH$_2$	Label of terminal phosphate of nucleotides	82	—
N_3—C$_6$H$_3$(NO$_2$)—NH$_2$	Useful synthetic intermediate	b	From fluoro compound with dry methanolic ammonia, room temperature, 1 wk

Compound	Use	Reference	Comments
N_3–C$_6$H$_4$–COOH	Coupling to ribose hydroxyl group in ATP	102	
N_3–C$_6$H$_4$–COCl	Useful synthetic intermediate	87, 114	Can be prepared from p-aminobenzoic acid with ^{14}C in carboxyl group[c] in high yield[114b]
N_3–C$_6$H$_4$–NHCOCH$_2$Br	Useful intermediate	114, b	^{14}C compound can be obtained from carboxylic acid in high yield with thionyl chloride[114b]
	Attachment to aminoacyl-tRNA, on 4-thio-U	82	—
O$_2$N,N_3–C$_6$H$_3$–NHCOCH$_2$Br	Attachment to reactive nucleophiles	b	From amine with BrCH$_2$COBr in dichloromethane

TABLE VI (*Continued*)

Reagent	Special use	References[a]	Comments
N-hydroxy succinimide esters of	Cleanly label free amino groups		
	Peptide synthesis	87	—
	Peptide synthesis	87	With radioactive glycine
	Peptide synthesis	87	—
	Peptide synthesis	87	With radioactive glycine
	Attachment to aminoacyl-tRNA on free amino group	78	With radioactive glycine
	Peptide synthesis	77	—

[a] Numbers refer to references cited in text footnotes.
[b] H. Bayley and J. R. Knowles, unpublished data.
[c] ICN Pharmaceuticals, Inc., Cleveland, Ohio.

The absorption characteristics have been used to estimate the number of nitroazidophenyl groups that are attached to protein derivatized with 4-azido-2-nitrofluorobenzene.[105] The ultraviolet-visible absorption or the characteristic infrared band at approximately 2100 cm^{-1}, often broad or a doublet, can be used to follow photolyses.

Stability of Aryl Azides

Most aryl azides are stable in the solid state or in solution. The nitroaryl azides are, in general, less stable to heat and light and should be recrystallized with care to avoid thermal decomposition. As with diazo compounds, they are best handled in dim incandescent light. Low-molecular-weight aryl azides are explosive when heated and should be handled carefully, especially in the solid state.[155]

Radioactive azides stored as solids often darken, but radiochemical purity is little affected. Nevertheless, they are best stored in solution to minimize radiolytic decomposition. Films containing a few micrograms per square centimeter of material are especially sensitive to light.

Photolysis

Apparatus

Irradiation in the ultraviolet, near-ultraviolet, or visible can be achieved satisfactorily by simple means although elaborate devices have been used.

For irradiation at 254 nm or "350 nm," the inexpensive Rayonet Minireactor[156] is ideal for studies on the microliter to tens of milliliters scale; a number of samples can be irradiated concurrently. The sample is contained in a quartz, jacketed cell of 1-cm pathlength through which ice water can be rapidly circulated, and which is clamped 3–4 cm from the center of the single lamp.

The 254-nm lamp (Rayonet RPR-2537 Å) emits little radiation below

---

[155] Further information on the chemistry of the azido group can be obtained from the following sources: "The Chemistry of the Azido Group" (S. Patai, ed.). Wiley (Interscience), 1971; "Houben-Weyl: Methoden der Organischen Chemie." Thieme, Stuttgart; Band IV/5a, p. 1260 (1975); Band X/3, p. 777 (1965).

[156] Rayonet Photochemical Mini-reactor, Model-RMR-400, The Southern New England Ultraviolet Company, Middletown, Connecticut. The RPR-3000 Å lamp, when appropriately filtered to remove 254-nm light, can be used to irradiate samples in the 275–325 nm range.

this wavelength. However, an acetic acid solution can be used to filter out the shorter wavelengths if required.

The "350 nm" lamp (Rayonet RPR-3500 Å) emits a broad band of radiation between 300 and 420 nm. Various Corning filters can be used to limit the light emitted at the lower end of this range. Details of these and other filters and further useful information on photolysis equipment may be found in the monograph by Calvert and Pitts.[157]

Photolysis at visible wavelengths may be necessary with sensitive materials or with preparations that are opaque to ultraviolet light. The use of Mazda 125 watt MB/V pearl glass lamps or a movie-projector lamp filtered through 0.1–1.0 M aqueous $NaNO_2$ will allow irradiation at >400 nm.[96,105,107,157]

A 1-cm pathlength cell is not ideal for the photolysis of samples with high absorbance, wherein the illumination of the back of the sample is considerably reduced. This problem can be overcome by using thin films[24,109] or a cell of shorter pathlength. For samples of large volume, thin films are not practical, and stirring may be necessary.

Before the design or synthesis of photolabile ligands, it is useful to determine the half-life of the biological system by whatever assays are available, under irradiation at 250, 350, and >400 nm. This can put limits on the molecules that might be used as reagents. However, in a number of systems it has been found (e.g., ref. 34) that the rate of ligand photolysis at 254 nm compares favorably with the rate of destruction of the biological system. Further, it has been observed that diazo and azido aryl compounds can show higher labeling efficiencies on irradiation at 254 nm than by illumination with light of longer wavelengths.[33,24,158] In contrast, Katzenellenbogen et al.[24] have shown that the efficiency of labeling by a triarylethylene derivative is higher at *longer* wavelengths.

Photolysis at 254 nm can be much faster than is often supposed. Thus, irradiation of the herbicide analog N-4-azidobenzoyl-3,4-dichloroaniline in methanol is half complete in 20 sec (70 μM, 4 cm from the Rayonet lamp), and on irradiation in chloroplast membranes the half-life increases only to 45 sec (150 μM, 3.4 cm from the lamp).[159]

---

[157] J. G. Calvert and J. N. Pitts, Jr., "Photochemistry." Wiley, New York, 1966.
[158] Some authors have also found that the extent of labeling depends on temperature. At present we can offer no guidance on this effect, other than the obvious statement that if K_d decreases at low temperatures, then both the extent and specificity of labeling will be improved by cooling the sample. The temperature at which photolysis of azides is carried out can dramatically affect the reactivity of the nitrene that is formed. See, e.g., J. M. Lindley, I. M. McRobbie, O. Meth-Cohn, and H. Suschitzky, *Tetrahedron Lett.*, p. 4513 (1976).
[159] H. Bayley, unpublished observations (1975).

Rapid photolysis is important in the case of chemically unstable ligands and can usefully minimize slow "dark" reactions undergone by some photochemically derived products. The ligand–receptor preparation can often be photolyzed in a few minutes and then be dialyzed or subjected to gel filtration immediately.

A standard photolysis reaction, which may simply be a solution of the reagent under study, should be used to monitor relative radiation doses in subsequent experiments under different photolysis conditions— for instance, for reactions in a different cell, or at a different distance from the center of the lamp. (The radiation dose can vary considerably along the length of a lamp and with the age of a lamp.)

Photolysis should be carried out under an atmosphere of nitrogen or argon to prevent chromophore-sensitized photoxidation of the biological receptor system.[125] Solutions should be purged with the gas at a rate of approximately 0.2 ml of gas per milliliter of solution per second for a minimum of 30 min before photolysis. Microliter samples (or preparations that do not survive the bubbling of gas) may be treated by degassing on a pump followed by flooding with nitrogen at least ten times. In some cases oxygen is required for coupling of the reagent, as with thiocarbonyl compounds,[120,132] and in many cases with other reagents, the presence of oxygen has not prevented the achievement of satisfactory results.

Reagent Concentration

The concentration of reagent is important. In cases in which the system can be considered as a simple equilibrium: receptor + ligand ⇌ receptor-ligand, labeling specificity at the expense of a high extent of labeling is maximized by a high receptor concentration (greater than the ligand dissociation constant) and a low ligand concentration. Repeated labeling or the use of a reagent of very high specific radioactivity may be required to obtain useful labeling levels.

Where a high extent of site labeling is required (e.g., in studying the effects of a permanently switched hormone receptor site), excess ligand (at a concentration greater than the ligand dissociation constant) over receptor is called for, and a repeated labeling may be necessary. In those cases in which there is tight binding or slow exchange, it is possible to saturate a receptor and then remove excess ligand, thereby producing the ideal 1:1 complex (see references cited in footnotes 36, 45, 105, 107, 110). These simple guidelines can be used as a basis for thinking about more complicated systems. For instance, labeling experiments with different concentrations of reagent, wherein receptor sites in complex assemblies are under study, can give useful information about the number and nature of these sites.[52]

Often the reagent must of necessity be added in organic solvent. It is essential that the quantity of solvent be very small, since subtle changes can be induced in biological preparations by organic solvents at concentrations so low that obvious denaturation is undetectable. The uptake of reagent into binding sites can be rather slow, and proper equilibration must be obtained before photolysis; equilibration of this sort may not be possible with reactive conventional affinity reagents.

Duration of Photolysis

Prolonged photolysis can result in such undesirable effects as slow nonspecific labeling by photolysis products and destruction or alteration of the binding site, as a result of which further photolysis can lead only to nonspecific labeling.

Rather than indulging in time-consuming determinations of quantum yields and radiation doses, the more empirical approach is almost always sufficient. However, when new types of photolabile reagents are being considered, it is advisable to check the literature on the quantum yields at different wavelengths.

A rough idea of a reasonable photolysis time can be determined by irradiation of the molecule in question at the appropriate concentration in a suitable solvent. The photolysis is usually followed by absorption spectroscopy or by monitoring of the appropriate infrared band. However, examples are known in which the sensitivity of the ligand to irradiation changes dramatically on binding to the receptor; the half-life of the ligand alone is only a very approximate guide to the behavior of the ligand:receptor complex. For example, p-azidocinnamoyl-α-chymotrypsin is considerably more sensitive to light than is the model compound p-azidocinnamoyl methyl ester; darkroom techniques are needed for the former, whereas the latter survives handling in full daylight.[160] Conversely, because of screening effects in the biological preparation due to light absorption by the receptor, the half-life of the reagent may be increased on binding. In complex systems the photolysis can often be followed by ultraviolet or visible difference spectroscopy or by quantitative thin-layer chromatography of the remaining radioactive ligand.

The addition of scavengers or protecting agents (see below) can cause increased absorption of the sample. This may be corrected for by the addition of an inert substance to give a constant absorbance[54] at the

---

[160] A. Bridges, unpublished observations (1972).

wavelength of irradiation or by the use of a split photolysis cell in the front compartment of which solutions of the scavengers or protecting agent can be placed as absorption controls.[93]

Repeated Treatment

A single period of irradiation of a ligand:receptor complex may not yield complete labeling. Repeated treatments can be effective,[41,84] but, since the noncovalently bound reaction products can block the receptor site, methods have been developed to remove these products before further treatment. These methods involve dialysis[34] (the most generally useful), gel filtration, or immobilization of the receptor on an ion-exchange column[34] through which the ligand and its photolysis products can flow. In certain cases, affinity columns may be used to separate specifically labeled material, the remainder being recycled. These methods involve the exposure of the sample to a larger total radiation dose, which may have deleterious effects on both biological activity and ligand-binding properties.

Identification of Products and Essential Control Experiments

Most of the remarks in this section are intended to be of general applicability. Assays and controls peculiar to a given system will not be discussed explicitly. Many of the methods mentioned are treated in detail elsewhere in this volume.

Extent of Labeling

Binding Studies

The normal method of determining whether labeling has occurred is to perform a standard assay on the preparation after dilution or removal of the remaining ligand and its photolysis products. If an enzyme active site has been blocked, it will no longer be active.[24] If a hormone binding site has been labeled, it will no longer bind hormone.[65] However, this approach can be misleading, and consideration must be given to the following points:

1. It may be difficult to remove unreacted ligand or photolysis products that are bound noncovalently to the receptor, without denaturing the receptor.

2. Even if irradiation of the receptor alone does not result in loss of binding capacity, it must be remembered that ligands can either enhance

the sensitivity of enzymes to photolytic inactivation, or protect them (see, e.g., Cysyk and Prusoff[124]). The bound label may also mediate photooxidation or other photosensitized destruction[125] of the receptor site. This type of inactivation will show saturation with respect to ligand, just as proper photoaffinity labeling should.

3. Because of the lack of chemical specificity of photoaffinity labeling, the labeled material will be a mixture of closely related products. Some of these products may retain partial biological activity.

It should be noted that, in certain kinds of experiments, one requires biological activity to be unaffected by the photochemical coupling.[56,77,78,101]

Radioactive Labeling

The binding method is a useful screening procedure for prospective photochemical labels, but must be accompanied by radioactive labeling studies if it is to be of real value. Radioactivity gives the most dependable measure of the extent of labeling, but great care must be taken to remove noncovalently bound label. Indeed, the success of a labeling experiment usually depends upon tight noncovalent binding of ligand analogs. In work on antibody labeling, it has been clearly demonstrated that prolonged dialysis, even in the presence of a nonradioactive hapten, is not completely effective in removing bound radioactive ligand.[107] In this case, the covalently bound label was measured by precipitation of the protein in trichloroacetic acid followed by extensive washing of the precipitate. Frequently, even this method gives misleading results, since ligand may be physically entrapped in the protein precipitate. A more reliable method is gel filtration in the presence of denaturing levels of sodium dodecyl sulfate. In many cases, suitable controls have shown that both the photolabile ligand and its noncovalently bound products are removed by less drastic procedures, including gel filtration or pelleting the labeled preparation in the presence of nonradioactive ligand. For macromolecular *ligands*, gel electrophoresis may be required to assess the extent of binding.

General Controls

The following is a list of controls that should normally be performed.

1. Photolyze the biological preparation alone. Test for biological activity. Will the receptor still bind ligand? Can the receptor still be photolabeled by the ligand analog? With relatively small single protein receptors, loss of sensitive amino acids, such as cysteine or tryptophan,

can be determined by amino acid analysis.[87] Cross-linking[28] or cleavage[84] of macromolecules can be detected in simple systems by gel electrophoresis. Cross-linking can be reduced in the case of monomeric proteins by labeling in more dilute solution.

2. Does the preparation react with the ligand analog in the *dark* during the time required for a typical photolysis? During prolonged photolyses, e.g., diazo compounds at 350 nm, a diazoacetyl derivative may label the receptor in a dark reaction by an electrophilic mechanism. Analysis of such incubation mixtures also tests for removal of noncovalently bound material in the work-up procedure.

3. Prephotolyze the reagent. Do the ligand photoproducts react with the receptor with or without photolysis? Do the products protect the receptor from fresh reagent? This information can be useful in evaluating the possible success of renewal experiments in which the label is renewed and photolysis continued.

4. The labeling site should be both saturable and show saturation kinetics for photochemical labeling. The half-maximally effective ligand concentration should be close to the binding or inhibition constant of the ligand in a biological assay (see, e.g., references cited in footnotes 24, 52, 60, 95, 96).

5. The labeling site should be protected by the natural ligand, if this is inert in light, and by nonphotolabile ligand analogs. For example, it would be desirable for a sugar binding site to be protected by the sugar itself from labeling with an azidonitrophenyl derivative, but not if the labeling was reduced on photolysis in the presence of free azidonitrophenol. Although a high concentration of protecting natural ligand will successfully compete with the photolabile analog for the receptor sites, irreversible attachment of the photolabile analog after relatively long irradiation times[161] can still lead to high levels of site labeling. The reversible binding equilibria are continuously perturbed by the irreversible blockage of the receptor by the photolabel.[53] Photolabeling with nonradioactive reagent should also protect the site against subsequent photolabeling by the radioactive species. Allosteric effectors and other molecules that affect ligand binding indirectly should also inhibit or stimulate photolabeling of the target locus.[124]

Nonspecific Labeling

Low extents of labeling, and nonspecific labeling, have been the two major difficulties encountered in photochemical labeling studies. Low

[161] Long, that is, with respect to the ligand residence time.

extents of labeling can often be dealt with by the design of the reagents and by the repetitive labeling experiments described above.

By nonspecific labeling, we mean labeling that occurs outside the receptor site.[93] This can be on the same macromolecule (a difficulty encountered in studies of the hapten binding site in antibodies, if precautions are not taken to remove excess reagent) or on different "irrelevant" macromolecules in more complex systems (a difficulty encountered in labeling receptors in membranes). In general, such labeling is not saturable.[24,30,95,96]

Alleviation of these difficulties may be approached in two ways:

1. A protection experiment is performed and the results are compared with those of the normal labeling experiment. This technique has been used so often in conventional labeling experiments that it need not be expanded upon here. The protecting agent does not prevent nonspecific labeling but indicates what proportion of the observed labeling is relevant.

2. The photolabeling is performed in the presence of scavengers, which are intended to destroy all photogenerated intermediates at places other than the ligand binding site. Random labeling by activated molecules created outside the receptor site, or even by those of long half-life initially created within the site, can to some extent be stopped by scavengers. Such molecules as p-aminobenzoic acid,[93,95,96] hydroquinone,[86] soluble proteins,[93] p-aminophenylalanine,[65] dithiothreitol,[84,96] β-mercaptoethanol,[84] and Tris[162] have been proposed as scavengers (Fig. 1).

There are three caveats. First, if the protecting molecule or scavenger absorbs at the wavelength required to activate the photolabile ligand, proper controls must be carried out to ensure that any observed reduction in labeling is not simply due to a lower incident radiation level. Second, the protecting molecules obviously should, but the scavengers should not, bind to the receptor site. This must be tested. Last, if proteins such as bovine serum albumin are added as scavengers, it must be ascertained that they are not themselves binding the ligand.

With tight binding sites, scavenging need not normally be considered; with very weak binding sites, it may be ineffective.

Location of the Site of Labeling

Identification of Receptor Macromolecules

The simplest method of characterizing an affinity-labeled macromolecular complex is by electrophoresis. Other techniques may be applicable,

[162] Photolysis in certain buffers (e.g., Tris) can cause inhibition of enzymes. Buffer effects should be checked, and inorganic buffers used where possible.

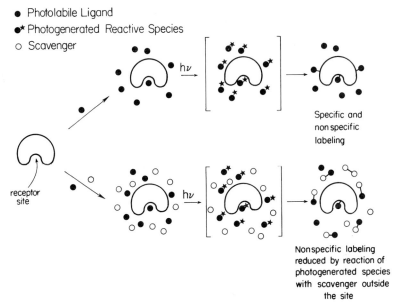

- ● Photolabile Ligand
- ●★ Photogenerated Reactive Species
- ○ Scavenger

hν

Specific and
nonspecific
labeling

receptor
site

hν

Nonspecific labeling
reduced by reaction of
photogenerated species
with scavenger outside
the site

FIG. 1. The use of scavengers to increase the proportion of specific receptor site labeling.

or cruder methods may be used when it is desired only to track the label down to a certain fraction of the complex. These methods are fully treated in other volumes of this series.

Before subjecting the photolysed sample to electrophoretic analysis, it is essential that all noncovalently bound ligand be removed. We have found that certain reagents give radioactive bands on gels in sodium dodecyl sulfate which do not appear only at the bottom of the gel. Photolysis products can also lead to a smear of radioactivity near the gel front; this has been attributed to labeled lipid. In the case of membranes, careful washing of the preparation with solvents appropriate for the dissolution of the reagent and its photoproducts can clear the background of the gels considerably. In other systems, extensive dialysis, gel filtration, or gradient ultracentrifugation may be more appropriate.

Autoradiography of samples run in parallel on slab gels is often the method of choice for comparing samples and controls.[52,53] With the advent of autofluorography,[163] even tritium-labeled samples are readily detected.

Finally, a number of investigators have observed that labeled proteins

[163] W. M. Bonner and R. A. Laskey, *Eur. J. Biochem.* **46,** 83 (1974); R. A. Laskey and A. D. Mills, *Eur. J. Biochem.* **56,** 335 (1975).

may not comigrate exactly with their unlabeled counterparts on electrophoresis.[30,76,77] This is particularly apparent on two-dimensional gels.

Localization of Binding Sites within Receptor Macromolecules

Often, the sole purpose of a photolabeling experiment has been to map a receptor site in order to identify the particular amino acids with which the reagent has reacted.[7,11,16-18,36-39,44,45,75,105,107] Since the usual techniques of protein chemistry are applicable, only those points particularly relevant to photolabeling are noted here.

1. Since levels of labeling by photoaffinity reagents are usually much lower than those by chemical reagents, such methods as affinity chromatography are especially important in the isolation of specifically labeled material.[107]

2. The exact nature of linkages formed to receptors by photolabile reagents has not been clarified fully. A few attempts to locate labeled amino acids have indicated that some of the linkages may be chemically labile, since a considerable amount of radioactivity is lost during the analysis. For instance, significant losses of nitrene-derived label have been shown to occur on treatment with cyanogen bromide.[107]

3. The electrophoretic mobilities and other physical properties of small peptides that are important in their separation may be greatly changed by the covalent binding of the label.

4. Since the labeling by the reagent is intentionally nonspecific toward receptor site residues, the analysis of smaller fragments can become very difficult. For a labeled protein, one can find reasonably coherent distributions of radioactivity for peptide fragments of 30 residues or so, but on further digestion a large number of only slightly different oligopeptides may result. This is an inevitable consequence of the randomness of the labeling process.[75,105,107]

5. As with chemical affinity labeling, the modification of an amino acid may interfere with protein fragmentation reactions, i.e., the Edman degradation, and cyanogen bromide cleavage.[39]

6. The indiscriminate nature of the photolabeling process can produce a range of chemically similar species. Gas-chromatography mass spectrometry is likely to prove a powerful method for the analysis of such multiple products.[44,45]

Despite these difficulties, particular residues at ligand binding sites have been successfully identified by photoaffinity labeling.

Acknowledgments

We are grateful to Professor F. H. Westheimer for copies of his work prior to publication, and to the National Institutes of Health for support.

[9] Design of Exo Affinity Labeling Reagents

By Michael Cory, Judith M. Andrews, and David H. Bing

In 1964, Baker distinguished two kinds of affinity labeling reagents, endo and exo.[1] Endo affinity labeling reagents are those that covalently bind to amino acids within the active center. They are of the type described by Wofsy *et al.*[2] and are discussed in detail in other sections of this volume. Exo affinity labeling reagents are those that covalently bind to an amino acid outside the active center and consist of three regions of activity, S-B-X. The group S is designed to be complementary with the enzyme active site. The chemically reactive group, X, covalently binds the label to the enzyme. The bridging group, B, links S to X and ensures that

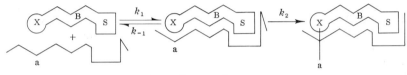

SCHEME 1

X reacts with an amino acid outside the enzyme active site. Endo affinity labeling reagents interfere directly with, and thus identify, amino acids involved in the catalytic process or in the binding of the specific ligand. In a few special cases, endo reagents have been used to distinguish enzymes with the same catalytic mechanism but different substrate specificities.[3] In contrast, the bulkier exo affinity labeling reagents offer greater opportunity for covalent binding to those amino acids that are away from the catalytic center but are involved in determining substrate specificity.

Excellent examples of exo affinity labeling reagents are the benzamidine sulfonylfluorides described by Baker and Cory.[4,5] In our laboratory we have used these reagents to distinguish the biological and biochemical activities of C1 esterase, a serine protease component of

[1] B. R. Baker, *J. Pharm. Sci.* **53**, 347 (1964).
[2] L. Wofsy, H. L. Metzger, and S. J. Singer, *Biochemistry* **1**, 1031 (1962).
[3] E. Shaw, *in* "Proteases and Biological Control" (E. Reich, D. B. Rifkin, and E. Shaw, eds.), p. 455. Cold Spring Harbor Laboratory, Cold Spring Harbor, New York, 1975.
[4] B. R. Baker and M. Cory, *J. Med. Chem.* **14**, 119 (1971).
[5] B. R. Baker and M. Cory, *J. Med. Chem.* **14**, 305 (1971).

TABLE I

BENZAMIDINE SULFONYL FLUORIDE INHIBITORS OF $C\bar{1}s^a$

			Percent inhibition		
R		n	Venule leakage[b]	Catalytic activity[c]	Esterolysis[d]
1	$C_6H_4SO_2F\text{-}m$	2	71	95	90
2	$NHC_6H_4SO_2F\text{-}p$	3	75	100	67
3	$NHC_6H_4SO_2F\text{-}m$	3	77	93	85
4	$C_6H_3\text{-}2\text{-}Cl\text{-}5\text{-}SO_2F$	3	46	100	14
5	$C_6H_2\text{-}2,4\text{-}Cl_2\text{-}5\text{-}SO_2F$	3	80	100	100
6	$NHC_6H_2\text{-}2,4\text{-}(CH_3)_2\text{-}5\text{-}SO_2F$	3	60	100	90
7	$NHC_6H_3\text{-}2\text{-}Cl\text{-}5\text{-}SO_2F^e$	3	75	92	87
8	$NHC_6H_3\text{-}2\text{-}Cl\text{-}5\text{-}SO_2F^e$	3	75	45	75
9	$NHC_6H_3\text{-}4\text{-}Cl\text{-}3\text{-}SO_2F$	3	46	83	25
10	$NHC_6H_2\text{-}2,4\text{-}(CH_3)_2\text{-}5\text{-}SO_2F$	3	50	0	60
11	$NHC_6H_3\text{-}2\text{-}Cl\text{-}5\text{-}SO_2F$	4	64	100	56
12	$NHC_6H_4\text{-}4\text{-}SO_2F$	4	82	92	67

[a] J. M. Andrews, F. S. Rosen, S. J. Silverberg, M. Cory, E. E. Schneeberger, and D. H. Bing, *J. Immunol.* **118**, in press (1977).

[b] Venule leakage in guinea pig skin as measured with Evans blue. Percent inhibition is average lesion diameter as compared to the average diameter of a positive control.

[c] Catalytic activity as measured by EAC42 formation.

[d] Esterolysis of the synthetic substrate, *N*-carbobenzoxy-L-tyrosine-*p*-nitrophenyl ester.

[e] 7 is the picrate salt and 8 is the TsOH salt.

complement.[6] Some of these reagents and their reactivity with C1 esterase are presented in Table I. The data compare the ability of these compounds to inhibit three known activities of C1 esterase: leakage from the post-capillary venules of guinea pig skin, catalysis of EAC42[7] formation, and

[6] J. M. Andrews, F. S. Rosen, S. J. Silverberg, M. Cory, E. E. Schneeberger, and D. H. Bing, *J. Immunol.* **118**, in press (1977).

[7] Terminology for the complement system is that suggested in the *Bull. W.H.O.* **39**, 935 (1968). Thus C1 esterase (C1s̄) is the enzymically active form of C1s, the third subunit of the first component (C1) of human complement. The other subunits of C1 are C1q and C1r; C2 and C4 are the second and fourth components of complement EA, sheep erthrocytes treated with antisheep erthrocyte antiserum.

the esterolysis of N-carbobenzyoxy-L-tyrosine-p-nitrophenyl ester. Some of the reagents tested inhibited all three activities, whereas others inhibited only one or two of the activities of the protease. Small changes in the structure of the compounds, e.g., compounds *8, 9, 11*, led to many changes in the nature of the inhibition, a result consistent with the hypothesis that these compounds are exo affinity labeling reagents.

Synthesis of Benzamidine Sulfonyl Fluorides

The synthesis of all the reagents in Table I follows the same route. Examples of the synthesis of two of the compounds in Table I are given below. The general procedure for converting aromatic nitriles (*13*) to benzamidines (*15*) depends on the classical Pinner reaction,[8,9] in which a nitrile is allowed to react with dry gaseous HCl and a slight excess of ethanol (1.5–10 moles/mole) in an inert solvent. The solvents that are most

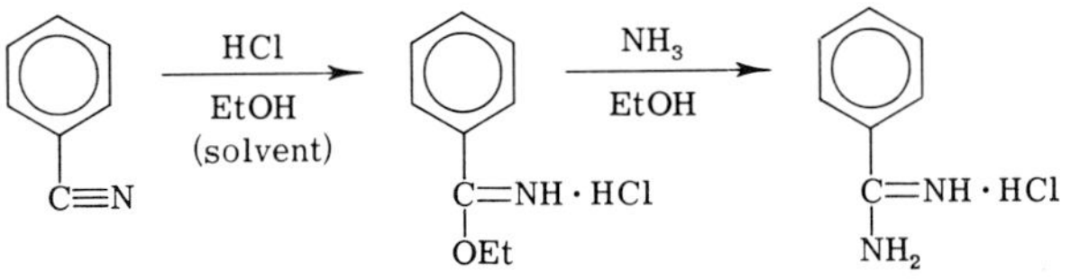

SCHEME 2

applicable are chloroform and ether. The technique is to dissolve or suspend the nitrile in the solvent at 0°, add ethanol, and bubble dry HCl gas through the mixture for at least 2 hr. The mixture is then stirred at ambient temperature for 15 hr and evaporated under reduced pressure to dryness. An infrared spectrum of the crude imino ether is obtained to determine whether the nitrile, $C\equiv N$, absorption band at 2200 cm^{-1} has been removed.[10] If this band is still present the imino ether (*14*) is again treated with dry HCl and excess ethanol (2–10 moles/mole). After complete conversion of the nitrile, the crude imino ether is treated with a large excess of dry ammonia at room temperature for 24 hr. The reaction mixture is filtered to remove precipitated ammonium chloride and evap-

[8] A. Pinner, A. Gradenwitz, and F. Gradenwitz, *Justus Liebigs Ann. Chem.* **298,** 47 (1897).

[9] A. W. Pox and F. C. Whitmore, *in* "Organic Synthesis Collective Vol. I" (H. Gilmand and A. H. Blatt, eds.), p. 5. Wiley, New York, 1941.

[10] R. M. Silverstein and G. C. Bassler "Spectrometric Identification of Organic Compounds," 2nd ed., p. 98. Wiley, New York, 1967.

orated under reduced pressure to dryness. The crude product is recrystallized from aqueous or aqueous-ethanol solution with an excess of hydrochloric, p-toluenesulfonic (TsOH), or benzenesulfonic (BzOH) acids.

A. *Synthesis of m-{2-[o-(m-Fluorosulfonylbenzamido)phenoxy]-ethoxy}benzamidine (1).*[4] To a solution of 10 mmoles of m-(o-aminophenoxyethoxy)benzamidine (*16*) in 2 ml of dry N,N-dimethylformamide (DMF)[11] are added 2.0 mmoles of pyridine. After stirring at ambient temperature for 1 hr, a solution of m-fluorosulfonylbenzoylchloride (*17*) in 2 ml of dry DMF is added. The mixture is stirred 1 hr at ambient temperature and poured into 75 ml of diethyl ether. The resulting mixture is chilled, and the ether is decanted. The residue is dissolved in hot 95% aqueous ethanol, and excess p-toluenesulfonic acid is added; upon cooling, a crystalline product is obtained. Recrystallization from 95% ethanol gives compound *1*.

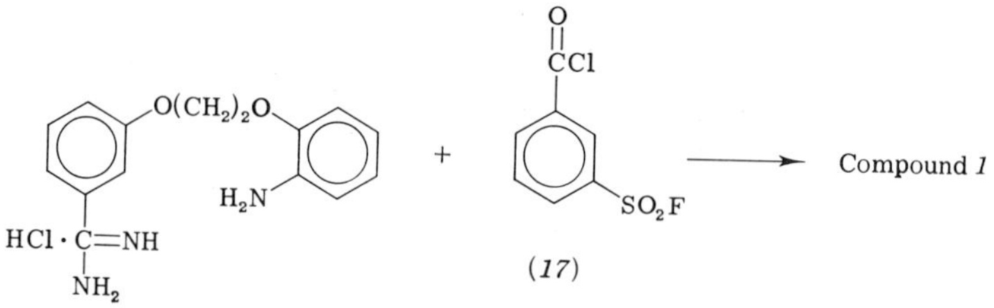

SCHEME 3

B. *Synthesis of m-{4-[o-(2-Chloro-5-fluorosulfonylphenylureido)-phenoxy]butoxy}benzamidine (11).*[5] To 5.2 mmoles of 3-amino-4-chlorobenzenesulfonylfluoride (*18*) (Aldrich Chemical Co.) are added 25 ml of benzene (dry reagent grade) and 5.5 mmoles of p-nitrophenylchloroformate (*19*) (Aldrich Chemical Co.). The mixture is stirred at reflux for 4 hr, at which point hydrochloric acid vapor is no longer detectable. The mixture is cooled, and N-(2-chloro-5-fluorosulfonyl-O-(4-nitrophenyl) carbamate (*20*) is collected. Recrystallization is from methylene chloride.[12] To a solution of m-[4-(o-aminophenoxy)butoxy]benzamidine p-toluenesulfonate (*21*) in 10 ml of DMF[11] are added 10.0 mmoles of pyri-

---

[11] N,N-Dimethylformamide can be dried by placing about 0.5 inch of 3A molecular sieves (Linde) in the bottom of a freshly opened bottle of reagent-grade solvent (MCB). Molecular sieves are natural zeolites with interconnecting cavities that can trap water molecules. See L. F. Fieser and M. Fieser, "Reagents for Organic Synthesis," Vol. 1, p. 703. Wiley, New York, 1967.

[12] B. R. Baker and N. M. J. Vermeulen, *J. Med. Chem.* **12**, 74 (1969).

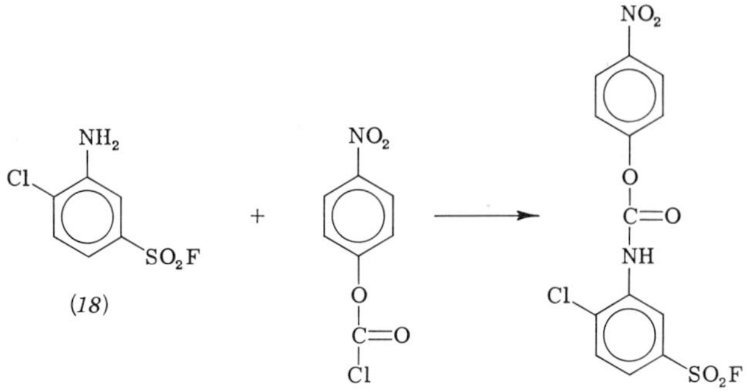

SCHEME 4

dine. The resulting mixture is stirred 1 hr at ambient temperature, and a solution of 5.5 mmoles of compound *20* in DMF[11] is added. The mixture is stirred at ambient temperature for 24 hr then poured into 300 ml of diethyl ether. The mixture is chilled, and the ether phase is decanted. Recrystallization from 60% ethanol gives compound *11*, m.p. 152°–154°.

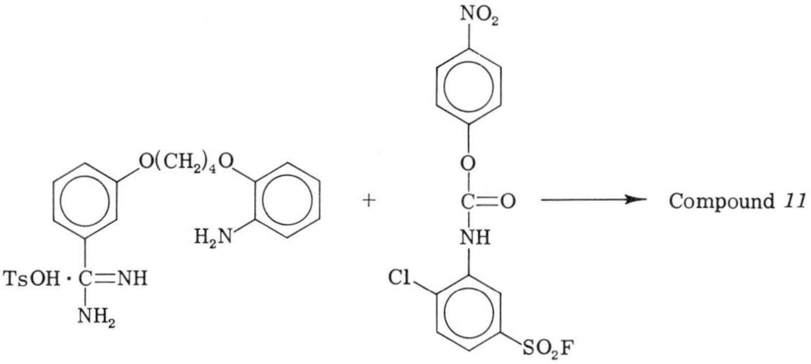

SCHEME 5

Primary Features of Exo Reagents

The degree of specificity observed in the compounds in Table I is attributed not to the reactive group, X, but to the affinity group, S. The specificity of the exo affinity labeling reagent is determined by that portion of the molecule which possesses an affinity for the binding site of the enzyme. An example of the importance of specificity is seen with C1 esterase and certain other serine proteases that are inhibited by the substituted benzamidines. Compound *11*, m-{4-[o-(2-chloro-5-fluorosulfonyl-phenylureido)phenoxy]butoxy}benzamidine, is a highly reactive affinity label causing 50% inhibition of C1 esterase, human plasmin and thrombin, and bovine trypsin at concentrations of 3–5 μM. However, a similar

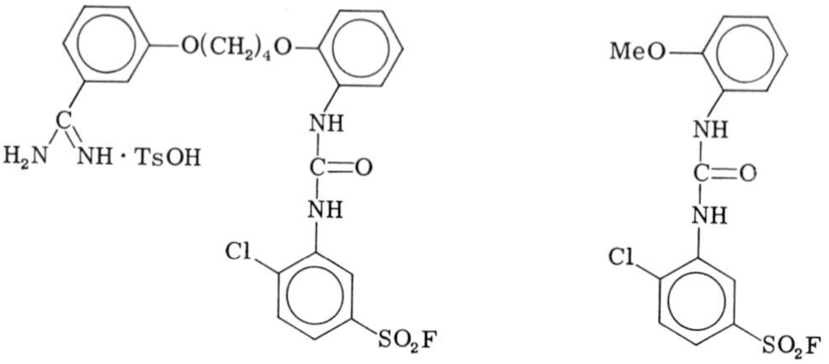

SCHEME 6

compound *22*, m-[o-(2-chloro-5-fluorosulfonylphenylureido)methoxy]-benzene, identical to *11* except for the deletion of the benzamidine and alkoxy side chain, shows no inactivation at 50–60 μM.[13] Another example of the importance of the benzamidine group is demonstrated by compound *23*. In 2-[m-(3-phenoxy)propoxyphenyl]-2-imidazoline hydrochloride hydrate (*23*), the cationic benzamidine group has been blocked by bridging with an ethyl group. This compound has no inhibitory activity toward guinea pig complement at 1.0 mM, whereas the unblocked benza-

(23)

SCHEME 7

midine, compound *25* (Table II), causes 60% inhibition of guinea pig complement at 0.5 mM.[13]

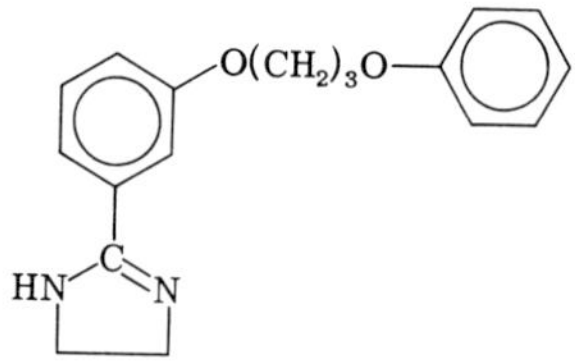

The specific binding moiety has been shown to be essential for an

---

[13] D. H. Bing and M. Cory, manuscript in preparation.

TABLE II
ω-PHENYL BRIDGED COMPLEMENT INHIBITORS

B		% Inhibition[a]
24	O—(CH₂)₂—O	42
25	O—(CH₂)₃—O	54
26	O—(CH₂)₄—O	14
27	O—(CH₂)₃	56
28	O—(CH₂)₄	51[b]
29	NHCO(CH₂)₂	0
30	O(CH₂)₃NHCO	22[b]
31	HC=CH	4[b]
32	(CH₂)₂	14
33	(CH₂)₄	40[b]

[a] Percent inhibition of whole guinea pig complement at 0.5 mM inhibitor concentration [B. R. Baker and E. H. Erickson, *J. Med. Chem.* **12**, 408 (1969)].
[b] Inhibitor concentration 0.25 mM [B. R. Baker and M. Cory, *J. Med. Chem.* **12**, 1049 (1969)].

exo affinity reagent. However, it is not solely responsible for the affinity of the compound. The bridging group, B, which links the chemically reactive X to S, does play a role in the activity of the affinity label. This interaction of the bridging group with the enzyme was not recognized in the early work on exo affinity labeling reagents. Baker and Cory,[4,5,14] in a series of studies on reversible and irreversible inhibitors of guinea pig C1, varied the nature of the bridging group. The effect of changes in the bridging group on the ability of the compound to inhibit the whole guinea pig complement is summarized in Table II. These compounds are inhibitory in a typical dose-response fashion.[4,5,14] We have interpreted these data to mean that this bridging group is interacting with secondary enzyme sites that play a role in the specificity of the biological function of complement. The synthesis of the bridging groups in Table II are presented in the following section.

[14] B. R. Baker and M. Cory, *J. Med. Chem.* **12**, 1049 (1969).

Synthesis of Bridging Groups

C. Synthesis[15] *of m-(ω-Phenoxyalkoxy)benzamidines* (*24, 25, 26*). The preparation of substituted m-phenoxyalkoxybenzamidines follows a general scheme in which an α,ω-dihaloalkane (*35*) in large excess is used to monoalkylate m-cyanophenol (*34*). The resulting α-bromo-ω-(3-cyanophenyl)alkane (*36*) is alkylated with an excess of phenol (*37*) to give an α-phenyl-ω-(3-cyanophenyl)alkane (*38*), which is converted to the m-substituted benzamidine (*24–26*) by the Pinner reaction. When $n = 3$, the synthesis is as follows. To 20 mmoles of m-cyanophenol (*35*) in 10 ml

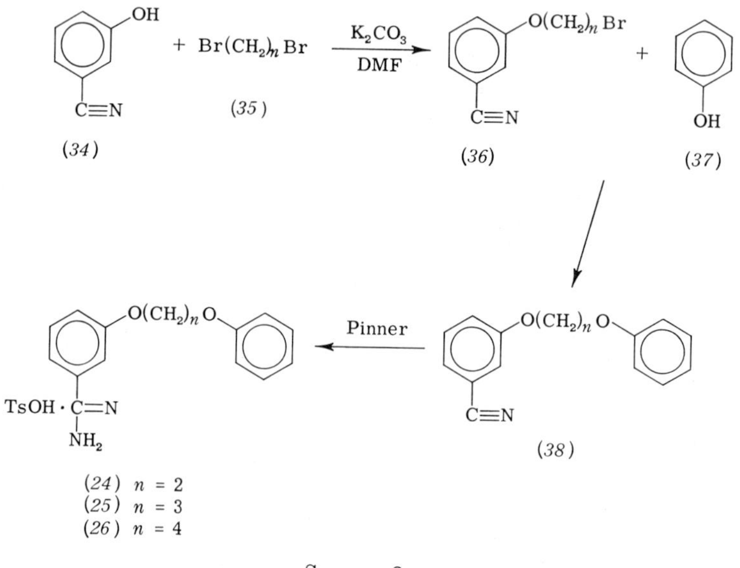

SCHEME 8

of dry DMF[11] are added 20 mmoles of anhydrous potassium carbonate and 100 mmoles of 1,3-dibromopropane (*35, n = 3*). The resulting mixture is stirred at 70° for 15 hr, poured into 100 ml of water, and extracted with three 30-ml portions of ethyl acetate. The extracts are combined, washed thoroughly with water, and evaporated under reduced pressure to dryness. The resulting oil is further evaporated on a high-vacuum pump to remove unreacted 1,3-dibromopropane. The residual oil is dissolved in 10 ml of dry DMF,[11] and 20 mmoles of anhydrous potassium carbonate are added. The mixture is treated with 20 mmoles of phenol (*37*). After stirring at 70° for 15 hr, the reaction is poured into 100 ml of water, extracted with three 30-ml portions of ethyl acetate, and washed thoroughly

[15] B. R. Baker and E. H. Erickson, *J. Med. Chem.* **12**, 112 (1969).

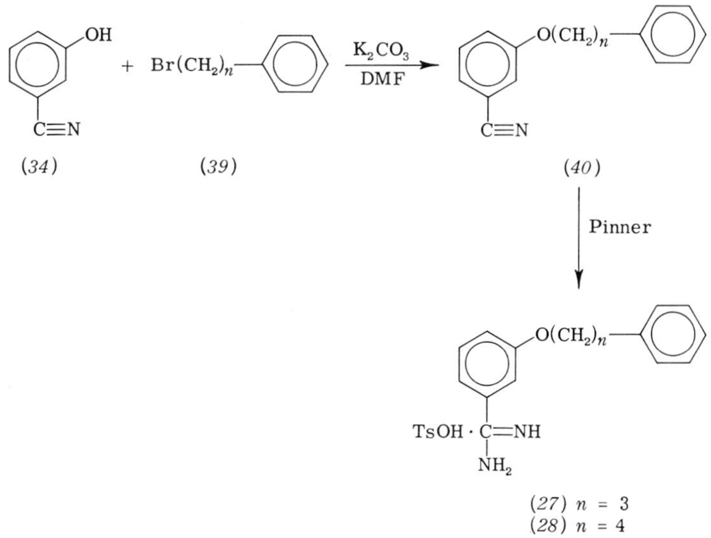

SCHEME 9

with water. The combined extracts are evaporated under reduced pressure to yield an oil that is used directly in the Pinner reaction. The product of the Pinner reaction, m-(3-phenoxypropoxy)benzamidine hydrochloride (25), melts at 141°–144°.

D. *Synthesis*[14] of *m-(ω-Phenylalkoxy)benzamidines (27, 28)*. To 20 mmoles of m-cyanophenol (34) in 10 ml of dry DMF[11] are added 10 mmoles of anhydrous potassium carbonate and 20 mmoles of ω-phenylalkylbromide (39). The resulting mixture is stirred at 70° for 15 hr, poured into 100 ml of water, and extracted with three 30-ml portions of ethyl acetate. The extracts are combined, washed thoroughly with water, dried over magnesium sulfate, and evaporated under reduced pressure to yield an oil, m-(ω-phenylalkoxy)cyanophenyl ether (40), which is used directly in the Pinner reaction. The resulting m-(4-phenylbutoxy)benzamidine p-toluenesulfonate (28) is recrystallized from water and melts at 147°–148°. The m-(3-phenylpropoxy)benzamidine p-toluenesulfonate (27) is recrystallized from ethanol–water and has a melting point of 203°–205°.

E. *Synthesis*[16] of *m-(3-Phenylpropionamidobenzene)benzamidine p-Toluenesulfonate (29)*. A solution of 5 mmoles of m-aminobenzamidine dihydrochloride (Aldrich) (41) in 5 ml of DMF[11] and 2 ml of pyridine is stirred at 0°. A solution of 5.3 mmoles of hydrocinnamic acid chloride

[16] B. R. Baker and E. H. Erickson, *J. Med. Chem.* **12**, 408 (1969).

(*42*) (Eastman) in 5 ml of DMF[11] is added dropwise to this 0° mixture. After 3 hr at ambient temperature, the mixture is diluted with 50 ml of ether, and the solution is decanted from the precipitated oil. The residual oil is recrystallized from aqueous toluenesulfonic acid solution to give *m*-(3-phenylpropionamidobenzene)benzamidine *p*-toluenesulfonate (*29*), m.p. 192°–194°.

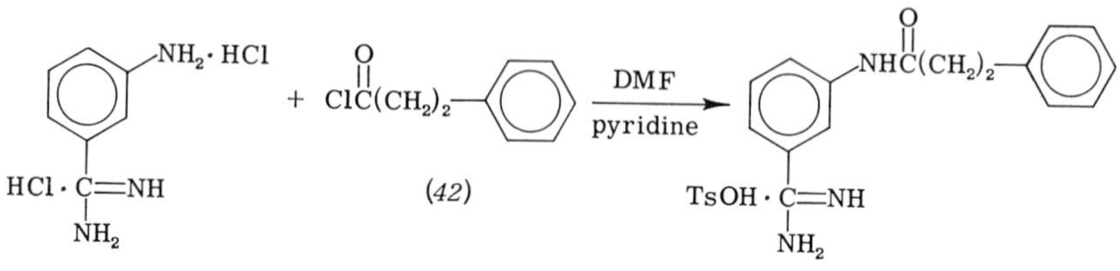

SCHEME 10

F. *Synthesis*[16] *of* *N-benzyl-3-(m-amidinophenoxy)propylamine* *p-Toluenesulfonate* (*30*). To a solution of 16 mmoles of *m*-cyanophenol (*34*) (Aldrich) in 10 ml of dry DMF[11] are added 16 mmoles of anhydrous powdered potassium carbonate and 20 mmoles of *N*-carbobenzoxy-3-bromopropyl ether (*43*). The resulting mixture is stirred at 70° for 15 hr, poured into 100 ml of 1.0 *N* NaOH, and extracted with ethyl acetate. The crude oily product, *N*-carbobenzoxy-3(3-cyanophenoxy)propylamine (*44*), is dissolved in 10 ml of 32% hydrobromic acid in acetic acid (Eastman) and stirred at room temperature for 2 hr. The solution is diluted with 90 ml of ether. The product is collected an recrystallized from acetone to yield white crystals of *m*-(3-cyanophenoxy)propylamine hydrobromide (*45*). To 8 mmoles of *45* are added 20 ml of chloroform and 20 mmoles of triethylamine followed by 8 mmoles of benzoylchloride (*46*). The resulting solution is stirred at ambient temperature for 24 hr, then washed successively with three 30-ml portions of 1.0 *N* NaOH and three 30-ml portions of water. The dried solutions (dried over magnesium sulfate) are evaporated under reduced pressure. Recrystallization from benzene gives *N*-benzyl-3-(3-cyanophenoxy)propylamine (*47*) with a melting point of 96°–98°. The nitrile (*47*) is treated in the Pinner reaction to give compound *30*, m.p. 208°–210°.

G. *Synthesis*[16] *of* *m*(-4-Phenylbutyl)benzamidine *p-Toluenesulfonate* (*33*). To a solution of 0.15 mmole of *m*-cyanobenzyl bromide (*48*) (Aldrich) in 150 ml of xylene is added a solution of 0.15 mmole of triphenylphosphine (*49*) (Aldrich) in 150 ml of xylene. The mixture is stirred at reflux for 1 hr and then chilled. The product, *m*-cyanobenzyl-

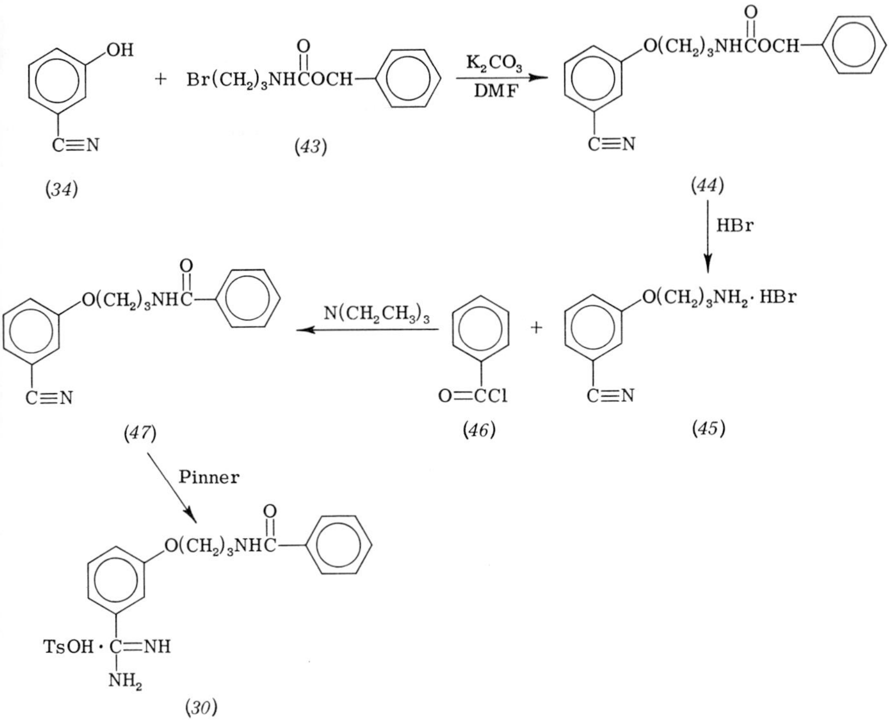

SCHEME 11

triphenylphosphonium bromide (50), is collected and recrystallized from
n-propanol, m.p. 311°–320°. To a solution of 4.3 mmoles of m-cyano-
benzyltriphenylphosphonium bromide (50) and 4.3 mmoles of hydrocinna-
maldehyde (51) (Aldrich) in 250 ml of methanol are added 4.6 mmoles

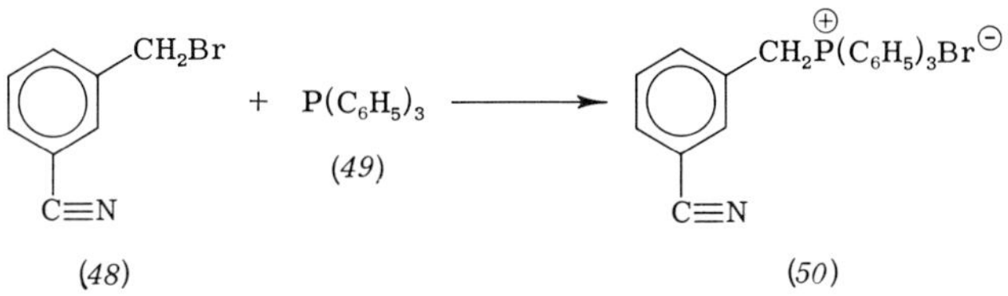

SCHEME 12

of sodium methoxide (Matheson) in 100 ml of methanol. The mixture is
stirred for 15 hr at room temperature and then evaporated under reduced
pressure to dryness. The residue is extracted four times with petroleum
ether (30°–60° boiling range). The combined extracts are evaporated
under reduced pressure to yield a crude oil, 1-(3-cyanophenylethyl)-2-

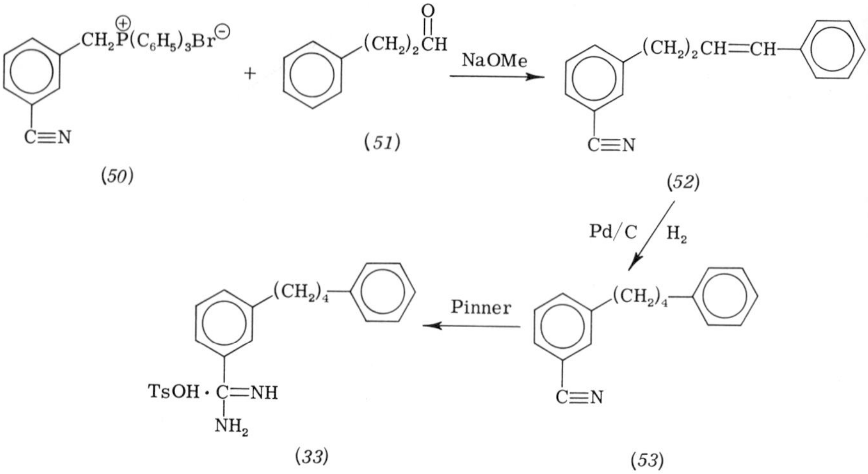

SCHEME 13

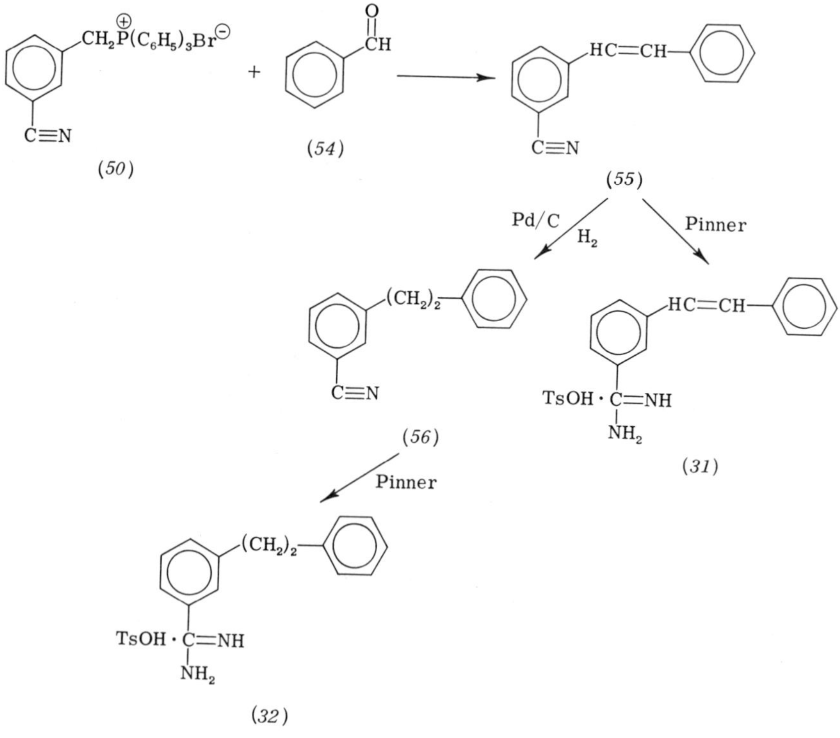

SCHEME 14

phenylethylene (*52*). The oil is dissolved in 100 ml of ethanol, and 0.1 g of 5% palladium on carbon is added. The mixture is shaken with hydrogen at 2–3 atmospheres in a Parr apparatus; reduction is complete in 3 hr. The filtered solution is evaporated to dryness under reduced pressure. The colorless residue, 1-(3-cyanophenyl)-4-phenylbutane (*53*), is used directly in the Pinner reaction; the product (*33*) has a melting point of 110°–112°.

H. Synthesis[16] *of m-Styrylbenzamidine p-Toluenesulfonate (31) and m-Phenylethylbenzamidine p-Toluenesulfonate (32).* By a procedure similar to that for compound *52*, equal molar amounts of *m*-cyanobenzyltriphenylphosphonium bromide (*50*) and benzaldehyde (*54*) are allowed to react with a slight excess of sodium methoxide in methanol. The resulting 2-(3-cyanophenyl)styrene is used directly in the Pinner reaction to yield compound *31*.

Compound *31* is reduced with hydrogen using a 5% palladium on carbon in a procedure analogous to that for the preparation of compound *33*. The resulting oily product, 2-(3-cyanophenyl)-1-phenylethane (*56*), m.p. 100°–112°, is used in the Pinner reaction to produce *32*, m.p. 176°–179°.

Secondary Considerations

All the compounds in Table II can be converted to potential exo affinity reagents by the addition of a reactive group to the terminal phenyl ring. Such groups are introduced by synthesizing the compound with the appropriate substituted terminal phenyl ring. At this point the choice of structure is often limited by synthetic considerations. For example, in the case of *m*-(3-phenoxypropoxy)benzamidine (*25*, synthesis C), this would mean that compound *37* would be a substituted phenol. Similarly, a substituted 4-phenylbutylbromide (*39*) would be used in the preparation of the *m*-(4-phenylbutoxy)benzamidine *p*-toluenesulfonate (*28*). Many substituted phenols are available commercially. Substituted 4-phenylbutylbromides are more difficult to obtain commercially or synthetically. This synthetic consideration and the inhibitory activity of *m*-(3-phenoxypropoxy)benzamidine *p*-toluenesulfonate (*25*) dictated that *25* would be the parent compound for further chemical modification.

Baker and Cory[4] found that the inhibitory activity of the reversible inhibitors in Table II could be further enhanced by the addition of a *p*-nitrophenyl amide on *p*-nitrophenyl urea groups to the terminal phenyl ring. This addition to the bridging group led to compounds such as *m*-{3-[*m*-(*p*-nitrophenylureido)phenoxy]propoxy}benzamidine (*57*) and *m*-{3-[*m*-(*p*-nitrobenzamido)phenoxy]propoxy}benzamidine (*58*). Posi-

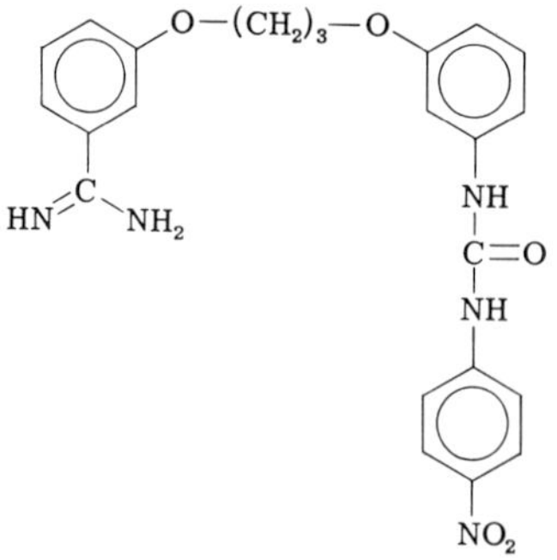

(57) 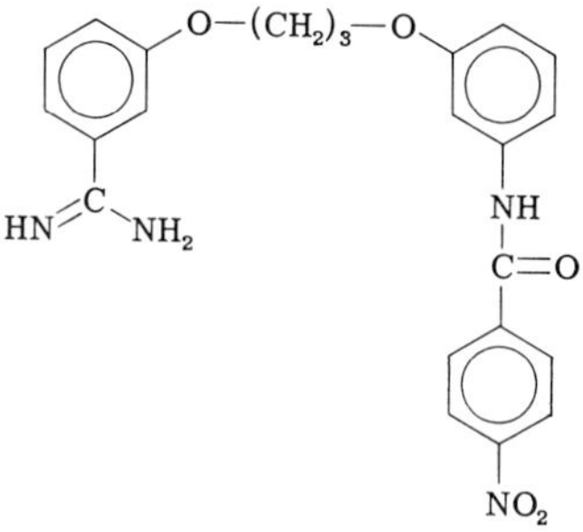   (58)

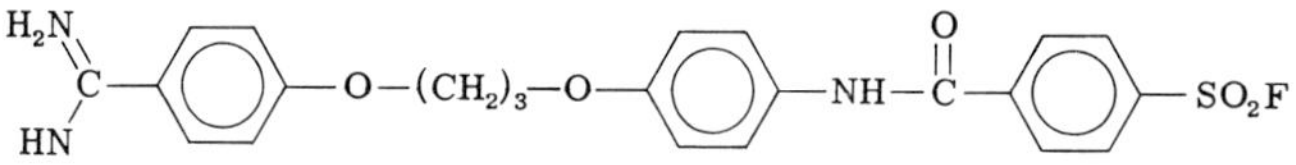

(59)

SCHEME 15

tion as well as chemical composition were found to be important considerations in the inhibitory activity of exo reagents. The affinity label for trypsin, p-{3-[p-(p-fluorosulfonylbenzamido)phenoxy]propoxy}benzamidine (59),[17] does not inactivate thrombin, whereas compound 11 (Table I) is a highly active affinity reagent for thrombin and other serine proteases.[13]

Compounds such as 57 and 58 can be converted to exo labeling reagents by the addition of a sulfonylfluoride to the terminal ring. Synthetically this is accomplished by the same basic routes described in syntheses A and B. The inhibitory activity of the exo label can be affected by the position of the amido or ureido linkage, the position of the SO_2F group on the terminal ring, and the presence of other substituents on the terminal ring. The effect of these structural changes on activity is illustrated in Table III. Compounds in which the ureido or amido group is ortho are better inhibitors than the corresponding meta-substituted benzamidines.

The examples of exo affinity labeling reagents in Table III have SO_2F as the reactive group, X. This group will react primarily with a serine or threonine and possibly a tyrosine hydroxyl.[18,19] Serine and threonine

[17] T. Chase and E. Shaw, *Biochemistry* 8, 2212 (1969).
[18] R. Kitz and I. B. Wilson, *J. Biol. Chem.* 237, 3245 (1962).
[19] D. E. Fahrney and A. M. Gold, *J. Am. Chem. Soc.* 85, 997 (1963).

TABLE III

EFFECT OF R ON INHIBITORY ACTIVITY

R	% Inhibition
60 m-NHCONH—C$_6$H$_4$—SO$_2$F-p	77[a]
61 m-NHCONH—C$_6$H$_4$—SO$_2$F-m	51[a]
62 m-NHCO—C$_6$H$_4$—SO$_2$F-p	54[a]
63 m-NHCO—C$_6$H$_4$—SO$_2$F-m	40[a]
64 o-NHCO—C$_6$H$_3$-2-Cl-5-SO$_2$F	8[b]
65 o-NHCO—C$_6$H$_3$-2-MeO-5-SO$_2$F	15[b]
66 o-NHCO—C$_6$H$_2$-2,4-Cl-5-SO$_2$F	23[b]
67 o-NHCO—C$_6$H$_4$—SO$_2$F-p	45[b]

[a] Percent inhibition of whole guinea pig complement at 0.062 M inhibitor concentration [B. R. Baker and M. Cory, *J. Med. Chem.* **14**, 305 (1971)].

[b] Percent inhibition of whole guinea pig complement at 0.031 M inhibitor concentration [B. R. Baker and M. Cory, *J. Med. Chem.* **14**, 119 (1971)].

comprise about 10% of the amino acid residues in proteases and are often on the surface because of their hydrophilic nature. Thus, SO$_2$F is a reasonable reactive group to attach to a reagent designed to label residues outside of the active site. Other types of reactive groups can be introduced into exo affinity labeling reagents by simply varying the nature of the terminal portion of the reagent. Elsewhere in this volume details of syntheses are presented for haloamides, epoxides, haloketones, nitrenes, and diazoniums as the reactive group, X, in affinity reagents. All these methods are applicable to the synthesis of exo affinity labeling reagents. The choice of the active group need be dictated only by the reactivity of the amino acids to be labeled.

Conclusion

Essential components of an exo affinity labeling reagent are a substrate analog with a high affinity for the enzyme binding site, a bridging group that has the potential for interaction with secondary sites on the enzyme, and a reactive group for covalent binding to the protein. We have discussed in detail the approach used initially by Baker and subsequently in our laboratories to design an exo affinity label for a series of serine proteases. Other enzymes will present different problems and the relative ease of obtaining an exo affinity labeling reagent will be determined by the accessibility of the substrate to chemical modification. Regardless of the nature of the enzyme or substrate, the general considerations should be the same.

Acknowledgments

Supported by NIH Grants AI 11231, AM 17351, and CA 17376. Judith M. Andrews is an NIH Postdoctoral Fellow AI 100264. David H. Bing is an Established Investigator of the American Heart Association. The authors acknowledge the excellent secretarial assistance of Ms. Dale Levine and Ms. Rachelle Rosenbaum.

[10] Haloketones as Affinity Labeling Reagents[1]

By Fred C. Hartman

General Considerations

One of the first affinity labels described was the chloromethylketone derived from N-tosyl-L-phenylalanine (TPCK), which provided convincing evidence for the catalytic involvement of an imidazole side chain in proteolytic enzymes.[2] This and other successes of Elliott Shaw[3,4] and his colleagues stimulated many investigators to design ligandlike α-haloketones as probes of the active sites of many different enzymes and nonenzymic proteins. α-Haloketones remain one of the more popular chemical classes of affinity labels.

Their principal advantages as affinity labels are that they are highly reactive and provide a means for introducing a radioactive marker, after the covalent modification has taken place, through reduction with sodium borotritride. Haloketones are more reactive as alkylating agents than are haloacetylated amines. Since haloketones are potentially reactive with most nucleophiles found in proteins, the chances are good that modification of some residue within the active site will occur, provided the reagent

[1] Research from the author's laboratory was sponsored by the U.S. Energy Research and Development Administration under contract with the Union Carbide Corporation.
[2] G. Schoellmann and E. Shaw, *Biochemistry* 2, 252 (1963).
[3] E. Shaw, *Physiol. Rev.* 50, 244 (1970).
[4] E. Shaw *in* "The Enzymes" (P. D. Boyer, ed.), Vol. 1, p. 91. Academic Press, New York, 1970.

has an affinity for the active site. The opportunity to introduce conveniently and economically an isotopic label into the reagent moiety of the derivatized protein removes the necessity of synthesizing the labeled reagent.

Haloketones also have some disadvantages. Their high reactivities can be a drawback in that the greater the reactivity, the greater the likelihood of undesired, nonselective modifications. It is possible to exert some control over reactivity by the choice of halogen; bromine is invariably a better leaving group than is chlorine (steric considerations aside), whereas fluorine is usually such a poor leaving group that it is of little use. Iodoketones are not always alkylating agents but instead can be oxidizing agents in which the reagent serves as a source of iodonium ion.[5] Haloketones (with the exception of the fluoro compounds) are certainly too reactive for general use as chemotherapeutic agents. Another disadvantage of haloketones is that the investigator faces a unique problem in identifying the derivatized amino acids for each new reagent used successfully. This contrasts with the characterization of proteins modified with bromoacetylated compounds, in which, irrespective of the entire structure, the derivatized amino acids are converted to easily identified carboxymethyl amino acids upon acid hydrolysis.[6,7] A third disadvantage is that the synthesis of a desired haloketone frequently becomes a research problem in organic chemistry. Again, this contrasts with the preparation of bromoacetylated compounds, in which well-defined, simple procedures are applicable to many alcohols and amines.[6,7]

Very recently a potential affinity label for the β-adrenergic receptor in erythrocytes has been prepared by treating an amine with dibromoacetone to form the corresponding alkylated amine containing a reactive bromoketone group.[8] This provides a synthetic procedure that should be readily adaptable to a multitude of amines.

Successful application of affinity labeling, with haloketones or any other class of reagent, typically requires four types of endeavors: (1) search for suitable reagents; (2) demonstration that inactivation results from an active-site-directed process; (3) characterization of the covalently modified protein; (4) determination of the relationship of the results to the mechanism of the enzyme-catalyzed reaction.

In rare instances, suitable reagents can be obtained from commercial sources; more often newly designed reagents are synthesized after

---

[5] F. C. Hartman, *Biochemistry* **9**, 1776 (1970).
[6] F. Naider, J. M. Becker, and M. Wilchek, *Isr. J. Chem.* **12**, 441 (1974).
[7] M. Wilchek and D. Givol, this volume [11].
[8] D. Atlas, M. L. Steer, and A. Levitzki, *Proc. Natl. Acad. Sci. U.S.A.* **73**, 1921 (1976). See also this volume [69].

thoughtful deliberations on the known structural requirements for substrate binding and on the most appropriate placement, with respect to available mechanistic information, of the reactive group within the reagent.

Many of the criteria for ascertaining whether a reagent is acting as a true affinity label are universally applicable and are described in a previous chapter of this volume.[9]

Methods for determining the kinds of amino acid residues modified depend somewhat upon the type of reagent used; the introduction of a carbonyl group into a protein provides a handle which can facilitate the identification of the labeled residue, as will be seen later in this chapter.

The degree to which data from affinity labeling will increase understanding of the enzymatic mechanism cannot be predicted *a priori*. In some cases nothing more is obtained than an indication of a particular residue's presence at the substrate binding site, whereas in other cases detailed information is obtained concerning a residue's role in a precisely defined catalytic step and its geometric relationship to other active-site residues.

Most of the methodologies used and problems encountered with haloketones as affinity labels are covered by considerations of bromopyruvate, haloacetol phosphates, and haloketone derivatives of pyridine nucleotides. Similar information is gained from numerous elegant studies in which halomethylketone derivatives of amino acids and peptides have been used as affinity labels for proteases.[3,4,10] These investigations are excluded from the present article, since they are considered elsewhere in this volume.[10]

Selected Examples

Bromopyruvate

Detailed investigations by H. Paul Meloche and his associates of the reaction of bromopyruvate with 2-keto-3-deoxyglucomate-6-phosphate (kdGtP) aldolase exemplify the wealth of information that can be gleaned from affinity labeling. Although this particular aldolase is not ubiquitous (in most studies the enzyme from *Pseudomonas putida* was used), it typifies a large number of aldolases, isomerases, and β-decarboxylases that activate a hydrogen atom (as a proton) α to the substrate carbonyl to generate an intermediate carbanion or enol. When Meloche initiated his studies, the kdGtP aldolase mechanism could be depicted as shown in

[9] F. Wold, this volume [1].
[10] J. C. Powers, this volume [16].

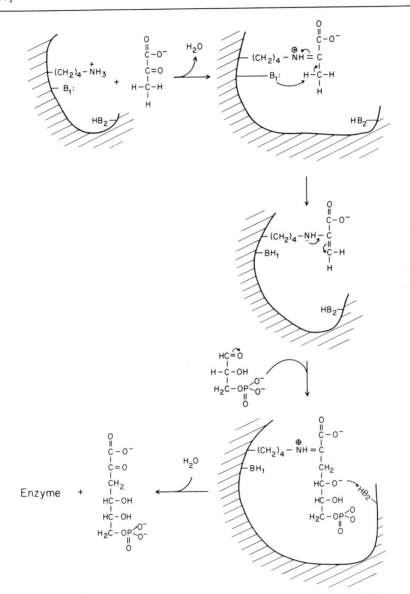

FIG. 1. Possible mechanism for 2-keto-3-deoxygluconate-6-phosphate aldolase.

Fig. 1. Schiff-base formation was detected by reduction with borohydride,[11] and the presence of a basic group (B_1) that abstracts a proton

[11] E. Grazi, H. Meloche, G. Martinez, W. A. Wood, and B. L. Horecker, *Biochem. Biophys. Res. Commun.* **10**, 4 (1963).

from C3 was assumed from the demonstration that the enzyme catalyzes the exchange of tritium from water into the methyl group of pyruvate.[12] Meloche[13,14] hypothesized that placement of a good leaving group at C3 would form an electrophilic agent capable of alkylating the unidentified basic group (B_1). Thus, on the basis of the probable mechanism, bromopyruvate appeared to be an ideal reagent for the selective modification of an active-site residue with a known catalytic function.

Bromopyruvate is a rare example of a reagent that not only appeared ideal from mechanistic considerations, but also was readily available. The compound can now be purchased from commercial sources. 1-[14C]Bromopyruvate can be obtained by direct bromination of commercial [14C]pyruvate.[14] [3H]Bromopyruvate has also been prepared from randomly tritiated pyruvate obtained by the action of kdGtP aldolase on pyruvate in the presence of T_2O.[15] Bromopyruvate is assayed by its conversion at pH 8.0 to hydroxypyruvate, which is quantitated either enzymically with lactate dehydrogenase or chemically as the semicarbazone.[12,14]

Inactivation of kdGtP aldolase by bromopyruvate is clearly an active-site-directed process.[14] Inactivation is first order with respect to the remaining native enzyme, and a rate-saturation effect is observed as the concentration of bromopyruvate is increased. Pyruvate protects against inactivation, and kinetic experiments in the presence of pyruvate show that reagent and substrate compete for the same site. The aldehyde substrate, D-glyceraldehyde 3-phosphate, affords no protection, indicating that only the pyruvate domain within the active site is alkylated. With enzyme that had prior treatment with nonradioactive bromopyruvate in the presence of pyruvate so that nonselective sites were blocked, the extent of specific incorporation as determined with [14C]bromopyruvate is close to 1 mole of reagent per mole of catalytic subunit inactivated. These observations meet the usual criteria indicative of affinity labeling.

To determine the kinds of residues modified, the aldolase inactivated with 1-[14C]bromopyruvate was first reduced, in the presence of a protein denaturant, with sodium borohydride to convert the ketone group of the incorporated pyruvyl moiety to a hydroxyl group. This was necessitated by the decarboxylation of the pyruvyl moiety (and therefore loss of the 14C label) during hydrolysis of the protein.[16]

Borohydride reduction of covalently incorporated ketones has become a widely practiced procedure. Not only does such reduction confer stability on the protein–reagent linkage and other labile groups α to the

[12] H. P. Meloche and W. A. Wood, *J. Biol. Chem.* **239**, 3511 (1964).
[13] H. P. Meloche, *Biochem. Biophys. Res. Commun.* **18**, 277 (1965).
[14] H. P. Meloche, *Biochemistry* **6**, 2273 (1967).
[15] H. P. Meloche, M. A. Luczak, and J. M. Wurster, *J. Biol. Chem.* **247**, 4186 (1972).
[16] H. P. Meloche, *Biochemistry* **9**, 5050 (1970).

carbonyl, but it aso provides an ideal method for introducing a stable radioactive isotope into any protein labeled by a haloketone and a means for determining the extent of reagent incorporation. These are important considerations, since in many cases labeled starting materials are either unavailable or prohibitively expensive. Even when labeled starting materials are available, lengthy synthetic routes for preparing a particular affinity label make introduction of label subsequent to the chemical modification more attractive.

Meloche[16] used approaches to identify the modified residues after reduction that are fairly representative of those used in characterization of new derivatives of amino acids: The elution positions of the radioactive components in acid hydrolyzates were determined by ion-exchange chromatography on an amino acid analyzer. These elution positions were then compared with those of synthetically prepared standards. Hydrolyzates of inactivated kdGtP aldolase contained two ^{14}C-labeled derivatives, which eluted from the long column with the front and just ahead of aspartic acid, respectively. The compound that was not retarded was ninhydrin negative and thus was assumed to be a decomposition or hydrolytic product. Subsequently, this compound was isolated and identified as glyceric acid. From this finding and studies of the stability of the protein-reagent bond, Meloche[16] concluded that bromopyruvate esterifies a carboxyl group of either a glutamyl or aspartyl residue. The corresponding tryptic peptide from inactivated enzyme has been recently shown to contain only glutamate as a potential esterification site (personal communication from H. P. Meloche).

The compound that emerges just ahead of aspartic acid cochromatographed with 1-carboxy-1-(DL)-hydroxyethylcysteine prepared by alkylation of glutathione with bromopyruvate followed by reduction and hydrolysis (Fig. 2). The alkylated glutathione could also be degraded to give carboxymethylcysteine by oxidation with hydrogen peroxide followed by hydrolysis (Fig. 2), but this approach with the inactivated enzyme was not reported. Peroxide treatment of cysteine thioethers presents the risk of sulfoxide formation. Sulfoxides undergo decomposition during hydrolysis leading to ambiguous results and must be reduced back to the thioether with HI prior to hydrolysis. Whenever labeled protein undergoes extensive loss of radioactivity subsequent to acid hydrolysis and drying thioether sulfoxide existence should be suspected. The conversion of an incorporated α-methylketone to a carboxymethyl group is an attractive method (which was used earlier to characterize histidine modified by bromopyruvate[17]) for identifying modified residues, since all carboxy-

[17] R. L. Heinrikson, W. H. Stein, A. M. Crestfield, and S. Moore, *J. Biol. Chem.* **240**, 2921 (1965).

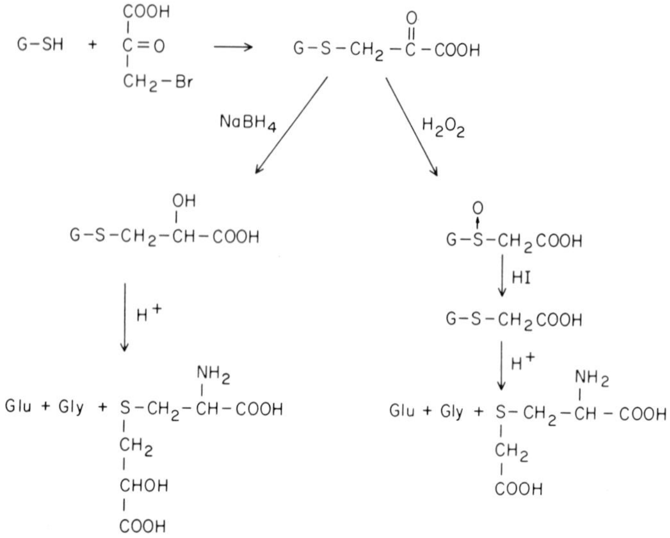

G-SH + C=O (COOH / C=O / CH₂-Br) ⟶ G-S-CH₂-C(=O)-COOH

NaBH₄ ↙   ↘ H₂O₂

OH
|
G-S-CH₂-CH-COOH

O
‖
G-S-CH₂COOH

↓ H⁺

↓ HI

G-S-CH₂COOH

↓ H⁺

NH₂
|
Glu + Gly + S-CH₂-CH-COOH
|
CH₂
|
CHOH
|
COOH

NH₂
|
Glu + Gly + S-CH₂-CH-COOH
|
CH₂
|
COOH

Fig. 2. Characterization of glutathione (G-SH) alkylated by bromopyruvate.

methyl amino acids are well characterized on the amino acid analyzer.[18,19] However, if carbon–carbon cleavage between the carbonyl carbon and the methyl carbon occurs, a carboxymethyl residue will not be formed.

Although the observed stoichiometry of the reaction of kdGtP aldolase with bromopyruvate was 1:1, both a sulfhydryl and a carboxyl group were modified. This result suggested that, within a given subunit (the enzyme is a trimer), modification of either residue prevents modification of the other and that both residues are in or near the active site. Subsequently, it was shown that at low ionic strength (20 mM citrate) 99% of the incorporated bromopyruvate was present as the ester, and at high ionic strength (250 mM citrate) 73% of the incorporated reagent was present as the thiol ether. The stoichiometry remained constant, even though at intermediate ionic strengths the incorporated label partitioned to varying degrees between two residues. The interpretation of these results offered by Meloche[16] is that the salt concentration alters an equilibrium between two conformers of the enzyme. In one conformer the active-site sulfhydryl is brought into position for alkylation, and in the other conformer the carboxyl is in proper orientation for esterification to occur. Consistent with this interpretation are the observed differential

[18] H. J. Goren, D. M. Glick, and E. A. Barnard, Arch. Biochem. Biophys. 126, 607 (1968).
[19] F. R. N. Gurd, this series, Vol. 25, p. 424.

effects of ionic strength on V_{max} for the exchange reaction and V_{max} for the cleavage reaction.[20]

An instructive aspect of this study is the demonstration that rather minor changes in experimental conditions can lead to substantive new findings about the identity of residues at the active site.

The close structural resemblance of bromopyruvate to the ketone substrate provides unusual opportunities for ascertaining whether the enzyme really recognizes bromopyruvate as substrate and whether bromopyruvate really alkylates the base that activates an α-hydrogen. Since the reagent contains both a ketone group and α-hydrogen atoms, one can ask if the reagent forms a Schiff base with the essential amino group and if one of the α-hydrogens in the bromopyruvyl–enzyme Schiff-base exchanges with solvent protons. In answering these questions, Meloche[15,21] has provided an additional criterion of affinity labeling that is applicable to certain other situations and has reinforced a general mechanism of aldolase catalysis.

$(3R,S)$-$[3$-$^3H_2]$Bromopyruvate is a substrate for kdGtP aldolase in the exchange reaction under conditions in which 80% of incorporated reagent is bound as an ester.[15] The enzyme is stereospecific for the Pro-R hydrogen. The kinetics of detritiation and inactivation are consistent with the two processes occurring at the same site according to the equation

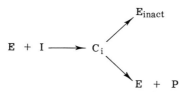

where E is free enzyme, I is bromopyruvate, C_i is the enzyme-bromopyruvate complex (a ketimine Schiff base), E_{inact} is inactivated enzyme, and P is detritiated bromopyruvate. Since the rates of both inactivation and exchange are proportional to the concentration of a common intermediate (C_i), a constant ratio of moles of bromopyruvate detritiated per mole of enzyme inactivated should be seen irrespective of the concentration of I. Furthermore, the concentration of bromopyruvate giving one-half the maximal rate of inactivation should also give one-half the maximal rate of exchange. Both predictions were verified[15]; the ratio of exchange to alkylation was about 50 over a wide range of reagent concentrations, and the half-maximal concentration of bromopyruvate for both processes was 1 mM. It has not been demonstrated directly that bromo-

[20] H. P. Meloche, unpublished data cited in ref. 16.
[21] H. P. Meloche and J. P. Glusker, *Science* **181**, 350 (1973).

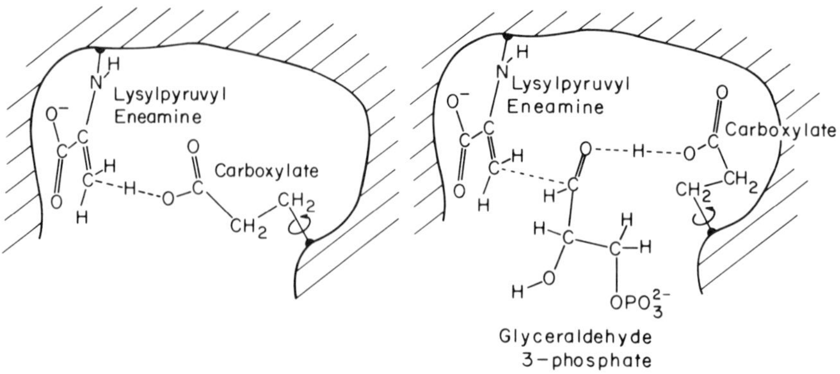

FIG. 3. Single-base mechanism for 2-keto-3-deoxygluconate-6-phosphate aldolase.

pyruvate forms a Schiff base with the enzyme as an obligatory precursor of alkylation. This appears, however, to be an inescapable conclusion, since Schiff-base formation precedes exchange and since both exchange and alkylation require the same complex.

Proof that the carboxylate, which is susceptible to esterification by bromopyruvate, is the base that activates a substrate hydrogen comes from results of borohydride reduction.[22] Reduction of the esterified aldolase (in the absence of protein denaturant) produces a N_6-lysyl secondary amine, demonstrating the presence of a Schiff base formed between the covalently fixed carboxyketomethyl group and the essential ϵ-amino group. Thus, when the carbonyl of bromopyruvate is bound at the active site as a Schiff base, C3 of the bound reagent is juxtaposed opposite the carboxylate so that either esterification or carboxylate-catalyzed exchange of protons occurs. In contrast a reagent bridge between the alkylated sulfhydryl and the essential amino group cannot be demonstrated by borohydride reduction.[22]

Realizing that a carboxylate is probably the base involved in activation of pyruvate hydrogens, Meloche and Glusker[21] have shown with models that a single base could serve in both proton-transfer steps (see Fig. 1) necessary to effect the condensation of pyruvate and D-glyceraldehyde 3-phosphate to the 6-carbon sugar phosphate. Rotation about the bond between C2 and C3 of a glutamyl side-chain positions the carboxylate so that it can be adjacent to either C3 of pyruvate or C4 of 2-keto-3-deoxygluconate-6-phosphate (Fig. 3).

Meloche and Monti[23] have recently extended these studies to the

[22] H. P. Meloche, *J. Biol. Chem.* **248**, 6945 (1973).
[23] H. P. Meloche and C. T. Monti, *Biochemistry* **14**, 3682 (1975).

TABLE I
EXAMPLES OF ACTIVE-SITE LABELING WITH BROMOPYRUVATE

Enzyme	Residue modified	References
3-Deoxy-D-arabino-heptulosonate 7-phosphate synthase	—	a
Isocitrate lyase	—	b
Carbonic anhydrase	His	c
N-Acetylneuraminate lyase	Cys (chloropyruvate used)	d
Pyruvate carboxylase	Cys	e
Glutamate decarboxylase	Cys	f
Malic enzyme	Cys	g, h

a M. Staub and G. Dénes, *Biochim. Biophys. Acta* **139**, 519 (1967).
b T. E. Roche, B. A. McFadden, and J. O. Williams, *Arch. Biochem. Biophys.* **147**, 192 (1971).
c P. O. Göthe and P. O. Nyman, *FEBS Lett.* **21**, 159 (1972).
d J. E. G. Barnett and F. Koliss, *Biochem. J.* **143**, 487 (1974).
e P. J. Hudson, D. B. Keech, and J. C. Wallace, *Biochem. Biophys. Res. Commun.* **65**, 213 (1975).
f M. L. Fonda, *J. Biol. Chem.* **251**, 229 (1976).
g G. Chang and R. Y. Hsu, *Biochem. Biophys. Res. Commun.* **55**, 580 (1973).
h G. Chang and R. Y. Hsu, *Biochemistry*, in press.

reaction of bromopyruvate with 2-keto-3-deoxygalactonate-6-phosphate aldolase to compare the geometry and chirality of the pyruvate portions of the active sites of two different aldolases.

Bromopyruvate has proved to be an exceedingly versatile affinity label; some of the more successful examples of its use as an active-site probe are listed in Table I. Especially interesting are recent studies by Chang and Hsu[24,25] in which the reagent is shown to be both an inactivator and substrate for pigeon liver malic enzyme, and half-of-sites stoichiometry is observed.

3-Haloacetol Phosphates

These compounds are close structural analogs of dihydroxyacetone phosphate, differing only in replacement of the hydroxyl with a halogen. The rationale for designing haloacetol phosphates was much the same as that which led Meloche[13,14] to believe that bromopyruvate would be a likely affinity label for kdGtP aldolase. Reactions catalyzed by both

[24] G. Chang and R. Y. Hsu, *Biochem. Biophys. Res. Commun.* **55**, 580 (1973).
[25] G. Chang and R. Y. Hsu, *Biochemistry,* in press.

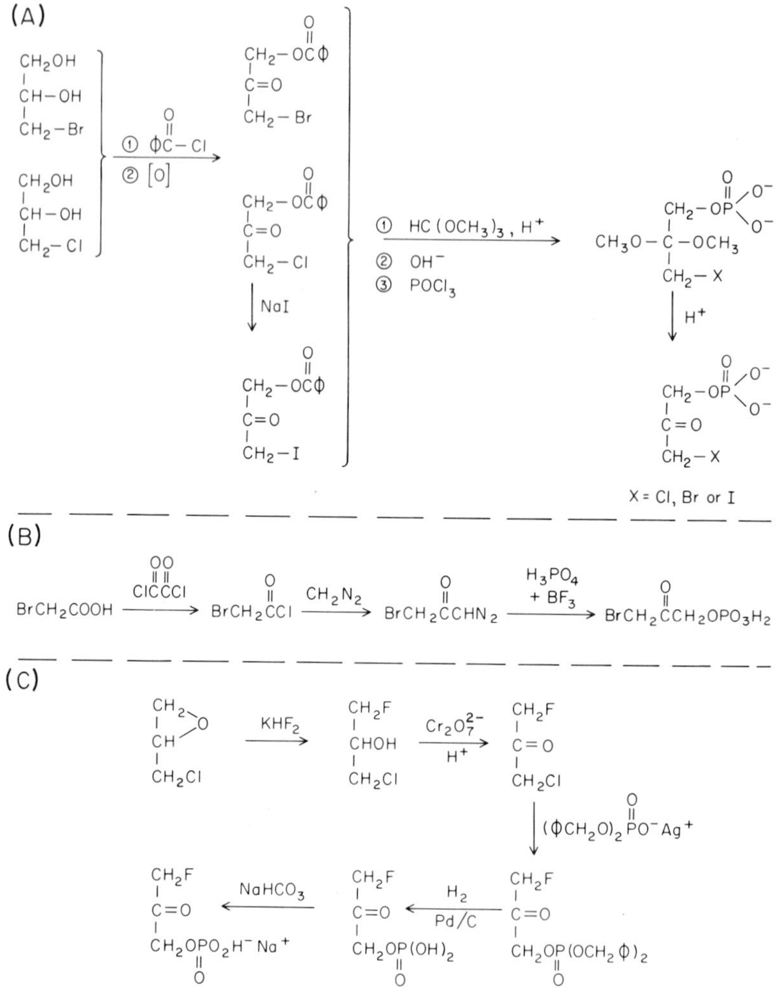

(A)

$$\left.\begin{array}{l} \text{CH}_2\text{OH} \\ | \\ \text{CH}-\text{OH} \\ | \\ \text{CH}_2-\text{Br} \\[2mm] \text{CH}_2\text{OH} \\ | \\ \text{CH}-\text{OH} \\ | \\ \text{CH}_2-\text{Cl} \end{array}\right\} \xrightarrow[\;\text{②}\;[\text{O}]\;]{\text{①}\;\Phi\overset{\text{O}}{\overset{\|}{\text{C}}}-\text{Cl}}$$

$$\begin{array}{l} \text{CH}_2-\text{OC}\overset{\text{O}}{\overset{\|}{}}\Phi \\ | \\ \text{C}=\text{O} \\ | \\ \text{CH}_2-\text{Br} \\[2mm] \text{CH}_2-\text{OC}\overset{\text{O}}{\overset{\|}{}}\Phi \\ | \\ \text{C}=\text{O} \\ | \\ \text{CH}_2-\text{Cl} \end{array}$$

$$\left.\right\} \xrightarrow[\;\text{②}\;\text{OH}^-\;]{\;\text{①}\;\text{HC(OCH}_3)_3,\;\text{H}^+\;} \xrightarrow{\;\text{③}\;\text{POCl}_3\;}$$

$$\begin{array}{l} \text{CH}_2-\text{O}\overset{\text{O}}{\overset{\|}{\text{P}}}\overset{\diagup \text{O}^-}{\diagdown \text{O}^-} \\ | \\ \text{CH}_3\text{O}-\text{C}-\text{OCH}_3 \\ | \\ \text{CH}_2-\text{X} \end{array} \downarrow \text{H}^+$$

↓ NaI

$$\begin{array}{l} \text{CH}_2-\text{OC}\overset{\text{O}}{\overset{\|}{}}\Phi \\ | \\ \text{C}=\text{O} \\ | \\ \text{CH}_2-\text{I} \end{array}$$

$$\begin{array}{l} \text{CH}_2-\text{O}\overset{\text{O}}{\overset{\|}{\text{P}}}\overset{\diagup \text{O}^-}{\diagdown \text{O}^-} \\ | \\ \text{C}=\text{O} \\ | \\ \text{CH}_2-\text{X} \end{array}$$

X = Cl, Br or I

(B)

$$\text{BrCH}_2\text{COOH} \xrightarrow{\overset{\text{O O}}{\overset{\|\;\|}{\text{ClCCCl}}}} \text{BrCH}_2\overset{\text{O}}{\overset{\|}{\text{C}}}\text{Cl} \xrightarrow{\text{CH}_2\text{N}_2} \text{BrCH}_2\overset{\text{O}}{\overset{\|}{\text{C}}}\text{CHN}_2 \xrightarrow[+\;\text{BF}_3]{\text{H}_3\text{PO}_4} \text{BrCH}_2\overset{\text{O}}{\overset{\|}{\text{C}}}\text{CH}_2\text{OPO}_3\text{H}_2$$

(C)

$$\begin{array}{l} \text{CH}_2 \\ | \quad\diagdown\text{O} \\ \text{CH} \\ | \\ \text{CH}_2\text{Cl} \end{array} \xrightarrow{\text{KHF}_2} \begin{array}{l} \text{CH}_2\text{F} \\ | \\ \text{CHOH} \\ | \\ \text{CH}_2\text{Cl} \end{array} \xrightarrow[\text{H}^+]{\text{Cr}_2\text{O}_7^{2-}} \begin{array}{l} \text{CH}_2\text{F} \\ | \\ \text{C}=\text{O} \\ | \\ \text{CH}_2\text{Cl} \end{array}$$

$$\downarrow (\Phi\text{CH}_2\text{O})_2\overset{\text{O}}{\overset{\|}{\text{P}}}\text{O}^-\,\text{Ag}^+$$

$$\begin{array}{l} \text{CH}_2\text{F} \\ | \\ \text{C}=\text{O} \\ | \\ \text{CH}_2\text{OPO}_2\text{H}^-\,\text{Na}^+ \\ \quad\quad\overset{\|}{\text{O}} \end{array} \xleftarrow{\text{NaHCO}_3} \begin{array}{l} \text{CH}_2\text{F} \\ | \\ \text{C}=\text{O} \\ | \\ \text{CH}_2\text{OP(OH)}_2 \\ \quad\quad\overset{\|}{\text{O}} \end{array} \xleftarrow[\text{Pd/C}]{\text{H}_2} \begin{array}{l} \text{CH}_2\text{F} \\ | \\ \text{C}=\text{O} \\ | \\ \text{CH}_2\text{OP(OCH}_2\Phi)_2 \\ \quad\quad\overset{\|}{\text{O}} \end{array}$$

FIG. 4. Syntheses of haloacetol phosphates.

fructose-bisphosphate aldolase and triosephosphate isomerase involve stereospecific removal of one of the prochiral protons from C3 of dihydroxyacetone phosphate. Thus, the possibility existed that a single reagent could be used to identify active-site residues with common catalytic functions in different enzymes.

Iodoacetol phosphate was the first haloacetol phosphate synthesized,[26,27] and the sequence of reactions used (Fig. 4A) also made avail-

[26] F. C. Hartman, *Fed. Proc., Fed. Am. Soc. Exp. Biol.* **27**, 454 (1968).
[27] F. C. Hartman, *Biochem. Biophys. Res. Commun.* **33**, 888 (1968).

able the bromo and chloro compounds.[5] Two other syntheses were independently devised, one for bromoacetol phosphate (Fig. 4B)[28] and one for the corresponding fluoro and chloro analogs (Fig. 4C).[29] The reagents can be labeled with ^{32}P (Hartman[30]) or ^{14}C (Coulson et al.[28]). Since both the halogen and phosphate groups of haloacetol phosphates are quite labile, their concentrations in solution should be carefully determined just prior to the chemical modification study. This is conveniently accomplished by quantitating base-labile phosphate content and reactive halogen content.[5] In the latter method a molar excess of glutathione, whose sulfhydryl group is readily alkylated by chloro- or bromoacetol phosphate, is added to a solution of the reagent, and the amount of glutathione remaining is determined by titration with p-chloromercuribenzoate[31] or 5,5'-dithiobis(2-nitrobenzoic acid).[32]

The most successful application of haloacetol phosphates as affinity labels has been in the partial characterization of the active site of triosephosphate isomerase. Very similar studies, carried out independently in the laboratories of F. C. Hartman and J. R. Knowles, demonstrated an essential glutamyl γ-carboxylate (esterified by the reagent) in the enzyme.[33-38] The recently determined primary structure of the enzyme from rabbit muscle places the glutamyl at position 165.[39] All the usual criteria of affinity labeling were satisfied and have been well documented. Certain aspects of these studies that either relate to the use of haloketones in general or have provided evidence of the carboxylate's intimate role in the catalytic process will be considered.

Characterization of the product of the reaction of chloroacetol phosphate with triosephosphate isomerase was facilitated by reduction of the derivatized protein with sodium borotritide. Initially the only isotopically labeled chloroacetol phosphate available was the ^{32}P-labeled material,

[28] A. F. W. Coulson, J. R. Knowles, and R. E. Offord, Chem. Commun. 1, 7 (1970).
[29] J. B. Silverman, P. S. Babiarz, K. P. Mahajan, J. Buschek, and T. P. Fondy, Biochemistry 14, 2252 (1975).
[30] F. C. Hartman, Biochemistry 9, 1783 (1970).
[31] P. D. Boyer, J. Am. Chem. Soc. 76, 4331 (1954).
[32] G. L. Ellman, Arch. Biochem. Biophys. 82, 70 (1959).
[33] F. C. Hartman, J. Am. Chem. Soc. 92, 2170 (1970).
[34] F. C. Hartman, Biochem. Biophys. Res. Commun. 39, 384 (1970).
[35] F. C. Hartman, Biochemistry 10, 146 (1971).
[36] F. C. Hartman, this series, Vol. 25, p. 661.
[37] A. F. W. Coulson, J. R. Knowles, J. D. Priddle, and R. E. Offord, Nature (London) 227, 180 (1970).
[38] S. De La Mare, A. F. W. Coulson, J. R. Knowles, J. D. Priddle, and R. E. Offord, Biochem. J. 129, 321 (1972).
[39] P. H. Corran and S. G. Waley, FEBS Lett. 30, 97 (1973).

which is not well suited for characterization of modified residues because of the lability of the phosphate group. Reduction with borotritride provided a stable isotopic label on the carbon chain of the reagent that survives conditions used to hydrolyze proteins and also stabilized the phosphate group so that it was not liberated as P_i during proteolytic digestion of the modified protein.[33,35]

In addition to providing a marker, the incorporated tritium can be quantitated, thereby revealing the amount of reagent covalently attached to the protein. The specific radioactivity of $NaBT_4$ must be measured indirectly, since lack of purity and stability preclude preparing stock solutions of known concentration based on weight. One approach is to reduce glutathione that has been alkylated with chloroacetol phosphate with a portion of the same borotritride solution used to reduce the modified protein. The glutathione derivative is easily purified on Dowex 50 (H^+) and quantitated on the amino acid analyzer, thus providing an accurate measurement of the specific radioactivity of the borotritride.[40,41] The extent of incorporation of chloroacetol phosphate into triosephosphate isomerase determined indirectly by reduction is identical to that determined with ^{32}P-labeled reagent.[42] This indirect method has also been used in the case of ribulose-bisphosphate carboxylase labeled with 3-bromo-1,4-dihydroxy-2-butanone 1,4-bisphosphate.[40,41]

Knowles and colleagues[38] discovered that if the inactivated isomerase was not reduced with borohydride, the phosphate group of the incorporated reagent was displaced by the phenolic hydroxyl group of an adjacent tyrosyl residue, thereby forming a cross-link. After proteolytic digestion, the reagent moiety was found attached to the tyrosyl residue through an ether linkage rather than to the glutamyl carboxylate, the initial site of reaction. Thus, with any haloketone containing an additional group that can be activated by the adjacent carbonyl, possibilities exist for cross-linking two active-site residues or for migration to a second site. As indicated by Knowles, this latter possibility raises the danger of an incorrect identification of the initial site of reaction and therefore an incorrect conclusion as to the identity of an active-site residue.

A difficult question to answer from chemical modification studies alone, even when a high degree of selectivity is achieved, is whether the residue modified plays an intimate role in the catalytic process. Two approaches have been used in the case of inactivation of triosephosphate isomerase by chloroacetol phosphate that provide rather convincing, albeit indirect,

[40] I. L. Norton, M. H. Welch, and F. C. Hartman, *J. Biol. Chem.* **250**, 8062 (1975).
[41] I. L. Norton and F. C. Hartman, this volume [42].
[42] F. C. Hartman, unpublished data.

evidence that the implicated carboxyl group is essential to catalysis. The rate of esterification of Glu-165 by chloroacetol phosphate is extremely rapid [at 2° and pH 6.5, the apparent k_{2nd} is 2300 M^{-1} sec^{-1} (Hartman[35]) ; at 25° and pH 7.0, k_{2nd} is 14,000 M^{-1} sec^{-1} (Davis et al.[43])]. In contrast, esterification of glutamic acid by chloroacetol phosphate has not been demonstrated. At 25° and pH 8.1, incubation of chloroacetol phosphate (20 mM) with glutamic acid (1 mM) for 12 hr resulted in no detectable loss of glutamic acid as measured with the amino acid analyzer.[5] If, for purposes of calculation, one assumes a 10% loss of glutamic acid (a 5% loss would have been readily detected) in the 12-hr period, k_{2nd} would be 9×10^{-5} M^{-1} sec^{-1}. Thus, the rate enhancement for the reaction of Glu-165 of triosephosphate isomerase with chloroacetol phosphate as compared with free glutamic acid is $>1.5 \times 10^8$. Such an enhancement is as large as the difference in rate of some enzyme-catalyzed reactions vs. their nonenzymic counterparts, so that the esterification of Glu-165 might be said to be "enzyme-catalyzed."

The other type of evidence that strongly suggests a catalytic role of Glu-165 comes from comparative studies. The isomerases from yeast,[44] chicken muscle,[38] rabbit muscle,[35] and human red blood cells[45] have all been subjected to modification with haloacetol phosphates; in each case inactivation results from esterification of a glutamyl γ-carboxylate. Furthermore, in each case the amino acid sequences of hexapeptides containing the essential residue are identical: -Ala-Tyr-Glu-Pro-Val-Trp-. This high degree of species invariance of the active-site glutamyl residue and the adjacent primary structure strongly implies that the carboxyl group is functional in catalysis. Banner et al.[46] have recently published the three-dimensional structure of chicken muscle triosephosphate isomerase. If it is assumed that the binding site for dihydroxyacetone phosphate is in the same region as the binding site for sulfate, which was visualized, Glu-165 is at the active site.

The interconversion of D-glyceraldehyde-3-phosphate and dihydroxy-acetone phosphate as catalyzed by triosephosphate isomerase involves proton abstraction from C3 of the ketonic substrate by an acid-base group of the enzyme to generate an enediol intermediate, followed by proton transfer from the acid-base group to C2 of the enediol to form the alde-

[43] R. H. Davis, Jr., P. Delaney, and C. S. Furfine, Arch. Biochem. Biophys. **159**, 11 (1973).
[44] I. L. Norton and F. C. Hartman, Biochemistry **11**, 4435 (1972).
[45] F. C. Hartman and R. W. Gracy, Biochem. Biophys. Res. Commun. **52**, 388 (1973).
[46] D. W. Banner, A. C. Bloomer, G. A. Petsko, D. C. Phillips, C. I. Pogson, I. A. Wilson, P. H. Corran, A. J. Furth, J. D. Milman, R. E. Offord, J. D. Priddle, and S. G. Waley, Nature (London) **255**, 609 (1975).

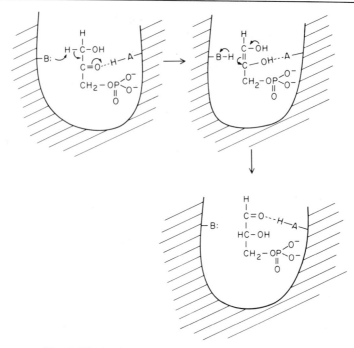

FIG. 5. Mechanism for triosephosphate isomerase.

hyde (Fig. 5).[47,48] Based on the observation of esterification of Glu-165, the carboxylate is considered a likely candidate for the essential base. Although this postulate has been generally accepted and also supported by results with glycidol phosphate[49,50] (an affinity label that also esterifies Glu-165), definitive proof will probably have to await description of the crystallographic structure of the enzyme–substrate complex. However, data are available that permit calculations of the maximal value for the pK_a of the acid-base group. During the enzyme-catalyzed conversion of dihydroxyacetone phosphate (stereospecifically tritiated at C3) to D-glyceraldehyde 3-phosphate, >98% of the tritium is exchanged for solvent protons and <2% is transferred to C2.[51] An explanation for this result is that the conjugate acid of the essential base ionizes (i.e., exchanges its proton for a water proton) at least 50 times more rapidly

[47] E. A. Noltmann, in "The Enzymes" (P. D. Boyer, ed.), 3rd ed., Vol. 6, p. 271. Academic Press, New York, 1972.

[48] I. A. Rose, Adv. Enzymol. 43, 491 (1975).

[49] J. C. Miller and S. G. Waley, Biochem. J. 123, 163 (1971).

[50] K. J. Schray, E. L. O'Connell, and I. A. Rose, J. Biol. Chem. 248, 2214 (1973). See also this volume [41].

[51] B. Plaut and J. R. Knowles, Biochem. J. 129, 311 (1972).

than it transfers the proton to C2 of the enediol intermediate. With this value and the value for k_{cat}, Rose[48] has calculated that the group involved in proton transfer must have a pK_a less than 5.0.

Based on the pH dependency of the inactivation rate of rabbit muscle triosephosphate isomerase with glycidol phosphate, the pK_a for Glu-165 was calculated to be less than 5.5 in one study[50] and 6.0 in another.[52] Since this difference could be due to the complication of a variable affinity of the reagent for the enzyme as a result of the change in ionization state of the reagent over the pH range examined, the pH dependency of inactivation rate was also studied with the strong, monoprotic acid chloroacetol sulfate, another compound which selectively esterifies Glu-165.[53] With this compound, the pK_a of Glu-165 in the rabbit muscle enzyme was found to be less than 5.0. An exact value could not be determined because of the instability of the enzyme to acid; however, with the more stable yeast enzyme, a pK_a of 3.9 was calculated.[53] Thus, the acidity of the essential carboxyl group is consistent with its postulated role in catalysis.

Haloketone Analogs of NAD

Reactive analogs of coenzymes are of special interest because they provide the opportunity to probe the active sites of a multitude of dehydrogenases and thus possibly answer questions concerning the degree of homology among nucleotide binding sites and the identity of basic residues involved in hydrogen transfer from the alcohol substrate to the coenzyme. Several α-haloketones that resemble pyridine nucleotides have been designed (primarily by C. Woenckhaus and colleagues) and used successfully to label active-site residues in a number of enzymes, including alcohol dehydrogenase, lactate dehydrogenase, and estradiol 17β-dehydrogenase (Table II).

A very simple derivative of nicotinamide, 3-bromoacetylpyridine, preferentially alkylates both an essential cysteinyl and an essential histidyl residue in pig heart lactate dehydrogenase.[54] The reagent is selective for the histidyl residue if the enzyme is pretreated with mercuric ions to protect the sulfhydryl group. Subsequent to these investigations the three-dimensional structures of dogfish lactate dehydrogenase and

[52] S. G. Waley, Biochem. J. 126, 255 (1972).

[53] F. C. Hartman, G. M. LaMuraglia, Y. Tomozawa, and R. Wolfenden, Biochemistry 14, 5274 (1975).

[54] C. Woenckhaus, J. Berghäuser, and G. Pfleiderer, Hoppe-Seyler's Z. Physiol. Chem. 350, 473 (1969).

TABLE II

NUCLEOTIDE ANALOGS THAT CONTAIN HALOKETONE GROUPS

Reagent	Enzyme	Residue modified	References
3-Bromoacetyl-pyridine	Lactate dehydrogenase NAD kinase	Cys-165 His-195 Unknown	a b
3-Bromoacetyl 1-carboxyethyl pyridinium (ion)	Lactate dehydrogenase	Cys	c

P¹-[3-(3-Bromoacetylpyridinium)propyl]-P²-5′-adenos-5′-yl diphosphate

Nicotinamide[5-(bromoacetyl)-4-methylimidazol-1-yl]dinucleotide

Yeast alcohol dehydrogenase	Cys-43	d–f
Yeast alcohol dehydrogenase	Cys-43	g–i
Liver alcohol dehydrogenase	Cys-174	g–i

TABLE II (Continued)

Reagent	Enzyme	Residue modified	References
3-Chloroacetylpyridinium adenine dinucleotide	Liver alcohol dehydrogenase	—	i
	Yeast alcohol dehydrogenase	—	i
	Estradiol 17β-dehydrogenase	Cys	k

[a] C. Woenckhaus, J. Berghäuser, and G. Pfleiderer, Hoppe-Seyler's Z. Physiol. Chem. **350**, 473 (1969).

[b] D. K. Apps, FEBS Lett. **5**, 96 (1969).

[c] M. Leven, G. Pfleiderer, J. Berghäuser, and C. Woenckhaus, Hoppe-Seyler's Z. Physiol. Chem. **350**, 1647 (1969).

[d] C. Woenckhaus, M. Zoltobrocki, and J. Berghäuser, Hoppe-Seyler's Z. Physiol. Chem. **351**, 1441 (1970).

[e] C. Woenckhaus, M. Zoltobrocki, J. Berghäuser, and R. Jeck, Hoppe-Seyler's Z. Physiol. Chem. **354**, 60 (1973).

[f] H. Jörnvall, C. Woenckhaus, E. Schättle, and R. Jeck, FEBS Lett. **54**, 297 (1975).

[g] C. Woenckhaus, M. Zoltobrocki, and J. Berghäuser, Hoppe-Seyler's Z. Physiol. Chem. **351**, 1441 (1970).

[h] C. Woenckhaus and R. Jeck, Hoppe-Seyler's Z. Physiol. Chem. **352**, 1417 (1971).

[i] H. Jörnvall, C. Woenckhaus, and G. Johnscher, Eur. J. Biochem. **53**, 71 (1975).

[j] J. Biellmann, G. Branlant, B. Y. Foucaud, and M. J. Jung, FEBS Lett. **40**, 29 (1974).

[k] J. Biellmann, G. Branlant, J. Nicolas, M. Pons, B. Descomps, and A. Crastes de Paulet, Eur. J. Biochem. **63**, 477 (1976).

its ternary complex with NAD-pyruvate were determined.[55,56] The histidyl residue (His-195), first implicated in catalysis by its selective modification with 3-bromoacetylpyridine,[54] appears to be involved in proton transfer to or from the substrate, since in the three-dimensional structure the imidazole ring of His-195 is oriented toward the carbonyl of pyruvate.[57] Although numerous chemical modification studies,[58-61] in addition to the one with 3-bromoacetylpyridine, have implicated Cys-165 in catalysis, the crystallographic work suggests that this sulfhydryl does not play a direct role in catalysis but occupies a position near the active site.[57]

Bromoacetylpyridine is prepared by bromination of commercially available [^{14}C]acetylpyridine (labeled in the carbonyl carbon). With the other compounds shown in Table II, the halogen is also inserted (usually by direct halogenation) in the last step of the synthesis. In the recently described[62] synthesis of 3-chloroacetylpyridine adenine dinucleotide, the immediate precursor is the corresponding 3-diazomethylketone. The diazo group is then exchanged for chlorine by treatment with lithium chloride in hydrochloric acid, a procedure similar to one used for the preparation of halomethylketone derivatives of amino acids.[2-4]

Amino acid residues of dehydrogenases modified by coenzyme analogs have been identified in several cases by the laborious but unequivocal sequential procedures of reduction with borohydride, proteolytic digestion, and characterization of purified peptides carrying the reagent moiety. If the primary sequence of the enzyme is known, the amino acid compositions of the pure peptides normally reveal the kind of residue modified (by the absence of one residue equivalent of a free amino acid and its replacement with an amino acid derivative) and its location in

[55] M. J. Adams, M. Buehner, K. Chandrasekhar, G. C. Ford, M. L. Hackert, A. Liljas, P. Lentz, Jr., S. T. Rao, M. G. Rossmann, I. L. Smiley, and J. L. White, *in* "Protein–Protein Interactions" (J. Jaenicke and E. Helmreich, eds.), p. 139. Springer-Verlag, Berlin and New York, 1972.

[56] M. G. Rossmann, M. J. Adams, M. Buehner, G. C. Ford, M. L. Hackert, P. J. Lentz, Jr., A. McPherson, Jr., R. W. Schevitz, and I. L. Smiley, *Cold Spring Harbor Symp. Quant. Biol.* **36**, 179 (1971).

[57] M. J. Adams, M. Buehner, K. Chandrasekhar, G. C. Ford, M. L. Hackert, A. Liljas, M. G. Rossmann, I. E. Smiley, W. S. Allison, J. Everse, N. O. Kaplan, and S. S. Taylor, *Proc. Natl. Acad. Sci. U.S.A.* **70**, 1968 (1973).

[58] A. H. Gold and H. L. Segal, *Biochemistry* **4**, 1506 (1965).

[59] J. J. Holbrook, *Biochem. Z.* **344**, 141 (1966).

[60] T. P. Fondy, J. Everse, G. A. Driscoll, F. Castillo, F. E. Stolzenbach, and N. O. Kaplan, *J. Biol. Chem.* **240**, 4219 (1965).

[61] J. J. Holbrook and R. A. Stinson, *Biochem. J.* **120**, 289 (1970).

[62] J. Biellmann, G. Branlant, B. Y. Foucaud, and M. J. Jung, *FEBS Lett.* **40**, 29 (1974).

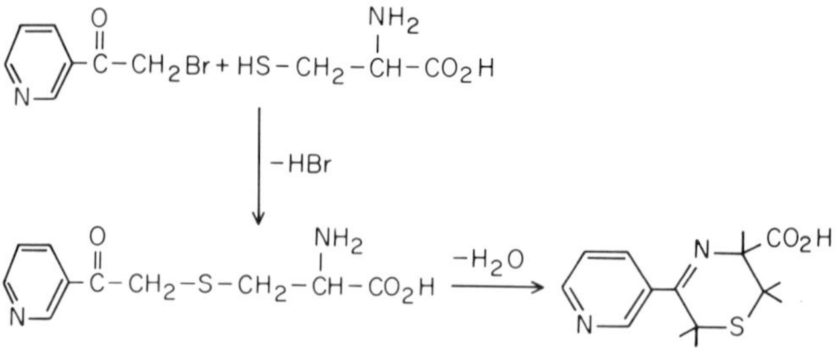

FIG. 6. Reaction of cysteine with bromoacetylpyridine.

the primary sequence. For example, in a study of the interaction of horse liver and yeast alcohol dehydrogenases by nicotinamide [5-(bromo-acetyl)-4-methylimidazol-1-yl] dinucleotide (which is active as a coenzyme), peptides were isolated whose amino acid compositions were consistent with Cys-43 (which corresponds to Cys-46 in the liver enzyme) as the site of alkylation in the yeast enzyme and Cys-174 as the site of alkylation in the horse liver enzyme.[63] Recent crystallographic data show that Cys-46 and Cys-174 in the horse liver enzyme function in binding of the catalytic zinc atom.[64]

Two other approaches have been used to identify the residues modi-fied by these nucleotide derivatives: (a) oxidation with hydrogen peroxide and subsequent identification of carboxymethyl amino acids in acid hydrolyzates, and (b) reduction with borohydride and comparison of derivatives in the hydrolyzates to synthetically prepared standards. The second approach was used in identifying a sulfhydryl and an imidazole group as the sites of modifiication of lactate dehydrogenase by bromo-acetylpyridine.[54] An interesting observation was the spontaneous forma-tion of a thiazine (by intramolecular condensation between the ketone group of the reagent moiety and the amino group of cysteine) from the initial S-alkylated product (Fig. 6). This serves to emphasize that small peptides are generally preferable to free amino acids in the synthesis of derivatives in which selective side-chain substitution is desired.

In a very nice example of affinity labeling in which 3-chloroacetyl-pyridine adenine dinucleotide (a compound that is active as a hydrogen acceptor) is shown to be active-site-specific for estradiol 17β-dehydro-genase from human placenta, Biellmann et al.[65] have identified the

[63] H. Jörnvall, C. Woencklaus, and G. Johnscher, Eur. J. Biochem. 53, 71 (1975).
[64] H. Eklund, B. Nordström, E. Zeppezauer, G. Söderlund, I. Ohlsson, T. Boiwe, and C.-I. Brändén, FEBS Lett. 44, 200 (1974).
[65] J. Biellmann, G. Branlant, J. Nicolas, M. Pons, B. Descomps, and A. Crastes de Paulet, Eur. J. Biochem. 63, 477 (1976).

TABLE III
SELECTED EXAMPLES OF SITE-SPECIFIC HALOKETONES

Reagent	Protein	Residue modified	References
 2-Bromo-4'-hydroxy-3'- nitroacetophenone	Transglutaminase	Cys	a
 (S)-N-[α-(Chloroacetyl)phenethyl]- p-toluenesulfonamide	Elongation factor Tu Luciferase	— Cys	b, c d
 (S)-3-Amino-1-chloro- 4-phenyl-2-butanone	Tyr/Phe transport protein in *Bacillus subtilis*	—	e
 1-Amino-3-chloro- 2-propanone	Neutral amino acid transport protein in *Trypanosoma brucei*	—	f

TABLE III (*Continued*)

Reagent	Protein	Residue modified	References

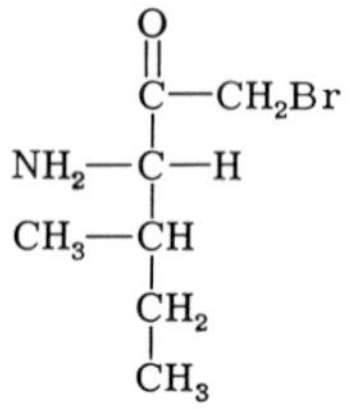

O
‖
C—CH₂Br
NH₂—C—H
CH₃—CH
CH₂
CH₃

(3*S*,4*S*)-3-Amino-1-bromo-
4-methyl-2-hexanone

L-Isoleucine: tRNA ligase   Cys   *g*

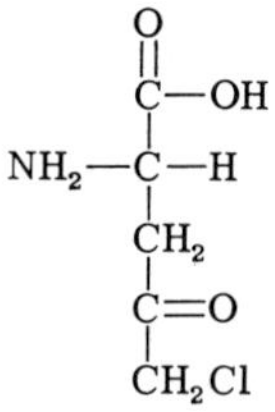

O
‖
C—OH
NH₂—C—H
CH₂
C=O
CH₂Cl

(*S*)-2-Amino-5-
chlorolevulinic acid

Carbamylphosphate synthetase   —   *h*

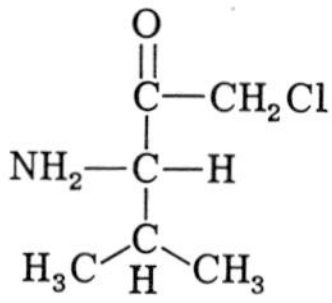

O
‖
C—CH₂Cl
NH₂—C—H
H₃C   H   CH₃

(*S*)-3-Amino-1-chloro-
4-methyl-2-pentanone

L-Valine: tRNA ligase   —   *i, j*

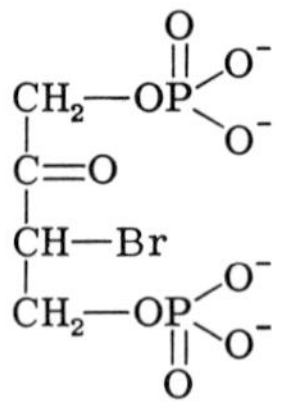

O   O⁻
‖   ╱
CH₂—OP
│   ╲O⁻
C=O
CH—Br
│   O⁻
CH₂—OP╱
‖   ╲O⁻
O

(*R,S*)-3-Bromo-1,4-
dihydroxy-3-butanone-
1,4-biphosphate

Ribulose-bisphosphate carboxylase   Lys   *k, l*

TABLE III (*Continued*)

a J. E. Folk and M. Gross, *J. Biol. Chem.* **246**, 6683 (1971).
b N. Richman and J. W. Bodley, *J. Biol. Chem.* **248**, 381 (1973).
c J. Sedlacek, J. Jonak, and I. Rychlik, *Biochem. Biophys. Acta* **254**, 478 (1971).
d R. Lee and W. D. McElroy, *Biochemistry* **8**, 130 (1969).
e S. M. D'Ambrosio and G. I. Glover, *Arch. Biochem. Biophys.* **167**, 754 (1975).
f M. J. Owen and H. P. Voorheis, *Eur. J. Biochem.* **62**, 619 (1976).
g P. Rainey, E. Holler, and M. Kula, *Eur. J. Biochem.* **63**, 419 (1976).
h E. Khedouri, P. M. Anderson, and A. Meister, *Biochemistry* **5**, 3552 (1966).
i L. Frolova, G. K. Kovaleva, M. B. Agalarova, and L. L. Kisselev, *FEBS Lett.* **34**, 213 (1973).
j J. Silver and R. A. Laursen, *Biochim. Biophys. Acta* **340**, 77 (1974).
k I. L. Norton, M. H. Welch, and F. C. Hartman, *J. Biol. Chem.* **250**, 8062 (1975).
l F. C. Hartman, this volume [10].

modified residue as carboxymethyl cysteine following oxidation with hydrogen peroxide; however, this approach may be fraught with danger. They reference the oxidation procedure used by Woenckhaus *et al.*[66] that indicated carboxymethyl histidine was formed from yeast alcohol dehydrogenase that was modified by P^1-[3-(3-bromoacetylpyridinium)propyl] P^2-5'-adenos-5'-yl diphosphate. A subsequent paper[67] stated that the material previously identified as carboxymethyl histidine was in fact a decomposition product as a result of harsh conditions and therefore in actuality the only modified residue was a cysteinyl.

[66] C. Woenckhaus, M. Zoltobrocki, J. Berghäuser, and R. Jeck, *Hoppe-Seyler's Z. Physiol. Chem.* **354**, 60 (1973).
[67] H. Jörnvall, C. Woenckhaus, E. Schättle, and R. Jeck, *FEBS Lett.* **54**, 297 (1975).

Other Examples

Additional successful applications of haloketones as affinity labels are listed in Table III.

[11] Haloacetyl Derivatives

By MEIR WILCHEK and DAVID GIVOL

A variety of haloacetyl reagents are being used for affinity labeling[1] although bromoacetyl is the predominant functional group.[2] The advantages of bromoacetyl reagents stem from the following features:

[1] E. Shaw, *Physiol. Rev.* **50**, 244 (1970).
[2] F. Naider, J. M. Becker, and M. Wilchek, *Isr. J. Chem.* **12**, 441 (1974).

1. *Spectrum of reactivity.* All protein nucleophiles can be alkylated by this type of reagent (Table I). Evidence for affinity labeling of cysteine,[3,4] histidine,[5] lysine[6-8] and α-amino groups,[8] tyrosine,[6,7] methionine[9,10] serine,[11] and glutamic acid[12] has been reported, and it is likely that aspartic acid and threonine can also be labeled. Such a wide spectrum of reactivity is important for successful labeling of residues at the reactive site.

2. *Relatively low reactivity.* Although cysteine can be readily labeled by haloacetyl reagents, most other residues are not being labeled significantly. Therefore a very low background and high specificity of labeling are present in affinity labeling of proteins by this type of reagent, whereas the predominant label is dictated by the structural requirement and concentration of the reagent at the site. This is illustrated by the difference in inactivation of various enzymes by haloacetyl reagents and their corresponding affinity label (Table II).

The low reactivity of the bromoacetyl derivatives is also sustained toward their reaction with the solvent. Thus, these reagents decompose slowly and are stable for long reaction periods; the low reactivity secures the presence of the reagent for many hours.

3. *Ease of preparation.* Bromoacetyl groups can be attached to amino groups (unstable compounds can also be prepared with compounds containing hydroxyl, sulfhydryl, and imidazole) by convenient procedures using bromoacetyl bromide,[13] bromoacetic anhydride,[14] or bromoacetyl-*N*-hydroxysuccinimide ester[6,7]; synthesis with bromoacetyl-*N*-hydroxy-

---

[3] A. Light, *Proc. Natl. Acad. Sci. U.S.A.* **52**, 1276 (1964).

[4] N. Sonenberg, M. Wilchek, and A. Zamir, *Proc. Natl. Acad. Sci. U.S.A.* **70**, 1423 (1973). See also this volume [73].

[5] S. L. Bradbury, *J. Biol. Chem.* **244**, 2002 (1969).

[6] P. Cuatrecasas, M. Wilchek, and C. B. Anfinsen, *J. Biol. Chem.* **244**, 4316 (1969). See also this volume [38].

[7] Y. Weinstein, M. Wilchek, and D. Givol, *Biochem. Biophys. Res. Commun.* **35**, 694 (1969).

[8] S.-Y. Cheng, M. Wilchek, M. J. Cahnman, and J. Robbins, *Biochemistry,* in press. See also this volume [48].

[9] F. Naider, Z. Bohak, and J. Yariv, *Biochemistry* **11**, 3202 (1972). See also this volume [43].

[10] W. B. Lawson and M. J. Schramm, *Biochemistry* **4**, 377 (1965).

[11] W. B. Lawson, M. D. Leafer, A. Tewes, and G. J. S. Rao, *Hoppe-Seyler's Z. Physiol. Chem.* **349**, 251 (1968).

[12] G. M. Hass and H. Neurath, *Biochemistry* **10**, 3535 (1971).

[13] P. H. Strausbauch, Y. Weinstein, M. Wilchek, S. Shaltiel, and D. Givol, *Biochemistry* **10**, 4342 (1971).

[14] D. Givol, P. H. Strausbauch, E. Hurwitz, M. Wilchek, J. Haimovich, and H. N. Eisen, *Biochemistry* **10**, 3461 (1971).

TABLE I

MODIFICATION OF AMINO ACID RESIDUES BY HALOACYL DERIVATIVES

Protein[a]	Amino acid modified	Product[b]
Papain[3b]	Cysteine	$-\text{NH}-\text{CH}-\text{CO}-$ $\quad\quad\quad\mid$ $\quad\quad\text{CH}_2$ $\quad\quad\quad\mid$ $\quad\quad\text{S}-\text{CH}_2-\text{COO}^-$
Carboxypeptidase[12]	Glutamic	$-\text{NH}-\text{CH}-\text{CO}-$ $\quad\quad\quad\mid$ $\quad\quad(\text{CH}_2)_2$ $\quad\quad\quad\mid$ $\quad\quad\text{CO}-\text{O}-\text{CH}_2-\text{COO}^-$
Carbonic anhydrase[5]	Histidine	$-\text{NH}-\text{CH}-\text{CO}-$ $\quad\quad\quad\mid$ $\quad\quad\text{CH}_2$ (imidazole ring) $\text{N}-\text{CH}_2-\text{COO}^-$
Staphylococcal nuclease[6] MOPC—315[13]	Lysine	$-\text{NH}-\text{CH}-\text{CO}-$ $\quad\quad\quad\mid$ $\quad\quad(\text{CH}_2)_4$ $\quad\quad\quad\mid$ $\quad\quad\text{NH}-\text{CH}_2-\text{COO}^-$
β-Galactosidase[9] chymotrypsin[10]	Methionine	$-\text{NH}-\text{CH}-\text{CO}-$ $\quad\quad\quad\mid$ $\quad\quad(\text{CH}_2)_2$ $\quad\quad\quad\mid$ $\text{H}_3\text{C}-\text{S}-\text{CH}_2-\text{COO}^-$
Trypsin[11]	Serine	$-\text{NH}-\text{CH}-\text{CO}-$ $\quad\quad\quad\mid$ $\quad\quad\text{CH}_2$ $\quad\quad\quad\mid$ $\quad\quad\text{O}-\text{CH}_2-\text{COO}^-$
Staphylococcal nuclease[6] MOPC—315[13]	Tyrosine	$-\text{NH}-\text{CH}-\text{CO}-$ $\quad\quad\quad\mid$ $\quad\quad\text{CH}_2$ (benzene ring) $\text{O}-\text{CH}_2-\text{COO}^-$

[a] The superscript numbers refer to text footnotes.

[b] Hydrolysis of alkylated protein gave the carboxymethylated derivative of the modified amino acid residue except for modified glutamic acid, which was reconverted to the starting amino acid.

TABLE II

COMPARISON OF THE EFFECT OF AFFINITY LABELS VERSUS NONAFFINITY
LABELS IN THE IRREVERSIBLE INACTIVATION OF ENZYMES

| | Enzyme activity (half-life) | | | |
Inactivating agent	Trypsin	Chymo-trypsin	Carboxy-peptidase	β-Galacto-sidase
Haloacyl acid or acetamide	24 hr	>2 wk	>3 days	24 hr
Bromoacetyl affinity label	3 hr	75 min	25 min	30 min

succinimide is described here. The preparation of highly purified reagents proceeds with high yield. This facile preparation is very important when radioactive reagents are required, and radioactive labeled reagents can be prepared from radioactive bromoacetic acid. Recommended procedure in this case will be to use succinimide ester of bromoacetic acid since this avoids waste of expensive radioactive reagent. An additional advantage of using bromoacetyl is that the introduction of the radioactive functional group is always the last step in the preparation of the reagent.

4. *Preparation of a homologous series of reagents.* Since the ligand can be prepared with side chains of different length, terminating in an amino group, affinity labeling reagents of systematically increasing size can be prepared and used as a ruler for measuring distances at the combining site.[13]

5. *Convenient procedures for identification of labeled residues.* Since the bromoacetyl moiety is attached by an amide bond to the ligand, acid hydrolysis of the protein will always result in a carboxymethyl derivative of the labeled amino acids (Table I). Carboxymethyl amino acids can be readily analyzed with an amino acid analyzer, since their position is known,[15] or by high voltage paper electrophoresis, where the additional charge will change the mobility of the labeled amino acid.[13] The synthesis of carboxymethyl derivatives of amino acids as markers is achieved by alkylation with haloacetate.[16] In some cases, the carboxymethyl derivatives are not stable and caution is necessary in their identification. With methionine, for example, acid hydrolysis will produce carboxymethyl homocysteine and homoserine as degradation products of methionine carboxymethyl sulfonium salt[15]; treatment with sulfhydryl reagents can also remove label from the modified methionine.[17] In the

[15] F. R. N. Gurd, this series Vol. 11, p. 532 (1967).
[16] H. J. Goren, D. M. Glick, and E. A. Barnard, *Arch. Biochem. Biophys.* **126**, 607 (1968).
[17] F. Naider and Z. Bohak, *Biochemistry* **11**, 3208 (1972).

case of glutamic or aspartic acids, the ester bond formed between the amino acid and the reagent is unstable in acid and alkali, and identification should proceed in other ways, e.g., by isolation of labeled peptides.

Synthesis of Reagents

Bromoacetyl-N-hydroxysuccinimide Ester. To a solution of bromoacetic acid (139 mg, 1.0 mmole) and N-hydroxysuccinimide (115 mg, 10 mmoles) dissolved in 5 ml of dioxane is added dicyclohexylcarbodiimide (206 mg, 1.0 mmole) in 2 ml of dioxane. After 1 hr at room temperature, the dicyclohexylurea is removed by filtration and washed with dioxane. The filtrate is evaporated to dryness and crystallized from isopropanol, yield, 75%; m.p., 117°.

Iodoacetyl-N-hydroxysuccinimde Ester. To a solution of recrystallized iodoacetic acid[18] (185 mg, 1.0 mmole) and N-hydroxysuccinimide (140 mg, 1.2 mmoles), dissolved in 5 ml of dioxane, is added dicyclohexylcarbodiimide (250 mg, 1.2 mmoles) in 2 ml of dioxane. After 1 hr at room temperature the dicyclohexylurea is removed by filtration and washed with dioxane. The filtrate is evaporated to dryness and crystallized from isopropanol. Yield, 70%; m.p., 148°.

Synthesis of Radioactive Labeled Reagents. The hydroxysuccinimide ester of radioactive bromoacetic and iodoacetic acid is prepared by a scaled-down modification of the procedure for the preparation of the corresponding nonradioactive reagent. Quantities of ^{14}C or ^{3}H reagents were prepared, varying from 0.1 μmole up to 100 μmoles. When the amount used is very small, one proceeds with the next step without isolation of the product or filtration of the dicyclohexylurea. The affinity labeling reagent prepared by this method is usually purified by thin-layer chromatography. Examples appear in this volume.[19]

Radioactive Bromoacetic Anhydride. This compound is prepared by coupling two equivalents of bromoacetic acid with one equivalent of dicyclohexylcarbodiimide in an organic solvent (dioxane, CH_2Cl_2, $CHCl_3$, CCl_4). To a solution of bromoacetic acid (13.9 mg, 100 μmoles) in 0.5 ml of dry dioxane is added dicyclohexylcarbodiimide (10.5 mg, 50 μmoles). The reaction mixture is left for 1 hr at room temperature. Whenever possible, dicyclohexylurea should be removed by filtration. The bromoacetic anhydride thus formed is ready for coupling to ligands.

---

[18] To recrystallize iodoacetic acid, dissolve 5 g in 80 ml of hot cyclohexane and cool to 5°. Collect the precipitate, wash it with petroleum ether, and dry.

[19] D. Givol and M. Wilchek, this volume [53].

[12] Acetylenic Irreversible Enzyme Inhibitors

By Robert R. Rando

The chemical basis of action of acetylenic inhibitors seems to reside mainly in the fact that acetylenes can be enzymically converted into conjugated allenes. The latter compounds are powerful Michael acceptors

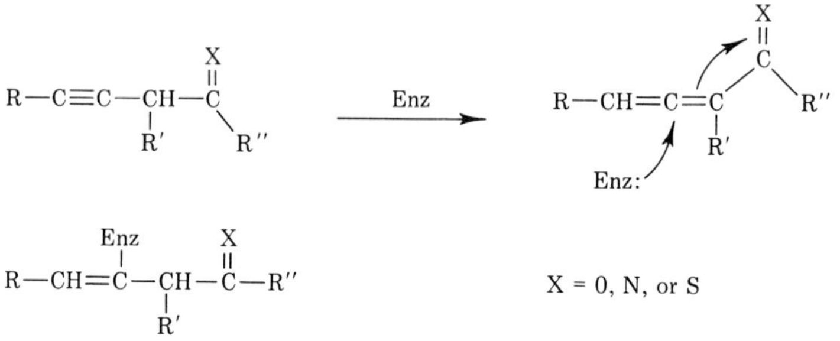

$$
R-C\equiv C-CH-\overset{\overset{X}{\parallel}}{C}\underset{R''}{\overset{R'}{\big\backslash}} \xrightarrow{\quad Enz \quad} R-CH=C=C\underset{R'}{\overset{\overset{X}{\parallel}}{C}}R''
$$

$$
Enz:
$$

$$
R-CH=\overset{\overset{Enz}{|}}{C}-CH-\overset{\overset{X}{\parallel}}{C}-R'' \qquad X = 0, N, \text{ or } S
$$
$$
\underset{R'}{}
$$

whereas the former are chemically unreactive. Inhibitors of this type have been developed for enzymes that catalyze isomerization, oxidation, elimination, and transamination reactions.

Isomerases

The first inhibitor of this series discovered, and indeed the best understood, is 3-decynoyl-N-acetylcysteamine (NAC).[1] This molecule irreversibly inactivates β-hydroxydecanoylthioester dehydrase, an enzyme that interconverts β-hydroxydecanoyl thioesters with their *trans* α, β- and *cis* β, γ-unsaturated counterparts. The enzyme is, then, capable of both dehydrase and isomerase activities.[2] When it isomerizes the acetylenic thioester, it produces a conjugated allenic thioester that alkylates an active site histidine residue before diffusing from the enzyme surface.[2]

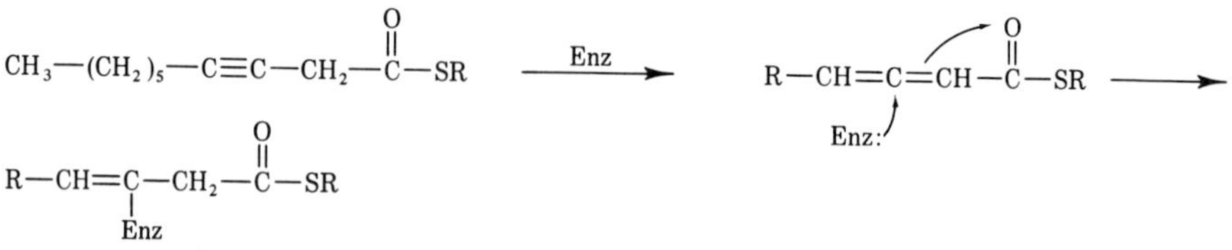

$$
CH_3-(CH_2)_5-C\equiv C-CH_2-\overset{\overset{O}{\parallel}}{C}-SR \xrightarrow{\quad Enz \quad} R-CH=C=CH-\overset{\overset{O}{\parallel}}{C}-SR \longrightarrow
$$

$$
Enz:
$$

$$
R-CH=\overset{}{C}-CH_2-\overset{\overset{O}{\parallel}}{C}-SR
$$
$$
\underset{Enz}{|}
$$

It is worthwhile considering how the mechanism of inhibition, presented above, was demonstrated. First of all, the chain length specificity

[1] K. Bloch, *Acc. Chem. Res.* **2**, 193 (1969).
[2] K. Endo, G. M. Helmkamp, and K. Bloch, *J. Biol. Chem.* **245**, 4293 (1970).

of the enzyme for substrates paralleled that for acetylenic inhibitors.[3] Furthermore, there is a kinetic isotope effect on the rate of conversion of α-deuterated substrates to products.[4] There is an essentially identical deuterium isotope effect on the rate of inactivation of the enzyme by α-D$_2$-decynoyl-NAC. Finally, the addition of exogenous allenic thioester results in the irreversible inhibition of the enzyme by an affinity labeling mode. And, indeed, only one of the enantiomeric pair of allenic thioesters is active as an inhibitor.[5] The great specificity of the inhibitor proved to be quite useful in studying the physiological role of β-hydroxydecanoyl thioester dehydrase *in vivo*.[6] This enzyme was shown to be absolutely required for unsaturated fatty acid biosynthesis in *Escherichia coli*.[6]

Recently, a very elegant example of an acetylenic substrate inhibitor of an isomerase has been reported. $\Delta^{3,5}$-Ketosteroid isomerase catalyzes the isomerization of Δ^5 to Δ^3 steroids. β, γ-Acetylenic analogs of the substrate irreversibly inhibit the enzyme by first being isomerized to the allene and then undergoing a reaction with an active-site residue.[7]

The bond between the inhibitor and the enzyme is stable to dialysis although it is broken with mild acid or base treatment. This suggests that

[3] G. M. Helmkamp, R. R. Rando, D. H. J. Brock, and K. Bloch, *J. Biol. Chem.* **243**, 3229 (1968).
[4] R. R. Rando and K. Bloch, *J. Biol. Chem.* **243**, 5627 (1968).
[5] M. Morisaki and K. Bloch, *Bioorg. Chem.* **1**, 188 (1971).
[6] L. R. Kass and K. Bloch, *Proc. Natl. Acad. Sci. U.S.A.* **58**, 1168 (1967); L. R. Kass, *J. Biol. Chem.* **243**, 3223 (1968).
[7] F. H. Batzold and C. H. Robinson, *J. Am. Chem. Soc.* **97**, 2376 (1975).

an ester linkage is involved that would require the proximity of a glutamate or aspartate residue to the substrate in the active-site complex.

A simple kinetic scheme was developed for this inhibitor that might
be useful in the analysis of the mechanism of other inhibitors of this type.[7]
This scheme for the inhibition is represented by Eq. (1). Here EI′ is the
enzyme-inhibitor derivative. Assuming that the rate of covalent bond

$$E + I \underset{k_2}{\overset{k_1}{\rightleftarrows}} E \cdot I \overset{k_3}{\rightarrow} EI' \tag{1}$$

formation between the enzyme and inhibitor is rate limiting, and that the
Michaelis assumption prevails, then with $(I) \gg (E)$, Eqs. (2)–(4) apply.
In

$$\frac{(E + E \cdot I)}{(E_t)} = \frac{-k_3}{1 + K_{I/I}} \tag{2}$$

where E_t = total enzyme concentration, by defining

$$k_{app} = \frac{-k_3}{1 + K_{I/I}} \tag{3}$$

one obtains

$$1/k_{app} = \frac{1}{k_3} + \frac{K_I}{k_3(I)} \tag{4}$$

Double reciprocal plots then permit the determination of K_I and k_3. Compound 1 had a K_I of 56 μM with a k_3 of 1.98×10^{-3} sec^{-1}.

With both isomerases, enzymically generated acetylenes appear to
react with an active-site residue before diffusing from the enzyme's surface. This suggests that the latter rate is slow relative to the rates of the
reactions occurring in the enzyme–substrate complex. It might be suggested that the rate of enzyme–substrate desorption is rate limiting in
these cases. At least with β-hydroxydecanoyl thioester dehydrase, this
seems to be true.[4]

Oxidases

Flavin- and nonflavin-linked amine oxidases carry out the oxidative
deamination of physiologically important amines.[8]

$$R—CH_2—NH_2 \overset{Enz}{\longrightarrow} R—CH=NH \overset{H_2O}{\longrightarrow} R—CHO$$

[8] "Advances in Biochemical Psychopharmacology" (E. Costa and P. Greengard,
eds.), Vol. 5. Raven, New York, 1972.

These enzymes are highly susceptible to inactivation by β,γ-acetylenic amines. Flavin-linked mitochondrial monoamine oxidase is irreversibly inhibited by the antidepressant drug pargyline. Structure activity studies have shown that the acetylenic unit is crucial and that it has to be β,γ to the nitrogen.[9]

$$\text{(phenyl)}-CH_2-\underset{\underset{CH_3}{|}}{N}-CH_2-C\equiv C-H \qquad\qquad H-C\equiv C-CH_2-NH_2$$

Pargyline   Propargyl amine

Furthermore, propargyl amine itself is an irreversible inactivator of the enzyme.[10] The mechanism of inhibition by these acetylenic amines appears to involve the following steps.[11] The reaction of the inactivated sub-

strate occurs with the flavin cofactor, not with active-site amino acid residues.[10,11] In this instance, it appears that the interconversion of acetylene and allene need not be postulated to account for the observed structure of the flavin-inhibitor complex, although this is not ruled out by the current evidence.[11] The enzyme remains irreversibly inactivated because the flavin is covalently linked to the enzyme. Plasma monoamine oxidase, a nonflavin-linked enzyme, is also irreversibly inhibited by propargyla-

[9] L. R. Swett, W. B. Martin, J. D. Taylor, G. M. Everett, A. A. Wykes, and Y. C. Gladish, *Ann. N. Y. Acad. Sci.* **107**, 891 (1963).
[10] L. Hellerman and V. G. Erwin, *J. Biol. Chem.* **243**, 5234 (1968).
[11] A. L. Maycock, R. H. Abeles, J. I. Salach, and T. P. Singer, *Biochemistry* **15**, 114 (1976).

mine.[12] In this case the mechanism can be rationalized as follows:

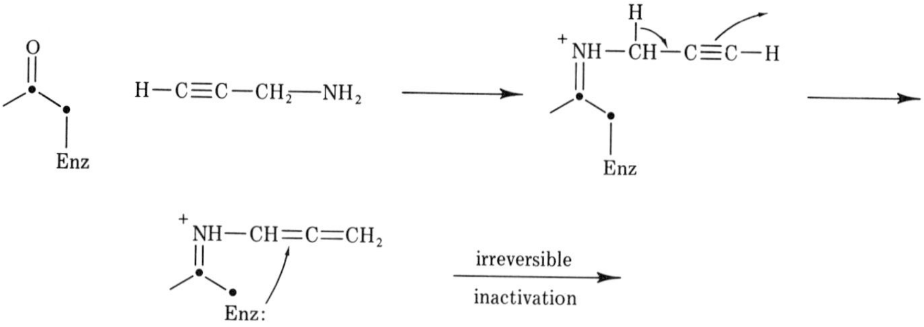

It is thought that the enzyme has a carbonyl acceptor which undergoes Schiff-base formation with substrate amines. Subsequent steps involve the abstraction of the α-C—H bond followed by a redox reaction. The enzyme also contains an essential cupric ion. Removal of this ion by chelating agents renders the apoenzyme catalytically inactive. The enzyme can be reactivated by the readdition of cupric ion. The apoenzyme is not in the least affected by incubation with high concentrations of propargylamine.[12] This is the case because a catalytically inert enzyme cannot activate the propargylamine, and propargylamine by itself cannot inhibit the enzyme. Propargyl amine will also irreversibly inhibit diamine oxidase, an enzyme similar in mechanism to the plasma monoamine oxidase.[12] Although this enzyme is maximally active with diamines, 1,4-diamino-2-butyne is not a better inhibitor of it than is propargylamine.

$$NH_2—CH_2—C \equiv C—CH_2—NH_2$$
1,4-Diamino-2-butyne

Although propargylamine will irreversibly inhibit both plasma and flavin linked monoamine oxidases, one can still arrange to inhibit one enzyme without affecting the other. Only the flavin-linked enzyme can oxidize tertiary amines, so that acetylenic amines that are tertiary, e.g., pargyline, will inhibit only the flavin-linked enzyme.

Other flavin-linked oxidase enzymes are also irreversibly inhibited by acetylenes. Lactate dehydrogenase is irreversibly inhibited by 2-hydroxy-3-butynoic acid.[13] As with flavin-linked monoamine oxidase the reaction takes place with the flavin, not with the protein. Finally, flavin-linked

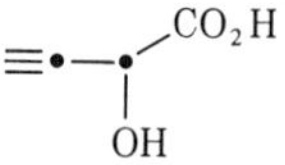

---

[12] R. R. Rando and J. DeMairena, *Biochem. Pharmacol.* **23**, 463 (1974).
[13] C. T. Walsh, A. Schonbrunn, O. Lockridge, V. Massey, and R. H. Abeles, *J. Biol. Chem.* **247**, 6004 (1972).

sarcosine demethylase is irreversibly inhibited by N-propargylglycine.[14] Presumably, the chemical reaction here also occurs with the flavin, not the protein.

Pyridoxal Phosphate-Linked Transaminases and Cleavage Enzymes

γ-Aminobutyric acid (GABA) transaminase is a pyridoxal-linked enzyme that degrades the neurotransmitter GABA. The enzyme functions by the usual transamination pathway that requires Schiff-base formation between the GABA and pyridoxal form of the enzyme followed by α-C—H bond abstraction and transamination. This enzyme is irreversibly inhibited by 4-amino-5-hexynoic acid.[15] The exact mechanism of this inhibition is not as yet known. Interestingly, through, 4-amino-2-butynoic acid is not an inhibitor of the enzyme. This is a surprising result, but one that reminds one of the complexities involved in dealing with enzyme inhibition.

$$H—C \equiv C—\underset{\underset{NH_2}{|}}{CH}—CH_2—CH_2—CO_2H \qquad\qquad NH_2—CH_2—C \equiv C—CO_2H$$

4-Amino-5-hexynoic acid   4-Amino-2-butynoic acid

Propargylglycine is an irreversible inhibitor of both pyridoxal-linked γ-cystathionase[16] and aspartate aminotransferase.[17] The mechanism of inhibition of γ-cystathionase involves the following sequence of steps.

[14] J. Kraus, J. Yaounans, and G. Stratz, *Eur. J. Med. Chem.*, p. 507 (1975).
[15] M. J. Jung and B. W. Metcalf, *Biochem. Biophys. Res. Commun.* **67**, 301 (1975).
[16] R. Abeles and C. Walsh, *J. Am. Chem. Soc.* **95**, 6124 (1973).
[17] P. Marcotte and C. Walsh, *Biochem. Biophys. Res. Commun.* **62**, 677 (1975).

This inhibition of aspartate aminotransferase by the inhibitor is of interest because it points up the uncontrollable aspects of these inhibition studies. A priori, one would not expect propargylglycine to be an inhibitor of this enzyme because it should not be able to activate the acetylene. Fortuitously, a basic amino acid is adjacent to the relevant C—B bond of the inhibitor and can, in fact, break it with the resulting activation of the acetylene.

Conclusion

Enzymes capable of generating a carbanion or a carbanion-like intermediate adjacent to an acetylene are susceptible to irreversible inhibition by these acetylenic reagents. Flavin-linked and pyridoxal-linked enzymes and isomerases have proved to be good candidates for inhibition by the acetylenic inhibitors. This is the case because these enzymes are generally involved in C—H bond abstraction with concomitant carbanion or carbanion-like intermediate formation. However, the critical factor is the formation of a free valance adjacent to the acetylenic unit. An enzyme that could do so is a candidate for inhibition by these inhibitors. Thymidylate synthethase, for example, should be irreversibly inhibited by 5-ethynouracil; the free valance would be generated adjacent to the acetylene by means of an enzyme-catalyzed nucleophilic addition to the uracil.[18]

[18] H. Sommer and D. Santi, *Biochem. Biophys. Res. Commun.* **57**, 689 (1974).

[13] Organic Isothiocyanates as Affinity Labels

By ROBERT G. LANGDON

Affinity labels may be regarded as being comprised of two parts: a portion, A, which associates specifically with a protein active site and a second reactive portion, C, which reacts to form a covalent linkage between a reactive group in the protein and the affinity probe itself.

$$A - C + \text{protein} \rightarrow A - C - \text{protein}$$

Ideally, the C group should not interfere with interaction between the A group and its binding site. It should react with some nearby protein functional group much more readily than it interacts with solvent, and it should form a stable product. Finally, it should be easily synthesized from readily available precursors.

In several ways, organic isothiocyanates satisfy these criteria rather well. The isothiocyanate group itself has no net charge and is not excessively bulky, so that it is not subject to electrostatic repulsion, and it is sufficiently small that it may avoid steric problems. Isothiocyanates react readily with amino groups in proteins to form disubstituted thioureas, which are stable under physiological conditions. On the other hand, reactivity with water is so slight that steam-volatile isothiocyanates may be purified by steam distillation. Finally, isothiocyanates are rather easily prepared by several methods.

The preparation, properties, and reactions of isothiocyanates have been the subject of valuable, comprehensive reviews, which should be consulted for details.[1,2]

Synthesis of Isothiocyanates

Preparative methods that are available include the following:

1. From alkyl halides and metal thiocyanates:

$$R - Cl + NH_4SCN \rightarrow R - NCS\ (+\ RSCN) + NH_4Cl$$

2. From salts of dithiocarbambic acid derivatives:

(a) $R-\overset{\overset{\displaystyle S}{\|}}{N}HCSNH_4 + Pb(NO_3)_2 \rightarrow RNCS + NH_4NO_3 + HNO_3 + PbS$

(b) $R-\overset{\overset{\displaystyle S}{\|}}{N}HCSNH_4 + COCl_2 \rightarrow RNCS + NH_4Cl + HCl + COS$

(c) $R-\overset{\overset{\displaystyle S}{\|}}{N}HCSNH_4 + 4NaOCl + NaOH \rightarrow$
$$RNCS + 3NaCl + NH_4Cl + Na_2SO_4 + H_2O$$

3. From disubstituted thioureas:

$$R-\overset{\overset{\displaystyle H}{|}}{N}-\overset{\overset{\displaystyle S}{\|}}{C}-\overset{\overset{\displaystyle H}{|}}{N}-R \xrightarrow{\text{HCl, heat}} RNCS + R-NH_3Cl$$

4. From primary amines and thiocarbonyl diimidazole[3]:

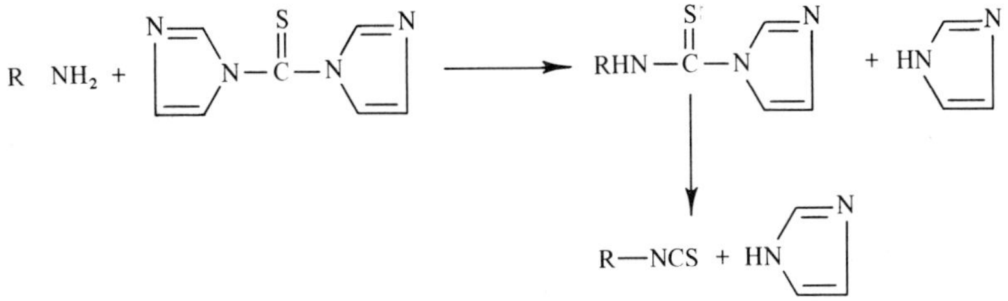

[1] M. Bögemann, S. Petersen, O. E. Schultz, and H. Söll, in Houben-Weyl "Methoden der Organischen Chemie" (E. Müller, ed.), Vol. IX, pp. 773–913. Thieme, Stuttgart, 1955.
[2] S. J. Assony, in "Organic Sulfur Compounds" (N. Kharasch, ed.), Vol. I, pp. 326–338. Pergamon, London, 1961.
[3] H. A. Staab and G. Walther, Justus Liebigs Ann. Chem. 657, 98 (1962).

5. From primary amines and thiophosgene:

$$R \cdot NH_2 + CSCl_2 \rightarrow [R{-}NH{-}\overset{\overset{\text{S}}{\|}}{C}{-}Cl] + HCl$$
$$\downarrow$$
$$R{-}NCS + HCl$$

Method 5, the reaction of thiophosgene with primary amines, is probably the most generally useful, since primary amines are frequently the most readily available of the starting materials, either by purchase or by synthesis, and thiophosgene is commercially available. Furthermore, the reaction may be carried out either in organic solvents or in aqueous solutions of water-soluble amines, because both thiophosgene and isothiocyanates react only slowly with cold water near neutrality; thiophosgene is sufficiently soluble in water so that the reaction proceeds well. The amine may be used as the free base or, if it is sufficiently basic, in its protonated form; the latter is usually preferable because it prevents undesired reaction of the synthesized isothiocyanate with its precursor amine and results in higher overall yields. The major disadvantage of this method is that thiophosgene, a liquid at room temperature (b.p. 73.5°), is very toxic and must be used in an efficient fume hood with precautions adequate to prevent skin contact or inhalation of its vapor. For storage, small screw-capped bottles of thiophosgene are placed in a desiccator, which is kept at −20°; prior to use, the desiccator is allowed to come to room temperature in the fume hood.

The following description of the preparation of D-glucosyl isothiocyanate[4] is illustrative of the method. To 71.2 mg (400 μmoles) of glucosylamine dissolved in 4.0 ml of 0.4 M NaHCO$_3$ (1600 μmoles) at room temperature in a glass-stoppered tube are added 50 μl (657 μmoles) of CSCl$_2$ with a microliter syringe to effect the transfer. The mixture, stirred rapidly with a magnetic stirrer, is allowed to react for 12 min. At the end of this period, evolution of CO$_2$ should have stopped or become quite slow, and droplets of orange-colored thiophosgene should have disappeared. To the mixture are added 15 ml of dichloromethane. The tube is stoppered, shaken, and centrifuged at low speed. The subnatant dichloromethane solution containing residual dissolved thiophosgene is removed with a syringe and transferred into a beaker containing ammonium hydroxide (*hood!*). The upper aqueous phase is extracted again with a second 15-ml portion of dichloromethane, and, after centrifugation, the upper aqueous phase is transferred to a fresh tube with care to avoid contamination with

[4] R. D. Taverna and R. G. Langdon, *Biochem. Biophys. Res. Commun.* **54,** 593 (1973).

dichloromethane. The aqueous glucosyl isothiocyanate solution is placed in an ice bath and used within 30 min.

Reactions of Isothiocyanates with Protein Functional Groups

The reactions of isothiocyanates with proteins have been extensively studied, particularly in connection with the Edman procedure.[5-7] The major initial reaction that occurs under mild conditions at neutral or slightly alkaline pH is the formation of thiourea derivatives with amino terminal residues and with ε-amino groups of lysyl residues.

$$R \cdot NCS + NH_2 - protein \rightarrow R—\underset{H}{\overset{}{N}}—\underset{S}{\overset{\parallel}{C}}—\underset{H}{\overset{}{N}} - protein$$

Although further modification, e.g., substituted thiohydantoin formation, may result under other circumstances, thiourea derivatives are ordinarily quite stable near neutrality.

Isothiocyanate derivatives of stilbene sulfonates have been utilized with marked success as affinity labels for an anion transport protein of the erythrocyte membrane by Rothstein and his associates,[8,9] and glucosyl isothiocyanate has been employed to label a glucose transport protein of the human erythrocyte membrane.[4] Both proteins fall in the 90,000–100,000 dalton range (band 3 in the terminology of Fairbanks *et al.*[10]). In both cases, transport has been reconstituted using phospholipid vesicles and a band 3 preparation[9,11] after affinity labeling with the appropriate isothiocyanate had provided preliminary evidence that this protein class might be involved in transport.

Because affinity labels may differ in the avidity with which they are bound, proximity of their binding site to an amino group on the protein, and in other respects, no specific directions can be given for interaction of isothiocyanate affinity labels with proteins which would be universally applicable.

In general, a substrate analog containing an isothiocyanate reactive group is incubated at different concentrations for varying periods of time

[5] P. Edman, *Acta Chem. Scand.* **10**, 761 (1956).
[6] W. A. Schroeder, this series, Vol. 25, p. 298 (1972).
[7] H. Fraenkel-Conrat, J. I. Harris, and A. L. Levy, *Methods Biochem. Anal.* **2**, 383 (1955).
[8] Z. I. Cabantchik and A. Rothstein, *J. Membr. Biol.* **10**, 311 (1972).
[9] A. Rothstein, Z. I. Cabantchik, and P. Knauf, *Fed. Proc., Fed. Am. Soc. Exp. Biol.* **35**, 3 (1976).
[10] G. Fairbanks, T. L. Steck, and D. F. H. Wallach, *Biochemistry* **10**, 2606 (1971).
[11] M. Kasahara and P. C. Hinkle, *Proc. Natl. Acad. Sci. U.S.A.* **73**, 396 (1976).

with the protein or membrane of interest in a neutral or alkaline buffer free of ammonia, amines, or mercaptans, and conditions are chosen such that approximately 50% inhibition of function is observed. Experiments are then conducted to ascertain whether a normal substrate protects competitively against the inhibitory action of the presumed affinity label. Finally, the isothiocyanate-containing affinity label may be prepared in radioactive form, and its specific interaction with a particular protein among a mixture of proteins may be examined; again, demonstration of competitive inhibition of reaction by a normal substrate or by a potent reversible competitive inhibitor is a valuable or essential criterion for specificity of interaction between the affinity label and the protein of interest.

As a specific example of methodolgy, reaction of D-glucosyl isothiocyanate with the erythrocyte membrane may be cited.[4] In preliminary experiments 50% suspensions of washed human erythrocytes in Krebs-Ringer phosphate buffer at pH 7.4 were incubated at 37° with 0–100 mM D-glucosyl isothiocyanate for periods of 0–60 min. After incubation, the cells were washed twice with 20 volumes of ice-cold buffer and the rates of influx of 30 mM [^{14}C]glucose were measured. It was found that prior treatment of the cells with 20 mM D-glucosyl isothiocyanate at 37° for 5 min resulted in 50% inhibition of the transport rate. The presence of 100 mM maltose or D-glucose during exposure to the isothiocyanate protected against inhibition whereas L-glucose was without effect. When 20 mM D-[^{14}C]glucosyl isothiocyanate was incubated with intact erythrocytes at 37° for 5 min, incorporation into a number of membrane proteins as well as into hemoglobin occurred. The simultaneous presence of 100 mM D-glucose inhibited incorporation of the labeled material, particularly into band 3, while L-glucose had no effect.

Limitations. Isothiocyanates react rather rapidly with exposed amino groups, whether or not they are attached to affinity groups. Therefore, nonspecific reaction can be expected to occur, and the investigator must demonstrate that the effects he observes with isothiocyanate affinity labels are indeed specific effects rather than the result of nonspecific interaction.

[14] Photo-Cross-Linking of Protein–Nucleic Acid Complexes

By PAUL R. SCHIMMEL and GERALD P. BUDZIK

In recent years the rapid expansion of biological research has resulted in an increasing interest in macromolecular complexes. This interest has been sparked by the growing awareness that many cellular processes, and many essential cellular structures, are expressed through macromolecule–macromolecule interactions. As a result, many investigations are aimed at elucidating the physical basis for, and the structural organization of, specific macromolecular complexes.

Because of limitations in applying physical techniques, alternative approaches have been devised to provide insight into the structural features of such complexes. One of the most promising developments has been the development of procedures to join covalently macromolecular surfaces that are in close contact. By careful determination of the location within each macromolecule of the segments that participate in the cross-linking reaction, one can draw a reasonable picture of the architectural design of the complex.

In the case of protein–nucleic acid complexes, which form a large part of the macromolecular systems currently under investigation, photochemical procedures that avoid the introduction of extraneous reagents or labels have been particularly successful. This approach is based on the fact that nucleotide bases and certain amino acid side chains can be joined directly through the action of ultraviolet light.[1,2] For example, there are data to indicate that ten of the amino acids can be joined to uracil.[3] These circumstances would appear to give a wide latitude for potential cross-linking sites in protein–nucleic acid complexes. Indeed, this direct photochemical approach has facilitated cross-linking in a number of complexes including DNA and RNA polymerase with double-stranded duplexes,[4–6] gene 5 protein with fd DNA,[7] aminoacyl tRNA synthetases with tRNAs,[8–10] and ribosomal proteins with ribosomal RNA.[11,12]

The discussion given below outlines many of the considerations and

[1] S. Y. Wang (ed.), "Photochemistry and Photobiology of Nucleic Acids." Gordon & Breach, New York, 1974.

[2] K. C. Smith (ed.), "Aging, Carcinogenesis and Radiation Biology: The Role of Nucleic Acid Addition Reactions." Plenum, New York, 1976.

[3] K. C. Smith, Biochem. Biophys. Res. Commun. 34, 354 (1969).

[4] A. Markovitz, Biochim. Biophys. Acta 281, 522 (1972).

[5] H. Weintraub, Cold Spring Harbor Symp. Quant. Biol. 38, 247 (1973).

[6] G. F. Strniste and D. A. Smith, Biochemistry 13, 485 (1974).

[7] E. Anderson, Y. Nakashima, and W. Konigsberg, Nucleic Acid Res. 2, 361 (1975).

[8] H. J. P. Schoemaker and P. R. Schimmel, J. Mol. Biol. 84, 503 (1974).

[9] G. P. Budzik, S. S. M. Lam, H. J. P. Schoemaker, and P. R. Schimmel, J. Biol. Chem. 250, 4433 (1975).

[10] H. J. P. Schoemaker, G. P. Budzik, R. C. Giegé, and P. R. Schimmel, J. Biol. Chem. 250, 4440 (1975).

[11] L. Gorelic, Biochim. Biophys. Acta 390, 209 (1975).

[12] L. Gorelic, Biochemistry 14, 4627 (1975).

pitfalls that must be taken into account when using the photochemical approach to obtain structural information on protein–nucleic acid complexes. Specific details and illustrations are given for aminoacyl tRNA synthetase–tRNA complexes.[13,14] However, many of these experimental details can be adapted and applied to a variety of other protein–nucleic acid systems.

Irradiation and Determination of Cross-Linking Efficiency

References to the preparation of specific *Escherichia coli* aminoacyl tRNA synthetases and tRNAs are given elsewhere.[8–10] For the purposes of the experiments described below, it is useful to have the nucleic acid radioactively labeled. This can be achieved in a straightforward manner by incubating the tRNA in ^3H_2O for 25 min at 90°.[15,16] After cooling the reaction mixture, free and fast exchanging tritium is removed by lyophilization and dialysis. This yields tritium stably incorporated into the purine C-8 positions; with a 20% counting efficiency, tRNA typically has a specific activity of 2 to 3×10^5 cpm/A_{260}.

Good cross-linking, and an accurate quantitative analysis of the data obtained, can be achieved with an extremely simple experimental setup. In a typical experiment, a 1-ml reaction mixture contains about 1 nmole of enzyme and 0.5 nmole of [^3H]tRNA, together with 10 mM MgCl$_2$ and 50 mM sodium cacodylate at pH 5.2–5.5. This pH range is chosen because synthetase–tRNA complexes have a smaller dissociation constant at about pH 5.5 than at higher pH values (see Lam and Schimmel[17]); in one system it was shown that the regions on a tRNA that cross-link to synthetase are the same at pH 7 as at the lower pH 10. The enzyme concentration is well above the dissociation constant of the complex.[17]

The reaction mixture is contained in a droplet which is placed on a parafilm-lined glass petri dish; the dish in turn is floated on an ice bath. The mixture is then irradiated for varying periods with two 15-watt mercury lamps (General Electric) at a distance of about 7 cm. The lamp yields radiation predominantly at 253:7 nm. Short wavelengths (<210 nm) can be cut off with a 2 mm-thick Vycor glass filter[18]; however, in our

[13] L. L. Kisselev and O. O. Favorova, *Adv. Enzymol. Relat. Areas Mol. Biol.* **40**, 141 (1974).

[14] D. Söll and P. R. Schimmel, *Enzymes* **10**, 489 (1974).

[15] R. C. Gamble and P. R. Schimmel, *Proc. Natl. Acad. Sci. U.S.A.* **71**, 1356 (1974).

[16] R. C. Gamble, H. J. P. Schoemaker, E. Jekowsky, and P. R. Schimmel, *Biochemistry* **15**, 2791 (1976).

[17] S. S. M. Lam and P. R. Schimmel, *Biochemistry* **14**, 2775 (1975).

[18] J. G. Calvert and J. N. Pitts, Jr., "Photochemistry," 2nd ed., p. 748. Wiley, New York, 1967.

studies, some experiments indicate that the same results are obtained with and without the filter.[9]

The irradiation dose from the lamp can be estimated by performing similar irradiations with solutions of uridine and monitoring the well-documented photohydration reaction. This serves as a uridine actinometer (for a general discussion of photochemical procedures, see Johns[19]). With this kind of actinometer, the dosage received by the droplet described above is equal to about 0.6 microeinstein/min/cm² or 3.6×10^{17} photons/min/cm². This value is in reasonable agreement with that estimated from the manufacturer's lamp specifications and the geometry of the experimental design. The actual "effective" intensity, received by the average molecule in solution, is less than expected owing to absorption by protein and tRNA. In general, if $I_{0,\lambda}$ is the incident intensity at wavelength λ (photons/unit time/unit area) and A_λ is the absorbance of the reaction mixture, then the "effective" intensity $\bar{I}_\lambda$ experienced by the average molecule in the solution is[19,20]

$$\bar{I}_\lambda = (2.3 A_\lambda)^{-1}(1 - e^{-2.3 A_\lambda}) I_{0,\lambda} \tag{1}$$

Note that A_λ is the absorbance of the reaction mixture with respect to the path length of the irradiation sample, which is not necessarily 1 cm. In the experiment given above, the effective intensity is about 60% of the incident intensity.

Stable joining of protein to nucleic acid can be demonstrated by several procedures. Perhaps the easiest approach is a simple phenol extraction method that gives accurate and reproducible results. This method is based on the observation that proteins and nucleic acids ordinarily can be separated by phenol extraction; protein migrates into the phenol layer whereas nucleic acids remain in the aqueous phase. However, upon linkage of a tRNA to a synthetase the nucleic acid migrates with the protein into the phenol layer. A similar observation was made earlier by Markovitz, who showed that DNA cross-linked to DNA polymerase is also extracted into phenol.[4]

To obtain a quantitative estimate of the extent of cross-linking, aliquots are removed from the reaction mixture at various times and extracted with an equal volume of phenol. The phenol layer is back-extracted twice with an equal volume of water. The amount of [³H]tRNA in the phenol phase is determined by scintillation counting on the aqueous phase, and then computing by that difference which is in the phenol.

[19] H. E. Johns, *Photochem. Photobiol.* 8, 547 (1968).

[20] In Eq. (1) we have assumed that the thickness of the sample is the same as the average path length through the sample for the beam of radiation. See Johns[19] for more discussion.

Upon irradiation of a synthetase-tRNA complex, the amount of cross-linking steadily increases for the first 20 min or so and then levels off with about 30–50% of the nucleic acid joined to enzyme.[8–10]

Characterization of Photo-Cross-Linked Complexes

Specific versus Nonspecific Cross-Linking

The first, most serious question is whether the observed cross-linking is due to a specific interaction between the macromolecules. It is conceivable, for example, that photoactivated species are generated in solution and that these in turn collide in various orientations to yield stable cross-linked species. Alternatively, spurious nonspecific complexes between proteins and nucleic acids may form and subsequently photo-cross-link. Therefore, before conducting a further characterization of the cross-linked material, it is critical to check whether it formed as a result of a specific interaction.

In the case of the Ile-tRNA synthetase–tRNA^{Ile} system, specificity of cross-linking was checked as follows.[9] Under conditions in which the enzyme gave good cross-linking to tRNA^{Ile}, none was observed when tRNA^{Ile} was replaced with tRNA^{Tyr} or tRNA^{Phe}. The lack of joining between Ile-tRNA synthetase and tRNA^{Tyr}, for example, is not due to an intrinsic inability of this nucleic acid to photo cross-link; when Tyr-tRNA synthetase is irradiated with tRNA^{Tyr}, good joining of tRNA^{Tyr} to the enzyme is achieved. Also, with a concentration of bovine albumin comparable to the Ile-tRNA synthetase concentration used, no joining of this nonspecific protein to tRNA^{Ile} was observed. These data, and the results of another type of control,[9] suggest that cross-linking under the conditions employed results from specific complex formation.

Of course, not all protein–nucleic acid interactions can be tested with respect to specificity in the same way as the synthetase–tRNA systems in which a variety of cognate and noncognate protein–nucleic acid pairs are possible. However, there is generally one or another means for taking advantage of the particular features of any given system so as to have a good control for specificity of cross-linking. For example, Lin and Riggs studied the cross-linking of the *lac* repressor to the *lac* operator.[21] In this case, the *lac* operator was carried on the DNA of a transducing phage. It was possible to obtain conditions under which cross-linking was achieved to this DNA, but not to the same DNA which lacked the *lac* operator segment. Thus, cross-linking was shown to be completely dependent on the presence of the *lac* operator region.

[21] S.-Y. Lin and A. D. Riggs, *Proc. Natl. Acad. Sci. U.S.A.* **71**, 947 (1974).

Structural Modifications Introduced by Irradiation

The question arises as to whether the irradiation introduced significant damage into the reacting macromolecules. The simplest procedure is to irradiate separately each reacting partner. Each can then be checked for effects on biological activity and for structural modifications.

In the case of Ile-tRNA synthetase, it has been shown that irradiation of the enzyme alone, with dosages comparable to those giving a maximal yield of cross-linking with tRNA^{Ile}, yields enzyme that retains a significant fraction of its tRNA binding activity.[9] Irradiation of tRNA^{Ile} alone produces no substantial alterations in the nucleic acid.[9] For example, the kinetics of and extent of aminoacylation of the irradiated tRNA are similar to that of an unirradiated control. Electrophoresis on polyacrylamide gels under dissociating conditions shows 5% or less chain breakage produced by irradiation. Finally, the T1 fingerprint of the irradiated tRNA was the same as that of an unirradiated control; this suggests that formation of intramolecular cross-links or base modifications is minimal.

These types of controls give some assurance that irradiation has not seriously distorted the reacting partners. This makes it more likely that the cross-linked molecules retain much of their native structural characteristics.

Identification of Macromolecule in the Complex That Is Responsible for Incomplete Cross-Linking

With all synthetase–tRNA complexes that have been studied, substantially less than complete joining of tRNA to synthetase is achieved, even though the enzymes are typically present in a molar excess and the concentrations used are well above the dissociation constant of the complex. This suggests that competing photoprocesses inactivate one or both reacting partners with respect to cross-linking.

The identification of which partner is inactivated with respect to cross-linking may be checked by a simple experiment. The reaction mixture is irradiated until the plateau level of cross-linking is achieved. At this point, additional enzyme or additional nucleic acid is added, and the irradiation is continued. If additional cross-linking is achieved with one of the additions, but not the other, this clearly identifies the species that is inactivated with respect to further cross-linking.

In the Ile-tRNA synthetase–tRNA^{Ile} system, addition of tRNA^{Ile} after reaching the plateau results in no further cross-linking; however, addition of Ile-tRNA synthetase gives an increase in the percent of tRNA cross-linked. This experiment clearly suggests that the plateau in cross-linking is due to competing photoinactivation processes that prevent many of the enzyme molecules from cross-linking to the tRNA. A detailed analysis of the enzyme's structure would be required in order to pinpoint the nature of the inactivation process.

Causes of Inefficient Cross-Linking

Although a systematic study of the causes for inefficient cross-linking in macromolecular systems has not been made, certain considerations apply.

It is well known that basic functions and α-carbon atoms can react readily with nucleotide bases. This is well demonstrated in the photo-cross-linking of 1-propylamine to 1,3-dimethyluracil; both the amino group and the α-carbon form adducts with the pyrimidine ring.[22] Irradiation of solutions of cysteine and uracil give 5-S-cysteine-6-hydrouracil;[23] 5-heteroadducts have also been identified after irradiation of thymine and cysteine.[24] With purines, the C-8 position of adenosine as well as of guanosine can photo-cross-link to 2-propanol.[25]

Photoprocesses between protein side chains and solvent molecules must also be considered. Although many side chains may have a very low extinction coefficient at the exciting wavelength, some compensation may be achieved if the quantum yield for reaction is relatively high. For example, the disulfide cystine is known to be relatively photoactive when irradiated at 254 nm, particularly because of its significant quantum yield for reaction.[26]

These considerations suggest that certain reagents should be excluded from buffers in which photo-cross-linking is to be carried out. For example, high concentrations of sulfhydryl reagents, such as mercaptoethanol, may inhibit protein–nucleic acid cross-linking by virtue of their potential to photoreact with nucleotides (to give products analogous to those formed between cysteine and uracil). In addition, buffer species that contain reactive functions should also be avoided. Moreover, since little is known about many photoprocesses, it is always advisable to try several different buffer conditions when attempting to achieve optimal cross-linking.

A more subtle variable that will influence the degree of cross-linking is the geometric and photochemical features of the complex itself. Unless

[22] L. S. Gorelic, P. Lisagor, and N. C. Yang, *Photochem. Photobiol.* **16**, 465 (1972).
[23] K. C. Smith and R. T. Alpin, *Biochemistry* **5**, 2125 (1966).
[24] A. J. Varghese, *Biochemistry* **12**, 2725 (1973).
[25] H. Steinmaus, I. Rosenthal, and D. Elad, *J. Am. Chem. Soc.* **91**, 4921 (1969).
[26] A. D. McLaren and D. Shugar, "Photochemistry of Proteins and Nucleic Acids." Macmillan, New York, 1964.

some amino acid side chains and nucleotide bases are properly positioned in the complex for good cross-linking, and unless the photochemical processes that occur lead to intermolecular cross-linking as opposed to other photoprocesses, cross-linking may not be observed in spite of good complex formation and the elimination of other interfering effects. An example in which these considerations may enter is provided by the *lac* repressor–*lac* operator system: good cross-linking of the repressor to *lac* DNA is not achieved unless the thymine residues in the DNA are substituted with the more photoreactive BrdU.[21] The substitution permits good cross-linking to be achieved between the repressor and *lac* DNA. This appears to be a clear example where geometric and, additionally or alternatively, photochemical constraints prevent cross-linking with the unsubstituted DNA.

Determination of Location of Cross-Linking Regions

For the purposes of obtaining detailed structural information on a macromolecular complex, the issue of most interest is the determination of the structural location of the cross-linked regions. This analysis can be pursued once there is sufficient confidence that the cross-linked product results from specific complex formation.

Ideally, one wishes to determine the location of the cross-linking sites within each partner of the complex. However, detailed structural information on both partners is rarely available. For example, although sequences of many tRNA species are known[27] and a crystallographic structural model for this nucleic acid is available,[28,29] the corresponding information for synthetases has not been obtained. In this case, attention is directed at the location in the tRNA of the cross-linking sites. From this, a picture can be drawn of the way in which the enzyme molecule lies across the nucleic acid.

The most obvious approach is to digest with a nuclease the cross-linked protein–nucleic acid complex so as to remove regions of the nucleic acid that are not cross-linked to the protein. However, discretion must be exercised in the choice of nuclease. For example, a nonspecific nuclease, such as T2 ribonuclease, is of little value since it may leave only mononucleotide units attached to the protein. Because the four common nucleotides are distributed throughout the nucleic acid, determination of

[27] B. G. Barrell and B. F. C. Clark, "Handbook of Nucleic Acid Sequences." Joynson-Bruvvers, Ltd., Oxford, 1974.
[28] S. H. Kim, F. L. Suddath, G. J. Quigley, A. McPherson, J. L. Sussman, A. H. J. Wang, N. C. Seeman, and A. Rich, *Science* **185**, 435 (1974).
[29] J. D. Robertus, J. E. Ladner, J. T. Finch, D. Rhodes, R. S. Brown, B. F. C. Clark, and A. Klug, *Nature (London)* **250**, 546 (1974).

which are cross-linked to the protein reveals nothing as to the position in the structure of the cross-linked nucleotides. A better approach is to use a highly specific nuclease, such as T1 ribonuclease (which cleaves after G's). This nuclease generates many unique fragments; if a particular one is found cross-linked to the protein, its location in the structure can usually be pinpointed unambiguously.

In choosing an approach for determining the regions on the nucleic acids that are cross-linked, two other important considerations apply. First, the direct structural characterization and identification of a T1 fragment that is cross-linked to the protein may be unusually difficult. This is particularly true because little is known about the types of photoproducts that are generated in these systems. Second, it became apparent in our work that the cross-linked complexes are heterogeneous on a microscopic basis; i.e., if three regions cross-link to the protein, some complexes may be cross-linked at only one or two of these sites, different permutations being achieved in different microscopic complexes. Thus, an approach is required that not only yields the locations of the cross-linked regions on the nucleic acid, but also gives quantitative information on the frequency of cross-linking at that site.

General Procedure For Identifying Cross-Linked Regions and for Determining Frequency of Cross-Linking of Each T1 Fragment. The procedure that was designed is outlined in Scheme 1.[9] Enzyme and [^{3}H]-

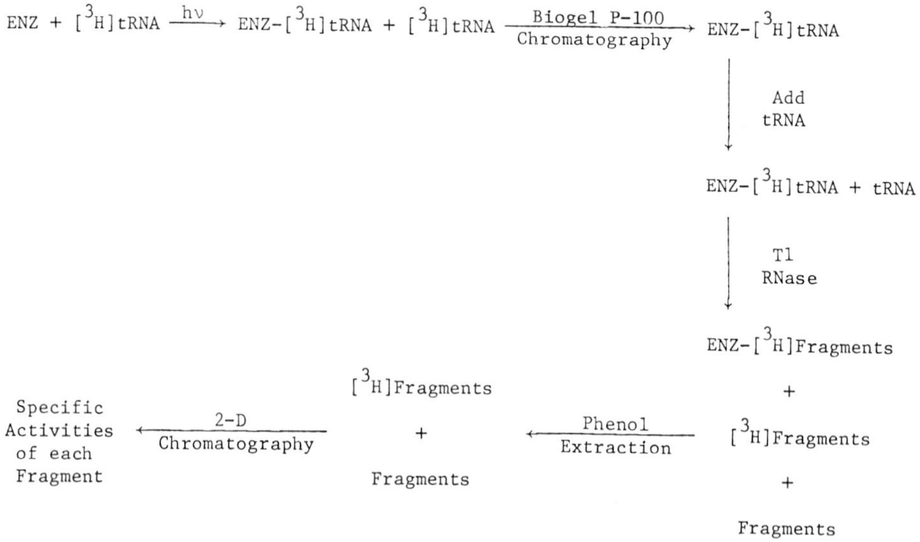

$$\text{ENZ} + [^3\text{H}]\text{tRNA} \xrightarrow{h\nu} \text{ENZ-}[^3\text{H}]\text{tRNA} + [^3\text{H}]\text{tRNA} \xrightarrow[\text{Chromatography}]{\text{Biogel P-100}} \text{ENZ-}[^3\text{H}]\text{tRNA}$$

$$\downarrow \begin{array}{c} \text{Add} \\ \text{tRNA} \end{array}$$

$$\text{ENZ-}[^3\text{H}]\text{tRNA} + \text{tRNA}$$

$$\downarrow \begin{array}{c} \text{T1} \\ \text{RNase} \end{array}$$

$$\text{ENZ-}[^3\text{H}]\text{Fragments}$$
$$+$$

Specific Activities of each Fragment $\xleftarrow[\text{Chromatography}]{\text{2-D}}$ [^{3}H]Fragments + Fragments $\xleftarrow[\text{Extraction}]{\text{Phenol}}$ [^{3}H]Fragments + Fragments

SCHEME 1. Scheme for determining the frequency of cross-linking for each T1 fragment. Reproduced by permission from G. P. Budzik, S. S. M. Lam, H. J. P. Schoemaker, and P. R. Schimmel, *J. Biol. Chem.* **250**, 4433 (1975).

tRNA are first cross-linked by irradiation. Uncross-linked [³H]tRNA is removed through molecular seive chromatography. At this point unlabeled tRNA is added to the cross-linked complex and the mixture is digested exhaustively with T1 ribonuclease. This gives enzyme-[³H]fragments plus [³H]fragments plus unlabeled fragments. A phenol extraction removes the cross-linked fragments, and the resulting free oligonucleotides are then separated by two-dimensional thin-layer chromatography. Finally, the specific activity of each fragment is determined.

The rationale of this procedure is that the cross-linked oligonucleotides will have the lowest specific activities. (The labeled cross-linked fragments are removed in the phenol extraction.) The percent joining of each T1 fragment in the cross-linked complex can be calculated (see below). This simple procedure gives a quantitative estimate of the frequency of cross-linking of each T1 fragment.

Two important controls must be performed. First, it must be shown that the nucleic acid does not lose radioactivity in specific locations as a result of the radiation per se. Second, the procedure should be carried out with the nucleic acid and enzyme mixed together after irradiation of the nucleic acid alone. This control checks whether T1 ribonuclease quantitatively liberates all noncovalently bound T1 fragments in the presence of enzyme. In the case of the Ile-tRNA synthetase–tRNA^Ile complex that has been used here for illustrative purposes, these controls demonstrate that irradiation does not release tritium from the nucleic acid and that the nuclease liberates all noncovalently bound fragments.[9]

The detailed procedures associated with Scheme 1 are outlined below.

Detailed Procedure. In all procedures given, it is assumed that a specific tRNA species is used.

A 1-ml synthetase-tRNA reaction mixture (see above) is irradiated as described above. The cross-linked [³H]tRNA is separated from uncross-linked [³H]tRNA by chromatography on a 1.2×100 cm Bio-Gel P-100 (100–200 mesh) column. The elution is performed with 10 mM NH_4HCO_3 (pH 8.5); at this pH non-cross-linked complex is dissociated so that only cross-linked species are isolated.

The fractions containing cross-linked species are pooled and lyophilized several times to remove the volatile salt. The residue is dissolved in approximately 100 μl of 10 mM ammonium carbonate at pH 7.5 and mixed with unlabeled tRNA. This mixture is then digested for 4–5 hr at 37° with T1 ribonuclease (about 5 units per A_{260} of total tRNA[30]) and bacterial alkaline phosphatase (about 5 μg per A_{260} of total tRNA). After digestion, the reaction mixture is extracted with an equal volume of

---

[30] K. Takahashi, *J. Biochem. Tokyo* **49**, 1 (1961).

buffer-saturated phenol. The phenol phase is back-extracted twice with equal volumes of buffer. The aqueous phases are pooled, extracted 4 times with ether, and lyophilized to dryness.

The residue is dissolved in about 10 μl of H_2O and applied to the origin of a cellulose thin-layer plate (Celplate-22; 20 $\times$ 20 cm; 0.1 mm thickness; Brinkmann). The origin is chosen as approximately 2.5 cm from either edge. Since each plate has a capacity for only 1.5–2.0 A_{260} units, it is necessary to spot the digest on several plates. The chromatograms are developed first by ascending chromatography with solvent I (n-propanol/concentrated NH_4OH/H_2O, 55:10:35, v/v). The plates are dried, rotated 90°, and developed by ascending chromatography with solvent II (isobutyric acid/concentrated NH_4OH/H_2O, 66:1:33, v/v). The locations of the separated fragments are visualized with an ultraviolet lamp.

The areas of cellulose containing the fragments are scraped and mixed with 1 ml of 0.1 M NaCl, 20 mM sodium cacodylate (pH 4.5). After thorough mixing of the slurry, the cellulose is removed by centrifugation. Elution of the small T1 fragments (1–3 bases in length) is quantitative, and about 60% efficiency is attained with the large fragments (10–15 bases in length).

The identity of each fragment is determined by standard methods.[9] The concentration of each fragment is determined by measurement of the absorbance at 260 nm; fragment extinction coefficients are calculated from the base compositions and the known extinctions of individual bases, together with hypochromicity corrections.[31] A portion, 0.8 ml, of each eluent is counted in 10 ml of Bray's scintillation fluid. Specific activities of the fragments are then calculated.

A similar procedure is followed with a control in which nonirradiated complex is used. Comparison of the fragment specific activities from the irradiated complex with those of the nonirradiated one allows determination of the cross-linking frequencies of the various fragments. For each T1 oligonucleotide, the percent joining is given by Eq. (2).

$$\% \text{ Joining} = 100 \times [(SA_0 - SA)/SA_0] \qquad (2)$$

Here, SA is the specific activity of the fragment from the irradiated complex and SA_0 is that of the same fragment from the nonirradiated complex.

Example of Data Obtained. Figure 1 gives the sequence and cloverleaf structure of *Escherichia coli* tRNA[IIe].[32] Dotted lines enclose the T1 fragments; these are numbered according to their positions on the two-

[31] R. C. Gamble, Ph.D. Thesis, Massachusetts Institute of Technology, Cambridge, Massachusetts, 1975.

[32] M. Yarus and B. G. Barrell, *Biochem. Biophys. Res. Commun.* **43**, 729 (1971).

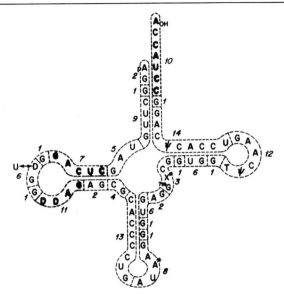

Fig. 1. Sequence and cloverleaf structure of *Escherichia coli* tRNA^{Ile}. T1 oligonucleotides are enclosed by dotted outline, and cross-linked fragments are indicated by shading. Reproduced by permission from G. P. Budzik, S. S. M. Lam, H. J. P. Schoemaker, and P. R. Schimmel, *J. Biol. Chem.* **250**, 4433 (1975).

dimensional chromatogram. Note that some of the fragments are unique, whereas others are redundant, i.e., they occur more than once in the structure.

Figure 2 is a bar graph representation of the percent joining of each

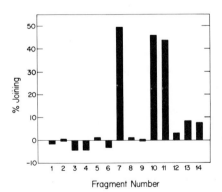

Fig. 2. Bar graph representation of percent joining of individual T1 fragments of tRNA^{Ile} to Ile-tRNA synthetase in a cross-linked Ile-tRNA synthetase–tRNA^{Ile} complex. Reproduced by permission from G. P. Budzik, S. S. M. Lam, H. J. P. Schoemaker, and P. R. Schimmel, *J. Biol. Chem.* **250**, 4433 (1975).

of the T1 fragments to Ile-tRNA synthetase and was determined by the procedure outlined. The error in the bars is about $\pm 10\%$ joining. Negative values for the percent joining simply reflect experimental error. The figure shows clearly that only three T1 fragments are cross-linked to the enzyme. These are fragments 7, 10, and 11. Their locations in the cloverleaf structure are indicated by shading in Fig. 1. In each case the fragments are cross-linked to the extent of 40–50%. (If a fragment has $x\%$ joining, this means that $x\%$ of the cross-linked complexes are joined at this fragment.) Since there are three fragments cross-linked to the extent of 40–50%, there are, on the average, 1.2–1.5 cross-links per cross-linked complex.

Since a three-dimensional structural model for tRNA is available,[28,29] the spatial locations of the cross-linked sites within this structure can be identified. This, in turn, gives a picture of the way in which the enzyme lies across the nucleic acid. Details of these types of analyses have been presented.[9,10]

Conclusions

The photochemical approach for cross-linking described here has obvious advantages. For example, no extraneous labels or chemicals are required to achieve cross-linking; the irradiations can be carried out under mild solution conditions; and there appears to be a large number of potential photoreactions that will yield cross-linked adducts.

An ambiguity associated with this approach is that little is known about the chemical structures of the cross-linked adducts that are produced. As research on the structures of the photoproducts continues, however, it is likely that the most commonly produced adducts will become well known and relatively easy to identify. And even without detailed knowledge of the chemical structures produced, the approach is nevertheless one of the best for mapping topological relationships in protein–nucleic acid complexes.[33]

[33] The photo-cross-linking work has been supported by Grant No. GM 15539 from the National Institutes of Health. We also thank Professor R. W. Chambers for advice on the photochemical procedures.

[15] Affinity Labeling of Multicomponent Systems

By Arthur E. Johnson and Charles R. Cantor

Affinity labeling is an extremely useful technique for the study of multicomponent systems, such as organelles or other complex assemblies of macromolecules. This approach offers the most direct means of identifying those components that are adjacent to the binding site of a particular ligand and may be helpful in the elucidation of structure–function relationships within the complex. Unfortunately, multicomponent systems introduce a number of complications into an affinity labeling experiment, primarily because of the large number of potential reactive sites, the possibility of multiple ligand binding sites, and the possibility of multiple system states. Hence, affinity labeling experiments with multicomponent systems must be strictly controlled to ensure that the labeling that occurs is true affinity labeling. In this chapter we shall discuss some general principles that must be considered when designing affinity label experiments for complex assemblies. We shall focus on our experience with the affinity labeling of ribosomes using tRNA derivatives, but most of the comments are applicable to other multicomponent systems, such as chromatin and membranes.

Binding of tRNAs to Ribosomes

A brief introduction to the ribosome system is necessary to acquaint the reader with some terminology. The ribosome is the site of protein biosynthesis. Its substrates are aa-tRNAs[1] that cycle on and off, transferring their amino acids into the nascent protein chain according to information encoded by mRNA. Transfer RNA binds to ribosomes in at least two functionally distinct sites, which are defined in terms of peptide bond formation. The peptide donor (a pep-tRNA) is bound in the P site during peptide bond formation; the peptide acceptor (an aa-tRNA) is bound in the A site. There is also some functional[2,3] evidence of a third tRNA-ribosome binding conformation.

A pep-tRNA or an aa-tRNA is considered bound to the P site if it reacts with the antibiotic puromycin. The structure of puromycin resembles the aminoacyl end of an aa-tRNA, and the ribosome treats it as an acceptor in peptide bond formation. One rarely obtains a puromycin reactivity of 100%. Assumptions are often made about the binding sites of the tRNAs that do not react with puromycin, but the actual binding sites are unknown. The puromycin-unreactive tRNAs may be binding to a non-P site, or to a P site on an inactive ribosome, or may simply

---

[1] Abbreviations used: aa-tRNA, aminoacyl-tRNA; pep-tRNA, peptidyl-tRNA; BrAcPhe-tRNA, N-bromoacetyl-Phe-tRNA; BrAcMet-tRNA, N-bromoacetyl-Met-tRNA$_1^{Met}$; ϵ-BrAcLys-tRNA, N^ϵ-bromoacetyl-Lys-tRNA; EF-T$_u$, elongation factor T$_u$; EF-G, elongation factor G; GMPPNP, guanylyl-imidodiphosphate.

[2] A. Skoultchi, Y. Ono, J. Waterson, and P. Lengyel, *Biochemistry* 9, 508 (1970).

[3] A. Haenni and J. Lucas-Lenard, *Proc. Natl. Acad. Sci. U.S.A.* 61, 1363 (1968).

react slowly with puromycin. Other assays that are supposedly site-specific, such as tetracycline inhibition of tRNA binding, are not reliable.[4] Some indications of tRNA bound to the A site can be obtained by using EF-G. This can transfer tRNA from the A to the P site, and the transfer can be monitored by the appearance of puromycin reactivity.

The *in vitro* ribosome binding characteristics differ for the various aa-tRNAs and pep-tRNAs. But site-specific binding to ribosomes is maximized under enzymic, i.e., factor-dependent, binding conditions. At 5 mM Mg^{2+}, the binding of fMet-tRNA or BrAcMet-tRNA is dependent on the presence of initiation factors and is nearly all puromycin reactive.[5] At 7 mM Mg^{2+}, in the presence of excess unacylated tRNA, the binding of ϵ-BrAcLys-tRNA is dependent on the presence of EF-T$_u$ and is nearly all puromycin-unreactive.[4,6] In the absence of factors, preferential binding to the P site is obtained at low Mg^{2+} concentrations (10 mM). Higher Mg^{2+} concentrations and the presence of unacylated tRNA favor the binding of pep-tRNA and aa-tRNA to puromycin-unreactive sites on the ribosome.

Choice of Chemically Reactive Groups

Specific reactive groups suitable for affinity and photoaffinity labeling are discussed elsewhere in this volume and in recent review articles[7,8] and will not be dealt with here. It should be noted, however, that varying the length or flexibility of the reactive group may alter the reaction pattern. This approach can be used in a multicomponent system to locate components that are different distances from the ligand binding site.[9]

The considerations that govern the choice of reactive group are the same as in simple systems, except in one respect. Multicomponent systems may exhibit a time-dependent change from one state into another,

[4] A. E. Johnson, R. F. Fairclough, and C. R. Cantor, *in* "Nucleic Acid–Protein Recognition" (H. Vogel, ed.), in press. Academic Press, New York, 1977.

[5] M. Sopori, M. Pellegrini, P. Lengyel, and C. R. Cantor, *Biochemistry* **13**, 5432 (1974).

[6] A. E. Johnson and C. R. Cantor, unpublished data.

[7] C. R. Cantor, M. Pellegrini, and H. Oen, *in* "Ribosomes" (M. Nomura, A. Tissières, and P. Lengyel, eds.), pp. 573–585. Cold Spring Harbor Laboratory, Cold Spring Harbor, New York, 1974. See also this volume [8].

[8] B. Cooperman, *in* "Aging, Carcinogenesis and Radiation Biology. The Role of Nucleic Acid Addition Reactions" (K. C. Smith, ed.), p. 315. Plenum, New York, 1976.

[9] D. Eilat, M. Pellegrini, H. Oen, Y. Lapidot, and C. R. Cantor, *J. Mol. Biol.* **88**, 831 (1974). See also this volume [9].

such as the movement of a probe from the A to the P site (or vice versa) on the ribosome. It is especially desirable to control the time at which the probe moiety reacts with the multicomponent complex. The time of reaction can be rigorously specified only by using photoaffinity labels with extremely short reactive lifetimes and exciting with a very brief pulse of light. The resulting reaction products will reflect the amount of the probe in each state of the system at the time of photolysis. However, a less intense light pulse of longer duration is often used. This offers no advantage over electrophilic affinity labels unless the system has reached equilibrium before the photolysis. In both chemical affinity labeling and long-duration nonequilibrium photoaffinity labeling experiments, the reaction products obtained reflect the reaction rates of the reactive group in the various states of the system as well as the rates of change from one state to another.

Whichever type of reactive group is chosen, it should be designed to attach covalently to the ligand in a position that maintains some ligand functions intact. Functional assays can then be used to determine the state of the ligand–receptor complex. This enables one to determine the functional state of the multicomponent system, and thus to correlate the affinity labeling structural information with functional data. For example, a bromoacetyl moiety attached to the *side chain* of the lysine of Lys-tRNA still allows the modified aa-tRNA to participate in peptide bond formation from either the A or the P site of the ribosome.[6] The widely used α-NH$_2$-derivatized aa-tRNAs can only function from the P site.

Macromolecular Ligands

Large multicomponent systems, such as ribosomes, offer the possibility of using macromolecules to carry the reactive label to the active site. As probes, modified macromolecules have several advantages over small label-carrying ligands. A macromolecular ligand can accept a modification with a large probe, such as an arylazide, with less chance of significantly changing the strength of the ligand–receptor association, or of altering the specificity of binding. This also enables one to use a greater variety of probes (short or long, rigid or flexible, hydrophobic or hydrophilic) in searching for reactive residues and in maximizing the efficiency of reaction.

A macromolecule will, in general, bind more strongly than a small molecule to a multicomponent receptor. This allows one to use a smaller concentration of ligand, with a concomitant reduction in nonspecific labeling. The stronger binding usually results in a longer ligand–receptor association time, thereby increasing the chance of receptor reaction with electrophilic affinity labels. A macromolecular ligand may allow placement of a probe at more than one position on the ligand, which will not affect its biological activity. One can therefore obtain information about different regions of the ligand binding site.

Complex receptors typically have a multiplicity of reactive residues outside the ligand binding site. Nonspecific reaction of affinity labels with these groups can be a serious problem. Some of these residues are likely to be accessible only to small molecules. Hence, the amount of nonspecific labeling will be less when a macromolecular probe is used.

A disadvantage in using a macromolecule as an affinity label is that chemical characterization of the reagent and its reaction products can be difficult. Considerable effort is often required to determine the number of covalently attached reactive groups per macromolecule, the sites of attachment of these groups, and the number of these groups that have reacted with the multicomponent receptor. Ideally, each of these quantities should be precisely one. It is sometimes very helpful to design the affinity label so that the reactive group is attached to a labile portion of the ligand. Cleavage of the reactive group moiety from the bulk of the ligand can then speed analysis. For example, reactive groups attached to the aminoacyl moiety of an aa-tRNA can easily be separated from the tRNA by a mild alkaline incubation.

Transfer RNA is the macromolecule which has most often been used as the label-carrying ligand in affinity and photoaffinity labeling studies of ribosomes. This is the result of the intimate involvement of tRNA in nearly all aspects of ribosomal function and the availability of convenient sites on the tRNA for the attachment of reactive groups. To date, the amino acid and the 4-thiouridine base (present in some tRNAs from *Escherichia coli*) have been the preferred modification sites, simply because the chemistry is straightforward and side reactions are minimal.

There are some particular advantages to attaching covalently the reactive group to the aminoacyl moiety of a tRNA. The amino acid must be near an active site (peptidyltransferase center) of the ribosome. A probe at this location is more likely to yield information about the structural effects of factors, factor-dependent GTP hydrolysis, and antibiotics than probes at most other places on the tRNA. If the aminoacyl-probe moiety can still participate in peptide bond formation, it is possible to correlate directly particular affinity-labeled products with ribosome function. The need for synthesizing a radioactive reactive group is avoided. Instead a radioactive amino acid is enzymically attached to a tRNA, and then this product is chemically modified to add the reactive group.

The many unusual bases found in tRNA can complicate the preparation of affinity labels. The X base, which is found in a number of tRNAs from different species,[10,11] contains a primary amino group.[12] Attempts to modify aa-tRNA at the α-amino group of the amino acid may therefore be accompanied by the reaction of a second label with the amino group of the X base of the tRNA. Similarly, using a primary alkyl halide reagent to modify 4-thiouridine may result in labels also being covalently attached to pseudouridine.[13] It is worth noting that the tRNA most often used in labeling studies of the ribosome, *E. coli* tRNA^Phe, contains X, 4-thiouridine, and pseudouridine.[14] It is very important to establish that the modified ligand contains only one reactive group, covalently attached to only one site on the ligand.

The problem of two or more reactive groups attached to the same ligand is particularly severe if the covalent reaction is monitored through a radioactive reactive group or a generally radioactive ligand. For example, the covalent reaction of a reactive group attached to an X base would be very difficult to distinguish from that of a reactive group attached to the amino acid. One can circumvent this problem by placing radioactivity in only one location on the ligand, i.e., either in the amino acid or in a strange base. After incubating the affinity label with the receptor, one can destroy most of the ligand, leaving only the reactive group and the amino acid (or strange base) attached to the receptor component. This can be accomplished either by nuclease digestion or alkaline hydrolysis. Then, only covalent adducts derived from reactive groups attached to the radioactive moiety will be detected. Destruction of the macromolecular ligand prior to analysis has a second advantage. Since it reduces the extent of chemical modification of the receptor components to a minimum, it simplifies the procedures necessary to analyze the pattern of covalently incorporated radioactivity.

Macromolecular ligands pose a further potential complication when they are converted to chemically reactive species. Intermolecular or intramolecular reactions between ligand sites is a real possibility. For example, BrAcPhe-tRNA, BrAcMet-tRNA, and ϵ-BrAcLys-tRNA all have sites, e.g., 4-thiouridine or pseudouridine, capable of reacting with the bromoacetyl moiety. We store ϵ-BrAcLys-tRNA in 1 mM potassium acetate (pH 5). At this pH essentially none of the N^ϵ-bromoacetyllysine reacts with tRNA. Under our usual affinity labeling conditions (pH 7.4, 37°,

[10] S. Friedman, *Biochemistry* **11**, 3435 (1972).
[11] S. Friedman, *Nature (London), New Biol.* **244**, 18 (1973).
[12] S. Friedman, H. J. Li, K. Nakanishi, and G. Van Lear, *Biochemistry* **13**, 2932 (1974).
[13] C. Yang and D. Söll, *Biochemistry* **13**, 3615 (1974).
[14] B. G. Barrell and F. Sanger, *FEBS Lett.* **3**, 275 (1969).

40 min), no tRNA aggregates are formed in a control incubation lacking ribosomes. However, when ε-BrAcLys-tRNA is incubated at pH 11.5 (0.05 M triethylamine, 37°, 90 min), alkylation of the tRNA does occur.

Another disadvantage of macromolecular affinity labels is ligand instability. If the label is attached to a part of the ligand that is chemically or enzymically labile, the covalent reactions observed in an experiment may not all result from true affinity labeling. Nonspecific reactions of the detached label with the receptor may be significant. As noted earlier, probes are often attached to the aminoacyl or peptidyl moieties of a tRNA. Aminoacyl-tRNAs are particularly susceptible to the chemical hydrolysis of the aminoacyl bond. Typically, only about 50% of the ε-BrAcLys-tRNA molecules added to an incubator remain aminoacylated after 50 min at pH 7.4 and 37°. (The percentage of intact aa-tRNAs obtained in a particular incubation depends on the amount of aa-tRNA bound to ribosomes, since this aa-tRNA is protected from deacylation.[15]) Pep-tRNAs are much less susceptible to chemical deacylation. However, all pep-tRNAs except peptidyl-tRNA$_f^{Met}$ are enzymically deacylated by an enzyme, peptidyl-tRNA hydrolase, which is found in some ribosome preparations.[16] (The enzyme does not deacylate aa-tRNAs.) It can be removed by washing the ribosomes in 1.0 M NH$_4$Cl.[16]

The Receptor

In the ideal case, the multicomponent receptors used experimentally are pure and fully active, and reproducibly so. In reality the individual particles in a solution of multicomponent receptors are usually heterogeneous with respect to composition, conformation, and activity. Furthermore, different preparations of the complexes are likely to differ in purity and activity. Typically, only 20–50% of the particles in a solution of salt-washed ribosomes are capable of exhibiting EF-T$_u$-dependent binding of an aa-tRNA. Moreover, many ribosomal proteins are present in the solution at an average of less than one copy per ribosome.[17] This heterogeneity is compounded by the presence of various factors and enzymes[16,18] that may copurify with the ribosomes.

The difficulty caused by a heterogeneous receptor population is most

[15] S. Pestka, *J. Biol. Chem.* **242**, 4939 (1967).

[16] J. R. Menninger, M. C. Mulholland, and W. S. Stirewalt, *Biochim. Biophys. Acta* **217**, 496 (1970).

[17] C. G. Kurland, *Annu. Rev. Biochem.* **41**, 377 (1972).

[18] J. S. Tscherne, I. B. Weinstein, K. W. Lanks, N. B. Gersten, and C. R. Cantor, *Biochemistry* **12**, 3859 (1973).

SOME YIELDS OF COVALENT REACTIONS BETWEEN 70 S
RIBOSOMES AND tRNA ANALOGS

Type of label[a]	Yield[b]	Primary reaction site
Electrophilic[19]	3–7%	Proteins L2, L27
Electrophilic[5]	5%	Protein L2
Photoreactive[20]	4%	23 S rRNA
Photoreactive[21]	15–20%	16 S rRNA
Photoreactive[22]	17–20%	23 S rRNA
Photoreactive[23]	5–6%	50 S proteins, 16 S rRNA
Photoreactive[24]	10%	23 S rRNA

[a] Numbers refer to text footnotes.
[b] The yield is defined as the percentage of ribosome-bound tRNA that reacts covalently with ribosomal components.

clearly appreciated after considering the data compiled in the table.[5,19-24] This table shows some of the yields of covalent reaction between probe and receptor that have been obtained in affinity and photoaffinity experiments with ribosomes. Such low yields raise the possibility that the affinity label has reacted with only a small subpopulation of the ribosomes, such as those missing a particular protein or those trapped in a nonfunctional conformation. Special controls are necessary to rule out this possibility.

One approach is to improve the homogeneity of the receptor preparation. For example, Noll et al. have described an improved purification procedure for ribosomes.[25] Unfortunately, it is not clear that purifications based on techniques such as centrifugation or conventional chromatography have the resolution to provide samples that are reproducibly homogeneous to greater than 90–95%. Affinity chromatography will undoubtedly be used in the near future to obtain ribosomes that are completely homogeneous in composition and conformation.[26]

A second approach is to design the affinity label experiment so that

[19] M. Pellegrini, H. Oen, D. Eilat, and C. R. Cantor, J. Mol. Biol. 88, 809 (1974).
[20] L. Bispink, and H. Matthaei, FEBS Lett. 37, 291 (1973).
[21] I. Schwartz and J. Ofengand, Proc. Natl. Acad. Sci. U.S.A. 71, 3951 (1974).
[22] A. S. Girshovich, E. S. Bochkareva, U. M. Kramarov, and Y. A. Ovchinnikov, FEBS Lett. 45, 213 (1974).
[23] I. Schwartz, E. Gordon, and J. Ofengand, Biochemistry 14, 2907 (1975).
[24] N. Sonenberg, M. Wilchek, and A. Zamir, Proc. Natl. Acad. Sci. U.S.A. 72, 4332 (1975).
[25] M. Noll, B. Hapke, and H. Noll, J. Mol. Biol. 80, 519 (1973).
[26] N. V. Belitsina, S. M. Elizarov, M. A. Glukhova, A. S. Spirin, A. S. Butorin, and S. K. Vasilenko, FEBS Lett. 57, 262 (1975).

one can selectively monitor only the covalent reactions that occur on functionally active receptors. This can be done by allowing the receptors to function normally on the ligand analogs, and then assaying the covalently bound ligands for receptor-mediated alterations.

Of course, it is quite possible that the system itself will perform the selection, in the sense that only active, intact receptors will bind the ligand analog and react covalently with the probe. One can presumably encourage this by using a low ligand:receptor ratio in the incubation. Alternatively, one can use conditions that require a nonreceptor molecule(s) to screen both the receptors and the ligand analogs for active particles. Examples of this are EF-T$_u$-dependent binding of ϵ-BrAcLys-tRNA[4] and initiation factor-dependent binding of BrAcMet-tRNA.[5]

Ligand–Receptor Association

A primary concern in affinity labeling experiments is whether the modified ligand, carrying the probe moiety, binds to the receptor in the same manner as the natural ligand. There are three ways of showing that the binding sites for the modified and unmodified ligands are the same.

One method is to demonstrate that the requirements for ligand binding are the same for the analog and the natural ligand. For example, the EF-T$_u$-dependent binding of ϵ-BrAcLys-tRNA is optimal under the same conditions that optimize the enzymic binding of Lys-tRNA to ribosomes.[6] The similarity in the binding characteristics of the two aa-tRNAs indicates that they are binding to the same site on the ribosome.

A second method is to show that the binding of the analog and the natural ligand is competitive. This approach is particularly useful in the case of small ligands, such as antibiotics.[27,28] However, in many instances it would be impractical to use this approach with macromolecular ligands. Because of its high affinity for ribosomes, tRNA can be used at a relatively low concentration in affinity labeling incubations, thereby minimizing nonspecific covalent reaction and maximizing binding site specificity (e.g., A or P). However, this leaves many tRNA binding sites empty. Even if ribosomes are not present in excess, the slow dissociation rate of tRNA from the ribosome and the unavoidable presence of unacylated tRNA in any aa-tRNA preparation make it difficult to devise an unambiguous experiment to demonstrate direct binding competition be-

[27] N. Sonenberg, M. Wilchek, and A. Zamir, *Proc. Natl. Acad. Sci. U.S.A.* **70**, 1423 (1973). See also this volume [73].

[28] O. Pongs, R. Bald, and V. A. Erdmann, *Proc. Natl. Acad. Sci. U.S.A.* **70**, 2229 (1973).

tween the modified and unmodified tRNAs under affinity labeling conditions.

The third and most convincing approach is to demonstrate that the ligand analog can participate in active site function on the multicomponent complex. For example, the binding of a modified tRNA to a ribosome can be considered normal if the tRNA participates in peptide bond formation.

Multicomponent receptors may have more than one functional binding site for a particular ligand, even for macromolecular ligands. This can lead to ambiguity in the interpretation of affinity labeling results because of the low yields of covalent reaction between the receptor and the ligand analog (see the table). For example, a particular tRNA analog may react much more efficiently with a ribosomal protein when bound in the A site than with a different protein when bound in the P site. In this case, the A site product may predominate even if 80% of the bound tRNA is puromycin reactive. Yet since 80% of the tRNA is bound to the P site, the A site product would probably be incorrectly designated as a P site protein. It is not sufficient, therefore, simply to determine whether the ligand analog binds to the receptor; one must distinguish between different receptor–ligand conformations. This is most easily accomplished by monitoring the functional status of the *same* molecule which covalently labels a receptor component.

Experimental Design

The key observation in affinity labeling experiments with ribosomes is that the ligand analog reacts with the receptor with a very low efficiency (see the table). This is crucial because the ribosomes are heterogeneous with respect to composition, activity, and ligand binding sites. Therefore, uncertainty exists as to which subpopulation of ribosomes actually reacts with a particular affinity or photoaffinity label. This situation is likely to exist, at least initially, in any multicomponent receptor system.

The low yields of covalent reaction dictate that one focus experimentally on the probe molecules that actually label receptor components. But at the same time, to resolve the heterogeneity problem, one must determine the functional status of the pertinent probe molecules. Hence, the basic feature of the optimal experimental design involves selective observation of the ligand analog molecules, which act both as affinity labels and as receptor substrates.

By designing experiments so that the same modified ligand molecule that reacts covalently with the receptor also participates in normal receptor function, one avoids most of the objections and uncertainties inherent in an affinity labeling experiment with a multicomponent system. A positive result enables one to conclude with confidence that the ligand analog has bound in a *functional manner* to a *functional site* on a *functional complex*. All three of these conclusions are important in view of the receptor heterogeneity problems and of the possibility of nonfunctional binding sites for the ligand on the receptor complexes.

This approach has been successful in a number of affinity labeling experiments.[4,5,19,29-31] The basic protocol requires using both a pep-tRNA and an aa-tRNA. One of them has a radioactive amino acid (alternatively, the two amino acid moieties may be labeled with different radioisotopes); the other has an electrophilic or photoactivatable group attached to the side chain or the α-amino nitrogen of the amino acid. The pep-tRNA is bound to ribosomes under conditions that favor binding to the P site. If this tRNA contains the affinity or photoaffinity label, reaction with the ribosomal components is allowed to proceed. Then the aa-tRNA is added to the incubation, and a peptide bond is allowed to form between the two aminoacyl moieties. Alternatively, the reactive probe may be attached to the aa-tRNA. Since peptide bond formation follows rapidly upon binding, the covalent reaction of the probe with ribosomal components presumably occurs after peptide transfer.

After both covalent reaction and peptide bond formation have occurred, the ribosomal components are separated and analyzed for the presence of radioactivity. Since the probe moiety is nonradioactive (or has a different radioisotope), the only way in which radioactivity can become covalently linked to a ribosomal component is through (i) a covalent reaction between a ribosomal component and an aminoacyl-probe moiety, and (ii) peptide bond formation between that *same* aminoacyl-probe moiety and a radioactive aminoacyl moiety. In this way analogs of peptidyl-tRNA,[5,19,29,30] aa-tRNA,[4] and puromycin[31] have been shown to function both as affinity labels and as peptide bond donors or acceptors.

One uncertainty that is difficult to resolve is whether the probe covalently reacts with the receptor while it is in the binding conformation indicated by functional analysis. The ligand analog may be translocated from one site to another between the time of function assay and the time of covalent reaction. It is extremely difficult to monitor a change in binding site which occurs subsequent to the functional assay in the ribosome system; it may be more easily monitored in other multicom-

[29] N. Hsiung, S. A. Reines, and C. R. Cantor, *J. Mol. Biol.* **88**, 841 (1974).
[30] N. Hsiung and C. R. Cantor, *Nucleic Acid Res.* **1**, 1753 (1974).
[31] P. Greenwell, R. J. Harris, and R. H. Symons, *Eur. J. Biochem.* **49**, 539 (1974).

ponent systems. Translocation of a probe from the A to the P site may be effected either by an intrinsic ribosomal mechanism[32] or by a contamination with a catalytic amount of EF-G. Thus, even though an aa-tRNA has been bound specifically to the A site with EF-T$_u$ and has formed a peptide bond with pep-tRNA, it is conceivable that the tRNA may move to the P site prior to covalent reaction. We have been unable to rule out this possibility in experiments that would otherwise demonstrate, unequivocally, affinity labeling from the A site[4] and photoaffinity labeling from the P site.[30]

The Covalent Reaction

It is well known that chemical affinity labels, such as the bromoacetyl moiety, react preferentially with certain amino acids. It is less well known that photoaffinity labels can also be quite selective.[8] Thus it is worthwhile to note that the absence of a covalent reaction does not necessarily mean that a particular component is not located at that binding site. The lack of reaction may mean that the reactive residues of the component in question are inaccessible to the probe, or that the rate of reaction with a different nearby component is much higher than with the component of interest.

The reactive moieties themselves may influence the reaction products obtained for reasons other than the reaction selectivity just discussed. For example, a hydrophobic probe will tend to bind to any hydrophobic region that is available to it. Thus, it will most likely react while in this region, and the reaction products will be biased. This is of particular concern in photoaffinity experiments that use substituted aromatic compounds as reactive groups.

The low efficiency of covalent reaction obtained in affinity and photoaffinity experiments with ribosomes (see the table) has been noted. The efficiency of photoaffinity labeling can be improved only by using a different label. However, the chemical affinity labeling efficiency may be increased by taking certain precautions. The reactivity of an electrophilic label, such as a bromoacetyl moiety, may be improved by decreasing the concentration of nucleophiles in the solution. Ribosomes and EF-T$_u$, for example, are usually stored in the presence of reducing agents, such as 2-mercaptoethanol and dithiothreitol, both of which are excellent nucleophilic reagents for the bromoacetyl group. Since complete removal of the reducing agent will inactivate the ribosomes and EF-T$_u$, one is

[32] L. P. Gavrilova, O. E. Kostiashkina, V. E. Koteliansky, N. M. Rutkevitch, and A. S. Spirin, *J. Mol. Biol.* 101, 537 (1976).

forced to determine empirically the concentration of reducing agent that will maximize retention of biological activity and minimize loss of reactive bromines.

We have not made a thorough study, but our experience in the EF-T$_u$-dependent affinity labeling of ribosomes with ϵ-BrAcLys-tRNA suggests some limits.[6] A dithiothreitol concentration of 5 mM abolishes alkylation of the 23 S rRNA by ϵ-BrAcLys-tRNA; a 1.5 mM concentration reduces alkylation by 90%. More efficient alkylation occurs when the reducing agent concentration is less than 1 mM. However, the total alkylation obtained in separate sets of experiments may vary considerably. It is not clear whether this variation results primarily from different amounts of EF-T$_u$ inactivation or from different amounts of unoxidized dithiothreitol. A reducing agent concentration of less than 1 mM can be achieved simply by dilution of the stock solutions into the affinity labeling incubation mixture. Alternatively, one can use a fast dialysis apparatus similar to that used in hydrogen exchange experiments[33] to reduce the dithiothreitol concentration in EF-T$_u$ or ribosome solutions immediately prior to use.

To minimize nonspecific covalent reaction during the product analysis procedures, we routinely terminate ϵ-BrAcLys-tRNA affinity labeling incubations by adding 2-mercaptoethanol to 0.1 M and incubating for 15 min at 37°. Nonspecific reactions can be minimized in photoaffinity labeling experiments by the use of "scavenger" molecules.[34]

The Reaction Products

The covalent attachment of a ligand analog to a receptor component will usually alter the properties of the component, particularly if the ligand is a macromolecule. As a result, affinity-labeled components may not copurify with their unmodified counterparts in whatever system is used to separate and identify the individual components of a multi-component receptor. Ribosomal proteins are routinely analyzed by two-dimensional gel electrophoresis. Any modification that alters a protein's net charge will cause it to migrate to a different position in the gel from that of the unmodified protein. Since the quantity of a particular modified protein is small (see the table), it is extremely difficult to locate its position by staining the protein in the gel. One is therefore forced to locate the radioactive modified proteins by examining sections of the gel surrounding the stained spots of unmodified protein. For example,

---

[33] S. W. Englander and D. Crown, *Anal. Biochem.* **12**, 579 (1965).
[34] A. E. Ruoho, H. Kiefer, P. E. Roeder, and S. J. Singer, *Proc. Natl. Acad. Sci. U.S.A.* **70**, 2567 (1973). See also this volume [59].

after alkylation by an electrophilic tRNA analog and nuclease treatment as described earlier, the electrophoretic mobility of protein L27 changes greatly.[6,19,35] Fortunately, other ribosomal proteins appear to exhibit little change in mobility after alkylation. Thus their identification is a problem only when two or more proteins are poorly resolved in the gel. Autoradiography or fluorography may be useful in locating the positions in the gel of those modified proteins that are present in sufficiently high concentrations to allow detection. One problem with two-dimensional gel electrophoresis is that a significant amount of the radioactive modified protein remains in the first-dimension disk gel. If some species are preferentially retained in the first dimension, an accurate quantitative analysis of the labeled proteins is impossible by this technique.

Identification of labeled receptor components by immunological means can circumvent the difficulties that may arise with physical techniques.[35] Another advantage in using antibodies is that very little of the modified protein is required for identification. The major limitation in this approach is the difficulty in obtaining a complete set of pure antibodies, each specific for a particular receptor component. Clearly, if there is significant cross-reaction between an antibody species and a "nonconjugate" protein, there is little advantage in this approach.

An unambiguous product identification can be made by determining the primary structure of the modified receptor component. Because of the difficulty in isolating sufficient quantities of the modified component, this method has been rarely used. However, since it can identify not only the component, but also its reactive site, this approach will be frequently employed in the future. It has particular advantages in the ribosome case because various techniques have indicated that a number of ribosomal proteins are elongated. In this situation, ligand analogs bound at different sites on the receptor may covalently react with different residues of the same receptor component. Protein L27, for example, is a primary target of electrophilic tRNA analogs as diverse as BrAc-(Gly)$_{12}$-Phe-tRNA,[9] BrAcPhe-tRNA,[19] and ϵ-BrAcLys-tRNA.[6] Sequence analysis of L27 will be needed to show whether this is due to a particularly reactive residue of L27 and flexibility in the aminoacyl ends of the tRNA analogs, or to separate reactive residues of L27 positioned at the A site, at the P site, and in the region occupied by the nascent protein chain.

To date, product analysis has not extended beyond identifying the receptor components adjacent to a ligand binding site. But once the reactive components (or residues) in a particular ligand binding site

[35] A. P. Czernilofsky, E. E. Collatz, G. Stöffler, and E. Kuechler, *Proc. Natl. Acad. Sci. U.S.A.* **71**, 230 (1974).

have been identified, affinity labeling can be used as a diagnostic assay to demonstrate the existence of different binding conformations. For example, the efficiency of ϵ-BrAcLys-tRNA alkylation of 23 S rRNA was reduced by more than 90% when the tRNA was bound to ribosomes with EF-T$_u$ and GMPPNP (a nonhydrolyzable analog of GTP) instead of EF-T$_u$ and GTP.[4] The nature and magnitude of the structural alteration that accompanies this change in alkylation are unknown, but the results corroborate functional evidence of a third tRNA-ribosome binding conformation.[2,3]

Quantitative analysis of affinity labeling data has its pitfalls. Most of the potential problems have been noted earlier and include the possible variation in alkylation yields resulting from different concentrations of active reducing agent in two separate sets of experiments. Appropriate control experiments are necessary to determine whether the multicomponent system is operating as expected. One control not mentioned above, which is necessary in photoaffinity labeling studies, is the dependence of covalent reaction on the photoactivatable group. This is necessary because photoreaction between unmodified nucleic acids and proteins has been observed.[36,37] Such reactions are interesting in themselves—assuming that they represent true affinity labeling of functional receptor sites—but their occurrence with reactive group-dependent reactions can lead to misinterpretations.

The possible perturbation of affinity labeling data by nonspecific small-molecule reactions must also be confronted experimentally. Pellegrini et al.[19] found that the contribution of small bromoacetyl compounds to the covalent labeling of ribosomal components was insignificant under the conditions used for BrAcPhe-tRNA affinity labeling. Alternatively, one can demonstrate that the covalent reaction is dependent upon the binding of the intact ligand. For example, both the enzymic binding of ϵ-BrAcLys-tRNA and the alkylation of ribosomal components by this analog are dependent on the presence of message [poly(rA)] and of EF-T$_u$.[6] The safest approach, mentioned before, involves assaying only the covalent reactions of probes which have also successfully participated in a normal receptor function requiring an intact ligand.

Acknowledgments

This work was supported by research grants GM 14825 and GM 19843 from the U.S. Public Health Service. A. J. is a fellow of the Helen Hay Whitney Foundation.

[36] H. J. P. Schoemaker and P. R. Schimmel, *J. Mol. Biol.* **84**, 503 (1974). See also this volume [14].

[37] I. Fiser, P. Margaritella, and E. Kuechler, *FEBS Lett.* **52**, 281 (1975).

Section II

Specific Procedures for Enzymes, Antibodies, and Other Proteins

A. Proteolytic Systems
Articles 16 through 22

B. Nucleotide and Nucleic Acid Systems
Articles 23 through 38

C. Carbohydrate Systems
Articles 39 through 44

D. Amino Acid Systems
Articles 45 through 49

E. Steroid Systems
Articles 50 through 52

F. Antibodies
Articles 53 through 58

G. Other Proteins
Articles 59 through 66

H. Receptors and Transport Systems
Articles 67 through 72

[16] Reaction of Serine Proteases with Halomethyl Ketones

By JAMES C. POWERS

The use of the chloromethyl ketones Tos-PheCH$_2$Cl (TPCK) and Tos-LysCH$_2$Cl (TLCK) as specific reagents for chymotrypsin and trypsin, respectively, is one of the classic demonstrations of the value of affinity labels for enzyme studies.[1] Subsequently, halomethyl ketones were used for characterization of functional groups and sequences in active sites, as probes of structure by providing labeled proteins that contain spectroscopic handles, in crystallographic studies of substrate binding sites, and in the study of the biological function of proteases. Today, halomethyl ketones are probably the most widely studied class of affinity label and have been discussed in numerous reviews.[2-6]

The reaction of a serine protease with most substrate-related haloketones probably first involves the formation of an enzyme inhibitor complex (Scheme 1) in which the inhibitor is recognized by specific interactions between the side chain of the P$_1$ amino acid residue and the S$_1$ or primary substrate binding site of the enzyme (nomenclature of Schecter and Burger[7]). Irreversible inhibition takes place within this complex by covalent bond formation between the active-site histidine residue and a methylene group of the inhibitor. In addition, crystallographic studies in a few cases indicate that the active-site serine oxygen has added to the carbonyl group of the inhibitor to give a tetrahedral

---

[1] The nomenclature used in this chapter for amino acid and peptide derivatives conforms to the 1971 Recommendations of the IUPAC–IUB Commission on Biochemical Nomenclature, *Biochemistry* **11**, 1726 (1972). Thus the chloromethyl ketone derived from tosyl-L-phenylalanine will be abbreviated as Tos-PheCH$_2$Cl instead of the more commonly used TPCK. The author recommends the use of systematic abbreviations such as Tos-PheCH$_2$Cl since they define the compound concisely, show the relationship between several related halomethyl ketones, and lead to less confusion for those not familiar with shorter abbreviations.

[2] See this volume [10].

[3] B. R. Baker, "Design of Active-Site-Directed Irreversible Enzyme Inhibitors." Wiley, New York, 1967.

[4] E. Shaw, *in* "The Enzymes" (P. Boyer, Ed.,), 3rd ed., Vol. 1, pp. 91–146. Academic Press, New York, 1970; *Physiol. Rev.* **50**, 244 (1970).

[5] S. J. Singer, *Adv. Protein Chem.* **22**, 1 (1967).

[6] J. C. Powers, *in* "Chemistry and Biochemistry of Amino Acids, Peptides, and Proteins" (B. Weinstein, ed.), Vol. 4, in press. Dekker, New York, 1977.

[7] I. Schechter and A. Berger, *Biochem. Biophys. Res. Commun.* **27**, 157 (1967).

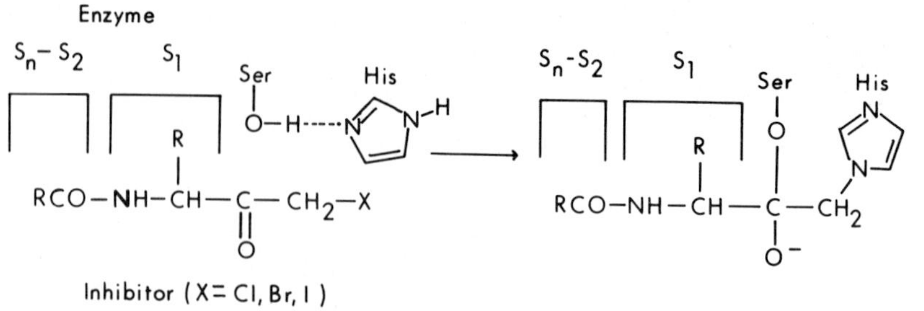

SCHEME 1. Reaction of a halomethyl ketone with the active site of a serine protease.

hemiketal.[8] It is likely that this is a general structural feature of serine proteases that are inhibited by haloketones.

In designing a halomethyl ketone affinity label for a new serine protease, some knowledge of the substrate specificity is required in order to choose the proper amino acid residue to place at the P_1 position of the inhibitor, i.e., the position that interacts with the enzymes S_1 subsite. The remainder of the inhibitor structure should optimally be free of nonpeptide functional groups. Tos-PheCH$_2$Cl, for example, will not inhibit subtilisin BPN' owing to the sulfonamide group, which would be forced into excessively close contact with the enzyme in any transition state leading to alkylation of the active-site histidine, whereas other chloromethyl ketones (RCO-PheCH$_2$Cl) are excellent inhibitors.[9-11] Even though the synthesis is more difficult, design of an inhibitor with an extended chain should be considered. In several crystallographic studies, interactions between the extended substrate binding site ($S_2 - S_n$ subsites) of the enzyme and an extended peptide chain of the inhibitor has been observed.[8,12,13] For enzymes such as elastase, inhibitors without extended chains react slowly or not at all.[14,15] There are now numerous

[8] T. Poulos, R. A. Alden, S. T. Freer, J. J. Birktoft, and J. Kraut, J. Biol. Chem. 251, 1097 (1976).

[9] J. D. Robertus, R. A. Alden, J. J. Birktoft, J. Kraut, J. C. Powers, and P. E. Wilcox, Biochemistry 11, 2439 (1972)

[10] K. Morihara and T. Oka, Arch Biochem. Biophys. 138, 526 (1970).

[11] J. C. Powers, M. O. Lively III, and J. T. Tippett, Biochim. Biophys. Acta. in press (1977).

[12] D. M. Segal, J. C. Powers, G. H. Cohen, D. R. Davies, and P. E. Wilcox, Biochemistry 10, 3728 (1971).

[13] J. D. Robertus, J. Kraut, R. A. Alden, and J. J. Birktoft, Biochemistry 11, 4293 (1972).

[14] L. Visser, D. S. Sigman, and E. R. Blout, Biochemistry 10, 735 (1971).

[15] A. Thomson and I. S. Denniss, Eur. J. Biochem. 38, 1 (1973).

examples reported in which extending the chain of the inhibitor has accelerated the rate of inhibition.[10,15-18] Increased reactivity is also obtained by proper choice of the amino acid residues at the various subsites.[11,18] In addition, selectivity within a group of related enzymes can be obtained by use of subsite inactions.[19,20]

The reactivity of haloketones is in the order I > Br > Cl.[21] However, increasing the reactivity of the haloketone in this way is dangerous because of the possibility of accelerating competing side reactions. For example, haloketones can react with nucleophilic side chains such as thiols[22] and thioethers. Therefore it is probably more advantageous to increase the reactivity of chloromethyl ketone by altering the peptide structure than by changing the leaving group.

Synthesis

Halomethyl ketone derivatives of blocked amino acids are readily prepared by the reaction of mineral acids (hydrochloric and hydrobromic) with the corresponding diazomethyl ketone (Scheme 2). Iodomethyl

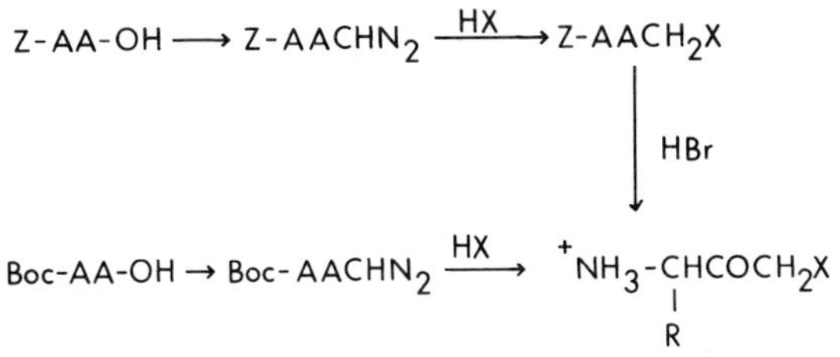

$$Z\text{-}AA\text{-}OH \longrightarrow Z\text{-}AACHN_2 \xrightarrow{HX} Z\text{-}AACH_2X$$

$$\downarrow HBr$$

$$Boc\text{-}AA\text{-}OH \rightarrow Boc\text{-}AACHN_2 \xrightarrow{HX} {}^+NH_3\text{-}\underset{\underset{R}{|}}{C}HCOCH_2X$$

SCHEME 2. Synthesis of halomethyl ketones.

ketones are prepared by reaction of a bromo- or chloroketone with NaI since reaction of HI with a diazomethyl ketone yields the methyl ketone. Usually the diazomethyl ketone is not isolated, but is generated *in situ* from the corresponding blocked amino acid. A number of blocking groups have been used including tosyl (Tos), benzyloxycarbonyl (Z), and *t*-butyloxycarbonyl (Boc); only the latter two should be used if the

[16] J. C. Powers and P. M. Tuhy, *Biochemistry* **12**, 4767 (1963).
[17] R. C. Thompson and E. R. Blout, *Biochemistry* **12**, 44 (1973).
[18] K. Kurachi, J. C. Powers, and P. E. Wilcox, *Biochemistry* **12**, 771 (1973).
[19] J. R. Coggins, W. Kray, and E. Shaw, *Biochem. J.* **138**, 579 (1974).
[20] C. Kettner and E. Shaw, *Fed. Proc., Fed. Am. Soc. Exp. Biol.* **35**, 1464 (1976).
[21] E. Shaw and J. Ruscica, *Arch. Biochem. Biophys.* **145**, 484 (1971).
[22] T. Rossman, C. Norris, and W. Troll, *J. Biol. Chem.* **249**, 3412 (1974).

goal is a peptide halomethyl ketone. Simple acyl groups should be avoided, since reaction with diazomethane will probably yield an oxazolone.[23,24] The diazomethyl ketone is prepared by reaction of diazomethane with the appropriate acid activated by means of dicyclohexylcarbodiimide (DCCI), by the mixed anhydride method[25] or via the acid chloride. The mixed anhydride procedure is probably the most convenient, avoids oxazolone formation, and allows the use of acid-labile protecting groups. Extreme caution must be exercised in using diazomethane since it is both toxic and explosive.[26]

Unblocked amino acid chloromethyl ketones are prepared by reaction of benzyloxycarbonyl blocked derivatives with HBr in HOAc,[12,27] trifluoroacetic acid,[21] or by hydrogenation.[10] The hydrogenolysis procedure can lead to difficulty in some cases since the chlorine is also susceptible to hydrogenolysis.[28-30] A very useful method for the synthesis of chloromethyl ketone derivatives of amino acids utilizes the ready removal of the t-butyloxycarbonyl (Boc) group with hydrochloric acid (Scheme 1). In one step the chloromethyl ketone moiety is generated from the diazomethyl ketone and the protecting group is removed.[11,31]

Synthesis of peptide chloromethyl ketones can be accomplished simply by coupling of an appropriate peptide or amino acid with an unblocked amino acid chloromethyl ketone.[10,12,27] A few dipeptides have been converted directly to the chloromethyl ketone using the mixed anhydride and CH_2N_2 followed by HCl.[32]

Various synthetic problems are encountered in the preparation of chloromethyl ketone derivatives of basic amino acids. The side chain usually must be blocked during synthesis, and difficulties are often encountered during removal of the blocking group. For example, deblocking of Tos-Lys(Z)CH_2Cl to Tos-LysCH_2Cl by several methods[28,33] gives a

[23] P. Karrer and R. Widmer, *Helv. Chim. Acta.* **8**, 203 (1925).

[24] P. Karrer and G. Bussman, *Helv. Chim. Acta.* **24**, 645 (1941).

[25] B. Penke, J. Czombos, L. Balaspiri, J. Petres, and K. Kovacs, *Helv. Chim. Acta.* **53**, 1057 (1970).

[26] J. A. Moore and D. E. Reed, *Org. Syn.* **41**, 16 (1961).

[27] J. C. Powers and P. E. Wilcox, *J. Am. Chem. Soc.* **92**, 1782 (1970).

[28] E. Shaw, M. Mares-Guia, and W. Cohen, *Biochemistry* **4**, 2219 (1965).

[29] L. Y. Frolova, G. K. Kovaleva, M. B. Agalarova, and L. L. Kisselev, *FEBS Lett.* **34**, 213 (1973).

[30] R. C. Thompson, *Biochemistry* **13**, 5495 (1974).

[31] P. L. Birch, H. A. El-Obeid, and M. Akhtar, *Arch. Biochem. Biophys.* **148**, 447 (1972).

[32] M. Tejima, M. Takeuchi, E. Ichishima, and S. Kobayashi, *Agr. Biol. Chem.* **39**, 1423 (1975).

[33] F. Sebestyen and J. Samu, *Chem. Ind.* (*London*), p. 1568 (1970).

low yield of product. Use of trifluoroacetic acid[34] or HF[35] was eventually found to give a good conversion to product. A number of peptide chloromethyl ketones with terminal LysCH$_2$Cl residues have been prepared via Lys(Z)CH$_2$Cl.[19] The chain was extended by coupling with Boc-blocked amino acids or peptides. Deblocking of the Boc group was accomplished with 5 M HCl in ethanol. Use of CF$_3$CO$_2$H in the case of dipeptides results in loss of the chloromethyl ketone group probably due to a cyclization reaction. In an analogous reaction Tos-OrnCH$_2$Cl slows cyclizes by reaction of the chloromethyl ketone moiety with the side-chain amino group.[34] Tos-LysCH$_2$Cl, on the other hand, is completely stable.

Arginine chloromethyl ketones, although of considerable interest, have proved to be difficult to synthesize. Z(NO$_2$)-ArgCH$_2$Cl was obtained as a contaminant (1.5–2%) of a cyclized product that was prepared by reaction of Z(NO$_2$)-Arg-Cl with CH$_2$N$_2$ followed by HCl.[34] Tos-ArgCH$_2$Cl has been successfully synthesized by using the nitro blocking group for the arginine side chain.[35] Peptide chloromethyl ketones with terminal Arg residues have recently been reported.[20]

The synthesis of several chloromethyl ketones have been reported in other volumes of this series. These include Tos-PheCH$_2$Cl, Z-PheCH$_2$Cl, Tos-LysCH$_2$Cl, PheCH$_2$Cl, and Z-Ala-Gly-PheCH$_2$Cl.[36,37]

Halomethyl or diazomethyl ketones derivatives of most amino acids (Ala, Arg, Asp, Cys, Glu, Gln, Gly, Ile, Leu, Lys, Phe, Pro, Thr, Trp, Tyr, Val, and several unusual amino acids) has been described in the literature. A recent comprehensive review lists all halomethyl and diazomethyl ketones reported in the literature through September 1975. Over a hundred derivatives of amino acids and approximately 60 peptide derivatives are listed.[6]

Two representative procedures are presented in detail. The first, preparation of Z-Gly-Leu-PheCH$_2$Cl, involves deblocking of Z-PheCH$_2$Cl and subsequent coupling of the deblocked chloromethyl ketone with Z-Gly-Leu-OH.[18] This compound is an excellent inhibitor of chymotrypsin[18] and cathepsin G.[38] The second procedure, that for Ac-Ala-Ala-Pro-ValCH$_2$Cl, illustrates the synthesis of a peptide chloromethyl ketone from Boc-Val-OH with a minimum of isolation and purification along the way[39]; this compound is an excellent inhibitor of porcine pancreatic and human leukocyte elastase.[39]

[34] E. Shaw and G. Glover, *Arch. Biochem. Biophys.* **139**, 298 (1970).
[35] K. Inouye, A. Sasaki, and N. Yoshida, *Bull. Chem. Soc. Jpn.* **47**, 202 (1974).
[36] E. Shaw, this series, Vol. 11, p. 677 (1967).
[37] E. Shaw, this series, Vol. 25, p. 655 (1972).
[38] J. C. Powers and B. F. Gupton, unpublished observations.
[39] J. C. Powers and R. J. Whitley, unpublished observations.

N-Benzyloxycarbonylglycyl-L-*leucyl*-L-*phenylalanine   Chloromethyl Ketone* (*Z-Gly-Leu-PheCH₂Cl*).[40] To a saturated solution of HBr in acetic acid are added 3.5 g (10.4 mmoles) of Z-PheCH₂Cl. After 20 min, ether is added and the resulting crystals are filtered. Recrystallization from ethyl acetate-methanol will allow an approximately 70% yield of L-phenylalanine chloromethyl ketone hydrobromide (m.p. 180°–181°, dec). It is not necessary to recrystallize the product; simply washing with dry ether provides material suitable for subsequent coupling. The hydrobromide must be kept anhydrous. In the present procedure, the intermediate was not recrystallized.

The mixed anhydride prepared from 3.2 g (10 mmoles) of Z-Gly-Leu-OH, 1.1 ml (10 mmoles) of N-methylmorpholine and 1.34 ml (10 mmoles) of isobutyl chloroformate in 60 ml of anhydrous tetrahydrofuran at $-10°$ to $-15°$ is allowed to react with L-phenylalanine chloromethyl ketone hydrobromide and 1.3 ml of N-methylmorpholine for 1.25 hr while the reaction mixture is allowed to warm to room temperature under anhydrous conditions. Work-up of the reaction mixture involves evaporation of tetrahydrofuran, extraction of the residue into ethyl acetate, washing with citric acid and sodium bicarbonate solutions, drying over magnesium sulfate, and evaporation. The product is recrystallized from ethyl acetate–cyclohexane to yield 3.41 g (68%) of product with m.p. 140.5°–143°.

N-Acetyl-L-*alanyl*-L-*alanyl*-L-*prolyl*-L-*valine   Chloromethyl   Ketone* (*Ac-Ala-Ala-Pro-ValCH₂Cl*).[40] The diazomethyl ketone Boc-ValCHN₂ is prepared from Boc-Val-OH (10 g, 45 mmoles) by the mixed anhydride method (N-methylmorpholine, 45 mmoles; isobutyl chloroformate, 45 mmoles) in tetrahydrofuran at $-10°$ to $-15°$. The mixed anhydride solution is filtered under anhydrous conditions to remove N-methylmorpholine hydrochloride and is then added to excess CH_2N_2 in ether.[26] After standing overnight at room temperature, anhydrous HCl is bubbled through the solution of Boc-ValCHN₂ at 5° for 15 min as the solution turns from yellow to colorless. The product ($HCl \cdot ValCH_2Cl$) is isolated by evaporation of the solvents and dried under reduced pressure to give 4.5 g (55%) of an extremely hydroscopic white solid.

Ac-Ala-Ala-Pro-OH (1.5 g, 5 mmoles) is dissolved in 50 ml of anhydrous tetrahydrofuran and stirred at $-15°$ while N-methylmorpholine (0.55 ml, 5 mmoles) and isobutyl chloroformate (0.65 ml, 5 mmoles) are

---

[40] Anhydrous conditions must be maintained during the synthesis of diazomethyl ketones, their reaction with HCl to form chloromethyl ketones, and coupling reactions. Solvents and most reagents can be purchased in anhydrous form. Prior to use, all apparatus is dried in an oven or flamed while passing a stream of dry nitrogen through glassware. Mixed anhydride solutions are filtered through a jacketed sintered glass filter, using dry nitrogen to exclude moisture.

added. After 10 min, the N-methylmorpholine salts were filtered and HCl · ValCH$_2$Cl (0.93 g, 5 mmoles) in 5 ml of DMF and N-methylmorpholine (0.55 ml, 5 mmoles) are added. Stirring is continued for 4 hr as the mixture warmed to 25°. Thereafter, the solvent is removed under reduced pressure and the resulting yellow oil is dissolved in methanol and chromatographed on Merck silica gel G (0.063–0.2 mm). The product, eluted with 8% methanol in chloroform, is recrystallized from methanol to give 1.05 g (49% yield, m.p. 195°–196°, dec.).

Kinetics of the Reaction

The irreversible reaction of a chloromethyl inhibitor with a protease may be represented by the overall reaction sequence in Eq. (1), where E·I represents a noncovalently bound complex of the enzyme with the inhibitor and E–I is the final product with the inhibitor irreversibly bound to the enzyme via a covalent linkage.

$$E + I \rightleftharpoons E \cdot I \xrightarrow{k_3} E - I \tag{1}$$

$$K_I = \frac{[E][I]}{[E \cdot I]} \tag{2}$$

If the inhibitor concentration is sufficiently greater than the total enzyme concentration ($[I]/[E] > 10$–20), it can be shown that the decrease in $E + E \cdot I$ concentration in the inhibition mixture follows pseudo-first-order kinetics at any fixed value of I. The pseudo-first-order rate constant can easily be measured by periodically drawing an aliquot from the inhibition mixture and assaying for residual enzyme activity. Dilution of the inhibitor concentration in the assay mixture will likely result in the absence of E·I in the assay mixture, so that the enzyme concentration obtained will represent the total enzyme ($E + E \cdot I$) present in the aliquot. A semilog plot of enzyme activity or concentration vs time will then give k_{obs} directly. It can be shown that k_{obs} is related to K_I by Eq. (2) and k_3 by Eq. (3).[18,41]

$$1/k_{obs} = (K_I/k_3[I]) + 1/k_3 \tag{3}$$

If data are available over a sufficient range of inhibitor concentrations, both K_I (dissociation constant of the E·I complex) and k_3 (the limiting rate of inactivation) may be evaluated by the use of a double-reciprocal plot ($1/k_{obs}$ vs $1/[I]$) and Eq. (3). When the K_I value is much greater than the chosen inhibitor concentrations, Eq. (3) reduces to Eq. (4) and

$$k_{obs}/[I] = k_3/K_I \tag{4}$$

[41] R. Kitz and I. B. Wilson, *J. Biol. Chem.* **237**, 3245 (1970).

predicts that a reciprocal plot of $1/k_{obs}$ vs $1/[I]$ will pass through the origin and that K_I cannot be evaluated. From a practical standpoint, this situation occurs often, since K_I values are in the millimolar range whereas inhibitor concentrations are frequently limited to lower values owing either to low solubility of the inhibitor or its high rate of reaction.[18] It should be pointed out that in a situation wherein k_{obs} is measured at one inhibitor concentration or where $k_{obs}/[I]$ does not vary in the range of inhibitor concentrations utilized, then it is not possible to demonstrate the existence of an enzyme–inhibitor complex.

The most appropriate parameter for comparison of the reactivity of various inhibitors is the inhibition parameter k_3/K_I. In the absence of sufficient data for determining k_3/K_I, $k_{obs}/[I]$ may be used. The differences in the magnitude of the numbers for related inhibitors reflect mostly the effect of structural changes on the binding of the inhibitor to the enzyme (K_I) and on the rate of reaction within the bound complex (k_3). It will be apparent that values of $k_{obs}/[I]$ are subject to distortion from nonlinear concentration effects when the inhibitor concentration is close to the K_I value of the inhibitor.

Since most halomethyl ketones are only slightly soluble in water, the kinetics of inhibition are usually measured in a mixed solvent. The inhibitors are dissolved in such solvents as methanol or dimethoxyethane and then diluted with buffer. If the solution becomes cloudy or the halo-ketone precipitates at this point, either the concentration of organic solvent must be increased or the concentration of inhibitor decreased. Solutions of chloromethyl ketones in organic solvents are fairly stable, of the order of a week or more, but aqueous solution slowly lose inhibitory activity owing to hydrolysis.[16,18] A solution of Ac-Gly-PheCH$_2$Cl was hydrolyzed 30% after standing for 20 hr in a buffer solution at pH 5.80.

The rates of inhibition with halomethyl ketone are quite variable and depend on the pH, the enzyme, and the structure of the inhibitor. The fastest reactions have half-lives of less than 1 min whereas poor inhibitors may take many hours. The optimum in most cases is at pH 7 to 8, but rates are often measured at lower values to slow the reaction with reactive inhibitors to a more manageable range.

Identification of the Site of Reaction. The alkylation of the active site histidine residue by a substrate related haloketone has been used often to locate a specific histidine in the primary sequence of a serine protease. However, haloketones are general alkylating agents, and if the reagent is too far removed in structure from a substrate, it is likely to alkylate nucleophilic sites other than the catalytic histidine.[42] In one

---

[42] K. J. Stevenson and L. B. Smillie, *Can. J. Biochem.* **46**, 1357 (1968).

case, alkylation of the catalytic serine residue of trypsin was observed when an inhibitor was synthesized with a decreased distance between the chloromethyl ketone group and a charged side chain relative to normal substrate or inhibitor.[43] However, this must be regarded as an exception, since it is likely that formation of a tetrahedral intermediate[8] by reaction of the catalytic serine with the carbonyl group of the inhibitor is a requirement for inhibition.

The site of reaction can be identified by isolation of a peptide fragment containing the modified amino acid residue. This process is quite efficient when a diagonal electrophoresis technique is utilized. This involves electrophoresis of a peptide mixture on some support, treatment with HCO_3H, and then electrophoresis in a direction perpendicular to the first. Oxidation of histidine alkylated by a halomethyl ketone yields carboxymethyl histidine. Disulfides are also oxidized to cysteic acid residues. All peptides that contained cystine residues or amino acid residues alkylated by the haloketone will lie off the diagonal after the second electrophoresis and can easily be isolated.[44-46] A modification of this technique allows separation of only the peptide containing the alkylated amino acid residue.[47] The alkylated protein is first reduced, then cyanoethylated, and the cyanoethylated cysteine residues are oxidized to the sulfone. This change is not sufficiently large to remove such peptides from the diagonal, and only the peptide with the alkylated amino acid residue is found off the diagonal.

Applications

Representative examples of halomethyl ketones which have been used as affinity labels for serine proteases are listed in the table. Several thiol proteases (papain, clostripain, cathepsin B,[1] and cocoonase) are also listed since the mechanism of the inhibition is probably similar for the two classes of enzymes. The inhibitors chosen for the table were either the most readily available or among the more reactive for a particular enzyme. For more information the reader is referred to a recent comprehensive review of all the enzymes that have been studied with halomethyl ketone inhibitors.[6]

Most of the potential applications of affinity labeling have been realized by the use of halomethyl ketones. They have been used to label

[43] D. D. Schroeder and E. Shaw, *Arch. Biochem. Biophys.* **142**, 340 (1971).
[44] K. J. Stevenson and L. B. Smillie, *J. Mol. Biol.* **12**, 937 (1965).
[45] L. B. Smillie and B. S. Hartley, *Biochem. J.* **101**, 232 (1966).
[46] K. J. Stevenson and L. B. Smillie, *Can. J. Biochem.* **46**, 1357 (1968).
[47] K. J. Stevenson, *Biochem. J.* **139**, 215 (1974).

INHIBITION OF PROTEOLYTIC ENZYMES WITH HALOKETONE DERIVATIVES OF
AMINO ACIDS AND PEPTIDES

Enzyme	Inhibitor	References[a]
Chymotrypsin	TosPheCH$_2$Cl	5
	Z-Gly-Leu-PheCH$_2$Cl	18
Cathepsin G	Z-Gly-Leu-PheCH$_2$Cl	38
Subtilisin	Z-PheCH$_2$Br	b
	Z-Ala-Gly-PheCH$_2$Cl	10
	Ac-Phe-Gly-Ala-LeuCH$_2$Cl	11
	Phe-Ala-LysCH$_2$Cl	19
Streptomyces griseus protease B	Boc-Gly-Leu-PheCH$_2$Cl	c
Elastase	Ac-AlaCH$_2$Cl	15
	Ac-Ala-Ala-AlaCH$_2$Cl	16, d
	Ac-Ala-Ala-Pro-AlaCH$_2$Cl	16, 17, e,f
Trypsin	Tos-LysCH$_2$Cl	34, g
	Tos-ArgCH$_2$Cl	g
	Lys-Ala-LysCH$_2$Cl	19
Acrosin	Tos-LysCH$_2$Cl	h–j
Thrombin	Tos-LysCH$_2$Cl	k
	Phe-Ala-LysCH$_2$Cl	19, l
	Glu-Ala-LysCH$_2$Cl	19, l
Plasmin	Tos-LysCH$_2$Cl	m, n
Kallikrein	Z-LysCH$_2$Cl	19, o
	Ala-Phe-LysCH$_2$Cl	19
Carboxypeptidase Y	Z-PheCH$_2$Cl	p, q
Carboxypeptidase C	Tos-PheCH$_2$Cl	r
Acid carboxypeptidases	Tos-PheCH$_2$Cl	s, t
Papain	Tos-PheCH$_2$Cl	u, v
	Z-Gly-Leu-PheCH$_2$Cl	w, x
Clostripain	Tos-LysCH$_2$Cl	y
Cathepsin B^1	Tos-PheCH$_2$Cl	z
	Ac-Ala-Ala-Ala-AlaCH$_2$Cl	
Cocoonase	Tos-LysCH$_2$Cl	aa

[a] Numbers refer to text footnotes.

[b] F. S. Markland, E. Shaw, and E. L. Smith, *Proc. Natl. Acad. Sci. U.S.A.* **61**, 1440 (1968).

[c] A. Gertler, *FEBS Lett.* **43**, 81 (1974).

[d] J. Travis and R. C. Roberts, *Biochemistry* **8**, 2884 (1969).

[e] J. C. Powers and P. M. Tuhy, *J. Am. Chem. Soc.*, **94**, 6544 (1972).

[f] P. M. Tuhy and J. C. Powers, *FEBS Lett.* **50**, 359 (1975).

[g] N. Yoshida, A. Sasaki, and K. Inouye, *Biochem. Biophys. Acta* **321**, 615 (1973)

[h] L. Zaneveld, K. Polakoski, and W. Williams, *Biol. Reprod.* **6**, 30 (1972).

[i] L. Zaneveld, B. Dragoje, and G. Schumacher, *Science* **177**, 702 (1972).

[j] K. L. Plaskoski and R. A. McRorie, *J. Biol. Chem.* **248**, 8183 (1973).

[k] T. M. Chulkova and V. N. Orekhovick, *Biokhimiya* **33**, 1222 (1968).

[l] E. Shaw, *in* "Proteinase Inhibitors" (H. Fritz, H. Tschesche, L. J. Green, and E. Truscheit, eds.), pp. 531–540. Springer-Verlag, Berlin and New York, 1974.

^{m} F. F. Buck, B. Hummel, and E. DeRenzo, *J. Biol. Chem.* **243**, 3648 (1968).

^{n} W. R. Groskopf, B. Hsieh, L. Summaria, and K. Robbins, *J. Biol. Chem.* **244**, 359 (1969).

^{o} C. Sampaio, S. Wong, and E. Shaw, *Arch. Biochem. Biophys.* **165**, 133 (1974).

^{p} R. Hayashi, Y. Bai, and T. Hata, *J. Biochem. (Tokyo)* **76**, 1355 (1974).

^{q} R. Hayashi, Y. Bai, and T. Hata, *J. Biol. Chem.* **250**, 5221 (1975).

^{r} R. W. Kuhn, K. A. Walsh, and H. Neurath, *Biochemistry* **13**, 3871 (1974).

^{s} E. Ichishima, S. Sonoki, K. Hirai, Y. Torii, and S. Yokoyama, *J. Biochem. (Tokyo)* **72**, 1045 (1972).

^{t} T. Nakadai, S. Nasuno, and N. Iguchi, *Agric. Biol. Chem.* **37**, 1237 (1973).

^{u} S. S. Husain and G. Lowe, *Chem. Commun.* **1965**, 345 (1965).

^{v} M. L. Bender and L. J. Brubacker, *J. Am. Chem. Soc.* **88**, 5880 (1966).

^{w} S. Roffman, M. Levy, and W. Troll, *Fed. Proc., Fed. Am. Soc. Exp. Biol.* **32**, 466 (1973).

^{x} S. L. Roffman, *Diss. Abstr. Int.* **35B**, 3218 (1975).

^{y} W. H. Porter, L. W. Cunningham, and W. M. Mitchell, *J. Biol.Chem.* **246**, 7675 (1971).

^{z} M. C. Burleigh, A. J. Barrett, and G. S. Lazarus, *Biochem. J.* **137**, 387 (1974).

^{aa} F. Kafatos, J. H. Law, and A. M. Tartakoff, *J. Biol. Chem.* **242**, 1488 (1967).

and locate the active site histidine in numerous serine proteases including chymotrypsin, trypsin, subtilisin, thrombin, and elastase. Heavy atom-containing halomethyl ketones have been synthesized for possible use in crystallographic phasing during the determination of the three-dimensional structure of proteins.[48] Indeed, the determination of the structures of chymotrypsin and subtilisin that had been treated with halomethyl ketones led to the discovery of the extended substrate binding site in these enzymes.[8,9,12] Numerous kinetic studies with chloromethyl ketones have shown the existence of extended binding sites in other proteases or have sought to determine the nature of the individual subsites.[10,11,15-20] Halomethyl ketones have been used to place either spectroscropic,[49] fluorescent,[50] or spin-labeled[51] reporter groups in the active sites of serine proteases. A chloromethyl ketone has been used in a nuclear magnetic resonance study of catalytic residues of a serine protease.[52] More recently, halomethyl ketones have been extensively used in studies of role of proteases in such diverse biological processes as fertilization, cell growth, protein synthesis, virus maturation, and diseases such as emphysema and cancer.[6] However, in many cases appropriate controls were not carried out and it is impossible to discern whether the effect of the haloketone is

^{48} A. Tulinsky, R. L. Vandlen, C. N. Morimoto, N. V. Mani, and L. H. Wright, *Biochemistry* 4185 (1973).

^{49} D. S. Sigman and E. R. Blout, *J. Am. Chem. Soc.* **89**, 1747 (1967).

^{50} G. Schoellmann, *Int. J. Peptide Protein Res.* **4**, 221 (1972).

^{51} D. J. Kosman, *J. Mol. Biol.* **67**, 247 (1972).

^{52} G. Robillard and R. G. Shulman, *J. Mol. Biol.* **86**, 519 (1974).

due to specific inhibition of a serine protease or to other alkylation reactions. In the future, there is an excellent possibility that some of these affinity labels may be used medicinally in the control of both normal and abnormal physiological processes involving serine proteases.

[17] Reaction of Serine Proteases with Aza-Amino Acid and Aza-Peptide Derivatives

By JAMES C. POWERS and B. FRANK GUPTON

Aza-amino acid residues are analogs of amino acids in which the α-CH has been replaced by a nitrogen atom.[1] This substitution has a profound effect on the reactivity of aza-peptides, i.e., those containing

$$-NH-CH-CO- \qquad\qquad -NH-N-CO-$$
$$\qquad\quad | \qquad\qquad\qquad\qquad\qquad | $$
$$\qquad\quad R \qquad\qquad\qquad\qquad\qquad R$$

aza-amino acid residues. In particular, it has been shown that aza-peptide p-nitrophenyl esters are inhibitors and active site titrants of several serine proteases.[2-4] Inhibition of these enzymes is believed to arise from the acylation of the active-site serine residue yielding an acylated enzyme (Fig. 1), which is substantially less reactive toward deacylation than a normal acylated enzyme owing to the influence of the adjacent nitrogen atom. At present, there is no evidence that the active-site serine residue is actually the site of reaction. However, the close structural resemblance of aza-peptides to normal peptides substrates of serine proteases points strongly to the serine residue as the one being labeled.

The only requirement for design of an aza-peptide reagent for a new serine protease is knowledge of the enzyme's substrate specificity so that an appropriate side chain may be placed on the aza-amino acid residue. The employment of p-nitrophenol as a product of the acylation step is advantageous since it is a good leaving group, and the release of p-nitro-

[1] The use of standard IUPAC nomenclature (2-substituted carbazic acid derivatives) for the naming of aza-amino acid residues is particularly disadvantageous since the names are cumbersome and give no indication of the structural relationship with amino acids. For this reason, the method of naming these analogs in this text will be to precede the name of the corresponding amino acid with "aza." For example, 2-benzylcarbazoic acid p-nitrophenyl ester will be referred to as aza-phenylalanyl p-nitrophenyl ester. Likewise, the method of abbreviating these analogs will be to prefix the abbreviation of the corresponding amino acid with an "A." For example, the abbreviation of aza-phenylalanine will be Aphe.
[2] A. N. Kurtz and C. Niemann, *J. Am. Chem. Soc.* **83**, 1879 (1961).
[3] D. T. Emore and J. J. Smyth, *Biochem. J.* **107**, 103 (1968).
[4] J. C. Powers and D. L. Carroll, *Biochem. Biophys. Res. Commun.* **67**, 639 (1975).

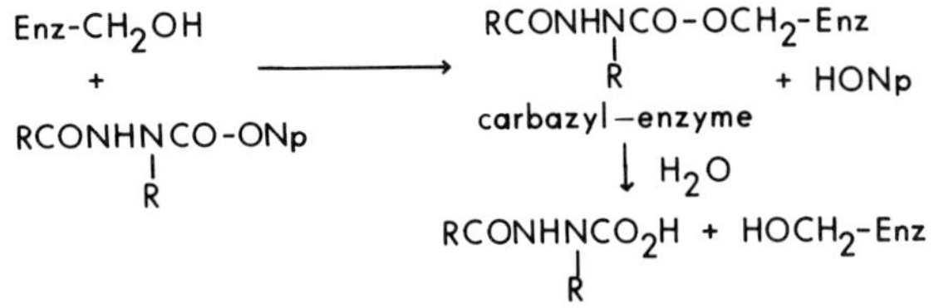

Enz-CH$_2$OH   RCONHNCO-OCH$_2$-Enz
                                           |
        + $\longrightarrow$ R   + HONp
                               carbazyl—enzyme
RCONHNCO-ONp
        | $\downarrow$ H$_2$O
        R
                          RCONHNCO$_2$H + HOCH$_2$-Enz
                                 |
                                 R

Fig. 1. Probable course for the reaction of a serine protease with an aza-amino acid derivative.

phenol serves as a useful means of monitoring the reaction between the aza-peptide and the enzyme.

Aza-peptide p-nitrophenol esters may also be used as active-site titrants of serine proteases[4] by measuring the initial burst of p-nitrophenol from the acylation step. Since the hydrolysis rate of the acylated enzyme is often very slow, but not zero, aza-peptides may also be employed as reversible blocking groups of the active site of serine proteases. The rate of deacylation is variable and depends on the pH, the enzyme itself, and the structure of the specific aza-peptide.

Synthesis of Aza-Peptides and Aza-Amino Acid Derivatives

Two general approaches are available for the synthesis of aza-peptides and aza-amino acid derivatives. The first method is illustrated in Scheme 1 and involves the reaction of an acyl hydrazide with an aldehyde fol-

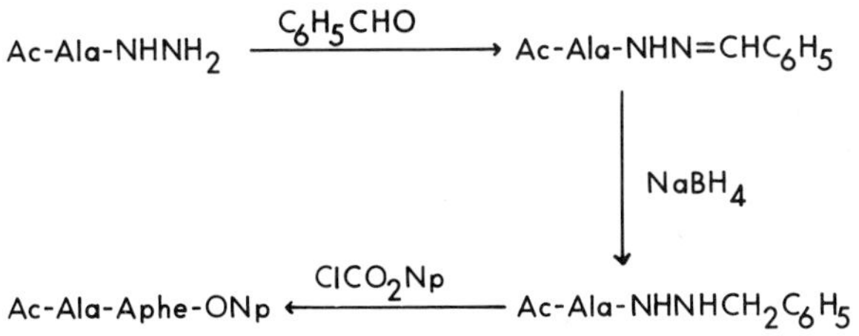

Ac-Ala-NHNH$_2$ $\xrightarrow{\text{C}_6\text{H}_5\text{CHO}}$ Ac-Ala-NHN=CHC$_6$H$_5$

                                          $\downarrow$ NaBH$_4$

Ac-Ala-Aphe-ONp $\xleftarrow{\text{ClCO}_2\text{Np}}$ Ac-Ala-NHNHCH$_2$C$_6$H$_5$

SCHEME 1

lowed by reduction with sodium borohydride and reaction with p-nitrophenyl chloroformate.[5] This route is advantageous in that the side chain of the aza-amino acid moiety is introduced on the proper nitrogen atom without any possibility of the formation of the alternate isomer. In addition, many aldehydes are commercially available or easily synthesized, making most aza-amino acid residues accessible by this synthetic route.

[5] A. S. Dutta and J. S. Morley, *J. Chem. Soc.* **1975**, 1712 (1975).

The second approach to aza-peptides involves the reaction of a mono-substituted hydrazine with an ester. Both isomers (*1* and *2*, Scheme 2)

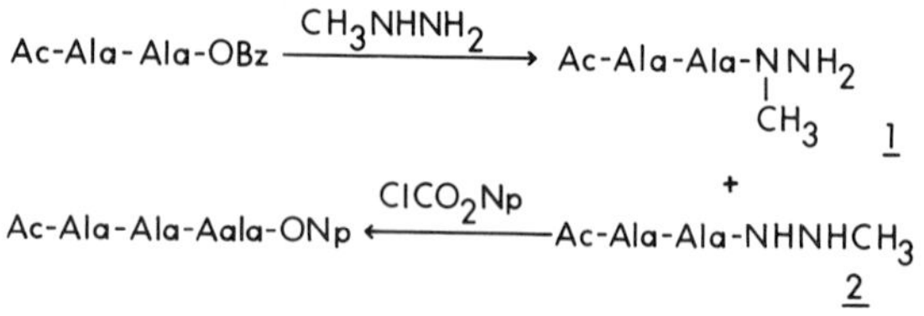

$$\text{Ac-Ala-Ala-OBz} \xrightarrow{\text{CH}_3\text{NHNH}_2} \text{Ac-Ala-Ala-N}\underset{\overset{|}{\text{CH}_3}}{\text{NH}_2} \quad \underline{1}$$

$$+$$

$$\text{Ac-Ala-Ala-Aala-ONp} \xleftarrow{\text{ClCO}_2\text{Np}} \text{Ac-Ala-Ala-NHNHCH}_3 \quad \underline{2}$$

SCHEME 2

are formed in the reaction and must be separated. The correct isomer, *2*, is usually formed in higher yield and can easily be converted to the final product. The isomeric hydrazides, *1* and *2*, can be readily distinguished by their PMR spectra. The alkyl group of the desired isomer *2* will appear at higher field (CH_3—, 2.42 ppm) than in the alternate isomer *1* (CH_3—, 3.20 ppm). This route has the advantage of being two steps shorter, and with higher yields, than the first method since peptide hydrazides are typically prepared from esters; however, relatively few monosubstituted hydrazines are commercially available, and the necessity for synthesizing the hydrazine may negate the advantages. Alternative methods for forming the monosubstituted hydrazine, e.g., coupling an acid bearing a monsubstituted hydrazine with dicyclohexylcarbodiimide or by a mixed anhydride procedure, should also be considered. However, the ratio of isomers formed upon acylation of a monosubstituted hydrazine is greatly dependent on conditions of synthesis.[6]

Synthesis of Ac-Ala-Aphe-ONp by Method I

Acetylalanyl hydrazine (2.8 mmoles) is coupled with benzaldehyde (3.5 mmoles) by refluxing the two reagents in 50 ml of methanol overnight. The product is readily crystallized from the reaction mixture by addition of 3.5 ml of ether. The hydrazone (2.0 mmoles) is subsequently reduced to the corresponding benzyl hydrazide with NaBH$_4$ (3.2 mmoles) in 40 ml of ethanol (24 hr, room temperature). Purification of this product by silica-gel chromatography (elution with 5% CH$_3$OH in CHCl$_3$) is required in order to separate inorganic starting materials and by-products. The aza-peptide Ac-Ala-Aphe-ONp is prepared by dissolving the substituted hydrazide (1.5 mmoles) in 50 ml of pyridine in the presence of triethylamine (0.8 mmole). A solution containing *p*-nitrophenyl chloroformate (1.8 mmoles) in 5 ml of anhydrous tetrahydrofuran is added

[6] F. E. Condon, *J. Org. Chem.* **37**, 3608 (1972).

dropwise to the pyridine solution at $0°$. The reaction mixture is stirred for 2 hr at $0°$ and an additional hour at room temperature. Upon evaporation of the reaction mixture, the crude product is chromatographed on silica gel to give a 50% yield of product, m.p. $156°–158°$ (recrystallized from acetone).

Peptide and amino acid hydrazides and hydrazones may easily be differentiated from other compounds on thin-layer chromatograms by treatment of the plates with ferric chloride followed by potassium ferricyanide[7]; a "royal" blue color is obtained. Likewise, the p-nitrophenyl moiety may be detected by addition of base to obtain the bright yellow color due to the p-nitrophenolate anion.

Synthesis of Ac-Ala-Ala-Aala-ONp by Method II

The methyl hydrazide is prepared by allowing Ac-Ala-Ala-OBz (6.9 mmoles) to react (24 hr, room temperature) with methyl hydrazine (170 mmoles). When one equivalent of water is added and dioxane (85 ml) is used as the solvent, the hydrazide crystallizes from the reaction mixture. The desired product, Ac-Ala-Ala-NHNHCH$_3$, is isolated from the isomeric hydrazide, Ac-Ala-Ala-N(CH$_3$)NH$_2$, by recrystallization from ethanol. The coupling of the substituted hydrazide with p-nitrophenyl chloroformate to form Ac-Ala-Ala-Aala-ONp is carried out by the procedure described in Method I to give a 66% yield of product with m.p. $200°–202°$ (ethanol).

Kinetic Considerations

The kinetics of the reaction of serine protases with substrate-related acylating agents are described by Eq. (1). The initial binding of the substrate to enzyme

$$\text{E} + \text{S} \rightleftarrows \text{E} \cdot \text{S} \xrightarrow{k_2} \text{E} - \text{S} + \text{P}_1 \xrightarrow{k_3} \text{E} + \text{P}_2 \tag{1}$$

is followed by formation of an acyl enzyme intermediate $(\text{E} - \text{S}) + \text{P}_1$ (p-nitrophenol in the case of p-nitrophenyl esters of aza-peptides). Deacylation (k_3) results in regeneration of active enzyme. An example of the type of results obtained by following the enzyme-catalyzed release of p-nitrophenol is shown in Fig. 2. This biphasic curve represents the initial stoichiometric burst of p-nitrophenol followed by the steady-state turnover of the aza-peptide ester. The background hydrolysis of the aza-peptide ester may be compensated for by having substrate in both sample and reference cells.

[7] H. Ertel and L. Horner, *J. Chromatogr.* **7**, 268 (1962).

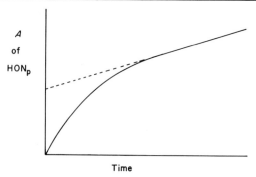

FIG. 2. A typical pattern for the enzyme-catalyzed release of p-nitrophenol from an aza-peptide p-nitrophenyl ester.

The equation[8-10] for the curve shown in Fig. 2 is represented by Eq. (2)

$$[P_1] = At + B(1 - e^{-bt}) \tag{2}$$

where

$$A = \frac{k_{cat}[E_0][S_0]}{[S_0] + K_m}$$

$$B = [E_0] \left(\frac{k_2}{k_2 + k_3}\right)^2 \bigg/ \left(1 + \frac{K_m}{[S_0]}\right)^2$$

$$b = \frac{(k_2 + k_3)[S_0] + k_2 K_s}{S_0 + K_s}$$

$$k_{cat} = \frac{k_2 k_3}{k_2 + k_3}$$

$$K_s = \frac{[E][S]}{[E \cdot S]}$$

$$K_m = \frac{K_s k_3}{k_2 + k_3}$$

As t approaches infinity, Eq. (2) reduces to Eq. (3), i.e., a straight line relationship represented by the steady-state portion of the curve. If $[S_0] \gg K_m$

$$[P_1] = At + B \tag{3}$$

[8] M. L. Bender, F. J. Kezdy, and F. C. Wedler, *J. Chem. Educ.* **44**, 84 (1967).

[9] M. L. Bender, M. L. Begue-Canton, R. L. Blakeley, L. J. Brukacher, J. Feder, C. R. Gunter, F. J. Kezdy, J. V. Kilheffer, T. H. Marshall, C. G. Miller, R. W. Roeske, and J. K. Stoops, *J. Am. Chem. Soc.* **88**, 5890 (1966).

[10] F. J. Kezdy and E. T. Kaiser, this series, Vol. 19, p. 1 (1970).

then the slope of this portion of the curve is $k_{cat} E_0$. The value of B may be obtained from Eq. (3) by extrapolating the steady-state portion of the curve back to $t = 0$. The value of $[P_1]$ at this point is directly proportional to $[E_0]$ if the acylation is stoichiometric. Under conditions wherein $k_2 \gg k_3$ and $[S_0] \gg K_m$, the value obtained from this treatment is equal to E_0. If the acylation rate (k_2) is greater than the deacylation rate (k_3) then $k_{cat} = k_3$. These requirements are readily met with several of the aza-peptide esters since these analogs rapidly acylate certain serine proteases at saturating substrate concentrations to form relatively stable acylated enzymes. An alternative method of measuring k_3 is to follow enzyme activity after acylation under conditions in which the enzyme is in excess. If the deacylation is extremely slow, then the acyl enzyme can be separated from excess reagent by dialysis or gel filtration and the acylation rate measured by following the reappearance of enzyme activity.

Two general procedures for monitoring the reaction of aza-peptide nitrophenyl esters with serine proteases will be discussed. The first (Ac-Ala-Aphe-ONp + chymotrypsin) illustrates the use of enzymes that are readily accessible to the investigator. It is also a situation in which the acylated enzyme has a long lifetime. The second (Ac-Ala-Ala-Aala-ONp + elastase) is designed so that only minimal amounts of enzyme need be employed and requires the use of a spectrophotometer with an expanded scale of 0.01 A. In this case, turnover of the acylated enzyme is fairly rapid, but the procedure is useful for titrating elastase.

Reaction of Chymotrypsin A_α with Ac-Ala-Aphe-ONp

Reagents

Ac-Ala-Aphe-ONp, 5.0 mM in acetonitrile. Care should be taken to avoid exposure to moisture. The stock solutions are stable for 1 week.

Chymotrypsin A_α, approximately 100 μM in 1.0 mM HCl. Approximately 10 mg of crystalline enzyme is measured out and dissolved in 5.0 ml of 1.0 mM HCl. The exact protein concentration is calculated from the absorbance at 280 nm [E(1%) = 20.5].

To each of two cuvettes is added 2.0 ml of 0.1 M citrate at pH 6.0, followed by 100 μl of the inhibitor stock solution. The two cuvettes are placed in the spectrophotometer, and a base line is recorded at 345 nm with a chart span of 0.1 A. To the reference cell, 100 μl of 1.0 mM HCl are added followed by the addition of a 100-μl aliquot from the enzyme stock solution to the sample cell. The recorder is immediately started upon addition of the enzyme. The solution is rapidly mixed, and the production of p-nitrophenol is measured (mixing time + machine equilibration time < 15 sec). Bursts of approximately 0.03 A are observed. The steady-state portion of the curve is indistinguishable from the background noise ($k_{cat} < 1.8 \times 10^{-4}$ sec^{-1}). Commercial enzyme preparations were found to be approximately 85–90% active.

The acylated enzyme may be isolated by gel filtration with Sephadex G-25 at pH 4.0 and 4° or by dialysis under similar conditions. The isolated enzyme derivative has a half-life of approximately 4 days at pH 5.8.[11]

Reaction of Porcine Elastase with Ac-Ala-Ala-Aala-ONp

Reagents

Ac-Ala-Ala-Aala-ONp, 0.5 mM in acetonitrile. Store in a cool dry place. Stock solutions are stable for 1 week.

Porcine elastase, approximately 10 μM in 1.0 mM HCl. The enzyme stock solution is prepared by dissolving approximately 1 mg of the crystalline enzyme in 5.0 ml of 1.0 mM HCl. The exact protein concentration is determined by its absorbance at 280 nm [E(1%) = 20.2].

To a 1-dram vial is added 1.0 ml of 0.1 M citrate at pH 6.0, followed by 50 μl of the inhibitor stock solution. After the solution is mixed, a 500-μl portion of the solution is added to each of two microcuvettes, and a base line is recorded at 345 nm with a chart span of 0.01 A. To the reference cell is added 25 μl of 1.0 mM HCl, followed by the addition of the enzyme stock solution to the sample cell. The solution is rapidly mixed, and the production of p-nitrophenol is measured. The burst of p-nitrophenol can be used to obtain the concentration of active sites (345 nm; $\epsilon = 6.25 \times 10^{3}$). Bursts of approximately 4×10^{-3} A are obtained; k_{cat} is 1×10^{-5} sec^{-1}.

In both experiments the burst of p-nitrophenol is stoichiometric (1:1) with respect to the enzyme concentration. Initial acylation rates are unobtainable, since this reaction is essentially complete before the mixing in the spectrophotometer cell is finished.

Use of Aza-Amino Acid and Aza-Peptide Derivatives

The table lists the aza-derivatives that have been allowed to react with serine proteases and the results obtained.[2–4,11,12,13] In most cases,

[11] J. C. Powers, N. Nishino, R. Boone, P. M. Tuhy, B. F. Gupton, and D. L. Carroll, to be published.

[12] S. A. Barber, C. J. Gray, J. C. Ireson, R. C. Parker, and J. V. McLaren, *Biochem. J.* **139**, 555 (1974)

[13] G. J. Gray and R. C. Parker, *Tetrahedron* **31**, 2940 (1975).

REACTION OF SERINE PROTEASES WITH AZA-AMINO ACID AND
AZA-PEPTIDE DERIVATIVES

Inhibitor	Enzyme	Comments	References[a]
Ac-Aphe-OEt	Chymotrypsin	Competitive inhibition	2
Ac-Aphe-OPh	Chymotrypsin	Stochiometric acylation, deacylation ($k_{deacyl} = 1.2 \times 10^{-4}$ sec^{-1})	12
Ac-Aphe-ONp	Chymotrypsin	Stochiometric (1:1) acylation, turnover ($k_{cat} < 1.8 \times 10^{-4}$ sec^{-1})	3, 11
	Subtilisin BPN'	Stochiometric (1:1) acylation, turnover ($k_{cat} < 1.8 \times 10^{-4}$ sec^{-1})	4, 11
	Trypsin	No reaction	4
	Porcine elastase	No reaction	4
	Human leukocyte elastase	No reaction	4
Ac-Aorn-OPh	Chymotrypsin	Slow inactivation	13
	Trypsin	Rapid acylation, deacylation ($k_{deacyl} = 2.1 \times 10^{-4}$ sec^{-1})	13
Ac-Ala-Aphe-ONp	Chymotrypsin	Stochiometric (1:1) acylation, turnover ($k_{cat} < 1.8 \times 10^{-4}$ sec^{-1})	4, 11
	Subtilisin BPN'	Stochiometric (1:1) acylation, turnover ($k_{cat} = 1.1 \times 10^{-2}$ sec^{-1})	4, 11
	Trypsin	No reaction	4
	Porcine elastase	No reaction	4
	Human leukocyte elastase	No reaction	4
Ac-Ala-Aala-ONp	Chymotrypsin	No reaction	4
	Subtilisin BPN'	Stochiometric (1:1) acylation, turnover ($k_{cat} = 8.8 \times 10^{-2}$ sec^{-1})	4, 11
	Trypsin	No reaction	4
	Porcine elastase	No reaction	4
	Human leukocyte elastase	No reaction	4
Ac-Ala-Ala-Aala-ONp	Chymotrypsin	Acylation ($k_{acyl} = 2.3 \times 10^{-2}$ sec^{-1})	11
	Porcine elastase	Stochiometric (1:1) acylation, turnover ($k_{cat} = 1.1 \times 10^{-5}$ sec^{-1})	11
Z-Ala-Ala-Pro-Aala-ONp	Chymotrypsin	Stochiometric (1:1) acylation, turnover ($k_{cat} < 1.8 \times 10^{-4}$ sec^{-1})	4, 11
	Subtilisin BPN'	Stochiometric (1:1) acylation, turnover ($k_{cat} = 1.7 \times 10^{-1}$ sec^{-1})	4, 11
	Trypsin	No reaction	4
	Porcine elastase	Stochiometric (1:1) acylation, turnover ($k_{cat} < 1.8 \times 10^{-4}$ sec^{-1})	4
	Human leukocyte elastase	Stochiometric (1:1) acylation, turnover ($k_{cat} < 1.8 \times 10^{-4}$ sec^{-1})	4

[a] Numbers refer to text footnotes.

reaction occurs only between an enzyme and those aza-peptides that are related to its normal substrates. For example, Ac-Ala-Aphe-ONp does not react with elastase, and Ac-Ala-Ala-Aala-ONp will react only slowly with chymotrypsin. These observations point to the importance of fulfilling the enzyme specificity requirements. This property is particularly important since it enables titration and/or inhibition of one enzyme in the presence of another. Ac-Ala-Ala-Aala-ONp, for example, may be used to titrate elastase in the presence of an equivalent concentration of chymotrypsin. Similarly, Ac-Ala-Aala-ONp may be used to titrate subtilisin BPN' and will not react with chymotrypsin or elastase.

These analogs may find further utility in the field of affinity chromatography. This method has already been used for the isolation of chymotrypsin by employing an aza-phenylalanine phenyl ester as a ligand.[12]

[18] Active Site-Directed Inhibition with Substrates Producing Carbonium Ions: Chymotrypsin

By EMIL H. WHITE, DAVID F. ROSWELL, IEVA R. POLITZER, and BRUCE R. BRANCHINI

Irreversible enzyme inhibition can result from compounds with little or no intrinsic derivatizing power but which carry "primed" functional groups. If at the active site this group can be uncoupled to yield an active species, derivatization can occur.[1,2] The first report of this type of specific inhibition was made by Bloch, who used the enzymic conversion of an acetylenic analog of the substrate to a reactive allenic form, which then rapidly alkylated the enzyme at the active site.

Our approach[3] involves the production of carbonium ions, as a result of enzyme action, at the active site of the enzyme; the carbonium ions are capable of alkylating the enzyme. Since N-nitroso amides have been shown to hydrolyze and produce carbonium ions,[4] as shown in Eq. (1)

$$R-\overset{\overset{\displaystyle O}{\|}}{\underset{\underset{\displaystyle N=O}{|}}{C}}-N-R' \rightarrow RCO_2H + \overset{\overset{\displaystyle N-R'}{}}{\underset{\underset{\displaystyle N-OH}{\|}}{}} \rightarrow R'-N_2^{\oplus} \rightarrow R'^{\oplus} + N_2 \qquad (1)$$

[1] K. Bloch, *Acc. Chem. Res.* **2**, 193 (1969).

[2] R. R. Rando, *Science* **185**, 320 (1974). See also this volume [3] and [12].

[3] E. H. White, D. F. Roswell, I. R. Politzer, and B. R. Branchini, *J. Am. Chem. Soc.* **97**, 2290 (1975)

[4] R. A. Moss, D. W. Reger, and E. M. Emery, *J. Am. Chem. Soc.* **92**, 1366 (1970); E. H. White and D. J. Woodcock, *in* "The Chemistry of the Amino Group" (S. Patai, ed.). Wiley, New York, 1968.

enzymatic catalysis of this hydrolysis by a protease such as chymotrypsin would result in the formation of carbonium ions at the active site [Eq. (2)].

$$R-\overset{\overset{\displaystyle O}{\|}}{C}-N-R' \xrightarrow[Enz]{} R-\overset{\overset{\displaystyle O}{\|}}{C}-O-(serine) \sim Enz + RN_2OH \rightarrow R^{\oplus} + N_2 \quad (2)$$
$$\underset{N=O}{|}$$

Carbonium ions derived from diazonium ions are very reactive species capable of alkylating any functional group on the enzyme.

The α-chymotrypsin catalyzed hydrolysis of the N-nitroso-N-benzyl amide of N'-isobutyrylphenylalanine (I) has been shown to result in active site-directed inhibition.[3]

(I)

Hydrolysis produced N-isobutyrlphenylalanine and a benzyl carbonium ion [Eq. (2)], which alkylates the enzyme.

Analogs of (I) in which the benzyl group was replaced by a methyl or ethyl group showed almost no inhibition although their hydrolysis is enzyme catalyzed. This lack of inhibition may result from the conversion of intermediate alkyl diazonium ions into diazoalkanes.

To prevent the enzymically produced carbonium ion from diffusing away from the active site and reacting with solvent or nonactive site located amino acids, cyclic analogs of (I) have been prepared. Compounds (II) and (III) have the ability to form carbonium ions while

(II)   (III)

still in the acyl enzyme form [Eq. (3)]. Thus alkylation of the enzyme can occur intramolecularly.

$$\text{(structure)} \xrightarrow{\text{Enz}} \text{(structure)} \quad O-\text{Enz} + N_2 + H_2O \quad (3)$$

The synthesis and properties of N-nitrosovalerolactam (II) and N-nitroso-1,4-dihydro-6,7-dimethoxy-3(2H)-isoquinolone (III) are reported here along with directions for their use in the active site-directed inhibition of α-chymotrypsin.

General Properties. Compounds (II) and (III) are crystalline solids that decompose on melting. They are reactive compounds, and we store them at $-20°$ in a dry atmosphere. They are quite soluble in organic solvents such as benzene, ether, chloroform, methylene chloride, and acetonitrile and are stable in these solvents for hours. Their gradual decomposition can be monitored by the intensity of the absorption band in the visible spectrum.

They decompose rapidly in alcohol or aqueous media. When acetonitrile solutions are added to buffer (Tris or phosphate) at pH 7.9 to give 10 mM to 10 nM solutions (9–10% in acetonitrile), the compounds have half-lives of 10–15 min. *Note:* These compounds may be carcinogenic based on their structural relationship to known carcinogens such as N-nitrosomethylurea, and precautions should be taken against exposure.

Synthesis: N-Nitrosovalerolactam (II). This compound is prepared according to the method of Huisgen.[5] δ-Valerolactam, 1.5 g (0.012 mole), is dissolved in 30 ml of diethyl ether, and the solution cooled in an ice bath. Nitrogen tetroxide gas is bubbled through until a green solution is obtained. The solution is extracted with ice water and with a 5% sodium bicarbonate solution and then dried over anhydrous sodium bicarbonate. The ether solution is decanted and cooled in an acetone–Dry Ice bath, and when crystallization occurs, the ether is removed by filtration. Recrystallization is repeated three times. The residue, after removal of solvent traces under reduced pressure, had a melting point of 42°–43° (lit[5] m.p. 45°–47°). It showed a single spot on thin-layer chromatography, $R_f = 0.7$ (silica gel, chloroform:acetonitrile 3:1); IR (CCL$_4$ 2900, 1750, 1520 cm^{-1}); ultraviolet (CH$_3$CN) 253 nm (log ϵ 3.83), 408 (sh) (1.65), 428(1.87), 450(1.88).

An alternative procedure involves dissolving 1.5 g of the lactam in 30 ml of chloroform and cooling this solution to $-10°$ in an ice/methanol bath. Anhydrous sodium acetate (7 g) is then added. The reaction mixture is stirred at $-10°$ for 1 hr while 700 ml of nitrogen tetroxide gas are

[5] R. Huisgen and J. Reinertshofer, *Justus Liebigs Ann. Chem.* **575**, 197 (1952).

added via a syringe. After warming to 0°, the mixture is filtered and the chloroform is removed under reduced pressure. Crystallization from ether is carried out as described above to give a 90% yield of pure product with the properties reported above.

1,4-Dihydro-6,7-dimethoxy-3(2H)-isoquinolone. 6,7-Dimethoxy-3-isochromanone,[6] 2.7 g (12.5 mmoles), is mixed with urea, 4.0 g (66.6 mmoles), in a 50-ml round-bottom flask and heated to 190°–200° for 0.5 hr. The reaction mixture is cooled, and 50 ml of water are added. The aqueous solution is extracted four times with chloroform (10 ml each), and the combined organic layers are washed with water and dried over sodium sulfate. The chloroform is removed under reduced pressure, and the resulting material is crystallized from 95% ethanol to yield 1.1 g (5.25 mmoles, 42%) of lactam in the form of white crystals, m.p. 198°–199.5° (lit[6] m.p. 201°–204°); IR (CHCl$_3$) 3420, 2930, 2960, 1675, 1650, 1660, 1618, and 1125 cm^{-1}; NMR (CDCl$_3$) δ 3.58 (2H, t), 3.93 (6H, s), 4.52 (2H, m,) 6.70 (2H, s); UV (CH$_3$OH) 284 nm (log ε 3.57) and 390 nm (sh) (1.03). This method proved to be superior to the literature method.

N-Nitroso-1,4-dihydro-6,7-dimethoxy-3(2H)-isoquinolone (III). 1,4-Dihydro-6,7-dimethoxy-3(2H)-isoquinolone, 0.2 g (0.96 mmole), is dissolved in methylene chloride (30 ml), and 1 g (12 mmoles) of anhydrous sodium acetate is added. The mixture is cooled in a Dry Ice–carbon tetrachloride bath at about −20°. Gaseous nitrogen tetroxide, 200 ml (10 mmoles) is added, and the reaction mixture is stirred for 15 min. The mixture is then washed once with ice water, three times with 1 N sodium bicarbonate, again with water, and then is dried over sodium sulfate. The volume of the dried organic layer is reduced to 5 ml and applied to a silica gel column (2 × 5 cm); the product is eluted with chloroform:acetonitrile (3:1, v/v). The fast-running yellow band is collected, the solvent is removed under reduced pressure, and the resulting material is recrystallized from methylene chloride–hexane to give 85 mg of bright yellow crystals (0.36 mmole, 37%), m.p. 124°–124.5° dec; IR (CHCl$_3$) 2910, 2840, 1730, 1600, 1450, and 995 cm^{-1}; UV (CH$_3$CN) 234 nm (log ε 4.05), 270 (sh) (3.69), 402 (sh) (1.85), 420 (2.03), and 442 (1.99); NMR (CDCl$_3$) δ 7.5 (2H, s), 5.2 (2H, s), 4.05 (2H, s), 4.15 (6H, d). The material may be stored at −20°.

Inhibition Studies. Compounds (II) and (III) can be dissolved in acetonitrile to give 0.1 to 1.0 mM solutions. Aliquots are withdrawn and added to solutions of α-chymotrypsin (10 to 50 μM) in 0.05 M sodium phosphate at pH 7.8 and room temperature (final concentration of CH$_3$CN ≤ 10%). During the incubation period, aliquots are withdrawn

---

[6] J. J. McCorkindale and A. W. McCullock, *Tetrahedron* **27**, 4653 (1971).

EFFECT OF INHIBITORS ON α-CHYMOTRYPSIN ACTIVITY

Compound	α-Chymo-trypsin $\times 10^{-5} M$	Inhib./α-Chy	Incubation time (min)	Percent inhibition
(II)	1.3	51	60	91
(II)	1.3	26	60	84
(III)	2.2	26	60	98
(III)	2.3	110	120[a]	94
(III)	38.0	8	10	75
(III)	38.0	8	60	50

[a] Incubated at 0° instead of room temperature.

and assayed for enzymic activity. In this case, the N-benzyltyrosine ethyl ester assay[7] was performed. Results from this procedure are shown in the table.

We have observed that at low ratios of inhibitor to chymotrypsin some reversible inhibition occurs. This may be due to alkylation of certain enzymes sites to yield groups that are labile to alkaline hydrolysis.

Substrates analogous to (I), (II), and (III) should function with trypsin, pepsin, papain, and other proteolytic enzymes. Furthermore, oxidases could be labeled by this technique with substrates such as hydrazine derivatives that yield diazonium salts on oxidation. Reductases and other enzymes could presumably be derivatized by other suitable substrates and active species. We have recently found that the D form of compound I effectively and fully inhibits chymotrypsin within 15 min when used at a 10/1 ratio to the enzyme.[8]

[7] B. C. W. Hummel, *Can. J. Biochem. Physiol.* **37**, 1393 (1959). This method is described *in* "Worthington Enzyme Manual," p. 129. Worthington Biochemical Corp., Freehold, New Jersey, 1972.
[8] *J. Am. Chem. Soc.*, submitted.

[19] Peptide Aldehydes: Potent Inhibitors of Serine and Cysteine Proteases

By ROBERT C. THOMPSON

The peptide acids generated by protease-catalyzed hydrolysis of proteins are generally good inhibitors of the enzymes that produce them. In the past few years it has become clear that derivatives of peptide acids in which the carboxylic acid group has been replaced by an aldehyde group are even more potent inhibitors of certain proteases than are the original acids. This phenomenon was first discovered by Umezawa and his colleagues while screening filtrates of actinomycetes for naturally occurring protease inhibitors.[1] It was independently discovered by Westerick and Wolfenden[2] and by Thompson[3] in a search for transition state analogs of cysteine and serine proteases. The latter workers have proposed that the peptide aldehydes bind to these proteases as enzyme-aldehyde hemithioacetals and hemiacetals, respectively. The enzyme-aldehyde adduct would then resemble in structure the transition state or high energy tetrahedral intermediate of the proteolytic reaction. It would therefore be stabilized by those forces that stabilize the transition state and confer on the enzyme its catalytic powers. The most important of these forces may be the stability of the covalent bond formed between the enzyme and the aldehyde.[4]

There is at present no direct evidence that the enzyme-aldehyde adduct has the proposed structure. However, indirect evidence that the adduct resembles the transition state for substrate hydrolysis comes from studies with elastase in which the relative binding energies of different peptide aldehydes were found to reflect k_{cat}/K_m and not the K_m for analogous substrates.[3] The proposed structure is also consistent with the finding that the enhanced binding due to the aldehyde group is more marked with the cysteine protease papain[2] than with any serine protease so far studied. This is in accord with the greater stability of hemithioacetals.[5]

Peptide aldehydes have been synthesized both by oxidation of peptide alcohols[1,3] and by reduction of peptide acids.[6] Both routes are restricted to peptides that do not contain other functional groups sensitive to the reducing and oxidizing reagents, and the oxidative route at least, subjects the whole peptide to a reaction which sometimes gives only poor yields.[1,3] To provide increased flexibility of synthesis of peptide aldehydes, I have explored another synthetic route, which involves the preparation from α-amino alcohols[7] of suitably protected α-amino aldehydes, the coupling of the protected aldehydes to preformed peptides, and the subsequent removal of protection from the aldehyde group. This type of synthesis has

[1] K. Kawamura, S. Kondo, K. Maeda, and H. Umezawa, *Chem. Pharm. Bull. (Tokyo)* **17**, 1902 (1969).
[2] J. Westerick and R. Wolfenden, *J. Biol. Chem.* **247**, 8195 (1972).
[3] R. C. Thompson, *Biochemistry* **12**, 47 (1973).
[4] R. C. Thompson, *Biochemistry* **13**, 5494 (1974).
[5] G. E. Lienhard and W. P. Jencks, *J. Am. Chem. Soc.* **88**, 3982 (1966).
[6] B. Shimizu, A. Saito, A. Ito, K. Tokawa, K. Maeda, and H. Umezawa, *J. Antibiot.* **25**, 515 (1972).
[7] S. Yamada, K. Koga, and H. Matsuo, *Chem. Pharm. Bull. (Tokyo)* **11**, 1140 (1963).

been used previously by Shimizu et al.,[6] who prepared the aldehyde group by reduction of a carbobenzoxy amino acid imidazolide and protected it as a semicarbazone during the coupling reaction. However, using the acidic conditions quoted by Shimizu et al., I have experienced difficulty in regenerating the aldehyde group from the peptide aldehyde semicarbazone without simultaneously degrading the peptide. I have therefore used the more acid-labile diethyl acetal of the aldehyde to mask the aldehyde function during the coupling reaction. The potential success of this method had been indicated previously by work of Westerick and Wolfenden,[2] who used the commercially available α-amino acetaldehyde dimethyl acetal.

The procedure described below has been used successfully to synthesize peptides containing phenylalaninal (Pheal), alaninal (Alaal), or glycinal (Glyal). It should be applicable to the preparation of peptides containing other α-amino aldehydes, including those with reactive side chains, provided that suitable acid-labile protecting groups are available for those side chains. For a list of acid-labile protecting groups for amino acids, the reader should consult a text or recent review of the synthesis of peptides, e.g., "Peptide Synthesis."[8]

N-Trifluoroacetyl L-Phenylalaninol

A mixture of 520 mg of L-phenylalaninol (Fluka AG, Switzerland) (3.4 mmoles) and 15 ml of 1 M aqueous sodium bicarbonate is stirred vigorously in an Erlenmeyer flask, and 2 ml of trifluoroacetyl ethanethiol (Pierce Chemical Co.) (2.6 g, 16 mmoles) are added. The flask is closed with a paper plug, and stirring is continued overnight. Nitrogen gas is blown through the reaction mixture to remove unreacted thiolester and thiol, and the solid product is removed by filtration. After drying under reduced pressure over phosphorus pentoxide, 750 mg of product (89% yield), m.p. 134°–136°, are obtained. The product is homogeneous by thin-layer chromatography (TLC) on silica gel, R_f 0.3 in chloroform: methanol (98:2). It is visualized by exposing the plate to chlorine vapor for 1 min, to air for 10 min, and then spraying with a 0.3% solution of tolidine in 1% aqueous potassium iodide 6% acetic acid.

N-Trifluoroacetyl L-Phenylalaninal

A mixture of 100 mg of N-trifluoroacetyl L-phenylalaninol (0.4 mmole), 500 mg of 1-ethyl(3,3-dimethylaminopropyl)carbodiimide hydrochloride

---

[8] M. Bodansky and M. A. Ondetti, "Peptide Synthesis." Wiley (Interscience), New York, 1966. Later work is summarized each year in "Specialist Periodical Reports. Amino Acids, Peptides, and Proteins." The Chemical Society, Letchworth, England.

(2.6 mmoles), and 2.0 ml of redistilled dimethyl sulfoxide is stirred for 10 min, and 200 μl of a 2 M solution of anhydrous phosphoric acid[9] in dimethyl sulfoxide are added. The mixture is stirred for 2.5 hr, then 5 ml of 1 M potassium phosphate at pH 7.5 and 5 ml of ethyl acetate are added, the mixture is shaken, and the layers are separated. The ethyl acetate solution is washed with an additional 3 ml of the phosphate buffer, dried over sodium sulfate, and evaporated. The residue is dissolved in 5 ml of ethanol and added to a solution of 4 g of sodium bisulfite in 7 ml of water while stirring vigorously. After standing for 10 min, the ethanol is removed by evaporation of the solution to approximately 7 ml, and the aqueous residue is extracted twice with 5 ml of ether. Solid sodium carbonate is added to the aqueous solution to bring the pH to between 8.5 and 9.0. The mixture is allowed to stand for 5 min and is then extracted twice with 5 ml of ether. The ether extracts from the basic solution are pooled, dried, and evaporated to yield 68 mg (69%) of product, homogeneous by TLC, R_f 0.5 on silica gel in chloroform:methanol(98:2).

The aldehyde is visualized either by spraying the plate with a solution of dinitrophenylhydrazine in phosphoric acid prepared according to Johnson,[10] or by the chlorine–tolidine reagent described above. A solution of the product in hexane kept at $-20°$ deposited 58 mg of product (m.p. 78°–80°) with the NMR spectrum anticipated for an N-acyl phenylalaninal.

We have found that 1-ethyl(3,3-dimethylaminopropyl)carbodiimide hydrochloride generally gives better yields of product than does another water-soluble carbodiimide, 1-cyclohexyl 3-morpholinoethyl carbodiimide hydrochloride. This is particularly true when the dimethyl sulfoxide is not redistilled before use. The use of dimethyl sulfoxide that has been redistilled under reduced pressure is recommended for all oxidation reactions. Prolongation of the oxidation reaction leads to a decreased yield of aldehyde product. After 24 hr, the yield of TFA Pheal is 36% before, and 27% after, crystallization from hexane.

L-*Phenylalaninal Diethylacetal*

N-Trifluoroacetyl L-phenylalaninal, 50 mg (0.20 mmole), is dissolved in 2 ml of ethanol. Then 100 μl of 1 M p-toluenesulfonic acid monohydrate in ethanol and 1 ml of triethylorthoformate are added and the mixture is allowed to stand for 30 min. The solvents are evaporated under high vacuum. The residue is dissolved in ethyl acetate and extracted twice with 5% aqueous sodium bicarbonate. The ethyl acetate solution

[9] R. E. Ferrell, H. S. Olcott, and H. Fraenkel-Conrat, *J. Am. Chem. Soc.* **70**, 2101 (1948).

[10] C. D. Johnson, *J. Am. Chem. Soc.* **73**, 5888 (1951).

is dried over sodium sulfate and evaporated. The residue is homogeneous by TLC on silica gel, R_f 0.8 in hexane:ether(4:1). The product is visualized either by the dinitrophenylhydrazine or chlorine–tolidine reagents described above. Unlike the parent aldehyde, the diethylacetal forms a yellow spot with dinitrophenylhydrazine solution only on heating the plate after spraying.

The residue is dissolved in 1.2 ml of methanol, and 1.6 ml of 0.5 M aqueous sodium hydroxide are added. After standing for 45 min, 5 ml of water are added, the solution is partially evaporated to remove the methanol, and is then extracted with ethyl acetate. The ethyl acetate solution is dried and evaporated to give 35 mg of an oil. TLC of the oil on silica gel showed the major product to have R_f 0.5 in chloroform:methanol(8:2). The product is visualized by spraying with a 0.3% solution of ninhydrin in acetone or with the dinitrophenylhydrazine reagent.

Occasionally a trace of a second product, R_f 0.4, was observed when the product was visualized with the ninhydrin reagent. The amount of this product increased with prolongation of the base hydrolysis.

Acetylprolylalanylprolylphenylalaninal

The α-amino aldehyde diethylacetal can be coupled to peptides using the mixed anhydride procedure of Anderson et al.[11] In the example given below there is no danger of racemization of the peptide during this process but, even were this possible, the procedure used would reduce the risk of racemization to an acceptable level.[11]

Acetylprolylalanylproline, 65 mg (0.2 mmole), and 22 μl of N-methylmorpholine (0.2 mmole) are dissolved in 2 ml of acetonitrile and the mixture is cooled to −20° in a bath of Dry Ice and carbon tetrachloride. Isobutyl chloroformate, 26 μl (0.2 mmole), is added with stirring, and 5 min later, the L-phenylalaninal diethylacetal prepared above is added as a solution in 0.5 ml of ethyl acetate. The stirred reaction mixture is allowed to warm to room temperature over a period of about 4 hr. The solvents are evaporated, and the residue is dissolved in 5 ml of water. The solution is then cooled to 0° and treated with REXYN I-300 ion exchange resin for 10 min and filtered. This procedure gives an aqueous solution of the peptide aldehyde diethylacetal which shows a single spot, R_f 0.9 on TLC on silica gel in chloroform:methanol(9:1). The product is visualized using the dinitrophenylhydrazine or chlorine–tolidine reagents. In the former case, it is necessary to heat the chromatography plate.

[11] G. W. Anderson, J. E. Zimmerman, and F. M. Callahan, J. Am. Chem. Soc. 89, 5012 (1967).

Dowex AG50W resin (H⁺ form), 100 mg, is added, and the solution is stirred for 2.5 hr at 25°. This results in complete conversion of the diethylacetal to peptide aldehyde. The product, 45 mg (47%), is isolated as an amorphous solid by evaporation of the aqueous solution and trituration under ether. It was homogeneous by TLC on silica gel, R_f 0.7 in chloroform:methanol (8:2). The product is visualized using the dinitrophenylhydrazine or chlorine-tolidine reagents.

Modifications Necessary to Prepare Aldehydes from
Water-Soluble Alcohols

A similar series of reactions has been used to prepare the peptide aldehyde AcProAlaProAlaal. However, in this case the intermediate, N-trifluoroacetyl alaninal (TFA Alaal), could not be separated from unreacted N-trifluoroacetyl alaninol (TFA Alaol) by purification of a water-soluble bisulfite addition compound since TFA Alaol is itself fairly soluble in water. Instead, the aldehyde is purified by direct conversion to the ether-soluble diethylacetal.

Potassium phosphate, 1 M at pH 7.5, 60 ml, is added to the carbodiimide-dimethyl sulfoxide reaction mixture, and the resulting solution is extracted first with 30 ml of ether and subsequently thrice with 60 ml each of ethyl acetate. The pooled ethyl acetate extracts are dried, evaporated, and treated with triethylorthoformate as described for the preparation of TFA Pheal diethylacetal. Any contaminating TFA Alaol is removed by extraction of an ethereal solution of the product with aqueous sodium bicarbonate. The procedure yields 48 mg (52%) of TFA Alaal diethylacetal which has the correct proton magnetic resonance spectrum and is homogeneous by TLC on silica gel, R_f 0.8 in chloroform. Crystallization from hexane at −20° gave 33 mg (36%) of white crystals.

The TFA Alaal diethylacetal is deprotected and coupled to acetylprolylalanylproline by the same method with approximately the same yields as those described from the synthesis of acetylprolylalanylprolylphenylalaninal.

[20] Carboxypeptidases A and B

By M. Sokolovsky

Carboxypeptidases A (EC 3.4.12.2) and B (EC 3.4.12.3) are pancreatic exopeptidases. Carboxypeptidase A preferentially hydrolyzes terminal peptide bonds in which the amino donor is a hydrophobic amino acid whereas the specificity of carboxypeptidase B is directed toward the

AFFINITY LABELING OF CARBOXYPEPTIDASES

Reagent	Enzyme[a]	K_I (M) of inhibitor or analog[b]	Site of modification	References[c]
N-Bromoacetyl-N-methyl-L-phenylalanine	CPA	—	Glu-270	5, 12
N-Bromoacetyl-N-methyl-L-phenylalanine	CPB	—	Glu-270	6
α-N-Bromoacetyl-D-arginine	CPB	5×10^{-4d}	Glu-270	1, 11
N-Bromoacetyl-p-aminobenzyl-L-succinic acid	CPB	6×10^{-7e}	Met	9
Bromoacetamidobutylguanidine	CPB	2×10^{-3d}	Tyr-248	14

[a] CPA, carboxypeptidase A; CPB, carboxypeptidase B.

[b] Values in this column refer to the inhibition constants of the inhibitors or substrate analogs without the reactive moiety.

[c] Numbers refer to text footnotes.

[d] N-Acetyl-D-arginine [E. C. Wolff, E. W. Schrimer, and J. E. Folk, *J. Biol. Chem.* **237**, 3094 (1962)].

[e] N-Acetyl-p-aminobenzylsuccinic acid (M. Sokolovsky, unpublished results).

terminal basic amino acids. The design of reagents for affinity labeling of these enzymes is based on their specificity, i.e., a free α-carboxylate group, a hydrophobic or basic side chain, and a reactive group (see the table). However, it must be kept in mind that carboxypeptidase B has, in addition to its known specificity toward basic substrates, an intrinsic activity with specificity similar to that of carboxypeptidase A.[1-4] This implies, in general, that a reagent designed for reaction with carboxypeptidase A may also be effective with carboxypeptidase B, whereas a nonbasic reagent used for carboxypeptidase B may also act with carboxypeptidase A. As shown in the table, this fact was used by Neurath's group[5,6] for the modification of both carboxypeptidases employing the same reagent. To achieve the required high degree of specificity of affinity labeling for the carboxypeptidases, it is very important that the reagent should react at a minimal number of alternate sites owing to the multiple

[1] M. Sokolovsky and N. Zisapel, in the Arieh Berger Memorial Issue of *Isr. J. Chem.* **12**, 631 (1974).

[2] J. W. Prahl and H. Neurath, *Biochemistry* **5**, 4137 (1966).

[3] E. Wintersberger, D. J. Cox, and H. Neurath, *Biochemistry* **1**, 1069 (1962).

[4] N. Zisapel and M. Sokolovsky, *Biochem. Biophys. Res. Commun.* **46**, 357 (1972).

[5] G. M. Hass and H. Neurath, *Biochemistry* **10**, 3535 (1971).

[6] G. M. Hass, M. A. Govier, D. T. Gahn, and H. Neurath, *Biochemistry* **11**, 3787 (1972).

modes of binding of substrates and inhibitors.[4,7,8] Furthermore, with the usual affinity labeling reagents, containing a moderately reactive chemical group, nonspecific labeling outside the active site might take place. Although individual nonspecific reactions may occur relatively slowly, the sum of these nonspecific reactions may be cumulatively rather large.

Synthetic Procedures

Bromoacetyl-p-aminobenzylsuccinic Acid.[9] Bromoacetic acid, 1.5 mmoles, and 1.5 mmoles of N-hydroxysuccinimide dissolved in 6 ml of dioxane are coupled with 1.5 mmoles of dicyclohexylcarbodiimide for 2 hr at 0°. The urea derivative is removed by filtration, and the active ester in the dioxane solution is coupled with 1.2 mmoles of p-aminobenzyl-succinic acid dissolved in 4 ml of 80% dioxane. The reaction mixture is kept at room temperature for 6 hr and then evaporated to dryness. The oily residue is dissolved in ethyl acetate. The organic phase is washed twice with H_2O, dried over Na_2SO_4, and evaporated to dryness. The compound crystallizes from ethyl acetate–petroleum ether; m.p. 126°.

α-N-Bromoacetyl-D-arginine. This reagent can be obtained by acylation of D-arginine[10] with the active ester as described above for the benzylsuccinate derivative or by coupling of D-arginine with bromo-acetylbromide[11]: D-Arginine·HBr (3 mmoles) is dissolved in 6 ml of ice-cold 0.1 N NaOH, and bromoacetylbromide (3 mmoles) is added slowly to the solution in the course of 20 min. The reaction mixture is stirred magnetically and maintained between pH 8.4 and 8.6 by addition of 5 N NaOH. After 30 min, the reaction mixture is adjusted to pH 3.0, and the solution is applied to a column (2.0 × 197 cm) of Sephadex G-10 equilibrated at 4° in 0.1 N acetic acid at a flow rate of 11 ml/cm² per hour. Effluent fractions (5 ml) are collected. The most retarded peak, which is the bromoacetyl-D-arginine, is collected and lyophilized. The material yields >98% of the theoretical quantity of arginine after acid hydrolysis.

N-Bromoacetyl-N-methyl-L-phenylalanine. This is prepared by acylating N-methyl-L-phenylalanine with bromoacetyl bromide in alkaline solution.[5] The product is crystallized from ethanol–water; m.p. 109°.

[7] N. Zisapel, N. Kurn-Abramowitz, and M. Sokolovsky, *Eur. J. Biochem.* **35**, 507 (1973).
[8] B. L. Vallee, J. F. Riordan, J. L. Bethune, T. L. Coombs, D. S. Auld, and M. Sokolovsky, *Biochemistry* **7**, 3547 (1968).
[9] N. Zisapel and M. Sokolovsky, *Biochem. Biophys. Res. Commun.* **58**, 957 (1974).
[10] M. Wilchek, unpublished results.
[11] T. H. Plummer, Jr., *J. Biol. Chem.* **246**, 2930 (1971); **247**, 7864 (1972).

Affinity Labeling of Bovine Carboxypeptidase A.[5,12] Bovine carboxypeptidase A (2 mg/ml in 2 N NaCl–0.2 M Tris-HCl, pH 7.5) is inactivated by incubation with an equal volume of 4 mM [^{14}C]bromoacetyl-N-methylphenylalanine at 25°. At suitable intervals samples of 2 ml are removed and the protein and small molecules are separated on a column (1.7 × 15 cm) of Sephadex G-25 equilibrated with 0.25 M NaCl–0.05 M Tris-chloride at pH 7.5. Reaction carried out for 24 hr in the dark yields a modified protein containing 2 moles of inhibitor per mole of carboxypeptidase A. If the reaction is performed in the presence of 0.05 M D-phenylalanine, only one inhibitor molecule is incorporated without loss of activity.

Applications: The Active Site of Carboxypeptidases

The residues modified in carboxypeptidases by the various affinity labels are given in the table. In most cases the enzymes are inactivated stoichiometrically with the incorporation of about 1 mole of reagent per mole of enzyme. With increasing reagent concentrations the rate of inactivation becomes maximal, indicating saturation kinetics, and implying the formation of a reagent–enzyme complex prior to inactivation, as required for active site-directed reagents. Further evidence for that requirement derives from the correlation observed between strength of binding of competitive inhibitors of carboxypeptidase A and carboxypeptidase B and their ability to protect the enzymes from inactivation by the various reagents employed. Bromoacetyl-N-methylphenylalanine alkylates the γ-carboxylate of Glu-270 at the active site of carboxypeptidase A and of carboxypeptidase B, forming a derivative in which the side chain of the N-methylphenylalanine moiety presumably occupies the binding pocket.[5,6,12] The γ-carboxylate of Glu-270 is believed to function as nucleophile in the hydrolysis of peptide and ester substrates.[13] Similarly, bromoacetyl-D-arginine[1,11] alkylated Glu-270 at the active site of carboxypeptidase B whereas 4-bromoacetamidobutylguanidine modifies the same enzyme at Tyr-248.[14] The two reagents, although similar in structure, do not bind in an identical manner and hence attack different residues. Alkylation of nitrocarboxypeptidase A and acetyl carboxypeptidase A with bromoacetyl-N-methylphenylalanine[15] indicated that Tyr-248 plays no role in the orientation of the affinity label but functions

[12] G. M. Hass and H. Neurath, *Biochemistry* **10**, 3541 (1971).
[13] W. N. Lipscomb, *Acc. Chem. Res.* **4**, 81 (1970).
[14] T. H. Plummer, Jr., *J. Biol. Chem.* **244**, 5246 (1969).
[15] G. M. Hass, B. Plikaytis, and H. Neurath, *FEBS Lett.* **46**, 162 (1974).

primarily in a catalytic role in the carboxypeptidase-catalyzed hydrolysis of peptide substrates.

Bromoacetyl-p-aminobenzyl succinic acid alkylates a methionyl residue in carboxypeptidase B.[9] As shown above, similar reagents modify amino acids other than methionyl residues, i.e., Tyr-248 and Glu-270. In those cases, the alkylating moiety was located in close proximity to the scissile bond of a normal substrate. Therefore, it would most likely interact with a residue that presumably functions in the catalytic step. The hydrophobic nature of bromoacetyl-p-aminobenzylsuccinate and the fact that the reactive side chain is carried on the aromatic part of the molecule suggests that the methionyl residue modified is part of the substrate recognition site in the hydrophobic pocket of carboxypeptidase B.[1,9,16] This possibility is reminiscent of the role of Met-192 in chymotrypsin, where it appears to function as a flexible hydrophobic lid on the substrate binding pocket.

[16] R. G. Reeck, K. A. Walsh, M. A. Hermodson, and H. Neurath, *Proc. Natl. Acad. Sci. U.S.A.* **68**, 1226 (1971).

[21] N-Substituted Arginine Chloromethyl Ketones

By B. KEIL

Although the early efforts to extend the chloromethyl ketones series to arginine derivatives met with chemical difficulties, two active site-directed inhibitors of trypsin and trypsinlike enzymes are now available: N^α-p-nitrobenzyloxycarbonyl-L-arginine chloromethyl ketone (p-NO$_2$-ZACK, Fig. 1a)[1] and N^α-tosyl-L-arginine chloromethyl ketone (TACK, Fig. 1b).[1] p-NO$_2$-ZACK is a highly efficient inhibitor of trypsin[2-4] and an extremely rapid inactivator of clostripain.[4] TACK is about three times more potent than TLCK[1] for the inactivation of trypsin, and it was found to be an efficient inhibitor of a trypsinlike enzyme from *Streptomyces erythreus*.[2]

[1] Abbreviations used: TACK, N^α-tosyl-L-arginine chloromethyl ketone (L-1-chloro-3-tosylamido-6-guanidinohexan-2-one); p-NO$_2$-ZACK, N^α-p-nitrobenzyloxycarbonyl-L-arginine chloromethyl ketone; TLCK, N^α-tosyl-L-lysine chloromethyl ketone; Tos- tosyl; Z-, benzyloxycarbonyl-; DTT, dithiothreitol; TLME, N^α-tosyl-L-lysine methyl ester; TAME, N^α-tosyl-L-arginine methyl ester.

[2] N. Yoshida, A. Sasaki, and K. Inouye, *Biochim. Biophys. Acta* **321**, 615 (1973).

[3] E. Shaw and G. Glover, *Arch. Biochem. Biophys.* **139**, 298 (1970).

[4] O. Siffert, I. Emöd, and B. Keil, *FEBS Lett.* **66**, 114 (1976).

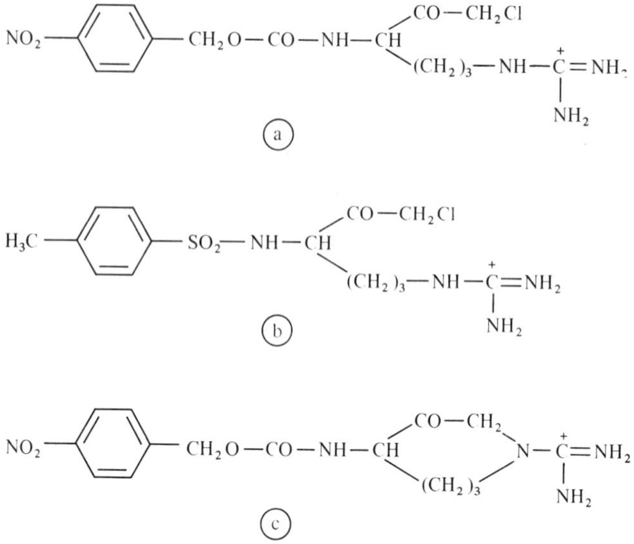

Fig. 1. Derivatives of arginine chloromethyl ketone: (a) p-NO$_2$-ZACK; (b) TACK; (c) inactive cyclic derivative of p-NO$_2$-ZACK.

Method of Preparation of p-NO$_2$-ZACK

Shaw and Glover[3] made the first attempt to synthesize an arginine chloromethyl ketone derivative for affinity labeling of the active sites of trypsinlike enzymes. Although the synthesis yielded a mixture containing less than 2% of the active chloromethyl ketone (Fig. 1a) and mainly an inactive cyclic product (Fig. 1c), they were able to show that this agent was a rapid inhibitor of trypsin.

By a modification of the synthesis and the introduction of additional purification steps Siffert et al.[4] prepared p-NO$_2$-ZACK in a pure form. Their procedure is described here.

$$p\text{-NO}_2\text{-Z-Arg-OH} \xrightarrow{\text{PCl}_5} p\text{-NO}_2\text{-Z-Arg-Cl} \xrightarrow{\text{CH}_2\text{N}_2} p\text{-NO}_2\text{-Z-Arg-CHN}_2 \xrightarrow{\text{HCl}}$$
$$p\text{-NO}_2\text{-Z-Arg-CH}_2\text{Cl}$$

α-(p-Nitrobenzyloxycarbonyl)arginine (706 mg) prepared according to Gish and Carpenter[5] is suspended in anhydrous tetrahydrofuran (25 ml), chilled in an ice–salt bath, and treated by adding phosphorus pentachloride (840 mg) in several portions with stirring. The mixture is stirred for 120 min during which time the temperature is allowed to rise to 0°. After addition of anhydrous ether (100 ml) the precipitate is

---

[5] D. T. Gish and F. H. Carpenter, *J. Am. Chem. Soc.* **75**, 5872 (1953).

filtered, washed with ether, and dried under reduced pressure over sodium hydroxide pellets at 4°. The yield of resulting α-(p-nitrobenzyloxycarbonyl)arginyl chloride hydrochloride is 700 mg (86%); m.p. = 120°; infrared spectrum (KBr) = 1785 cm^{-1}.

To the product dissolved in anhydrous tetrahydrofuran (15 ml) is added at 0° an excess of ethereal solution of diazomethane, until a yellow color persists. After an hour of stirring, 80 ml of anhydrous ether are added. The light yellow precipitate is filtered, washed several times with anhydrous ether, and dried under reduced pressure. The yield is 500 mg (77%) of the diazo derivative.

To 380 mg (1 mmole) of this product is added 1 N HCl in acetic acid (15.2 ml) at room temperature. After an hour, evolution of nitrogen ceases. The solvent is evaporated under reduced pressure at 35°, the last traces being removed with an oil pump.

The resulting crude p-NO$_2$-ZACK is purified consecutively by two procedures: after passage through a column of silica gel (Kieselgel, methanol–chloroform 9:1 v/v) the product is taken to dryness, dissolved in water, and freed from salts by passage through a column of Bio-Gel P-2 (400 mesh). Lyophilization of the eluate gives 150 mg of the desired p-NO$_2$-ZACK, m.p. 98–105° (38% yield). Thin-layer chromatography (Kieselgel GE 254, Merck; 1-butanol–acetic acid–water 18:2:5 by volume) followed by the Sakaguchi reaction reveals a single spot. Infrared spectrum (KBr) = 1739, 1700, 1675, 1520 cm^{-1}.

A ^{14}C-labeled reagent can be synthesized by this procedure using labeled diazomethane for the conversion of α-(p-nitrobenzyloxycarbonyl)arginyl chloride to the diazoderivative.[6]

Method of Preparation of TACK

Inouye *et al.*[7] developed two syntheses for TACK, starting with either N^G-nitro- or N^G-tosylarginine. The first variant, yielding pure and stable products, will be described here.

$$H\text{-}Arg(NO_2)\text{-}OH \xrightarrow{Tos\text{-}Cl} Tos\text{-}Arg(NO_2)\text{-}OH \xrightarrow{PCl_5} Tos\text{-}Arg(NO_2)\text{-}Cl \xrightarrow{CH_2N_2}$$

$$Tos\text{-}Arg(NO_2)\text{-}CHN_2 \xrightarrow{HF} Tos\text{-}Arg(NO_2)\text{-}CH_2Cl \xrightarrow{HCl} Tos\text{-}Arg\text{-}CH_2Cl$$

To a solution of N^G-nitroarginine (5.5 g) in 2 N NaOH (37.5 ml) is added sodium carbonate (2.65 g) and acetone (25 ml) followed by a dropwise addition of p-toluenesulfonyl chloride (7.15 g) in acetone

---

[6] O. Siffert, in preparation.

[7] K. Inouye, A. Sasaki, and N. Yoshida, *Bull. Chem. Soc. Jpn.* **47**, 202 (1974).

(20 ml) at 0°. After stirring at 0° for 3.5 hr and subsequent removal of acetone under reduced pressure, the aqueous solution is acidified with 4 N HCl. The crystalline precipitate is filtered, washed with water, and dried under reduced pressure. Pure tosylnitroarginine is obtained after recrystallization from methanol–water (7.39 g, 79%), m.p. 167°–169.°C.

To the chilled solution of 750 mg of tosylnitroarginine in anhydrous tetrahydrofuran (10 ml) in an ice–salt bath is added phosphorus pentachloride (840 mg) in several portions. The temperature is allowed to rise to 0° during an hour of stirring. Addition of 80 ml of anhydrous ether gives a syrupy precipitate, which, after decantation of the supernatant fluid and trituration with ether, is dried under reduced pressure over sodium hydroxide pellets at 4°. The resulting amorphous tosylnitroarginyl chloride (470 mg, 60%) is dissolved in anhydrous tetrahydrofuran (10 ml) and an excess of an ethereal solution of diazomethane is added at 0° until the yellow color persists. After 1 hr of stirring at 0°, the diazo derivative is precipitated by addition of anhydrous ether (80 ml), filtered, washed with ether, and dried under reduced pressure (360 mg, 45%).

To 250 mg of the diazo derivative is added 1 N HCl in acetic acid (10 ml) at room temperature. After an hour the evolution of nitrogen ceases. The solvent is evaporated under reduced pressure at 35° and, after addition of acetic acid, the product is lyophilized.

Crude tosylnitroarginine chloromethyl ketone is purified by passage through a column of silica gel (25 g, mesh 0.05–0.2 mm; Merck) in methanol–chloroform (1:9 by volume). Fractions containing the desired product (as followed by thin-layer chromatography) give, after evaporation of the solvent, a syrup that crystallizes upon addition of chloroform (120 mg, 49%). Pure nitroderivative is obtained after recrystallization from methanol (86% yield).

This intermediate, 180 mg, and 0.2 ml of anisole are placed in a reaction vessel made of fluorinated polyethylene and, after chilling the mixture in a Dry Ice–acetone bath, about 5 ml of hydrogen fluoride are introduced. The mixture is stirred at 0° for 30 min, and hydrogen fluoride is evaporated under reduced pressure. The residue is dissolved in water, washed with ethyl acetate, and passed through a column (0.9 × 10 cm) of Amberlite CG-400 in acetate form with subsequent elution by water. To the combined eluates is added 0.5 ml of 1 N HCl; subsequent lyophilization yields 180 mg of the desired N^α-tosylarginine chloromethyl ketone (98%). Infrared spectrum (KBr) = 1737 cm^{-1}, $[\alpha]_D^{25}$ = $-21.2°$ ± 0.6° (c 1.0, 0.1 N HCl). Thin-layer chromatography (cellulose F, Merck) in 1-butanol–acetic acid–water (18:2:5 v/v/v) followed by the Sakaguchi reaction reveals a single spot.

Affinity Labeling of Trypsin and Trypsinlike Enzymes

Inhibition of β-Trypsin by p-NO₂-ZACK and TACK

In a preliminary study on the inactivation of β-trypsin by a preparation of 2% p-NO$_2$-ZACK and 98% of the inactive cyclic compound (Fig. 1c), Shaw and Glover[3] were able to show the high efficiency of the inhibitor. They estimated that pure p-NO$_2$-ZACK should be orders of magnitude more rapid than the lysine derivative, TLCK. From their data it could be deduced that the impure inhibitor preparation totally eliminates the activity of trypsin within 10 min at pH 7 and 25° at an inhibitor–enzyme ratio of 50:1. They have also shown that a histidine residue is replaced by a 3-carboxymethylhistidine residue in the total hydrolyzate of the inactivated trypsin, which proves the active site-oriented action of the inhibitor.

Siffert *et al.*[6] studied the inhibition of β-trypsin by pure p-NO$_2$-ZACK. The enzyme concentration was 10 μM, inhibitor concentrations varied between 6 and $2.5 \times 10^{-4} = 0.25$ mM, and the pH was maintained at 7.0 at 25°. Within 10 min, activity was completely lost at an inhibitor–enzyme ratio of 17:1. The study of the saturation kinetics of inactivation showed that the line could be drawn through the origin, indicating that essentially no reversible intermediary complexes were formed.

Yoshida *et al.*[2] studied the trypsin–TACK system and found values of $K_i = 9$ mM and $K_3 = 0.2$ min^{-1}, i.e., TACK is 3.4 times more rapid than TLCK in inactivating trypsin. On the other hand, it is much less effective than p-NO$_2$-ZACK. The authors[2,3] agree, however, that the increased effectiveness of p-NO$_2$-ZACK over TLCK may be due, in part, to the difference of side chains (Tos- and p-NO$_2$-Z-), not only to the advantage of arginine over lysine. The same effect has been observed in the case of inhibition of chymotrypsin and subtilisin by the phenylalanine chloromethyl ketone derivative.[8,9]

Inhibition of the Serine Protease from Streptomyces erythreus by TACK

Streptomyces erythreus serine protease has a substrate specificity similar to that of bovine trypsin. A comparative study of its inactivation by TACK and TLCK made by Yoshida *et al.*[2] has shown that the rates of inactivation of the protease by the two inhibitors are not different, in contrast with the higher efficiency of TACK toward trypsin (see the table).

---

[8] E. Shaw and J. Ruscica, *J. Biol. Chem.* **243**, 6312 (1968).
[9] E. Shaw and J. Ruscica, *Arch. Biochem. Biophys.* **145**, 484 (1971).

KINETIC DATA FOR THE SUBSTRATES AND IRREVERSIBLE INHIBITORS OF
Streptomyces erythreus PROTEASE AND TRYPSIN[a]

	Substrate $K_m \cdot 10^{-4}$		Inhibitor			
			$K_i \cdot 10^{-4} M$		$k_3 \cdot min^{-1}$	
Enzyme	TLME	TAME	TLCK	TACK	TLCK	TACK
S. erythreus protease	0.6	0.2	0.7	0.5	0.14	0.11
Trypsin	1.0	0.1	2.5	0.9	0.16	0.22

[a] From N. Yoshida, A. Sasaki, and K. Inouye, *Biochim. Biophys. Acta* **321**, 615 (1973).

Inhibition of Clostripain by p-NO₂-ZACK

Clostripain (EC 3.4.22.8) from the culture filtrate of *Clostridium histolyticum* is a thiol proteinase with a highly limited specificity directed at the carboxyl bond of arginyl residues in proteins and in synthetic substrates. Porter *et al.*[10] have described the inactivation of clostripain by TLCK. Their enzyme preparation (specific activity 1.1 μkat/mg) incorporated approximately 4 moles of TLCK per mole of enzyme of molecular weight 50,000 instead of the expected 1:1 molar ratio. Because they found by independent means that the enzyme was homogeneous, they concluded that a large fraction of the enzyme was in an inactive state.

Recently Siffert *et al.*[4] obtained complete inactivation of clostripain by pure p-NO₂-ZACK. The enzyme they used was four times more active (4.5 μkat/mg) than that of Porter. It was activated for 2 hr at room temperature prior to the assay at a concentration of 20 μM in 50 mM Tris chloride at pH 7.4, containing 2.5 mM DTT and 50 mM CaCl₂. In inhibition experiments at an initial enzyme concentration of 10 μM, the rate of inactivation is extremely rapid: a 4M excess of the reagent removes the activity completely in less than 2 min at room temperature.

Evidence that the reaction is oriented in both cases to the active site of clostripain is given by the results of experiments effected in the presence of a competitive inhibitor. Benzamidine ($K_i = 42.9$ mM) is effective in protecting the active site of clostripain from reaction with both TLCK[10] and p-NO₂-ZACK.[4]

At present, there is no direct experimental evidence as to which residue in clostripain is substituted by the action of the chloromethyl ketones.

[10] W. H. Porter, L. W. Cunningham, and W. M. Mitchell, *J. Biol. Chem.* **246**, 7675 (1971).

Another question that remains open is the number of inhibitor residues incorporated in the enzyme. Although Siffert et al.[4] used a different reagent from that of Porter et al.,[10] and a highly active enzyme, the results of the two studies indicate a molar ratio of 4 moles of the chloromethyl ketone per mole of enzyme.

Evaluation

Enzymes cleaving polypeptide substrates at the carboxyl group of arginine residues can be irreversibly inhibited by two arginine chloromethyl ketones, p-NO$_2$-ZACK and TACK. The reaction is specifically oriented to their active sites.

Both inhibitors can be prepared in pure and stable form. In all cases studied, p-NO$_2$-ZACK was found more efficient than TACK. The highest rate of inactivation was observed with clostripain.

[22] Renin

By TADASHI INAGAMI and YASUNOBU SUKETA

Renin (EC 3.4.99.19), which plays a key role in renovascular hypertension and blood pressure regulation by producing angiotensin I, is a protease with an extremely limited substrate specificity: it will cleave only the unique leucylleucine peptide bond of angiotensinogen and the tetradecapeptide renin substrate H-Asp-Arg-Val-Tyr-Ile-His-Pro-Phe-His-*Leu-Leu*-Val-Tyr-Ser-OH.[1] This unique substrate specificity is reflected in stringent structural requirements for an extensive segment of substrate molecules.[2]

Results of competitive inhibition experiments using various peptides possessing partial structures of the substrate indicate that only the C-terminal portion, Leu-Leu-Val-Tyr, of the tetradecapeptide substrate contributes significantly to the binding affinity of the substrate to the enzyme.[3,4]

α-Bromoisocaproic acid has a hydrocarbon structure identical with that of leucine, its α-bromine atom substituting for the α-amino group

[1] L. T. Skeggs, J. R. Kahn, K. E. Lentz, and N. P. Shumway, *J. Exp. Med.* **106**, 439 (1957).

[2] L. T. Skeggs, K. E. Lentz, J. R. Kahn, and H. Hochstrasser, *J. Exp. Med.* **128**, 13 (1968).

[3] K. Poulsen, J. R. Burton, and E. Haber, *Biochemistry* **14**, 3892 (1975).

[4] T. Kokubu, E. Ueda, S. Fujimoto, K. Hiwada, A. Kato, H. Akutsu, and Y. Yamamura, *Nature (London)* **217**, 456 (1968).

INACTIVATION OF MOUSE SUBMAXILLARY GLAND RENIN A BY
α-BROMOIOSCAPROYL PEPTIDES

	Pseudo-first-order rate constant of inactivation ($k \times 10^2$, hr^{-1})	
Inactivators[a]	pH 5.6	pH 4.0
L-BIC-Leu-Val-Tyr-Ser-OH	8.3	28
D-BIC-Leu-Val-Tyr-Ser-OH	1.5	5.2
L-BIC-Val-Tyr-Ser-OH	8.2	25
D-BIC-Val-Tyr-Ser-OH	3.2	4.9
L-BIC-Leu-Val-OCH$_3$	11	—
L-BIC-Leu-Val-OH	3.7	—
L-BIC-Val-Tyr-NH$_2$	20	22
D-BIC-Val-Tyr-NH$_2$	0.9	1.0
L-BIC-Leu-OCH$_3$	9.5	9.7

[a] BIC, α-bromoisocaproyl.

of the leucine residue. The bromine atom should have an elevated reactivity due to the adjacent carboxyl group. Since the leucyl-leucine structure should be bound at the catalytically active region of the enzyme, oligopeptides (see the table for structure) with an α-bromoisocaproyl (BIC) group substituting for one of the two leucine residues were prepared as active site-directed inactivators of renin.[5]

D- *and* L-α-*Bromoisocaproic Acid.* The α-bromo-substituted isocaproic acids can be prepared by a slight modification of the method of E. Fischer.[6] To an ice-cold suspension of L-(or D-) leucine (5.0 g) in 15 ml of cold 48% HBr (Fisher Scientific Co.) is added 1 ml of Br$_2$ with stirring. Nitric oxide (NO) gas[7] is passed through the solution for 1 hr. After another addition of Br$_2$ (1 ml) the NO bubbling is continued for an additional hour. The mixture is extracted with ethyl ether, and the ether layer is washed successively with 1% sulfuric acid, 10% sodium carbonate, and water; the solution is dried over anhydrous sodium sulfate overnight. After evaporation of the ether, α-bromoisocaproic acid is obtained by distillation under reduced pressure (boiling point 160° ∼ 162° at 6.7 mm Hg) as a colorless liquid with $[\alpha]_D^{20} = -6.10°$ (c 5.06, methanol) for the D enantiomer. The acid chloride of α-bromoisocaproic acid is prepared by directly treating the acid (200 mg) with 2 ml of thionyl chloride. After allowing reaction for 2 hr at room temperature, thionyl chloride is evaporated under reduced pressure with a rotary evaporator.

[5] Y. Suketa and T. Inagami, *Biochemistry* **14**, 3188 (1975).
[6] E. Fischer, *Chem. Ber.* **39**, 2893 (1906).
[7] Toxic! Use a fume hood and trap the waste gas in alkali-permanganate solution.

Traces of thionyl chloride are removed by reevaporation after addition of dry benzene. The purity of the peptides can be checked by thin-layer chromatography on silica gel using two solvent systems: (1) 1-butanol–acetic acid–water, 18:2:5 (v/v) ; and (2) toluene–pyridine–water, 80:10:1 (v/v). All preparations should give a single spot.

α-Bromoisocaproyl (BIC) Peptides. Peptides, listed in the table, are prepared by dropwise addition of the acid chloride (1.5 mmoles) of α-bromoisocaproic acid to ice-cooled solutions of peptides or amino acids (1.5 mmoles in 2 ml). The addition continues over a period of 1.5 hr, the pH being held between 10 and 11 with the use of 1 N NaOH. After an additional 30 min of reaction, the pH is lowered to 2.5–3.0 with ice-cold 1 N HCl. The product is extracted from the solution with ethyl acetate. The ethyl acetate is evaporated under reduced pressure, and the residue is recrystallized from methanol–petroleum ether. Methyl esters of BIC peptides are prepared by esterifying parent BIC peptides in methanol–thionyl chloride.

Reaction of Submaxillary Renin and Bromoisocaproyl Peptides. To 160 μl of 0.05 M sodium pyrophosphate at pH 5.6 in a glass-stoppered centrifuge tube are added 20 μl of a 5 mM solution of BIC peptide in ethylene glycol monomethyl ether. This solvent does not cause irreversible inactivation of mouse submaxillary gland renin[8,9] whereas other organic solvents, e.g., 1-propanol and dioxane, resulted in essentially complete loss of enzyme activity under the same conditions for 20 hr. After equilibration at 37°, the reaction is started by the addition of 20 μl of a 0.1% solution of mouse submaxillary renin in the same buffer that also contains 0.1 M NaCl. For reactions at pH 4, 0.05 M sodium acetate may be used instead of the pyrophosphate buffer.

Renin Activity Determination. With crude renin preparations it is best to use a specific radioimmunoassay determination of angiotensin I produced by renin.[10] However, for the modification of pure renin, the fluorometric method of Roth and Reinharz[11] or the radioisotope method of Gregerman[12] will be convenient. In the fluorometric method, the renin containing solution (10 μl) from the above reaction mixture is mixed with 230 μl of 0.05 M sodium pyrophosphate (pH 5.6), and the reaction

[8] S. Cohen, J. M. Taylor, K. Murakami, A. M. Michelakis, and T. Inagami, *Biochemistry* **11**, 4286 (1972).
[9] Stability of renin also depends on protein (renin) concentration. At a much lower level of concentration, inactivation will take place even in this solvent system.
[10] E. Haber, T. Koerner, L. B. Page, B. Kliman, and A. Purnode, *J. Clin. Endocrinol. Metab.* **29**, 1349 (1969).
[11] A. Reinharz and M. Roth, *Eur. J. Biochem.* **7**, 334 (1966).
[12] N. M. Bath and R. J. Gregerman, *Biochemistry* **11**, 2845 (1972).

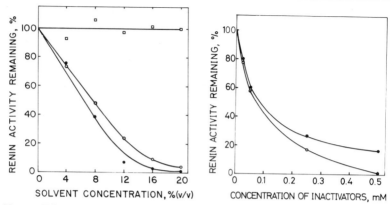

FIG. 1. Effect of the concentration of α-bromoisocaproyl (BIC) peptides on the inactivation of submaxillary renin. Mouse submaxillary renin A (0.1 mg in 1 ml) was incubated with L-BIC-Val-TyrNH₂ (○) or L-BIC-Leu-Val-Tyr-Ser-OH (●) in 0.04 M sodium pyrophosphate (pH 5.6) at 37° for 10 hr. The inactivators were added as solution in ethylene glycol monoethyl ether. The concentration of the organic solvent was 10% (v/v) throughout this experiment.

is started by addition of 10 μl of substrate (the fluorogenic octapeptide renin stubtrate,[13] Bachem, Inc., Marina Del Ray, California) solution in dimethylformamide (1 mg/ml). After incubation at 37° for 1 hr the reaction is stopped by heating in a boiling water bath for 10 min. The β-naphthylamine is released by a second reaction at 37° for 2 hr after the addition of 10 μl of 0.1 M zinc acetate, 10 μl of $2M$ Tris chloride at pH 8.6, and 10 units of aminopeptidase M (Sigma Chemical Co.). The concentration of β-naphthylamine released is determined by spectrofluorometry with excitation at 340 nm and emission at 410 nm using a β-naphthylamine solution (1 μg per 1 ml of water) as a standard.

The rate of inactivation is strongly dependent on the concentration of BIC-peptides as shown in Fig. 1. In 10% ethylene glycol monomethyl ether the least soluble peptide, BIC-Val-Tyr-NH₂, can be maintained at a concentration of 0.5 mM peptides. The rate of reaction is also strongly dependent on pH as shown in Figs. 2 and 3. A pH below 4 seems to favor the inactivation reaction. However, renin tends to denature rapidly below pH 4. Thus for the mouse submaxillary gland renin, pH 4.0 seems to be the best compromise.

Among the inactivators listed in the table, the best seems to be L-BIC-Val-Tyr-NH₂, which is uniformly efficient throughout the entire pH range (Fig. 3). However, its solubility is the lowest among the inhibitors

[13] M. Roth and A. Reinharz, *Helv. Chim. Acta* **99**, 1903 (1966).

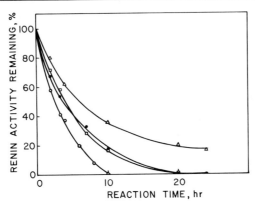

FIG. 2. Time course of the inactivation of submaxillary renin by L-BIC-Leu-Val-Ser-OH at different pH. Inactivation reactions were carried out in a manner similar to that described for Fig. 1 at 37° and with 0.5 mM of the inactivator. Aliquots of reaction mixtures were assayed directly using a fluorogenic octapeptide substrate. Buffers (0.04 M) used included acetate at pH 4 (○), pyrophosphate at pH 5 (●) and pH 6 (△), and phosphate at pH 7 (□). At pH 4 and 7 additional 0.04 M pyrophosphate was added for the stabilization of renin.

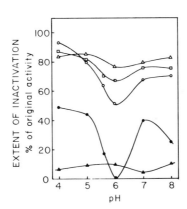

FIG. 3. Extent of inactivation of submaxillary renin by different L- and D-bromo-isocaproyl peptides as a function of pH. Inactivators used are L-BIC-Leu-Val-Tyr-Ser-OH (○), L-BIC-Val-Tyr-Ser-OH (□), L-BIC-Val-Tyr-NH₂ (△), D-BIC-Leu-Val-Tyr-Ser-OH (●), and D-BIC-Val-Tyr-NH₂ (▲). Reaction conditions and buffers are identical with those given for Fig. 2. Sodium pyrophosphate (40 mM) was used for pH 8. Inactivators were added as solutions in ethylene glycol monomethyl ether to a final concentration of 0.5 mM; the final ethylene glycol monomethyl ether concentration was 10% (v/v). After 10.5 hr at 37° and the noted pH, the inacti-vators were dialyzed out immediately against the same buffer containing the 10% solvent in the cold.

synthesized. L-BIC-Leu-Val-Tyr-Ser-OH and L-BIB-Val-Tyr-Ser-OH when used at pH 4.0 or at a slightly higher pH appear to be satisfactory. Since these compounds seem to be somewhat more soluble than L-BIC-Val-Tyr-NH₂, it may also be possible to reduce the concentration of ethylene glycol monomethyl ether.

Renin can be inactivated more readily by aliphatic diazo compounds such as diazoacetyl-DL-norleucine methyl ester or diazoacetylglycine methyl ester in the presence of cupric ion.[14,15] These reagents are soluble in water and hence do not require organic solvents. However, the reaction is less selective, since these latter reagents can react with other acid proteases.

[14] T. Inagami, K. Misono, and A. M. Michelakis, *Biochem. Biophys. Res. Commun.* **56,** 503 (1974).
[15] M. M. McKown and R. J. Gregerman, *Life Sci.* **16,** 71 (1975).

[23] Adenosine Derivatives for Dehydrogenases and Kinases

By ROBERTA F. COLMAN, PRANAB K. PAL, and JAMES L. WYATT

Among the coenzymes and regulatory compounds involved in biochemical reactions, adenine nucleotides are present in a large proportion of cases. Adenosine triphosphate and the corresponding diphosphate are required for many kinase reactions, in which a phosphoryl group is transferred to a variety of substrates; and adenosine is part of the coenzymes, DPN and TPN, that participate in most dehydrogenase reactions. In addition, these compounds are important in activating or inhibiting the activity of a number of regulatory enzymes. In trying to identify by means of chemical modification the amino acid residues in the region of the protein binding sites of the adenine nucleotides, affinity labeling is a particularly suitable approach since such amino acid residues are not necessarily expected to be particularly reactive. Therefore specific modification can most reasonably be accomplished by a reagent that simulates the structure of the naturally occurring adenine nucleotide and thereby forms a reversible enzyme–inhibitor complex at a purine nucleotide site; a functional group of the reagent, which should be capable of reacting broadly with different types of amino acids, may then form a covalent bond within the binding site during the existence of this complex. A few scattered reports have appeared in which purine derivatives have been used to modify enzymes catalyzing reactions in which purines are sub-

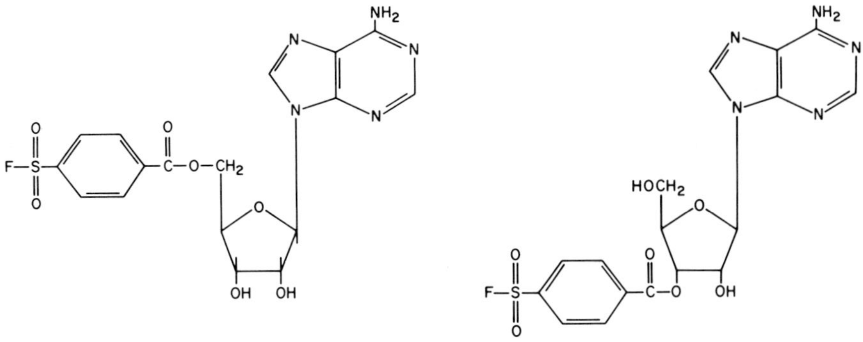

FIG. 1. Structures of 5'-p-fluorosulfonylbenzoyladenosine and 3'-p-fluorosulfonyl-benzoyladenosine.

strates. Examples of these include the use of 6-(purine 5'-ribonucleotide)-5-(2-nitrobenzoic acid)-thioether to label the AMP site of phosphorylase b,[1] and the application of the photolabile compounds diazomalonyl cAMP,[2] and 8-azidoadenosine 3',5'-monophosphate[3] to label the nucleotide sites of a cAMP binding protein of erythrocytes and of a partially purified brain protein kinase, respectively. However, the first of these reagents[1] contains a bulky functional group directly linked to the purine ring; such a bulky group may either interfere with the binding of the compound or necessitate its binding at a different location on the enzyme from the usual adenine site. Its covalent reactions may be limited to those involving cysteine residues. Photoaffinity labels, while exhibiting broad reactivity, also have certain inherent difficulties, since upon irradiation the label will frequently tend to react with the solvent as well as with any amino acid adjacent to the compound. The reagent may therefore react incompletely with several amino acid residues, making it difficult to ascertain which residues are actually involved in binding the purine nucleotide.

Figure 1 shows the structures of two synthetic adenosine analogs which contain alkylating agents capable of reacting covalently with proteins. The 5'-p-fluorosulfonylbenzoyladenosine (5'-FSBA) may reasonably be considered as an analog of ADP, ATP, or DPN(H). In addition to the adenine and ribose moieties, it has a carbonyl group at the 5'-

[1] F. W. Hulla and H. Fasold, *Biochemistry* **11**, 1056 (1972).

[2] C. C. Guthrow, H. Rasmussen, D. J. Brunswick, and B. S. Cooperman, *Proc. Natl. Acad. Sci. U.S.A.* **70**, 3344 (1973).

[3] A. H. Pomerantz, S. A. Rudolph, B. E. Haley, and P. Greengard, *Biochemistry* **14**, 3858 (1975).

position that is structurally similar to the first phosphoryl group of the naturally occurring purine nucleotides. If the molecule is arranged in an extended conformation, the sulfonyl fluoride moiety may be located in a position analogous to the terminal phosphate of ATP or to the ribose proximal to the nicotinamide ring of DPNH. This sulfonyl fluoride moiety can act as an electrophilic agent in covalent reactions with several classes of amino acids, including serine, tyrosine, lysine, and histidine.[4] Thus, it might reasonably be expected that 5'-FSBA would react with amino acid residues directly within adenine nucleotide binding sites in proteins. In the case of several proteins, including glutamate dehydrogenase and pyruvate kinase, it has been found to produce stoichiometric labeling of specific sites.

In contrast, the 3'-p-fluorosulfonylbenzoyladenosine (Fig. 1) is not as close an analog of ATP, ADP, and DPN(H) as is 5'-FSBA. The 3'-hydroxyl group of ribose is generally free in purine nucleotides that bind to the substrate and regulatory sites of enzymes and thus, for those enzymes that can tolerate the added bulk in this region, 3'-FSBA may function in accordance with Baker's[5] definition of an exo-alkylating agent: one that participates in covalent bond formation with an enzymic nucleophilic group located in a position immediately adjacent to if not actually within the purine nucleotide binding site. It is thus anticipated that 3'-FSBA may prove complementary to 5'-FSBA as a general affinity label for adenine nucleotide sites of dehydrogenases and kinases.

Preparation of Adenosine Analogs

5'-p-Fluorosulfonylbenzoyladenosine. The 5'-p-fluorosulfonylbenzoyladenosine is most readily prepared by reaction of adenosine with p-fluorosulfonylbenzoyl chloride by a modification of the procedure of Pal, Wechter, and Colman[6]: Adenosine (1.13 g, 4.2 mmoles) is dissolved in 10 ml of hexamethyl phosphoric triamide by warming to 50° in a water bath. Upon cooling, p-fluorosulfonylbenzoyl chloride (1.32 g, 6 mmoles) is added, and the mixture is allowed to stand at room temperature for 18 hr. The reaction mixture is extracted once with 30 ml of petroleum ether. After separation of the two layers, the product is precipitated from the lower layer by the slow addition of 40 ml of ethyl acetate:diethyl ether (1:1). The product is collected by filtration and air dried. This procedure yields between 1.6 (2.5 mmoles) and 2.0 g (3.1 mmoles) of

---

[4] T. L. Paulos and P. A. Price, *J. Biol. Chem.* **249**, 1453 (1974).

[5] B. R. Baker, "Design of Active-Site-Directed Irreversible Enzyme Inhibitors." Wiley, New York, 1967. See also this volume [9].

[6] P. K. Pal, W. J. Wechter, and R. F. Colman, *J. Biol. Chem.* **250**, 8140 (1975).

product with a melting point of 149°–150°. Elemental analysis of the compound revealed the presence of 1 mole of hexamethyl phosphoric triamide per mole of 5'-p-fluorosulfonylbenzoyl adenosine. The ultraviolet absorption spectrum exhibits maxima at 259 nm ($\epsilon = 1.35 \times 10^4$ cm^{-1} M^{-1}) and 232 nm ($\epsilon = 1.88 \times 10^4$ cm^{-1} M^{-1}) when measured in ethanol. Radioactive 5'-FSBA can be prepared by the addition of either 8-[^{14}C]adenosine or 2-[^{3}H]adenosine to the nonradioactive adenosine; reaction with p-fluorosulfonylbenzoyl chloride is then conducted as described above. The radioactive 8-[^{3}H]adenosine cannot be used for this preparation since tritium exchanges from the 8-position of the purine ring under the basic conditions used in this synthesis.

It is possible (although not necessary) to recrystallize 5'-FSBA from dimethylformamide, as described by Pal et al.[6] The resultant product contains 1 mole of dimethylformamide in the crystal and exhibits a melting point of 159°–160°. This form of the compound exhibits the same features for the ultraviolet absorption spectrum, but with somewhat different extinction coefficients ($\epsilon_{259\text{ nm}} = 1.58 \times 10^4$ cm^{-1} M^{-1}; $\epsilon_{232\text{ nm}} = 2.17 \times 10^4$ cm^{-1} M^{-1}).

The purity of the 5'-p-fluorosulfonylbenzoyl adenosine can be assessed by thin-layer or descending paper chromatography. With EM silica gel F-254 (fast running) thin-layer plates on aluminum and a solvent system composed of methyl ethyl ketone:acetone:water (60:20:15), the R_f for 5'FSBA is 0.76, whereas that for 3'-FSBA is 0.80. With a solvent consisting of methanol:chloroform (15:85) the R_f for 5'-FSBA is 0.56, whereas that for 3'-FSBA is 0.64. In descending paper chromatography (Whatman No. 3 MM), with n-butanol:acetic acid:water (4:1:5) as solvent, the 5'-FSBA exhibits an R_f of 0.76, whereas the 3'-FSBA has an R_f of 0.86.

3'-p-Fluorosulfonylbenzoyladenosine. The 3'-FSBA is prepared by reaction of adenosine with p-fluorosulfonylbenzoyl chloride in the presence of the base, 1,5 diazabicyclo[3.4.0]nonene-5, to accept the mole of HCl from the reaction of the acid chloride with the alcohol, in a modification of the procedure of Pal, Wechter, and Colman.[7] The initial crude product is a mixture in which 3'-FSBA predominates; substantial amounts of 5'-FSBA as well as smaller amounts of 2',3'-bis-p-fluorosulfonylbenzoyladenosine and 2',3',5'-Tris-p-fluorosulfonylbenzoyladenosine are also detected. The 3'-FSBA is purified from the mixture. A typical preparation is as follows: Adenosine (0.80 g, 3 mmoles) is dissolved in 15 ml of dimethylformamide and 1,5-diazabicyclo[3.4.0]nonene-5 (0.37 g, 3 mmoles) is added. After the mixture has been stirred for 2 hr at room temperature,

[7] P. K. Pal, W. J. Wechter, and R. F. Colman, *Biochemistry* **14**, 707 (1975).

a clear solution results to which is added p-fluorosulfonylbenzoyl chloride (0.67 g, 3 mmoles). The solution is allowed to stand for 48 hr, after which the solvent is evaporated under reduced pressure to yield a glassy residue. This material is dissolved in 1.5 ml of dimethylformamide, and the crude product is precipitated as an oil by the addition of 4 ml of ethyl acetate:diethyl ether (1:1). The oil is converted to a solid by the slow addition of 5 ml of water. After being allowed to stand at 0° for 2 hr, the material is filtered, washed with cold water, and dried.

The crude product (0.75 g) is initially purified by column chromatography on silica gel (28–200 mesh, grade 12) at 4°. A column of 2.6 cm $\times$ 39 cm has been used with methyl ethyl ketone:acetone:water (72:20:8) as solvent. The crude product is dissolved in 7 ml of the solvent and applied to the column. The effluent is collected in 1.9-ml fractions which are monitored by thin-layer chromatography, as described for the preparation of 5′-FSBA. After 74 ml of effluent, the ultraviolet absorbing material begins to emerge. The next 23 ml (pool I) contain most of the 2′,3′,5′-Tris-p-fluorosulfonylbenzoyl adenosine and 2′,3′-bis-p-fluorosulfonylbenzoyl adenosine as indicated by spots of R_f higher than that of 3′-FSBA on thin-layer chromatograms. The 3′-FSBA predominates in the next 36 ml (pool II) although it continues to be contaminated with the other products. Final purification is achieved by preparative descending paper chromatography of pool II on Whatman 3 MM for 15 hr using as solvent n-butanol:acetic acid:water (4:1:5). The compounds are detected by examining the chromatogram under ultraviolet light; 3′-FSBA is eluted from the paper with the same solvent. The 3′-p-fluorosulfonylbenzoyladenosine has a melting point of 201°–202° and exhibits an ultraviolet absorption spectrum with maxima at 258 nm ($\epsilon = 1.63 \times 10^4$ cm^{-1} M^{-1}) and 232 nm ($\epsilon = 2.01 \times 10^4$ cm^{-1} M^{-1}).

Conditions for Reaction of 3′-FSBA and 5′-FSBA with Proteins. Because of their limited water solubility, it is convenient to prepare solutions of the compounds in organic solvents and to add small volumes of these solvents to a reaction mixture containing enzyme. The compounds are relatively stable in absolute ethanol, and solutions of about 7–8 mM can be prepared in this solvent and stored at −12° up to several weeks. If more concentrated solutions are required, it is necessary to use a solvent in which the compounds are more soluble, e.g., dimethylformamide; concentrations of about 50 mM reagent can be prepared with dimethylformamide. However, since the compounds are less stable in dimethylformamide, it is advisable to prepare fresh solutions each day. Reactions involving 3′- and 5′-FSBA have been successfully conducted in reaction mixtures containing 1–10% ethanol and 1–15% dimethylformamide.

In addition to reacting with enzymes, as described below, both 3'- and 5'-FSBA have been observed to react with certain commonly used buffers, with the concomitant release of fluoride ion in accordance with first-order kinetics. In the case of 3'-FSBA in buffers at pH 8 and 25°, the half-life is about 37 min for 0.01 M potassium phosphate and 0.01 M Tris acetate, whereas triethanolamine chloride at the same pH reacted more vigorously. In 0.01 M sodium barbital at pH 8, the half-life for 3'-FSBA was about 63 min; this buffer has proved to be satisfactory for reaction with proteins. The 5'-FSBA does not react as readily with buffers, and at 30° in 0.01 M sodium barbital at pH 7.6, containing 0.2 M KCl and 15% dimethylformamide, its half-life was found to be about 8.4 hr. The reaction with the 3'- and 5'-FSBA might be expected to involve the unprotonated form of susceptible amino acids, and therefore the rate of reaction in many cases may proceed more rapidly at pH values that are on the alkaline side of neutrality. However, it must be kept in mind that the ester linkage of both 3'- and 5'-FSBA has limited stability below pH 6 and above pH 9.

Determination of Extent of Reaction of 3'- and 5'-FSBA with Proteins. There are three relatively simple methods to detect the extent of reaction of these adenosine analogs with proteins. First, the release of fluoride ion from the compound can be followed with a specific fluoride electrode during incubation with a known amount of protein. Such data can be used to estimate the number of sites on the protein that have been modified, providing appropriate corrections are made for any reaction between the compounds and the buffer under the same conditions. Second, the extent of incorporation of sulfonylbenzoyladenosine groups can be approximated from the difference in absorbance at 259 nm between the isolated modified enzyme and native enzyme, using the extinction coefficient of free 3'- or 5'-FSBA at that wavelength. Third, and most directly, the number of moles of reagent incorporated per mole of protein can be determined with radioactive 3'- or 5'-FSBA.

Examples of Reactions of 3'-FSBA and 5'-FSBA with Particular Proteins

Glutamate Dehydrogenase. Native glutamate dehydrogenase is activated by ADP and inhibited by high concentrations of DPNH binding at a regulatory site distinct from the active site. Both of these effects are irreversibly decreased upon incubation of the enzyme with 3'-FSBA, whereas the intrinsic enzymic activity as measured in the absence of regulatory compounds (Fig. 2A) remains unaltered.[7] The decrease in extent of ADP activation (Fig. 2B) obeys pseudo first-order kinetics, as shown in the inset of Fig. 2, allowing the calculation of a rate constant for the reaction. A plot of the rate constant for modification as a function of the 3'-FSBA concentration is not linear, but rather exhibits saturation kinetics, suggesting that the adenosine derivative binds to the enzyme ($K_I = 0.10$ mM) prior to the irreversible modification. This interaction requires the adenosine portion of the molecule since p-fluorosulfonyl benzoic acid, which lacks the adenosine moiety, does not modify the enzyme under conditions in which 3'-FSBA reacts with a half-life of 25 min. Protection against modification by 3'-FSBA is provided by ADP and by high concentrations of DPNH, but not by the normal substrates of the enzyme. The isolated modified enzyme contains approximately 1 mole of sulfonylbenzoyladenosine per peptide chain, indicating that a single specific regulatory site has reacted with 3'-FSBA. The implication is that the activating ADP site and the inhibitory DPNH site are at least

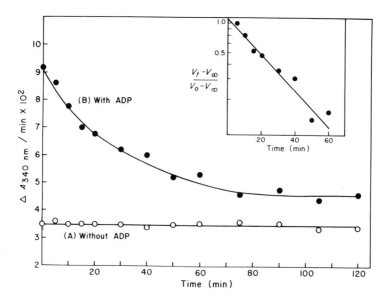

FIG. 2. Reaction of 3'-p-fluorosulfonylbenzoyladenosine with bovine liver glutamate dehydrogenase. Glutamate dehydrogenase (0.21 mg/ml) was incubated with 3'-FSBA (0.496 mM) at 24° in 0.01 M sodium barbital buffer (pH 8) containing 0.43 M KCl and 5% ethanol. At each indicated time, an aliquot was withdrawn, diluted 20-fold with Tris-0.1 M acetate buffer (pH 8) at 0°, and assayed (A) in the absence and (B) in the presence of 100 μM ADP. *Inset:* Determination of the pseudo first-order rate constant from the decrease in activation by ADP. (V_t and V_0 are the enzymic velocities measured in the presence of ADP and the given and zero time, respectively, and V_∞ is the constant velocity at the end of the reaction. The pseudo first-order rate constant calculated is 0.0351 min^{-1}.) Data are taken from P. K. Pal, W. J. Wechter, and R. F. Colman, *Biochemistry* **14,** 707 (1975).

partially overlapping and that 3'-FSBA reacts covalently in the region of these regulatory sites.

Differentiation between the two sites, however, is afforded by reaction of glutamate dehydrogenase with 5'-FSBA.[6] This compound also reacts covalently with glutamate dehydrogenase with incorporation of about 1 mole of 5'-sulfonylbenzoyladenosine per peptide chain. However, in the case of the reaction with 5'-FSBA neither the catalytic activity as measured in the absence of ADP (Fig. 3A) nor the activation by added ADP (Fig. 3B) is affected. The major change in the kinetic characteristics of this modified enzyme is a total loss of the inhibition by DPNH. These results demonstrate that the catalytic as well as the ADP regulatory sites are distinct from the inhibitory DPNH site and are consistent with the postulate that the 5'-p-fluorosulfonylbenzoyladenosine attacks exclusively the second inhibitory DPNH site. Enzyme that has combined stoichiometrically with 5'-p-fluorosulfonylbenzoyladenosine is still able to react with 3'-FSBA; thus, the two adenosine analogs appear to react at distinct sites on glutamate dehydrogenase. Indeed, it has been demonstrated that a tryptic peptide isolated from radioactive 3'-p-sulfonylbenzoyladenosine-modified glutamate dehydrogenase is chromatographically distinct from a

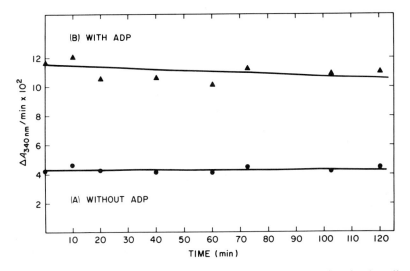

Fig. 3. Reaction of 5'-p-fluorosulfonylbenzoyl adenosine with bovine liver glutamate dehydrogenase. Glutamate dehydrogenase (0.21 mg/ml) was incubated with 5'-FSBA (0.43 mM) at 30° in 0.01 M sodium barbital buffer, pH 8, containing 0.4 M KCl and 5% ethanol. At the indicated times, aliquots were withdrawn, diluted 20-fold with Tris-0.1 M acetate buffer pH 8 at 0°, and assayed (A) in the absence and (B) in the presence of 100 μM ADP using 100 μM TPNH as coenzyme.

tryptic peptide derived from a 5'-SBA-modified enzyme.[8] These results indicate not only that both 3'- and 5'-FSBA can function as affinity labels of purine nucleotide regulatory sites, but that comparison between the effects of the two compounds may provide additional information about these sites in enzymes.

Pyruvate Kinase. Rabbit muscle pyruvate kinase is inactivated upon incubation with 5'-FSBA at 30° in 0.02 M sodium barbital at pH 7.4, containing 15% dimethylformamide, whereas enzyme incubated under the same conditions in the absence of the reagent does not lose activity.[9] A plot of log E/E_0 versus time reveals a biphasic rate of inactivation which is consistent with a rapid reaction to form partially active enzyme (54%) followed by a slower reaction to yield totally inert enzyme. At 2 mM 5'-FSBA, $k_1 = 4.5 \times 10^{-2}$ min^{-1} and $k_2 = 0.35 \times 10^{-2}$ min^{-1}. The rate of incorporation of radioactive 5'-FSBA is also biphasic and indicates that 2 moles of reagent per peptide chain are bound when the enzyme is completely inactive. The pseudo first-order rate constant for the rapid rate is linearly dependent on 5'-FSBA concentration whereas the constant for the slow rate is a nonlinear function of the reagent concentration, suggesting that the reagent binds reversibly to the second site prior to its covalent modification. Protection against both phases of inactivation is provided by phosphoenol pyruvate, magnesium ion, and Mg^{2+}-ADP. It appears that 5'-FSBA labels two residues in the binding region for metal ion and the phosphoryl group of phosphoenol pyruvate that is transferred during the catalytic reaction of pyruvate kinase. Thus, 5'-FSBA not only can react at a regulatory site (as in the case of glutamate dehydrogenase), but may also label the active site of an enzyme in which an adenine nucleotide participates as a substrate.

ADP Binding Proteins of Human Platelets. In addition to affinity labeling of purified proteins, the adenosine analogs described here may be useful in identifying adenine nucleotide binding proteins in intact cells. For example, it is known that ADP initiates a change in the shape of human blood platelets, followed by aggregation of the platelets; and it has been postulated that ADP may bind to a specific receptor protein on the platelet membrane.[10,11] The compound 5'-FSBA has been found to competitively inhibit the initial event of ADP-platelet interaction, the change in shape, with a K_I of 11 μM. Incubation of whole platelets or

[8] P. K. Pal, W. J. Wechter, and R. F. Colman, *Abstr. Int. Congr. Biochem. 10th* (Hamburg, Germany), 1976.

[9] J. L. Wyatt and R. F. Colman, *Fed. Proc., Fed. Am. Soc. Exp. Biol.* **35,** 1434 (1976); *Biochemistry* **16,** in press (1977).

[10] G. V. R. Born, *Nature (London)* **20,** 1121 (1965).

[11] R. L. Nachman and B. Ferris, *J. Biol. Chem.* **249,** 704 (1974).

platelet membrane preparations with radioactive 5'-FSBA leads to irreversible binding of this ADP analog as indicated by observations that the labeled complex is stable to extensive washing and prolonged dialysis as well as to exposure to 8 M urea, 10% sodium dodecyl sulfate, or dithiothreitol. The simultaneous incubation of ADP and radioactive 5'-FSBA with whole platelets or platelet membranes markedly decreases the incorporation of the radioactive analog, suggesting that 5'-FSBA is acting as an affinity label for the platelet by being incorporated into ADP binding proteins.[12]

Radioactivity can be shown to be associated with protein upon electrophoresis in polyacrylamide gels containing sodium dodecyl sulfate.[12] Since 5'-FSBA inhibits the ADP-induced platelet shape change, it is possible that one of the labeled proteins is responsible for the initial ADP-membrane interaction by serving as an ADP platelet membrane receptor.

Conclusions. These three examples indicate that 3'- and 5'-FSBA may be useful in the affinity labeling of receptor proteins as well as of the regulatory and active sites of enzymes. Within the constraints imposed by the limitations of solubility and stability of these compounds, the fluorosulfonylbenzoyladenosines thus may have broad applicability in the labeling of adenine nucleotide sites in proteins.

[12] J. S. Bennett, R. W. Colman, W. Figures, and R. F. Colman, *Abstr. Int. Congr. Biochem. 10th* (Hamburg, Germany), 1976.

[24] Alcohol Dehydrogenases

By Christoph Woenckhaus and Reinhard Jeck

NAD is the coenzyme of alcohol dehydrogenase. It serves in the reversible enzymic reaction as an acceptor of the hydrogen from the substrate and is converted to NADH. This occurs in a ternary complex of enzyme, pyridine nucleotide, and substrate, in which coenzyme and substrate interact closely with amino acid residues of the enzyme, and thereby are activated sufficiently to allow reaction.

The coenzyme substrate, NAD$^+$, is composed of two nucleotide moieties: the nonfunctional adenosine monophosphate and the functional nicotinamide mononucleotide, which are connected by a pyrophosphate bridge. X-Ray structure analysis of the complex of enzyme and coenzyme have indicated that the interactions with the protein are hydrophobic ones for adenine, hydrophilic for the riboses of the two nucleotides, and ionic for the pyrophosphate group. In the neighborhood of the nicotina-

TABLE I
COENZYME FUNCTION OF 3-ACETYLPYRIDINE-n-ALKYL ANALOGS
OF NAD$^+$ AND NADH WITH ALCOHOL DEHYDROGENASES[a]

n-Alkyl chain	Oxidized coenzyme analogs[b]		Dihydroforms[b]	
	Yeast	Horse liver	Yeast	Horse liver
Ethyl	−	−	−	−
Propyl	−	−	(+)	(+)
Butyl	(+)	+	(+)	+
Pentyl	−	−	(+)	(+)
Hexyl	−	−	−	−

[a] Ethanol oxidation was tested in 0.2 M glycine-NaOH at pH 9.5, and acetaldehyde reduction was tested in 0.2 M potassium phosphate at pH 7.

[b] +: Active as hydrogen donor or acceptor, respectively; (+): very weak coenzymic activity as hydrogen donor or hydrogen acceptor; −: no coenzyme activity.

mide and at the binding site for the other substrate are the side chains that must be responsible for activation of the two substrates. Modifications of amino acid residues of the enzyme that prevent binding and activation of substrates result in a reduction of enzymic activity.

By the introduction of reactive groups in NAD, analogs are obtained that complex with the enzyme at its active center and then form covalent bonds with corresponding amino acid residues in a manner similar to NAD itself. For this purpose, halogen ketones, azo groups, and diazo ketones were found to be suitable reactive groups.[1]

In enzymic assays, 3-acetylpyridine adenine dinucleotide (3-APAD) can replace NAD as substrate.[2] No one has as yet succeeded in the direct bromination to form 3-bromo-APAD, although 3-chloro-APAD has been synthesized.[3] If the ribose of the functional moiety of 3-APAD is replaced by a hydrocarbon chain, the redox potential is altered to −320 mV, a magnitude close to that of NAD$^+$. Compounds of this structure can be easily labeled by using methyl or [^{14}C]carbonyl-labeled acetylpyridine for synthesis.

The coenzyme function of the 3-acetylpyridine-alkyl analogs of NAD$^+$ depends upon the length of the hydrocarbon chain (Table I). If the

[1] C.-I. Brändén, H. Jörnvall, H. Eklund, and B. Furugren, in "The Enzymes" (P. D. Boyer, ed.), 3rd ed., Vol. 11, p. 176. Academic Press, New York, 1975.

[2] N. O. Kaplan, M. M. Ciotti, and F. E. Stolzenbach, J. Biol. Chem. 221, 833 (1956).

[3] J.-F. Biellmann, G. Branlant, B. Y. Foucaud, and M. J. Jung, FEBS Lett. 40, 29 (1974).

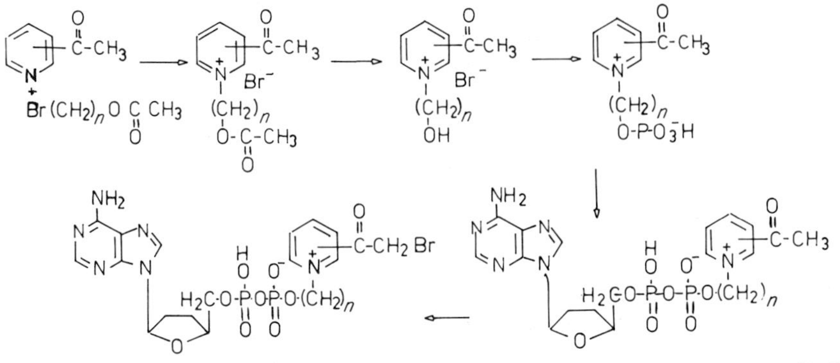

FIG. 1. Preparation steps for the synthesis of [ω-(bromoacetylpyridinio)-n-alkyl] adenosine pyrophosphates.

3-acetylpyridine is replaced by 2-acetylpyridine or 4-acetylpyridine, the resultant analogs will not function enzymically but will act as competitive inhibitors of NAD⁺ or NADH with certain dehydrogenases.[4,5]

The Synthesis of Acetyl Pyridinio-n-alkyl Monophosphates (Fig. 1)

A mixture of 10 mmoles of acetylpyridine and 10 mmoles of ω-halogen-alkylacetate is stored for 7 days at room temperature. 2-Acetylpyridine is alkylated to a satisfactory degree only at 100° for 30 hr. To hydrolyze the acetyl group, the residue is dissolved in 10 ml of methanol, containing 0.5 ml of concentrated HCl, and maintained for 24 hr at room temperature. The solvent is evaporated under reduced pressure at 30°. Assay for esters by the hydroxamate method performed on the residue should be negative.[6] The yields are about 40–90% and depend upon the length of hydrocarbon chain and the position of the acetyl group of the pyridine ring. Hydroxybutyl and hydroxypentyl derivatives of 3-acetylpyridine, as well as quaternary 2-acetylpyridinio salt, have been obtained in lower yields (40–60%). Acetyl-1-(ω-hydroxy-n-alkyl)pyridinio halide, 3 g, is mixed with 15 g of metaphosphoric acid; the latter is freshly prepared according to the procedure of Karrer by heating concentrated ortho-phosphoric acid until it just becomes turbid and white smoke is developed.[7] The reaction mixture is stirred for 3 hr in a water bath at 60° as gaseous hydrochloric acid escapes from it.[7] To the syrupy product, 10 ml

[4] R. Jeck and C. Woenckhaus, Z. Naturforsch. Teil C 29, 180 (1974).
[5] C. Woenckhaus, R. Jeck, E. Schättle, and J. Berghäuser, Hoppe Seyler's Z. Physiol. Chem. 353, 559 (1972).
[6] R. E. Buckles and C. J. Thelen, Anal. Chem. 22, 676 (1950).
[7] M. Viscontini, C. Ebnöther, and P. Karrer, Helv. Chim. Acta 34, 1834 (1951).

of ice water are added, and the solution is slowly poured into 2 liters of ethanol at −20°.[8] After 20 hr at −20°, a precipitate of acetylpyridinio-*n*-alkyl polyphosphoric acid is collected by centrifugation and dissolved in 60 ml of 1 *N* HCl.

To hydrolyze the anhydride bonds of the polyphosphoric acid esters, this solution is refluxed for 20 min. After cooling it is filtered, and the filtrate is concentrated to 10 ml under reduced pressure. The solution is applied to a Dowex 1 × 8 column (formate form, 100–200 mesh, 2 × 50 cm) and eluted with water. After 100 ml of water have run through the column, acetylpyridinio-*n*-alkyl phosphates are collected in the following 160 ml. The solution is brought to pH 2 with dilute HCl, concentrated under reduced pressure to a volume of 10 ml, and placed on a Dowex 50 W × 8 column (H⁺ form, 2 × 20 cm, 200–400 mesh). With water elution, acetylpyridinio-*n*-alkyl monophosphates generally appear after about 1.5 liters and are collected in the following 2–5 liters. Only those nucleotide analogs having long alkyl chains, e.g., pentyl or hexyl, are retarded more strongly by the acidic ion exchange resin; they are eluted after 10–20 liters of water have run through the column and in a volume of about 20 liter. The aqueous solution of the product is evaporated under reduced pressure at 30° to 5 ml. The phosphoric acid ester is precipitated by addition of cold acetone. The yields are generally about 50%, based on the starting material.

Synthesis of Dinucleotide Analogs

To prepare the [ω-(acetylpyridinio)-*n*-alkyl]adenosine pyrophosphates, 10 mmoles of acetylpyridinio-*n*-alkylphosphoric acid and 10 mmoles of adenosine 5′-phosphoromorpholidate (as salt of 4-morpholine *N,N′*-dicyclohexylcarboxamidine) are dissolved in 20 ml of freshly distilled *o*-chlorophenol.[9] The mixture is stored at room temperature for 7 days. Progress of the condensation reaction is checked daily by paper electrophoresis (0.1 *M* Tris chloride at pH 8.1, 30 V/cm). After the reaction is completed, 60 ml of water are added to the mixture and the suspension is extracted 3 times, each with 200 ml of ethyl ether. The first ether extract obtained is reextracted with 30 ml of water. The combined aqueous phases are reduced to a volume of 5 ml under reduced pressure and are charged on to a Dowex 1 × 8 column (formate form, 2 × 50 cm, 100–200 mesh). The column is washed with 6.5 liters of water, and the coenzyme analogs are eluted from the resin with a convex gradient of formic acid;

---

[8] In the preparation of ¹⁴C-labeled compounds, ethanol precipitation is not used so as to avoid large losses of radioactively labeled material.

[9] J. G. Moffatt and H. G. Khorana, *J. Am. Chem. Soc.* **83**, 649 (1961).

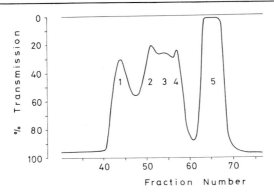

FIG. 2. Chromatography of [5-(3-acetyl-1,4-dihydro-1-pyridyl)-*n*-pentyl]adenosine pyrophosphate on Sephadex G-10 (200 × 2 cm). The column was eluted with 50 mM glycine/NaOH buffer, pH 9.5; fractions of 4.3-ml volume were collected, and the flow rate was 13 ml/hr. The dihydrocoenzyme analog was eluted in peak 5.

the mixing chamber contains 500 ml of water, and the reservoir vessel is filled with 2 liters of 25 mM formic acid. The compound appears after 700 ml of eluent and is collected in a volume of about 1 liter. The solution is evaporated under reduced pressure at 30° to a volume of 7 ml, from which the coenzyme analog is precipitated with cold acetone. Yields of 30–70% were obtained for the different analogs. The electrophoretic mobilities are identical to that of NAD+. The analog [3-(2-acetylpyridinio)propyl]adenosine pyrophosphate shows an $\epsilon^{pH5.6}_{261\,nm} = 17,200$, and [3-(4-acetylpyridinio)propyl]adenosine pyrophosphate an $\epsilon^{pH5.6}_{259\,nm} = 16,300$.

Preparation of the Dihydrocoenzyme Analogs

The reduced forms of the coenzyme analogs can be obtained only from the 3-acetylpyridinio-*n*-alkyl analogs of NAD. The pyridine ring of these analogs is transformed to the dihydro form by treatment with sodium hyposulfite.[10] The coenzyme analog, 25 mg, together with 62.5 mg of sodium bicarbonate and 25 mg of sodium hyposulfite are dissolved in 1.75 ml of water. The solution is placed in a boiling water bath for 90 sec. The cooled solution is subjected to a stream of air for 30 min in order to destroy excess hyposulfite. The mixture is placed on a Sephadex G-10 column (2 × 200 cm) and eluted with 50 mM glycine-NaOH at pH 9.5. The dihydro form of the coenzyme analogs is eluted immediately after the oxidized form and its decomposition products (Fig. 2). It is characterized by its absorption maxima at 260 nm and 380 nm (Table II).

[10] S. Gutcho and E. D. Stewart, *Anal. Chem.* **20**, 1185 (1948).

Alkyl chain	Oxidized form[a]		Dihydro form[a]			
	λ_{max}	ϵ	λ_{max_1}	ϵ	λ_{max_2}	ϵ
Ethyl	260	17,500	260	14,000	380	9,900
n-Propyl	260	18,100	260	14,500	380	10,300
n-Butyl	260	17,200	260	14,600	380	10,600
n-Pentyl	260	18,400	260	14,500	380	10,300
n-Hexyl	260	17,900	260	14,600	380	10,800

[a] λ_{max} (nm), $\epsilon(cm^{-1} \times M^{-1})$.

Preparation of Inactivating Reagents

Bromine reacts with acetylpyridinio-n-alkyl analogs of NAD in acidic medium, forming unstable bromoacetyl derivatives. To prepare the inactivating reagent, 0.1 mmole of [ω-(acetylpyridinio)-n-alkyl]adenosine pyrophosphate is dissolved in 2 ml of 3% aqueous HBr, to which 16 mg (0.1 mmole) of bromine are added. The solution is either irradiated from a distance of 20 cm with a Shandon lamp (2 × 6 watt, white light) or stored in sunlight. Complete discoloration is reached after 10–20 hr. Hydrogen bromide and a small amount of residual bromine are removed by extracting 5 times with 10-ml volumes of ethyl ether. To the aqueous phase, cold acetone is added until the coenzyme analog is completely precipitated. The precipitate, collected by centrifugation, is dried under reduced pressure. Since bromoacetyl compounds are not very stable in a solid state, even at −20°, it is advisable to use the reagent only in the freshly prepared form. The rates of hydrolysis of several of these inactivating reagents, determined at 37°, are noted in Table III.

Assay of Inhibitor

Enzymic activity of alcohol dehydrogenases is determined spectrophotometrically at 25° at 340 nm in a cuvette containing 3 ml of 0.2 M glycine–NaOH at pH 9.5 with 0.25 mM NAD$^+$ and 50 mM ethanol. The reaction is started by addition of 1 μg of the yeast enzyme or 0.1 mg of the liver enzyme. The decrease in enzymic activity is determined using the relation, 100 v/v_0, where v is the rate of NADH formation catalyzed

TABLE III
RATE OF HYDROLYSIS OF THE [BROMOACETYLPYRIDINIO-
PROPYL]ADENOSINE PYROPHOSPHATES AT 37°, REPORTED
AS HALF-LIFE, $\tau/2$ IN MINUTES

Substituent of the pyridinio ring	pH			
	7	8	9	10
2-Bromoacetyl	50	8	1	—
3-Bromoacetyl	300	150	20	3
4-Bromoacetyl	30	5	—	—

in presence of enzyme and inactivator and v_0 is the rate of control sample without inhibitor.

Preparation of the Covalently Linked Enzyme–Coenzyme Compounds

To 100 mg of alcohol dehydrogenase in 20 ml of 0.3 M potassium phosphate at pH 6.5 and 25°, inactivator is added to a final concentration of 0.7 mg/ml. The decrease in enzymic activity is checked enzymically. Upon reaching the desired degree of inactivation, cysteine is added up to a concentration of 0.01 M to terminate the inactivation reaction. Excess inactivators and cysteine are removed by gel chromatography on a Sephadex G-25 fine column (2×100 cm), and washed with 0.3 M potassium phosphate at pH 6.5.

In the presence of 1 mM ethanol, the phosphate buffer, and [4-(3-bromoacetylpyridinio)-n-butyl]adenosine pyrophosphate, inactivation and the redox reaction take place simultaneously: the coenzyme analog is covalently bound partly in the reduced and partly in the oxidized form. The complex, in which the coenzyme analog is incorporated in the oxidized form, can be completely transformed into the dihydroform by treatment with sodium hyposulfite.

To 10 mg of inactivated enzyme in 3 ml of the same phosphate buffer is added 1 mg of sodium hyposulfite. This mixture is allowed to stand at room temperature for 15 min. Excess reducing agent is removed by passage through a Sephadex G-25/fine column (1×30 cm) and washing with the buffer that will be used subsequently.

Kinetics of Inactivation

When the enzymes are incubated with the inactivating reagents in presence of NAD+ or NADH, the rate of inactivation is decreased, e.g., no significant decrease in enzymic activity can be measured if 10 mM NAD$^+$ is present in the inactivation medium. The coenzyme analog, 3-APAD, and the acetylpyridinioalkyl analogs of NAD also decrease the rate of inactivation.

Inactivation of both alcohol dehydrogenases by bromoacetylpyridinio-n-alkyl analogs of NAD$^+$ follows saturation kinetics, indicating that formation of covalent enzyme-analog complexes are preceded by reversible binding of the inactivators to the enzymes. A kinetic expression for such a two-step inactivation has been given by Gold and Fahrney.[11]

$$E + I \underset{k_{-1}}{\overset{k_{+1}}{\rightleftharpoons}} E \cdots I \overset{k_2}{\to} E - I$$

The dissociation constants of the first-formed reversible enzyme–coenzyme analog complex can be obtained by plotting $1/k_{obs}$ ($= \tau/2 : \ln 2$) versus $1/c_I$, assuming that $k_2 \ll k_{-1}$. The K_I values of the complexes are found to be of the order of magnitude determined for the Michaelis constants. For those analogs that are not effective as hydrogen acceptors in the enzyme assay, they are comparable in magnitude to the competitive inhibitor constants of the compounds when not brominated. With ^{14}C-labeled inactivators, complete loss of activity is observed when 1 mole of inactivator has reacted with 1 mole of enzyme subunit in the case of both alcohol dehydrogenases (Table IV).

Optical Properties of Enzyme–Coenzyme Analog Compounds

The optical properties of inactivated dehydrogenases correspond to those of the binary complexes between coenzyme analog and the alcohol dehydrogenases. Upon excitation at 280 nm, protein fluorescence measured at 340 nm is reduced as compared with the apoenzyme; such quenching of fluorescence is increased after reduction of covalently bound 3-acetylpyridinio-n-alkyl analogs. The absorption spectrum of these dihydrocoenzyme-enzyme compounds shows an increase in extinction at 260 nm and a band at 380 nm that corresponds to the absorption of the dihydropyridine moiety. In the fluorescence excitation spectrum there is an energy transfer band at 290 nm, the emission maximum of which lies in the region of coenzyme fluorescence at about 480 nm. As in the reversible alcohol dehydrogenase–coenzyme complexes, the intensity of

[11] A. M. Gold and D. Fahrney, *Biochemistry* **3**, 783 (1964).

TABLE IV

INHIBITOR CONSTANTS OF THE COENZYME ANALOGS WITH ALCOHOL DEHYDROGENASE[a]

Coenzyme analog	Yeast enzyme			Liver enzyme		
	pH	K_1 (mM)	k_2 (min^{-1})	pH	K_1 (mM)	k_2 (min^{-1})
2-Bromoacetylpyridiniopropyl-ADP	6.5	20	10	No inactivation		
2-Acetylpyridiniopropyl-ADP	9.5	17	—	—	—	—
3-Bromoacetylpyridiniopropyl-ADP	6.5	10	0.9	No inactivation		
3-Acetylpyridinopropyl-ADP	7.1	14	—	—	—	—
	9.5	12	—	—	—	—
3-Bromoacetylpyridinio-n-butyl-ADP	6.6	4	1.7	6.6	2	0.1
3-Acetylpyridinio-n-butyl-ADP	9.5	10	—	9.5	4	—
				(9.5 K_m = 7 mM; turn. = 50 min^{-1})		
4-Bromoacetylpyridiniopropyl-ADP	6.5	4	1	6.5	7	0.2
4-Acetylpyridiniopropyl-ADP	7.1	2	—	7.1	4	—
	9.5	3	—	9.5	7	—

[a] ADP = adenosine pyrophosphate; K_m = Michaelis constant; turn. = maximum turnover number; the inhibitor constants, K_1, of the unbrominated compounds are defined according to A. M. Gold and D. Fahrney [*Biochemistry* **3**, 783 (1964)] as $K_1 = (k_{-1} + k_2)/k_{+1}$ and are corresponding in magnitude to dissociation constants provided that $k_2 \ll k_{-1}$.

the coenzyme fluorescence is enhanced and the emission maximum is shifted to shorter wavelength.[12]

Identification of Amino Acid Residues

The inactivation of yeast alcohol dehydrogenase by carbonyl-[^{14}C] [3-(3-bromoacetylpyridinio)-n-propyl]adenosine pyrophosphate and methylene-[^{14}C][3-(4-bromoacetylpyridinio)-n-propyl]adenosine pyrophosphate was found to be solely the result of ketoalkylation of cysteine residue 43.[13] With horse liver alcohol dehydrogenase, cysteine-174 reacted nearly exclusively with the inactivator [3-(4-bromoacetylpyridinio)-n-propyl]adenosine pyrophosphate.[13] Carbonyl-[^{14}C][4-(3-bromoacetyl-pyridinio)-n-butyl]adenosine pyrophosphate modifies cysteine residue

[12] R. Jeck, P. Zumpe, and C. Woenckhaus, *Justus Liebigs Ann. Chem.* **1973**, 961.
[13] H. Jörnvall, C. Woenckhaus, E. Schättle, and R. Jeck, *FEBS Lett.* **54**, 297 (1975).

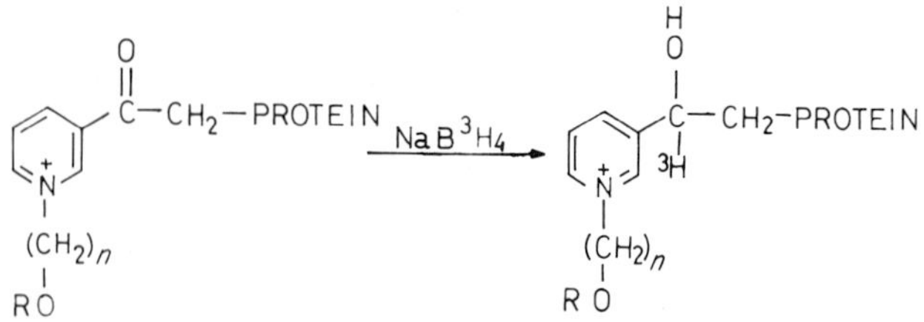

FIG. 3. Stabilization of enzyme-analog compound by reduction of carbonyl to a carbinol group.

38 of the alcohol dehydrogenase of *Bacillus stearothermophilus*[14]; this residue, unlike its counterparts in the yeast and mammalian enzymes, is not reactive with thiol group reagents.[15]

Oxidation of the Inactivator–Enzyme Compound

By oxidation of the modified protein, the coenzyme can be removed leaving a carboxymethylated amino acid associated with protein. With a carbonyl or methylene-^{14}C-labeled coenzyme analog, the label remains partly with the protein. The procedure involves dialysis of inactivated enzyme for 6 hr at 25° against 1 liter of 0.1 M Tris chloride at pH 8.2, containing 5 mmoles of hydrogen peroxide. Excess hydrogen peroxide is destroyed by dialysis for 12 hr against a solution containing 5 mg of catalase in 1 liter of 10 mM potassium phosphate at pH 7. About 50% of the radioactivity initially incorporated by covalent binding of the inactivator is lost. After complete hydrolysis of the proteins, ^{14}C-labeled CM-Cys and CM-His have been detected with an amino acid analyzer.[16,17]

Stabilization of the Bond between Coenzyme and Protein

After covalent binding of the inactivator, the enzyme–analog compound is stabilized by reduction of its carbonyl to a carbinol group (Fig. 3). To 100 mg of inactivated enzyme in 0.15 M sodium pyrophosphate (pH 6.0) at 0° are added 10 mg of sodium borohydride. At this pH, formation of dihydro or tetrahydropyridine derivatives could not be observed. Excess sodium borohydride is removed by dialysis. With tritium-labeled sodium borohydride this method yields a tritiated protein complex.

[14] C. Woenckhaus, R. Jeck, and J. I. Harris, in preparation.
[15] J. Bridgen, E. Kolb, and J. I. Harris, *FEBS Lett.* 33, 1 (1973).
[16] C. Woenckhaus, *in* "Topics in Current Chemistry" (F. L. Boschke ed.), Vol. 52, p. 209. Springer-Verlag, Berlin and New York, 1974.
[17] C. Woenckhaus, M. Zoltobrocki, J. Berghäuser, and R. Jeck, *Hoppe Seyler's Z. Physiol. Chem.* 354, 60 (1973).

[25] Arylazido Nucleotide Analogs[1] in a Photoaffinity Approach to Receptor Site Labeling

By RICHARD JOHN GUILLORY and STELLA JYHLIH JENG

The ATP molecule contains three distinct structural groupings: the adenine base, the ribose sugar, and the triphosphate component. On the basis of the interaction of actomyosin with ATP analogs modified in these positions, a model of the enzyme substrate complex has been postulated.[1a,2] This model is based upon evidence suggesting that the amino grouping at the 6 position of the adenine base and the 3'-hydroxyl of the ribose ring are hydrogen bonded to the protein. The nucleotide binding is thought to involve an NH group at the trinitrobenzene sulfonate interacting site[3] as well as conjugation of triphosphate with asparagine via metal interaction[4] and binding of the terminal pyrophosphate to a sulfhydryl and guanidinium group at the active site.[5] An important aspect of these actomyosin–nucleotide analog studies is the appreciation that the ribose ring can be drastically modified without affecting markedly the rate of myosin-catalyzed hydrolysis of the analog.

In view of the above facts it was decided to develop methods by which an active photosensitive adjunct could be attached to the ribose portion of nucleoside triphosphates. An additional restraint that we impose upon ourselves was that the photogenerated species be formed at wavelengths of light remote from regions of protein damaging radiation.

This chapter describes the successful chemical coupling of arylazido groupings to adenine nucleotides and the extension of this methodology

[1] Abbreviations most commonly used: arylazido ATP or arylazido-β-alanine ATP, 3'-O-{3-[N-(4-azido-2-nitrophenyl)amino]propionyl}ATP; arylazido-β-alanine NAD$^+$, 3'-O-{3-N-(4-azido-2-nitrophenyl)amino]propionyl}NAD$^+$; arylazido-β-alanine, N-(4-azido-2-nitrophenyl)β-alanine; NMN, nicotinamide mononucleotide; CoQ$_1$, coenzyme Q with one isoprenoid unit in the side chain; F$_1$, coupling factor 1 ATPase.

[1a] Y. Tonomura, S. Kubo, and K. Imamura, in "Molecular Biology of Muscular Contraction" (S. Ebashi, F. Oosawa, T. Sekine, and Y. Tonomura, eds). Igaku Shoin, Tokyo, 1965.

[2] Y. Tonomura, "Muscle Proteins, Muscle Contraction and Cation Transport," Chapter 9. Univ. Park Press, Baltimore, Maryland, 1973.

[3] S. Kubo, S. Tokura, and Y. Tonomura, J. Biol. Chem. 235, 2835 (1960).

[4] N. Asuma, M. Ikegara, E. Otsuka, and Y. Tonomura, Biochim. Biophys. Acta 60, 104 (1962).

[5] T. Yamashita, Y. Soma, S. Kobayashi, T. Sekine, K. Titani, and K. Narita, J. Biochem. (Tokyo) 55, 576 (1964).

to the pyridine nucleotides. The synthesis of arylazido analogs of CoA and tetrodotoxin are mentioned.

The use of the adenine analogs in photodependent inhibition studies of myosin ATPase (EC 3.6.1.3) fragments and F_1 (mitochondrial) ATPase as well as the interaction of the pyridine nucleotide analogs with alcohol dehydrogenase (EC 1.1.1.1) and complex I (NADH–CoQ reductase) are illustrated. The synthesis of radiolabeled analogs is described together with experiments illustrating the use of these labeled compounds in evaluating the covalent binding of the analogs to myosin fragments and F_1 ATPase.

The general utility of our procedure for the synthesis of nucleotide analogs by substitution at the ribose hydroxyl group is indicated by the synthesis of spin-labeled analogs of the adenine and pyridine nucleotides having 3-carboxyl-2,3,5,5-tetramethylpyrroline-1-oxyl coupled to this position. The use of photoaffinity analogs with a variation in the distance of the azido group from the 3'-hydroxyl position on the nucleotide ribose is discussed as a tool in the mapping of nucleotide binding regions of proteins.

Because of their potential application to other systems, a number of reactions describing the reactivity of the 3'-hydroxyl of the nucleotide ribose are initially outlined together with attempted synthesis of diazo derivatives of ATP as potential precursors of active carbene analogs.

Arylazido Adenine Nucleotide Analogs

Carbodiimidazole-Catalyzed Chemical Synthesis of Arylazido Adenine Nucleotide Analogs. The utilization of carbodiimidazole to facilitate the formation of activated carboxylic acids was elegantly applied by Gottikh and his group[6,7] to the synthesis of a wide spectrum of aminoacyl nucleosides, nucleotides, nucleoside di- and triphosphates, and tRNA derivatives. The formation of an imidazolide intermediate has been used in our work in the synthetic schemes leading to the esterification of adenosine and diphosphopyridine nucleotides.

It is known that amino acylation can take place at the 2'- or 3'-hydroxy position or at the 6-amino group of the adenine ring structure. The solvent systems, varying in polarity and solvation capacity, play an important role in orienting precisely where the reaction occurs. The

[6] B. P. Gottikh, A. A. Kraevskii, P. P. Peuygin, T. L. Tsilevich, Z. S. Belova, and L. N. Rudzite, *Izv. Akad. Nauk SSSR, Ser. Khim.* (*Engl.*) 2453 (1967).
[7] B. P. Gottikh, A. A. Krayevsky, N. B. Tarussova, P. P. Purygin, and T. L. Tsilevich, *Tetrahedron* 26, 4419 (1970).

dimethyl formide:water (1:5) solvent system we have adopted has been shown by Gottikh *et al.*[7] to be most advantageous in restricting the reaction center to the ribose hydroxyl groups rather than at the 6-amino group.

Chemical Reactivity of the 3′-Hydroxyl Group of Adenosine Triphosphate

Condensation with p-Nitrophenyl Ester of 3-Chloropropionic Acid

Our initial experiments utilized 2′-deoxyadenosine in attempting to acquire knowledge of the reactivity of the 3′-hydroxyl group for condensation reactions with the *p*-nitrophenyl ester of 3-chloropropionic acid with free imidazole as catalyst.[8] The choice of the 3-chloropropionic acid rested upon the feasibility of converting the halide first into an azide, then an amine, and finally a diazo derivative.

When this reaction was carried out according to the procedure outlined below, three products were detected by paper chromatography using ultraviolet (UV) light as a detecting agent. The infrared (IR) and nuclear magnetic resonance (NMR) spectrum of the major component indicated that during the reaction imidazole displaced the chlorine atom from the 3-chloropropionyl moiety. Structure (I) was tentatively assigned to this product. It was thus concluded that esterification of the 3′-OH of the ribose using imidazole catalysis can take place with good yields. On the other hand, our expectation that chlorine would be present in the product for subsequent transformation into the diazo compound was frustrated.

(I)

[8] J. A. Secrist, J. R. Barris, and N. J. Leonard, *Chem. Ind.* (*London*) p. 1960 (1967).

Condensation with N-t-Butoxycarbonyl-α-alanylimidazole

An alternative procedure attempted was the utilization of N-t-butoxycarbonyl amino acids for the esterification of the 3'- (or 2'-) hydroxyl group of ATP with carbodiimidazole catalysis. Products of a similar esterification of the hydroxyl groups at the 2'- or 3'-positions of nucleosides and nucleotides by amino acids have been investigated as analogs for an intermediate step in protein synthesis.

Experimentally, N-t-butoxycarbonyl-α-alanylimidazole formed in anhydrous dimethylformamide was allowed to react with ATP in aqueous solvent at room temperature. Descending preparative paper chromatography using n-butanol:water:acetic acid (5:3:2) as the migrating solvent afforded three well defined, UV-absorbing bands with R_f values of 0.37, 0.45, and 0.69. The fractions isolated by eluting each band from the paper with water were stored in powder form desiccated at reduced pressure at —20°. Fractions with R_f values of 0.45 and 0.69 were reexamined after storage and were found to have been partially converted to that fraction with an R_f value of 0.37. Acyl migration to the thermally more stable 3'-OH substituted structure[9] is felt to be responsible for this interchange reaction. It was also noticed that, during storage in an aqueous medium at room temperature, all three fractions produced ATP.

A potassium bromide pellet prepared from the major component indicated the formation of an ester linkage, the presence of a phosphate, and an N-t-butoxycarbonyl group. A UV spectrum with a maximum absorption at 257.5 nm identical to that of ATP eliminated the possibility that aminoacylation may have taken place at 6-amino group of nucleotide. Nuclear magnetic resonance spectroscopy demonstrated the presence of tertiary butyl protons. On the basis of the above evidence, structure (II) has been assigned to this product.

The cleavage of the N-t-butoxycarbonyl group from the N-t-butoxycarbonylamino-α-alanyl ATP (II) was performed by treating with trifluoroacetic acid briefly at ice bath temperature (Eq. 1).

$$(1)$$

[9] C. S. McLaughlin and V. M. Ingram, *Biochemistry* **4**, 1442, 1448 (1965).

Diazotization of the aminoacyl derivative of ATP (III) using sodium nitrite in 5% sulfuric acid was not successful. Successful diazotization is known to depend upon being able to extract the product into a water-immiscible phase immediately upon formation. The water solubility of the nucleotide would thus appear to limit the formation of the diazotization product.

Condensation with N-t-Butoxycarbonyl α-Alanine p-Nitrophenyl Ester and N-t-Butoxycarbonyl β-Alanine p-Nitrophenyl Ester

An additional method attempted for the synthesis of diazo derivatives of ATP was via the condensation of diazotized α-alanine p-nitrophenyl ester with ATP under imidazole catalysis. The N-protecting group from N-t-butoxycarbonyl α-alanine p-nitrophenyl ester was removed by acid and the product diazotized prior to attempted condensation with ATP.

The first step worked out with quantitative yields. However, yields for the formation of the diazo ester vary from 8% to near 0 depending to a great extent upon how rapidly and how thoroughly the methylene chloride extraction of the diazo ester can be managed. The diazo ester once isolated from the reaction mixture by preparative thin-layer chromatography can be stored indefinitely at 0° without decomposition.

The condensation of the diazo ester and the ATP triethylammonium salt[10] was attempted by allowing the reaction to proceed at room temperature for 4 hr in dimethyl sulfoxide under imidazole catalysis. However, no products other than ATP and unchanged diazo ester were recovered. The lower activity of the diazo p-nitrophenyl esters is explained by the electron distribution resulting from the presence of the α-diazo group. This arrangement makes it difficult for the carbonyl carbon to assume the partial positive charge required for the nucleophilic attack.

The difficulty encountered in carrying out the final step in the synthesis of the precursors of active carbene analogs of ATP directed us toward investigations of the synthesis of the diazo derivative of β-alanine p-nitrophenyl ester, a compound in which the diazo group is not conjugated to the active ester group. Diazotization of β-alanine p-nitrophenyl ester hydrochloride salt has, however, not been accomplished. Since the diazo group is not conjugated with a carbonyl, it is very reactive and tends to decompose and undergo further reaction as soon as it is formed.

Preparation of Arylazido ATP Analogs and Their Utilization

In view of the relative stability of nitrene precursors and the difficulties encountered in the final stages of synthesis of the diazo deriva-

[10] J. Zemlicka and S. Chladek, *Collect. Czech. Chem. Commun.* **33**, 3293 (1968).

tives of ATP, it was decided to investigate the synthesis of ATP analogs containing nitrene precursor groups. ATP was reacted with p-azidobenzoic acid in the presence of carbodiimidazole. The reaction mixture was chromatographed and was shown to contain, in addition to ATP and p-azidobenzoic acid, a new UV-absorbing material. This band with an R_f of 0.35 in the solvent system n-butanol:water:acetic acid (5:3:2) was eluted with water. The UV spectrum of the material recovered after lyophilization showed a maximum at 266 nm and a shoulder at 285 nm. A similar reaction carried out with NADP replacing ATP showed that the azidopyridine nucleotide analog could be formed.

The successful synthesis of these compounds showed the feasibility of using the carbodiimidazole coupling reaction in the formation of arylazido ATP analogs. For the general synthesis of the arylazidocarboxylic acids, 4-fluoro-3-nitroaniline is first diazotized in a concentrated hydrochloric acid medium in the presence of sodium nitrite [Eq. (2)]. The diazonium salt is subsequently treated with sodium azide, which results in the formation of a light-sensitive 4-fluoro-3-nitrophenyl azide (IV) [Eq. (3)]. The structure of (IV) has been confirmed by its melting point of 52°–52.5° (recrystallized from petroleum ether); its NMR spectrum, 7.47 ppm (2 protons), 7.78 ppm (1 proton); and its mass spectrum, m/e 182.

$$F-\!\!\bigcirc\!\!-NH_2 \xrightarrow[NaNO_2]{HCl} F-\!\!\bigcirc\!\!-N_2^+ \ Cl^- \tag{2}$$
$$\overset{|}{NO_2} \qquad\qquad \overset{|}{NO_2}$$

$$F-\!\!\bigcirc\!\!-N_2^+ \ Cl^- \xrightarrow{NaN_3} F-\!\!\bigcirc\!\!-N_3 \tag{3}$$
$$\overset{|}{NO_2} \qquad\qquad\qquad \overset{|}{NO_2}$$
$$\text{(IV)}$$

The resulting azide is then condensed with an amino carboxylic acid in an ethanolic aqueous solution containing sodium carbonate. N-4-azido-2-nitrophenylaminocarboxylic acids with varying hydrocarbon chain lengths separating the amino and carboxylic group have been synthesized [Eq. (4)], and their structures were identified via infrared and mass spectrometry.

$$N_3-\!\!\bigcirc\!\!-F \ + \ NH_2(CH_2)_n \, COOH \xrightarrow{Na_2CO_3} N_3-\!\!\bigcirc\!\!-NH(CH_2)_n \, COOH$$
$$\quad\ \overset{|}{NO_2} \qquad\qquad\qquad\qquad\qquad\qquad\qquad\qquad \overset{|}{NO_2}$$

$$n = 1 \ (V); \ 2 \ (VI); \ 3 \ (VII); \ 4 \ (VIII); \ 5 \ (IX); \ 10 \ (X); \ 11 \ (XI) \tag{4}$$

The N-(4-azido-2-nitrophenyl)aminocarboxylic acids are then condensed with ATP by means of carbodiimidazole catalysis [Eq. (5)].

$$n = 2 \text{ (XII)}; \quad 3 \text{ (XIII)}; \quad 5 \text{ (XIV)}$$

Attempts to prepare an ATP analog by condensation of the azido derivative of glycine with ATP under our standard conditions were unsuccessful. Presumably the nucleophilic reaction center, i.e.,

$$-\overset{\overset{\text{O}}{\|}}{\text{C}}-\text{OH}$$

is sufficiently deactivated by its proximity to the α-amino group. The condensation reaction with the azido derivative of β-alanine, 4-aminobutyric acid, or 6-aminocaproic acid, in which the reaction center is not α to the amino-deactivating group, was found to proceed satisfactorily. In the case of the synthesis of arylazido-β-alanine ATP (XII), following the condensation with ATP, the crude reaction product was first dried under reduced pressure and then subjected to preliminary purification by repeated washing with acetone. This removes the excess carbodiimidazole and the azido derivative of the β-alanine (VI). The solid residue was taken up in water and applied to Whatman No. 3 MM paper. The material from the band migrating at an R_f of 0.38 was recovered by elution with water and concentrated by lyophilization. This material (XII) gave a UV spectrum with maxima at 260, 290 (sh), and 480 nm. The increase in 260 nm peak intensity as compared with the starting material is due to the presence of the adenine ring structure.

Methods

Esterification of ATP with N-t-Butoxycarbonyl-α-alanine

Carbodiimidazole (71 mg, 0.43 mmole) and N-t-BOC-α-alanine (70.8 mg, 0.40 mmole) dissolved in 0.2 ml of dimethylformamide (dried over molecular sieves) are stirred for 10 min at room temperature.[11] The solu-

[11] R. Paul and G. W. Anderson, *J. Am. Chem. Soc.* **82**, 4596 (1960).

tion is added to ATP (24.2 mg, 0.04 mmole in 1 ml of water), and the reaction mixture is allowed to react at room temperature for 3.5 hr. Solvent is evaporated under reduced pressure, and the resulting oil is washed with ether to remove imidazole and excess N-t-BOC-α-alanine. The residue from ether extraction (87.8 mg) is dissolved in water and chromatographed on 20 $\times$ 20-cm Whatman No. 42 paper using n-butanol: water:acetic acid (5:3:2) as eluent. A resulting, well defined, UV absorbing band at R_f 0.37, a diffused band at R_f 0.45, and a narrow band at R_f 0.69 are eluted from the paper with water and lyophilized, resulting in the collection of 7.3 mg, 1.0 mg, and 2.9 mg, respectively. The fractions with R_f values of 0.45 and 0.69 are the 2′ isomer and 2′,3′-disubstituted compound. The fraction with R_f 0.37 is the 3′ isomer (II).

Esterification of ATP at the 3′ Position with p-Azidobenzoic Acid

The 3′-p-azidobenzoyl ATP is prepared with p-azidobenzoic acid and ATP using carbodiimidazole as a catalyst as indicated above for the esterification of ATP with N-t-BOC-α-alanine. The reaction mixture is subjected to chromatographic resolution with n-butanol:water:acetic acid (5:3:2) as solvent. The reaction product has a UV absorption band with an R_f of 0.37. The material is eluted from the paper with water and found to have absorption peaks at 266 and 285 (sh) nm. Irradiation of the aqueous solution with UV light shifts the 266 nm absorption maximum to 257 nm. The starting p-azidobenzoic acid has a maximum absorption peak at 271 nm which is broadened by UV irradiation.

Preparation of 4-Fluoro-3-nitrophenyl Azide (IV)[12,12a]

4-Fluoro-3-nitroaniline (4.38 g, 0.028 mole) dissolved in 30 ml of concentrated HCl and 5 ml of water at 45° is filtered and then chilled to −20° in a Dry Ice–acetone bath. Sodium nitrite (2.4 g, 0.034 mole) in 5 ml of water is added slowly to the well stirred acid medium while the flask temperature is kept at −15° to −20°. The reaction mixture after a 10-min stirring is filtered into a flask at −20°. To this filtrate, sodium azide (2.2 g, 0.032 mole) dissolved in 8 ml of water is added dropwise while the reaction mixture is stirred and kept in the dark at −15°

---

[12] Azido derivatives of this nature are known to be light sensitive. They are explosive in large quantities when dry, and caution is urged in their preparation. We always utilized a ventilated hood, and reactions were carried out in the dark.
[12a] G. W. J. Fleet, J. R. Knowles, and R. R. Porter, *Biochem. J.* **128**, 499 (1972).

to $-20°$. The light brown solid obtained by filtration is washed with ice water and dried in a vacuum desiccator. Recrystallization from petroleum ether yields straw-colored needles. Yield: 3.2 g, 63%; m.p.: $52°-52.5°$; MS: m/e 182; NMR ($CDCl_3$): 7.47 ppm (2 H), 7.78 ppm (1 H).

Preparation of N-4-Azido-2-nitrophenyl-β-alanine (VI)[12a]

To 5.4 ml of an aqueous solution of β-alanine (534 mg, 6 mmoles) and sodium carbonate (1.08 g, 10 mmole) is added 4-fluoro-3-nitrophenyl azide (900 mg, 4.9 mmoles). Ethanol (6.75 ml), water (5.4 ml), and another portion of ethanol (13.5 ml) are added subsequently to enhance the homogeneity of the reaction mixture. The reaction mixture suspension is stirred at $52°$ overnight with an attached cooling condenser. The resulting dark red mixture is first concentrated under reduced pressure to about one-third of its volume and then diluted with 18 ml of water. Two extractions with 45 ml of ether remove all the excess starting azide. The aqueous layer is acidified with 3 N HCl to pH 2 and extracted with three 90-ml portions of ether. The combined ether extract is washed three times, each with 50 ml of saturated NaCl solution, and is dried over sodium sulfate and then evaporated to dryness. The residue is recrystallized from hot ethanol. Yield: 736 mg (59%); m.p.: $142.5°-145°$; MS: m/e 251; UV: $\lambda_{max}^{CH_3OH}$ 260 (molar extinction coefficient 27.2 $\times$ 10^3), 280 (sh) and 460 (molar extinction coefficient 5.9 $\times 10^3$) nm; IR: λ_{mox}^{KBr} 2.96, 3.43, 4.74, 4.80, 5.83, 6.13, and 6.37 μm; NMR (chemical shift is expressed in ppm, J in cps, solvent = CD_3OD): 7.79 (aromatic H-3, d, $J = 2$), 7.28 (aromatic, quartet, H-5, $J_{56} = 10$, $J_{35} = 2$), 7.12 (aromatic H-6, d, J = 10), 3.66 (CH_2 β to the carboxylic proton, t, $J = 6$).

Preparation of N-4-Azido-2-nitrophenyl Amino Carboxylic Acid Derivatives

The N-4-azido-2-nitrophenyl derivatives of glycine (VI), α-alanine (VIa), 4-aminobutyric acid (VII), 5-aminovaleric acid (VIII), 6-amino-caproic acid (IX), 11-aminoundecanoic acid (X), and 12-aminodode-canoic acid (XI) were prepared following the procedure as outlined for the preparation of N-4-azido-2-nitrophenyl-β-alanine (VI), and the physical characteristics of these derivatives are indicated below. The R_f values given are those obtained on silica-gel plates with n-butanol saturated with water as the solvent.

N-4-Azido-2-nitrophenylglycine (*V*). UV: $\lambda_{max}^{CH_3OH}$ 260, 280 (sh), and 460 nm; IR: λ_{max}^{KBr} 3, 3.5, 4.7, 5.70, and 6.12 μm; R_f: 0.31; MS: m/e 237.

N-4-Azido-2-nitrophenyl-α-alanine (*VIa*). UV: $\lambda_{max}^{CH_3OH}$ 263, 284 (sh), and 460 nm; IR: λ_{max}^{KBr} 3.0, 3.5, 4.71, 4.78, 5.79, 6.12, and 6.38 μm; R_f: 0.41; MS: m/e 251.

N-4-Azido-2-nitrophenyl-4-aminobutyric Acid (*VII*). UV: $\lambda_{max}^{CH_3OH}$ 260, 280 (sh), and 460 nm; IR: λ_{max}^{KBr} 2.96, 3.36, 4.71, 5.88, 6.12, and 6.38 μm; R_f: 9.63; MS: m/e 265.

N-4-Azido-2-nitrophenyl-5-aminovaleric Acid (*VIII*). UV: $\lambda_{max}^{CH_3OH}$ 257, 282 (sh), and 460 nm; R_f: 0.68.

N-4-Azido-2-nitrophenyl-6-aminocaproic Acid (*IX*). UV: $\lambda_{max}^{CH_3OH}$ 259, 280 (sh), and 460 nm; IR: λ_{max}^{KBr} 2.96, 3.42, 4.75, 5.88, 6.13, and 6.38 μm; MS: m/e 293; R_f: 0.68.

N-4-Azido-2-nitrophenyl-11-undecanoic Acid (*X*). UV: $\lambda_{max}^{CH_3OH}$ 261, 285 (sh), and 460 nm; IR: λ_{max}^{KBr} 2.96, 3.43, 4.72, 5.90, 6.13, and 6.39 μm; MS: m/e 363; R_f: 0.77.

N-4-Azido-2-nitrophenyl-12-aminododecanoic Acid (*XI*). UV: $\lambda_{max}^{CH_3OH}$ 259, 282 (sh), and 460 nm; IR: λ_{max}^{KBr} 2.96, 3.44, 4.77, 5.87, 6.14, and 6.40 μm; R_f: 0.77; MS: m/e 377.

Preparation of 3'-O-{3-[N-(4-Azido-2-nitrophenyl)amino]propionyl}-adenosine 5'-Triphosphate (XII), "Arylazido-β-alanine ATP"

Carbodiimidazole (270 mg, 1.67 mmoles) and *N*-4-azido-2-nitrophenyl-β-alanine (125.5 mg, 0.5 mmole) dissolved in 0.5 ml of dimethylformamide (dried over molecular sieves) are stirred for 15 min prior to introducing 60.5 mg of ATP (0.1 mmole) dissolved in 2.5 ml of water. The reaction is allowed to proceed overnight in the dark at room temperature. After the evaporation of the solvent to dryness, the residue is repeatedly washed with acetone followed by centrifugation. The acetone-washed residue obtained after drying under reduced pressure is dissolved in about 0.2 ml of water. This sample is then applied to Whatman No. 3 MM paper and eluted with n-butanol:water:acetic acid (5:3:2). Two major orange-colored UV-absorbing bands of R_f values 0.91 and 0.38, in addition to a colorless UV-absorbing band with an R_f value 0.05, are obtained. The front moving band (R_f = 0.91) and the slow moving band (R_f = 0.05) have been identified as *N*-4-azido-2-nitrophenyl-β-alanine and ATP, respectively. The material (19.3 mg) recovered by eluting the band of R_f value 0.38 with water followed by lyophilization gives a spectrum with maxima at 480 (molar extinction coefficient 4.2×10^3), 290 (sh), and 260 nm (molar extinction coefficient 35.4×10^3) attributed to the addition of the ATP and azido-β-alanine molecules. NMR (chemical shift is expressed in ppm, J in cps, solvent = D_2O): 8.29 (H-8), 7.63 (H-2), 6.54 (H-1', d, $J = 7$), 5.00 (H-2', buried under water peak), 6.08 (H-3', quartet $J = 7$), 4.80 (H-4'), 4.64 (H-5'), 4.00 (CH$_2$ β to the carboxylic group, 5, $J = 6$), 3.30 (CH$_2$ α to the carboxylic group, $J = 6$).

Azido-4-aminobutyric ATP (XIII) and azido-6-aminocaproic ATP (XIV) were prepared according to the procedure described for the preparation of azido-β-alanine ATP (XII). The R_f values given refer to paper chromatography. Attempts to prepare azidoglycine ATP and azido-α-alanine ATP by identical procedures gave no indication of product formation.

Azido-4-aminobutyric ATP (XIII). $R_f = 0.40$ with n-butanol:water: acetic acid (5:3:2) as the solvent; $R_f = 0.64$ with isobutyric acid:0.5 N NH$_3$ (5:3) as the solvent; UV: $\lambda_{H_2O}^{max}$ 260, 290 (sh), and 480 nm.

Azido-6-aminocaproic ATP (XIV). $R_f = 0.49$ (solvent, n-butanol: water:acetic acid, 5:3:2); $R_f = 0.65$ (solvent, isobutyric:0.5 N NH$_3$, 5:3); UV: $\lambda_{H_2O}^{max}$ 260, 290 (sh), 480 nm.

Preparation of Azido-[1-¹⁴C]β-alanine-ATP

(*i*) The [1-¹⁴C]β-alanine (0.05 mC, in 0.2 ml of 0.01 N HCl, 0.5 mg, 6.2 μmoles) is concentrated under reduced pressure dissolved in 0.2 ml of water. Sodium carbonate (3.9 mg, 37.2 μmoles) is added followed by 0.2 ml of an ethanol solution containing 3.4 mg of 4-fluoro-3-nitrophenyl azide (18.6 μmoles). The resulting clear solution is maintained at 50° in a sand bath for 5.5 hr.

The reaction mixture, after evaporation to dryness, is extracted with acetone and then methanol and the combined solvents are concentrated to yield a solid residue. The residue is dissolved in a minimal amount of methanol and chromatographed on a 20 × 20-cm silica gel plate with n-butanol saturated with water as the solvent. Three colored bands are observed (orange to yellow color), and radioactivity can be detected at the band with an R_f value of 0.41 (azido-[1-¹⁴C]β-alanine). The labeled azido-β-alanine is recovered from the plate by elution with methanol.[13]

(*ii*) The azido-[1-¹⁴C]β-alanine diluted with 61.5 mg (0.25 mmole) of cold N$_3$-β-alanine is allowed to react with ATP (30.25 mg, 0.05 mmole) in the presence of carbodiimidazole (60.5 mg. 0.37 mmole) as a catalyst. The azido-[1-¹⁴C]β-alanine ATP obtained by paper chromatography separation had a specific activity of 20,000 cpm/μmole with UV maxima at 260, 290 (sh), and 474 nm.

[13] R. A. G. Smith and J. R. Knowles, *Biochem. J.* **141**, 51 (1974).

Preparation of Tritium and ^{32}P-Labeled Analogs of Arylazido-β-alanine ATP

Tritium-labeled arylazido-β-alanine ATP (1.6×10^7 cpm/μmole) is prepared by condensing 3-[N-(4 azido-2-nitrophenyl)]aminopropionic acid with [2,8-^{3}H]adenosine 5-triphosphate tetrasodium salt (1 mCi) obtained from New England Nuclear. The ^{32}P-labeled ATP analog is prepared in the identical manner utilizing γ-labeled [^{32}P]ATP. The synthesis of a number of labeled arylazido-β-alanine nucleotides is schematically outlined in Fig. 1.

The esterification of ATP by N-4-azido-2-nitrophenyl-β-alanine could result in the formation of either the 2′ or 3′ isomer. In the case of amino acid esters of nucleotides, such isomers can be resolved chromatographically. In the esterification of ATP with N-t-butoxycarbonyl-α-alanine, the reaction products could be shown to be resolved into three components

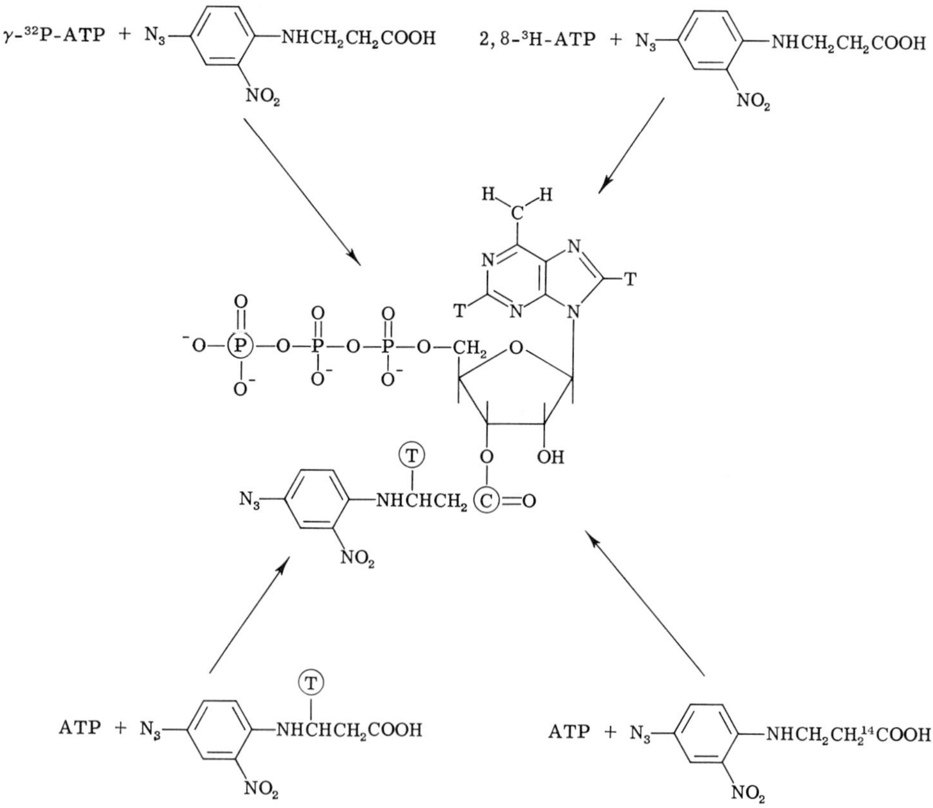

Fig. 1. Synthesis of radioactive arylazido-β-alanine ATP analogs.

by paper chromatography using n-butanol:water:acetic acid (5:3:2) as eluent. The two minor components with R_f values of 0.45 and 0.69 were converted to the major component (R_f 0.37) upon standing at room temperature. We assume that the component with R_f 0.69, because of its lesser polarity, represents the 2′,3′-disubstituted ATP analog. The component with R_f 0.45 represents the 2′ isomer, and the component with R_f 0.37 the 3′ isomer. The possibility of conversion of the 2′ to the 3′ isomer by acyl migration is well supported by a number of studies.[9,14] Evidence in the literature indicates that the 2′-hydroxyl groups of both adenosine and uridine are more open to electrophilic attack than the 3′-hydroxyl group.[15] Thus the 2′-hydroxyl group is kinetically more reactive for substitution. However, such substitution is relatively less stable in comparison to the thermally more favorable esterification at the 3′-hydroxyl group.[14]

Our NMR studies indicate that in the case of arylazido-β-alanine ATP a proton shift from 4.98 ppm, where H-3′ normally absorbs to a position substantially down field at 6.08 ppm, occurs owing to esterification at the 3′ position.

The accumulation of data indicates the 3′ isomers to be the specific components isolated under our synthetic conditions. Further support for this conclusion is seen in the recent observation that the esterification of adenosine with fluorosulfonylbenzoyl chloride results in the preferential formation of 3′-p-fluorosulfonylbenzoyladenosine.[16]

Hydrolysis of Arylazido-β-alanine ATP (XII)

The stability of {3′-O[3-N(4-azido-2-nitrophenyl)amino]propionyl}-adenosine 5′-triphosphate (arylazido-β-alanine ATP) is tested at 20° in neutral and basic solutions. Small amounts of the reagent are dissolved in (a) 0.5 M NH$_4$HCO$_3$; (b) 1.0 M NH$_4$HCO$_3$; (c) H$_2$O; and (d) 5 mM Tris chloride pH 6.7. Incubation is allowed to continue, and at 0, 1, and 24 hr a small aliquot of the reaction mixture is spotted on Whatman No. 3 MM paper. Chromatography is conducted with N-butanol:water:acetic acid (5:3:2) as solvent. The paper is dried and examined for UV absorbing material. All four samples immediately upon mixing show a single clear UV-absorbing spot with an R_f of 0.38 corresponding to the arylazido-β-alanine ATP analog and indicating that no hydrolysis has taken place. One hour later, sample (a) gives one major spot at R_f 0.90

[14] P. C. Zamecnik, *Biochem. J.* **85**, 257 (1962).
[15] B. E. Griffin, M. Jarman, C. B. Reese, J. E. Sulston, and D. R. Trenthan, *Biochemistry* **5**, 3638 (1966).
[16] P. K. Pal, W. J. Wechter, and R. F. Colman, *Biochemistry* **14**, 707 (1975).

corresponding to arylazido-β-alanine and two spots at R_f 0.38 and 0.05, which correlate with the arylazido-β-alanine ATP and ATP. Sample (b) is completely hydrolyzed, only ATP and arylazido-β-alanine being detected. Overnight storage of sample (a) at room temperature results in complete hydrolysis to ATP and arylazido-β-alanine. Samples (c) and (d) retain resistance to degradation following overnight incubation at 20° or at 4°C.

Spectral Characteristics and Photochemical Reactivity of Arylazido-β-alanine ATP

In all the experiments described, photolysis was carried out in an air-cooled unit kept below 15°. The sample in either 13 × 100 mm culture tubes or a glass cuvette was located equidistant between two tungsten-halogen projector lamps, 650 watts, DVY 3400°K, spaced 20 cm apart. The volume of the irradiated mixture varied from 0.1 to 3 ml. The sample

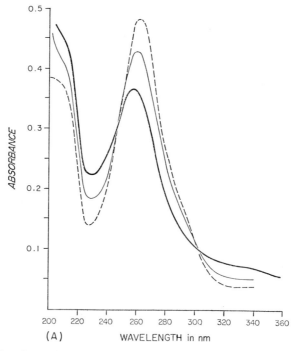

Fig. 2. The absorption spectra of arylazido-β-alanine ATP as a function of the time of photoirradiation. Arylazido-β-alanine ATP was dissolved in water to 14.25 μM and subjected to photoirradiation for periods of 0 min (----), 6 min (——), and 10 min (▬▬). (A) Absorbance from 200 to 360 nm; (B) absorbance from 350 to 600 nm.

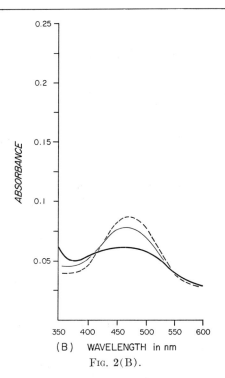

(B)   WAVELENGTH in nm

FIG. 2(B).

was removed from the photolysis apparatus, chilled, and mixed well at 30-sec intervals during the irradiation in order to prevent overheating.

Figure 2 shows the absorption spectra of arylazido-β-alanine ATP (XII) as a function of the time of photoirradiation. The ATP analog has maxima at 260 and 480 nm. The increase in the 260 nm peak intensity above that of ATP is due to the combined presence of the adenine ring structure and (XI). The absorption band at 470–480 nm is a consequence of the 4-azido-2-nitrophenyl adjunct. As can be seen when (XII) is subjected to photoirradiation in an aqueous solution, there occurs a light-dependent decrease in the intensity of both the 260 nm and 470–480 nm absorption band.

Biochemical Interactions

Inhibition of Myosin Subfragment 1 ATPase with Arylazido-β-alanine ATP (XII)

When subfragment 1[17,18] was irradiated in the presence of arylazido-β-alanine ATP, a 67% loss in ATPase activity was observed within 60 sec,

[17] S. Lowey, H. S. Slater, A. G. Weeds, and H. Baker, *J. Mol. Biol.* **42**, 1 (1969).
[18] P. Liu-Osheroff and R. J. Guillory, *Biochem. J.* **127**, 419 (1972).

while the control dark sample with arylazido ATP remained stable. The extent of photoinactivation has been found to approach maximal at irradiation of 2-min duration. A variable degree of inhibition has been obtained with different protein preparations; a 2-min photolysis of myosin subfragment 1 in the presence of azido-β-alanine ATP resulted in a 45–75% decrease in ATPase activity. There are indications that the age of the protein preparation influences the degree of photoinactivation. One sample was shown to be initially inhibited 75%, whereas 16 days later, under otherwise identical conditions, only 46% inhibition was obtained.

The effectiveness of arylazido-β-alanine ATP as a photoinactivator of ATPase activity is influenced greatly by the presence of the natural substrate ATP. When equivalent amounts of ATP and arylazido-β-alanine ATP were added to the subfragment, irradiation induced no inhibition. The additional fact that arylazido-β-alanine (VI) without the nucleotide adjunct is not an effective photoinactivator strongly indicates that the arylazido ATP photoinactivation is due to its active site-directing ability.

The Dephosphorylation of Arylazido-β-alanine ATP by Myosin Subfragment 1 ATPase

The possibility that the arylazido analog might act as a substrate for subfragment 1 ATPase was tested initially with the EDTA-stimulated ATPase activity. The control assay system contained in 0.5 ml, 71 mM EDTA, 0.41 M NH$_4$OH, 86 mM Tris chloride pH 8, and 10 mM ATP.[19] In the experimental vessel 3 mM arylazido-β-alanine ATP replaced the 10 mM ATP. Dephosphorylation was allowed to proceed for 20 min at 30° in the dark, and the degree of ATPase activity was determined by the amount of inorganic phosphate liberated. The arylazido analog of ATP was found to be dephosphorylated at a rate equivalent to 17% of the control ATPase activity. (Control ATPase activity was 230 μmoles of phosphate liberated per milligram of protein per hour.)

Subsequent kinetic studies on the Ca^{2+}-dependent ATPase of subfragment 1 were accomplished by using the ATP analog containing ^{32}P in the phosphate position. The ATPase activity was determined by measuring the amount of radioactivity extracted into an isobutanol phase in the presence of acid molybdate.[20] A K_m of 6.15 μM was evaluated in this way for the nucleotide analog, a value remarkably similar to that for the natural adenine nucleotide substrate.

[19] R. Yount, J. S. Frye, and K. R. O'Keefe, *Cold Spring Harbor Symp. Quant. Biol.* **37**, 113 (1972).

[20] R. Yount, D. Ojala, and D. Babcock, *Biochemistry* **10**, 2490 (1971).

When ATP hydrolysis is measured in the presence of different concentrations of arylazido-β-alanine ATP in the dark, the analog is observed to be an effective competitive inhibitor. This is in contrast to the irreversible inhibition of ATPase activity following photoirradiation in the presence of the analog.

Covalent Labeling of Myosin Subfragment 1 with
Arylazido-1-[^{14}C]β-alanine ATP

Photoinactivation of myosin subfragment 1 with arylazido-1-[^{14}C]β-alanine ATP resulted in the labeling of the protein with approximately stoichiometric amounts of radioactivity. Subfragment 1 (2.6 mg, 2×10^{-2} μmole) in 5 mM Tris HCl, pH 6.7, was incubated with arylazido-1-[^{14}C]β-alanine ATP (0.925 mg, 2.2×10^4 cpm, 1.1 μmoles) for 15 min at 0° followed by 3 min of irradiation. The protein was dialyzed against 5 mM Tris HCl, pH 6.7, and then against 25 mM Tris chloride, pH 7.6, for 24 hr. The dialyzate was then treated with 2.6 mg of Norit. The Norit-treated preparation (1.5 ml) was passed through a Sephadex G-75 column (1 $\times$ 15 cm), and 0.35-ml fractions were collected. Protein eluted at the void volume contained radioactivity. These fractions were combined, and the total protein was determined at 6.8 mmoles. A molecular weight of 1.2×10^5 was used for subfragment 1.[21] Radioactivity measurements indicated 7.1 mmoles of the arylazido compound to be covalently bound to the protein. Such measurements indicate a label:protein ratio of 1.04.

A Comparison of the Arylazido-β-alanine ATP Labeling of
Myosin Subfragment 1 and Heavy Meromyosin after Photolysis

Different concentrations of tritium-labeled arylazido-β-alanine ATP were incubated with either subfragment 1 or heavy meromyosin.[22] After photolysis the protein was subjected to chromatography on Sephadex G-50. When the specific radioactivity of the protein eluting at the void volume was evaluated together with the remaining enzymic activity, total photodependent inhibition of subfragment 1 was accomplished at a ratio of 1 mole of arylazido-β-alanine bound per mole of protein. In the case of heavy meromyosin, the ratio was 2 moles of nucleotide analog bound per mole of enzyme. Covalent binding of the arylazido-β-alanine

[21] D. M. Young, S. Himmelfarb, and W. F. Harrington, *J. Biol. Chem.* **240**, 2428 (1965).
[22] P. D. Wagner and R. Yount, *Biochemistry* **14**, 1900 (1975).

ATP is thus associated in a very precise manner with Ca^{2+}-dependent ATPase activity.

Reactivity of Other Arylazido-ATP Analogs

Two additional arylazido ATP analogs, arylazido-4-aminobutyric ATP (VII) and arylazido-6-aminocaproic ATP (IX), in which there are three and five methylene groups, respectively, bridging the distance between the ribose portion of the adenine nucleotide and the potential aryl nitrene, have been tested on myosin subfragment 1 ATPase activity. While the specificity of only the arylazido-β-alanine ATP analog has been investigated extensively, all three arylazido ATP analogs inhibit subfragment 1 enzymic activity in a photodependent reaction. In each case the extent of photoinactivation is dependent upon the concentration of the arylazido ATP analog and shows saturation kinetics.

Interaction of Arylazido-β-alanine ATP with Mitochondrial ATPase[23,24]

Arylazido-β-alanine ATP brings about a sizable photodependent inhibition of the soluble mitochondrial F_1 (ATPase). That the photodependent inhibition by the ATP analog is not a characteristic unique to the soluble enzyme can be seen from Table I, where the ATP analog is shown to bring about a greater than 80% photodependent inhibition of ATPase activity associated with a number of mitochondrial particles, all of varying specific ATPase activity.[25,26] The membrane-bound ATPase activity appears to be much more sensitive to the arylazido-β-alanine ATP-dependent photoinhibition than the soluble ATPase. In separate experiments the presence of p-aminobenzoic acid during photolysis and up to 15 μM did not affect ATPase activity and had only a slight influence upon the degree of photoinactivation due to arylazido-β-alanine ATP. The presence of the natural substrate (ATP) prevents the photodependent inhibition by arylazido-β-alanine ATP.

When arylazido-β-alanine ATP containing ^{32}P in the γ phosphate is incubated with F_1 (ATPase) in the dark, there is a sizable turnover of the enzyme. Within 15 sec, in excess of 40% of the added 2 mmoles of the analog one is hydrolyzed by 6.37 μg of F_1 (ATPase). Assay of the enzyme under these conditions, i.e., in the absence of a nucleotide re-

[23] A. F. Knowles and H. S. Penefsky, J. Biol. Chem. 247, 6617 (1972).
[24] M. F. Pullman, H. S. Penefsky, A. Datta, and E. Racker, J. Biol. Chem. 235, 3322 (1960).
[25] J. M. Fessenden and E. Racker, J. Biol. Chem. 241, 2483 (1966).
[26] E. Racker and L. L. Horstman, J. Biol. Chem. 242, 2547 (1967).

TABLE I
ARYLAZIDO-β-ALANINE ATP-DEPENDENT PHOTOINHIBITION
OF THE ATPASE ACTIVITY OF SUBMITOCHONDRIAL
PARTICLES[a]

Arylazido-β-alanine ATP[c]	ATPase, % control[b]		
	ETP[d]	ETP$_A$[d]	Sephadex[e]
1.28	38	—	22
2.53	—	33	—
3.85	31	—	16
6.40	29	—	—
7.60	—	22	—
12.70	—	15	—
25.60	—	—	10
50.70	—	12	—

[a] Submitochondrial particles (ETP, 1,015 mg; ETP$_A$, 0.514 mg, and Sephadex particles, 1.01 mg) were incubated with various amounts of arylazido-β-alanine ATP (6.5–118.2 μM) in a total volume of 0.2 ml containing 0.25 M sucrose and 10 mM Tris HCl, pH 7.5. Following a 4-min irradiation period, the ATPase activity was assayed with the ATP regenerating system described by M. F. Pullman, H. S. Penefsky, A. Datta, and E. Racker, *J. Biol. Chem.* **235**, 3322 (1960).

[b] The specific activity (micromoles of ATP hydrolyzed per milligram of protein per minute) was for ETP (0.82), ETP$_A$ (2.0), Sephadex particles (4.2); the data are recorded as percentage of these values.

[c] The concentration of the arylazido-β-alanine ATP is presented as nanomoles of analog per milligram of protein.

[d] J. M. Fessenden and E. Racker, *J. Biol. Chem.* **241**, 2483 (1966).

[e] E. Racker and L. L. Horstman, *J. Biol. Chem.* **242**, 2547 (1967).

generating system, but as saturating levels of [^{32}P]ATP (5 mM), resulted in an estimated activity of 61 μmoles per milligram of protein per minute. The low specific activity observed (0.51 μmole per milligram protein per minute at 15 sec) is felt to represent a limitation due to the low level of substrate.

Photodependent Labeling of F_1(ATPase) by Arylazido-β-alanine ATP

That the arylazido-β-alanine ATP is bound covalently to the F_1 (ATPase) following photolysis was indicated from experiments utilizing a tritium-labeled nucleotide analog.[27]

The labeled arylazido-β-alanine ATP associated with the F_1 (ATPase) protein following photolysis is stable to Norit treatment and extensive

[27] J. Russell, S. J. Jeng, and R. J. Guillory, *Biochem. Biophys. Res. Commun.* **70**, 1225 (1976).

dialysis at neutral pH. Dissociation of the tritium-labeled protein by incubation with sodium dodecyl sulfate in the presence of mercapto-ethanol followed by extensive dialysis did not result in the liberation of a significant portion of the label. When the labeling was carried out as outlined by Russell *et al.*[27] assuming a molecular weight of 3.6×10^5 for F_1 (ATPase),[28] 0.8 μmole of arylazido-β-alanine ATP was bound per micromole of F_1 (ATPase) protein.

Preliminary experiments indicate that the β subunit of the F_1 (ATPase) is the major site for labeling by arylazido-β-alanine when photolysis is carried out as described.[28a]

Influence of Arylazido-β-alanine ATP on Reactions Associated with the Mitochondrial Membrane

Energy-Linked Transhydrogenation. The energy-dependent transhy-drogenation reaction associated with the mitochondrial membrane can be driven independently by ATP or by respiratory energy.[29] Presumably the succinate-driven pathway utilizes the energized state of the mitochondrial membrane, independent of ATP-dependent reactions and thus presumably independent of F_1 (ATPase) involvement. From Table II it can be seen that, after light irradiation of ETP in the presence of varying concentra-tions of arylazido-β-alanine ATP, the ATP-driven transhydrogenation appears to be preferentially inhibited over that of transhydrogenation supported by succinate oxidation. No inhibition of transhydrogenation was observed for dark controls. The relatively constant and low-level inhibition of the succinate-driven reaction by irradiation in the presence of arylazido-β-alanine ATP may be due to the presence of a small amount of endogenous nucleotide in ETP. It would appear that arylazido-β-alanine ATP is a specific reagent in its interaction with the mitochondrial membrane, restricting itself to reaction with the bound F_1 (ATPase). Although ATPase activity of the ETP particle in inhibited some 50% at 30.6 μM arylazido-β-alanine ATP, there is no major inhibition of either respiration or the phosphorylation reaction. We believe it to be highly significant that those particles whose ATPase is most sensitive to photoinhibition in the presence of arylazido-β-alanine are as well

---

[28] A. R. Senior, *Biochim. Biophys. Acta* **301**, 249 (1973).

[28a] Ching-san Lai, unpublished observations.

[29] L. Ernster and C.-P. Lee, *Annu. Rev. Biochem.* **33**, 729 (1964); C.-P. Lee and L. Ernster, *in* "Regulation of Metabolic Processes in Mitochondria" (J. M. Tager, S. Papa, E. Quagliariello, and E. C. Slater, eds.), pp. 218–234. Am. Elsevier, New York, 1966.

TABLE II

INFLUENCE OF ARYLAZIDO-β-ALANINE ATP ON THE
ENERGY-DEPENDENT TRANSHYDROGENATION OF ETP[a]

Arylazido-β-alanine ATP present during photoirradiation[b]	Transhydrogenation (% control)	
	ATP dependent	Succinate dependent
1.6	51	70
4.9	34	85
8.2	21	71

[a] Electron transport particles (ETP), 1.0 mg in 0.25 ml of solution containing 0.25 M sucrose and 10 mM Tris chloride pH 7.5, were subjected to 4-min photoirradiation in the presence of varying amounts of arylazido-β-alanine ATP. After photoirradiation, an aliquot containing 400 μg of protein was taken for the fluorometric analysis of transhydrogenation in a total volume of 3 ml consisting of 140 mM sucrose, 40 mM Tris HCl pH 7.5, 50 mM sucrose, 40 mM Tris HCl pH 7.5, 50 mM ethanol, 10 mM MgSO$_4$, 0.020 mM NAD$^+$, 0.20 mM NADP$^+$, 1 μg of rotenone, and 0.25 mg of yeast alcohol dehydrogenase. Control transhydrogenation driven by either 5 mM succinate or 3 mM ATP was 19.5 and 8.8 nmoles of NADPH formed per milligram per minute, respectively. The results are presented as percentages of these control activities.

[b] Nanomoles of analog per milligram of protein.

particles that have the lowest phosphorylation potential and highest specific ATPase activity.

Arylazido-β-alanine Pyridine Nucleotide Analogs

Chemical Synthesis and Structural Elucidation

The successful synthesis of ATP analogs containing a photoactive arylazido grouping esterfied to the 3'-hydroxyl of ATP offers the possibility for the synthesis of a wide variety of different nucleotide photoactive probes. In our search for reagents to assist in a comparative study of nucleotide binding regions of a variety of enzyme systems, we decided to attempt the synthesis of a photoactive arylazido pyridine nucleotide. The NAD$^+$ molecule contains four potentially reactive hydroxyl groups within the nucleotide ribose positions.

Preparation of Arylazido-β-alanine NAD$^+$

The N-(4-azido-2-nitrophenyl)-β-alanine is coupled to NAD$^+$ under the same conditions as those described for the synthesis of arylazido-β-

alanine ATP except for the substituting of NAD[+] for the adenine nucleo-tide.[30] Chromatography of the reaction product on Whatman No. 3 MM paper as described for the adenine analog indicates the formation of two major orange-colored UV-absorbing bands of R_f values 0.62 and 0.47 in addition to a band of R_f 0.91 corresponding to arylazido-β-alanine and one of 0.12 corresponding to NAD[+].

Characterization of the Material with R_f Value of 0.47

The product with the R_f value 0.47 is eluted from the Whatman paper with H_2O, the solvent is removed by lyophilization, and the residue is redissolved in H_2O. Rechromatography indicates that this reaction prod-uct if free of NAD[+] and arylazido-β-alanine.

This material has an absorption spectrum with maxima at 475, 290 (sh), and 260 nm in H_2O. The 475 nm band is characteristic of arylazido-β-alanine analogs. The concentration of the material is determined from the absorbance at 275 nm by utilizing the molar extinction coefficient, 5.9×10^3, as previously determined for arylazido-β-alanine at 460 nm.[30]

The compound was found to undergo reaction with yeast alcohol dehydrogenase. The spectrum obtained following the reaction showed the development of a 340-nm absorption band in addition to the initial two maxima at 475 and 260 nm. The increment of absorbance at 340 nm was linearly related to the concentration of this material as measured by its absorption at 475 nm. In view of these experimental facts, the material has been designated arylazido-β-alanine NAD[+]. The above experiments indicate as well that arylazido-β-alanine NAD[+] can substitute for NAD[+] as a substrate for yeast alcohol dehydrogenase.

In the case of arylazido-β-alanine ATP an arylazido-β-alanine adjunct is attached to the hydroxyl group at the 3'-OH of the ribose moiety of ATP.[30] For NAD[+] there are two ribose units and thus four hydroxyl groups that are potential esterification sites. Our results indicate that but a single arylazido residue has been esterified to NAD[+] under the conditions used for its preparation.

The Structure of Arylazido-β-alanine NAD[+]

The arylazido-β-alanine derivatives of NAD[+] and AMP[31] are well separated from NAD[+] and AMP by paper chromatography on Whatman

[30] S. J. Jeng and R. J. Guillory, J. Supramol. Struct. 3, 448 (1975).

[31] Arylazido-β-alanine AMP was synthesized using the same procedure as that described for the synthesis of arylazido-β-alanine ATP except for the substitution of AMP for the ATP. One product with an R_f value of 0.63 was obtained with the solvent system n-butanol–water–acetic acid (5:3:2).

TABLE III
CHROMATOGRAPHY OF NAD$^+$, AMP, NMN,
AND ARYLAZIDO-β-ALANINE DERIVATIVES IN
SOLVENT SYSTEM n-BUTANOL–WATER–
ACETIC ACID (5:3:2)

Compound	R_f
NAD$^+$	0.12
Arylazido-β-alanine NAD$^+$	0.47
AMP	0.30
Arylazido-β-alanine AMP	0.62
NMN	0.20
Arylazido-β-alanine NAD^{+a}	
Product A	0.18
Product B	0.60

^{a} Chromatography following incubation with nucleotide pyrophosphatase; see text for incubation conditions.

No. 1 paper with n-butanol-water-acetic acid (5:3:2). Their R_f values are shown in Table III.

To determine the structure of arylazido-β-alanine NAD$^+$, the analog is treated with nucleotide pyrophosphatase isolated from potatoes as described by Kornberg and Pricer.[32] After incubation of the arylazido-β-alanine NAD$^+$ with pyrophosphatase, the products are separated on Whatman No. 1 filter paper with n-butanol–water–acetic acid (5:3:2). A single UV absorbing spot with R_f value of 0.18 and two orange spots with R_f values of 0.60 and 0.91 were evident (Table III). The orange material with the highest R_f value (0.91) was characteristic of arylazido-β-alanine.

The UV-absorbing material (R_f 0.18) is considered to be NMN owing to the similarity of their R_f value in this solvent system. When the material is eluted from the paper with H$_2$O, it is found to have an absorption peak at 260 nm. An additional absorption band develops at 325 nm upon the addition of 1 M KCN, indicating the presence of a nicotinic base.[33,34] Further characterization of this material has not yet been attempted.

The orange material with an R_f of 0.60, when eluted from the paper with H$_2$O, was found to have an absorption band at 475–480 nm as well

[32] A. Kornberg and W. E. Pricer, *J. Biol. Chem.* **182**, 763 (1950).
[33] Y. Nishizuka and O. Hayaishi, *J. Biol. Chem.* **238**, 3369 (1963).
[34] R. M. Burton and N. O. Kaplan, *Arch. Biochem. Biophys.* **101**, 139 (1963).

as one at 260 nm. This material could not, however, be reduced by yeast alcohol dehydrogenase in the presence of ethanol. On the basis of a similar R_f value to arylazido-β-alanine AMP (0.62) in this solvent, it was considered to be arylazido-β-alanine AMP. Since the ester bonding of arylazido-β-alanine to the ribose hydroxyl of ATP is known to be susceptible to rapid and complete cleavage when incubated at basic pH,[30] the yellow material with R_f 0.60 was eluted from the paper and the aqueous solution was incubated for 12 hr at room temperature in 1 M NH$_4$HCO$_3$ (pH 9). The reaction products were examined by chromatography on Whatman No. 1 paper with a solvent system consisting of 1 M ammonium acetate–ethanol (3:7). One of the products is a UV-absorbing material with an R_f of 0.10. In this solvent this material cochromatographs with AMP while NMN has an R_f value of 0.27.[33] The other reaction product of the alkaline hydrolysis is orange and has an R_f of 0.68, identical to that of arylazido-β-alanine.

Together with nuclear magnetic resonance spectroscopy studies indicating that the 3′-OH of ATP is esterified during the synthesis of arylazido-β-alanine ATP, the above data indicate that the esterification of NAD$^+$ by arylazido-β-alanine occurs on the 3′-OH of the ribose associated with the adenine portion of the nicotinamide adenine dinucleotide (see Fig. 3).

The reaction product with R_f value of 0.62 has not as yet been completely characterized; it is not, however, reduced by yeast alcohol dehydrogenase.

Biochemical Interactions

Alcohol Dehydrogenase

In addition to acting as a substrate for yeast alcohol dehydrogenase, arylazido-β-alanine NAD$^+$ can bring about a photodependent inhibition of this enzyme. In order to assure ourselves that the photoinhibition resulting from irradiation of alcohol dehydrogenase in the presence of arylazido-β-alanine NAD$^+$ was due to a true photoaffinity effect of the NAD$^+$ analog, irradiation was carried out in the presence of arylazido-β-alanine. Incubation of arylazido-β-alanine at molar excess of 291-fold over that of alcohol dehydrogenase resulted in only a minor inhibition of the enzyme during photoirradiation, whereas the photolysis of the enzyme in the presence of a 59-fold molar excess of arylazido-β-alanine NAD$^+$ resulted in a 63% photodependent inhibition of the dehydrogenase activity (Table IV).

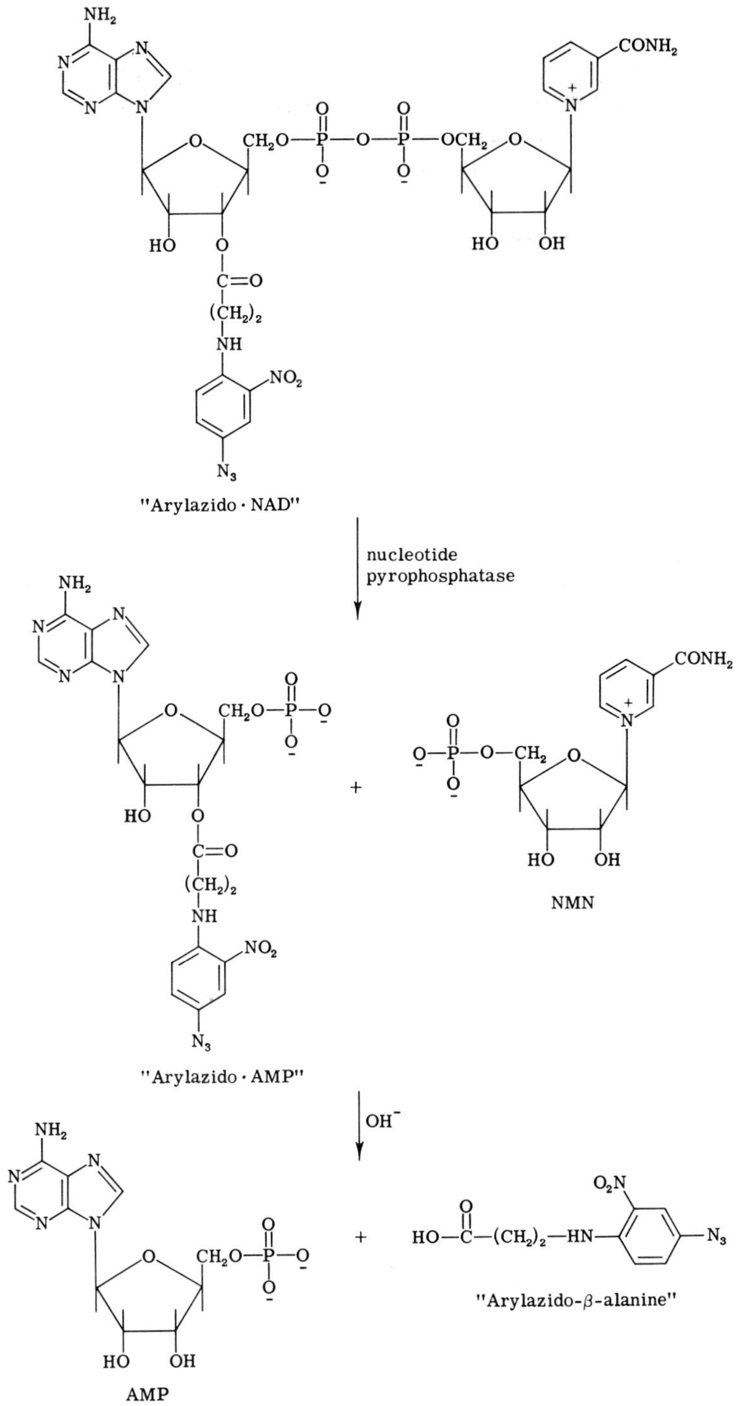

Fig. 3. An illustration of the procedure used to determine the structure of arylazido-β-alanine NAD.

TABLE IV

A COMPARISON OF THE PHOTODEPENDENT INHIBITION OF
YEAST ALCOHOL DEHYDROGENASE BY ARYLAZIDO-β-
ALANINE NAD$^+$ AND ARYLAZIDO-β-ALANINE[a]

Additions	Molar ratio[b]	Percent activity
Arylazido-β alanine NAD$^+$	59	36.6
Arylazido β-alanine	291	94.8

[a] For photolysis, a mixture containing 0.5 M phosphate buffer, pH 7.0, 0.2 mg of yeast alcohol dehydrogenase, and 0.93 mM arylazido-β-alanine NDA$^+$ or 1.94 mM arylazido-β-alanine in 0.21 ml was irradiated for 10 min at 15°. After irradiation, 10 μl of the above mixture was diluted to 1 ml with 0.5 M phosphate buffer, pH 7.0. The assay was initiated by the addition of 5 μl of the diluted irradiated sample to a mixture containing 0.15 M phosphate buffer, pH 7.0, 1.34 mM NAD$^+$, 17 mM ethanol, and 0.59 mM glutathione in 1 ml at 30°. The rate of NAD$^+$ reduction was assayed by the increase in absorbance at 340 nm. The activity of a nontreated control was taken as 100% and is equal to 422 μmoles of NADH produced per minute per milligram of protein.

[b] The molar ratio of the arylazido-β-alanine NAD$^+$ and arylazido-β-alanine to yeast alcohol dehydrogenase is based upon an enzyme molecular weight of 141,000 [Y. Hatefi, in "Comprehensive Biochemistry" (M. Florkin and E. Stolz, eds.), Vol. 14, p. 199. Elsevier, Amsterdam, 1966].

Reaction with NADH-CoQ Reductase[35]

When alcohol and alcohol dehydrogenase are incubated together with arylazido-β-alanine NAD$^+$, they act as a regenerating system capable of maintaining the analog in its reduced state. Under these conditions the analog can bring about a reduction of CoQ$_1$ catalyzed by complex I at a rate of some 75% that of the natural substrate. The reduction of CoQ$_1$ by arylazido-β-alanine NADH catalyzed by complex I is as well rotenone sensitive.

Photodependent Inhibition of NADH-CoQ Reductase by Arylazido-β-alanine NADH

The fact that the reduced form of arylazido-β-alanine NAD$^+$ in the presence of complex I catalyzes the reduction of CoQ$_1$ in a rotenone-sensitive reaction indicates clearly that the reduced analog is acting as a substrate for the NADH dehydrogenase of complex I. The pyridine nucleotide analog is as well a potent photodependent inhibitor of the enzyme complex. In fact the reduced form of the arylazido-β-alanine NAD$^+$ is a more effective inhibitor with greater than 95% inhibition of

[35] Y. Hatefi, D. G. Haavik, and D. E. Griffiths, J. Biol. Chem. 182, 1676 (1962).

NADH-CoQ reductase after a 4-min period of photoirradiation, and an equal concentration of arylazido-β-alanine NAD results in 75% inhibition. The photodependent inhibition is a function of both the time of irradiation and the concentration of arylazido-β-alanine NADH.

Competitive and Noncompetitive Inhibition of NADH-CoQ Reductase by Arylazido-β-alanine NAD$^+$

Arylazido-β-alanine NAD$^+$ can bring about a sizable inhibition of NADH-CoQ reductase activity when incubated with the enzyme complex in the dark without prior photoirradiation. In this case the addition of higher substrate concentrations (NADH) results in a reversal of this inhibition. Reduced NAD at 53 μM is able to completely reverse the inhibitory influence of 11 μM arylazido-β-alanine NAD$^+$. On the other hand, when complex I is subjected to photoirradiation in the presence of 11 μM arylazido-β-alanine NAD$^+$, the addition of higher concentrations of the natural substrate (NADH) is not able to reverse the inhibition. It would thus appear that the arylazido-β-alanine NAD$^+$ acts as a competitive inhibitor of NADH-CoQ reductase in the dark and is converted to a noncompetitive irreversible inhibitor upon light irradiation in the presence of the protein complex.

Other Applications of the Carbodiimidazole-Catalyzed Esterification

Preparation of Spin-Label ATP, ADP, AMP, and NADP

3-Carboxyl-2,2,5,5-tetramethylpyrroline-1-oxyl (184 mg, 1 mmole) and carbodiimidazole (173 mg, 1.1 mmoles) dissolved in 0.5 ml of anhydrous dimethylformamide are allowed to react at room temperature for 10 min to form the imidazolide. ATP (60.5 mg, 0.1 mmole) in 2.5 ml of water is then added to the DMF solution, and the mixture is stirred at room temperature for 10 hr. The liquid sample obtained following lyophilization is washed with ether and methanol to remove the starting material. The solid residue (37.5 mg) is chromatographed with isobutyric acid:0.5 N NH$_3$ (5:3) as the solvent. Material (XV) recovered from a UV-absorbing band appearing at R_f 0.58 weighed 12.3 mg and had UV maxima at 259 and 208.5 nm, corresponding to adenine nucleoside derivatives and an α,β-unsaturated carbonyl group, respectively. The electron spin resonance spectrum of this material gave three sharp N-oxide peaks. Spin label ADP, spin-label AMP, and spin-label NADP are prepared according to the same procedure. Products isolated utilizing the same chromatography system were found with R_f values of 0.74, 0.84, and 0.63, respectively, all three compounds having strong ESR signals. The UV spectrum of the spin-label ADP showed two maxima, one at 259 and the other at 208 nm.

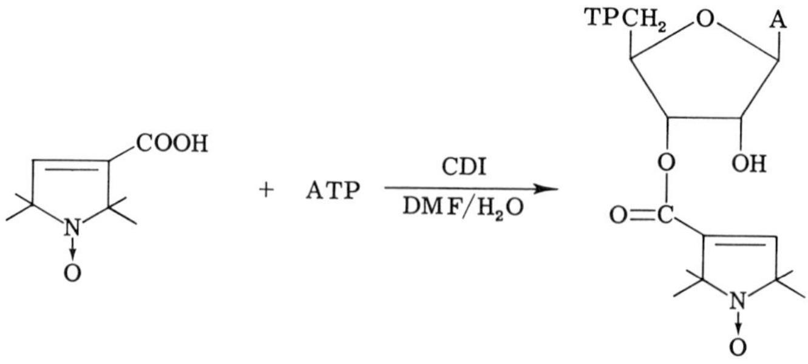

(XV)

Recently Westheimer and his group[36] have reported the synthesis of 2-diazo-3,3,3-trifluoropropionyl chloride from trifluorodiazoethane and phosgene. The reagent is a powerful acylating agent and *should* react with ATP at the 2'- or 3'-hydroxyl group (XVI).

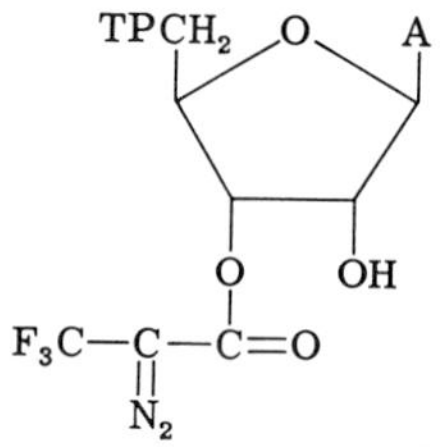

(XVI)

The diazo derivative itself is known to be stable to acid pH and readily undergoes photolytic decomposition generating a carbene.

Discussion

The carbodiimidazole-catalyzed formation of activated carboxylic acids containing a photosensitive arylazido has been utilized in the synthesis of a number of photoreactive adenine and pyridine nucleotide analogs. These analogs have properties consistent with the requirements of photoactive site-directed reagents. Their usefulness has been demon-

[36] V. Chowdhry, R. Vaughan, and F. H. Westheimer, *Proc. Natl. Acad. Sci. U.S.A.* **73**, 1406 (1976).

strated in the investigation of the nucleotide binding regions of a number of enzyme systems. The photoactive species arylazido-β-alanine does not inhibit myosin or mitochondrial F_1 (ATPase) in comparison to the extensive photodependent inhibition observed in the presence of arylazido-β-alanine ATP. The ATP moiety is thus required to direct the nitrene precursor to the active site region of the enzyme. This conclusion is supported by the competitive nature of the interaction of the analogs in the presence of the natural ligands. Similar conclusions can be made for the pyridine nucleotide photoactive analogs and their interaction with alcohol dehydrogenase and with NADH dehydrogenase of complex I of the mitochondrial electron transport chain.

The arylazido-β-alanine analog has been found to be as well a potent inhibitor of yeast hexokinase and firefly luciferase.[37]

The variation in the distance separating the photoactive azido group from the 3'-hydroxyl position using azido-β-alanine, azido-4-aminobutyric, and azido-6-aminocaproic ATP and other potential analogs is anticipated to become a powerful test in the chemical mapping of the topographical components of nucleotide binding regions of proteins. Such analogs can be used for the scanning of the region about the binding site, revealing the three-dimensional alignment of the region without the necessity of a complete knowledge of the primary structure of the protein.

Because of the chemical nature of the analogs discussed in this paper, one must assume that the covalent insertion reaction does not occur directly at the ligand binding site. This may be one of the reasons for the less than 100% inhibition of myosin ATPase activity with nevertheless reasonable stoichiometry of analog labeling following photolysis. Labeling occurs within the vicinity of the active site with inhibition of enzymic activity being due to the irreversibly bound nucleotide blocking access of substrate to this region in a flexible manner. Under these conditions there is no reason to assume that the inhibition of enzymic activity should be directly proportional to active site binding. This would be especially true for those analogs with long spacers between the photoactive azido and the nucleotide (i.e., ligand) binding site. Thus while one might not be able to specify precisely that covalent insertion occurs directly at the active site, since the reactive reagent is at a distance from the binding site, one does have an exact knowledge of the distance separating the point of insertion from the point of ligand binding.

The preparatory procedures outlined have great potential use in that they may be used as the basis for the synthesis of a wide variety of

---

[37] Unpublished observations of R. J. Guillory.

different nucleotides[38] as well as nonnucleotide probes. In addition to the photoaffinity labels, spin labels and fluorescent adjuncts may also be coupled to the ribose while still allowing the ligand to retain its enzymic function.

That the point of esterification need not be the hydroxyl group of a nucleotide can be seen from some recent experimental findings with tetrodotoxin.[39] This toxin is a specific inhibitor of the transient conductance of the muscle (and nerve) membrane.[40] The specificity of action (at nanomole concentrations) has raised the possibility of tetrodotoxin being used in the identification and characterization of membrane sodium channels. While the binding of tetrodotoxin to these channels is rapid, it is reversible, and attempts at the further characterization of the channel components are limited. Tetrodotoxin when subjected to a carbodiimidazole-catalyzed esterification by arylazido-β-alanine results in the formation of a number of analogs of tetrodotoxin containing the azido ligand. Testing of these materials revealed that one in particular had the biological characteristics of the natural tetrodotoxin. While this material could be adequately washed from the muscle tissue to restore the membrane's normal transient Na⁺ conductance, this reversibility was completely prevented when the muscle bathed in the tetrodotoxin analog-containing medium was subjected to photoirradiation. The material has all the characteristics of a photoaffinity active site directed reagent for the labeling of the specific Na⁺ channels of excitable tissue. Carried to its reasonable extension, the use of the arylazido-β-alanine tetrodotoxin analog shall enable us to characterize the components within the vicinity of the Na⁺ channel of excitable tissue.

The availability of the described photoaffinity reagents represent chemical tools for the examination of the ligand binding regions of a wide variety of biologically active systems.

Acknowledgments

Portions of this work were supported by a grant from the Hawaii Heart Association. The authors would like to express their appreciation for the collaboration of Mr. S. Chen and Mr. J. Russell in a number of the experiments detailed in this work. A portion of this work was accomplished during the tenure of R. J. G. as an Established Investigator of the American Heart Association and of S. J. J. as a recipient of NIH Research Fellowship 1FO 2 GM 57060.

[38] The carbodiimidazole-catalyzed esterification reaction with arylazido-β-alanine has recently been successfully applied to the synthesis of a photoaffinity analog of coenzyme A. Unpublished observations of S. J. Jeng.

[39] Unpublished observations of R. J. Guillory in collaboration with Dr. M. Rayner and Dr. J. S. D'Arrigo.

[40] T. Narahashi, *Physiol. Rev.* **54**, 813 (1974).

[26] Aromatic Thioethers of Purine Nucleotides

By Hugo Fasold, Franz W. Hulla, Franz Ortanderl,
and Michael Rack

Modification of the SH-group of 6-mercaptopurine by suitable substituents results in strong activation of the thioether bonds. A family
of substances (Fig. 1) has been used to attach a covalent label to amino
acid side chains, preferentially to the —SH group. The reaction proceeds
to a cleavage of the aromatic thioether with liberation of a thiophenol.
The carbon-6 of the purine is bound to the amino acid side chain although,
in some cases, the nitrophenyl moiety is transferred onto the protein.

Preparation of S-(Dinitrophenyl)-6-mercaptopurine
Riboside Triphosphate

2′,3′-O-Isopropylidene-S-(dinitrophenyl)-6-mercaptopurine   Riboside.
Isopropylidated 6-mercaptopurine riboside, 2.88 g (9.6 mmoles), prepared
from the riboside with the aid of 2,2-dimethoxypropane,[1] is suspended
in 300 ml of ethanol and dissolved by the addition of 10 ml of a 1 N
solution of sodium hydroxide in 50% ethanol. A solution of 1.24 ml (10
mmoles) of 2,4-dinitro-1-fluobenzene in 60 ml of ethanol is added in several portions over a period of 30 min, and the mixture is stirred for

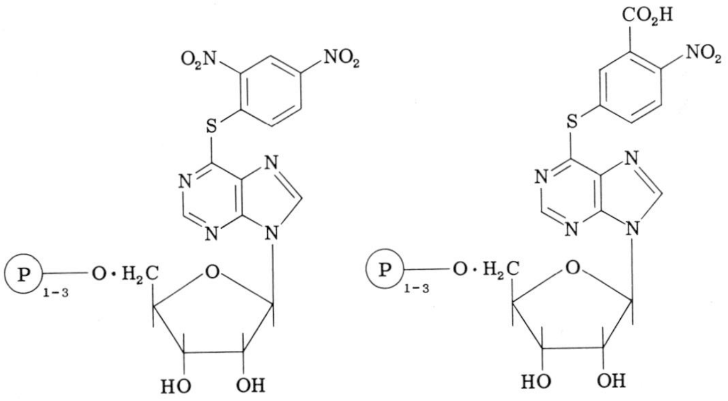

Fig. 1. Substances used to attach a covalent label to amino acid side chains.

[1] J. A. Zderic, J. G. Moffat, D. Kan, K. Gerson, and W. E. Fitzgibbon, *J. Med. Chem.* **8**, 275 (1965).

another 30 min. After evaporation to a turbid syrup, the residue is taken up in 300 ml of methylene chloride, and the mixture is extracted three times with 50 ml of water. After removal of the solvent, the product is sufficiently pure for phosphorylation; it is only slightly contaminated by dinitrophenol. Purity may be checked by thin-layer chromatography on silica gel in methanol or dibutyl ether with R_f of 0.6 and 0.5, respectively. The compounds carrying the aromatic thioether bonds can be detected on thin-layer chromatograms or by paper electrophoresis by spraying with alkaline 1% mercaptoethanol. The slight yellow color of the bands deepens perceptibly after a few minutes. The yield at this stage is 95%. The riboside thioether may be completely purified by dissolving a sample of about 100 mg in 4 ml of warm ethanol, adding the same amount of ethyl acetate and 200 mg of dry silica gel powder. The mixture is applied to a silica gel column in ethyl acetate, and the column is developed with the same solvent.[2]

S-(*Dinitrophenyl*)-6-mercaptopurine *Riboside Monophosphate.* The isopropylidated thioether riboside, 3 g (6.16 mmoles), is dissolved in 12 ml of dry triethyl phosphate, and the solution cooled to —20°. A solution of 0.85 ml (9.3 mmoles) of freshly distilled $POCl_3$ in 2 ml of triethyl phosphate is added in small portions, and the reaction mixture is maintained below —16° with the aid of a cryostat. Moisture should be strictly excluded during the addition. The solution is stirred for 2 hr at room temperature. A mixture of 70 ml of a freshly prepared 10% barium acetate solution, 35 ml of acetone, and 40 g of ice is stirred vigorously, and the phosphorylation mixture is added from a pipette. The pH is brought to 7.0 by addition of 4 N NaOH and kept constant until the excess of $POCl_3$ is hydrolyzed. Then ethanol is added to a final concentration of 80% (v/v). The slightly yellow precipitate is centrifuged, and extracted four times with 200 ml each of 80% ethanol and twice with 200 ml each of acetone to remove salts and dinitrophenol. The barium salt of the monophosphate may be extracted from the precipitate with portions of 200 ml of water at pH 6 until the absorbance at 260 nm of the supernatant liquids has decreased to 5% of the value of the first extract (approximately 6 extractions are required). The barium salt may be stored in the cold for several months.

To isolate the monophosphate, the barium salt is mixed with an equal weight of Dowex 50 × 2 resin in the H⁺ form. The mixture is applied to a Dowex 50 × 2 column (2 × 35 cm for 2 g of barium salt) in the H⁺ form in water. Upon elution of the column with water, the monophosphate emerges as the second peak after a small peak of impurities. Fractions may be monitored by electrophoresis at pH 3.5 in 0.1 M am-

[2] U. Faust, H. Fasold, and F. Ortanderl, *Eur. J. Biochem.* **43**, 273 (1974).

monium citrate. Yield: 45%. The solution of the purified monophosphate is lyophilized, and the product is used immediately for protein reactions or for the synthesis of the triphosphate.

Triphosphate I. The monophosphate, 425 mg (0.8 mmole), is dissolved in 6 ml of dry picoline and 0.19 ml of tributylamine. The solution is evaporated to dryness under reduced pressure, and the tributylammonium salt is taken up in 5 ml of dry dioxane. Tributylamine, 0.36 ml, and 0.24 ml (1.16 mmoles) of freshly distilled diphenyl chlorophosphonate are added, and the mixture is stirred for 2 hr. Evaporation under reduced pressure leaves an oily residue, which is extracted with 10 ml of diethyl ether for 15 min and usually is transformed into a fluffy yellow precipitate adhering to the glass wall. The ether is decanted, the residue is dissolved in 2 ml of dry dioxane, and the solvent is again evaporated under reduced pressure. The gum is suspended in a few milliliters of picoline and dissolved by the addition of 0.5 ml of dimethylformamide. The solution is added to well-dried tributylammonium pyrophosphate, prepared from 356.8 mg (0.8 mmole) of tetrasodium pyrophosphate after removal of the cations over a small Dowex 50 column in the usual manner.[3] Pyrophosphorylation is begun by the addition of 0.8 ml of dry pyridine. After 30 min, the solution is concentrated under reduced pressure, the syrup is shaken for a few minutes with 5 ml of diethyl ether, and this solvent is decanted. The syrup is taken up in 10 ml of water, and the turbid solution is extracted four times with 10 ml each of ether. The pH of the water phase is brought to 7.0 and the solution is applied to a 2×20 cm column of DEAE-Sephadex A-25 in the bicarbonate form that has been well washed with water. The column is developed with a linear gradient of 2000 ml total volume from 0.02 to 0.5 M LiCl or from 4.0 to 1 M triethylammoniumbicarbonate at pH 7.2. The elution diagram shows three major peaks with several shoulders; the triphosphate is found in the third peak. The column fractions are again monitored by electrophoresis at pH 3.0 and the fractions containing the triphosphate are lyophilized. The fluffy material is extracted twice with 5 ml of absolute ethanol to remove salts, dried well, and used for labeling experiments within a few days. It may be stored as a concentrated aqueous solution for a few weeks. Yield: 28%.

Synthesis of S-(Nitro-4-carboxyphenyl)-6-mercaptopurine Riboside 5'-Monophosphate

S-(2-Nitro-4-carboxyphenyl)-6-mercaptopurine Riboside. 5,5'-Dithio-bis-(2-nitrobenzoic acid) (DTNB, Ellman's reagent) 3.96 g (10 mmoles)

<hr>

[3] J. G. Moffat, *Can. J. Chem.* **42**, 599 (1964).

is suspended in 50 ml of water and neutralized with 4 N NaOH. Isopropylidated 6-mercaptopurine riboside (see above) 300 mg (1 mmole) is added, and the pH is adjusted to 8.0. The reaction mixture is stirred for 7 days, the pH being readjusted each day. This somewhat unusual reaction proceeds to the desired thioether with elimination of the SH group of the nucleoside.[4] The reaction is monitored by occasional electrophoresis in 0.1 M pyridine acetate at pH 6.5. The resultant solution is applied to a DEAE-Sephadex A-25 column (3 × 35 cm) in the bicarbonate form, and the product is eluted with a linear gradient of 2000 ml total volume from water to 0.2 M annonium bicarbonate at pH 6.5 (adjusted with CO_2). It appears as a second peak after a small amount of unreacted mercaptopurine riboside. Purity is checked by electrophoresis at pH 6.5, and the thioether is detected with a mercaptoethanol spray as described above. The solution of the thioether is lyophilized several times to remove almost all the bicarbonate. Yield: 80%.

Monophosphate II. The thioether riboside, 489.5 mg (1 mmole), is dissolved in 8 ml of dry triethyl phosphate, and 1 ml of freshly distilled $POCl_3$ is added at temperatures below 0° in small portions; the solution is maintained at 0° for 20 hr. Thereafter, 20 ml of 1 M barium acetate solution are added with vigorous stirring while the mixture is cooled to below 20°. After addition of 80 ml of water, the pH is held between 7 and 8.5 by the addition of triethylamine. When all the $POCl_3$ is hydrolyzed, the pH is adjusted to 8.5 and the precipitated barium phosphate is removed by centrifugation. The precipitate is redissolved at pH 2.0 and reprecipitated at pH 8.5 several times to wash out the coprecipitated barium salt of the desired nucleotide. The combined supernatant fluids are mixed with 3 volumes of ethanol. After standing at 0° for 3 hr, the precipitated barium salt of the monophosphate is isolated by centrifugation and washed 3 times with 80% ethanol and with ether.

The free monophosphate may be obtained by passage of the solution of its barium salt over a Dowex 50 × 2 column in the H⁺ form, as described above. The isopropylidene group is also removed during this operation. Fractions containing the compound are monitored by electrophoresis at pH 3.0, as described above, or by paper chromatography in isopropanol–0.5 M ammonium acetate pH 6.0 (5:2), with the aid of alkaline mercaptoethanol spray. The solution is lyophilized twice. The compound may be stored as a concentrated aqueous solution at pH 4 for a few weeks whereas the barium salt is quite stable. Yield: 65%.

Triphosphate. The procedure described above for the S-(dinitro-

---

[4] F. W. Hulla and H. Fasold, *Biochemistry* 11, 1056 (1972).

phenyl)-6-mercaptopurine riboside triphosphate may also be applied to the prophosphorylation of this monophosphate. The yield in this case is about 55%.

These compounds show rapid reaction with sulfhydryl groups at pH values above 7.5. Of all other amino acid side-chain functional groups tested, only primary amines and phenols are able to react with the aromatic thioether at elevated temperatures and over incubations of several hours' duration.[2,4]

Related Compounds

The compounds described above may be regarded as analogs of adenosine nucleotides. The guanosine nucleotide analogs of the same group have been synthesized from 2-amino-6-mercaptopurine nucleoside using very similar procedures. A 3',5'-cyclic monophosphate was synthesized from the 5'-monophosphate (I) after the method of Smith et al.[5]

Examples of Affinity Labeling of Proteins

Phosphorylase b. Phosphorylase *b* is dependent in its activity upon the allosteric effector AMP, and thus AMP analog (II) could be used to label the effector binding site.[4] The protein was first freed from AMP by filtration with Sephadex G-50 and charcoal columns. Portions of the enzyme, 60 mg, were incubated in 2 ml of 0.01 glycerophosphate-HCl at pH 8.0 with 40 mg of (II) for 0.5 to 2.0 hr. The analog in this case had been labeled by ^{32}P. The pH was then lowered to 6.5, thereby terminating the modification reaction. Excess nucleotide was removed by filtration over Sephadex G-25, and specific enzymic activity as well as covalently bound ^{32}P-labeled nucleotide was determined. The activation of phosphorylase *b* by the attached label is shown in Fig. 2. It was not possible to activate the enzyme to a higher degree since the nonspecific reaction of the nucleotide with superficial SH groups became prevalent. The peptic fingerprint of a preparation covalently activated to 20% (see Fig. 2) gave only a single labeled peptide spot in an autoradiogram.

Rabbit Muscle Actin. The monomeric G form of actin carries an ATP firmly bound, which is hydrolyzed in a stoichiometric reaction to ADP during polymerization to the fibrillar F form under the influence of inorganic cations, especially Mg.$^{2+}$ By removal of the triggering cations, and displacement of the ADP by fresh ATP, the protein depolymerized again to the G form.

[5] M. Smith, G. I. Drummond, and H. G. Khorana, *J. Am. Chem. Soc.* **83**, 698 (1961).

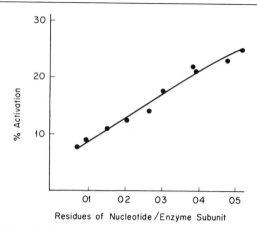

FIG. 2. Activation of phosphorylase *b* by covalently bound nucleotide after reaction with monophosphate (II). Ordinate: Percentage of maximal activation by AMP.

Prior to labeling by the ATP analog, one easily accessible SH group of the protein was allowed to react with *N*-ethylmaleimide. It had previously been shown that this modification does not change the polymerization reaction and interaction with myosin in any of its properties.[6,7] The label was attached by incubation of F-actin at a concentration of 10 mg/ml with a 100-fold molar excess of the ATP analog at pH 7.5 for 2 hr in 1 mM Tris chloride containing 0.2 mM ascorbic acid. The depolymerization, as measured by loss of viscosity, proceeded as in the case of ATP. After the removal by centrifugation of nondepolymerized material, approximately 20% of the total, the protein was repolymerized and centrifuged down as modified F-actin. Samples for determinations of bound nucleotide, inorganic $^{32}P_i$, and specific viscosity were taken before and after repolymerization, as well as after the attempt at depolymerization of the modified F-actin. Only 20% of the protein could be brought into the monomeric state again; the major portion was "frozen" in the F form. During repolymerization, the nucleotide was split to the ADP analog and inorganic phosphate. Tryptic fingerprint of this protein revealed only one labeled peptide spot on autoradiography.[2]

Other Proteins. Whereas a specific binding site was labeled in the examples cited above, a very different reaction ensued with phosphoribosyl pyrophosphate–ATP-ligase. After incubation of the enzyme in 0.05 M sodium bicarbonate solution with AMP analog (I), label from

[6] C. J. Lusty and H. Fasold, *Biochemistry* **8**, 2933 (1969).
[7] W. W. Kielley and C. B. Bradley, *J. Biol. Chem.* **218**, 653 (1956).

both ^{32}P and ^{14}C in the dinitrophenyl group[8] was attached to protein. The dinitrophenyl moiety was attached stoichiometrically to its subunits; a tryptic fingerprint revealed only two labeled peptides on autoradiography.

With Na$^+$,K$^+$-ATPase from nerve tissue, on the other hand, the nucleotide moiety was able to label the enzyme, although extremely long incubation periods of several days were necessary.[9]

[8] T. Dall-Larsen, H. Fasold, C. Klungsøyr, H. Kryvi, C. Meyer, and F. Ortanderl, *Eur. J. Biochem.* **60**, 103 (1975).

[9] P. Patzelt, H. Pauls, E. Erdmann, and W. Schoner, *Hoppe-Seyler's Z. Physiol. Chem.* **355**, 758 (1974).

[27] N^6-o- and p-Fluorobenzoyladenosine 5'-Triphosphates

By ALEXANDER HAMPTON[1] and LEWIS A. SLOTIN

The title compounds (I) and (II) were prepared as potential ATP-site-directed reagents.[1] The fluorobenzoyl substituents were selected because (a) the smallness of fluorine lessens steric hindrance to reaction of (I) and (II) with enzymic groups, (b) fluorine undergoes nucleophilic displacement from aromatic systems under milder conditions than other halogens, (c) o- and p-fluorobenzamides are stable in physiological buffers, and (d) (I) and (II) could be attacked by a nucleophilic group of an enzyme not only at their ortho and para carbons, respectively, but also at their amide carbons, thus approximately doubling the probability that (I) or (II) could act as an affinity label for the ATP site of any given enzyme.

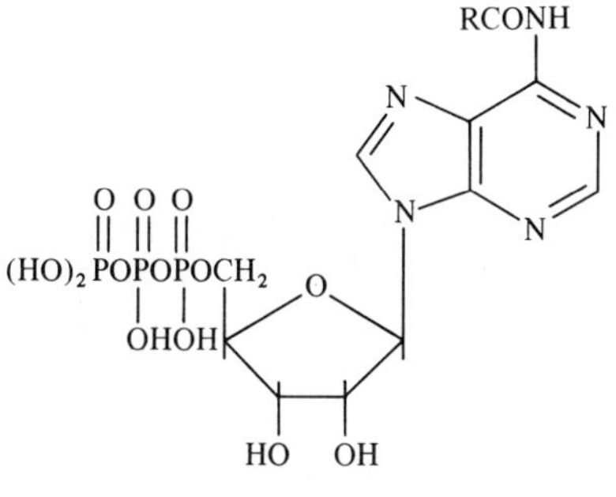

(I): R = o-fluorophenyl

(II): R = p-fluorophenyl

[1] A. Hampton and L. S. Slotin, *Biochemistry* **14**, 5438 (1975).

Synthesis of N^6-o- and p-Fluorobenzoyladenosine 5'-Triphosphates

General. Pyridine is distilled from p-toluenesulfonyl chloride and then from calcium hydride and stored over potassium hydroxide. N,N-Dimethylformamide is distilled from calcium hydride and stored over molecular sieves. Dioxane is distilled from phosphorus pentoxide. Tri-n-butylamine, triethylamine, diphenyl phosphorochloridate, and o- and p-fluorobenzoyl chlorides are distilled before use. Tri-n-butylammonium pyrophosphate is prepared at room temperature according to the method of Moffatt and Khorana[2] and stored at 5°. Paper chromatography is carried out by the descending technique on Whatman No. 1 paper in (A) isobutyric acid–1 M NH$_4$OH (60:40) and (B) 1-propanol–water (7:3). Electrophoresis is carried out on Whatman No. 1 paper at pH 3.5 in 0.035 M citric acid–0.0148 M sodium citrate (1:1). Evaporations are carried out under reduced pressure at bath temperatures below 30°. Phosphate analyses of nucleoside triphosphates are performed by the method of Lowry and Lopez[3] after treatment of approximately 1 μmole of these compounds for 60 min at 22° in 1 ml of Tris chloride at pH 10.4 containing 0.02 mg of alkaline phosphatase of calf intestinal mucosa (type VII, Sigma Chemical Co.).

N^6-o-Fluorobenzoyladenosine 5'-Phosphate. An aqueous solution of AMP (1.44 mmoles) is percolated through a column of Dowex 50 ion-exchange resin (pyridinium form, 20 ml). The eluate and aqueous washings are combined and evaporated under reduced pressure, and the residue is dried by repeated evaporation from it of pyridine (5 × 30 ml). To the residual oil is added pyridine (30 ml) and o-fluorobenzoyl chloride (30 mmoles), and the light yellow solution is stirred under anhydrous conditions in the dark. After 1.5 hr at room temperature it is poured into cold (<5°) water–chloroform (1:1, 200 ml) and the mixture is maintained at 0° for 15 min. The aqueous layer is washed with chloroform (2 × 30 ml), and the chloroform solutions are combined and evaporated. A solution of the residue in pyridine–water (2:1, 60 ml) is cooled in an ice bath. Aqueous 2 N NaOH (50 ml) at 0° is added, and the mixture is stirred in an ice bath for 4 min. Excess of pyridinium Dowex 50 is added to remove sodium ions, and the resin is removed by filtration and washed with water. The eluates are concentrated under reduced pressure to a gum which is extracted with anhydrous ether (5 × 50 ml). The residue is treated with anhydrous methanol (10 ml) and triethylamine (0.5 ml, 3.6 mmoles), and volatiles are removed under reduced pressure. The residue is dissolved in methanol (10 ml) with gentle warming, and

[2] J. G. Moffatt and H. G. Khorana, *J. Am. Chem. Soc.* **83**, 649 (1961).
[3] O. H. Lowry and J. A. Lopez, *J. Biol. Chem.* **162**, 421 (1946).

3 ml of 1 M NaI in acetone are added. An excess of acetone is added and the precipitate thereby obtained is dried over P_2O_5 under reduced pressure at room temperature to give the disodium salt of the product as a white amorphous powder (671 mg, 79% yield) that contains 4.5 molecules of water. The product was found to be homogeneous in solvents A (R_f 0.73) and B (R_f 0.34) and on paper electrophoretograms. The UV maximum at pH 7 was 282 nm (ϵ 19,800).

N^6-o-Fluorobenzoyladenosine 5'-Triphosphate (I). The procedure is based on that of Michelson[4] for the synthesis of nucleoside 5'-triphosphates. An aqueous solution of the foregoing AMP derivative (0.18 mmole) is passed through a column of Dowex 50 (pyridinium form, 10 ml). The combined eluate and washings are concentrated to dryness under reduced pressure. Pyridine (2 ml) and tri-n-butylamine (0.14 ml, 0.6 mmole) are added, and the solution is concentrated again. N,N-Dimethylformamide (2 ml) is added, and the solution is again evaporated. Dioxane (0.5 ml), N,N-dimethylformamide (0.1 ml), diphenyl phosphorochloridate (0.06 ml), and tri-n-butylamine (0.07 ml) are added, and the solution is stirred at room temperature under anhydrous conditions. After 3 hr the solvents are evaporated and the residue is extracted several times with anhydrous ether. The residue is dissolved in dioxane (0.2 ml), and the solution is concentrated to dryness. Tri-n-butylammonium pyrophosphate in pyridine (2 ml, 1 mmole) is added, and the solution is stirred at room temperature for 45 min. Volatiles are removed, and the residue is extracted with ether and then dissolved in water and chromatographed on 6 sheets of Whatman No. 3 MM paper (each 23 cm wide) in solvent A. The band at R_f 0.50 is eluted with water; yield, 55%. The solution is concentrated to dryness and treated with methanol (5 ml) and triethylamine (0.2 ml). The mixture is warmed to effect dissolution, and the solution is evaporated to dryness and treated with methanol (1 ml) and 1 M NaI in acetone (0.3 ml). An excess of acetone is added, and the precipitate is collected and dried over P_2O_5 under reduced pressure at room temperature to give the product (86 mg) as a white amorphous powder. Electrophoretic mobility is 1.80 relative to AMP. The R_f value on paper in solvent B is 0.18. The UV spectrum in water (pH 7) shows a maximum at 282 nm (ϵ 20,300, calculated for a trihydrate of the tetrasodium salt). The action of alkaline phosphatase liberates 2.89 times more inorganic phosphate from the product than from an equivalent amount of AMP.

N^6-p-Fluorobenzoyladenosine 5'-Triphosphate (II). This is obtained in similar overall yield starting from AMP and p-fluorobenzoyl chloride

---

[4] A. M. Michelson, *Biochim. Biophys. Acta* **91**, 1 (1964).

by the procedures described above for its o-fluoro isomer. The product was homogeneous as judged by paper electrophoresis (mobility 1.78 relative to AMP) and chromatography in systems A and B (R_f values 0.5, 0.17, respectively). The UV maximum at pH 7 was at 282 nm (ϵ 20,900; calculated for a trihydrate of the tetrasodium salt). The product produced 2.92 times as much inorganic phosphate as an equivalent amount of AMP after treatment with alkaline phosphatase.

Reaction of (I) and (II) with Adenylate Kinases of Pig, Rabbit, and Carp Muscle

The reactions catalyzed by each enzyme were followed by the changes in absorbance at 340 nm, which occurred in 1 ml of 0.1 M Tris-HCl (pH 7.6) containing $MgSO_4$ (0.92 mM), KCl (0.11 M), PEP cyclohexylammonium salt (0.31 mM), NADH (0.38 mM), AMP (0.39 mM), ATP (0.24 mM), pyruvate kinase (10 μg), and lactate dehydrogenase (10 μg).

Studies of the rates of inactivation by (I) were carried out in solutions that included all the assay components except AMP and ATP at levels such that addition of 400–500 μl of the Tris buffer and 100 μl of 3.9 mM AMP and 2.4 mM ATP gave the assay concentrations listed above. Rates of inactivation by (II) were studied under the same conditions except that 2.64 mM of ATP was present and that the assay levels of Mg^{2+} and ATP were 4.5 mM and 1.06 mM, respectively. Mixtures lacking (I) or (II) were utilized as controls to monitor denaturation of the enzymes; this did not exceed 11% over a 6-hr period. Inactivation mixtures and their controls were maintained at 0° and assayed at 22°.

The three adenylate kinases were inactivated by (I) with first-order kinetics during the period of observation (6 hr); the apparent first-order rate constants for inactivation by four levels (0.88, 1.22, 2.22, and 2.93 mM) of (I) were calculated from the times for 50% inactivation and their inverses were plotted against the inverse of the (I) levels to give for the rabbit, pig, and carp enzymes, respectively, rate constants for inactivation of 0.91, 0.13, and 0.19 min^{-1} and apparent dissociation constants of the (I)–enzyme complexes of 7.15 mM, 0.47 mM, and 1.96 mM. The above values have been recomputed and differ from those previously published.[1] Since (I) is a good substrate of the three enzymes and since ATP markedly reduced the rate of inactivation, it was concluded that (I) is acting in each case as an ATP-site-directed reagent.

The N^6-p-fluorobenzoyl derivative of ATP (II) in the presence of 2.64 mM ATP inactivated the rabbit and carp adenylate kinases by pseudo-first-order kinetics; the half-life of the rabbit kinase was 210 min in the presence of 50 μM (II), and the half-life of the carp kinase was 130 min in the presence of 100 μM (II). Retardation of the inactivations by ATP indicated that the effects were ATP-site directed.

Interaction of (I) and (II) with Other ATP-Utilizing Enzymes

Rabbit muscle pyruvate kinase was not inactivated when treated at 0° for 5 hr with N^6-p-fluorobenzoyl-ATP (100 μM), N^6-p-fluorobenzoyl-ADP (260 μM), or N^6-o-fluorobenzoyl-ADP (3 mM).[1] Yeast hexokinase was apparently not inactivated by (I) or (II)[1] although searching tests could not be made because of the excellent substrate properties of (I) and (II) in the presence of glucose and because of the marked instability of this enzyme in the absence of glucose.

[28] 6-Chloropurine Ribonucleoside 5′-Phosphate

By ALEXANDER HAMPTON

The title compound (I) can be viewed as an enzyme affinity label produced by attachment of a relatively small leaving group (chlorine) to a hydrophilic substrate (inosine 5′-phosphate) in the expectation that the enzymic substrate binding sites will contain a preponderance of hydrophilic and hence potentially derivatizable amino acid residues. Evidence indicates that (I) forms covalent bonds at the nucleotide binding sites of two of four purine nucleotide utilizing enzymes which were examined.

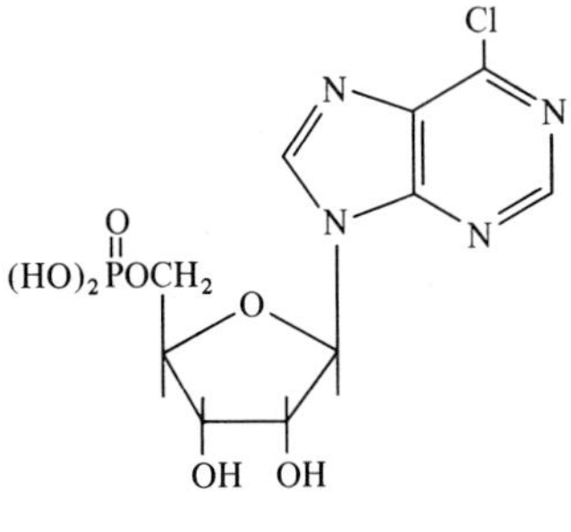

(I)

Synthesis of (I)

The compound can be prepared by phosphorylation of 2′,3′-O-isopropylidene-6-chloropurine ribonucleoside.[1] On a small scale it is con-

---

[1] A. Hampton and M. H. Maguire, *J. Am. Chem. Soc.* 83, 150 (1961).

veniently obtained in a single step from the commercially available barium salt of 6-mercaptopurine ribonucleoside 5'-phosphate[2] utilizing a chlorination procedure described by Robins[3] for the conversion of 6-mercaptopurine ribonucleoside to 6-chloropurine ribonucleoside.

A suspension of 50 mg of barium 6-thioinosinate in 2 ml of methanol is cooled to below $-10°$ in an acetone–solid CO_2 bath with exclusion of moisture. Chlorine gas is slowly bubbled into the mixture for 15 min during which the temperature is maintained below $-5°$. The suspended material dissolved within 7 min. The yellow solution is maintained at the same temperature for an additional 15 min, after which cooling is continued and nitrogen is bubbled through to remove excess chlorine. Saturated methanolic ammonia, precooled to $-10°$, is added to adjust the pH of the solution to 7. A heavy white precipitate forms; water is added to dissolve it (final volume of the mixture, 8 ml), and 0.5 ml of 1 M barium acetate is added. After 2 hr at 4° the solution is centrifuged to remove small amounts of insoluble material. Ethanol (32 ml) is added, and the mixture is stored at 4° overnight. The white precipitate collected by centrifugation is washed once with 8 ml of aqueous 80% ethanol. The product is dried over NaOH under reduced pressure, then dissolved in 2 ml of water and applied to a column containing about 20 equivalents of Dowex 50 (Na^+) ion-exchange resin. The column is washed with water, and 10 ml of effluent are collected. The spectral characteristics of this solution in 0.05 M potassium acetate at pH 4.8 are λ_{max} 264 nm, λ_{min} 226.5 nm; absorbancy ratio $A_{264}:A_{226.5} = 4.3$, $A_{250}:A_{260} = 0.76$, and $A_{280}:A_{260} = 0.19$. The values reported for 6-chloro-9-β-D-ribofuranosylpurine 5'-phosphate are 263 and 226 nm, and absorbancy ratios are 4.0, 0.82, and 0.175, respectively.[1] Chromatography of the solution in 1-butanol–acetic acid–water (5:3:2), isopropyl alcohol–1% $(NH_4)_2SO_4$ (2:1), and isoamyl alcohol–0.3 M phosphate (pH 6.9) (1:1) reveals one fluorescent spot with the same R_f as previously reported.[1] The yield of 6-chloropurine nucleotide by the present procedure was determined spectrophotometrically to be 92%.

The barium salt is precipitated from the solution of the sodium salt with barium acetate and ethanol under the conditions given above. When dried over P_2O_5 for 3 hr at 0.1 mm and 100°, it showed λ_{max} 264 nm (ϵ 8260) at pH 4.8 (reported previously, ϵ 8400 at 263 nm); this extinction coefficient corresponds to the anhydrous form of the barium salt of this nucleotide, and this form has been shown[1] to be produced by the above drying conditions. (I) is stable for at least several days in aqueous solutions at pH 4.5–7.5.

[2] L. W. Brox and A. Hampton, *Biochemistry* 7, 398 (1968).
[3] R. K. Robins, *Biochem. Prepn.* 10, 145 (1963).

Reaction of (I) and Inosine 5'-Phosphate (IMP)
Dehydrogenase of *Aerobacter aerogenes*[4,5]

The enzyme was purified to near homogeneity[5,6] and dissolved in 0.02 M potassium phosphate at pH 7.4 containing 2 mM glutathione and 0.1 M KCl to give a protein concentration of 2 mg/ml (measured by absorption at 280 nm). To 1 ml of this solution was added 5 μl of a neutral 3.3 mM solution of the sodium salt of (I); 100 min later, a second 5-μl portion was added; and 20 min later a third 5-μl portion was added. A progressive loss in enzyme activity was observed during this time, and 0.5 hr after the third addition of (I) only 11% of the initial activity remained.[5] Activity was not restored by increasing the level of IMP; the rate of inactivation was decreased by IMP[4] and GMP,[6] accelerated by glutathione,[4] and unaffected by NAD.[6] Changes in the ultraviolet absorption of the mixture indicate that a cysteine residue was alkylated and that the enzyme could become inactivated after reaction with as little as 2 molar equivalents of (I).[5] The rate of inactivation at different levels of (I) followed saturation kinetics from which the apparent dissociation constant of the enzyme–(I) complex at pH 7.0 and 24° was found to be 260 μM and the apparent first-order rate constant for inactivation was 0.125 sec^{-1}.[5]

Reaction of (I) and Guanosine 5'-Phosphate (GMP) Reductase of *Aerobacter aerogenes*[2]

In 1 ml of 0.04 M Tris chloride pH 7.5, containing 2 mM glutathione, about 50-fold purified GMP reductase (21 μg of protein) was progressively inactivated by (I): at 23° 10 μM (I) caused 80% inactivation after 40 min and 100 μM (I) caused 95% inactivation after 35 min. During the 40-min period, inactivation by 10 μM (I) was significantly decreased by 200 μM GMP and essentially prevented by 1 mM GMP; in addition, it was markedly decreased by 50 μM GMP in combination with 60 μM NADPH (the second substrate) although the same concentrations of either GMP or NADPH singly had no effect on the rate of inactivation. The loss in activity, once established, could not be reversed by 1 mM GMP. Various lines of indirect evidence indicate that C(6) of (I) probably forms a thioether linkage with an enzymic sulfhydryl group located near the GMP binding site.[2]

[4] A. Hampton, *J. Biol. Chem.* **238**, 3068 (1963).

[5] L. W. Brox and A. Hampton, *Biochemistry* **7**, 2589 (1968).

[6] A. Hampton and A. Nomura, *Biochemistry* **6**, 679 (1967).

Interaction of (I) with Other Mononucleotide-Utilizing Enzymes

IMP dehydrogenase of mouse sarcoma 180 is inactivated by (I)[7] in a manner similar in many respects to its inactivation of the same enzyme from *A. aerogenes*. Yeast adenylosuccinate AMP-lyase was not appreciably inactivated by 1 mM (I) under the usual assay conditions, nor was *Escherichia coli* IMP:L-aspartate ligase (GDP) appreciably inactivated by 200 μM (I).[8]

Affinity Labels Related to (I)

6-Thio-IMP and 6-thio-GMP under certain conditions inactivate IMP dehydrogenase[4] and GMP reductase[2] of *A. aerogenes*, and evidence suggests that these effects could be associated with formation of disulfide bonds between the substrate analogs and a sulfhydryl group located near the substrate site of each enzyme.[2,4,6] 6-Thio-GMP and 6-thio-IMP do not appear to inactivate yeast adenylosuccinate AMP-lyase or *E. coli* IMP:L-aspartate ligase (GDP).[8]

[7] J. H. Anderson and A. C. Sartorelli, *Biochem. Pharmacol.* **18**, 2737 (1969).
[8] A. Hampton, *Fed. Proc., Fed. Am. Soc. Exp. Biol.* **21**, 370 (1962).

[29] Carboxylic-Phosphoric Anhydrides Isosteric with Adenine Nucleotides

By ALEXANDER HAMPTON, PETER J. HARPER, TAKUMA SASAKI, and PAUL HOWGATE

Adenine nucleotide analogs of this type are of interest as potential affinity labels because (1) they are structurally closely similar to the parent nucleotides, (2) after adsorption to the substrate site they can either acylate or phosphorylate amino acid residues, and (3) it is likely that binding of these analogs to the substrate site will frequently be accompanied by partial neutralization of negative charges on the phosphoryl groups that would increase the reactivity of the mixed anhydride function in the enzyme-bound analogs and thereby tend to increase the specificity of labeling of the substrate site. The present article describes the synthesis of carboxylic-phosphoric mixed anhydride analogs of AMP (I)[1,2] and ATP (II)[2] and conditions under which they appear to react covalently at the adenine nucleotide sites of three enzymes.

[1] A. Hampton and P. J. Harper, *Arch. Biochem. Biophys.* **143**, 340 (1971).
[2] A. Hampton, P. J. Harper, T. Sasaki, P. Howgate, and R. K. Preston, *Biochem. Biophys. Res. Commun.* **65**, 945 (1975).

(I)   (II)

Synthesis of (I)

A solution of sodium phosphate in aqueous 50% methanol is passed through a column containing a 15-fold excess of tri-n-butylammonium Dowex 50 ion-exchange resin. The solvent is removed under reduced pressure, and the residue is thrice coevaporated under reduced pressure with anhydrous pyridine; the product is dissolved in anhydrous pyridine to give a 0.5 M solution of tri-n-butylammonium phosphate.

In order to obtain a homogeneous product it is necessary to carry out the following operations in a glove box in a dry nitrogen atmosphere with freshly prepared anhydrous reagents and solvents. To the sodium salt of 9-(β-D-ribofuranosyluronic acid)adenine (0.5 mmole)[3] is added N,N-dimethylformamide (2 ml). Diphenyl phosphorochloridate (0.48 mmole) is added, and the suspension is stirred for 5 hr. The mixture is centrifuged to quantitatively remove unreacted sodium salt together with sodium chloride. The supernatant fluid is treated with diethyl ether (50 ml), and the precipitated diphenyl ester of (I) is collected by centrifugation and dissolved in dioxane (2 ml); the solution is clarified by centrifugation, and the supernatant is treated with diethyl ether (50 ml). The precipitate is collected and freed of diethyl ether under reduced pressure. The purity and identity of this diphenyl ester can, if desired, be checked by chromatography after treating a portion with anhydrous ammonia-N,N-dimethylformamide.[1] The diphenyl ester of (I) is not stable and is therefore immediately dissolved in 2 ml of the above 0.5 M tri-n-butylammonium phosphate solution. After 5 hr, the solution is clarified by centrifugation and treated with diethyl ether (50 ml). The precipitate is collected, dissolved in pyridine-N,N-dimethylformamide (4 ml of 1:1), and reprecipitated with ether (50 ml). That the product

[3] P. J. Harper and A. Hampton, *J. Org. Chem.* **35**, 1688 (1970).

is free of the diphenyl ester of (I) is shown by the absence of IR peaks at 690 and 782 cm^{-1} that would be due to monosubstituted phenyl groups; furthermore, the amination described below does not produce diphenyl phosphate as judged by chromatography in previously described systems.[1]

The white precipitate (0.31 g), which contained a 90% yield (calculated from UV absorption) of the tri-n-butylammonium salt of (I) in admixture with about an equimolar amount of tri-n-butylammonium phosphate, is freed of ether under reduced pressure; IR 3250, 3150 sh, 1742, 1642, 1602 sh, 1200, and 1090 cm^{-1}; UV spectra are identical with those of the starting acid.[3] The product is dissolved in the minimum of anhydrous N,N-dimethylformamide, and the solution [about 1% of (I)] is stored at $-25°$ under nitrogen. The identity and homogeneity of the product may be further established by treatment of a portion of the solution under anhydrous conditions with dry ammonia, after which paper chromatography reveals a single ultraviolet-absorbing component of R_f 0.45 in ethanol–1 M ammonium acetate (7:3) and R_f 0.70 in n-butanol–acetic acid–water (5:2:3), corresponding to the authentic carboxamide[1] of the starting material.

The acyl phosphate (I) hydrolyzes in Tris buffer at pH 7.7 at least 100-fold faster than does acetyl phosphate, as shown by the inability of (I) to inactivate adenylosuccinate-lyase after a 15-sec exposure to the buffer before addition of the enzyme. Compound (I) also lost its ability to inactivate AMP aminohydrolase after 15-sec hydrolysis at pH 6.5.[2] After the solution of (I) in N,N-dimethylformamide had been stored at $-25°$ for 2 days, it showed UV spectral changes (upon dilution into aqueous solutions) indicative of N^6-acylaminopurine nucleoside formation. In the case of AMP aminohydrolase, this change is associated with a marked reduction in the degree of enzyme inactivation,[2] and the use of freshly prepared solutions of (I) in studies of enzyme inactivation is therefore indicated.

Synthesis of (II)[2]

A solution of (II) in N,N-dimethylformamide is prepared by the same procedure as described for (I) by substituting sodium tripolyphosphate for sodium phosphate. The identity and homogeneity of the product were established by treating a portion of the solution with dry ammonia gas when the only UV-absorbing paper chromatographic component obtained was the amide of the starting carboxylic acid. In addition, the reaction mixture was subjected to thin-layer chromatography on PEI-cellulose using two developments with 4 M sodium formate of pH 3.4, and the plates were exposed to the vapor of concentrated HCl for 3 min and then sprayed with a molybdate–perchloric acid solution.[4] Phosphate-containing components were seen as white spots on a yellow background; this showed that the reaction mixture contained much tripolyphosphate (a streak at R_f 0.1–0.3) but no pyrophosphate (R_f 0.5) and a trace of phosphate (R_f 0.9). The latter was found to be formed in the same proportion by the action of ammonia on the stock solution of tri-n-butylammonium tripolyphosphate.

Compound (II) is extensively hydrolyzed in 15 sec at pH 7.6 and room temperature.[2]

Reaction of (I) with Adenosine 5′-Phosphate (AMP) Aminohydrolase[2]

In these experiments initial reaction rates were measured by the decrease in absorbance at 265 nm and calculated as nanomoles per minute from $\Delta\epsilon = 6600$. The final volume of 0.90 ml contained 0.01 M potassium citrate (pH 6.5), 0.016 μg of enzyme (Sigma Chemical Co., grade IV, from rabbit muscle), 0.005 M KCl, 60 μM (I), and 50 μM AMP. The order of addition was (a) buffer, (b) enzyme, (c) (I), followed 1 min later by AMP. In control assays carried out in conjunction with the inactivation experiments, the order of addition was (a) buffer, (b) (I), followed 1 min later by (c) the enzyme and (d) AMP. These control assays showed that hydrolyzed (I) was not inhibitory; the observed rates varied between 1.11 and 1.13 nmoles/min.

A nominal initial level of 60 μM of freshly prepared (I) caused 32% inactivation of the enzyme; when 50 μM AMP (12% the K_m of AMP) were present prior to the addition of (I) no inactivation occurred. The same amount of (I) was added to the enzyme in less buffer (about 0.1 ml) to give a nominal initial level of 540 μM (I) and after 1 min the mixture was assayed by the standard procedure. Under these conditions, (I) caused 40% inactivation when added in one portion and 47% inactivation when added in 3 portions at 20-second intervals. The inactivation by (I) was abolished by as little as 15-second hydrolysis of (I) in the assay buffer prior to contact with the enzyme, thus indicating that inactivation is the result of acylation or phosphorylation of the enzyme by the mixed anhydride and that the reaction between (I) and the enzyme is extremely rapid.

Reaction of (I) with Adenylosuccinate AMP-Lyase[1]

The above solution of (I) was mixed at 22° with partially purified *Escherichia coli* adenylosuccinate AMP-lyase in 40 mM Tris chloride–10

---

[4] C. S. Hanes and F. A. Isherwood, *Nature (London)* **164**, 1107 (1949).

mM sodium ethylenediamine tetraacetate (pH 7.7) to give 80 μM (I) (and 80 μM tri-n-butylammonium phosphate); about 5 sec later ammonium adenylosuccinate (120 μM final concentration) was added. UV absorbance measurements[5] showed 95–99% reduction in the rate (0.3 nmole/min) of formation of AMP. This inhibition was abolished (a) by 30-sec hydrolysis of (I) in the buffer prior to successive addition of enzyme and substrate, (b) by prior addition to the enzyme of 120 μM adenylosuccinate ($K_m = 20$ μM), and (c) by substituting 1 mM tri-n-butylammonium phosphate or 2 mM acetyl phosphate (20 min interaction with enzyme) for (I). This and other findings indicated that (I) either acylates or phosphorylates the enzyme and that the reaction occurs at the nucleotide binding site.[1]

Reaction of (II) with Pyruvate Kinase

Initial reaction velocities were measured at 340 nm. For all experiments (except where noted) the final volume of 1.00 ml of 0.1 M Tris chloride (pH 7.6) contained 13 μg of rabbit muscle lactic dehydrogenase, 0.05 μg of rabbit muscle pyruvate kinase, 0.1 M KCl, 0.025 M MgSO$_4$, 1.5 mM phosphoenolpyruvate (sodium salt), 0.25 mM ADP (sodium salt), and 0.25 mM NADH. The order of addition of the components was the same as described above for AMP aminohydrolase.

When a nominal initial level of 100 μM of (II) was added to the pyruvate kinase in three increments, about 50% inactivation occurred (see the table). The relationship between the degree of inactivation and the initial concentration of (II) was not examined further. Inactivation was prevented by 15-sec hydrolysis of (II) in the buffer prior to addition of enzyme. Pyruvate kinase inactivated by (II) did not regain activity when stored in the assay medium for 16 hr at 22°. Protection of the enzyme from a 100 μM nominal level of (II) was afforded by 100 μM ATP, 2.5 mM ADP, or 1.5 mM phosphoenolpyruvate, the levels of the last two compounds being selected so as to be in excess of their enzyme–substrate dissociation constants (0.8 mM and 0.08 mM, respectively). Protection by these three substrates was concluded to imply that the action of (II) is probably ATP-site-directed.[2]

Studies with Other Enzymes and Attempted Syntheses of Anhydride Analogs of Other Nucleotides

AMP kinase of rabbit muscle was not inactivated by 1 mM nominal initial levels of (I) (a 2-day-old sample) or of (II) (freshly prepared).

[5] C. E. Carter and L. H. Cohen, *J. Biol. Chem.* **222**, 17 (1956).

INACTIVATION OF RABBIT MUSCLE PYRUVATE KINASE BY (II)[a]

Additions prior to (II)	Rate × 10² (ΔA_{340}/min)		Inactivation (%)
	Enzyme plus hydrolyzed (II)	Enzyme plus (II)	
None	3.10	1.80	43
None[b]	3.00	1.45	52
None[c]	3.06	1.81	41
0.1 mM ATP	3.00	2.95	2
1.5 mM PEP	3.10	2.95	5
2.5 mM ADP[d]	3.02	2.96	2

[a] The nominal initial concentration of (II) was 100 μM.
[b] (II) added in 3 equal increments at 20-second intervals.
[c] After treatment with (II) the enzyme solution was stored at 22° for 16 hr before the velocity determination.
[d] The enzyme was exposed to 2.5 mM ADP, then 0.1 mM (II), in 0.1 ml of buffer and assayed after dilution to the standard volume of 1 ml.

The method used to convert adenosine-5'-carboxylic acid to (I) and (II) did not convert uridine- or thymidine-5'-carboxylic acids[6] to the analogous UMP or TMP carboxylic-phosphoric anhydrides.

[6] G. P. Moss, C. B. Reese, K. Schofield, R. Shapiro, and A. R. Todd, *J. Chem. Soc.* **1963**, 1149 (1963).

[30] Active-Site Labeling of Thymidylate Synthetase with 5-Fluoro-2'-deoxyuridylate[1]

By YUSUKE WATAYA and DANIEL V. SANTI

Thymidylate synthetase catalyzes the reductive methylation of 2'-deoxyuridylate (dUMP) to thymidylate with concomitant conversion of 5,10-methylenetetrahydrofolate (CH₂-H₄folate) to 7,8-dihydrofolate. A number of studies of the thymidylate synthetase reaction have led to the proposal that a primary event in the catalytic sequence involves the addition of a nucleophilic group of the enzyme to the 6-position of the substrate dUMP, a required step for subsequent condensation of the 5-posi-

[1] This work was supported by U.S. Public Health Service Grant CA-14394 from the National Cancer Institute. D. V. S. is a recipient of a National Institutes of Health Career Development Award.

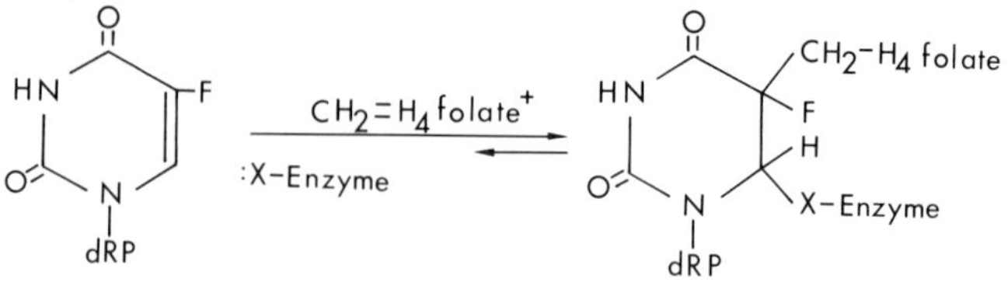

Fig. 1. Reaction of FdUMP, CH$_2$-H$_4$folate, and thymidylate synthetase to form the reversible covalent complex. dRP = 5'-phospho-2'-deoxyribosyl.

tion of nucleotide with the one carbon unit of the cofactor; details of this mechanism have appeared in a recent review.[2] To verify this mechanism, we sought an affinity labeling agent which might react with the proposed nucleophilic catalyst of thymidylate synthetase.

It had been known for some time[3,4] that FdUMP is an extremely potent inhibitor of thymidylate synthetase, but the nature of inhibition was the topic of considerable controversy.[5] Since the 6-position of 1-substituted 5-fluorouracils is quite susceptible toward nucleophilic attack,[6-8] it was suspected that FdUMP might exert its inhibitory effect by reaction with the proposed nucleophilic catalyst of thymidylate synthetase. It is now well established that, in the presence of CH$_2$-H$_4$folate, 5-fluoro-2'-deoxyuridylate (FdUMP) behaves as a quasi-substrate for thymidylate synthetase[9-12] and is, in effect, an affinity labeling agent for the enzyme. Whereas FdUMP binds relatively poorly to free enzyme, in the presence of the cofactor, CH$_2$-H$_4$folate, a covalent bond is formed between an amino acid residue of the enzyme and the 6-position of the nucleotide to give the complex depicted in Fig. 1. Although covalent bonds are involved in linking the components of the complex, the reaction is slowly reversible. Nevertheless, the complex is sufficiently stable ($K_d \simeq 10^{-13}$ M) to permit isolation and characterization.

[2] A. L. Pogolotti, Jr. and D. V. Santi, "A Survey of Contemporary Bioorganic Chemistry." Academic Press, New York, in press.

[3] S. S. Choen, J. G. Flaks, H. D. Barner, M. R. Loeb, and J. Lichenstein, *Proc. Natl. Acad. Sci. U.S.A.* **44**, 1004 (1958).

[4] C. Heidelberger, G. Kaldor, K. L. Mukherjee, and P. B. Danenberg, *Cancer Res.* **20**, 903 (1960).

[5] R. L. Blakley, "The Biochemistry of Folic Acid and Related Pteridines," pp. 245–247. American Elsevier, New York, 1969.

[6] J. J. Fox, N. C. Miller, and R. J. Cushley, *Tetrahedron Lett.* p. 4927 (1966).

[7] B. A. Otter, E. A. Falco, and J. J. Fox, *J. Org. Chem.* **34**, 1390 (1969).

[8] E. J. Reist, A. Benitez, and L. Goodman, *J. Org. Chem.* **29**, 554 (1964).

[9] D. V. Santi and C. S. McHenry, *Proc. Natl. Acad. Sci. U.S.A.* **69**, 1855 (1972).

[10] R. J. Langenbach, P. V. Danenberg, and C. Heidelberger, *Biochem. Biophys. Res. Commun.* **48**, 1565 (1972).

[11] D. V. Santi, C. S. McHenry, and H. Sommer, *Biochemistry* **13**, 471 (1974).

[12] P. V. Danenberg, R. J. Langenbach, and C. Heidelberger, *Biochemistry* **13**, 926 (1974).

Dissociation of the complex depicted in Fig. 1 is enzyme catalyzed, and denaturation stabilizes the covalent bonds. Thus, peptides covalently bound to both FdUMP and CH_2-H_4folate are sufficiently stable to permit isolation and investigation. Such peptides have thus far been obtained by proteolytic digestion[11-15] or cyanogen bromide cleavage[15] of denatured FdUMP-CH_2-H_4folate-thymidylate synthetase complexes. Utilizing Pronase digestion and TEAE-cellulose chromatograhy,[13,14] good yields (about 50%) of an FdUMP-CH_2-H_4folate peptide, presumably the active site peptide, may be obtained which is suitable for structural studies; the peptide has six amino acids with the sequence Leu-Pro-Pro-Cys-His-Thr and the structure of covalently bound ligands shown in Fig. 1.

Most important, the isolable FdUMP-CH_2-H_4folate-thymidylate synthetase complex is analogous to a steady-state intermediate of the normal enzymic reaction, and studies of its formation and structure are receiving considerable attention in investigations of the mechanism of thymidylate synthetase. Such studies are facilitated by dramatic changes in the ultraviolet,[11,12,16] fluorescence,[16,17] and circular dichroic spectra[16,18] which accompany formation of the complexes. In addition, since FdUMP and CH_2-H_4folate act as active site titrants of thymidylate synthetase, utilization of radioactive ligands permit direct quantitation of the enzyme at levels limited only by the specific activity of the ligand.[11,19]

We describe here the general procedures used in this laboratory for preparation and assay of the FdUMP-CH_2-H_4folate-thymidylate synthetase complex. It is emphasized that procedures have been optimized for thymidylate synthetase from *Lactobacillus casei* and may require modification for use with enzymes from other sources. Details of the experiments which led to the protocols have been reported.[11,19]

Materials

The thymidylate synthetase used here is obtained from an amethopterin-resistant strain of *Lactobacillus casei*[20] purified by the method of

[13] H. Sommer and D. V. Santi, *Biochem. Biophys. Res. Commun.* **57**, 689 (1974).
[14] A. L. Pogolotti, Jr., K. M. Ivanetich, H. Sommer, and D. V. Santi, *Biochem. Biophys. Res. Commun.* **70**, 972 (1976).
[15] R. L. Bellisario, G. F. Maley, J. H. Galivan, and F. Maley, *Proc. Natl. Acad. Sci. U.S.A.* **73**, 1848 (1976).
[16] H. Donato, J. L. Aull, J. A. Lyon, J. W. Reinsch, and R. B. Dunlap, *J. Biol. Chem.* **251**, 1303 (1976).
[17] R. K. Sharma and R. L. Kisliuk, *Biochem. Biophys. Res. Commun.* **64**, 648 (1975).
[18] J. H. Galivan, G. F. Maley, and F. Maley, *Biochemistry* **14**, 3338 (1975).
[19] D. V. Santi, C. S. McHenry, and E. R. Perriard, *Biochemistry* **13**, 467 (1974).
[20] T. C. Crusberg, R. Leary, and R. L. Kisliuk, *J. Biol. Chem.* **245**, 5292 (1970).

Galivan et al.[18]; we omit thiols from all stages of purification and perform recrystallizations from 30% $(NH_4)_2SO_4$ at pH 7.0 and 4°. The enzyme (about 6 mg/ml) is stored in 60 mM potassium phosphate (pH 6.9) containing 1 mM EDTA at —80°. Once thiols have been added, solutions are stored at 0°, or at —20° in 40% glycerol; freezing solutions of enzyme that have been treated with thiols leads to loss of activity. With exception of labeled FdUMP, all other materials may be obtained from commercial sources.

Preparation of Radioactive FdUMP. [³H]- and [¹⁴C]FUdR are commercially available and conveniently converted to the 5′-monophosphate using thymidine kinase[21] found in the 20–40% $(NH_4)_2SO_4$ fraction of the cell-free supernatant from *Escherichia coli* B. A solution (1.0 ml) containing about 1 mM [³H]- or [¹⁴C]FUdR, 7.5 mM ATP, 7.5 mM MgCl$_2$, 30 mM KF, 0.5 mg of bovine serum albumin, 70 mM Tris·HCl (pH 7.8), and approximately 0.1 mg of the crude thymidine kinase preparation is incubated at 37°. The reaction is monitored by application of aliquotes to a micro DEAE-cellulose column (0.5 ×0.8 cm) and stepwise elution with 3 ml of 5 mM ammonium formate (pH 4.5) to elute unreacted nucleoside and 3 ml of 100 mM of the same buffer to elute the nucleotide; radioactivity is determined in each fraction to ascertain the extent of conversion. After the reaction is 80–90% complete (usually 60 min), the mixture is diluted to 20 ml with water, and purified on a DEAE-cellulose column (1 × 5 cm) by the method of Rustum and Schwartz.[22]

Preparation of Complex and Filter Assay

In the presence of CH_2-H_4folate, FdUMP forms a specific, stable complex with thymidylate synthetase in which all components are covalently bound as depicted in Fig. 1. The affinity constant of this complex is sufficiently high that, with typical concentrations of components used in most experiments, the limiting reagent (FdUMP or enzyme) is completely bound. Using [³H]FdUMP of high specific activity, low levels of complexes present in solution may conveniently be assayed by retention on nitrocellulose filter membranes under conditions in which the free nucleotide is readily removed. The radioactivity remaining on the filter is determined to quantitate the complex. Other conventional methods (e.g., gel filtration, charcoal adsorption of free FdUMP, protein precipitation) may be used for this purpose, but are more tedious and apparently less efficient. There are expectedly few proteins that will form isolable

---

[21] R. Okazaki and A. Kornberg, *J. Biol. Chem.* **239**, 269 (1964).
[22] Y. M. Rustum and H. S. Schwartz, *Anal. Biochem.* **53**, 411 (1973).

complexes with FdUMP at the low concentrations of components usually used; even if such proteins were to exist, the probability that binding would also require the cofactor, CH_2-H_4folate, is negligible. Thus, controls are performed by omission of cofactor. In this regard, it should be cautioned that crude preparations may contain sufficient amounts of CH_2-H_4folate to give erroneously high controls. This difficulty can be circumvented by pretreatment of controls with unlabeled FdUMP or passage of the crude preparation through Sephadex G-15 prior to assay. If the filter assay is to be used for exact quantitation of thymidylate synthetase, it is necessary to determine the filtration efficiency.[19,23] This reflects the probability that the protein–ligand complex will survive the filtration and washing procedure and is used to convert the amount of filter-bound complex to the amount of complex actually present in solution prior to filtration. We determine this parameter by using enzyme in excess of [³H]FdUMP; the filtration efficiency is the ratio of bound radioactivity to the total applied to the filter. Values are constant within an experiment, but may vary with the lot of filters and purity of components. Details of parameters influencing the assay have been published.[19]

Procedure

Formation of Complex. The standard mixture contains 50 mM N-methylmorpholine-HCl (pH 7.4), 25 mM $MgCl_2$, 1 mM EDTA, 75 mM 2-mercaptoethanol, 6.5 mM formaldehyde, about 0.2 mM CH_2-H_4folate, [6-³H]FdUMP,[24] and thymidylate synthetase. The reaction is initiated by the addition of enzyme; CH_2-H_4folate is omitted in controls. If small amounts of highly purified enzyme are used, bovine serum albumin (50 μg/ml) may be added to avoid adsorption to containers, although this appears not to be necessary for most purposes. After 30 min, an aliquot (about 50 μl) is removed for determination of total [³H]FdUMP concentration and 100-μl duplicates are assayed as described below. Although 30 min incubation is sufficient for even the most dilute solutions, an additional 100-μl aliquote is usually assayed after 1 hr to ensure that equilibrium has been obtained.

[23] M. Yarus and P. Berg, *Anal. Biochem.* 35, 450 (1970).
[24] When larger quantities of enzyme are used, [¹⁴C]FdUMP is more convenient because of its greater stability. [³H]FdUMP may be obtained in specific activities exceeding 5 Ci/mmole, but has a tendency to decompose upon storage; we recommend storage in 70% EtOH at $-20°$ and periodic checks of purity. For large-scale preparations of the purified complex, concentrations of 0.15 mM FdUMP, 0.05 mM *L. casei* thymidylate synthetase, and 0.3 mM CH_2-H_4folate may be used without modification of the described procedure.

For assay of preparations containing unknown quantities of enzyme, we generally use 40 nM [³H]FdUMP (about 5 Ci/mmole), vary the enzyme, and assay 100-μl aliquots. With crude cytosol from rat liver hepatoma cells, 50 μg of protein contains sufficient enzyme to be easily detectable.[19]

Filtration Assay. Nitrocellulose membranes are soaked before use in 25 mM potassium phosphate (pH 7.4) and 25 mM MgCl$_2$; filters which are not wetted within 2 min are discarded. The filter disks are placed on a filter manifold (Hoeffer Scientific) and a gentle vacuum is applied (about 2 ml/min filtration rate) to remove excess moisture. After removal of vacuum, 100-μl aliquots of the reaction mixture are applied to each disk and allowed to permeate the membrane. The filters are washed at about 2 ml/min filtration rate with seven 0.5-ml portions of a solution containing 25 mM phosphate buffer (pH 7.4) and 25 mM MgCl$_2$. The damp filters are dissolved in 10 ml of Bray's solution[25] and counted.

[25] G. A. Bray, *Anal. Biochem.* 1, 279 (1960).

[31] Inert Co(III) Complexes as Reagents for Nucleotide Binding Sites

By Moshe M. Werber and Antoine Danchin

Principle

About one-third of all currently purified enzymes require a metal cation for their activity. In some cases, as with nucleotide binding sites, the metal ion is acting as a bridge between the substrate and the enzyme. It would therefore be useful to devise techniques for labeling the metal binding sites of enzymes. Among the transition metal ions, those that form d³ or d⁶ low spin complex ions in octahedral fields are the most inert with regard to substitution of their ligands; they have dissociation times of the order of 10⁷ sec. The cobaltic cation (d⁶ low spin) binds its ligand in an octahedral ligand field and the distances of Co(III)–nitrogen or Co(III)–oxygen resemble those observed for corresponding Mg(II) complexes because of the similarity in ionic radii (about 0.65 Å). In addition, Co(III) complexes are frequently stabilized, in spite of their high redox potential, by reason of the low probability of exchange of the ligands by external groups.[1]

It is therefore possible to synthesize Co(III) analogs of metal sub-

[1] F. Basolo and R. G. Pearson, "Mechanisms of Inorganic Reactions," 2nd ed. Wiley, New York, 1967.

strate complexes, by oxidation of the corresponding Co(II) complexes with biological ligands, and to try to use them as affinity labels of the active sites of enzymes. When a cobaltic complex possessing the correct stereoconformation interacts with a binding site on an enzyme, the barrier of free energy of activation for a substitution reaction is considerably lowered. Under these conditions, a liganding group from the enzyme will displace a "nonessential" ligand from the complex and result in affinity labeling.

Synthesis of Co(III) Complexes

Method I: Electrolytic Preparation of Nucleotide Complexes

The various nucleotides can be mixed with cobaltous chloride at the anode of an electrolysis apparatus; cobalt(II) is oxidized to cobalt(III), a species with nonexchangeable ligands.

The electrolysis chamber should contain at least 50 mM KCl in order to conduct the current, and is connected by an agar bridge (1% agar) saturated in KCl to a chamber containing 1 N HCl and saturated KCl. The cathode is fixed in the saturated KCl chamber, and the anode, preferably rotating, is in the electrolysis chamber. Any kind of platinum anode of large surface can be used provided that strong stirring is applied.

Synthesis of Co(III)-AMP

Cathode chamber: HCl 1 N; KCl saturated
Anode chamber: KCl 50 mM; CoCl$_2$ 10 mM; AMP 10 mM adjusted to pH 3.5

A voltage difference of about 1.1 volt is established, and the electrolysis is allowed to proceed for 24–72 hr. The pH should be readjusted to 3.5 from time to time as it slowly becomes acidic. The yield is 10%.

The same procedure can be applied to ADP, ATP, GDP, IDP, and ITP. With these complexes, the experiment should be performed at low temperature and the pH should be carefully checked: pHs lower than 2.5 and higher than 4.0 must be avoided. With GMP and IMP one obtains insoluble complexes. With the pyrimidine nucleotides, no complex can be formed under these conditions.

Purification

The complexes are somewhat unstable; one should especially avoid pHs higher than 7.2, and separation on many solid supports causes considerable degradation. Purification is therefore performed by ethanol precipitation. The electrolytic mixture is brought to 150 mM in KCl, adjusted to pH 6.0, and put on ice; cold ethanol up to 0.5 volume of ethanol per volume of solution is added with stirring. The mixture is allowed to stand at −20° for 30 min and centrifuged. The pellet is redissolved in 150 mM KCl, and the procedure is repeated at least three times. The pH should be maintained.

Method II: Air Oxidation

Synthesis

Co(III) complexes can be obtained from the Co(II) complexes by vigorous aeration at alkaline pH. Thus, a complex containing ATP, 1,10-phenanthroline (abbreviation: phen), and cobalt(III) may be synthesized as follows:

ATP, 11 mM, and phenanthroline, 10 mM, are dissolved and brought to neutrality in ice; CoCl$_2$, 10 mM, is then added. The pH is raised to 10 and vigorous bubbling of air (or oxygen) is initiated. The solution is magnetically stirred, and the pH is maintained between 9.5 and 10 with 0.1 or 1 N NaOH. The reaction is allowed to continue until no color increase is observed or A_{590} = 0.85 to 0.90 (usually 3–5 hr). The yield is variable between 25 and 75%.

The complex is purified by precipitating twice with ethanol at −20° as in Method I. In this case, however, the pH is maintained at 10 and KCl may be omitted from the solution. The complex is stored in either frozen or lyophilized state.

Characterization

The molecular weight of the complex was 850 as determined by gel filtration on a Sephadex G-10 column. The composition of the complex was determined as Co:phen:ATP 1:1:1 after reduction to the Co(II) complex at low pH and separation of the reduced complex on a G-10 column. The cobalt content was calculated from the absorbancy of its complex with thiol groups (see below), and the ATP and phenanthroline concentrations from their respective absorbancies at 260, 264, and 287 nm. That the complex also contained the anion O$_2^-$ was deduced both from the fact that it is paramagnetic and that O$_2^-$ can be displaced by CN$^-$; O$_2^-$ may be monitored by its inducing effect on the photoreactivity of luminol.

Parent Complexes

The same procedure can be applied to all purine di- and trinucleotides. Mononucleotides, such as AMP, yield insoluble complexes under these conditions.

Similar ligands, other than phenanthroline, have been used, including ethylenediamine, dimethylethylenediamine, and bipyridyl.

Method III: Oxidation by Hydrogen Peroxide

Method III is very similar to Method II, except that oxidation is performed with a 4-fold molar excess of H_2O_2. The synthesis of Co(III)–phen–ATP is performed as follows: Na_2ATP, 11 mM, phen and $CoCl_2$, 10 mM, are mixed at 0°; the pH is raised to 10, and H_2O_2 is added at a final concentration of 40 mM. The pH is kept between 9.5 and 10 with NaOH until almost no pH decrease occurs (about 20 min). At this stage, the complex is purified by precipitation with cold ethanol as in Method II.

Affinity Labeling of Enzymes

Labeling Time Course

The complex is incubated with the enzymes under conditions of their optimal stability. Usually 30- to 300-fold excess of labeling reagent is necessary. The kinetics of labeling are followed by assaying aliquots of the incubation mixture for activity. It is advisable to run a blank enzyme sample without labeling reagent under the same conditions. The concentration of labeling reagent should be sufficiently high as to enable the labeling to be completed within 2–5 hr.

Reversal of Labeling by Thiolysis

Many Co(III) complexes are sensitive to thiols at slightly alkaline pH values. Under these conditions, strongly colored thiol complexes of Co(III) are formed. These complexes will also form spontaneously with Co(II) in the presence of O_2.

Thus, incubation for 15 min at 37° of Co(III) complexes or of enzymes labeled with Co(III) complexes with 25 mM of dithiothreitol (DTT) or mercaptoethanol causes conversion to the Co(III)–thiol complex, thereby usually completely restoring enzyme activity.

Determination of Stoichiometry of Labeling

The labeled enzymes are first freed from excess reagent, either by dialysis, by precipitation and centrifugation, or by column chromatography.

Since the labeling is performed on high molecular-weight compounds, the proportion of the label is low. In some cases it is possible to use ^{57}Co, ^{58}Co, and ^{14}C-, ^{3}H-, or ^{32}P-labeled ATP for the labeling and express the stoichiometry in terms of radioactivity that remains bound to the enzyme after it has been freed from excess reagent.

An approximate stoichiometry can also be determined by thiolysis of the labeled enzymes with DTT or mercaptoethanol. The Co(III)–thiol polynuclear complexes exhibit a visible absorption spectrum composed of two distinct peaks, usually centered around 405 and 480 nm with molar absorbancies around 9000. However, these figures are correct only at low dilutions of the complexes (lower than 50 μM in cobalt), since at higher concentration the absorbancy does not follow the Beer–Lambert law. Since the thiolysis reaction requires an exchange of ligands, it must heavily depend on the nature of the ligands, temperature, pH, and reagents concentration. In the presence of o-phenanthroline, for instance, in complexes with a ratio phen:Co 1:1, such as Co(III)–phen–ATP, the molar absorbancy of thiolized complexes is slightly altered at 400 nm ($\epsilon \sim 8500$), and strongly diminished at 480 nm ($\epsilon \sim 3000$, i.e., about one-third of the value in the absence of phenanthroline).

A third method of determination of the stoichiometry of labeling is based on atomic absorption evaluation of the cobalt content of the labeled enzymes.

Affinity Labeling of the AMP Allosteric Site of Rabbit Muscle Glycogen Phosphorylase b

Several complexes of AMP, ADP, ATP, and Co(III), prepared as described in Method III, have been assayed for their specific activity toward phosphorylase b (5 mM of the complexes in solution containing 1 mg/ml of the enzyme); it was observed that there is inhibition of enzyme activity and binding of Co(III) to the protein. A thorough investigation has also been conducted using the Co(III)–AMP complex prepared by Method I. Since incubation of the complex with solubilized enzyme yielded irreversible denaturation, probably because of the oxidation of an essential thiol group, labeling was performed with about 5 mg of microcrystals of the enzyme per milliliter suspended in a medium containing 0.1 mM AMP, 0.1 mM MgCl$_2$, 50 mM Tris-acetate at pH

TABLE I
DETERMINATION OF THE STOICHIOMETRY OF COBALTIC COMPLEXES[a]

| Method | Complex A | | | Complex B | | | Phosphorylase b–Co(III)–AMP | | |
	Co	AMP	Co–AMP ratio	Co	AMP	Co–AMP ratio	Co	AMP	Phosphorylase b
Double labeling[b] (mM)	0.08	0.21	2.65:1	0.86	0.82	1.05:1	—	—	—
Simple labeling (mM)				0.63	0.64	0.98:1	—	—	—
Absorbancies[c] (μM)	12.2	24.3	1.99:1	27.7	27.9	1.01:1	6.4	6.0	6.1

[a] Reprinted with permission from A. Danchin and H. Buc, *J. Biol. Chem.* **248**, 3241 (1973). Copyright by the American Chemical Society of Biological Chemists.

[b] Based on the radioactivity of a doubly labeled complex [^{14}C]AMP (2 μCi/mmole)–^{58}Co (10 μCi/mmole).

[c] The concentration of the components was determined spectroscopically after dissociation of the Co(III)–AMP complex with DTT. The following extinction coefficients were used: Co(III)–DTT, $\epsilon_{480} = 9000$; AMP, $\epsilon_{260} = 15,400$; phosphorylase, $\epsilon\, b$, $\epsilon_{280}^{1\%} = 13.2$. Phosphorylase *b* is labeled with complex B. For each of the two Co(III)–AMP complexes, a stock solution of known absorbancy at 260 nm was added to a 10 mM solution of DTT in Tris buffer (pH 7.5) and read after evaluation of the Co(III) content at 480 nm against the same DTT solution at the same final concentration in cobalt, in the presence of oxygen [thus containing the same amount of Co(III)–DTT].

7.0.[2] The concentration of the label varied from 10 μM to 100 μM and yielded, after 24 hr at 4°, an enzyme that carried up to one AMP residue and one Co(III) ion per subunit, with a parallel loss in its phosphorolytic activity.

Table I summarizes the results, which show that labeling occurs at one site per subunit. Tentatively, this is assumed to be the AMP binding site, since equilibrium dialysis experiments with ^{14}C-labeled AMP show a decrease in binding of complementary AMP to the irreversible binding of the Co(III)–AMP.

The effect of the labeling with this complex could be reversed after incubation with mercaptans, such as 2-mercaptoethanol, DTT, or glutathione. Not only is the binding site of AMP abolished by labeling with Co(III)–AMP but the cooperative behavior of this allosteric enzyme is also altered. Cooperativity disappears at high degrees of labeling with a concomitant decrease in the affinity for AMP. Nevertheless, AMP is still required for activity.

Hybrid molecules of phosphorylase *b*, half-labeled with Co(III)–AMP, can be separated from nonlabeled molecules and from totally

[2] A. Danchin and H. Buc, *J. Biol. Chem.* **248**, 3241 (1973).

labeled ones on an AMP-coupled Sepharose column, after elution with increasing concentrations of the allosteric inhibitor, glucose 6-phosphate (20 mM). It is necessary to use a slightly alkaline pH (8.0) for their separation in order to prevent rapid rearrangement between subunits.

Affinity Labeling of ATPase: Rabbit Muscle Myosin and
  Chloroplast Coupling Factor 1

Several Co(III)–ATP derivatives have been tested as potential affinity labels of myosin. Whereas electrolytically prepared Co–ATP was inefficient as a labeling reagent, the ternary complexes prepared by air oxidation displayed some labeling properties. However, both classes of ATP complexes, the electrolytically prepared Co–ATP as well as the ternary complexes prepared by air oxidation, could *inhibit* the Ca ATPase activity of myosin by about 50% when present in the assay medium at a concentration of 0.1–0.2 mM ([ATP] = 2.5 mM).

The labeling of myosin by Co–(dimethylethylenediamine)–ATP could not be reversed by DTT. With Co–(phen)–ATP as labeling reagent, both labeled myosin and coupling factor 1 ATPases were reactivated by thiolysis of the Co(III) label with DTT.[3,4]

The time course of the labeling of myosin by Co–(phen)–ATP (Fig. 1) occurs in two phases: at first ATPase activity (Ca^{2+}-, EDTA-, and actin-dependent activities) are enhanced, and in the second phase the activities are abolished in a pseudo-first-order process. In the cases of the myosin subfragments—"double-headed" heavy meromyosin (HMM) and "single-headed" subfragment 1 (S-1)—no enhancement phase is observed and the labeling occurs as a pseudo-first-order inactivation process.[3,5] The kinetic parameters of the Co–(phen)–ATP affinity labeling of myosin and its subfragments are summarized in Table II.

The labeling of coupling factor 1 (Fig. 2) occurs also as a single inactivation process. However, a residual activity remains after the labeling process and is shown to depend on the ATP concentration in the assay medium.[4] On the other hand, ($Na^+ + K^+$)-ATPase from sheep kidney is inactivated with biphasic kinetics, each phase representing approximately 50% of the initial activity.[6]

The labeling of myosin (and of its subfragments) was performed at

[3] M. M. Werber, A. Oplatka, and A. Danchin, *Biochemistry* **13**, 2683 (1974).
[4] M. M. Werber, A. Danchin, Y. Hochman, C. Carmeli, and A. Lanir, *in* "The Jerusalem Symposium on Quantum Chemistry and Biochemistry" (B. Pullman and N. Goldblum, eds.), Vol. 9, in press. Reidel Publ., Dordrecht, Holland, 1976.
[5] A. Oplatka, M. M. Werber, and A. Danchin, *FEBS Lett.* **47**, 7 (1974).
[6] S. Karlish and M. M. Werber, unpublished results.

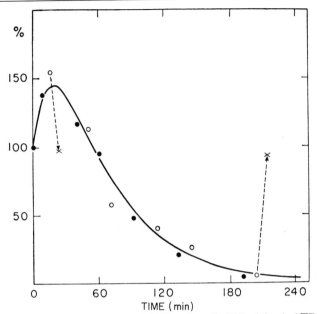

Fig. 1. Time course of myosin labeling by Co(III)–(phen)–ATP. ●——●, EDTA–ATPase activity: ○——○, Ca²⁺–ATPase activity; ×, Ca²⁺–ATPase activity after reactivation with 10 mM DTT. The solid line is drawn on the basis of the kinetic scheme for a two-step labeling (see Table II), developed in the Appendix of the article in which this figure was presented. Reprinted with permission from M. M. Werber, A. Oplatka, and A. Danchin, *Biochemistry* **13**, 2683 (1974).

TABLE II

KINETIC PARAMETERS OF THE Co–(PHEN)–ATP AFFINITY LABELING OF MYOSIN AND ITS SUBFRAGMENTS[a]

	Myosin			HMM	Sub-fragment 1
$[KCl]^b$ M	$k_1/2^t$ $(M^{-1} sec^{-1})$	s^c	k_2^b $(M^{-1} sec^{-1})$	k^b $(M^{-1} sec^{-1})$	k^b $(M^{-1} sec^{-1})$
0.60	0.297	1.18	0.148	0.222	—
0.15	0.305	1.40	0.107	0.126	0.084
0.12	0.311	1.75	0.119	0.110	—
0.022	—	—	—	0.044	—

[a] Reprinted in adapted form with permission from A. Oplatka, M. M. Werber, and A. Danchin, *FEBS Lett.* **47**, 7 (1974).

[b] $M + C \xrightarrow{k_1} MC$; $MC + C \xrightarrow{k_2} MC_2$, where M denotes myosin or its subfragment, and C the complex. In the case of HMM, $k_1 = k_2 = k$.

[c] s is defined as the enhancement in the activity of a myosin molecule labeled at one head only.

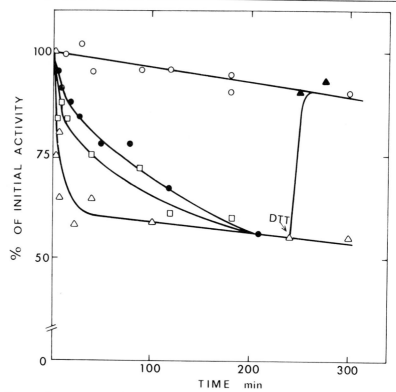

Fig. 2. Time course of CF$_1$-ATPase inactivation in the presence of various concentrations of Co(III)–(phen)–ATP: ●——●, 8.8 mM; □——□, 4.5 mM; △——△, 2.5 mM; ○——○, no complex; ▲——▲, after addition of DTT. Trypsin-activated coupling factor 1 (CF$_1$) was incubated with the Co(III) complex at 15°. At various times 1-μl aliquots were assayed for ATPase activity by following the decomposition of highly labeled [^{32}P]ATP at 37° as described by Y. Hochman, A. Lanir, and C. Carmeli, *FEBS Lett.* **61**, 255 (1976). The reaction mixture contained 10 μg of CF$_1$, 40 mM HEPES buffer, pH 8, and 5.4 mM Ca ATP in 1.5 ml. Reprinted from M. M. Werber, A. Danchin, Y. Hochman, C. Carmeli, and A. Lanir *in* "The Jerusalem Symposium on Quantum Chemistry and Biochemistry" (B. Pullman and N. Goldblum, eds.), Vol. 9, in press. Reidel Publ., Dordrecht, Holland, 1976.

pH 7.0 in phosphate buffer (5 mM) at 0°. Increasing the ionic strength seems to cause more rapid labeling (Table II). In the case of coupling factor 1, a cold-sensitive enzyme, incubation was performed at 15° and in 40 mM HEPES. Tricine at 10 mM is also effective. Buffers and reagents that possess the ability to chelate metal ions should be avoided because they can compete with the enzymes for Co–(phen)–ATP. Because Co–(phen)–ATP has a tendency to decompose spontaneously below pH 10, a correction for the concentration of destroyed complex is necessary when long (12 hr) incubations are employed. For example, at pH 7 and 0°, the half-life of Co–(phen)–ATP was found to be 20 hr.

The stoichiometry obtained both in the case of myosin and its sub-fragments[3,4] and in the case of coupling factor 1[4,7] indicates that the labeling occurs only at specific sites on the enzymes. Moreover, the fact that it is possible to separate the labeled enzymes from excess reagent is by itself a proof that the association with the complex is irreversible.

Co–(phen)–ATP was also shown to behave as a competitive inhibitor of myosin[3] and of coupling factor 1 when added directly into the assay medium.[8] Moreover, in the case of heavy meromyosin, both Mg ATP and Mg ADP could protect against labeling by Co–(phen)–ATP.[4] Experiments with thiol reagents (p-hydroxymercuribenzoate and N-ethylmaleimide) seem to indicate that Co–(phen)–ATP either binds to or protects some essential SH groups in the active site of myosin.

The structure of Co–(phen)–ATP has been investigated, and the following features emerge from this study[9]: the cobaltic ion is bound in the plane of the metal ion to phenanthroline and to the β- and γ-phosphate groups; the apical positions are occupied by the N-7 of the adenine ring and by the O_2^- anion, which is an exchangeable ligand that can be replaced by a protein ligand.

---

[7] Y. Hochman, C. Carmeli, A. Lanir, and M. M. Werber, to be published (1977).
[8] It has been established that Co–(phen)–ATP is not hydrolyzed by myosin or by coupling factor 1. This is probably due to the inertness of the complex, which precludes liberation of ligands in the medium.
[9] A. Danchin and M. M. Werber, to be published (1977).

[32] The Active Site of Ribonucleoside Diphosphate Reductase[1]

By Lars Thelander, John Hobbs, and Fritz Eckstein

Ribonucleotide reductase of *Escherichia coli*, which catalyzes the reduction of ribonucleoside 5'-diphosphates to 2'-deoxynucleoside 5'-diphosphates, consists of two nonidentical subunits, proteins B1 and B2. In the presence of Mg^{2+}, the two subunits form a 1:1 complex of active enzyme.[2] When separated, neither subunit has any known biological activity. Protein B1 has a molecular weight of 160,000, contains the active dithiols, is capable of interacting with thioredoxin, and contains

---

[1] L. Thelander, B. Larsson, J. Hobbs, and F. Eckstein, *J. Biol. Chem.* **251**, 1398 (1976).
[2] L. Thelander, *J. Biol. Chem.* **248**, 4591 (1973).

binding sites both for the ribonucleoside diphosphate substrates[3] and for the nucleoside triphosphate effectors.[4] Protein B2 has a molecular weight of 78,000 and contains bound nonheme iron and an organic free radical essential for activity.[2,5] The radical gives rise to a characteristic light absorption at 410 nm. B2 is inactivated by removal of the iron or by destruction of the radical with hydroxylamine. No binding of substrates or effectors to B2 can be demonstrated.

Here, the inactivation of ribonucleotide reductase by 2'-deoxy-2'-chlorocytidine and by 2'-deoxy-2'-chlorouridine 5'-diphosphate, as well as by 2'-deoxy-2'-azidocytidine 5'-diphosphate, is described.[1] The results indicate that both B1 and B2 contribute to the active site of the enzyme and that the radical present in B2 directly participates in the catalytic process together with the redox active dithiols of B1.

Synthesis of Substrate Analogs

2'-Chloro-2'-deoxycytidine[6]

2'-Chloro-2'-deoxy-4-thiouridine[7] (278.5 mg; 1 mmole) is dissolved in water (200 ml), and a 2-ml aliquot of a solution containing 4.725 g of anhydous sodium sulfite and 1.188 g of sodium hyposulfite in 50 ml of water is added. A brisk stream of air is drawn through the solution via a sintered-glass disk. Additional 2-ml aliquots are added at hourly intervals up to 4 hr. Examination of the UV spectrum of the reaction solution shows λ_{max} changing from 329 to 318 nm. After 5.5 hr, bubbling is stopped and 1 M ammonium acetate (10 ml) is added to the solution with sufficient concentrated ammonia to bring the pH to 8.7. The solution is stirred magnetically. After about 1 hr the pH is readjusted to 8.7 with a little more ammonia, and the solution is allowed to stir overnight. A large new maximum at about 270 nm is evident, the $\epsilon_{270}:\epsilon_{318}$ being about 50. The aqueous solution is evaporated, traces of water being removed by addition and evaporation of pyridine using an oil pump. The temperature throughout the reaction is not allowed to rise above 25°. The residue is thoroughly triturated with dry pyridine (3 × 15 ml), and the inorganic salts are removed by filtration. The pyridine is evaporated, and traces of

[3] U. V. Döbeln, *J. Biol. Chem.* **251**, 3616 (1976).

[4] N. C. Brown and P. Reichard, *J. Mol. Biol.* **46**, 39 (1969).

[5] C. L. Atkin, L. Thelander, P. Reichard, and G. Lang, *J. Biol. Chem.* **298**, 7464 (1973).

[6] J. Hobbs, H. Sternbach, M. Sprinzl, and F. Eckstein, *Biochemistry* **11**, 4336 (1972).

[7] I. L. Doerr and J. J. Fox, *J. Org. Chem.* **32**, 1463 (1967).

pyridine are removed from the residual gum by addition and evaporation of water. The gum is dissolved in a little methanol and applied to thin-layer chromatography (TLC) plates (2 mm thickness), which are developed with methanol–chloroform (40:60, v/v). The major band (R_f 0.74) is excised, and the product is eluted with methanol. The methanolic solution is evaporated to give a gum that does not crystallize, but yields a single spot on TLC (R_f 0.62) in the above system, and on paper chromatography (R_f 0.72) in system A. The yield may be estimated spectrophotometrically as 6.48×10^3 A_{269} units in H_2O (81%). Upon thawing a frozen concentrated aqueous solution of the product (about 31 mg/ml), white crystals are obtained. These show a melting phase, 109°–115°, followed by formation of a new crystalline phase in the range 118°–130°. This sinters at 190°–220°, decomposing at 230°–240°. [cf.7]

2′-Azido-2′-deoxyuridine

2′-Azido-2′-deoxyuridine is prepared by a slight modification of a procedure described by Verheyden et al.[8] Uridine (10 g) and diphenyl-carbonate (12 g) are stirred in hexamethylphosphotriamide (80 ml) in an oil bath at 140°, and $NaHCO_3$ (0.24 g) is added. After cessation of bubbling, approximately 30 min, LiN_3 (8 g) is added. After about 2 hr at 140° the solution is cooled, diluted with H_2O (16 ml), and extracted with $CHCl_3$ (2 × 200 ml). The combined $CHCl_3$ extracts are extracted with H_2O (2 × 16 ml), and the combined aqueous solution is again extracted with $CHCl_3$ (3 × 200 ml). The $CHCl_3$ extracts from this last step are evaporated under reduced pressure. The residue is triturated with a mixture of acetone (160 ml) and MeOH (60 ml) and filtered, and the filtrate is evaporated. The remaining oil is chromatographed on SiO_2 (200 g) which has been equilibrated with acetone; the column is eluted with acetone. Fractions containing the product are combined and further purified by preparative TLC (6 plates, 20 × 40 cm, 2-mm layer of SiO_2). The plates are developed with acetone/ethyl acetate (1:1). The product containing bands were scraped out and eluted with acetone. The acetone solution is evaporated, and the residue is taken up in pyridine (45 ml) to remove SiO_2 and filtered; the filtrate is evaporated and reevaporated several times with H_2O to remove traces of pyridine. The remaining slightly yellow oil is homogeneous by TLC (acetone/ethyl acetate, 1:1) with a yield of 50%. This material may be used without further purification for the synthesis of 2′-azido-2′-deoxycytidine. On standing at room temperature, the oil crystallizes.

[8] J. P. H. Verheyden, D. Wagner, and J. G. Moffatt, J. Org. Chem. 36, 250 (1971).

To remove the yellow color, the material (approximately 2 mmoles) may be applied to a Dowex 1 × 4 ion-exchange column (1.7 × 17 cm, OH⁻ form) and the column is washed sequentially with H_2O (500 ml) and with 50% aqueous MeOH (500 ml); the compound is eluted with 0.1 M triethylammonium bicarbonate. The eluate is evaporated, and the buffer is removed by repeated evaporations with MeOH. The residue is triturated with acetone (20 ml) and filtered; the acetone solution is evaporated, the residue is applied to a SiO_2 column (1.7 × 17 cm), and the column is eluted with acetone. The acetone solution is evaporated; the residue crystallizes on standing: white needles, m.p. 139°–147°, with darkening and decomposition which became very rapid above 180°. Yield 95%.

2'-Azido-2'-deoxycytidine[9,10]

2'-Azido-2'-deoxy-3',5'-diacetyluridine[8] (1.0 g) is dissolved in 13.9 ml of ethanol-free chloroform, and dry dimethylformamide (0.139 ml) and thionyl chloride (2.2 ml) are added. The solution is heated under reflux for 6.5 hr, cooled to room temperature, evaporated, and dissolved in methanol (60 ml) that is 50% saturated with ammonia. The solution is stirred at room temperature for 5 days. TLC on SiO_2 (methanol–chloroform, 40:60, v/v) shows the major product to have R_f 0.62. The solution is evaporated, and the product is separated by preparative TLC in the above system; the required band elutes with methanol. The eluate is evaporated and the residue is dissolved in water and applied to a column of Dowex 1 × 2 (OH⁻) (1.7 × 21.5 cm). The column is washed with water, and the required product is eluted with methanol–water (30: 70, v/v). The solvent is evaporated; the remaining gum, dissolved in a little ethanol, yields white crystals when stored at 5°. The yield (determined spectrophotometrically) is 47%, m.p. 215° (decomposition).

Phosphorylation of Nucleosides

2'-Chloro-2'-deoxycytidine 5'-Phosphate; 2'-Chloro-2'-deoxycytidine 5'-Monophosphate

2'-Chloro-2'-deoxycytidine (0.5 mmole) is dissolved in triethylphosphate (2 ml) and cooled to 0°; redistilled phosphoryl chloride (0.3 ml), also cooled to the same temperature, is added. The reaction mixture is

[9] J. Hobbs and F. Eckstein, *in* "Nucleic Acid Chemistry, New and/or Improved Synthetic Procedures" (L. B. Townsend, ed.), Vol. I, in press. Wiley, New York, 1977.

[10] J. Hobbs, H. Sternbach, M. Sprinzl, and F. Eckstein, *Biochemistry* **12**, 5138 (1973).

maintained at ice temperature. After 80 min, when the reaction appears to be far advanced as judged by TLC (on SiO_2 plates, developing with methanol/chloroform, 1:1, v/v), the reaction flask is placed on a vacuum rotary evaporator for 15 min to remove excess phosphoryl chloride, and then cooled to 0° with a small piece of ice added. After 15 min, the solution is again evaporated to remove traces of water and subsequently cooled to 0°, and triethylamine (2.5 ml) is added. A granular precipitate forms at once. On addition of a couple of drops of water, the precipitate becomes gummy, and the solution is decanted and discarded. The gum is dissolved in a little water and applied to a column of DEAE-cellulose (32 × 2.1 cm), which is washed thoroughly with water and then eluted with a linear gradient of triethylammonium bicarbonate (0–0.15 M in 3 liters). The major UV-absorbing peak is collected and evaporated; traces of triethylamine are removed by addition of methanol and evaporation, to obtain 2′-chloro-2′-deoxycytidine 5′-monophosphate (3150 A_{270} units, 79%) homogeneous on electrophoresis at pH 7.5 (30 volts/cm, 90 min). 2′-Azido-2′-deoxycytidine 5′-phosphate is prepared in the same manner. Preparation of 5′-diphosphates from the nucleoside 5′-phosphates was carried out according to Michelson.[11]

Protein Inactivation

Inactivation of Protein B1 by 2-Deoxy-2′-chlorocytidine 5′-Diphosphate (CclDP)

Addition of CclDP to a solution containing ribonucleoside diphosphate reductase resulted in inactivation of the B1 subunit. When increasing amounts of CclDP were incubated with proteins B1 and B2 in the presence of Mg^{2+} and a positive effector, there was progressive inactivation of B1; complete inactivation was attained with 4 moles of CclDP per mole of B1. Inactivation of B1 showed an absolute requirement for active protein B2. Protection of B1 against inactivation by CclDP was observed when increasing amounts of CDP were added to B1 assay mixtures together with a fixed amount of CclDP. The inactivation of B1 was much faster in the presence of a positive effector (ATP or dTTP) than in the presence of a negative one (dATP or dGTP).

As reaction products of the inactivation, free base (cytosine), chloride ion, and a compound which behaved electrophoretically and chromatographically like 2-deoxy-5-diphosphate could be identified. As a result of the reaction with CclDP a loss of titratable SH-group in protein B1 was observed. Addition of an excess of CclDP to a B1–B2 mixture resulted

[11] A. M. Michelson, *Biochim. Biophys. Acta* **91**, 1 (1964).

in the loss of about 6 titratable SH groups per mole of B1 at complete inactivation. A similar loss of titratable sulfhydryls occurred on addition of the normal substrate CDP but without inactivation.[12] In the latter case the SH groups were easily regenerated by addition of dithiothreitol, whereas neither sulfhydryl nor B1 activity could be regenerated after inactivation by CclDP, even in the presence of 6 M guanidinium hydrochloride. The nature of the sulfhydryl modification remains unknown.

When CclDP was added to a B1–B2 mixture a slow increase in absorbance at 320 nm was observed, which was removed with B1 after chromatography on dATP-Sepharose.[2] This modification is tentatively believed to be one of tryptophan.

Inactivation of Protein B2 by 2'-Deoxy-2'-azidocytidine 5'-Diphosphate (CzDP)

Addition of increasing amounts of CzDP to B1–B2 mixtures gave a progressive decrease in B2 activity until 100% inactivation was attained when approximately 0.2 mole of CzDP had been added per mole of B2. In the absence of B1 no inactivation was observed. Only B1 with redox active dithiols could fulfill this function. This indicated that reduction of CzDP was required for inactivation of B2. Increasing amounts of CDP together with a constant amount of CzDP resulted in a decreased inactivation of B2 indicating competition between the analog and substrate for the same site on the reductase. The reductase had to be in an active conformation in order to be inactivated by CzDP. Addition of the negative effector dATP slowed down the inactivation while further addition of ATP increased it.

When the B1–B2 complex was incubated with an excess of [β-^{32}P] CzDP, analyses of the reaction mixture after complete inactivation showed the loss of 2.3 moles of sulfhydryls in B1. Chromatography on Sephadex showed that no radioactivity was bound to protein. The characteristic 410 nm absorbance peak of protein B2[5] had been lost completely after inactivation. Iron analyses showed that this was not due to loss of iron. Removal of iron by dialysis and readdition of Fe^{2+} resulted in a regaining of activity and the reappearance of the absorption at 420 nm. It is concluded that inactivation of B2 by CzDP is due to the selective destruction of the free radical necessary for B2 activity.

Comments

Both the 2'-chloro- and the 2'-azido-2'-deoxyribonucleoside diphosphates are bound to the substrate binding site of the B1–B2 complex as

[12] L. Thelander, J. Biol. Chem. 249, 4858 (1974).

shown by the influence of allosteric effectors on the inactivation, protection of the enzyme by substrate, the inability of the monophosphates to act as inhibitors, and the requirement for the presence of both B1 and B2 to achieve inactivation. Neither CclDP nor CzDP reacted with protein B1 or B2 alone or with dithiothreitol. We suggest that the substrate analogs were converted to reactive species that, in turn, inactivated either B1 or B2 by modification of functional groups of the active site participating in the catalytic process. In B1 the redox active dithiols were modified and in B2 the radical was destroyed. In this activity, the chloro and azido derivatives behaved as irreversible enzyme inhibitors, i.e., k_{cat} inhibitors. These are characterized as unreactive compounds (proinhibitors) that are converted to a highly reactive form (inhibitor) within an active site by the specific action of a particular enzyme.[13]

CzDP is a very effective inhibitor, since it inactivated protein B2 stoichiometrically. The value of about 0.2 mole of CzDP required for total inactivation of 1 mole of B2 agrees with the known variable and low content of free radical in the B2 preparations.[14] CclDP is somewhat less effective, providing about 50% inactivation of B1 on addition of stoichiometric amounts of chloro derivative (2 moles of CclDP per mole of B1 based on two binding sites for ribonucleoside diphosphates per mole of B1[3]).

Earlier data[15] and those derived from these inhibition studies can be summarized in a model of ribonucleoside diphosphate reductase in which the active site is formed both from B1 and B2. It contains active dithiols contributed by B1 and a free radical contributed by B2. The active dithiols donate the electrons required for ribonucleotide reduction while participating in catalysis; the function and nature of the free radical remain unknown.

---

[13] R. R. Rando, *Science* **185**, 320 (1974). See also this volume [3] and [12].

[14] A. Ehrenberg and P. Reichard, *J. Biol. Chem.* **247**, 3485 (1972).

[15] N. C. Brown, Z. M. Canellakis, B. Lundin, P. Reichard, and L. Thelander, *Eur. J. Biochem.* **9**, 561 (1969).

[33] Adenosine Deaminase

By Giovanni Ronca, Antonio Lucacchini, and Carlo Alfonso Rossi

Adenosine deaminase plays a key role in adenosine metabolism. This nucleoside has some important pharmacological and toxic effects. An adenosine deaminase deficiency observed in some cases of severe congenital immunodeficiency represents the first link between an immunological and enzymic defect.[1]

The enzyme is widely distributed in animal tissues and microorganisms and has been purified by classical methods from several sources.[2-4] Multiple forms, which differ for electrophoretic mobility and molecular weight, have been observed in animal tissues; in some cases the origin of these forms is due to the binding of catalytically inactive peptides or proteins to adenosine deaminase molecules.[5] Recently, we have prepared a specific adsorbent, 9-(p-aminobenzyl)adenine bound to Sepharose, which allows the rapid and quantitative purification of the enzyme from several sources independently from the physicochemical properties of the multiple forms.[6]

Adenosine deaminase from calf intestinal mucosa, the most widely studied enzyme, hydrolytically deaminates, in addition to adenosine, some other natural and synthetic nucleosides and also dehalogenates 6-halopurine ribosides.[7] Many studies have been carried out to define the contribution of purine and ribose moieties in the binding of substrate and to help prepare effective inhibitors of the enzyme.[7-10]

One of these compounds, 9-(p-bromoacetamidobenzyl)adenine, has been used for labeling the amino acid residues of the active site. In fact the functional groups of adenosine deaminase are particularly unreactive.[11] The sulfhydryl groups present in the catalytic site react with p-mercuribenzoate and phenylmercuriacetate, but not with haloacetates, haloacetamides, N-ethylmaleimide, or 5,5'-dithiobis-(2-nitrobenzoic acid).[11] The enzyme is inactivated by 1-fluoro-2,4-dinitrobenzene and by haloacetates at pH values over 7.5 and 9.0, respectively; more than one lysine residue is modified under these conditions.[11]

[1] I. H. Porter, in "Combined Immunodeficiency Disease and Adenosine Deaminase Deficiency" (H. J. Meuwissen, B. Pollara, R. J. Pickering, and I. H. Porter, eds.), p. 3. Academic Press, New York, 1975.

[2] T. G. Brady and W. O'Connell, Biochim. Biophys. Acta 62, 216 (1962).

[3] M. K. Sim and M. H. Maguire, Eur. J. Biochem. 23, 17 (1971).

[4] V. D. Hoagland and J. R. Fisher, J. Biol. Chem. 242, 4341 (1967).

[5] H. Akedo, H. Nishihara, K. Shinkai, K. Komatsu, and S. Ishikawa, Biochim. Biophys. Acta 276, 257 (1972).

[6] C. A. Rossi, A. Lucacchini, U. Montali, and G. Ronca, Int. J. Peptide Protein Res. 7, 81 (1974).

[7] J. C. Cory and R. J. Suhadolnik, Biochemistry 4, 1729 (1965); Biochemistry 4, 1733 (1965).

[8] G. Ronca and G. C. Zucchelli, Biochim. Biophys. Acta 159, 203 (1968).

[9] H. J. Schaeffer and P. S. Bhargava, Biochemistry 4, 71 (1965).

[10] H. J. Schaeffer and E. Odin, J. Med. Chem. 9, 576 (1966).

[11] G. Ronca, C. Bauer, and C. A. Rossi, Eur. J. Biochem. 1, 434 (1967).

The low reactivity of the functional groups is due not only to the conformation of the active site, which is designed to receive the flat structure of the purine moiety of the substrate and may make access of some reagents difficult, but also to the presence of hydrophobic regions, as demonstrated by the inhibitory effectiveness of 9-alkylated adenines[12] and alkylureas.[13] The presence of positively charged groups may also influence the accessibility of reagents to the active site. Only the unprotonated forms of amidine derivatives are competitive inhibitors of adenosine deaminase.[13]

On the other hand, 9-(p-bromoacetamidobenzyl)adenine behaves as an affinity labeling reagent.[14] One equivalent reacts per mole of protein with loss of enzyme activity, and competitive inhibitors protect against inactivation. The acid hydrolysis of inactivated protein yields stoichiometric amounts of ϵ-monocarboxymethyllysine. During acid hydrolysis the amide bond between 9-(p-aminobenzyl)adenine and the acetyl moiety is broken; the resulting carboxymethylamino acid is easily determined with an amino acid analyzer.[14]

The Sepharose-bound 9-(p-aminobenzyl)adenine is very useful for the separation of the inactive alkylated from the active unreacted protein.

All the multiple forms of calf intestinal adenosine deaminase as well as those of calf thymus, calf thymocytes, and human erythrocytes and lymphocytes are inactivated by this reagent.[15]

Synthesis of 9-(p-Bromoacetamidobenzyl)adenine[16]

For the synthesis of 9-(p-bromoacetamidobenzyl)adenine we used the following method, which is rapid and convenient; it consists of the direct alkylation of adenine with p-nitrobenzylbromide (Fig. 1).

Synthesis of 9-(p-Nitrobenzyl)adenine (I). Sodium hydride (0.55 g) is added to a suspension of adenine (1.45 g) in 40 ml of dry N,N-dimethylformamide. After 1 hr of stirring at room temperature, p-nitrobenzylbromide (2.0 g) is added slowly and the mixture is stirred for 8 hr or more until the halide disappears as evaluated by carrying out a chromatography on Kieselgel 254 (Merck) with a solvent system of acetone–light petroleum 60°–80° (1:1). After cooling, the solid is

[12] H. J. Schaeffer and D. Vogel, *J. Med. Chem.* **8**, 507 (1965).
[13] G. Ronca and S. Ronca-Testoni, *Biochim. Biophys. Acta* **178**, 577 (1969).
[14] G. Ronca, M. F. Saettone, and A. Lucacchini, *Biochim. Biophys. Acta* **206**, 414 (1970).
[15] G. Ronca, A. Lucacchini, and C. A. Rossi, unpublished observations.
[16] G. Giovanninetti, A. Chiarini, L. Garuti, and A. Lucacchini, *Boll. Chim. Farm.* **113**, 91 (1974).

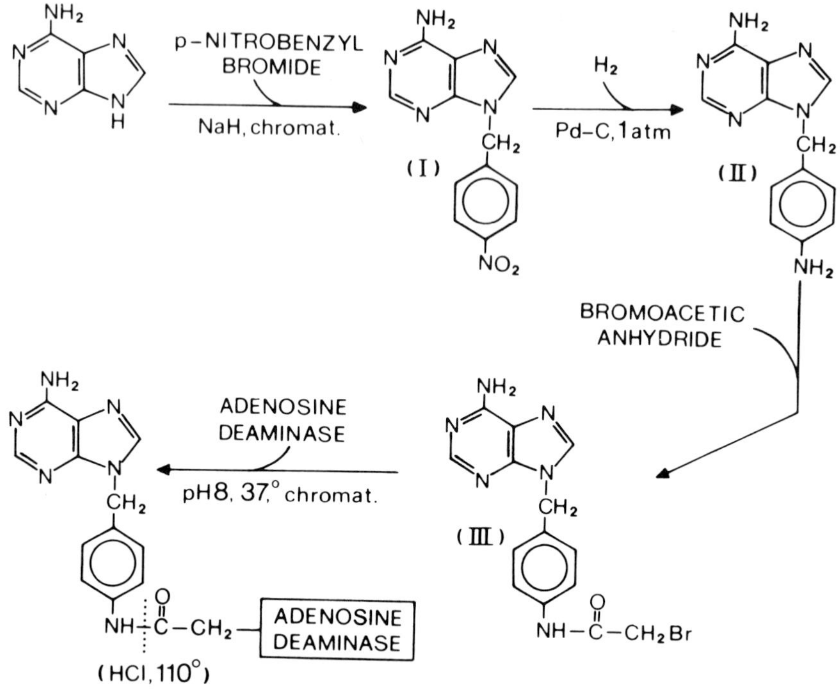

Fig. 1. Synthesis of 9-(p-bromoacetamidobenzyl)adenine, affinity labeling of lysine in adenosine deaminase active site, and acid hydrolysis of reagent amide bond (dotted line) to produce carboxymethylamino acids.

filtered and exhaustively washed with N,N-dimethylformamide. The filtrate and washings are evaporated under reduced pressure at 60°. The solid residue (4.3 g) is dissolved in chloroform–methanol (90:10) and chromatographed on a column (4 × 60 cm) of Kieselgel 60, 70–230 mesh (Merck). The elution of the column with chloroform–methanol (90:10) yields the following in sequence: deeply colored impurities, 9-(p-nitrobenzyl)adenine (light yellow); 7-(p-nitrobenzyl)adenine. 9-(p-Nitrobenzyl)adenine, crystallized from methanol, gives 1.373 g (54.9%); m.p. 256°–258°; IR (KBr) 3300, 1660, 1590, 1565, 1500, 1335 cm⁻¹.

Synthesis of 9-(p-Aminobenzyl)adenine (II). Pd–C powder, 210 mg, is added to a solution of 940 mg of (I) in 150 ml of glacial acetic acid, and the mixture is hydrogenated at room temperature and at atmospheric pressure. After filtration the solvent is evaporated and the solid residue is recrystallized from methanol to yield 0.8 g (96%) of 9-(p-aminobenzyl)adenine; m.p. 273°–275°; IR (KBr) 3350, 1675, 1590, 1570, 1510 cm⁻¹.

Synthesis of 9-(p-Bromoacetamidobenzyl)adenine (III). A solution of 0.35 g of bromoacetic anhydride in 2 ml of tetrahydrofuran is added to a cold solution of 200 mg of (II) in 5 ml of tetrahydrofuran and 0.8 ml of 10% aqueous acetic acid. The solution is stirred for 1.6 hr at 0° and for 0.5 hr at room temperature. After filtration, 15 ml of chloroform are added to the filtrate. The precipitated crude product is collected by filtration. Two precipitations from tetrahydrofuran–hexane give 140 mg of 9-(p-bromoacetamidobenzyl)adenine (46.7%) which begins to decompose at 230° but does not melt even at 380°; IR (KBr) 1700 (amide I) and 1530 (amide II) cm⁻¹.

Synthesis of 9-(p-Acetamidobenzyl)adenine. Acetic anhydride, 460 mg, in 4 ml of tetrahydrofuran is added to 0.15 g of (II) in 5 ml of tetrahydrofuran and 0.6 ml of 10% acetic acid. The solution is stirred for 2 hr at 0°, and the resulting precipitate is collected by filtration to yield 110 mg of 9-(p-acetamidobenzyl)adenine (59.6%); m.p. 235°, dec; IR (KBr) 1690 (amide I) and 1540 (amide II) cm⁻¹.

The preparation was carried out several times with consistent results. 9-(p-Aminobenzyl)adenine (II) was used for the synthesis of the selective adsorbent for the enzyme.

The synthesis of 9-(p-bromoacetamidobenzyl)adenine described by Schaeffer and Odin[15] consists of alkylation of 6-chloropurine with p-nitrobenzylbromide and substitution of —Cl with —NH₂. The synthesis is carried out as follows.

Synthesis of 6-Chloro-9-(p-nitrobenzyl)purine. 6-Chloropurine, 5 g, and 7.56 g of p-nitrobenzylbromide are dissolved in N,N-dimethylformamide (60 ml) containing triethylamine (3.54 g), stirred for 67 hr, and poured into 180 g of ice. The insoluble material (8.32 g; 89.7%) is dissolved in 200 ml of chloroform and chromatographed on a column of neutral alumina to yield 5.8 g of 6-chloro-9-(p-nitrobenzyl)purine (62.5%).

Synthesis of 9-(p-Nitrobenzyl)adenine (I). 6-Chloro-9-(p-nitrobenzyl)purine, 600 mg, in 50 ml of 20% methanolic NH₃ is heated at 73° for 45 hr in a stainless steel bomb. The precipitate, recrystallized from methanol, yields 400 mg (71%).

The hydrogenation of 9-(p-nitrobenzyl)adenine is carried out with an initial pressure of H₂ of 4 atm.

9-(p-Bromoacetamidobenzyl)adenine is obtained by treatment of (II) with bromoacetylbromide. Bromoacetylbromide, 35 μl, is added to 70 mg of (II) in 125 ml of 1,2-dimethoxyethane and the mixture is stirred for 1 hr in ice, for 2 hr at room temperature, and finally for 3 hr in ice. The material dissolved in warm methanol is precipitated with diethyl ether to give 80 mg (75%); a second precipitation from methanol–

diethyl ether provides an additional 66 mg of (III) (62%) as the hydro-bromide salt.

Inactivation of Enzyme and Stoichiometry of Reaction

Adenosine deaminase (0.5–200 μM), purified from calf intestinal mucosa, in 0.2 M sodium phosphate at pH 8, containing 10% dimethyl sulfoxide (v/v), is treated at 37° in the dark with 9-(p-bromoacetamido-benzyl)adenine (0.01–5 mM). The alkylating reagent is added dissolved in dimethyl sulfoxide. When necessary, the pH of the reaction mixture is maintained constant by addition of 1.0 N NaOH. At intervals 5–10 μl samples are removed and diluted in 0.1 M sodium phosphate at pH 7.5; adenosine deaminase activity is measured. The enzyme activity is deter-mined by the method of Kalckar[17] with adenosine as substrate by follow-ing the decrease in absorbance at 265 nm in 0.1 M sodium phosphate at pH 7.5.

To remove the unreacted 9-(p-bromoacetamidobenzyl)adenine the reaction mixture is passed through a column (0.9 × 10 cm or 2.5 × 10 cm) of Sephadex G-25 medium equilibrated with 0.1 M sodium phosphate at pH 8, or 0.1 M NH$_4$HCO$_3$. The major part of the reagent precipitates on the column. Alternatively, the reaction mixture is maintained at 0° for 60–120 min before column separation; during this period no further in-activation of the enzyme is observed and a great part of the reagent precipitates. The latter procedure is preferred when tritiated 9-(p-bromo-acetamidobenzyl)adenine is used. In this case Sephadex G-25 columns (0.9 × 20 cm or 2.5 × 20 cm) equilibrated with 10% dimethyl sulfoxide are far more useful.

The spectrophotometric determination of the reagent equivalents bound to the enzyme is based on the assumption that the spectrum of the alkylated enzyme results from the addition of adenosine deaminase and 9-(p-acetamidobenzyl) adenine spectra. The spectra have two isosbestic points: at 244 nm ($\epsilon = 21.1 \times 10^3$ M^{-1} cm^{-1}) and at 265 ($\epsilon = 21 \times 10^3$ M^{-1} cm^{-1}). The adenine derivative exhibits maximum absorbance at 255 nm ($\epsilon = 25 \times 10^3$ M^{-1} cm^{-1}), where the molar extinction coefficient of the enzyme is 13×10^3 M^{-1} cm^{-1}. From the absorbance of the enzyme eluted from Sephadex G-25 column at 244 nm and 265 nm, the sum of the molar concentrations of adenosine deaminase and the bound reagent can be calculated. The readings obtained at both wavelengths, which agree within a range of 2%, are averaged and used to calculate the en-

---

[17] H. M. Kalckar, *J. Biol. Chem.* **167**, 445 (1947). See also N. O. Kaplan, this series, Vol. 2 [69] (1955).

zyme concentration and the equivalents of the reagent bound. The following equations are used:

$$y = (A_{255} - x_2)/(\epsilon_1 - \epsilon_2) \tag{1}$$
$$x = y + z \tag{2}$$

where y is the molar adenosine deaminase concentration; A_{255} is the experimental absorption of 9-(p-bromoacetamidobenzyl)adenine-treated enzyme after Sephadex chromatography; x is the total molar concentrations (reagent bound plus adenosine deaminase); ϵ_1 and ϵ_2 are the molar extinction coefficients at 255 nm of adenosine deaminase and 9-(p-acetamidobenzyl)adenine, respectively; z is the molar concentration of the bound reagent.

When the enzyme is mixed with 9-(p-acetamidobenzyl)adenine a difference spectrum is observed; nevertheless the differences are small (lower than 1% at the considered wavelengths) in comparison to the absorbance of the reagent and the enzyme and are therefore disregarded.

A good relationship exists between inactivation and equivalents of reagent bound to the enzyme calculated by spectrophotometric method.

More recently tritiated 9-(p-bromoacetamidobenzyl)adenine has been used. The protein concentration is determined by the spectrophotometric method above described or with a microburette. With this method as well, a strict relationship is observed between the equivalents of reagent bound and residual activity.[14]

To separate inactive adenosine deaminase from the remaining active enzyme, the protein, after the gel filtration on Sephadex G-25, is passed through a 9-(p-aminobenzyl)adenine-sepharose column (1.0×1.5 cm) equilibrated with 0.1 M sodium phosphate at pH 8, or with 0.1 M NH$_4$HCO$_3$. The active adenosine deaminase binds to the column while the inactive enzyme passes through. The elution of the active enzyme is obtained by using 0.1 M sodium phosphate at pH 8, containing 4 mM guanylurea as a competitive enzyme inhibitor.

Identification of Alkylated Amino Acid Residues

During acid hydrolysis the amide bond between 9-(p-aminobenzyl)adenine and acetyl moiety of the alkylating reagent is broken (dotted line in Fig. 1) and carboxymethylamino acids are produced. To obtain the highest recovery of ϵ- and α-monocarboxymethyllysine (78% and 88%, respectively) from poly-L-lysine alkylated with 9-(p-bromoacetamidobenzyl)adenine, it is useful to carry out the hydrolysis in 5.7 N HCl at 110° in evacuated sealed vials for 48 hr. A more prolonged hydrolysis (72 hr) gives lower recoveries (67% and 82%) whereas hydrolysis for 24 hr is incomplete (72% and 47%). When the experiment is carried out with adenosine deaminase inactivated by 9-(p-bromoacet-amidobenzyl)adenine, the amount of ε-monocarboxymethyllysine after 48-hr hydrolysis accounts for 85% (not corrected for a possible destruction during hydrolysis) of the expected carboxymethylamino acids.[14]

Kinetics of the Reaction

Enzyme inactivation follows pseudo-first-order kinetics although the pseudo-first-order rate constants are not proportional to 9-(p-bromoacet-amidobenzyl)adenine concentration and, at high reagent concentration, become independent. This behavior is due to the formation of a reversible enzyme–reagent complex followed by the slow alkylation of lysine. The formation of the complex and the alkylation reaction are kinetically distinguishable, and therefore it is possible to calculate the equilibrium constant (K_{EI}) of the reversible complex and the rate constant (k_a) of the alkylation of the amino acid by plotting the reciprocal value of the experimental pseudo-first-order rate against the reciprocal value of the reagent concentration. K_{EI} and k_a values are 16 μM and 0.013 min^{-1}, respectively. As a consequence of the slowness of the irreversible reaction, the alkylating reagent behaves as a competitive inhibitor with respect to adenosine; the K_i value calculated from Lineweaver–Burk plots is 9 μM.

The competitive inhibitor of adenosine deaminase, purine riboside, strongly protects against inactivation.

Other Reagents

Schaeffer *et al.*[18,19] have synthesized and studied two other compounds that irreversibly inactivate commercial preparation of adenosine deaminase from bovine intestinal mucosa and show a kinetic behavior similar to that reported above. The compounds are 9-(o-bromoacetamidoben-zyl)adenine $(K_{EI} = 430 \ \mu M; \ K_i = 440 \ \mu M; \ k_a = 0.077 \ \text{min}^{-1})$ and 9-(m-bromoacetamidobenzyl) adenine $(K_{EI} = 720 \ \mu M; \ k_a = 0.28 \ \text{min}^{-1})$. 9-(m-Bromoacetamidobenzyl)adenine is only a good reversible inhibitor $(K_i = 36 \ \mu M)$. No indication exists that these reagents alkylate the lysine residue that reacts with 9-(p-bromoacetamidobenzyl)adenine, and the stoichiometry of the reaction has not been studied. However, it has been suggested that ortho and para derivatives alkylate different residues.[18]

[18] H. J. Schaeffer, M. A. Schwartz, and E. Odin, *J. Med. Chem.* **10**, 686 (1967).
[19] H. J. Schaeffer and R. N. Johnson, *J. Med. Chem.* **11**, 21 (1968).

The hydrophobic portion of the reagent, but not adenine, appears to be essential for the alkylation of the adenosine deaminase active site since benzylbromoacetate irreversibly inactivates the enzyme whereas the site is inaccessible to haloacetate and its amide. Benzylbromoacetate and 9-(p-bromoacetamidobenzyl)adenine are only 4 and 2.5 more reactive, respectively, than bromoacetate toward 4-(p-nitrobenzyl)pyridine.[14] The oxidation of ribose of purine riboside with periodic acid in order to produce carbonyl groups that may react with lysine residue destroys the inhibitory effectiveness of the nucleoside.[15]

[34] Direct Photoaffinity Labeling with Cyclic Nucleotides[1]

By Ross S. Antonoff, Tom Obrig, and J. J. Ferguson, Jr.

Photoaffinity labeling with chemically modified ligands has proved to be a useful technique for the analysis of biological receptors. Potentially even more advantageous, however, are photoreactions in which unmodified ligands are stably linked to their binding sites, a technique that we term *direct photoaffinity labeling*. In this technique UV irradiation of binding proteins in the presence of their radioactive ligand results in covalent attachment of the two moieties. This method obviates the possibility that a chemically modified ligand might bind with insufficient affinity and specificity, or in a topological orientation incongruent with that of the native ligand. In addition, it avoids the complexities of synthesizing a modified ligand, with attendant problems of yield and purification. These advantages have been discussed by Martyr and Benisek.[2] Direct photoaffinity labeling is an extremely simple technique. Many criteria for the specificity of labeling can rapidly be tested. Labeled receptors can be analyzed by a number of standard methods at various levels of resolution. Several uses of direct photoaffinity labeling have been described, utilizing Δ^5-ketosteroids,[2] puromycin,[3] DNA,[4] tRNA,[5] and cyclic nucleotides[6] as probe ligand. In addition, photoincorporation of

[1] This work was supported by National Institutes of Health Grant No. HD 05507.

[2] R. J. Martyr and W. F. Benisek, *Biochemistry* **12**, 2172 (1973). See also this volume [52].
[3] B. S. Cooperman, E. N. Jaynes, D. J. Brunswick, and M. A. Luddy, *Proc. Natl. Acad. Sci. U.S.A.* **72**, 2974 (1975). See also this volume [85].
[4] A. Markovitz, *Biochim. Biophys. Acta* **281**, 522 (1972).
[5] H. J. P. Schoemaker and P. R. Schimmel, *J. Mol. Biol.* **84**, 503 (1974).
[6] R. S. Antonoff and J. J. Ferguson, Jr., *J. Biol. Chem.* **249**, 3319 (1974).

chloramphenicol into ribosomes[7] and norepinephrine into particulate catecholamine binding sites[8] have been observed.

We here describe a procedure for using unmodified radioactive cyclic nucleotides as direct photoaffinity labels for high-affinity macromolecular receptors in tissue extracts. By UV irradiation of tissue extracts and [³H]cAMP[9] or [³H]cGMP, one can achieve specific covalent labeling of a class of peptides that bind cyclic nucleotides with high affinity.[6,10] We have used this photoreaction to label cyclic nucleotide binding sites in messenger ribonucleoprotein-like particles.[11] Several other applications of direct photoaffinity labeling with cyclic nucleotides have recently appeared. These include the labeling of renal plasma membrane receptors[12] and of binding proteins in rat liver and hepatoma subcellular compartments.[13]

Procedure

Preparation of Cytosol Extracts and Polysomes

Lamb testes freed of capsular tissue are homogenized in a Waring blender in 3 volumes of 0.01 M Tris chloride (pH 7.4) for 2 min at 4°. The homogenate is centrifuged for 30 min at 16,000 g and further for 1 hr at 106,000 g. The resulting supernatant solution (cytosol) is dialyzed against TMGE. Other tissues can be treated similarly[6,10] to obtain soluble extracts for direct photoaffinity labeling with cyclic nucleotides, but our studies on these tissues have not been extensive. Polysomes are prepared by homogenization of testis parenchyma in appropriate buffers, followed by differential centrifugation.[11]

[7] N. Sonenberg, A. Zamir, and M. Wilchek, *Biochem. Biophys. Res. Commun.* **59**, 693 (1974).

[8] R. S. Antonoff and J. J. Ferguson, Jr., *Fed. Proc., Fed. Am. Soc. Exp. Biol.* **35**, Abstr. 1491 (1976).

[9] Abbreviations: cAMP, adenosine-3',5'-monophosphate; cGMP, guanosine-3',5'-monophosphate; TMGE, 10 mM Tris chloride (pH 7.4), 6 mM β-mercaptoethanol, 10% glycerol, 1 mM EDTA; TCA, trichloroacetic acid.

[10] R. S. Antonoff, J. J. Ferguson, Jr., and George Idelkope, *Photochem. Photobiol.* **23**, 327 (1976).

[11] T. Obrig, R. S. Antonoff, K. S. Kirwin, and J. J. Ferguson, Jr., *Biochem. Biophys. Res. Commun.* **66**, 437 (1975).

[12] R. J. Walkenbach and L. R. Forte, *Fed. Proc., Fed. Am. Soc. Exp. Biol.* **34**, Abstr. 258 (1975).

[13] J. Kallos, *Fed. Proc., Fed. Am. Soc. Exp. Biol.* **34**, Abstr. 1822 (1975).

Photochemical Labeling

Photoactivation is performed in a cold room at 4°. Cytosol or polysomes are mixed with high specific activity [³H]cAMP or [³H]cGMP. Mixtures are transferred to a quartz 1-cm spectrophotometric cuvette positioned 1 cm from a Mineralite UVSL-25 lamp and irradiated with the source having peak emission at 253.7 nm. Radiant flux at the cuvette is approximately 400 μW/cm² in this configuration. The 366-nm source is not effective in inducing significant incorporation. Negligible incorporation is seen in a glass cuvette. Effective incorporation is also obtained by irradiating the cyclic nucleotide–protein mixture in a spot plate.[14] In addition, we have obtained significant incorporation of label using a Bausch & Lomb UV Monochromator (Model 33-86-75-01) with a high-intensity xenon lamp (Model 33-86-20). This instrument can be used for evaluating the wavelength dependence of the photoincorporation reaction.[15] Covalent incorporation of label is measured by counting a washed TCA precipitate.[6] The precipitate is dissolved in 0.2 ml of 88% formic acid and counted by liquid scintillation spectrometry in Bray's solution or Hydromix (Yorktown Research), with appropriate corrections for quenching by the formic acid. Incorporated label can also be detected after gel electrophoresis in sodium dodecyl sulfate[10,11] and after equilibrium centrifugation in CsCl.[11] With the quartz cuvette–Mineralite system, incorporation of label is linear for about 3 hr with 0.1 μM [³H] cAMP[6] and for 2 hr with 3.3 μM [³H]cGMP.[10] A 3-hr irradiation of testicular cystosol (2 mg of protein per milliliter) with 0.1 μM [³H]cAMP routinely produces photoincorporation of approximately 0.1 pmole of tritium per milligram of protein. The rate of incorporation into testicular cytosol varies as a function of concentration for both cyclic nucleotides and has the features of a saturatable system.[6,10]

Evaluation

Specificity of covalent labeling can be checked using the following criteria: (1) incorporation of label at a constant protein concentration should be linearly proportional to the amount of cyclic nucleotide bound noncovalently; (2) unlabeled cyclic nucleotide analogs should inhibit noncovalent binding and covalent incorporation in parallel fashion; (3) labeling of discrete peptides in a crude mixture should be observed; and

[14] A. H. Pomerantz, S. A. Rudolph, B. E. Haley, and P. Greengard, *Biochemistry* **19**, 3858 (1975).
[15] R. S. Antonoff and J. J. Ferguson, Jr., unpublished observations.

(4) proteins which do not bind a cyclic nucleotide should not incorporate it significantly. Noncovalent binding can be measured by several standard methods.[6,10,11] These criteria have been used to demonstrate the specificity of labeling of high-affinity cAMP receptors by both [³H]cAMP and [³H]cGMP in lamb testis.[6,10]

In cytosol (at 2 mg of protein per milliliter) 2 hr of irradiation with the Mineralite produces incorporation of 1–2% of the total noncovalently bound [³H]cAMP. This compares with 20% using [³H]cGMP.[10] With testicular polysomes incorporation of approximately 10% of total bound [³H]cAMP occurs after 2–3 hr of irradiation.[11] We have seen incorporation of up to 20% of bound cAMP using L-cell polysome receptors.[16] These differences in efficiency of incorporation are not yet explained.

Comments

The technique described here has several limitations. Photoincorporation in this system is a slow process, and, as a consequence, binding proteins may be denatured during irradiation. Indeed, we have found that up to 50% of total cAMP binding activity can be lost after 4 hr of irradiation. Of interest is the fact that this inactivation is prevented if cAMP is present during irradiation. This points to the advantage of obtaining incorporation sufficient for analytical purposes in the shortest possible exposure time. An important factor determining the minimal required irradiation time is the specific activity of cyclic nucleotide binding protein in an extract, longer irradiation being required with lower binding activity.

As indicated above, irradiation with saturating [³H]cGMP is useful in obtaining sufficient labeling of cAMP receptors in testis cytosol in a relatively short time (i.e., on the order of 0.5 hr). A number of other variables should also be considered in trying to minimize exposure time. We have observed that at protein concentrations greater than 2 mg/ml there is a decrease in the efficiency of cAMP incorporation.[6] We presume that this is due to absorbance of the incident radiation. Maximal specific incorporation is obtained with a protein concentration less than 2 mg/ml. We have also found that incorporation during irradiation at 4° is at least twice as rapid as with irradiation at 22°. When using cyclic nucleotides as ligands, appropriate measures must be taken to suppress phosphodiesterase activity during incubation of mixtures. We have found that dialysis of testis cytosol against 1 mM EDTA minimizes cyclic nucleotide degradation and have used 1 mM cAMP phosphodiesterase in-

---

[16] T. Obrig, unpublished observation.

hibitor Ro20-1724/1 (Hoffman-La Roche) routinely in photolabeling experiments with polysomal preparations.[11] Successful use of a phosphodiesterase inhibitor requires that it does not itself interfere with incorporation by a light filtration effect; i.e., it should absorb minimally at the wavelength of irradiation. This criterion is met by 1 mM Ro 20-1724/1 at 254 nm whereas theophylline at a concentration required for efficient phosphodiesterase inhibition is not as effective.

Action spectral studies with [³H]cAMP indicate that the optimum wavelength for photoincorporation is near 280 nm, this wavelength producing about 35% more incorporation than irradiation at 254 nm.[15] These studies suggest that photoincorporation of [³H]cAMP results from a primary photoactivation of aromatic amino acids of the receptor protein rather than of the cyclic nucleotide. Thus direct photoaffinity labeling using this technique may be possible with other ligands that are not photosensitive in themselves, but interact with receptors containing photosensitive aromatic amino acids.

[35] Adenosine 3',5'-Cyclic Monophosphate Binding Sites[1]

By Boyd E. Haley

The use of photoaffinity reagents to investigate the binding of ligands to macromolecules is a relatively new technique and is being rapidly developed. This technique has been successfully used to identify ATP and cAMP binding proteins of cells, cell membranes, and some purified protein kinases.[2-6] The reagent successfully used to photolabel cAMP binding sites was 8-azidoadenosine 3',5'-cyclic monophosphate (8-N₃-cAMP). In the absence of activating light, 8-N₃-cAMP is a good biological substitute for cAMP. Earlier published results have shown that 8-N₃-cAMP binds to cAMP sites with high affinity and activates cAMP-stimulated protein

[1] This research was supported in part by Grant No. GM 21998-02 from the National Institutes of Health. Journal Paper No. JA-856 of the Wyoming Experiment Station.

[2] B. E. Haley, *Biochemistry* **14**, 3852 (1975).
[3] A. H. Pomerantz, S. A. Rudolph, B. E. Haley, and P. Greengard, *Biochemistry* **14**, 3858 (1975).
[4] J. Owens and B. E. Haley, *Proc. ICN-UCLA Conf. Cell Shape* (1975).
[5] A. M. Malkinson, B. K. Krueger, S. A. Rudolph, J. E. Casnellie, B. E. Haley, and P. Greengard, *Metab. Clin. Exp.* **24**, 331 (1975).
[6] B. E. Haley and J. F. Hoffman, *Proc. Natl. Acad. Sci. U.S.A.* **71**, 3367 (1974).

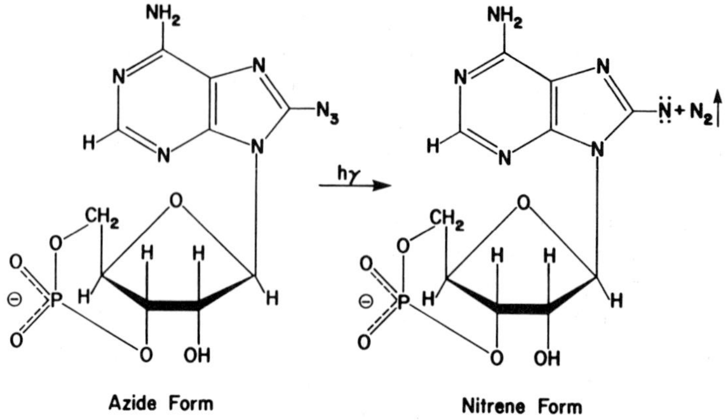

Azide Form   **Nitrene Form**

Fig. 1. Photolysis of 8-azidoadenosine 3′,5′-cyclic monophosphate.

kinases.[2,7,8] The ability to measure the degree of reversible interaction at the binding site by obtaining K_A, K_I, or K_M values in the absence of activating light is one of the advantages of photoactivated probes. After the kinetic values have been obtained, they may be used to optimally design the photoincorporation experiments. For example, if a K_A value of 10 nM is observed, it would be an error to photolabel with 10 μM analog since the latter amount would result in an increase in nonspecific labeling and would not substantially increase the labeling of the specific binding sites.

The arylazide, 8-N$_3$-cAMP, is converted to the nitrene form by photolysis with ultraviolet light in the 240–320 μm range (Fig. 1). Nitrenes are very reactive and will form a covalent bond with any amino acid R group. The possible reactions of azides via photogenerated nitrenes have been reviewed.[9,10] The bond formed when 8-N$_3$-cAMP is photoincorporated onto membrane proteins is stable to acid treatment at least 52° and probably higher and to boiling at pH 9.0. We have found nothing that will remove photoincorporated, covalently bound label without also destroying the primary protein chain.

8-N$_3$-cAMP is an analog of cAMP that has strong affinity for cAMP binding sites (i.e., it is biologically specific); on photolysis, it becomes 8-nitrene-cAMP, which rapidly reacts with any amino acid residue (i.e.,

---

[7] L. N. Simon, D. A. Sherman, and R. K. Robins, *Adv. Cycl. Nucleotide Res.* **3**, 294 (1973).

[8] K. Muneyama, R. J. Bauer, D. A. Sherman, R. K. Robins, and L. N. Simon, *Biochemistry* **10**, 2390 (1971).

[9] W. Lwowski, "Nitrenes." Wiley (Interscience), New York, 1970.

[10] J. R. Knowles, *Acc. Chem. Res.* **5**, 155 (1972).

it is chemically nonspecific). The advantages of biological specificity and chemical nonspecificity make 8-N_3-cAMP a potent tool for both identifying cAMP binding proteins and investigating the amino acid residues of the binding site peptides.

For 8-N_3-cAMP to be of optimal value it must be synthesized in radioactive form without loss of the high specific activity of the starting compound. The synthesis of [^{32}P]8-N_3-cAMP has been previously published.[2] Detailed herein is a modification of the earlier synthesis, which is easier to perform and consistently gives yields of 60–80% without dilution of the specific radioactivity. This procedure has also been used to prepare ^{32}P- and ^{14}C-labeled 8-azidoadenosine monophosphate. It cannot be used on compounds containing tritium-labeled adenosine without loss of the ^{3}H on the 8 position.

Synthesis of Nanomolar Quantities of High Specific Radioactive Photoaffinity Analogs of Adenosine-Containing Compounds

Adenosine (0.075 mmole, 20 mg) is dissolved in 2 ml of 1 M sodium acetate at pH 3.8. To the adenosine solution is added [^{32}P]cAMP (0.030 μmole, 10.0 Ci/mmole) (ICN) in 0.093 ml of aqueous solution at pH 7.4. To this solution is added 1 ml of bromine water containing 5 μl of Br (0.097 mmole). The bromination reaction is maintained at room temperature in a 4–8 ml screw-top test tube sealed with a Teflon-lined cap. The bromination proceeds rapidly at first but usually requires 4 hr for completion. The reaction may be followed by observing the shift in the λ_{max} of the UV spectra from 259 to 264 nm and by thin-layer chromatography. Adenosine is added to act as a carrier molecule and to allow UV monitoring of the bromination of all adenosine moieties in the reaction mix. It also allows the addition of sufficient Br_2 that can be easily measured and assures a ratio of Br_2:adenosine which results in the maximum yield of the 8-Br-adenosine product. For consistent yields, a container with minimal air space and a seal that is unreactive with Br_2 must be used.

After 4 hr, the excess Br_2 is removed by flushing with a stream of air. The resulting light bronze mixture is treated with 4 M NH_4OH, added dropwise, until the color becomes a reddish purple. This indicates that the solution is no longer acidic. (Note: Failure to raise the pH will cause the DEAE-cellulose-HCO_3 in the column to release CO_2 and ruin the column separation.)

The solution is applied to a column of DEAE-cellulose (2 $\times$ 25 cm) in the bicarbonate form. The column is washed with water until the 8-bromoadenosine is completely eluted; approximately 225 ml are required at a flow rate of 2.7 ml per minute. The column is then eluted with a linear gradient of 2 liters of 0–0.2 M triethylammonium bicarbonate (pH 7.5), and the [^{32}P]8-Br-cAMP is located by monitoring UV absorption and radioactivity. With this method only one major, somewhat skewed, peak is usually observed. In a model separation experiment, a 1:1 mixture of cAMP and 8-Br-cAMP revealed that cAMP is eluted only slightly faster than 8-Br-cAMP. By discarding the first 5% of radioactivity eluted, any trace of [^{32}P]cAMP that is not brominated will be eliminated; this is based on the assumption that the UV spectrum indicates approximately 100% bromination of all adenine groups.

Thin-layer chromatography (TLC) using Eastman TLC cellulose sheets 13254 and developing solvent systems of system A (isobutyric acid–NH$_4$OH–H$_2$O, 66:1:33 v/v), and system B (1-butanol–acetic acid–H$_2$O, 5:2:3 v/v), will serve to confirm the degree of bromination. The R_f values of 8-Br-cAMP, cAMP, AMP, ADP, and ATP, respectively, are system A, 0.78, 0.76, 0.70, 0.64, and 0.54; system B, 0.65, 0.63, 0.50, 0.35, and 0.18.

The bromination reaction appears to be the most sensitive step in the synthesis. If too little Br$_2$ is added, bromination will be incomplete and cAMP could contaminate the final product. Adding too much Br$_2$ allows the formation of an undesirable side product that migrates on both TLC systems more slowly than ATP with R_f values of 0.45 and 0.14 for systems A and B, respectively.

The tubes containing [^{32}P]8-Br-cAMP are pooled and evaporated to dryness at 30° under reduced pressure with a Buchi rotovaporator. The material is dissolved and coevaporated four times with methanol to remove excess triethylammonium bicarbonate. An ultraviolet spectrum of the product has a λ_{max} of 264 as expected for 8-Br-cAMP. The [^{32}P]8-Br-cAMP is placed in a 25-ml round-bottomed flask with 25 ml of dimethylformamide containing 10 mmoles of triethylammonium azide. The last solution is prepared by passing 500 mg of LiN$_3$ (10 mmoles), dissolved in dimethylformamide, through a 20-ml Dowex 50 column in the triethylammonium form and also in dimethylformamide as solvent.

For the Br-N$_3$ exchange reaction to proceed optimally, water contamination must be kept at a minimum in both the dimethylformamide and the Dowex 50 column. Dimethylformamide was dried over molecular sieve (Linde MS-1086, 4A, 1/16-inch pellets) 1 week before use.

The Dowex resin is prepared from Dowex-50W (Sigma) in the hydrogen form, and washed three times with twice the resin volume of anhydrous methanol and then dried in an oven at 85° for 6 hr. The resin is cooled in a desiccator and then covered with anhydrous methanol. Thiethylamine (Baker) is added slowly to the methanol–resin mixture until heat is no longer released. The resin is then washed two times with anhydrous methanol and four times with dry dimethylformamide. Each wash consists of twice the resin volume and is allowed to equilibrate for at least 1 hr. The resin is stored in dimethylformamide. On the day of use, the column is poured and washed with fresh, dry dimethylformamide before applying LiN₃; the resin discolors the covering dimethylformamide if stored in it for a long period of time, but discoloration does not appear to affect the exchange reaction.

The flask containing [³²P]8-Br-cAMP and triethylammonium azide is tightly sealed and heated at 75° for 10–12 hr. Thereafter, the reaction mixture is evaporated to dryness at 1 mm pressure and 18°. Although the azidoadenosine compounds are relatively stable to high temperature in dimethylformamide, they are very unstable when heated dry or in water. It is critical that the temperature be near 20° when the last few milliliters of dimethylformamide are evaporated. After evaporation to near dryness, the products are dissolved in 3–4 ml of water and chromatographed as described for 8-Br-cAMP with an initial water wash lasting until all traces of dimethylformamide are eluted from the column, as observed by UV monitoring. After dimethylformamide is eluted, the UV monitor is shut off since passing such small amounts of 8-N₃-cAMP through a UV beam will result in undesirable photolytic destruction of some of the compound. The products are located by the presence of radioactivity. Usually, two minor peaks are observed on each side of the major peak containing 90–96% of the total radioactivity. The three peaks are pooled separately and evaporated to dryness at 1 mm pressure and 18°. Excess triethylammonium bicarbonate is removed by four coevaporations with absolute methanol. The three products are isolated and stored at −20° in absolute methanol. The major product, [³²P]8-N₃-cAMP, migrates almost identically to [³²P]8-Br-cAMP with solvent systems A and B. However, if, after spotting on Eastman TLC cellulose sheets, the product is photolyzed for 5 min with a UVS·11 Mineral-Light (Ultra-violet Products, Inc.) at a distance of 2 cm, 15–25% of the radioactivity will usually bind irreversibly to the cellulose at the origin and not migrate on development. A huge portion of the material will exist as a smear trailing behind the unphotolyzed 8-N₃-cAMP. [³²P]8-Br-cAMP and the minor products are not affected by the preceding photolysis process, and their migration on TLC remains the same with or without photolysis.

The ultraviolet spectrum of the major product had a λ_{max} at 281 nm. Photolysis of this material in absolute methanol both decreases the optical density and shifts the λ_{max} as shown in Fig. 2. No changes in the spectra of cAMP and 8-Br-cAMP were observed after identical treat-

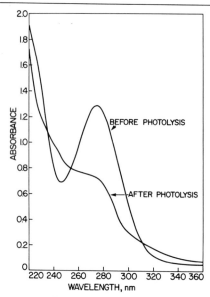

Fig. 2. Ultraviolet spectra of 8-azidoadenosine 3′,5′-cyclic monophosphate, before and after 4 min photolysis through quartz at 4 cm distance using UVS·11 Mineral-Light (UV Products, Inc.).

ment. The shape of the ultraviolet spectrum after photolysis may be quite different, depending mainly on the solvent and solutes present with which the nitrene form may react.

The 8-azidoadenosine derivatives are stable for at least several weeks if stored in absolute methanol at −20°. Short periods of exposure to normal room light, through glass, caused no noticeable breakdown of the compounds.

This synthetic method has also been used for making 8-N_3-cAMP of high specific radioactivity. The only difference in preparation is that the compound was eluted with a slightly higher concentration of triethylammonium bicarbonate. Its R_f values were 0.73 and 0.55 in solvent systems A and B, respectively, i.e., nearly identical to 8-Br-cAMP.

Photolabeling Procedure

[^{32}P]8-N_3-cAMP has been used to photolabel cAMP binding sites in freeze-fractured whole cells (Chinese hamster ovary, sarcoma 37, and *Dictyostelium discoideum*), in tissue homogenates (brain, heart, corpus luteum), in purified membranes (human erythrocyte and sarcoma 37), and in purified cAMP-activated protein kinases.[3] The only requirement for any of these systems is an ultraviolet light source such as a hand-held mineral-light. The time of photolysis depends upon the nature and optical density of the solution being photolyzed, and the watts/cm^2 output of the ultraviolet lamp. Photolabeling of freeze-fractured whole cells, tissue homogenates, or membranes has been successfully accomplished by photolyzing for 1–2 min at a distance of 2 cm from a lamp (UVS·11 Mineral-Light) that gives 240 μW/cm^2 at 6 inches. The solutions being photolyzed are usually no more than 2 mm thick. For best results with each system, a different determination of photolysis time versus amount of photolabel incorporated must be obtained.

Specific Considerations

Photolysis at 0° with constant stirring has given the highest degree of photoincorporation of [^{32}P]8-N$_3$-cAMP with minimal membrane protein cross-linking. Photolysis at 37° is also effective, and very little protein cross-linking of membrane protein occurs when exposures at 2 cm of 5 min or less to the UVS·11 Mineral-Light are used. Each system appears to be affected differently by photolysis. Some are very sensitive to ultraviolet light, and destruction of the cAMP binding sites occurs simultaneously with photolabeling. Obviously such behavior makes quantitative measurements of the number of binding sites with 8-N$_3$-cAMP tenuous.

To verify that proteins photolabeled with [^{32}P]8-N$_3$-cAMP are truly cAMP binding proteins, it must be demonstrated that addition of increasing amounts of cAMP reduces or abolishes covalent incorporation in correlation with the K_A value of cAMP. Unfortunately, this may be experimentally difficult. Both cyclic nucleotides bind so tightly at 0° that the one which is added first does not easily exchange with cyclic nucleotide added afterward. Addition of cAMP at 0° before addition of [^{32}P]8-N$_3$-cAMP completely prevents photolabeling. The same experiment at 37° decreases photoincorporation by 73%. Addition of [^{32}P]8-N$_3$-cAMP first, at 0°, followed by the addition of cAMP, does not measurably decrease the amount of photoincorporation. It does appear that, at 37°, the two cyclic nucleotides will exchange at the binding site. However, even this exchange appears to be slow and is incomplete within 8 min. Therefore, to obtain accurate measurement of the relative affinities of cAMP and 8-N$_3$-cAMP for specific binding sites, the two cyclic nucleotides should be added simultaneously.

Another problem that may lead to inconsistent results is the action of proteases on photolabeled proteins. Experiments have shown that the major 8-N$_3$-cAMP photolabeled protein in human erythrocytes is proteolytically cleaved to yield other cAMP binding proteins. This occurs before any general proteolysis can be detected in the Coomassie Brilliant

Blue staining profile of membrane proteins separated by sodium dodecyl sulfate gel electrophoresis.

Additional problems include the screening of ultraviolet light by high nucleotide concentration or possible enzymic conversion of 8-N_3cAMP to 8-N_3-AMP with subsequent labeling of AMP binding sites. Each protein photolabeled with a reagent, such as with [^{32}P]8-N_3-cAMP, must be checked with several experiments to ensure that it is indeed specific for both the natural compound and its biological mimic.

[36] 5-Formyl-UTP for DNA-Dependent RNA Polymerase

By Vic Armstrong, Hans Sternbach, and Fritz Eckstein

In order to determine which subunits of DNA-dependent RNA-polymerase contain the catalytic site, affinity labeling has been employed by various groups.[1-5] We have synthesized 5-formyluridine 5'-triphosphate (fo^{5}UTP) for this purpose. Such a derivative should be able to function as an affinity label by reaction with an amino group at the active

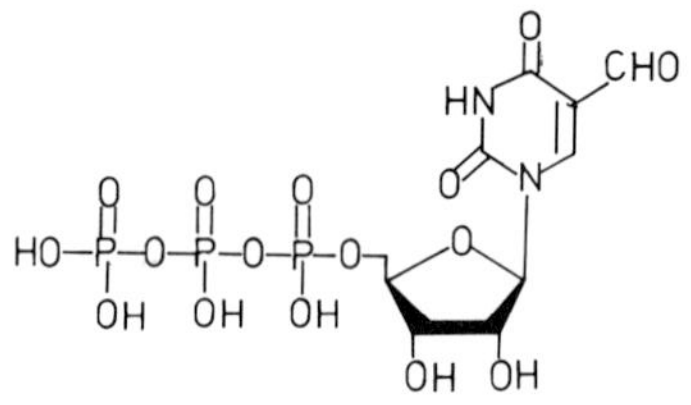

(I)

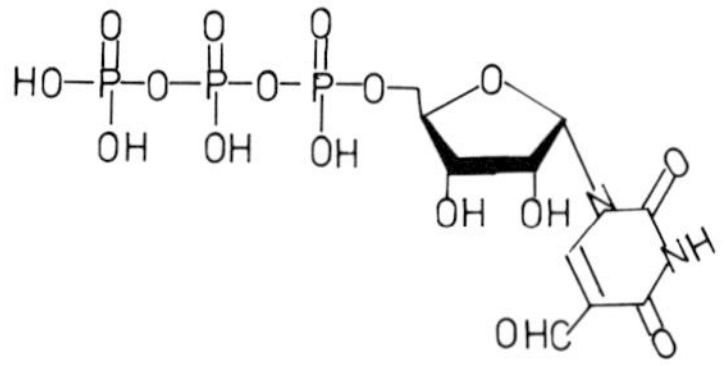

(II)

[1] V. W. Armstrong, H. Sternbach, and F. Eckstein, Biochemistry 15, 2086 (1976).
[2] A. M. Frischauf and K. H. scheit, Biochem. Biophys. Res. Commun. 53, 1227 (1973).
[3] J. Nixon, T. Spoor, J. Evans, and A. Kimball, Biochemistry 11, 4570 (1972).
[4] F. Y.-H. Wu and C.-W. Wu, Biochemistry 13, 2562 (1974).
[5] P. Bull, J. Zaldivar, A. Venegas, J. Martial, and P. Valenzuela, Biochem. Biophys. Res. Commun. 64, 1152 (1975).

site of the enzyme to form a Schiff base. Subsequent reduction with sodium borohydride would then covalently link the triphosphate to the enzyme.

However, during the synthesis of this triphosphate, a base-catalyzed anomerization[6,7] occurred such that the final product was a mixture of the α-(1) and β-(2) anomers. The anomers could be separated chromatographically with DEAE-Sephadex. The α-anomer has been used successfully to affinity-label RNA polymerase whereas the β-anomer is a substrate for the enzyme.

Synthesis of α- and β-fo⁵UTP

The required functionality is introduced into the 5 position of the uracil ring by hydroxymethylation of 2′,3′-isopropylideneuridine[8] followed by oxidation of the hydroxymethyl derivative with active MnO_2[9] to yield 5-formyl-2′,3′-isopropylideneuridine. The latter is converted into the mono- and triphosphate by conventional procedures. During the synthesis of the monophosphate by the method of Tener,[10] a base-catalyzed anomerization occurs so that a mixture of the α- and β-anomers of 5-formyluridine monophosphate is produced. After conversion to the triphosphate, the two anomeric triphosphates may be separated.

5-Hydroxymethyl-2′,3′-O-isopropylideneuridine[8]

2′,3′,-Isopropylideneuridine (24 g, 0.10 mole) is dissolved in a mixture of 200 ml of 0.5 N potassium hydroxide and 76 ml of (35%) formaldehyde. The solution is heated at 50° until reaction is complete (usually 4–6 hr). After allowing the solution to cool to room temperature, it is neutralized with Merck-I (H⁺ form) ion-exchange resin. The resin is removed by filtration and washed well with methanol–water (1:1, v/v); the filtrate and washing are combined and evaporated under reduced pressure. The resulting syrup is dissolved in chloroform, and the solution is applied to a silica-gel column (50 × 8 cm). The column is first washed with chloroform–methanol (19:1, v/v) until excess formaldehyde is eluted. It is then washed with chloroform–methanol (8:2, v/v), and the fractions

[6] V. W. Armstrong and F. Eckstein, *Nucl. Acids Res.* Suppl. 1, 97 (1975).
[7] V. W. Armstrong, J. K. Dattagupta, F. Eckstein, and W. Saenger, *Nucl. Acids Res.* 3, 1791 (1976).
[8] K. H. Scheit, *Chem. Ber.* 99, 3884 (1966).
[9] J. Attenburrow, A. F. B. Cameron, J. H. Chapman, R. M. Evans, B. A. Hems, A. B. A. Jansen, and T. Walker, *J. Chem. Soc.* 1952, 1094 (1952).
[10] G. M. Tener, *J. Am. Chem. Soc.* 83, 159 (1961).

containing 5-hydroxymethyl-2′,3′-isopropylideneuridine, as determined by thin-layer chromatography (TLC), are collected and evaporated under reduced pressure to yield 22.5 g of the product as a foam. Since this material is homogeneous by TLC and nuclear magnetic resonance (NMR), it may be used directly in the next step.

5-Formyl-2′,3′-O-isopropylideneuridine

To 5-hydroxymethyl-2′,3′-isopropylideneuridine (20 g, 64 mmoles) in dichloromethane (500 ml) is added active manganese dioxide[9] (100 g), and the suspension is stirred at room temperature for 48 hr. It is filtered over Celite with the aid of a suction pump and washed well with chloroform–methanol (1:1, v/v). The filtrate and washings are combined and evaporated under reduced pressure. The residue is dissolved in chloroform and applied to a silica gel column (50 × 6 cm) which is eluted with 1 liter of chloroform, 1.5 liters of chloroform–methanol (98:2, v/v), and then with chloroform–methanol (96:4, v/v). Fractions of 200 ml are collected and are monitored by TLC. Those fractions containing pure 5-formyl-2′,3′-isopropylideneuridine are combined and evaporated to dryness under reduced pressure to yield 7.2 g of the desired compound as a white crystalline residue, m.p. 157°–159°, $\lambda_{max}^{H_2O}$ 280 nm (13,600) and 235 nm (9000). This material is sufficiently pure for most synthetic purposes. It may be recrystallized from ethanol–water (95:5, v/v), m.p. 160°–161°.

Synthesis of the Mixture of α- and β-5-Formyluridine 5′-Monophosphates

5-Formyl-2′,3′-isopropylideneuridine (2 mmoles) are treated with 4 mmoles of β-cyanoethyl phosphate as described by Tener[10] but using 6 mmoles of triisopropylbenzenesulfonyl chloride instead of dicyclohexylcarbodiimide; this change reduces the reaction time to 4 hr. Hydrolysis of the cyanoethyl protecting group is carried out in 4 N NaOH–MeOH (1:1, v/v) for 30 min. Finally, the isopropylidene protecting group is cleaved with 50% aqueous acetic acid at 100° for 2 hr. The monophosphate is purified over a Dowex 1 × 4 (100–200 mesh, Cl⁻) column (40 × 4 cm), eluted with a linear gradient of 0.05 M LiCl–0.01 M HCl and 0.15 M LiCl–0.01 M HCl (2 liters of each). The product elutes between 0.07 and 0.09 M LiCl and is detectable by the ratio of its UV absorption at 280 nm to that at 260 nm (approximately 2.00). The fractions containing 5-formyluridine 5′-monophosphate are pooled and neutralized with 1 M LiOH. They are concentrated under reduced pressure to approximately 20 ml, and the solution is divided between two 50-ml centrifuge tubes. Saturated BaCl₂ solution, 5 drops, is added to each tube followed by 40 ml of ethanol. After centrifugation, the supernatant liquid is decanted and the residue is washed with 30 ml of 70% aqueous ethanol. The barium salt of 5-formyluridine 5′-monophosphate is converted into its Na^+ salt by stirring with Merck-I (Na^+ form) ion-exchange resin. After filtering the resin and washing with water, the filtrate and washings are evaporated under reduced pressure to yield 10,500 A_{280} units of 5-formyl-uridine 5′-monophosphate. The product gives a single spot on electrophoresis at pH 7.5 with a mobility of 12.8 cm (UMP, 13.2 cm). A single spot ($R_f = 0.21$) is obtained after paper chromatography in ethanol/1 M ammonium acetate (7:3, v/v).

Synthesis of α- and β-fo⁵UTP

Of the anomeric mixture of fo⁵UMP, 10,500 A_{250} units are converted to their triphosphates by a standard procedure.[11] The crude reaction product is charged onto a column of DEAE-Sephadex and eluted with a linear gradient of 2.5 liters each of 0.05 M and 0.45 M triethylammonium bicarbonate. The α-fo⁵UTP (856 A_{280} units) elutes between 0.39 and 0.42 M buffer and the β-anomer (1190 A_{280} units) between 0.35 and 0.38 M buffer. These structures have been assigned after comparison of their CD and NMR spectra to spectra of the corresponding nucleosides. For enzymic studies, these compounds may be purified further by passage over Dowex ion-exchange resin as described for [γ-³²P]-α-fo⁵UTP below.

Synthesis of [γ-³²P]α-fo⁵UTP

This derivative is prepared essentially according to the method of Glynn and Chappell.[12] α-fo⁵UTP, 150 A_{280} units, is incubated in a final volume of 1.2 ml containing 0.2 ml 1 M Tris-chloride at pH 8.0, 12 μl of 1 M $MgCl_2$, 0.2 ml of 0.1 M NaOH, 3 μl of 0.1 M Na_2HPO_4, 20 μl of 3-phosphoglycerate (cyclohexylammonium salt, 20 mg/ml), 20 μl of glyceraldehyde-3-phosphate dehydrogenase (10 mg/ml), 12 μl of phosphoglycerate kinase (10 mg/ml), and 0.1 ml of ³²P_i (10 mCi/ml, carrier free). After 16 hr at room temperature, the mixture is diluted with 1 ml of water and chromatograhed over a Dowex 1 × 4 (200–400 mesh, Cl⁻) ion-exchange column (10 × 0.08 cm) with a linear gradient of 0.01 M HCl–0.05 M LiCl and 0.01 M HCl–0.45 M LiCl (180 ml of each). The fractions containing α-fo⁵UTP are collected and neutralized with 1 M LiOH. After concentration under reduced pressure to approximately 5 ml, the solution is transferred to a centrifuge tube. Saturated $BaCl_2$ solution,

[11] A. M. Michelson, *Biochim. Biophys. Acta* **91**, 1 (1964).
[12] I. M. Glynn and J. B. Chappell, *Biochem. J.* **90**, 147 (1974).

5 drops, is added followed by 30 ml of EtOH to precipitate the barium salt of the triphosphate. After centrifugation, the supernatant liquid is decanted and the residue is washed twice with 2 ml of water by centrifugation. Finally, the triphosphate is converted to its Na^+ salt with Merck-I (Na^+ form) ion-exchange resin, and, after evaporation under reduced pressure, the α-fo⁵UTP is dissolved in 1 ml of H_2O and stored as a frozen solution at $-20°$. Yield $= 92$ A_{280} units. $280/260 = 1.98$. Specific activity $= 2.34 \times 10^8$ cpm/mole.

Enzyme Purification and Assay

Escherichia coli RNA-polymerase holo- and core enzymes[13] were 95% pure by sodium dodecyl sulfate (SDS) gel electrophoresis. Enzyme activity was measured as the amount of [¹⁴C]AMP or [¹⁴C]UMP incorporated into acid-insoluble material[14] after a 10-min incubation at 37°. The assay mixture contained in 0.1 ml 40 mM Tris-chloride at pH 8.0, 8 mM $MgCl_2$, 5 mM dithioerythritol, 0.2 A_{260} unit of poly[d(AT)], 0.05 M KCl, 1 mM ATP, and 1 mM [¹⁴C]UTP. For kinetic studies, a fixed concentration of ATP (0.4 mM) was used and the concentration of [¹⁴C]UTP was varied. Enzyme activity was measured as the amount of [¹⁴C]UTP incorporated into acid-insoluble material after 5 min.

Inhibition of RNA Polymerase by α-fo⁵UTP

α-fo⁵UTP is a noncompetitive inhibitor of the poly[d(AT)]directed synthesis by RNA polymerase with a $K_i = 0.54$ mM. The β-anomer, on the other hand, proved to be a substrate for the enzyme with a $K_m = 0.12$ mM [$K_m(UTP) = 0.05$ mM].

The inhibition of RNA polymerase by a α-fo⁵UTP is dependent upon the α-fo⁵UTP concentration, and for maximum inhibition at any particular concentration a short preincubation of enzyme and α-fo⁵UTP is required (Fig. 1). It is important that the α-fo⁵UTP concentration in the assay solution is the same as that in the preincubation solution, since on dilution of the α-fo⁵UTP the enzyme regains activity with time (Fig. 1) indicating that the inhibition is reversible.

Sodium Borohydride-Induced Irreversible Inhibition

Upon reduction of a preincubated enzyme/α-fo⁵UTP solution with an equal volume of 0.1 M sodium borohydride solution, the enzyme was irreversibly inhibited. The extent of the inhibition was dependent upon

[13] R. R. Burgess, *J. Biol. Chem.* **244**, 6168 (1969).
[14] F. J. Bollum, *Proc. Nucl. Acid Res.* **1**, 296 (1966).

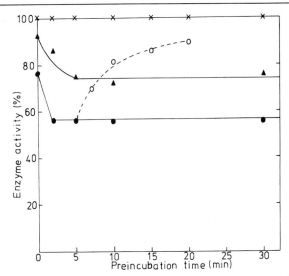

Fig. 1. The inhibition of RNA polymerase by α-fo⁵UTP. The enzyme was pre-incubated at 37° in a reaction mixture containing 40 mM Tris-chloride, pH 8.0, 8 mM MgCl₂, 0.05 M KCl and (a) H₂O (×——×), (b) 0.23 mM α-fo⁵UTP (▲——▲), and (c) 0.46 mM α-fo⁵UTP (●——●). After various times, 10-μl aliquots were removed and assayed. In a further experiment (d) (○---○) 10-μl aliquots were removed from (c) after 5-min preincubation and diluted with a solution (70 μl) containing 57.1 mM Tris-chloride, pH 8.0, 11.4 mM MgCl₂, 7.1 mM DTE, 0.071 M KCl, 1 mM ATP, and 1 mM [¹⁴C]UTP. The incubation was continued at 37°, and after various times the enzyme was assayed by the addition of a solution (20 μl) containing 0.2 A_{260} unit of poly[d(AT)].

the α-fo⁵UTP concentration, reaching 80% inhibition at 2 mM α-fo⁵UTP. To test the specificity of this inhibition, the nucleoside α-fo⁵U was also used. The inhibition in this case was about half that produced by the triphosphate, reaching only 42% at 2 mM α-fo⁵U. The β-analogs were also tested and found to be inferior to α-fo⁵UTP. At 2 mM, β-fo⁵UTP produced 35% and β-fo⁵U 24% inhibition after NaBH₄ reduction.

Further evidence for the specificity of the binding of α-fo⁵UTP was provided by protection experiments with ATP, UTP, and GTP. Addition of these triphosphates to the preincubation mixture of α-fo⁵UTP and RNA polymerase reduced the extent of the inhibition.

Stoichiometry of Binding of α-fo⁵UTP and Location of
   Its Binding Site(s)

The amount of [γ-³²P]-α-fo⁵UTP, prepared by a previously published procedure,[12] covalently linked to the enzyme after sodium borohydride reduction was determined by retention of the protein on nitrocellulose membrane filters. The amount of label retained on the filters increased with the time of the preincubation of enzyme and $[\gamma\text{-}^{32}P]\alpha\text{-fo}^5\text{UTP}$. However, during the same period no significant increase in the inhibition of the enzyme occurred. After a 20-sec preincubation the stoichiometry of binding was 1.1:1 ($[\gamma\text{-}^{32}P]\alpha\text{-fo}^5\text{UTP}$ bound:inactivated enzyme) and after 10 min it had risen to 2.42:1.

The location of the label after a 20-sec preincubation was determined by separating the enzyme into its different subunits by electrophoresis in 6 M urea on cellulose acetate plates.[15] Most of the label was found to run with the β subunit. This subunit has been shown to be the site of rifampicin action,[16,17] and other affinity labeling studies[2-4] have implicated this subunit as containing at least part of the catalyic center of RNA polymerase.

Comments

The data presented here were obtained with holoenzyme. Core enzyme gave similar results indicating that σ factor has no influence on the binding of $\alpha\text{-fo}^5\text{UTP}$. Poly$[d(AT)]$ also had no effect.

It had originally been our intention to use $\beta\text{-fo}^5\text{UTP}$ as an affinity label for RNA polymerase. However, this triphosphate was a substrate for the enzyme and only produced weak inhibition on reduction of an enzyme–$\beta\text{-fo}^5\text{UTP}$ mixture. On the other hand, $\alpha\text{-fo}^5\text{UTP}$ proved to be a potent inhibitor of the enzyme, and we therefore concentrated our effort on this analog. Although this no longer has the usual β configuration of those triphosphates that are substrates for RNA-polymerase, Rhodes and Chamberlin[18] have shown that the elongation site in the ternary complex has a general affinity for the triphosphate moiety.

The inhibition of RNA polymerase by the binding of $\alpha\text{-fo}^5\text{UTP}$ to the β subunit has been shown to satisfy several criteria for affinity labeling: (1) it forms a noncovalent complex prior to covalent attachment; (2) the inhibition and covalent attachment of $\alpha\text{-fo}^5\text{UTP}$ can be suppressed by the presence of nucleoside triphosphates; (3) the triphosphate $\alpha\text{-fo}^5\text{UTP}$ is a more potent inhibitor than the nucleoside $\alpha\text{-fo}^5\text{U}$; and (4) the $\alpha\text{-fo}^5\text{UTP}$ stoichiometrically labels RNA polymerase (after a 20-sec preincubation).

[15] D. Rabussay and W. Zillig, *FEBS Lett.* 5, 104 (1969).

[16] W. Zillig, K. Zechel, D. Rabussay, M. Schachner, V. Sethi, P. Palm, A. Heil, and W. Seifert, *Cold Spring Harbor Symp. Quart. Biol.* 35, 47 (1970).

[17] W. Stender, A. A. Stütz, and K. H. Scheit, *Eur. J. Biochem.* 56, 129 (1975).

[18] G. Rhodes and M. J. Chamberlin, *J. Biol. Chem.* 249, 6675 (1974).

[37] DNA-Dependent RNA Polymerase[1]

By A. P. KIMBALL

An essential lysine residue in the active center of the catalytic subunit of DNA-dependent RNA polymerase can be affinity labeled with methylthioinosinedicarboxaldehyde (MMPR-OP).[2,3] The ε-amino group of the lysine forms a Schiff base with an aldehyde group of MMPR-OP, which can be converted to the stable amine by mild reduction with sodium borohydride. The [35S]MMPR-OP can be prepared easily for radioactive affinity labeling of the enzyme.

MMPR-OP

Preparation of Methylthioinosinedicarboxaldehyde (MMPR-OP)

Synthesis of Unlabeled MMPR-OP

One millimole of 9-β-D-ribofuranosyl-6-methylthiopurine (298 mg) (methylthioinosine, MTI) is allowed to react with 1.1 mmoles of periodic acid (H_5IO_6, 251 mg) in 11 ml of distilled water in the dark for 30 min. The solution is rapidly pipetted onto a Dowex 1-formate column (10 × 30 mm), and the self-eluate and two water washes are collected in a 100-ml round-bottom flask. The solution of MMPR-OP, approximately 30 ml, is frozen in a thin film around the sides of the flask by rotation in a Dry Ice–acetone mixture, and the frozen solution is immediately lyophilized to give a white fluffy powder. A yield of 70–80% can be expected. The MMPR-OP should be stored dry at −20°. Solutions of

[1] This research was supported by research grants from the Robert A. Welch Foundation (E-321) and from the NCI (CA-12327) of National Institutes of Health.
[2] J. Nixon, T. Spoor, J. Evans, and A. P. Kimball, *Biochemistry* 11, 4570 (1972).
[3] T. C. Spoor, F. L. Persico, J. E. Evans, and A. P. Kimball, *Nature (London)* 227, 57 (1970).

MMPR-OP should be prepared immediately before use. Any excess solution should be discarded since the MMPR-OP slowly degrades in solution to 6-methylthiopurine with a half-life of about 4 days. The MMPR-OP is moderately difficult to dissolve in water, but heat may not be used to aid solution because this rapidly degrades the MMPR-OP.

Synthesis of [^{35}S]MMPR-OP

One millimole (284 mg) of 9-β-D-ribofuranosyl-6-thiopurine (thioinosine, mercaptopurine riboside, MPR) is allowed to react with 1 mmole (32 mg) of [^{35}S]rhombic sulfur, 20 mCi/mmole, in 25 ml of dry pyridine under reflux for 4 hr.[4] Care should be taken to keep the MPR from accumulating around the wall of the round-bottom flask at the solution surface if a heating mantle is used, since charring may occur. After 4 hr of reflux, 25 ml of distilled water is added to the pyridine solution, and the solution is taken to dryness under reduced pressure at 40°. Water, 25 ml, is added to the flask and gently heated in a water bath to dissolve the [^{35}S]MPR. A yellow-gray film that does not dissolve is [^{35}S]rhombic sulfur. The flask contents are again taken to dryness under reduced pressure at 40°. This last procedure is repeated twice more to strip the [^{35}S]MPR and [^{35}S]rhombic sulfur of residual pyridine. The stripping procedure is important because [^{35}S]rhombic sulfur dissolves in pyridine, and any pyridine remaining will carry along the sulfur in subsequent purification steps. The [^{35}S]MPR is dissolved in the flask with 15 ml of hot water and filtered while hot through a sintered-glass funnel with a fine filter and a suction flask. The [^{35}S]MPR which crystallizes from solution on cooling is redissolved with gentle heating and again filtered through a clean, fine sintered-glass funnel. The [^{35}S]MPR is again dissolved and dried in the same manner. The dried [^{35}S]MPR is taken up with 10 ml of hot water and filtered while hot through a clean, fine sintered-glass funnel. The solution in a 20-ml beaker is allowed to crystallize in the refrigerator overnight. Crystals of [^{35}S]MPR are collected and dried under reduced pressure to yield about 220 mg (77%) of >99% radiopure compound.

For preparation of the 6-methyl[^{35}S]thio derivative (6-methyl[^{35}S]-thioinosine, [^{35}S]MTI), 0.775 mmole (220 mg) of [^{35}S]MPR is allowed to react with 0.787 mmole (0.049 ml, 1.015 equivalents) of methyliodide in 1.67 ml of 0.4 N NaOH with constant stirring for 10 min. Then 0.27 ml of 0.4 N NaOH and 0.049 ml of methyliodide are added and constantly stirred for an additional 10 min. The reaction mixture is maintained at

[4] L. L. Bennett, Jr., R. W. Brockman, H. P. Schneibli, S. Chumley, G. J. Dixon, F. M. Schabel, Jr., E. A. Dulmadge, H. E. Skipper, J. A. Montgomery, and H. J. Thomas, *Nature (London)* **205,** 1276 (1965).

room temperature for several hours, seeded with a few crystals of un-labeled MTI, and placed in the refrigerator overnight to crystallize. The crystals of [³⁵S]MTI (190–200 mg) are collected, dried, and refluxed with 1.0 ml of *absolute* ethanol for several minutes and filtered through a fine sintered-glass funnel with suction while hot. It is important here to maintain anhydrous conditions since [³⁵S]MTI is very soluble in ethanol containing traces of water. The crystals are dried under reduced pressure. The yield of [³⁵S]MTI is 140–150 mg.

Periodate oxidation to give the final product, [³⁵S]MMPR-OP, is carried out as follows. [³⁵S]MTI, 0.475 mmole (142 mg), is allowed to react with 0.52 mmole (120 mg, 1.1 equivalents) of periodic acid in 5.20 ml of water in the dark for 30 min. The reaction mixture is pipetted onto a Dowex 1-formate column, 10 × 30 mm, and the eluate and two washes are collected in a 50-ml round-bottom flask and immediately lyophilized to yield about 100 mg of pure [³⁵S]MMPR-OP. The initial specific activity is about 7000 cpm/nmole. All these reactions may be scaled up or down, as long as the correct proportions described here are held constant.

Affinity Labeling DNA-Dependent RNA Polymerase

An affinity label has an affinity for the most reactive site on the enzyme; however, other sites will be bound at high label concentrations. Equation (1) can be used to determine the precise concentration [³⁵S]-MMPR-OP that binds to the active center of RNA polymerase in a 1:1 molar ratio. Here,

$$(AL \cdot E)/E = KAL \tag{1}$$

AL·E is the molar concentration of the [³⁵S] MMPR-OP-RNA poly-merase complex, E is the total molar enzyme concentration, AL is the molar concentration of [³⁵S]MMPR-OP, and K is the binding constant at equilibrium. Taking the log of both sides

$$\log (AL \cdot E)/E = \log KAL \tag{2}$$

and rearranging to give

$$\log (AL \cdot E)/E = \log K + \log AL \tag{3}$$

a plot of log AL·E/E vs. log AL gives a straight line. The point at which the line passes through zero is the concentration of [³⁵S]MMPR-OP that binds to RNA polymerase in a 1:1 molar ratio since

$$\log (AL \cdot E)/E = \log 0 \tag{4}$$

and the antilog of zero is 1.

DETERMINATION OF THE CONCENTRATION OF METHYLTHIOINOSINEDICARBOX-
ALDEHYDE WHICH BINDS IN A 1:1 MOLAR RATIO WITH RNA POLYMERASE[a]

Concentration of affinity label	Log AL concentration	AL:Enzyme (ratio)	Log AL·E ratio
1×10^{-3}	-3.00	5.96	$+0.78$
7.5×10^{-4}	-3.12	4.01	$+0.60$
5×10^{-4}	-3.30	2.65	$+0.42$
2.5×10^{-4}	-3.60	1.43	$+0.16$
1×10^{-4}	-4.00	0.56	-0.25
5×10^{-5}	-4.30	0.27	-0.57

[a] RNA polymerase (49 μg per reaction mixture) was incubated for 15 min at 37° with varying concentrations of [^{35}S]MMPR-OP in Tris at pH 7.9, containing 50 μmoles of Tris, 4 μmoles of MgCl$_2$, and 0.2 M with respect to KCl. No substrates or template were present. The reaction mixtures were cooled to 4°, and the reactions were stopped by the addition of 5% TCA. This treated mixture was filtered through 2.4-cm Whatman GF/C glass fiber filters, and the unbound radiolabeled compound was washed through with 50 ml of 2% TCA and 3 ml of 95% ethanol. The filters were allowed to air dry for 1 hr. Finally, they were placed directly into scintillation vials containing 10 ml of a toluene fluor and counted in a liquid scintillation spectrometer.

An application of this method and of Eq. (3) is shown in the table using 0.1 nmole of enzyme and varying concentrations of [^{35}S]MMPR-OP. A plot of these data gave a concentration of 0.18 mM [^{35}S]MMPR-OP, which would bind 0.1 nmole of enzyme in a 1:1 ratio. Higher concentrations of [^{35}S]MMPR-OP begin to bind to lysines not in the active center. Pragmatically, concentrations of [^{35}S]MMPR-OP less than 0.18 mM would have been used to label 0.1 nmole of enzyme if the stable amine bond were to be formed between [^{35}S]MMPR-OP and the enzyme by sodium borohydride reduction. For instance, 1.36 nmoles of enzyme and 0.5 mM [^{35}S]MMPR-OP were used to affinity label the lysine in the initiation subsite in the active center of RNA polymerase[2] (see below).

Enzyme Purification

Homogeneous preparations of *Escherichia coli* DNA-dependent RNA polymerase are required for affinity labeling. The method of Burgess and Jendrisak can be used to prepare the core enzyme.[5]

[5] R. R. Burgess and J. J. Jendrisak, *Biochemistry* **14**, 4634 (1975).

*Binding of [³⁵S]MMPR-OP to RNA Polymerase: Formation of the
Schiff Base*

Binding to DNA-dependent RNA polymerase has been carried out as
follows. Enzyme, 1.36 nmoles (679 μg), was incubated for 45 min at 37°
with 0.5 mM [³⁵S]MMPR-OP in 1.0 ml of 1.0 M KHCO₃ at pH 7.9,
which was 8 mM in MgCl₂ and 0.2 M in KCl. Tris buffer does not work
well here because [³⁵S]MMPR-OP reacts with the amino group of Tris
upon sodium borohydride reduction (see below). This prohibition probably holds for any buffer containing reactive amino groups if sodium
borohydride reduction is to be used.

Reduction of the Schiff Base to the Stable Amine

The Schiff base is reduced to the amine bond by mild reduction with
sodium borohydride (NaBH₄). Immediately after formation of the Schiff
base in the above reaction, the reaction mixture is cooled to 4° in ice and
15 mg of NaBH₄ in 1.0 ml of KHCO₃ at pH 7.9 is added. The reduction
is allowed to occur in a cold-room at 4° for 12 hr. Since the reaction
liberates hydrogen gas, reaction tubes are loosely capped to avoid a
buildup of pressure. Unreacted [³⁵S]MMPR-OP and NaBH₄ are removed
by dialyzing the reaction mixture against excess KHCO₃ at pH. 7.9 until
the radioactivity in the buffer drops to background levels.

Disk electrophoresis on polyacrylamide gels of the denatured enzyme–
[³⁵S]MMPR-OP complex is carried out to determine the subunit bound
by the affinity label.[2]

Comments

The [³⁵S]MMPR-OP has recently been used to affinity label the active
center of *E. coli* DNA-dependent DNA polymerase I.[6] The compound
has also been used to label the lysine near the active center of
ribonuclease A.[7]

In all studies carried out to date on enzymes of nucleic acid metabolism, MMPR-OP has shown specificity for lysine residues. Inhibition
of other enzymes of nucleic acid metabolism by MMPR-OP can be
taken as presumptive evidence for the presence of lysines in their active

[6] R. A. Salvo, G. F. Serio, J. E. Evans, and A. P. Kimball, *Biochemistry* **15**, 493
(1976).
[7] T. C. Spoor, J. L. Hodnett, and A. P. Kimball, *Cancer Res.* **33**, 856 (1973).

centers, and the procedures described above could be modified for their study.

Dowex 1-formate columns should be washed with concentrated formic acid and then eluted with distilled water until neutral before use for purifying MMPR-OP. The formic acid wash produces a "milky" solution at the solvent front that would otherwise contaminate the MMPR-OP.

The 6-methyl-[¹⁴C]MMPR-OP could be prepared by substituting [¹⁴C]methyliodide in the methylation step. This compound has the advantage of a long half-life; [³⁵S]MMPR-OP has a half-life of about 3 months.

[³⁵S]Rhombic sulfur is usually supplied in a solution of benzene. The solution is placed in the round-bottom flask in which the sulfur exchange is to be carried out, then the benezene is removed by distillation. Care should be exercised to avoid polymerizing or subliming the [³⁵S]rhombic sulfur.

[38] Staphylococcal Nuclease

By PEDRO CUATRECASAS and MEIR WILCHEK

Micrococcal nuclease is an extracellular enzyme of *Staphylococcus aureus* that hydrolyzes specific phosphodiester bonds of both RNA and DNA.[1] The protein is a convenient model for studies of conformational stability, effects of ligand binding, chemical modification, hydrolytic mechanism, and X-ray crystallographic analyses.[1]

Two groups of reagents were used for affinity labeling of this enzyme, bromoacetyl (Fig. 1, reagents I–III),[2] and diazonium salts (Fig. 1, reagents IV–VI[3,4]) of different thymidine-p-aminophenylphosphate derivatives.

Reagents I and IV are derived from pdTp-aminophenyl,[5] which is a potent competitive inhibitor of the enzyme, having a dissociation con-

---

[1] P. Cuatrecasas, H. Taniuchi, and C. B. Anfinsen, *Brookhaven Symp. Biol.* **21**, 172 (1969).

[2] P. Cuatrecasas, M. Wilchek, and C. B. Anfinsen, *J. Biol. Chem.* **244**, 4316 (1969).

[3] P. Cuatrecasas, *J. Biol. Chem.* **295**, 574 (1970).

[4] M. Wilchek, *FEBS Lett.* **7**, 161 (1970).

[5] Abbreviations: pdTp, deoxythymidine 3′,5′-diphosphate; pdTp-nitrophenyl, deoxythymidine 3′-p-nitrophenylphosphate 5′-phosphate; nitrophenyl-pdT, p-nitrophenyl ester of deoxythymidine 5′-phosphate; dTp-nitrophenyl, p-nitrophenyl ester of deoxythymidine 3′-phosphate; dTp-aminophenyl, deoxythymidine 3′-p-amino-phenylphosphate; pdTp-aminophenyl, deoxythymidine 3′-p-aminophenylphosphate 5′-phosphate; aminophenyl-pdT, deoxythymidine 5′-p-aminophenylphosphate.

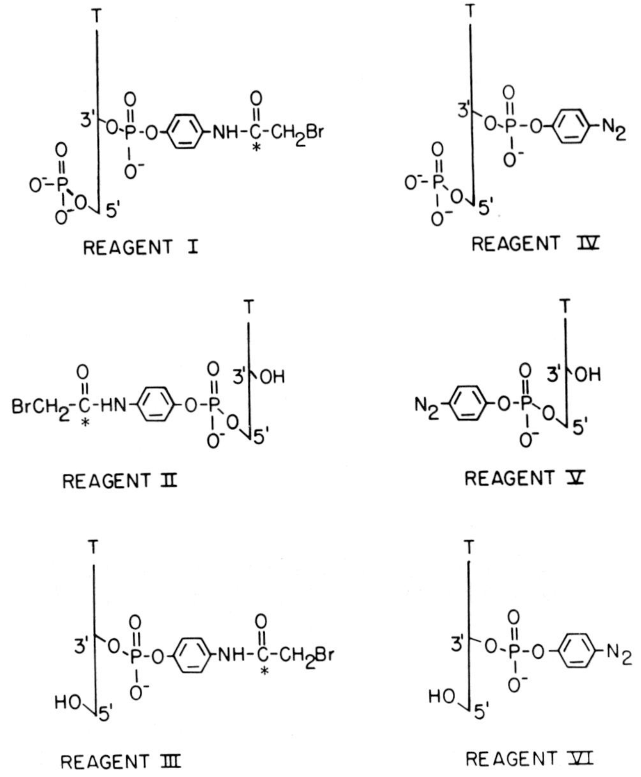

FIG. 1. Reagents used for affinity labeling of staphylococcal nuclease.

stant of 1.1 μM.[6,7] Reagents II and V are derived from 4-aminophenyl-pdT, a weak substrate with a dissociation constant of 2.2 mM. Reagents III and VI are derivatives of the weak inhibitor, dTp-aminophenyl, with a dissociation constant of 22 mM.

The reaction of nuclease with these reagents results in inactivation of enzyme activity. The specificity of the labeling reaction was determined (a) by following the stoichiometry of inactivation; (b) by studies of the prevention of inactivation (as well as reaction of the specific residue) by the addition of a strong, reversible, competitive inhibitor, pdTp[1,8]; (c) by studies of protection that results by omitting the divalent cation Ca^{2+}, which is required for the binding of substrates and inhibitors[1,9]; (d) by contrasting the qualitative patterns of chemical reaction,

[6] P. Cuatrecasas, M. Wilchek, and C. B. Anfinsen, *Biochemistry* 8, 2277 (1969).

[7] P. Cuatrecasas, M. Wilchek, and C. B. Anfinsen, *Science* 162, 1491 (1968).

[8] P. Cuatrecasas, S. Fuchs, and C. B. Anfinsen, *J. Biol. Chem.* 242, 1541 (1967).

[9] P. Cuatrecasas, F. Fuchs, and C. B. Anfinsen, *J. Biol. Chem.* 242, 3063 (1967).

i.e., by peptide maps and peptide isolation found with structurally different inhibitors having identical reactive groups (bromoacetyl, diazonium); and (e) by purification of the affinity-labeled enzymes followed by identification of their uniquely labeled residues.

Synthesis of Affinity Labeling Reagents

Preparation of [14C]Bromoacetyl Hydroxysuccimide. In 3 ml of dioxane, 87 mg (630 μmoles) of [1-14C]bromoacetic acid (1.54 mCi/nmole) and 86 mg (750 μmoles) of N-hydroxysuccinimide are dissolved. To this solution, 132 mg (700 moles) of dicyclohexylcarbodiimide are added. Urea precipitates immediately. After 1 hr the urea is removed and the solution is brought to 5 ml with dioxane. The compound is used without further purification.

Preparation of Reagent I. pdTp-Aminophenyl, 50 μmoles,[6,10] is dissolved in 0.3 ml of water, diluted with 0.7 ml of dioxane, and 1 ml of the [14C]bromoacetyl-N-hydroxysuccinimide ester solution (125 μmoles) is added. After 10 min, 3 equivalents of triethylamine are added, and the reaction is left for 1 hr. One volume of water is then added, and the mixture is passed through Dowex 50 (hydrogen form) and lyophilized. The product is dissolved in methanol, precipitated with ether, and the precipitate is dissolved in a small volume of water. After adjusting the pH to 3.5 with NaOH, the solution is lyophilized. The yield is 90%. The compound is pure on thin-layer chromatography using 2-propanol–NH$_4$OH–H$_2$O (7:1:2) and 1-butanol–acetic acid–water (5:2:4). The λ_{max} is 258 nm; the E_M at 258 nm is 14,900, and at 267 nm, 13,900. The specific activity is 1.10 μCi/μmole.

Preparation of Reagents II and III. These are prepared from nitrophenyl-dTp and dTp-nitrophenyl after reduction to the corresponding aminophenyl derivatives by catalytic hydrogenation using palladium on charcoal, as previously described.[6] The bromoacetyl derivatives are prepared essentially as described for reagent I. Of the aminophenyl ester, 50 μmoles are dissolved in 0.3 ml of water and 0.7 ml of dioxane, and 1 ml (125 μmoles) of the stock solution of [14C]bromoacetyl-N-hydroxysuccinimide ester is added. After an hour, 1 volume of water is added, and the solution is extracted 5 times with ether and concentrated to dryness. The product is dissolved in methanol and precipitated with ether. Yields are 90%, and the compounds are chromatographically pure on thin-layer chromatography with the solvents described for reagent I. The λ_{max} of reagent II is 260 nm, the E_M (260 nm) is 14,600, and the specific radio-

---

[10] M. Wilchek and M. Gorecki, this series, Vol. 34, p. 492 (1974).

activity is 1.30 μCi/μmole. For reagent III λ_{max} is 264 nm, the E_M (264 nm) is 14,500, and the specific radioactivity is 0.44 μCi/μmoles.

Preparation of Diazonium Reagents and Reactions with Enzyme. The diazonium reagents depicted in Fig. 1 are prepared from the corresponding aminophenyl derivatives by treatment with nitrous acid. The aminophenyl compounds are prepared from the respective nitrophenyl derivatives by catalytic hydrogenation with palladium on charcoal, as previously described.[6] Aminophenyl nucleotide, 10 μmoles, is dissolved in 150 μl of 2 N HCl. The subsequent reactions are carried out at 4°. Fifty microliters (containing 0.8 mg) of sodium nitrite are added over a 1-min period to the continuously stirred solution containing the nucleotide. After 7 min, 2.8 ml of cold 0.1 M sodium borate at pH 8.1 are added. Aliquots of the diazonium compound are used immediately in the modification reactions.

Affinity Labeling with Bromoacetyl Derivatives Reagents I, II, III

A solution of nuclease (6.2 μM) in 0.05 M sodium borate at pH 9.4, containing 10 mM CaCl$_2$, is treated with a 3- to 12-fold molar excess of reagent I. Protection from inactivation in the presence of a 12-fold molar excess of reagent I is provided by addition of 29 μM pdTp or by omitting Ca^{2+} from the incubation mixture. Affinity labeling with reagents II and III is performed in the same buffer with a 6- to 26-fold molar excess of the reagents. Incubations are performed for 30 hr at room temperature.

Affinity Labeling with Diazonium Salt Derivatives IV, V, VI

A solution of nuclease (3.2 μM) in a buffer containing 10 mM CaCl$_2$, 0.05 M bis-Tris chloride (pH 7.0, 7.5), or 0.05 M sodium borate (pH 8.2, 8.8, 9.4) is treated with 2- to 10-fold molar excess of reagent IV. Complete inactivation is achieved with a 2-fold molar excess. Incubation time is 90 min at room temperature. With reagents V and VI, larger amounts of reagent, longer incubation period, and higher enzyme concentrations are required to achieve similar degrees of inactivation.

Results

Localization of the amino acid residues of staphylococcal nuclease with which the bromoacetyl and diazonium affinity labeling reagents specifically react is summarized in Fig. 2. Reagent 1 is the most specific and attaches to Lys-48 and Lys-49 in 80% and to Tyr-115 in 15% yield. Reagent II reacts with Tyr-85. Reagent III probably reacts with Lys-24

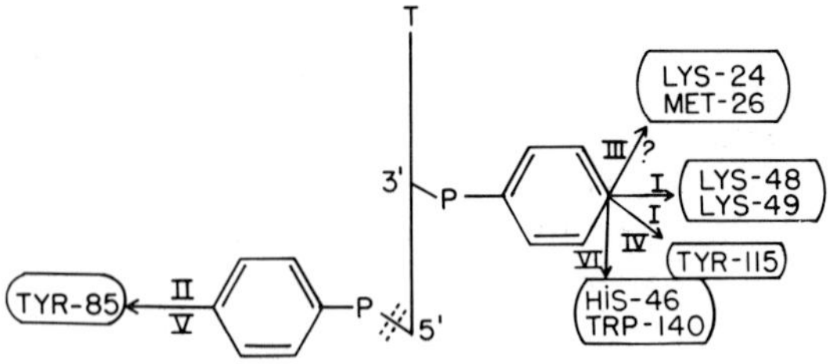

FIG. 2. Schematic representation of the residues of staphylococcal nuclease that were found to react with the affinity labeling reagents depicted in Fig. 1.

and possibly with Met-26. Reagent IV reacts exclusively with Tyr-115, and reagent V with Tyr-85. Reagent VI reacts specifically with His-46 (55%) and Trp-140(45%). Such studies have been useful in determining the topography of the active site of this enzyme in solution and in relating catalytic and binding functions to specific amino acid residues within this region.

Comments

Ribonuclease A also has been affinity labeled with the diazonium salt of adenosine 5'-p-aminophenylphosphate 2'(3')phosphate. Using methods similar to that described in this chapter, 1 mole of the reagent was found to be covalently bound to the enzyme. In this case Tyr-72 was modified.[11]

[11] M. Gorecki, M. Wilchek, and P. Patchornik, *Biochim. Biophys. Acta* **229**, 590 (1971).

[39] Carbohydrate Binding Sites

By E. W. THOMAS

Several carbohydrate analogs incorporating haloacetyl[1-4] or epoxide[5-10] groups have been synthesized and evaluated as affinity labels for

[1] D. H. Buss and I. J. Goldstein, *J. Chem. Soc. C* p. 1457 (1968).

[2] E. W. Thomas, *J. Med. Chem.* **13**, 755 (1970).

[3] F. Naider, Z. Bohak, and J. Yariv, *Biochemistry* **11**, 3202 (1972).

[4] S. Otieno, A. K. Bhargava, E. A. Barnard, and A. H. Ramel, *Biochemistry* **14**, 2403 (1975).

[5] E. W. Thomas, *Carbohydr. Res.* **13**, 225 (1970).

[6] G. Legler and E. Bause, *Carbohydr. Res.* **28**, 45 (1973).

[7] E. M. Bessel and J. H. Westwood, *Carbohydr. Res.* **25**, 11 (1972).

[8] G. Legler, *Biochem. Soc. Trans.* **3**, 847 (1975).

[9] M. L. Shulman, S. D. Shiyan, and A. Ya. Khorlin, *Carbohydr. Res.* **33**, 229 (1974).

[10] J. E. G. Barnett and A. Ralph, *Carbohydr. Res.* **17**, 231 (1971).

carbohydrate binding sites in proteins. Analogs incorporating photosensitive functions (photoaffinity labels) have also been described.[11-16] This article deals with the synthesis and properties of representative compounds of the above type, in particular those in which the reactive function is attached to the anomeric carbon atom of the carbohydrate.

Bromoacetylglycosylamines

The glycosylamines (1-amino sugars) are easily accessible by treatment of the appropriate aldose with anhydrous methanol saturated with ammonia.[17] The synthesis of β-D-galactosylamine is given here, since a published method[18] was not satisfactory.

β-D-Galactosylamine[19] (I)

Methanol (250 ml) at 4° is saturated with gaseous ammonia from a cylinder. D-Galactose (15 g) in hot water (20 ml) is added, and the reaction vessel is stoppered. Compound (I) crystallizes after storage at 23° for about 14 days; yields may be increased by more prolonged storage at 4°. The crystals are collected, washed with methanol, and recrystallized by dissolving water (2 parts), adding methanol (4 parts) followed by n-propanol to turbidity. The product has a melting point of 135°–136°.

The pH of an aqueous solution of (I) is about 9. The α-D-galactosylamine–ammonia complex, formed from D-galactose and ammonia in anhydrous methanol,[18] gives an aqueous solution of pH 11.

N-Bromoacetylation[2]

The preferred reagent is bromoacetic anhydride; the following procedure is convenient if radioactive reagent is required.

Bromoacetic Anhydride. Dry bromoacetic acid (1 mM) in dry CCl₄ (3 ml) is treated with dicyclohexylcarbodiimide (0.5 mM) at 4° with

[11] E. W. Thomas, *Carbohydr. Res.* **31**, 101 (1973).
[12] M. B. Perry and L. W. Heung, *Can. J. Biochem.* **50**, 510 (1972).
[13] E. Saman, M. Claeyssens, H. Kersters-Hilderson, and C. K. de Bruyne, *Carbohydr. Res.* **30**, 207 (1973).
[14] A. E. Burkhardt, S. O. Russo, C. G. Rinehardt, and G. M. Loudon, *Biochemistry* **14**, 5465 (1975).
[15] M. Beppu, T. Terao, and T. Osawa, *J. Biochem. (Tokyo)* **78**, 1013 (1975).
[16] G. Rudwick, H. R. Kaback, and R. Weil, *J. Biol. Chem.* **250**, 6847 (1975).
[17] H. Isbell and H. L. Frush, *J. Org. Chem.* **23**, 1309 (1958).
[18] H. L. Frush and H. I. Isbell, *J. Res. Natl. Bur. Stand.* **47**, 239 (1951).
[19] L. A. Lobry de Bruyn and F. H. Van Leent, *Recl. Trav. Chim. Pays-Bas* **14**, 134 (1895).

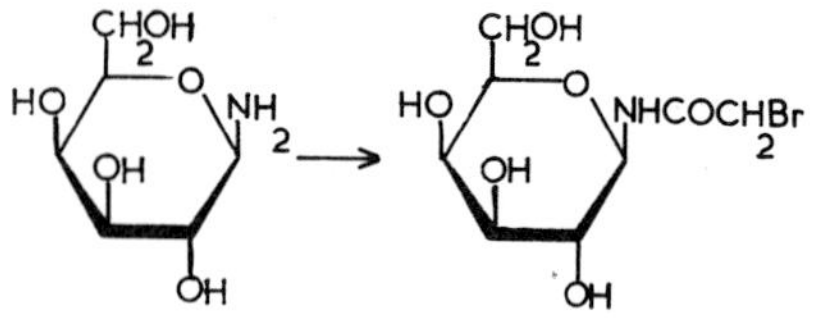

(I)   (II)

Fig. 1. Formation of N-bromoacetyl-β-D-galactosglamine (II) from galactosyl-amine.

stirring. After 5 min at 23°, dicyclohexylurea is removed by centrifuga-tion; the supernatant liquid either is used directly or concentrated under reduced pressure.

N-*Bromoacetyl-β-D-Galactosylamine* (II) (*Fig. 1*). Galactosylamine (I) (2 mM), suspended in dry dimethylformamide (2 ml), is treated with bromoacetic anhydride (2.5 mM) with stirring at 23°: (I) dissolves as reaction proceeds. A large excess of anhydride should be avoided as O-acylation can occur. After 6 hr, crude (II) is precipitated by adding the reaction mixture dropwise with stirring to ice-cold diethyl ether (200 ml). The resultant solid is collected, washed well with ether, and crystal-lized by dissolving in methanol and adding ether to turbidity. Compound (II), formed in 80% yield, has m.p. 192° (dec) and is homogeneous by TLC (Silica gel G; methanol:acetone 1:10 v/v; detected by charring with H_2SO_4). Other bromoacetylglycosylamines have the properties[2] in the tabulation.

N-Bromoacetyl-	Melting point	$\{\alpha\}_D^{20}(C = 1, H_2O)$
D-Glucosylamine	187°–190°	−13.5°
L-Glucosylamine	178°	+14°
L-Fucosylamine	175°	+2.8°
α-Lactosylamine	158°	—

Compound (II) inhibits β-galactosidase from *Escherichia coli* irreversi-bly; the inactivation shows saturation kinetics, with $K_I = 1.13$ mM and a first-order rate constant 0.063 min^{-1}. Enzymic activity reappears on treating with β-mercaptoethanol[3] and is ascribed to thiolysis of a sul-fonium salt formed at a methionine in the active site. N-Bromoacetyl-L-fucosylamine reacts with β-galactosidase, giving a covalent 1:1 complex which is enzymically active.[20]

[20] J. Yariv, K. J. Wilson, J. Hildesheim, and S. Blumberg, *FEBS Lett.* **15,** 24 (1971).

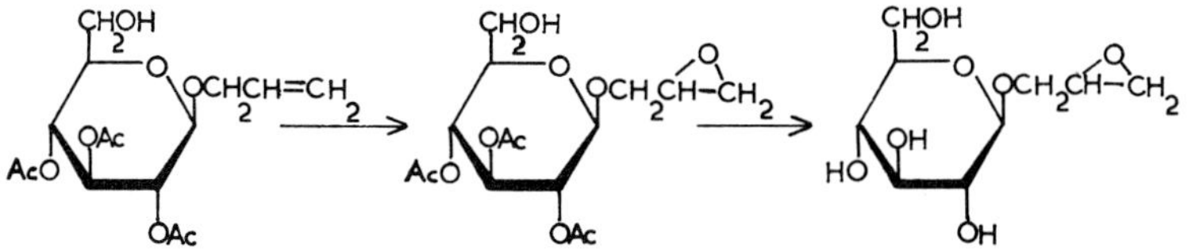

(III)   (IV)

FIG. 2. Formation of 2,′3′-epoxypropyl-2-acetamido-2-deoxy-β-D-glucopyranoside (IV) via 2′,3′-epoxypropyl-2-acetamido-3,4,6-tri-O-acetyl-2-deoxy-β-D-glucopyrano-side (III).

β-Galactosidase from *E. coli* is also inhibited *in vivo* by (II), but only if lac permease is functional.[21] Presumably, (II) is actively accumulated by cells and is not acting as a site-directed inhibitor of permease in short-term experiments.

Most of the bromoacetylglucosylamines show marked effects on glucose-stimulated insulin release from isolated pancreatic islet cells.[22]

Epoxypropyl Glycosides

These analogs are accessible[5] by epoxidation of the appropriate acetylated allyl glucoside (or vinyl C-glycoside[9]), followed by deacetylation. The following procedure for synthesis of epoxypropyl 2-acetamido β-D-glycoside is applicable to the corresponding glucose, cellobiose, chitobiose, and chitotriose analogs. The allyl glycosides are themselves prepared from the appropriate glycosyl halide and allyl alcohol.[e.g.,5]

2′,3′-Epoxypropyl-2-acetamido-3,4,6-tri-O-acetyl-2-deoxy-β-D-glucopyranoside (III) (Fig. 2)

Allyl 2-acetamido-3,4,6-tri-O-acetyl-2-deoxy-β-D-glucopyranoside (1 mM) in methylene chloride (10 ml) containing peroxyphthalic acid (2 mM; peroxybenzoic acid is also satisfactory) is refluxed gently for 3 hr. After cooling to 4°, the solution is filtered (if peroxyphthalic acid is used), washed with cold aqueous potassium bicarbonate solution until free of peroxy acid (as tested with acidified KI), dried with Na_2SO_4, and evaporated to dryness. Compound (III) is crystallized from methanol–ether. Yield, 85%; m.p., 162°–163°; $\{\alpha\}_D^{23}$, −34.4° (C = 0.4, $CHCl_3$).

[21] W. Daring and E. W. Thomas, unpublished data (1974).
[22] B. Hellman, L-A Idahl, A. Lernmark, I-B Taljedal, and E. W. Thomas, *Mol. Pharmacol.* **12**, 208 (1976).

2′,3′-Epoxypropyl-2-acetamido-2-deoxy-β-D-glucopyranoside (IV)

Compound (III), suspended in dry methanol (final concentration 10% w/v), is treated with methanolic barium methoxide solution to give a final concentration of 20 mM in base. After 12 hr at 4°, the reaction mixture is partly neutralized with a few chips of solid CO_2 and evaporated to dryness. The residue is triturated with methanol and filtered, and dioxane is added dropwise to the filtrate until a permanent turbidity obtains. Compound (IV) crystallizes as needles. Yield, 60%; m.p., 175°–178°; $\{\alpha\}_D^{23}$, −37.5° ($C = 2$, H_2O).

Compound (IV) is homogeneous by thin-layer chromatography (TLC) under identical conditions used to evaluate (II). Solutions of (IV) in 0.1 N sodium thiosulfate become alkaline within 5 min at 23°, indicating the presence of the epoxide group.

Epoxypropyl-β-D-glycosides of N-acetylglucosamine (Glc NAc), di-N-acetylchitobiose (Glc NAc)$_2$, and tri-N-acetylchitotriose (Glc NAc)$_3$ irreversibly inhibit lysozyme from hen's egg-white.[23] Although the kinetics of inactivation have not been rigorously examined, inactivation has been shown to result from selective esterification[24] of Asp-52, in the case of the (Glc NAc)$_2$ analog.

The covalent enzyme inhibitor complex has been crystallized, and X-ray diffraction studies[25] show that the two pyranose rings of the analog occupy subsites B and C as originally predicted. The esterified enzyme cannot be reactivated with alkaline hydroxylamine. Successful cleavage of the complex with $NaBH_4$ is possible only after prior chemical reduction at all disulfide linkages of the enzyme.[26]

Compound (IV) reacts with α-lactalbumin to give a 1:1 covalent complex.[27] The site of attachment has not been determined, but the α-lactalbumin is not protected from (IV) by either Glc NAc or (Glc NAc)$_2$.

The 3′,4′-epoxybutyl-β-D-glycoside of (Glc NAc)$_2$ inhibits hen's egg-white lysozyme at a similar rate to the epoxypropyl analog, but the 5′,6′-epoxyhexyl analog is much less effective.[28]

Photoaffinity Labels

The synthesis of a number of 2-nitro-4-azidophenyl[12,16] and 4-azidophenyl-β-D-glycosides[13] has been reported. Nitro-substituted analogs

[23] E. W. Thomas, J. F. McKelvy, and N. Sharon, Nature (London) 222, 485 (1969).
[24] Y. Eshdat, J. F. McKelvy, and N. Sharon, J. Biol. Chem. 248, 5892 (1973).
[25] J. Moult, Y. Eshdat, and N. Sharon, J. Mol. Biol. 75, 1 (1973).
[26] Y. Eshdat, A. Dunn, and N. Sharon, Proc. Natl. Acad. Sci. U.S.A. 71, 1658 (1974).
[27] Y. Eshdat and E. W. Thomas, unpublished data (1971).
[28] E. W. Thomas, unpublished data (1972).

have the advantage of being activable with light of wavelength >350 nm. Access to 2-nitro-4-azidophenyl-α-D-glycosides is not easy, except in the case of the mannoside, where reaction of α-acetochloromannose with the appropriate phenol proceeds with retention of configuration.[11]

4-(Benzyloxycarbonylamino)-2-nitrophenyl-α-D-mannopyranoside Tetraacetate (V) (Fig. 3)

4-(Benzyloxycarbonylamino)2-nitrophenol (3.39) in 1 M sodium hydroxide (17 ml) and acetone (30 ml) at 5° is treated with tetra-O-acetyl-α-D-mannopyranosyl chloride (3.7 g) dissolved in acetone (20 ml). After 12 hr at 5°, the mixture is poured into water (200 ml) and extracted five times with chloroform. Unreacted phenol is recovered from the chloroform extracts by six extractions with 1M sodium hydroxide. The chloroform layer is washed with dilute acid, dried, and evaporated to give a mixture of (V) and mannopyranosyl chloride. Elution of the mixture from neutral alumina (Woelm) with chloroform gives the unreacted mannosyl chloride. Compound (V), seen as a pale yellow band on the column, is eluted with chloroform–ethylacetate (1:1). Yield, 0.7 g (11%); m.p., 188° (from n-propanol); $\{\alpha\}_D^{23}$, +68° (C = 0.2, CHCl$_3$).

4-Azido-2-nitrophenyl-α-D-mannopyranoside (VI)

Compound (V) (2.3 g) in dry chloroform (10 ml) is treated with a 45% solution (30 ml) of hydrogen bromide in glacial acetic acid with vigorous stirring. After 15 min at 23° with stirring, the solution is poured into chloroform (200 ml), which is in turn poured onto crushed ice (200 ml). The chloroform layer is separated, freed from acid by several extractions with ice-cold aqueous sodium bicarbonate, dried, and evaporated. The residue obtained is washed with a small volume of ice-cold chloroform to remove benzyl bromide. In addition to removal of the benzyloxycarbonyl group, partial de-O-acetylation occurs during this step, the product being predominantly the tri-O-acetyl derivative. An

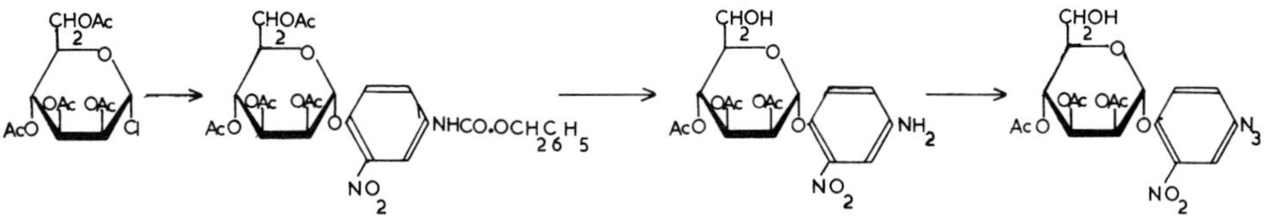

FIG. 3. A synthetic approach to 4-azido-2-nitrophenyl-α-D-mannopyranoside (VI).

improvement here would be to remove the benzyloxycarbonyl group by limited hydrogenolysis over palladium black.[12]

To the amine triacetate (500 mg) in methanol–acetic acid (1:1 v/v, 10 ml) at 0°, isopentylnitrite (0.5 ml) is added. After 30 min, 1 M hydrochloric acid (2 ml) is added, followed by solid sodium azide (1 molar equivalent). After stirring for 60 min, the mixture is poured into excess water and extracted with chloroform. The extract is washed, dried, and evaporated to yield a cream colored solid (Va) having $\nu_{max} = 2120$ cm^{-1} (N_3). Conventional deacetylation (a small chip of sodium added to a solution in dry methanol) results in precipitation of (VI): 100 mg, 26% yield. After recrystallization from hot water, (VI) has m.p. 132°–135°; $\{\alpha\}_D^{23}$, $+ 130°$ (C = 0.22, H_2O); $\lambda_{max}^{H_2O}$, 355 nm (ϵ1900); ν_{max}, 2100 cm^{-1} (N_3).

Irradiation of 4-azido-2-nitrophenyl-α-D-mannopyranoside with light of wavelength >350 nm gives the corresponding azo compound in high yield.[11] No evidence has been obtained[29] for specific labeling of concanavalin A by irradiating mixtures of the protein and (VI), although one report[15] claims specific labeling of concanavalin A with 4-azidophenyl-α-D-mannopyranoside. In the writer's opinion, the relatively long lifetimes of photogenerated nitrenes, and their propensity to undergo dimerization and intramolecular insertion reactions, make them unattractive canavalin A by irradiating mixtures of the protein and (Va), although one ing sites which do not contain nucleophilic side chains is still unsolved.

---

[29] E. W. Thomas, unpublished data (1973).

[40] Glucosidases

By G. LEGLER

Preparation of active site-directed inhibitors for glucosidases by the introduction of reactive groups into substrate analogs has been handicapped by the high specificity of these enzymes for the intact glucose molecule.[1,2] Modification of a hydroxyl group by a reactive acyl or alkyl residue usually prevents binding to the substrate binding site. Incorporation of such groups into the aglycon part is tolerated, but labeling will occur at functional groups of the enzyme not directly involved in cataly-

---

[1] G. Legler, *Mol. Cell. Biochem.* **2**, 31 (1973).
[2] G. Legler, *Biochem. Soc. Trans.* **3**, 847 (1975).

sis. With the exception of the reaction of β-galactosidase with N-bromo-acetyl-β-galactosylamine,[3] no successful labeling experiment of this type has been reported.

The most suitable inhibitors have been compounds derived from (3,5/4,6)-tetrahydroxycyclohex-1-ene[4] (conduritol B). Its epoxide has a low intrinsic reactivity, but it is activated by an acidic group of the enzyme, which is probably the same as that is necessary for the protonation of the substrate during hydrolysis. The epoxide can, therefore, be regarded as a k_{cat} inhibitor as defined by Rando.[5]

The specificity of conduritol B epoxide is such that β-glucosidases from widely differing sources have been found to react with loss of enzymic activity: from various *Aspergillus* species,[6,7] yeast,[8] snail (*Helix pomatia*),[8] sweet almonds,[9] and mammals.[10] The only exceptions have been the β-glucosidases from garbanzo plants (*Cicer arietum L.*) and from *Alocasia macrorrhiza*.[11] The only other enzymes that have been found to be covalently inhibited are α-glucosidase from yeast (*Saccharomyces cerevisiae*)[12] and the sucrase–isomaltase complex from rabbit small intestine.[13]

Other epoxides that have been reported to give a time-dependent inhibition are 2-methyleneoxide-(1,3,5/4,6)-pentahydroxy cyclohexane with sucrase–isomaltase[13] and 1,2-epoxy-(β-glucopyranosyl)ethane with β-glucosidase from sweet almonds.[14] Preparation of these two compounds in radioactive form has not been reported.

Another derivative of conduritol B, 6-bromo-3,4,5-trihydroxycyclohex-1-ene (bromoconduritol), was found to be effective for covalent inhibition of glucosidases.[14] It reacts rapidly with an α-glucosidase from yeast (maltase) at equimolar concentrations of enzyme and inhibitor. It also reacts with β-glucosidases from *Cicer arietum* and *Alocasia macrorrhiza*.[11]

[3] F. Naider, Z. Bohak, and J. Yariv, *Biochemistry* **11**, 3202 (1972).

[4] The fraction in parentheses denotes the position of substituents above and below the plane of the cyclohexane ring as specified by the IUPAC–IUB Commission on Biochemical Nomenclature.

[5] R. R. Rando, *Science* **185**, 320 (1974). See also this volume [3] and [12].

[6] G. Legler, *Hoppe-Seyler's Z. Physiol. Chem.* **349**, 767 (1968).

[7] G. Legler and L. M. Omar Osama, *Hoppe-Seyler's Z. Physiol. Chem.* **349**, 1488 (1968).

[8] G. Legler, *Hoppe-Seyler's Z. Physiol. Chem.* **345**, 197 (1966).

[9] G. Legler and S. N. Hasnain, *Hoppe-Seyler's Z. Physiol. Chem.* **351**, 25 (1970).

[10] J. N. Kanfer, G. Legler, J. Sullivan, S. S. Raghavan, and R. A. Mumford, *Biochem. Biophys. Res. Commun.* **67**, 85 (1975).

[11] W. Hösel, personal communication (1975).

[12] W. Lotz, Doctoral Thesis, University of Bonn, 1970.

[13] A. Quaroni, E. Gershon, and G. Semenza, *J. Biol. Chem.* **249**, 6424 (1974).

[14] M. L. Shulman, S. D. Shiyan, and A. Y. Khorlin, *Carbohydr. Res.* **33**, 229 (1974).

The following observations show that the reaction of conduritol B epoxide and of bromoconduritol with enzyme is specifically at the active site: (1) epoxides and allyl bromides not structurally related to glucose do not react; (2) the rate of inactivation is decreased in the presence of competitive inhibitors; (3) in all cases in which the stoichiometry of the reaction between inhibitor and glucosidase has been investigated, a 1:1 molar ratio was found[6,9,13,15]; (4) incorporation of 1 mole of inhibitor per mole of active sites results in complete loss of enzymic activity.

In all reported cases,[6,9,13,16] labeling with conduritol B epoxide occurs by nucleophilic attack of a carboxylate anion at the active site on the protonated epoxide to give an ester of inositol. The latter was identified as (+)-*chiro*-inositol with β-glucosidase from *Aspergillus wentii*.[6] Since racemic conduritol B epoxide was used for the inactivation, this observation indicates that the enzyme reacts specifically with that enantiomer corresponding to D-glucose. The other enantiomer, prepared from an optically active inositol derivative, was inactive both with α- and β-glucosidases.[17] The carboxylate group and the amino acid sequence around it have been identified in β-glucosidases from *A. wentii*[18] and from bitter almonds[16] and in the sucrase–isomaltase complex from rabbit small intestine.[19] In all cases, the side chain of an aspartic acid residue had reacted with the epoxide.

In the reaction of α-glucosidase from yeast with bromoconduritol, it appears that a histidine residue is modified.[15] There are indications that the reaction goes via an epoxide formed at the active site by dehydrobromination of the inhibitor. The enzyme contains two essential SH groups that are masked during the reaction but are freely accessible after denaturation of the modified enzyme.

Additional inhibitors have been tested to label other groups at the active site of β-glucosides from *A. wentii* and from bitter almonds.[20] These include *N*-bromoacetyl-β-glucosylamine, β-D-glucosyl azide, 2-azido-4-nitrophenyl- and 4-azido-2-nitrophenyl-β-D-thioglucopyranoside. The last three were conceived as photoaffinity labels. The compounds showed the expected competitive inhibition but a time-dependent loss of activity was not observed on irradiation at the appropriate wavelength. *N*-Bromoacetyl-β-glucosylamine was inactive against the almond enzymes but did

[15] G. Legler and W. Lotz, *Hoppe-Seyler's Z. Physiol. Chem.* **354**, 243 (1973).

[16] A. Harder, Doctoral Thesis, University of Cologne, 1976.

[17] D. Mercier, A. Olesker, S. D. Gero, and J. E. G. Barnett, *Carbohydr. Res.* **18**, 227 (1971).

[18] E. Bause and G. Legler, *Hoppe-Seyler's Z. Physiol. Chem.* **355**, 438 (1974).

[19] A. Quaroni and G. Semenza, *J. Biol. Chem.* **251**, 3250 (1976).

[20] G. Legler, unpublished results, 1975.

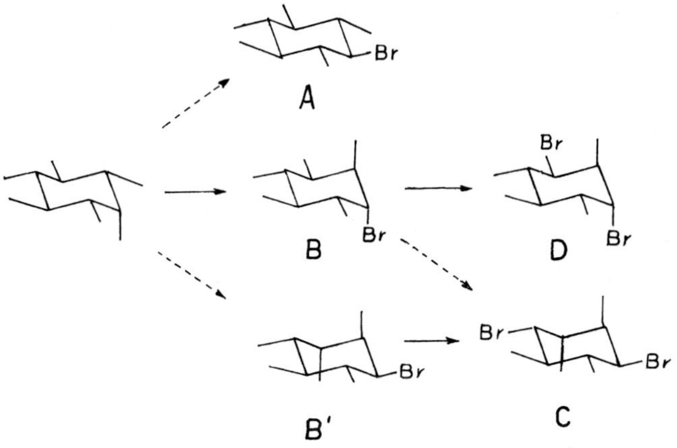

FIG. 1. Bromination of *myo*-inositol. Unmarked substituents denote hydroxyl or acetoxy groups.

inhibit in a time-dependent fashion with the enzyme from *Aspergillus*. However, the degree of inhibition did not reach 100% but leveled off at 85% and was partly reversible.

Preparation of Conduritol B Epoxide

Conduritol B epoxide can be prepared from *myo*-inositol by two different routes[8,21] and from benzene.[22] Owing to technical problems only the first two are suitable for the preparation of radioactive epoxide. For gram quantities, we prefer inositol as starting material.

Route A involves the protection of the hydroxyl groups of *myo*-inositol at C-1 and C-2 as the cyclohexanon ketal, and acetylation of the remaining hydroxyl groups followed by removal of the protecting group from C-1 and C-2. The resulting 3,4,5,6-tetra-*O*-acetyl-myo-inositol is converted to the 1,2-thionocarbonate, from which tetra-*O*-acetyl conduritol B is obtained by the removal of sulfur and carbon dioxide with trimethyl phosphite. Deacetylation and epoxidation with peroxy acids give the desired product.

In *route B*, more suitable for the preparation of less than a milligram of radioactive epoxide, it is prepared by dehydrobromination of (1,2,4/3,5,6)-6-bromocyclohexanepentol. The latter is obtained by bromination of *myo*-inositol with HBr in acetyl bromide and acetic acid (Fig. 1).

Depending on reaction conditions either B or D is obtained as the

[21] T. L. Nagabushan, *Can. J. Chem.* **48**, 383 (1970).
[22] M. Nakajima, L. Tomida, N. Kurihara, and S. Takei, *Chem. Ber.* **92**, 173 (1959).

main product. The latter can be converted to 6-bromo-6-deoxyconduritol B epoxide, which is more reactive with some β-glucosidases.[1,9]

Route A

1,2-O-Cyclohexylidene myo-Inositol (I).[23] A 1-liter, two-necked flask with stirrer, reflux condenser, and water trap is charged with 500 ml of cyclohexanone and 130 ml of benzene. Powdered *myo*-inositol, 50 g, is added and the mixture is stirred under reflux until no more water is separated. After the addition of 200 mg of *p*-toluenesulfonic acid, heating is continued for 4–5 hr. Another 200 mg of *p*-toluenesulfonic acid are added, and heating is continued until most of the inositol has dissolved. About 17–20 ml of water are separated.

The mixture is stirred at room temperature overnight, then unreacted inositol and a small amount of (I) are filtered and washed with 50 ml of benzene. About 7 g of (I) are extracted from the residue by boiling with 250 ml of ethanol containing 0.5 ml of triethylamine and crystallization at −20°.

To the first filtrate and benzene washing are added 50 ml of ethanol and 0.5 g of *p*-toluenesulfonic acid; the solution is stirred for 1 hr at room temperature. Biscyclohexylidene-*myo*-inositol is thereby converted into (I), which begins to crystallize after about 10 min; 2 ml of triethylamine are added to prevent further ethanolysis of (I), and crystallization is completed at −20° overnight. Yield, 60 g; m.p., 155°–156° or 176°–180° with change in crystal modification at 160°–165°. The material is sufficiently pure for the next step.

1,2-O-Cyclohexylidene Tetra-O-acetyl-myo-inositol (II)[24] Sixty grams of (I) are dissolved in 340 ml of pyridine and 400 ml of acetic anhydride and heated for 2 hr at 100°. After cooling to room temperature, the mixture is slowly poured into 1.5 liters of ice water. A sticky precipitate is formed that crystallizes within a few hours. The material is filtered and recrystallized from 300 ml of hot methanol. Yield, 75 g; m.p., 113°–115°.

3,4,5,6-Tetra-O-acetyl-myo-inositol (III).[24] For 3 hr, 75 g of (II) are heated in 150 ml of 80% acetic acid/water to 100°. The solution is concentrated under reduced pressure to a syrup; 100 ml of water are added, and the sample is again concentrated to a syrup. (III) is crystallized from 200 ml of methanol as hemihydrate. Yield, 40 g; m.p., 90°–91°.

3,4,5,6-Tetra-O-acetyl-myo-inositol-1,2-thionocarbonate (IV).[21] Forty

---

[23] D. E. Kiely, G. J. Abriscato, and V. Baburao, *Carbohydr. Res.* **34**, 307 (1974).
[24] S. J. Angyal, M. E. Tate, and S. D. Gero, *J. Chem. Soc. London* **1961**, 4116.

grams of (III) are dried under reduced pressure over phosphorus pentoxide for 24 hr in a large drying pistol heated to 70°.

Imidazol, 50 g, is dissolved in 320 ml of dry acetone and cooled to 0°. Thiophosgene, 14.5 ml, is added dropwise at such a rate that the temperature does not rise above 20°. After 90 min at room temperature imidazol hydrochloride is filtered and washed with 50 ml of acetone. Dried (III), 38 g is added to the combinded filtrates and heated under reflux for 3 hr (drying tube on reflux condenser). Acetone is removed under reduced pressure and the residue is dissolved in 400 ml of ethyl acetate. The solution is washed twice with 250 ml of 3 N HCl and once with 100 ml of 10% NaCHO$_3$ and concentrated to a syrup. The residue is crystallized from 350 ml of methanol. Yield, 32 g; m.p., 165°–170°.

Conduritol B Tetraacetate (V).[21] Compound (IV), 30 g, dissolved in 150 ml of trimethyl phosphite, is heated under reflux for 5 hr under nitrogen. The solution is concentrated under reduced pressure to a syrup and triturated with 100 ml of water. The sticky material crystallizes after a few hours. It is recrystallized from 150 ml of methanol/water 3:2. Yield, 18 g; m.p., 85°–89°.

Conduritol B (VI). Compound (V), 15 g, is suspended in 70 ml of methanol to which is added 1 ml of 1 M sodium methoxide. The suspension is stirred for 30 min at room temperature, and the methoxide is neutralized with 0.5 ml of 2 N acetic acid. Compound (V) sometimes dissolves completely before (VI) starts to crystallize. To complete the crystallization, 30 ml of ethanol are added and the suspension is left overnight at −20°. Yield, 6.4 g; m.p., 201°–203°.

Conduritol B Epoxide (VII).[8,21] Compound (VI), 5 g, (finely powdered) is suspended in 10 ml of water. Acetic acid (99%), 500 ml, and 20 g of p-nitroperbenzoic acid[25,26] are added and the suspension is stirred at 8°–10° overnight and then for 5 hr at room temperature. The mixture is concentrated under reduced pressure to a semisolid mass, which is stirred with 100 ml of water. p-Nitrobenzoic acid and unreacted peracid are filtered, and the filtrate is taken to a syrup. Methanol, 100 ml, is added, the preparation is again concentrated, and the residue is taken up in 100 ml of ethanol. The epoxide usually begins to crystallize during the first concentration. Crystallization is completed during 2 days at −20°. Yield, 4.2 g; m.p., 157°–159°.

Purity is checked by thin-layer chromatography on silica gel G (Merck) impregnated with sodium tetraborate. Plates are developed twice

[25] M. Vilkas, *Bull. Soc. Chim. Fr.* **1959**, 1401.
[26] The equivalent amount of commercial (85%) *m*-chloroperoxybenzoic acid (Aldrich) can also be used. The UV data for its detection are λ_{max} 280 nm, $\epsilon = 1.3 \times 10^4$ M^{-1} cm^{-1}.

in ethyl acetate/pyridine/water 20:5:2. Unreacted (VI) is revealed by spraying with 0.5% $KMnO_4$ in 5% Na_2CO_3 solution, (VI), (VII), and inositols with 30% H_2SO_4 in acetic acid and heating to 120°.

Purity can also be checked by descending paper chromatography in ethyl acetate/pyridine/water 3.6:1:1.15 (upper phase). The compounds are visualized by dipping the paper in 1% $AgNO_3$ in acetone, drying it, and spraying with 1 N NaOH in ethanol.

Traces of p-nitroperbenzoic acid can be detected by UV absorption (λ_{max} 270 nm, $\epsilon = 1.07 \cdot 10^4$ M^{-1} cm^{-1}). If found, they are removed by stirring the material with 100 ml of ethanol overnight at 5° and subsequent filtration.

Determination of Epoxide Content (Adapted from Kerkow).[27] About 15 mg of epoxide are weighed into a 10-ml vial with plastic cap and dissolved in 2 ml of 0.1 N HCl in saturated $CaCl_2$, which are weighed accurately (weighing of the $CaCl_2$ solution is necessary, because its viscosity and surface tension make accurate pipetting difficult). The vial is closed and left overnight at room temperature. Unreacted acid is titrated with 0.1 N NaOH using a 2-ml microburette. The amount of acid consumed by the epoxide is calculated using a blank of 2 ml of 0.1 N HCl/$CaCl_2$ solution. 1 ml of NaOH corresponds to 16.2 mg of epoxide.

6-Deoxy-6-bromoconduritol B Epoxide[9]

(1,2,4/3,5,6)-4,6-Dibromocyclohexanetetrol (Dibromotetrol D, VIII). *myo*-Inositol hexaacetate, 3.8 g [prepared by acetylation of *myo*-inositol in analogy to (II)] 10 ml of 30% HBr in acetic acid, and 1 ml of acetic anhydride are sealed in a 30-ml thick-walled Pyrex tube and heated to 120° for 5 hr. The mixture is poured into 100 ml of water and extracted three times with 50 ml of benzene. The combined extracts are washed with water and 10% $NaHCO_3$ solution, dried with sodium sulfate, and concentrated under reduced pressure to a viscous mass. This is dissolved in 30 ml of methanol and left to crystallize at room temperature for 45 min. The crystals (mostly tetraacetate of dibromotetrol C, m.p., 216°–222°) are filtered, and the mother liquor is left to crystallize at 4° overnight. (VIII)-tetraacetate crystallizes in coarse prisms that are recrystallized from 20 ml of methanol. Yield, 1.6 g; m.p., 130°–132°.

The tetraacetate is hydrolyzed by boiling in 1 N HCl in 50% aqueous ethanol for 6 hr. The solution is concentrated under reduced pressure to a syrup; the remaining water is removed by adding 30 ml of anhydrous ethanol and 30 ml of benzene and evaporating to dryness. The residue is

---

[27] F. W. Kerkow, *Z. Anal. Chem.* **108**, 249 (1937).

crystallized from 15 ml of ethanol plus 20 ml of benzene. Yield, 0.9 g; m.p., 214°–216°.

6-Deoxy-6-bromoconduritol B Epoxide (IX). Compound (VIII), 0.8 g, dissolved in 8 ml of water is dehydrobrominated with 2 N triethylamine in ethanol at pH 8.8, the triethylamine solution being added with a pH-stat burette. The reaction is completed after about 30 min. The solution is concentrated to 2 ml, the water is removed by repeated addition and evaporation of ethanol and benzene, and the residue is crystallized from 4 ml of ethanol to which 8 ml of chloroform are added at −10° overnight. It is recrystallized from 40 ml of hot ethyl acetate. Yield, 0.4 g; m.p., 174°–177°.

Determination of epoxide content as described above gives values corresponding to the calculated equivalent weight of 225.

Route B[6,9]

Bromination of myo-Inositol. The starting material is 2-[³H]*myo*-inositol which is usually supplied in ethanol/water solution with a specific activity of 2–5 mCi/mmole (∼10–25 mCi/mg). To bring the specific activity to about 10^7 cpm/μmole, approximately 2 mg of cold *myo*-inositol (weighed accurately) are added to 0.25 mCi of radioactive material. The solution is taken to dryness, and the residue is dissolved in about 2 g of anhydrous acetic acid. The exact specific activity is calculated from the amount of inositol (radioactive plus added inactive material) and the exact volume of acetic acid (calculated from its weight), and from the counts measured in a suitable aliquot.

Since the bromination rate is critically dependent on trace amounts of water, it is advised to carry out a few trial runs to find the optimal conditions for the formation of bromopentol B or dibromotetrol D. *myo*-Inositol, 1 mg, and 10 μl of the above solution of [³H]*myo*-inositol are added to 1 ml of bromination mixture (acetyl bromide/acetic acid/48% HBr in acetic acid 2:2:1, v/v) and heated in a sealed tube to 110° for 2, 4, and 6 hr. If dibromotetrol D is the desired product, 5% water is added before sealing.

The reaction mixture is transferred to a 20-ml pear-shaped flask and taken to dryness under reduced pressure. The residue is dissolved in 2 ml of 1 N HCl in 50% aqueous ethanol and heated under reflux for 5 hr. The mixture is again taken to dryness, and trace amounts of acid are removed in a desiccator over solid NaOH overnight. The residue is dissolved in 0.1 ml of methanol and an aliquot (30–50%) is investigated by descending paper chromatography in ethyl acetate/pyridine/water 3.6:1:1.2 (v/v, upper phase). Relative amounts of reaction products are obtained by measuring the radioactivity. R_f values of myo-inositol and its various bromination products are given in the table.

The main portion of [³H]myo-inositol is transferred to a reaction tube and taken to dryness. The residue is dissolved in 2 ml of bromination mixture and heated in the sealed tube to 110° for the time found to be optimal in a trial run. The material is worked up as before and the whole product is chromatographed on two strips of 8×56 cm Schleicher and Schüll 2043 b. Zones corresponding to bromopentol B and/or dibromotetrol D are cut out, eluted with water, and taken to dryness.

Conduritol B Epoxide. The above bromopentol B is dissolved in 2 ml of water and adjusted to pH 10.5–11 with a few microliters of triethylamine. The solution is kept at room temperature for 60 min at this pH, readjusted with triethylamine, if necessary, diluted with 10 ml of water, and concentrated to 1 ml to remove excess triethylamine. The solution is stable at 4° for several weeks. It is not advisable to concentrate to dryness since the mixture of triethyl ammonium bromide and epoxide may show a partial reversion to bromopentol.

Completeness of epoxide formation can be checked by paper chromatography as above with added inactive epoxide as control (R_f value in the table).

6-Bromo-6-deoxyconduritol B Epoxide.[9] This epoxide is prepared from dibromotetrol D by the same procedure as given for conduritol B epoxide in route B. However during dehydrobromination the pH is kept between 9.0 and 9.5. Since the epoxide and its precursor cannot be separated by paper chromatography, completeness of epoxide formation must be checked either by cocrystallization with inactive epoxide[9] or by conversion of the epoxide to bromopentol B' by heating it with 1 N H_2SO_4 for 2 hr under reflux, removal of H_2SO_4 with $BaCO_3$, and paper chromatography of the concentrated solution in the same solvent as above. Dibromotetrol D is not changed under these conditions.

PAPER CHROMATOGRAPHY OF INOSITOL AND ITS BROMO DERIVATIVES

Compound	R_f
myo-Inositol	0.04
Bromopentol A	0.20
Conduritol B epoxide	0.24
Bromopentol B	0.29
Bromopentol B'	0.35
Dibromtetrol D	0.72
Dibromotetrol C	0.81

Active-Site Labeling with Conduritol B Epoxide or Its 6-Deoxy-6-bromo Derivative

β-Glucosidases differ considerably in their reactivity with conduritol B epoxide and its 6-bromo derivative. It appears that enzymes in which the activity against normal β-glucosides is higher than against the corresponding 6-deoxyglucosides, e.g., from *Aspergillus* species, react up to 30 times faster with the former; enzymes with a higher activity against 6-deoxyglucosides (from sweet almonds and *Helix pomatia*) react up to 18 times faster with the bromo epoxide. The choice of epoxide will, therefore, depend on the type of enzyme.

As mentioned in the introduction, the epoxide oxygen has to be protonated by an acidic group at the active site in order to facilitate attack by a nucleophile of suitable orientation. The rate of inhibition is strongly pH-dependent and is usually governed by the concentration of the acidic group in its protonated form, resulting in a sigmoid pH-rate profile with maximal rate at low pH. The only exceptions are a β-glucosidase from *A. oryzae*[7] and the intestinal sucrase–isomaltase complex.[13] In these cases the pH-rate profile is determined by the concentration of a basic group, probably the attacking carboxylate ion.

The time dependence of the inhibition usually follows a first-order rate law. Epoxide concentrations necessary to achieve more than 98% inhibition in about 8 hr are in the range of 0.2 to 2 mM (40 mM with sucrase–isomaltase[13]). Deviations may occur that are due to an enzyme-catalyzed reaction of the epoxide with halide ions in the medium[6]; high concentrations of these ions should be avoided. If salt is necessary to keep the enzyme in solution, sodium perchlorate or another salt with a nonnucleophilic anion should be used.

Experimental suggestions. The pH-rate profile and concentration dependence of the rate of inhibition are determined in preliminary experiments with the cold epoxide. These should also reveal whether complete inhibition can be achieved and whether competitive inhibitors protect in accordance with their inhibition constants.

Labeling with radioactive epoxide is carried out with 0.01–1 μmole of enzyme under conditions adapted from the exploratory experiments. The amount of enzyme depends on whether stoichiometry of binding and inhibition is to be studied or whether the amino acid sequence of labeled peptides is to be determined. Loss of enzymic activity is followed in suitable aliquots. The inhibited enzyme is freed from excess epoxide by exhaustive dialysis or gel filtration over Sephadex G-25. Binding ratios are obtained from protein concentration and radioactivity measurements.

Unreacted epoxide can be recovered and used for further experiments by vacuum concentration of the solutions containing the low-molecular-weight material, provided that dialysis or gel filtration has been done against water. An aliquot should be tested by paper chromatography for decomposition of the epoxide.

Removal of Bound Epoxide and Identification of Reaction Product

In the cases investigated, the label could be removed by prolonged incubation of the labeled enzyme with 0.5 M hydroxylamine at pH 9.4 and 35°. With β-glucosidase from almonds, it was necessary to denature the enzyme with guanidinium chloride before complete cleavage with hydroxylamine was possible.[9] This stability of the label in the native, or only partially denatured, enzyme stands in marked contrast to the lability observed with labeled peptides obtained from these enzymes (see below).

The radioactive material is separated from the enzyme by dialysis against 20 volumes of deionized water. Salts are removed from the dialyzate by filtration over Dowex 50 × 8 (H+ form) and neutralization of the acid in the filtrate with Dowex 1 × 8 (OH− form). If denaturation is necessary, it is advised to remove the denaturant by dialysis before the hydroxylamine treatment. The desalted solution containing the released label is concentrated to dryness, and the reaction product is identified by descending paper chromatography in acetone/water 6:1 and n-butanol/acetic acid/water 4:1:1[28] and cocrystallization with authentic reference substances. Three reaction products can be expected from racemic conduritol B epoxide, provided the reacting group on the enzyme is a carboxylate ion: (+)-*chiro*-inositol, (−)-*chiro*-inositol, and *scyllo*-inositol.

(+)- and (−)-*chiro*-Inositol have been commercially available (from Calbiochem), but the supplier has discontinued this item and to the knowledge of the author there are no other suppliers. The racemate (±)-*chiro*-inositol can be prepared from conduritol B epoxide as described under 6-bromo-6-deoxy conduritol B epoxide (m.p. 253°–255° after crystallization from water/ethanol). The (−)-enantiomer is obtained by demethylation of its methyl ether quebrachitol[29] (available from Uniroyal Plantation Co, Baltimore, Maryland). (+)-*chiro*-Inositol is obtained in the same way from pinitol,[30] which does not, however, seem to be a commercial product.

[28] S. J. Angyal, D. J. McHugh, and P. T. Gilham, *J. Chem. Soc. London* 1957, 1432.
[29] E. P. Clark, *J. Am. Chem. Soc.* 78, 1009 (1936).
[30] L. Maquenne, *Ann. Chim. Phys.* 22, 264 (1891).

scyllo-Inositol (scyllitol) can be prepared from myo-inositol by converting it first to scyllo-inosose (myo-inosose-2)[31,32] and subsequent reduction of inosose pentaacetate with sodium borohydride.[33]

For cocrystallization, the radioactive material (about 20,000 cpm) is added to about 50 mg of reference substance dissolved in the minimum amount of water, and crystallized by the addition of ten volumes of anhydrous ethanol. The crystals are recrystallized five to ten times until the specific radioactivity of the crystals and that of the material in the mother liquor is constant or has been diluted out.

Isolation and Sequence Analysis of an Active-Site-Derived Peptide

For experimental details of peptide isolation and sequence analysis, the reader is referred to Volume XXV of this series. However, standard procedures for disulfide reduction, thiol blocking, and tryptic hydrolysis have to be slightly modified, owing to the extreme sensitivity of the bound label in the completely denatured protein or isolated peptides at pH values above 7. This also puts a limit on the conditions of ion-exchange chromatography and electrophoresis. With peptides from β-glucosidase A_3 from A. wentii, the following half-lives (at 25°) were observed.[18]: pH 7.0, 30 hr; pH 8.5, 5.4 hr; pH 9.0, 0.6 hr. However, the label is very stable under acidic conditions, e.g., pH 2.0, aqueous acetic acid up to 25%, 70% formic acid. A similar observation was made with β-glucosidases from sweet almonds[16] and sucrase–isomaltase.[19]

With the following modifications the losses of label could be kept below 5%: Reduction of disulfide bonds with dithiothreitol is performed at pH 6.7 in 0.2 M potassium phosphate in 6 M guanidinium chloride (4 hr at room temperature, 100-fold excess with respect to disulfide bonds). Cyanoethylation is carried out at pH 8 for 15 min with a 5-fold excess of acrylonitrile. After readjustment to pH 4, the protein solution is dialyzed and lyophilized.

No problems arise during cleavage with cyanogen bromide or pepsin. Tryptic hydrolysis may be performed at pH 7.3 with a trypsin concentration of 0.5 mg/ml and incubation times ranging from 0.5 to 3 hr at 25°. Hydrolysis is stopped by acidification to pH 5.0 or 10% acetic acid. The amount of label released during the longer incubation times is about 5%.

[31] T. Posternak, Biochem. Prep. 2, 57 (1952).
[32] K. Heyns and H. Paulsen, Chem. Ber. 86, 833 (1953).
[33] N. L. Stanzer and M. Kates, J. Org. Chem. 26, 912 (1961).

With β-glucosidase A_3 from *A. wentii,* difficulties arise from the tendency of the large cyanogen bromide and trypsin peptides to form insoluble aggregates. These could be overcome by carrying out the first chromatographic separation in 10% acetic acid on Sephadex G-50.

The site of attachment of conduritol B epoxide follows from the distribution of radioactivity during sequence determination by the dansyl-Edman procedure as described by Hartley.[34] Up to the amino acid residue bearing the label, negligible amounts of radioactivity are found in the butyl acetate phase containing the phenylthiazolinones from the degradation. When the amino acid bearing the label is cleaved, radioactivity is extracted into the organic phase. Owing to the hydrophilic inositol derivative, the partition is incomplete and the radioactivity tails over the next four to five degradation cycles.

Preparation of Bromoconduritol[15]

Conduritol B, 1 g, is dissolved in 5 ml of 48% hydrobromic acid and kept for 20 hr at room temperature in the dark. The solution, transferred to a flat dish, is taken to dryness in an evacuated desiccator over potassium hydroxide pellets. The brownish residue is dissolved in 10 ml of 99% ethanol and maintained at 0° for 4–5 hr so as to allow unreacted conduritol B to crystallize (about 0.25 g). To the filtrate, concentrated to about 5 ml, are added 20 ml of benzene. Bromoconduritol, 0.5–0.6 g, crystallizes after 3 days at 0° and is recrystallized from 99% ethanol. Yield, 0.45 g; m.p., 117°–119°.

Paper chromatography in ethyl acetate/pyridine/water 3.6:1:1.2 (upper phase) and thin-layer chromatography on silica gel G in ethyl acetate/acetonitrile/water 20:8:0.3 shows the presence of two compounds, A and B (detection with 5% $KMnO_4$ in 10% Na_2CO_3). The R_f values of A and B are 0.67 and 0.72 (paper chromatography) and 0.59 and 0.68 (thin-layer).

A and B are isomeric bromoconduritols as shown by elemental analysis and NMR spectra of their acetylation products. Both have the bromine substituent in an allylic position. Relative rates of dehydrobromination with sodium ethylate lead to the following tentative assignments: A: (3,5/4,6)- and B: (3,5,6/4)-6-bromo-3,4,5-trihydroxycyclohex-1-ene.

Preparative separation of A and B could not be achieved in a reproducible manner. Fractional crystallization from 99% ethanol sometimes leads to an enrichment of A, but there are indications that an isomerization A $\rightleftharpoons$ B may take place during recrystallization. Owing to the sensi-

[34] B. S. Hartley, *Biochem. J.* **119**, 805 (1970).

tivity of A and B to hydrolysis, column chromatography or preparative paper chromatography are not successful. The half-life in aqueous solution is about 30 min at pH 7.0 and decreases rapidly at pH levels above 8.

Inhibition of Glucosidases with Bromoconduritol. α-Glucosidase from yeast[15] and β-glucosidases from *Cicer arietum* and *Alocasia macrorrhiza*[11] show a time-dependent, irreversible inhibition with very low concentrations of bromoconduritol. From experiments with preparations partially enriched with respect to A or B, it was found that isomer A reacts with α-glucosidase whereas the above β-glucosidases react preferentially with B.

The contents of a reactive isomer in a given preparation can be determined from the degree of inhibition observed when a known molar excess of enzyme is allowed to react with the inhibitor. Based on the concentration of this isomer, we observed second-order kinetics for the reaction between α-glucosidase and one enantiomer of bromoconduritol A. The rate constants ranged from 8.7×10^4 M^{-1} min^{-1} at pH 6 to 2.2×10^4 M^{-1} min^{-1} at pH 5.0 and 7.4.

[41] Glycidol Phosphates and 1,2-Anhydrohexitol 6-Phosphates

By EDWARD L. O'CONNELL and IRWIN A. ROSE

The preparation of epoxide analogs was undertaken in the expectation that inactivation of an enzyme would signify that a nucleophile was positioned for a back-side attack on one of the ring carbons and that an acid group might be required for protonation of the ring oxygen. Such an antarafacial positioning of basic and acidic functional amino acids had been postulated in the aldose-ketose isomerase reactions for which the epoxide substrate analogs proved to be successful active site reagents in two cases tested: triosephosphate isomerase[1,2] and phosphoglucose isomerase.[3] In addition glycidol phosphate (glycidol-P) was effective against enolase.[1,2] In each case, a carboxyl oxygen on the enzyme was the nucleophile, leading to formation of an ester.

That the inactivators show *enantio*meric and diastereomeric specificities is seen in the table.

[1] I. A. Rose and E. L. O'Connell, *J. Biol. Chem.* **244**, 6548 (1969).
[2] K. J. Schray, E. L. O'Connell, and I. A. Rose, *J. Biol. Chem.* **248**, 2214 (1973).
[3] E. L. O'Connell and I. A. Rose, *J. Biol. Chem.* **248**, 2225 (1973).

STEREOSPECIFICITIES OF INACTIVATION

Enzyme	Compound	K_{inact} (mM)	V_{inact} (min^{-1})
Triosephosphate isomerase	R-Glycidol P	4	3.8×10^{-1}
	S-Glycidol P	5.2	3.8×10^{-2}
Enolase	R-Glycidol P	9	2.35×10^{-2}
	S-Glycidol P	15	6.70×10^{-3}
Phosphoglucose isomerase	$(2R)$-1,2-Anhydro hexitol-6-P	0.27	2.5×10^{-2}
	$(2S)$-1,2-Anhydro hexitol-6-P	0.47	5.0×10^{-4}

Preparation of Glycidol-P[1], -^{32}P, or -^{3}H Labeled

The phosphorylation reaction was carried out under anhydrous conditions. Use of N,N-dimethylaniline rather than pyridine gave a much purer and higher yield of glycidol-P by preventing acid catalyzed ring opening of the epoxide by Cl$^-$.

Reagents. These were purified as described below and stored under anhydrous conditions.

Trimethyl phosphate, dried over magnesium sulfate, filtered, and distilled at atmospheric pressure using a vacuum-jacketed distilling column; the fraction boiling at 193°–195° was collected

Dimethylaniline, dried over calcium hydride, filtered, and distilled from fresh calcium hydride at atmospheric pressure; b.p. 193°–195°

Phosphorus oxychloride, redistilled at atmospheric pressure before use; b.p. 106°–107°

Glycidol, fractionally distilled[4]

In an ice bath are mixed 0.5 ml of trimethylphosphate, 0.25 ml of N,N-dimethylaniline, and 0.1 ml of POCl$_3$ or ^{32}POCl$_3$ (1 mmole). To the stirred solution is added a mixture of 0.5 ml of trimethyl phosphate containing 0.07 ml (1 mmole) of glycidol. After 1 hr at 0°, the reaction is added dropwise to 10 ml of cold water while maintaining the pH between 5 and 9 with 2 N NaOH. After final adjustment to pH 7.0, the dimethylaniline is extracted twice with 6 ml of ether. The aqueous solution is adjusted to pH 7.5, 2 ml of 1 M barium acetate are added, and

---

[4] The glycidol is freshly distilled at 11 mm of pressure and 56° [J. C. Sowden and H. O. L. Fischer, *J. Am. Chem. Soc.* **64**, 1291 (1942)]. The tritiated form is prepared by reduction of glycidaldehyde (see following) with NaB^{3}H$_4$ [modified to a small scale from P. H. Williams, G. B. Payne, W. J. Sullivan, and P. R. Van Ess, *J. Am. Chem. Soc.* **82**, 4883 (1960)]. R,S-glycidol is available commercially, or the two epimers may be prepared as indicated next.

the resulting precipitate of barium phosphate is discarded. Ethanol (2.5 volumes) is added, and the precipitate of barium glycidol phosphate formed after 30 min in ice is collected. The precipitate is dissolved in 10 ml of water, 2.5 volumes of ethanol are added, and the mixture is kept at 30° until shiny crystals begin to form. After standing overnight at 4°, the barium glycidol-P salt is obtained by filtration (180 mg, 63% yield).

R,S-Glycidol-P was assayed by heating a $\sim$0.1 mM solution in 1 N HClO$_4$ at 100° for 20 min, neutralizing with KOH, and assaying the L-glycerol-P formed.[5] Correcting for formation of glycerol-2-P, 14% at equilibrium, only 43% of R,S-glycidol-P can be expected as L-glycerol-P.

Preparation of R,S-Glycidaldehyde for Synthesis of R,S-[³H] Glycidol-P and 1,2-Anhydro-D,L-hexitol-P's

R,S-Glycidaldehyde is available commercially (Aldrich) and should be vacuum-redistilled just before use. A simple preparation from acryl aldehyde with H$_2$O$_2$ at pH $\sim$8 has been described.[6]

Preparation of Mixed 1,2-Epoxides of Hexitol-6-P's by Condensation of Glycidaldehyde with Dihydroxyacetone-P and Muscle Aldolase[3]

Use of R,S-glycidaldehyde yields a mixture of four diastereoisomers, epimers at C-2 and C-5. R-Glycidaldehyde gives 1,2-anhydro-D-mannitol-

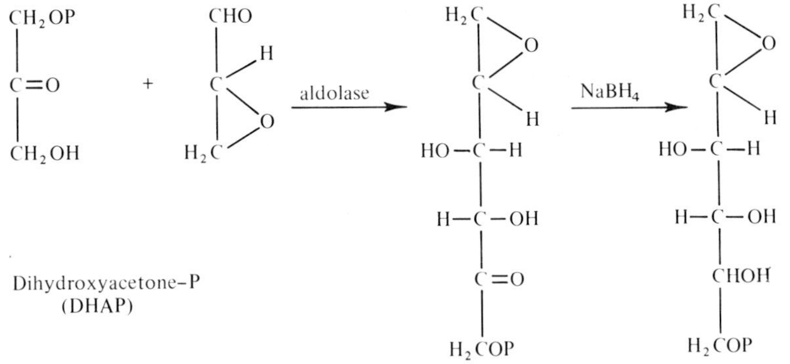

Dihydroxyacetone–P
(DHAP)

6-P and 1,2-anhydro-L-gulitol-6-P (the 2R mixture). S-Glycidaldehyde gives the D-glucitol and L-iditol-6-P's (the 2S mixture).

One hundred micromoles of DHAP, 200 μmoles of triethanolamine-

---

[5] H. U. Bergmeyer, ed., "Methods of Enzymatic Analysis." Academic Press, New York, 1965.

[6] G. B. Payne, $J. Am. Chem. Soc.$ **81**, 4901 (1959).

HCl, pH 7.5, and 2500 μmoles of glycidaldehyde are incubated at 25° in a final volume of 5.0 ml with 10 units of aldolase that is free of triosephosphate isomerase.[7] After 30 min, less than 3% of DHAP remains as assayed by glycerol-P dehydrogenase.[5] Aldolase is inactivated and precipitated by the addition of an equal volume of absolute alcohol. After 10 min at 25°, denaturated protein is removed by centrifugation. The supernatant liquid is concentrated to its original volume and chromatographed on a Sephadex G-10 column (2.5 × 92 cm) equilibrated with 0.05 M sodium acetate. The condensation product appears at 1.1 times the column void volume in approximately 20 ml. The condensation product is detected by the formation of dihydroxyacetone-P with adolase and its coupled reduction with glycerol-P dehydrogenase and DPNH. Attempts to avoid the Sephadex step and to purify the condensation product directly by barium precipitation led to an impure product.

The condensation product is adjusted to pH 9.0 and maintained at that pH during the stepwise addition of NaBH₄. If the hexitols are to be isotopically labeled, 5 μmoles of pyruvate are added as an internal standard for determination of tritium specific activity. An appropriate amount of carrier-free NaBT₄ is added and, after 10 min at 20°, is followed by 100 μmoles of NaBH₄. The loss of condensation product was followed by the coupled assay system described above. The reaction is terminated by dropping the pH to 4.0 with glacial acetic acid to destroy excess NaBH₄. After pH readjustment to 7.5, the volume is reduced to 4.0 ml in a rotary flash evaporator at 25° or less, and 1 ml of 1 M barium acetate and 16 ml of absolute ethanol are added. A precipitate forms slowly in an ice bath and is collected by centrifugation. The supernatant liquid is used for determination of specific activity of the lactate. The precipitate is dissolved in 2 ml of water; insoluble material is removed, and the precipitate is reformed by the addition of an equal volume of absolute ethanol. The precipitate is dried by washing with ethanol and ether and is weighed (34 mg; 75% yield containing 75 μmoles of organic phosphate). The specific activity that is used for further experiments is based on phosphorus content. The product is stored over desiccant at −70° to prevent decomposition.

Preparation of L-Glycidaldehyde[2] (for Synthesis of L-Glycidol-P and 2S Epoxihexitol-6-P's)

D-Mannitol to 1,6-di-O-methane sulfonyl-D-mannitol, (1) → (3) consists of three steps: (i) mannitol to the 1,2:3,4:5,6-tri-O-isopropylidine,[8]

---

[7] C. Richards and W. J. Rutter, J. Biol. Chem. 236, 3185 (1961).
[8] L. J. Wiggins, J. Chem. Soc. London 1946, 13.

(1)   (2)   (3)   (4)   (5)

(ii) partial hydrolysis to 3:4-isopropylidine mannitol,[9] and (iii) mesylation and hydrolysis to give (3).[10,11]

(i) Dry powdered mannitol (1), 160 g, is suspended in 2 liters of dry acetone with magnetic stirring. Concentrated H_2SO_4, 16 ml, is added, and the suspension is stirred for 48 hr; all the solid dissolves and the solution turns yellow. The stirred solution is neutralized by adding solid Na_2CO_3 until a sample is neutral to moist pH paper. After filtration, the residue is washed with acetone, and the combined filtrate plus wash is concentrated under reduced pressure with a flash evaporator. The solid residue is dissolved in acetone and dried twice to remove residual water. The solid is dissolved in absolute ethanol and redried twice to remove acetone. The solid is dissolved in 500 ml of 100% ethanol, filtered, and added slowly with stirring to 3 liters of water. The mixture is allowed to remain in the cold overnight, filtered, washed with H_2O until free of alcohol, and air dried. Yield, 140 g (53%); m.p. 69°.

(ii) The triacetone, 120 g, is selectively hydrolyzed according to Stern and Wasserman[9] to the 3,4-acetonyl mannitol in 72% yield.

[9] R. Stern and H. H. Wasserman, J. Org. Chem. 24, 1689 (1959).
[10] L. Vargha, J. Kuszman, and B. Dumbovich, Chem. Abstr. 55, 15365b (1959).
[11] L. Vargha and J. Kuszman, Naturwissenschaften 46, 84 (1959).

(*iii*) To 55 g of 3,4-acetone-D-mannitol (0.25 mole) (*2*), dissolved in 300 ml of anhydrous pyridine, is added dropwise with stirring at −10°, 57.5 g (0.51 mole) of methanesulfonyl chloride over a period of 60 min. The mixture is stirred at room temperature for 4 hr and then poured into 500 ml of ice water. After three extractions with 400 ml of chloroform each, the combined chloroform extracts are washed with 400 ml of ice-cold H_2O, twice with 400 ml each of ice-cold 5 N H_2SO_4, twice with 400 ml of ice-cold saturated $NaHCO_3$, and twice with 400 ml of ice-cold H_2O. The chloroform is dried over $MgSO_4$ and evaporated under reduced pressure to yield a syrup. The syrup is dissolved in 100 ml of anhydrous dioxane, and 8.9 ml of 50% aqueous methanesulfonic acid are added. The reaction is maintained at room temperature for 24 hr, after which it is heated to 60° for 2 hr, cooled, and allowed to crystallize. The product(s) are recrystallized from warm absolute ethanol and have yielded 16.8 g (18%) of 1,6-di-*O*-methanesulfonyl-D-mannitol, m.p. 129°–131°.

(*3*) → (*5*) *Epoxidation of (3)*[12]

Compound (*3*), 6.8 g (20 mmoles), is suspended in 5 ml of H_2O with stirring, warmed to 37°, and carefully titrated to the phenolphthalein end point with 5 N NaOH until no additional alkali is consumed (40 mmoles NaOH are used). This solution is added dropwise to a stirred mixture of 150 ml of ethyl acetate and 30 g of anhydrous Na_2CO_3 and more Na_2CO_3 is added until the solid returns to an easily suspended solid. The residue is washed on the filter with ethyl acetate, and the filtrate is concentrated under reduced pressure to about 25 ml at a bath temperature of 30°. This material is redried with $MgSO_4$ and concentrated to a volume of 5 ml. A gummy solid is obtained upon cooling in an ice bath. The solid is washed with a small amount of ethyl acetate, and 1 g is dissolved in 4 ml of warm ethyl acetate. The insoluble material is removed by centrifugation and the product is allowed to crystallize and is dried under reduced pressure. The 0.7 g (4.8 mmoles) of slightly gummy solid (m.p. 59°–63°) is dissolved in 4 ml of water. Methyl red is added as indicator and the solution is adjusted with NaOH to the acid side of the indicator end point. Add 5 mmoles $NaIO_4$ (1.07 g) in small portions, adjusting the pH with NaOH to the indicator end point until no additional periodate is consumed (by starch potassium iodide paper indicator). The pH of the mixture is readjusted to the indicator end point. The solution is transferred to a small stopcock-sealable distillation apparatus, frozen in a Dry Ice–acetone bath, and evacuated under high vacuum; the distillate is

[12] M. Jarman and W. C. J. Ross, *Carbohydr. Res.* 9, 139 (1969).

collected in a receiver kept in a Dry Ice–acetone bath while the reaction solution is allowed to warm to 25°. The condensate contained 6 mmoles (62%) of glycidaldehyde as determined by the thiosulfate titration method of Ross.[13] The glycidaldehyde distillate is neutralized and redistilled. The distillate is stored at −70°.

Preparation of D-Glycidaldehyde[2] (for Synthesis of D-Glycidol-P and 2R Epoxihexitol-6-P's)

The procedure differs from that for L-glycidaldehyde because L-mannitol is not available commercially. Instead, 1,6-di-O-methanesulfonyl-L-mannitol is prepared from quebrachitol (6) by the following steps:

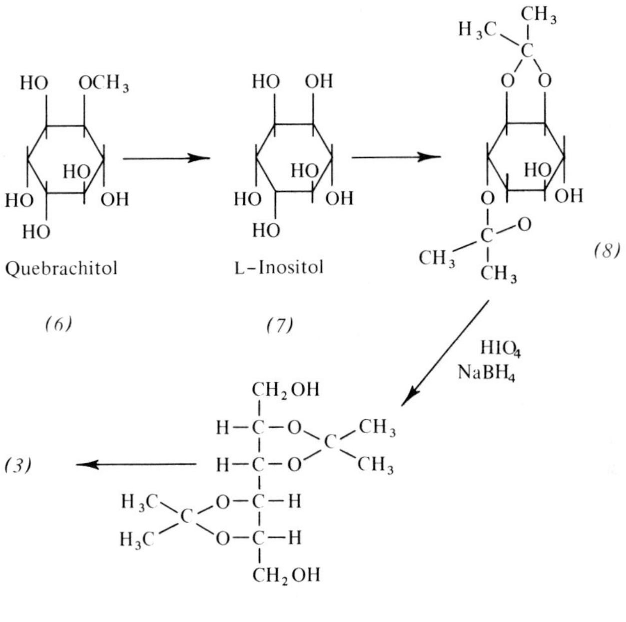

A crude sample of quebrachitol (Uniroyal Rubber Company, Plantation Division) was purified as reported.[2] The compound, 161 g, was converted to 76 g of (9) following the procedure of Gillett and Ballou.[14] The method for preparation of 1,6-di-O-methanesulfonyl-L-mannitol in 43% yield is given by Schray et al.[2]

[13] W. C. J. Ross, J. Chem. Soc. London 1950, 2257.
[14] J. W. Gillett and C. E. Ballou, Biochemistry 2, 547 (1963).

Acknowledgments

This work was supported by U.S. Public Health Service Grants GM-20940, CA-06927, and RR-05539 and also by an appropriation from the Commonwealth of Pennsylvania.

[42] Ribulosebisphosphate Carboxylase[1]

By I. LUCILE NORTON and FRED C. HARTMAN

Ribulosebisphosphate carboxylase[2-4] (EC 4.1.1.39), synonymous with fraction I protein,[5] is responsible for fixation of atmospheric CO_2 by plants and photosynthetic microorganisms.[6] The reaction catalyzed by this component of the reductive pentosephosphate cycle is shown in Reaction (1).

$$ \text{(1)} $$

D–Ribulose 1,5–bisphosphate   D–3–Phosphoglycerate

The divalent cation requirement can be served by Mn^{2+}, Co^{2+}, Fe^{2+}, or Ni^{2+} as well as Mg^{2+}.[2,7] Carbon dioxide, not bicarbonate, is the carbon species active in carboxylation.[8] Ribulosebisphosphate carboxylase also catalyzes the oxygenation of ribulosebisphosphate to yield phosphoglycolate and phosphoglycerate.[9-14] Thus, a single enzyme (termed ribu-

[1] Research from the authors' laboratory was sponsored by the Energy Research and Development Administration under contract with the Union Carbide Corporation.

[2] A. Weissbach, B. L. Horecker, and J. Hurwitz, J. Biol. Chem. 218, 795 (1956).

[3] W. B. Jakoby, D. O. Brummond, and S. Ochoa, J. Biol. Chem. 218, 811 (1956).

[4] E. Racker, Arch. Biochem. Biophys. 69, 300 (1957).

[5] S. G. Wildman and J. Bonner, Arch. Biochem. Biophys. 14, 381 (1947).

[6] M. Calvin, Science 135, 879 (1962).

[7] J. M. Paulsen and M. D. Lane, Biochemistry 5, 2350 (1966).

[8] T. G. Cooper, D. Filmer, M. Wishnick, and M. D. Lane, J. Biol. Chem. 244, 1081 (1969).

[9] G. Bowes, W. L. Ogren, and R. H. Hageman, Biochem. Biophys. Res. Commun. 45, 716 (1971).

[10] T. J. Andrews, G. H. Lorimer, and N. E. Tolbert, Biochemistry 12, 11 (1973).

[11] G. H. Lorimer, T. J. Andrews, and N. E. Tolbert, Biochemistry 12, 18 (1973).

[12] T. Takabe and T. Akazawa, Biochem. Biophys. Res. Commun. 53, 1173 (1973).

[13] F. J. Ryan and N. E. Tolbert, J. Biol. Chem. 250, 4229 (1975).

[14] F. J. Ryan and N. E. Tolbert, J. Biol. Chem. 250, 4234 (1975).

losebisphoshate carboxylase/oxygenase by Tolbert[13]) can account for the photosynthetic assimilation of CO_2 and photorespiration.

General Considerations

In view of the crucial metabolic role played by ribulosebisphosphate carboxylase and the ease with which this enzyme can be isolated (it constitutes about 16% of the total protein in spinach leaf[7]), it is not surprising that a multitude of studies have been reported concerning its isolation, chemical and physical characterization, mechanism of action, and mode of regulation. Excellent reviews[15-18] on these topics are available, and no attempt to duplicate or summarize them will be made here. However, properties of the enzyme that are pertinent to affinity labeling studies will be described briefly.

The most throughly studied ribulosebisphosphate carboxylase is from spinach, but carboxylases from many other species—including tobacco;[19] *Chromatium* strain D,[20] a purple sulfur bacterium; *Chlorella ellipsoidea*,[18] a green alga; and *Rhodospirillum rubrum*,[21,22] a purple nonsulfur bacterium—are sufficiently well characterized to be of use in comparative work that can be helpful in ascertaining the essentiality of a given residue.

Although information as to the identity of groups at the active site is limited, a consideration of structural requirements for binding of substrates to the carboxylase and a consideration of the reaction pathway (see footnote 15 and references therein for a detailed description) give some clues about the nature of active-site residues. As with most enzymes whose substrates are sugar phosphates, the anionic phosphates are crucial to binding, presumably interacting electrostatically with either protonated lysyl or arginyl residues.[7,23] Lobb et al.[24] recently pointed out that

[15] M. I. Siegel, M. Wishnick, and M. D. Lane, in "The Enzymes" (P. D. Boyer, ed.), 3rd ed., Vol. VI, p. 169. Academic Press, New York, 1972.

[16] B. A. McFadden, *Bacteriol. Rev.* **37**, 289 (1973).

[17] B. B. Buchanan and P. Schürmann, *Curr. Top. Cell. Regul.* **7**, 1 (1973).

[18] T. Akazawa, in "Protein Structure and Function" (M. Funatsu, K. Hiromi, K. Imahori, T. Murachi, and K. Narita, eds.), Vol. 2, p. 203. Kodansha Ltd., Tokyo, and Wiley, New York, 1972.

[19] P. H. Chan, K. Sakano, S. Singh, and S. G. Wildman, *Science* **176**, 1145 (1972).

[20] T. Akazawa, H. Kondo, T. Shimazue, M. Nishimura, and T. Sugiyama, *Biochemistry* **11**, 1298 (1972).

[21] F. R. Tabita and B. A. McFadden, *J. Biol. Chem.* **249**, 3453 (1974).

[22] F. R. Tabita and B. A. McFadden, *J. Biol. Chem.* **249**, 3459 (1974).

[23] G. Bowes and W. L. Ogren, *J. Biol. Chem.* **247**, 2171 (1972).

[24] R. R. Lobb, A. M. Stokes, H. A. O. Hill, and J. F. Riordan, *FEBS Lett.* **54**, 70 (1975).

arginyl residues are more frequently involved in binding of phosphate esters than had been realized. Therefore, an inspection of the reaction of arginyl-specific diketones with ribulosebisphosphate carboxylase might be fruitful. Subsequent to binding, the initial catalytic step is thought to be isomerization of ribulosebisphosphate to the corresponding C2–C3 enediol (or 3-keto compound), thereby developing a nucleophilic center (carbanion) at C2 to which either carbon dioxide or oxygen could add. Lorimer and Andrews[25] proposed that the carbanion-like transition state dictates a reactivity with molecular oxygen. The assumption that the existence of such a reactive transition state is a property of all ribulose-bisphosphate carboxylases is consistent with the lack of correlation between oxygenase activity and the sensitivity of the organism to oxygen (i.e., ribulosebisphosphate carboxylases from both faculative aerobes[26,27] and obligate anaerobes[12] possess oxygenase activity). The isomerization involves loss of the C3 hydrogen as a proton, and it is reasonable to predict, based on comparisons with other enzymes that catalyze similar isomerizations,[28] the presence at the active site of a basic group that abstracts the C3 hydrogen. The nucleophilicity of C2 or the electrophilicity of the CO_2 could be enhanced by an appropriately positioned acid–base group in the active site. Based on proton relaxation rate data and ^{13}C nuclear magnetic resonance (NMR) data with ^{13}C-labeled carbon dioxide, Miziorko and Mildvan[29] concluded that the divalent metal ion is probably directly coordinated to the C3 hydroxyl (which would facilitate deprotonation at C3), but that the distance between metal and carbon dioxide is great enough to indicate that a water molecule or a functional group of the protein is interposed between them.

Strong support for the existence of the postulated six-carbon carboxylated intermediate was the finding that a structural analog of the intermediate, 2-carboxy-D-ribitol 1,5-P_2, is a potent inhibitor of the enzyme.[30] Furthermore, the proposed intermediate, 2-carboxy-3-ketoribitol 1,5-P_2, has been synthesized and is a substrate for the carboxylase.[31] An interesting observation is that 2-carboxy-3-ketoribitol 1,5-P_2 also hydrolyzes spontaneously (in the presence of platinum oxide) to yield 3-phosphoglycerate. In contrast to the enzyme-catalyzed carboxylation of ribulose-

[25] G. H. Lorimer and T. J. Andrews, *Nature (London)* 243, 359 (1973).

[26] B. A. McFadden, *Biochem. Biophys. Res. Commun.* 60, 312 (1974).

[27] F. J. Ryan, S. O. Jolly, and N. E. Tolbert, *Biochem. Biophys. Res. Commun.* 59, 1233 (1974).

[28] E. A. Noltmann, *in* "The Enzymes" (P. D. Boyer, ed.), 3rd ed., Vol. VI, p. 271. Academic Press, New York, 1972.

[29] H. M. Miziorko and A. S. Mildvan, *J. Biol. Chem.* 249, 2743 (1974).

[30] M. I. Siegel and M. D. Lane, *Biochem. Biophys. Res. Commun.* 48, 508 (1972).

[31] M. I. Siegel and M. D. Lane, *J. Biol. Chem.* 248, 5486 (1973).

bisphosphate, which gives only D-3-phosphoglycerate, the phosphoglycerate derived from C1, C2, and the carboxyl group of the intermediate during spontaneous hydrolysis is of L-configuration. Thus, the enzyme directs the stereochemistry of the hydrolytic cleavage.[31]

No definitive identifications are available of residues in ribulosebisphosphate carboxylase involved intimately in binding or catalysis. There are a number of reports[23,32–36] concerning the essentiality of sulfhydryl groups, but their potential role in maintenance of tertiary and quaternary structure and their potential role in catalysis have not been distinguished. Based on inactivation studies with iodoacetamide and 5,5′-dithiobis-(2-nitrobenzoic acid), Trown and Rabin[32,33] proposed that a catalytically essential —SH condensed with the carbonyl of ribulosebisphosphate to form a thiohemiacetal. Although this remains a viable possibility, it is unproved. Akazawa's[36–38] group has thoroughly studied the reaction of the carboxylase with p-chloromercuribenzoate. At slightly alkaline pH, disruption of quaternary structure results, so there is no doubt that modification of sulfhydryls can lead to conformational changes.[37] In their most recent paper, this group concluded that —SH groups of both spinach and *Chromatium* carboxylases are important in maintaining the structural integrity of the enzyme and that the possibility exists of a —SH group at the active site.[38]

Cyanide combines reversibly with the carboxylase in the presence of ribulosebisphosphate to form an inactive ternary complex.[39] Although the exact chemical nature of the cyanide adduct is unknown, it is unlikely to have arisen by condensation with a Schiff base, because attempts to detect a Schiff base (which conceivably could be formed by the reaction of a protein amino group with the substrate carbonyl) by borohydride reduction were negative. An enzyme–CO_2 complex can be stabilized by treatment with diazomethane.[40] Lorimer *et al.*[41] suggested recently that this might be a consequence of esterification of a carbamate formed by

[32] B. R. Rabin and P. W. Trown, *Proc. Natl. Acad. Sci. U.S.A.* **51**, 497 (1964).

[33] P. W. Trown and B. R. Rabin, *Proc. Natl. Acad. Sci. U.S.A.* **52**, 88 (1964).

[34] J. H. Argyroudi-Akoyunoglou and G. Akoyunoglou, *Nature (London)* **213**, 287 (1967).

[35] T. Sugiyama, T. Akazawa, N. Nakayama, and Y. Tanaka, *Arch. Biochem. Biophys.* **125**, 107 (1968).

[36] T. Akazawa, T. Sugiyama, N. Nakayama, and T. Oda, *Arch. Biochem. Biophys.* **128**, 646 (1968).

[37] M. Nishimura and T. Akazawa, *J. Biochem.* **76**, 169 (1974).

[38] T. Takabe and T. Akazawa, *Arch. Biochem. Biophys.* **169**, 686 (1975).

[39] M. Wishnick and M. D. Lane, *J. Biol. Chem.* **244**, 55 (1969).

[40] G. Akoyunoglou, J. H. Argyroudi-Akoyunoglou, and H. Methenitou, *Biochim. Biophys. Acta* **132**, 481 (1967).

[41] G. H. Lorimer, M. R. Badger, and T. J. Andrews, *Biochemistry* **15**, 529 (1976).

the reaction of CO_2 with an amino group at a regulatory site. They showed that, in addition to serving as a substrate, carbon dioxide in the presence of magnesium activates ribulosebisphosphate carboxylase to give a species with a low K_m for carbon dioxide that is predicted from the observed *in vivo* rates of photosynthesis. The pH-dependence of the activation process is consistent with carbon dioxide combining with an unprotected amino group.

3-Bromo-1,4-dihydroxy-2-butanone 1,4-bisphosphate (Br-butanone-P_2), a reactive structural analog of ribulose 1,5-bisphosphate, inactivates the carboxylase by the preferential alkylation of two lysyl residues.[42] Although functionality of these residues in binding or catalysis is not proved by these studies, all available data are consistent with their presence at the binding site for ribulosebisphosphate. The synthesis of this reagent and its use as an affinity label for the spinach carboxylase are described.

Synthesis of Br-Butanone-P_2[43]

cis-1,4-Di-O-benzoyl-2-butene-1,4-diol. To a solution of 500 ml of $CHCl_3$ containing 44 g (0.5 mole) of *cis*-2-butene-1,4-diol (Aldrich Chemical Co.) and 121 ml (1.5 moles) of pyridine that has been cooled to $-5°$ in an ice–salt bath are added 140 ml (1.2 moles) of benzoyl chloride. This mixture is left in the ice–salt bath for 2 hr, then transferred to room temperature for 3 hr, at which time 20 ml of water are added. After remaining overnight at room temperature, the solution is washed in succession with two 500-ml portions of 1 N H_2SO_4, saturated sodium bicarbonate, and water. The chloroform layer is dried over anhydrous sodium sulfate and concentrated to dryness at 40° on a rotary evaporator. Crystallization of the residue from 200 ml of ethyl alcohol results in a 90% yield of the dibenzoate, m.p. 64°–65°.

1,4-Di-O-benzoyl-3-bromo-1,2,4-butanetriol. The above dibenzoate (100 g, 0.34 mole) is dissolved in dioxane (300 ml), and to this solution are added 71.2 g (0.4 mole) of N-bromosuccinimide and 50 ml of water. The resulting mixture is stirred until homogeneous (about 5 min). Within 1.5 hr the temperature of the reaction mixture rises from 24° to 33°. After the temperature decreases to 30° (total reaction time of 3.5 hr), the dioxane is removed by concentration.The residual liquid is mixed with 200 ml of $CHCl_3$, and this solution is washed, dried, and concentrated as described above. The product is crystallized from 100 ml of isopropyl ether to give 56 g (42%) of material, m.p. 97°–99°.

[42] I. L. Norton, M. H. Welch, and F. C. Hartman, *J. Biol. Chem.* **250**, 8062 (1975).
[43] F. C. Hartman, *J. Org. Chem.* **90**, 2638 (1975).

1,4-Di-O-benzoyl-3-bromo-2-butanone. A solution containing 500 ml of ether, 20 ml (0.28 mole) of dimethyl sulfoxide, 3 ml (0.037 mole) of pyridine, 50 g (0.24 mole) of dicyclohexylcarbodiimide, and 48 g (0.122 mole) of the dibenzoylbromobutanetriol is cooled to 8°. The oxidation (a method of Pfitzner and Moffatt[44]) is initiated by the addition of 3 ml (0.04 mole) of trifluoroacetic acid. The temperature of the reaction mixture rises rapidly to 33°. Fifteen minutes after initiation, the reaction is terminated by the addition of powdered oxalic acid (15 g, 0.12 mole). The insoluble dicyclohexylurea is removed by suction filtration, and the filtrate is washed, dried, and concentrated as described above. Crystals (30 g, 83%) with m.p. 83°–84° are obtained from 200 ml of isopropyl alcohol.

3-Bromo-2-butanone-1,4-diol Diethyl Ketal. A solution of the above ketone (5 g), ethanol (16 ml), freshly distilled triethyl orthoformate (33 ml), and concentrated sulfuric acid (1.4 ml) is incubated in the dark at room temperature for 7 days. At this time, thin-layer chromatography (MN-polygram Sil N-HR containing a fluorescent indicator, Brinkmann Instruments, Inc.) with ether–petroleum ether (1:1 v/v) as solvent should show an approximately 80% conversion of the ketone ($R_f = 0.40$) to the ketal ($R_f = 0.56$) (both compounds visualized under ultraviolet light). The reaction mixture is neutralized with 15 g of sodium bicarbonate. After the addition of 100 ml of ether, the mixture is filtered through Celite and concentrated to dryness at 60°. The residual liquid is dissolved in 150 ml of methyl alcohol, and to the solution is added 35 ml of 1 N NaOH. After 1 hr the methyl alcohol is removed by concentration, and the resulting aqueous mixture is extracted twice with 50-ml portions of ether. The extracts are dried and concentrated at 30° to yield 3.0 g of a slightly yellow, thin syrup. Thin-layer chromatography (same solvent) shows this material to contain a fluorescent substance ($R_f = 0.61$), tentatively identified as methyl benzoate, and two 2,4-dinitrophenyl-hydrazine-positive[45] (after heating the sprayed sheet at 100° for 5 min) components, a major one with $R_f = 0.24$ and a minor one with $R_f = 0.47$. The mixture is dissolved in 5 ml of cyclohexane and fractionated on a 2.5×23 cm column of Florisil (Floridin Co., Tallahassee, Florida) packed in cyclohexane. The column is washed in succession with 300 ml of cyclohexane (which elutes the methyl benzoate), 500 ml of cyclohexane–benzene (1:1) (which elutes the material with $R_f = 0.47$), and 225 ml of benzene–ether (1:1) (which elutes the material with $R_f = 0.24$ that is assumed to be the 3-bromo-2-butanone-1,4-diol diethyl ketal).

[44] K. E. Pfitzner and J. G. Moffatt, *J. Am. Chem. Soc.* **87**, 5661 (1965).
[45] R. A. Gray, *Science* **115**, 129 (1952).

Concentrations of the benzene–ether washings at 30° should yield 1.6 g (49% based on 5 g of the crystalline ketone) of chromatographically pure material as a colorless, slightly viscous liquid.

Br-Butanone-P₂ Diethyl Ketal. To an ice-cold solution of 3-bromo-2-butanone-1,4-diol diethyl ketal (1.5 g, 5.9 mmoles) in a mixture of pyridine (5 ml) and $CHCl_3$ (10 ml) is added ³²P-labeled diphenyl chlorophosphate (3.8 ml, 18 mmoles) (Amersham/Searle Corp.). The reaction mixture is maintained overnight at 4°, and then a few chips of ice are added. After an additional 12 hr at 4°, more $CHCl_3$ (100 ml) is added to the mixture, which is washed (with 1 N H_2SO_4 and saturated $NaHCO_3$), dried, and concentrated. The resulting viscous syrup is dissolved in 80 ml of ethyl alcohol; the solution is filtered through Celite and hydrogenated in the presence of platinum black (0.5 g) at 50 psi with a Parr apparatus. Consumption of hydrogen is completed within 3 hr, at which time the catalyst is removed by filtration through Celite. The filtrate is adjusted to pH 8.0 with cyclohexylamine and concentrated to dryness. The residue is slurried in 100 ml of acetone, and the insoluble triscyclohexylammonium salt of Br-butanone-P₂ diethyl ketal (3.4 g, 71%) is collected by filtration. Paper chromatography (descending method with Whatman No. 1 paper and a solvent composed of *n*-butyl alcohol–glacial acetic acid–water 7:2:5) revealed a single organic phosphate ester ($R_f = 0.53$) and a slight contamination with P_i ($R_f = 0.36$).[46] The salt is recrystallized by dissolving it in 4 ml of 20% (v/v) aqueous cyclohexlamine followed by the addition to this solution of 300 ml of isopropyl alcohol. Crystallization occurs during 24 hr at room temperature to yield 2.6 g (3.8×10^5 cpm/μmole) of the triscyclohexylammonium salt. Based on a molecular weight of 714, phosphate assays[47] revealed 1.91 molar equivalents of organic phosphate and 0.028 molar equivalents of P_i.

Br-Butanone-P₂. An aqueous solution (10 ml) containing 179 mg (0.025 M) of the triscyclohexylammonium salt of the diethyl ketal is swirled with 2 g of Dowex 50 (H⁺) to remove cyclohexylammonium ions and then filtered. The resulting acidic solution is incubated at 40° for 3 hr. At this time paper chromatography (same system as above) followed by visualization with the spray for phosphate esters[46] revealed an essentially complete conversion of the ketal ($R_f = 0.53$) to the ketone ($R_f = 0.23$). The latter component gave an immediate positive response with a silver nitrate dip,[48] whereas the ketal was not detected. Solutions of

[46] C. S. Hanes and F. A. Isherwood, *Nature (London)* **164**, 1107 (1949).
[47] B. B. Marsh, *Biochim. Biophys. Acta* **32**, 357 (1959).
[48] W. E. Trevelyan, D. P. Proctor, and J. S. Harrison, *Nature (London)* **166**, 444 (1950).

Br-butanone-P_2 (the free-acid form) may be stored at $-20°$ without appreciable decomposition during several months.

Labeling of Essential Lysyl Residues in Spinach Ribulosebisphosphate Carboxylase[42,49]

Ribulosebisphosphate carboxylase (isolated by a published procedure[50]) (250 mg, 3.57 μmoles of protomeric unit[51,52]) in 50 ml of metal-free 0.1 M Bicine/60 mM potassium bicarbonate/0.1 mM EDTA (pH 8.0) was treated at 25° with four successive 0.25-ml additions, at 20-min intervals, of 20 mM Br-butanone-$^{32}P_2$. Twenty minutes after the fourth addition, less than 10% of the initial activity (as determined by the method of Racker[53]) remained, and the reaction was terminated by the addition of 2-mercaptoethanol (10 mM). A duplicate enzyme solution containing ribulosebisphosphate (1 mM) was treated with reagent in an identical manner; 95% of the initial enzymic activity was retained. A third enzyme solution under the same conditions but lacking both the reagent and substrate served as control. The three protein solutions were dialyzed against 0.1 M sodium chloride at 4°; after dialysis the protein samples were made 0.1 M in sodium bicarbonate and 0.01 M in sodium [^{3}H]borohydride in order to reduce the carbonyl of the protein-bound reagent. The mixtures were maintianed in an ice bath for 30 min and then dialyzed exhaustively at 4° against 50 mM sodium chloride. The samples then were assayed for protein concentration, radioactivity (^{3}H and ^{32}P), and sulfhydryl content (see the table).

The reduced, dialyzed carboxylase was carboxymethylated with sodium iodoacetate in the presence of guanidinium hydrochloride by standard procedures and then dialyzed exhaustively. Aliquots of the dialyzed samples, containing about 2 mg of protein, were hydrolyzed for amino acid analyses, and the remaining samples were digested with trypsin.

To determine the specific radioactivity of sodium [^{3}H]borohydride, glutathione (30 μmoles) and chloroacetol phosphate[54] (20 μmoles) were incubated at room temperature for 30 min in 0.2 M sodium bicarbonate.

[49] F. C. Hartman, M. H. Welch, and I. L. Norton, *Proc. Natl. Acad. Sci. U.S.A.* **70**, 3721 (1973).

[50] See this series, Vol. 23 [53].

[51] A. C. Rutner, *Biochem. Biophys. Res. Commun.* **39**, 923 (1970).

[52] T. Sugiyama and T. Akazawa, *Biochemistry* **9**, 4499 (1970).

[53] E. Racker, *in* "Methods of Enzymatic Analysis" (H. U. Bergmeyer, ed.), p. 188. Academic Press, New York, 1963.

[54] See this series, Vol. 25 [59].

EXTENT OF INCORPORATION OF BR-BUTANONE-P$_2$ INTO RIBULOSEBISPHOSPHATE
CARBOXYLASE AND NUMBER OF SULFHYDRYL GROUPS MODIFIED

Sample	Enzymic activity (% remaining)	$^{32}P^a$ (moles reagent/ mole enzyme)	^3H (moles reagent/ mole enzyme[b])	Number of residues of car- boxy- methyl cysteine[c]	Number of sulfhydryl groups modified
Inactivated	1	5.2	13.8	90	8
Substrate-protected	95	3.0	12.2	87	11
Control	100	—	—	98	—

[a] Based on the specific radioactivity of the reagent, which is a bisphosphate.
[b] A correction has been made for the radioactivity found in the control sample.
[c] From amino acid analyses.

(Chloroacetol phosphate alkylates the sulfhydryl group.[55]) The reaction mixture was cooled in an ice bath and then treated with a portion of the same stock solution of sodium [^3H]borohydride that was used in the reduction of the modified carboxylase (see above). The final borohydride concentration in the reaction mixture was 0.1 M. After 1 hr the mixture was acidified to pH 2.0 with 1 N HCl and chromatographed on a column (1.2 × 22 cm) of Dowex 50 (H$^+$) equilibrated and eluted with 50 mM HCl. The concentration of the glutathione derivative in the peak fraction was determined with the amino acid analyzer[55]; the radioactivity in the same fraction was also determined, thereby providing the specific radio-activity of the [^3H]borohydride.

Analysis of acid hydrolyzates of the substrate-protected carboxylase and comparison of the elution position of tritium from the amino acid analyzer with standards[43] prepared from glutathione, whose —SH group had been modified with Br-butanone-P$_2$, showed that under protective conditions only —SH groups of the carboxylase were modified.[42] In contrast, identical experiments with the inactivated carboxylase coupled with characterization of tryptic peptides showed that lysyl residues, in addition to cysteinyl residues, were modified. Thus, inactivation is a consequence of modification of lysyl residues.

With the knowledge that under protective conditions only —SH groups are modified, one can begin to interpret the incorporation data shown in the table. In the substrate-protected sample the moles of the reagent incorporated per mole of enzyme (560,000 daltons) based on tritium agree reasonably well with the number of —SH groups modified. In the absence of ribulosebisphosphate, fewer —SH groups are modified,

[55] F. C. Hartman, *Biochemistry* 9, 1776 (1970).

so that the difference between tritium incorporation in the inactivated and protected samples is not a true reflection of the number of residues protected by ribulosebisphosphate from modification. However, in the inactivated sample the difference between the incorporation based on tritium (13.8 moles of reagent per mole of enzyme) and the number of —SH groups modified (8) should represent the number of lysyl residues (5.8) that react with Br-butanone-P_2 during inactivation. The extent of incorporation based on ^{32}P is low because of the instability of the phosphate group α to the carbonyl.[43]

The peptides containing the modified, essential lysyl residues were purified from a tryptic digest. The profiles of radioactive peptides in the digests of inactive and protected carboxylase are very similar with the exception of one major peak from the inactivated enzyme that is virtually lacking in the protected carboxylase. Based on radioactive compounds found in total acid hydrolyzates, the only peptides that contain modified lysyl residues are represented by the peak unique to the digest of inactivated carboxylase. Therefore, this material was further purified by chromatographic procedures to yield two distinct peptides, each containing a modified lysyl residue, arising from different regions of the primary sequence.[42] The identity of the derivatized amino acid residues as lysyl was based on their conversion, in acid hydrolyzates, to free lysine upon periodate oxidation.[42]

Comments

A variety of data indicate that the lysyl residues of ribulosebisphosphate carboxylase that are modified by Br-butanone-P_2 occupy positions within the binding domain for ribulosebisphosphate. The inactivation is subject to substrate protection by ribulosebisphosphate. Another substrate, carbon dioxide, whose binding apparently induces conformational changes,[41] increases the rate of inactivation of the enzyme.[49] The lysyl residues that are sites of reaction are unusually reactive compared with model systems in which the alkylation of —SH groups by Br-butanone-P_2 proceeds more than 100 times faster than the alkylation of amino groups.[43] The same lysyl residues in carboxylase, which has lost its enzymic activity during storage, are unreactive toward Br-butanone-P_2.[49] Both lysyl residues that are preferentially modified by the reagent are located in the large subunit of the carboxylase,[42] as is the active site.[56] None of these data specify the functionality or lack thereof of the two lysyl residues. In a broad sense they may be classified as essential in that their modification results in loss of carboxylase activity.

Br-butanone-P_2 is not an ideal affinity label for the carboxylase. The

[56] M. Nishimura and T. Akazawa, *Biochem. Biophys. Res. Commun.* **59**, 584 (1974).

reagent is unstable; the C1 phosphate and C3 bromine are lost at rates that are significant with respect to the rate of inactivation of the enzyme.[43] Since careful kinetic studies cannot be done, we do not know if the inactivation is subject to a rate-saturation effect as is required for an active-site-directed process. Also because of the instability of the C1 phosphate, the possibility exists that the reactive reagent is not Br-butanone-1,4-P_2 but is instead Br-butanone-4-P. There are also difficulties in the interpretation of the stoichiometric data. Ideally, with an affinity labeling reagent the loss of activity should correlate with the modification of one residue per catalytic subunit. In the case of the carboxylase, complicating factors are (1) the enzyme used may not be fully functional; (2) the nonselective alkylation of —SH groups occurs to different extents under inactivation and protective conditions; and (3) uncertainties exist in measuring the precise amount of covalent incorporation, since the functional group of the reagent that carries the radioactive marker (^{32}P-labeled phosphate) is partially lost. We can say only that inactivation results from the modification of about 0.6 residues of lysine per catalytic subunit, which is not unreasonable in view of the enzyme containing only four binding sites for ribulosebisphosphate per mole of octomer under the conditions we use for inactivation.[57] Since the incorporated reagent is found associated with two different lysyl residues and the average extent of lysyl modification is only 0.6 residues per catalytic subunit, it is likely that in a given subunit only one of the two residues can be modified, and that modification of either results in loss of enzymic activity.

Clearly, additional affinity labels with greater selectivity than, and different specificities from, Br-butanone-P_2 will be needed to characterize the active site of ribulosebisphosphate carboxylase.

[57] M. Wishnick, M. D. Lane, and M. C. Scrutton, *J. Biol. Chem.* **245**, 4939 (1970).

[43] β-Galactosidase

By JOSEPH YARIV

β-Galactosidase of *Escherichia coli* (MW 540,000) is a tetramer of identical subunits; it is obtained pure but is not of uniform activity.[1-3]

[1] I. Zabin and A. V. Fowler, *in* "The Lactose Operon" (J. R. Beckwith and D. Zipser, eds.), p. 27. Cold Spring Harbor Laboratory, Cold Spring Harbor, New York, 1970.

[2] G. R. Craven, E. Steers, Jr., and C. B. Anfinsen, *J. Biol. Chem.* **240**, 2468 (1965).

[3] M. E. Goldberg, *in* "The Lactose Operon" (J. R. Beckwith and D. Zipser, eds.), p. 273. Cold Spring Harbor Laboratory, Cold Spring Harbor, New York, 1970.

The highest reported activity is 910 units/mg.[3] The evidence from affinity labeling of enzyme of intermediate activity (427 units/mg) by N-bromoacetylgalactosylamine and its subsequent reactivation by treatment with mercaptoethanol indicates that not all subunits of the enzyme are catalytically active. This is at variance with the results of phenylethyl thiogalactoside binding to enzyme (also of intermediate activity), which demonstrated that the subunits of the enzyme are homologous.[4] However, binding of isopropyl thiogalactoside to enzyme is more complex and demonstrates that enzyme subunits are not equivalent in the tetramer.[5]

The chemical basis of the inactivation of β-galactosidase by N-bromoacetylgalactosylamine is the alkylation of one of the 21 methionyl residues of the enzyme, also accomplished, though less efficiently, by bromoacetamide.[6] The identity of this residue, in the polypeptide chain, has not been established, nor is the mechanism of inactivation understood. There is reason to believe that the site-directed reagent binds to a site other than the substrate-binding site since N-bromoacetylglucosylamine also shows a much enhanced activity in alkylation of this same methionyl residue.[7]

Reactivation of enzyme to full activity by treatment with a mercaptide ion is the result of the regeneration of the methionyl residue from its alkyl sulfonium salt.[6,8] However, it is not understood why the presence of the glycosyl residue in the alkyl side chain enhances the rate of the reactivation of the enzyme by more than 10-fold.[6] Nor is it understood why this enhancement in the reactivation rate is restricted to catalytically active subunits. This phenomenon is the basis of the method for determining the turnover number of the catalytically active subunit in the not fully active enzyme and will be described.

The other useful application of the affinity label derives from the fact that the sulfur in the methionyl residue that is alkylated is not essential for enzyme activity.[6] Therefore in enzyme modified by incorporation of norleucine, the fraction of enzyme which is resistant to alkylation by the reagent is a measure of methionine replacement by norleucine in this locus and in each and every methionine locus in simultaneously synthesized bacterial protein[9] (cf. Cowie et al.[10]).

[4] M. Cohn, Bacteriol. Rev. **21,** 140 (1957).

[5] J. Yariv and M. Yariv, Abstr. Meet. Fed. Eur. Biochem. Soc. 10th, Paris, 1975, Abstr. 942 (1975).

[6] F. Naider, Z. Bohak, and J. Yariv, Biochemistry **11,** 3202 (1972).

[7] O. Viratelle, Doctoral Dissertation, Université de Paris-Sud, Orsay, 1976.

[8] F. Naider and Z. Bohak, Biochemistry **11,** 3208 (1972).

[9] J. Yariv and P. Zipori, FEBS Lett. **24,** 296 (1972).

[10] D. B. Cowie, G. N. Cohen, E. T. Bolton, and H. de Robichon-Szulmajster, Biochim. Biophys. Acta **34,** 39 (1959).

Synthesis of Reagent

N-bromoacetyl β-D-galactopyranosylamine is prepared by acetylation of β-galactopyranosylamine with bromoacetic anhydride as described by Thomas.[11]

β-D-Galactopyranosylamine. This compound is prepared by dissolving 50 g of D-galactose in 400 ml of 30% ammonia in methanol (w/w, prepared by passing ammonia gas through absolute methanol) at room temperature. The solution is allowed to stand for 1 week at room temperature, by which time crystals of α-galactopyranosylamine separate. The supernatant liquid containing the product is decanted and allowed to crystallize at room temperature for 4 days. Crystalline β-D-galactopyranosylamine (7.5 g) is collected by filtration, washed with absolute methanol, and dried over sodium hydroxide (m.p. 151°; $[\alpha]_D^{20} + 62.7°$, H_2O).

Bromoacetic Anhydride. To prepare the bromoacetyl derivative it is best to start with bromoacetic acid, since tritium and ^{14}C-labeled bromoacetic acids are commercially available. The quality of the acid is of prime importance because it affects the ease with which the final product crystallizes.

Dry acid, 1.0 mmole, is dissolved with exclusion of moisture in dry carbon tetrachloride (3 ml). To this solution, 0.5 mmole of solid dicyclohexyl carbodiimide is added with rapid stirring at 0°. Stirring is continued for 15 min at 0° and for an additional 5 min at 25°. The mixture is then filtered directly into a suspension of β-D-galactosylamine in dimethylformamide, and the filter is washed with dry carbon tetrachloride.

N-Bromoacetyl β-D-Galactopyranosylamine. β-Galactosylamine, 2 mmoles, is suspended in 2 ml of dimethylformamide and 2.5 mmoles of the anhydride are added. After 3 hr at 25°, ethyl ether is added and the precipitated product is filtered and washed well with ethyl ether. For recrystallization, dry product is dissolved in a minimal volume of cold water. To this mixture are added 10 volumes of methanol, followed by ethyl ether to turbidity.

The purity of the compound is checked by thin-layer chromatography on silica gel with acetone–methanol in the ratio of 10:1 and by paper chromatography with isopropanol–water in the ratio of 4:1 as solvents.

Radioactive Reagent. In the synthesis of ^{14}C-labeled reagent, [2-^{14}C]bromoacetic acid (Radiochemical Centre, Amersham) of specific activity 55 mC/mmole was used. This was diluted 60-fold with cold bromoacetic acid. The radiochemical purity of product was determined with a Packard Radiochromatographic Scanner. Its concentration was

[11] E. W. Thomas, *J. Med. Chem.* **13,** 755 (1970). See also this volume [39].

determined by the phenol–sulfuric acid test[12] using an authentic sample of N-bromoacetyl-β-galactosylamine as standard.

Enzyme

Sources. β-Galactosidase of *E. coli* was prepared according to Craven *et al.*[2] (specific activity: 300–480 units/mg protein) or purchased from Worthington or Sigma.

When working with enzyme present in a bacterial extract, this was prepared by sonication of a thick suspension of *E. coli* cells in 0.1 M sodium phosphate at pH 7.5 containing 1 mM magnesium chloride; the clear supernatant liquid obtained with a refrigerated high speed centrifuge was used.

Inactivation and Labeling of Enzyme

The rate of enzyme inactivation by N-bromoacetylgalactosylamine is strongly dependent on temperature, and therefore inactivation is usually carried out in the temperature range of 30°–37° in a thermostated bath. Inactivation follows first-order kinetics and leads to complete inactivation of enzyme. For low residual activity ($\sim$1%), it is necessary to carry out the assay in the absence of mercaptoethanol since reactivation of enzymes under the standard assay conditions may confuse the result. The rate of ONPG hydrolysis is then markedly decreased and a conversion factor is used to bring the result to the standard unit of activity (cf. Viratelle and Yon[13]).

Inactivation of enzyme is best carried out with about 2 mM reagent in 0.1 M sodium phosphate, pH 7.5, containing 1 mM magnesium chloride. The alkylating reagent is added from a relatively concentrated stock solution ($\sim$0.1 M) in water at zero time. The course of inactivation is followed by transferring adequately small samples ($\sim$10 μl) of the reaction mixture (which contains the alkylating reagent) directly into the assay mixture.

In a labeling experiment the excess radioactive N-bromoacetylgalactosylamine is removed by extensive dialysis against many changes of cold buffer containing magnesium ions. When correlation between enzyme inactivation and incorporation of label is sought, the inactivation is performed in the presence of a large excess of iodoacetic acid (0.1 M). This is known to alkylate enzyme cysteine without inactivating the enzyme.[2]

---

[12] M. Dubois, K. A. Giles, J. K. Hamilton, P. A. Rebers, and F. Smith, *Anal. Chem.* **28**, 350 (1956).

[13] O. M. Viratelle and J. M. Yon, *Eur. J. Biochem.* **33**, 110 (1973).

It also prevents alkylation of cysteine by the specific reagent (cf. Yariv *et al.*[14]).

Reactivation of Enzyme

Reactivation of enzyme, previously inactivated by N-bromoacetyl-galactosylamine, is optimally carried out at pH 8.5 where native enzyme is stable even in the absence of mercaptoethanol. Mercaptoethanol is added to 0.1 M, and the temperature is maintained at 30°. Under these conditions, 90% reactivation is achieved in 5–6 hr.

The time course of the reactivation can be determined directly by transferring adequately small volumes of the reaction mixture for assay. To terminate the reactivation reaction, the sample is dialyzed against a relatively large volume of cold 0.1 M sodium phosphate at pH 7.0, containing 1 mM magnesium chloride. When the stoichiometry of the label lost on reactivation is important, dialysis is performed against many changes of buffer during 24 hr and, thereafter, the sample is dialyzed against a relatively small volume of buffer for at least 6 hr. An aliquot of the last buffer is used both to check whether the sample is essentially free of reagent and for correction of the counting rates of the protein solution.

Determination of the Turnover Number of the Enzyme

The procedure introduced here is based on the assumption that the label released from alkylated enzyme by treatment with a mercaptide ion corresponds numerically to regenerated active sites. Therefore the method allows, in principle, measurement of catalytic activity of a site in a preparation containing inactive protein whether homologous or foreign. The label released can also be used to calculate the equivalent weight of a site when the quantity of the label corresponds to that released on reactivation of enzyme from zero activity to 100% activity. The method is illustrated by description of an actual determination.

A commerical preparation of β-galactosidase by Sigma was dialyzed overnight at 4° against 0.1 M sodium phosphate at pH 7.5, containing 1 mM magnesium chloride and clarified in a refrigerated centrifuge. Protein concentration was 4.84 mg/ml and activity was 427 units/mg under standard conditions of assay.

To 1.98 ml of enzyme solution 0.22 ml of a 10 mM solution of ^{14}C-labeled N-bromoacetylgalactosylamine was added and incubated at 20° for 2 hr. A portion, 0.75 ml, of the reaction mixture was dialyzed extensively at 4° against 100 ml of 0.1 M sodium phosphate, pH 7.0, with 1

[14] J. Yariv, K. J. Wilson, J. Hildesheim, and S. Blumberg, *FEBS Lett.* **15**, 24 (1971).

mM magnesium chloride (six changes of buffer every 4 hr, the last change being 15 ml of buffer). This is the inactivated enzyme, E_I.

To 1.0 ml of the reaction mixture, 10 μl of 2-mercaptoethanol and 20 μl of 2 M sodium carbonate were added (pH 8.5) and incubated for 6 hr. Then, 0.75 ml was transferred for dialysis as described for the inactivated enzyme. This is the reactivated enzyme, E_R.

Determinations were carried out on the two preparations of the enzyme after clarification in the centrifuge: protein concentration, enzyme activity, and quantity of label incorporated. The results are tabulated below.

Enzyme	Protein (mg/ml)	K_M (mM)	V_{max} (μmoles/min/mg)	Label incorporated (nmoles/mg)
E_I	3.8	—	2	15.8
E_R	3.4	0.11	246	12.4

From these experimental values, the turnover number is obtained by the following calculation: Enzyme activity recovered is 244 μmoles/min/mg [$(V_{max} \, E_R) - (V_{max} \, E_I)$] and label released is 3.4 nmoles/mg (label E_I — label E_R); therefore, the turnover number is 244 μmoles/3.4 nmoles min^{-1}, or 1200 sec^{-1}.

In order to convert this number to standard conditions of enzyme assay, namely activity in the presence of 0.1 M mercaptoethanol, this is multiplied by 1.8 (a factor found empirically) to give 2160 sec^{-1}.

[44] Lysozyme

By YUVAL ESHDAT and NATHAN SHARON

The lysozymes are a group of enzymes characterized by their ability to lyse bacterial cells, notably those of *Micrococcus luteus* (previously known as *Micrococcus lysodeikticus*), by hydrolyzing the $\beta(1 \rightarrow 4)$ glycosidic bonds between the N-acetylmuramic acid (MurNAc) residues and N-acetyl-D-glucosamine (GlcNAc) residues in the cell wall peptidoglycan.[1-3] Of these enzymes, the most thoroughly investigated is hen egg-white lysozyme.[2-5] This and closely related enzymes, such as the

[1] P. Jollès, *Angew. Chem. Int.* **8**, 227 (1969).

[2] D. Chipman and N. Sharon, *Science* **165**, 454 (1969).

[3] T. Imoto, L. N. Johnson, A. C. T. North, D. C. Phillips, and J. A. Rupley, *in* "The Enzymes" (P. D. Boyer, ed.), 3rd ed., Vol. VII, p. 665. Academic Press, New York, 1972.

[4] A. Fleming, *Proc. R. Soc. London Ser. B* **93**, 306 (1922).

[5] D. C. Phillips, *Sci. Am.* **215** (5), 78 (1966).

duck and human lysozymes, will also hydrolyze oligosaccharides isolated from the peptidoglycan, such as the cell wall tetrasaccharide [GlcNAc-$\beta(1 \rightarrow 4)$-MurNAc-$\beta(1 \rightarrow 4)$-GlcNAc-$\beta(1 \rightarrow 4)$-MurNAc] and the corresponding hexasaccharide, as well as oligosaccharides derived from chitin [$\beta(1 \rightarrow 4)$ linked oligomers of N-acetyl-D-glucosamine, designated as (GlcNAc)$_n$]. In addition to hydrolysis, hen egg-white lysozyme and the related enzymes will also catalyze transglycosylation.[6,7] The cell wall oligosaccharides, as well as chitin oligosaccharides, can act as competitive inhibitors of these enzymes.

Hen egg-white lysozyme is readily obtained in crystalline form. It has a molecular weight of 14,500 and consists of a single polypeptide chain made up of 129 amino acids, with 4 intramolecular S-S bonds.[8-10] The three-dimensional structure of hen egg-white lysozyme and of several of its complexes with saccharides have been elucidated by Phillips and his co-workers.[5,11] Knowledge of these structures, and of the substrate and inhibitor specificity of hen egg-white lysozyme, made it possible to design affinity labeling reagents for the enzyme. These reagents, the 2',3'-epoxypropyl β-glycosides of N-acetyl-D-glucosamine and of its $\beta(1 \rightarrow 4)$ linked di- and trisaccharides, bind reversibly to the enzyme with affinity constants similar to those of the saccharides themselves.[12] While such reversible binding occurs, the epoxy group is located in the area of the active site which contains the side chains of Asp-52 and Glu-35, presumed to be involved in the catalysis.[2,5,13] The binding facilitates the chemical reaction between the epoxy group and the carboxylic group of Asp-52, leading to the formation of an ester bond between the amino acid and the affinity label.

Synthesis of 2',3'-Epoxypropyl β-Glycosides of N-Acetyl-D-Glucosamine and of Its Oligomers

Method A (Scheme 1) describes the synthesis of the 2',3'-epoxypropyl β-glycoside of (GlcNAc)$_2$ [designated as (GlcNAc)$_2$-Ep]. Both the octa-

[6] N. Sharon and S. Seifter, *J. Biol. Chem.* **239**, PC2398 (1964).

[7] N. Sharon, Y. Eshdat, I. Maoz, Y. Bernstein, E. M. Prager, and A. C. Willson, *Isr. J. Chem.* **12**, 591 (1974).

[8] J. Jollès, J. Jauregui-Adell, I. Bernier, and P. Jollès, *Biochim. Biophys. Acta* **78**, 668 (1963).

[9] R. E. Canfield, *J. Biol. Chem.* **238**, 2698 (1963).

[10] R. E. Canfield and A. K. Liu, *J. Biol. Chem.* **240**, 1997 (1965).

[11] C. C. F. Blake, L. N. Johnson, G. A. Mair, A. C. T. North, D. C. Phillips, and V. R. Sarma, *Proc. R. Soc. London Ser. B* **167**, 378 (1967).

[12] E. W. Thomas, J. F. McKelvy, and N. Sharon, *Nature (London)* **222**, 485 (1969).

[13] N. Sharon and Y. Eshdat, *in* "Lysozyme" (E. F. Osserman, R. E. Canfield, and S. Beychok, eds.), p. 195. Academic Press, New York, 1974.

acetate of $(GlcNAc)_2(I)$ and its glycosyl chloride (II) were prepared according to the procedure described by Osawa.[14] The synthesis of $(GlcNAc)_2$-Ep from (II) was adapted from the method developed by Thomas.[15] The same method, with slight modifications, was also applied successfully for the synthesis of the epoxy glycosides of N-acetyl-D-glucosamine and of $(GlcNAc)_3$. Method B (Scheme 2) describes the preparation of radioactive $(GlcNAc)_2$-Ep by introduction of the isotope into the N-acetyl group of the sugar moiety adjacent to the epoxy group. It is based on the procedure described by Horton et al.[16] for isotope labeling of N-acetyl-D-glucosamine. Method C (Scheme 3) describes a shorter procedure used for the preparation of the epoxypropyl β-glycoside of N-acetyl-D-glucosamine[17] and should also be suitable for the synthesis of the affinity labels of the higher oligomers.

Method A (Scheme 1)

O - (2 - Acetamido - 3,4,6 - tri - O - acetyl - 2 - deoxy-β-D - glucopyranosyl) - (1 → 4) -2-acetamido-1,3,6-tri-O-acetyl-2-deoxy-α-D-glucopyranose   (I).

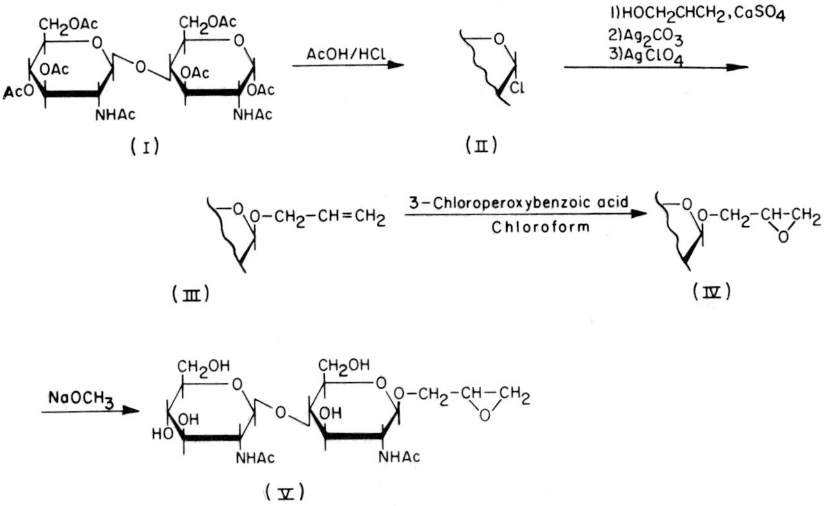

SCHEME 1

Chitin (Fluka, 20 g) is added to a cooled mixture of acetic anhydride (100 ml) and concentrated sulfuric acid (13 ml). After continuous stirring

[14] T. Osawa, *Carbohydr. Res.* **1**, 435 (1966).
[15] E. W. Thomas, *Carbohydr. Res.* **13**, 225 (1970). See also this volume [39].
[16] D. Horton, W. E. Mast, and K. D. Phillips, *J. Org. Chem.* **32**, 1471 (1967).
[17] Y. Eshdat, H. Flowers, and N. Sharon, unpublished observations (1971).

for 15 hr at room temperature followed by 8 hr at 55°, the solution is poured into a cooled solution of sodium acetate (80 g) in water (520 ml). After centrifugation, the supernatant is extracted with chloroform (3 × 300 ml), and the combined organic layers are washed once with water, several times with cold saturated sodium bicarbonate solution to neutrality, and again with water. The crystalline residue (13.5 g), obtained after evaporation of the chloroform, is dissolved in ethyl acetate and chromatographed on a silica gel (Merck No. 7734) column (4 × 150 cm). The column is eluted with 4:1 ethyl acetate–acetone, and the fractions containing (I) [R_f = 0.6 on silica gel G thin-layer chromatography (TLC)[18] plates, 1:1 ethyl acetate–acetone] are collected and evaporated. The dry material (5.1 g) contains residual amounts of acetylated N-acetyl-D-glucosamine and acetylated (GlcNAc)₃ and is recrystallized from methanol to give pure (I) in a yield of 2.0 g (6%); m.p. 302°(dec.), [$α$]²_D 55° (c 0.47, acetic acid).

O - (2 - Acetamido - 3,4,6 - tri - O - acetyl - 2 - deoxy-β-D - glucopyranosyl) - (1 → 4) -2-acetamido-3,6-di-O-acetyl-2-deoxy-α-D-glucopyranosyl chloride (II). A freshly prepared solution of glacial acetic acid, presaturated at 0° with dry hydrochloric acid (12 ml), is added to 2 g of (I), and the solution is kept at room temperature for 16 hr. Methylene chloride (200 ml) is added, and the solution is poured onto a mixture of water (40 ml) and ice (160 g). After short and vigorous stirring, the organic solution is removed and neutralized by washing twice with an ice-cold saturated solution of sodium bicarbonate. After a final washing with water, the organic phase is dried over sodium sulfate and then evaporated. The crystalline residue obtained gives one spot on TLC (R_f = 0.67; 1:1 ethyl acetate–acetone) with 75% yield (1.45 g), and is usually used directly for the synthesis of the allyl glycoside (III). Crystallization of (II) from a mixture of ethyl acetate, chloroform, and pentane gave 1.2 g (62% yield) of white needles, m.p. 208° (dec.), [$α$]²³_D 40° (c 0.39, chloroform). The product is stored in a desiccator under reduced pressure at 4°.

Allyl O - (2 - Acetamido - 3,4,6 - tri - O - acetyl-2-deoxy-β-D-glucopyranosyl) - (1 → 4) -2-acetamido-3,6-di-O-acetyl-2-deoxy-β-D-glucopyranoside (III). The acetylated glycosyl chloride of (GlcNAc)₂ (II) (1.1 g) and silver perchlorate (20 mg) are added to a mixture of dry allyl alcohol (15 ml), silver carbonate (425 mg), and powdered anhydrous calcium sulfate (1.5 g) prestirred for 2 hr in the dark under anhydrous conditions. After 6 hr, the mixture is filtered through Celite; the filtrate is evaporated to dryness and dissolved in 25 ml of methylene chloride. The solution

---

[18] Thin-layer chromatography was carried out on silica gel G precoated plates (Analtech Inc.).

is washed sequentially with cold 0.1 M hydrochloric acid, with saturated sodium bicarbonate solution to neutrality, and with water. The organic layer is dried with sodium sulfate and evaporated to a colorless syrup. Crystals of the product can be obtained from the syrup by triturating with isopropanol. Either the crystals or the syrup are dissolved in a minimal amount of 1:1 ethyl acetate–acetone; the solution is applied to a silica gel column (1.5 × 55 cm) and eluted with the same solvent. Fractions containing (III) (R_f = 0.55 on TLC) are collected and evaporated to dryness. Yield, 650 mg (59%); m.p. 244°, $[\alpha]_D^{23}$ −39° (c 2.1, chloroform).

2′,3′-Epoxypropyl O-(2-Acetamido-3,4,6-tri-O-acetyl-2-deoxy-β-D-glucopyranosyl)-(1 → 4)-2-acetamido-3,6-di-O-acetyl-2-deoxy-β-D-glucopyranoside (IV). Epoxidation of the allyl glycoside (III) is performed according to the procedure of Fieser and Fieser.[19] A solution of 600 mg of (III) in methylene chloride (10 ml) is well stirred, and 0.5 M m-chloroperbenzoic acid (Eastman Kodak Co.) in the same solvent (10 ml) is added dropwise during 30 min, keeping the temperature at about 25°. After 16 hr a solution of sodium sulfite (10% w/v) is added until the excess of the peracid is destroyed, as observed by a negative test on starch–iodide paper. The organic layer is washed successively with 5% sodium bicarbonate solution, with water, and with saturated sodium chloride solution, and dried with sodium sulfate. The solvent is evaporated, and the residual material is dissolved in a minimal volume of 7:3 ethyl acetate–acetone. Chromatography on a silica gel column (1.5 × 55 cm) with the latter solvent as eluent yields 450 mg of (IV) (73%), R_f = 0.28 on TLC with the same solvent, m.p. 261°, $[\alpha]_D^{23}$ −53° (c 2.0, chloroform).

2′,3′-Epoxypropyl O-(2-Acetamido-2-deoxy-β-D-glucopyranosyl)-(1 → 4)-2-acetamido-2-deoxy-β-D-glucopyranoside (V). Deacetylation of the O-acetylated epoxy-glycoside (IV) is performed according to the procedure of Zemplen et al.[20] The glycoside (400 mg) is suspended in dried methanol (20 ml) and stirred vigorously; freshly prepared 0.1 M methanolic sodium methoxide (~10 ml) is added dropwise to pH 8. After 15 hr at 23°, dry CO_2 (about 2 g) is added to neutralize the solution, which is then evaporated. The solid residue is dissolved in water and is additionally purified by gel filtration on a Sephadex G-25 (fine) column (0.9 × 200 cm) with water as eluent, to yield 180 mg of pure (GlcNAc)$_2$-Ep (65% yield), m.p. 284° (dec.), $[\alpha]_D^{23}$ −33° (c 2.5, water). The material was homogeneous by TLC (R_f = 0.17 in 4:1 acetone–

[19] L. F. Fieser and M. Fieser, *in* "Reagents for Organic Synthesis," p. 135. Wiley, New York, 1967.

[20] G. Zemplen, Z. Csuros, and S. Angyal, *Ber. Deut. Chem. Ges.* **70**, 1848 (1937).

methanol), and contained two glucosamine residues per mole (MW 480) as found by amino acid analysis.

Method B (Scheme 2)

O-(2-Acetamido-3,4,6-tri-O-acetyl-2-deoxy-β-D-glucopyranosyl)-(1 → 4)-2-amino-1,3,6-tri-O-acetyl-2-deoxy-α-D-glucopyranose Hydro-

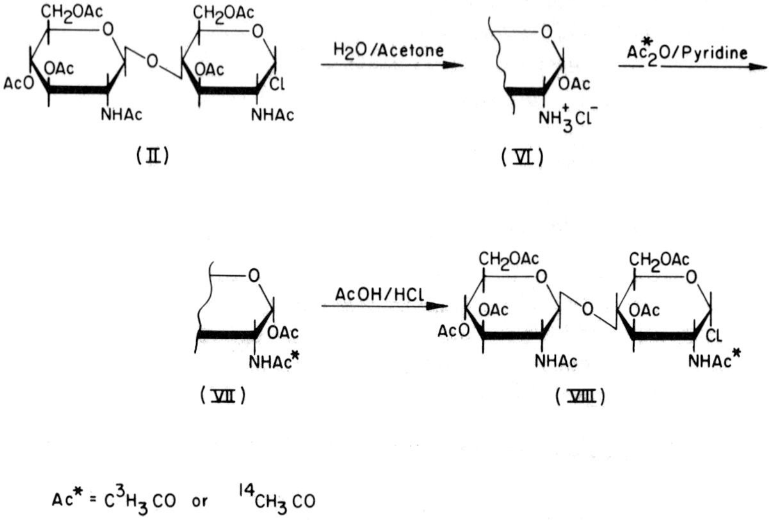

CH₂OAc ... CH₂OAc

AcO ... OAc ... O ... OAc ... Cl

NHAc ... NHAc

(II)

H₂O/Acetone →

O ... OAc ... NH₃⁺Cl⁻

(VI)

Ac*₂O/Pyridine →

(VII) ... O ... OAc ... NHAc*

AcOH/HCl →

CH₂OAc ... CH₂OAc

AcO ... OAc ... O ... OAc ... Cl

NHAc ... NHAc*

(VIII)

Ac* = C³H₃CO or ¹⁴CH₃CO

SCHEME 2

chloride (VI). A solution of 480 mg of (II) in acetone (5 ml) containing 15 μl of water is refluxed gently for 40 hr. After cooling to 0° for 2 hr, the resulting suspension is filtered and washed twice with cold actone. Upon drying under reduced pressure, 320 mg of (VI) (65% yield) are obtained, m.p. 215° (dec.), $[\alpha]_D^{23}$ 66° (c 0.7, water).

O-(2-Acetamido-3,4,6-tri-O-acetyl-2-deoxy-β-D-glucopyranosyl)-(1 → 4)-2-[³H]acetamido-1,3,6'-tri-O-acetyl-2-deoxy-α-D-glucopyranose (VII). A suspension of (VI) (320 mg) in dry pyridine (10 ml) is poured into an ampoule containing [³H]acetic anhydride (25 mCi, 5.6 Ci/mmole; The Radiochemical Center, Amersham), with its bottom preimmersed at −40° in a cold bath. The solution is stirred at room temperature, acetic anhydride (25 μl) is added with rapid stirring, and the ampoule is stoppered. After 2 hr, another dose (30 μl) of acetic anhydride is added with stirring and the solution is maintained for 18 hr at room temperature. To ensure completed acetylation, 100 μl of acetic anhydride are added. Four hours later a mixture of ice (30 g) and water (10 ml) is added. After

30 min, the resulting mixture is extracted with five 20-ml portions of chloroform, and the combined fractions are washed with water and with 5% sodium bicarbonate solution. The organic phase is dried with sodium sulfate and evaporated, and the residue is codistilled several times with toluene and dried under reduced pressure over P_2O_5. Recrystallization from methanol–ether gave 300 mg of (VII) (93% yield), identical in its migration on TLC and its analysis to the nonradioactive analog (I).

Tritium-labeled $(GlcNAc)_2$-Ep was obtained from (VII) by the procedure described in Method A, starting with the synthesis of the glycosyl chloride (VIII). The specific activity of the $[^3H](GlcNAc)_2$-Ep was 1.1×10^8 cpm/mmole. $[^{14}C](GlcNAc)_2$-Ep was prepared by the same method, using $[^{14}C]$acetic anhydride instead of the corresponding tritium-labeled compound.

Method C (Scheme 3)

2-Acetamido-3,4,6-tri-O-acetyl-2-deoxy-α-D-glucopyranosyl Chloride (X). The acetylated glycosyl chloride (X) is prepared from 50 g of N-

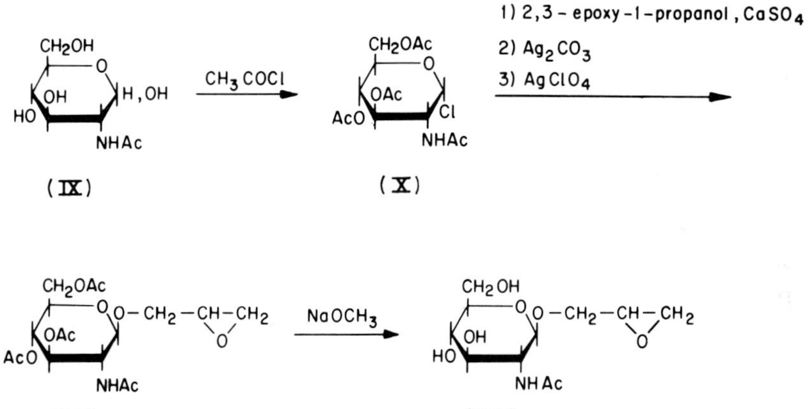

SCHEME 3

acetyl-D-glucosamine (IX, a gift from Pfizer Inc.) by following the procedure described by Lis et al.[21] The yield is 58 g (70%), m.p. 130°, $[\alpha]_D^{23}$ 109° (c 1.0, chloroform).

2',3'-Epoxypropyl 2-Acetamido-3,4,6-tri-O-acetyl-2-deoxy-β-D-glucopyranoside (XI). The acetylated glycosyl chloride of N-acetyl-D-glucosamine (X) (5 g) and 1.7 g of silver perchlorate are added to a mixture of

---

[21] H. Lis, R. Lotan, and N. Sharon, this series, Vol. 34, p. 341 (1974).

2′,3′-epoxy-1-propanol (Fluka, 175 ml), silver carbonate (3.9 g), and powdered anhydrous calcium sulfate (13.8 g), precooled to 10°, and stirred in the dark under anhydrous conditions for 2 hr. Care should be taken that the temperature does not exceed 25°, especially during the first hour of the reaction. After 4 hr of continuous stirring at 23° the mixture is filtered through celite, and methylene chloride (200 ml) is added. The solution is washed once with an equal volume of a saturated solution of sodium bicarbonate and 5 times with water to remove excess of the epoxy propanol. The organic solution obtained is dried over sodium sulfate and evaporated to dryness, and the product is redissolved in the minimal volume of ethyl acetate. Chromatography with the same solvent on a silica gel column gave (XI); $R_f = 0.15$ on TLC. Crystallization from methanol–ether yielded 2.4 g of (XI) (44%), m.p. 164°, $[\alpha]_D^{23} - 32°$ (c 0.72, chloroform).

2′,3′-Epoxypropyl 2-Acetamido-2-deoxy-β-D-glucopyranoside (XII). The O-acetylated epoxy glycoside (XI) (400 mg) is O-deacetylated with methanolic sodium methoxide, by the procedure used for the preparation of (GlcNAc)$_2$-Ep (V), to yield 190 mg of (XII) (69%); $R_f = 0.3$ on TLC (9:1 acetone–methanol), m.p. 177°, $[\alpha]_D^{23} - 37.1°$ (c 2.0, water).

Affinity Labeling of Hen Egg-White Lysozyme with (GlcNAc)$_2$-Ep

Hen egg-white lysozyme (0.125 mM) was incubated at 37° in water (pH 5.5) with (GlcNAc)$_2$-Ep (1.0 mM), and aliquots were taken at various times to determine the residual activity of the enzyme at pH 6.7 and 26° using cells of *M. luteus* as substrate.[22] The inactivation was exponential with time, and the enzyme lost 50% of its initial activity in 3 hr under the experimental conditions used. The rate of inactivation was markedly decreased when (GlcNAc)$_2$ (41 mM) was added to the incubation mixture. In the absence of (GlcNAc)$_2$-Ep and under the same incubation conditions, the enzymic activity remained constant. Incubation of the enzyme with an epoxide without the sugar moiety, such as propylene oxide (up to 0.5 M), did not affect its activity.[13]

For the quantitative examination of the extent of incorporation with time of the affinity label into hen egg-white lysozyme, the enzyme was incubated with ³H-labeled (GlcNAc)$_2$-Ep, and the excess of reagent was separated from the protein by TLC in 1:1 methanol–acetone. The silica layer at the origin, containing the protein only, was transferred to a scintillation vial, water and Bray's scintillation solution[23] were added,

---

[22] Y. Eshdat, J. F. McKelvy, and N. Sharon, *J. Biol. Chem.* **248**, 5892 (1973).
[23] G. A. Bray, *Anal. Biochem.* **1**, 279 (1960).

and the sample was counted in a scintillation counter. A 1:1 correlation was found between the molar incorporation of the radioactive affinity label and the inactivation of the enzyme.[13]

Isolation of the Affinity Labeled Hen Egg-White Lysozyme

Hen egg-white lysozyme (100 mg, 1.8 mg/ml) was incubated with the affinity label [³H] (GlcNAc)₂-Ep (26.7 mg) under the conditions described in the preceding section. After 24 hr the affinity labeled enzyme was separated from excess labeling reagent by dialysis against water. After lyophilization, the protein was dissolved in 5 ml of 0.1 M ammonium acetate and chromatographed on a column (2 × 150 cm) of Sephadex G-25 (fine) with the same solvent as eluent. Only one radioactive peak, which showed absorbance at 280 nm and had constant specific radioactivity corresponding to 1 mole of (GlcNAc)₂ per mole of enzyme, was obtained. The fractions containing the protein were combined and dialyzed against water. After lyophilization, the isolated protein was homogeneous by polyacrylamide gel electrophoresis at pH 4.5. Its enzymic activity, when assayed with cells of $M.$ $luteus$, was found to be less than 2% of that of an equal weight of the native enzyme. Analysis of the acid hydrolyzate (HCl 6 M, 110°, 22 hr) of the inactive lysozyme on an amino acid analyzer revealed, in addition to the expected amino acids, the presence of glucosamine in a ratio of 1.9 moles per mole of enzyme.[22]

Identification of the Site of Attachment

Proteolytic digestion was chosen as a suitable method to obtain peptide fragments from the affinity-labeled enzyme. Inactive radioactive-labeled enzyme was reduced by dithiothreitol and carboxymethylated by iodoacetic acid. All the radioactivity of [³H] (GlcNAc)₂-Pr-lysozyme was recovered in the reduced carboxymethylated affinity-labeled enzyme. Total enzymic digestion with subtilism and aminopeptidase M yielded two radioactive compounds that were further purified by paper electrophoresis and gel filtration chromatography. One compound, which composed 6% of the total radioactivity, contained glucosamine and no amino acid, and was most probably 2′,3′-propanediol β-glycoside of (GlcNAc)₂, which has been released from the affinity-labeled enzyme during the proteolytic digestion. The second material, with the recovery of 91% of total original radioactivity, was found to contain aspartic acid and glucosamine in a molar ratio 1:2.1, as revealed by amino acid analysis. This finding shows that the affinity label was bound solely to aspartic acid.

To establish the position of this aspartic acid residue in the polypeptide chain of the enzyme, the reduced and carboxymethylated affinity labeled enzyme was digested for 16 hr with pepsin in 5% formic acid. Most of the radioactivity recovered was associated with a single peptic peptide isolated by paper electrophoresis and gel filtration chromatography. Based on its amino acid and end group analysis, it was concluded that the peptide is located between Asn-39 and Tyr-53 in the polypeptide chain of the enzyme. Digestion of the labeled peptic peptide with an aminopeptidase obtained from *Clostridium histolyticum*[24] afforded a single radioactive peptide carrying the affinity label. The composition, partial structure, and other properties of this peptide correspond to those of the peptide located between Gly-49 and Tyr-53 in hen egg-white lysozyme, with the sequence: H_2N-Gly-Ser-Thr-Asp-Tyr-COOH. Mild treatment of the labeled peptide with triethylamine resulted in removal of the radioactivity and formation of a negative charged peptide at pH 6.5. Thus, in the inactivated lysozyme the affinity label is covalently bound via an ester bond to the carboxyl group of Asp-52.[22]

Further evidence on the site of attachment of the affinity label in hen egg-white lysozyme was obtained by X-ray crystallography[25] and immunochemical[26] and physicochemical studies.[13] These studies have conclusively demonstrated that the carbohydrate moiety of the affinity label is located in the active site of the enzyme, and that the aglycon is neighboring the side chain of Asp-52.

Affinity Labeling of Lysozymes from Different Sources

The affinity labeling reagents described above have proved to be useful in studies of the active sites of lysozymes isolated from different organisms.[7] Many of these lysozymes exhibit close similarity in their amino acid sequences, fluorescence spectra, and enzymic properties including not only the ability to digest cell walls of *M. luteus*, but also to catalyze transglycosylation reactions with a variety of saccharide acceptors when the cell wall tetrasaccharide or chitin oligosaccharides are used as substrates.[1,3,6,27]

All the bird lysozymes examined, except for the egg-white lysozyme from geese, were inactivated by $(GlcNAc)_2$-Ep at pH 5.5, 37°. The goose egg-white lysozyme was not affected by $(GlcNAc)_2$-Ep even at pH 4.7, at which it attacks N-acetyl-D-glucosamine oligosaccharides better than

[24] E. Kessler and A. Yaron, *Biochem. Biophys. Res. Commun.* **50**, 405 (1973).

[25] J. Moult, Y. Eshdat, and N. Sharon, *J. Mol. Biol.* **75**, 1 (1973).

[26] E. Maron, Y. Eshdat, and N. Sharon, *Biochim. Biophys. Acta* **278**, 243 (1972).

[27] A. Tsugita, in "The Enzymes" (P. D. Boyer, ed.), 3rd ed., Vol. V, p. 361. Academic Press, New York, 1971.

at higher pH.[28] In addition to the hen lysozyme, the enzymes that were found to interact with the affinity label were the lysozymes from the egg-white of the turkey, the bobwhite quail, the Japanese quail, the ring-necked pheasant, the chachalaca, and the duck lysozyme II. The rate of inactivation varied with each enzyme, but in all cases decreased significantly upon the addition of an excess of $(GlcNAc)_2$. The fact that these bird lysozymes, except the goose egg-white enzyme, were all affinity labeled by the same reagent suggests that the architecture of the active site region of these enzymes is similar. However, the active site of the goose enzyme is different from those of the other bird lysozymes tested. This conclusion is in accord with the results of the study of the primary structure of that enzyme,[29] which showed an amino terminal peptide of 30 residues completely different from that of the hen egg-white lysozyme.

Lysozymes derived from sources other than bird egg-whites were also tested for their susceptibility to inactivation by the epoxypropyl β-glycosides. Among these, human leukemic urine lysozyme was inactivated by $(GlcNAc)_2$-Ep at pH 5.5, whereas the lysozymes from papaya latex and the bacteriophage T4 were not inactivated at all. Similar results were also obtained at pH 4.6, at which the papaya lysozyme acts on chitin most efficiently. These results show that the active sites of human and the hen lysozymes are similar, whereas the active sites of the T4 phage and the papaya lysozymes differ from them. Indeed, the hen and the human lysozymes are similar in their amino acid sequences[29] and three-dimensional structures,[30] whereas the papaya and the T4 phage lysozymes differ from the hen egg-white lysozyme in their amino acid composition, molecular weight, and substrate specificity.[27,31]

Concluding Remarks

The affinity labeling of hen egg-white lysozyme by 2′,3′-epoxypropyl β-glycoside of $(GlcNAc)_2$ clearly demonstrates that Asp-52 forms part of the active site of the enzyme. Additional investigation performed with this reagent gave further insight into the nature of the active site of the enzyme and its mechanism of action. Thus, fluorescence polarization studies[32] of the affinity-labeled enzyme provided information about the binding of saccharides to subsites E and F of the active site.

[28] D. Charlemagne and P. Jollès, *Bull. Soc. Chim. Biol.* **49**, 1103 (1967).
[29] R. E. Canfield, S. Kammerman, J. H. Sobel, and F. J. Morgan, *Nature* (*London*), *New Biol.* **232**, 16 (1971).
[30] C. C. F. Blake and I. D. A. Swan, *Nature* (*London*), *New Biol.* **232**, 12 (1971).
[31] J. B. Howard and A. N. Glazer, *J. Biol. Chem.* **244**, 1399 (1969).
[32] V. I. Teichberg and M. Shinitzky, *J. Mol. Biol.* **74**, 519 (1973).

Insight into the role of Asp-52 was obtained by specific reduction of the ester bond which links the affinity label to this residue and enables the chemical conversion of Asp-52 to homoserine.[33] The modified lysozyme showed the same fluorescence spectrum as the native enzyme. With both proteins the fluorescence maximum shifted to the blue to a similar extent upon the addition of the saccharide inhibitors $(GlcNAc)_3$ and $(GlcNAc-MurNAc)_2$. Although the modified enzyme binds these two saccharides with nearly the same binding constants as those found for the native lysozyme, it lacks the catalytic ability to hydrolyze lysozyme substrates. The loss of the enzymic activity due to the modification of Asp-52 establishes the importance of this residue in lysozyme catalysis. In addition, the conversion of the catalytic residue Asp-52 to homoserine provides a novel approach by which functional carboxyl groups located in binding sites of proteins can be "mutated" specifically to their alcohol (CH_2OH) analogs.

[33] Y. Eshdat, A. Dunn, and N. Sharon, *Proc. Natl. Acad. Sci. U.S.A.* **71**, 1658 (1974).

[45] Glutamine Binding Sites

By LAWRENCE M. PINKUS

Glutamine plays a central role in the nitrogen metabolism of both prokaryotic and eukaryotic organisms.[1] In Fig. 1 are illustrated several compounds with structural similarity to glutamine but containing different functional groups in place of the normal side-chain amide. Some of these compounds might be expected to interfere with the utilization of glutamine only by inhibition in a competitive and reversible manner whereas others contain an alkylating group, such as a diazoketone (I, IV), chloroketone (III), diazoacetyl ester (II), or chloroisoxazole (V), which confers the potential of reacting with an amino acid in or near the glutamine binding site. Such reaction would result in covalent attachment of the inhibitor and, hence, irreversible inhibition. This chapter describes methods involving the potentially irreversible glutamine antagonists which have proved to be versatile in studies on glutamine utilization by enzymes.

Preparation and Properties of Glutamine Analogs

6-Diazo-5-oxo-L-norleucine (I) and O-diazoacetyl-L-serine (II) were originally isolated from *Streptomyces* fermentations, shown to have anti-

[1] "The Enzymes of Glutamine Metabolism" (S. Prusiner and E. R. Stadtman, eds.). Academic Press, New York, 1973.

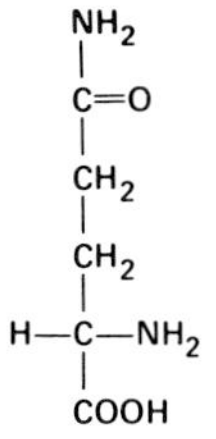

L-glutamine

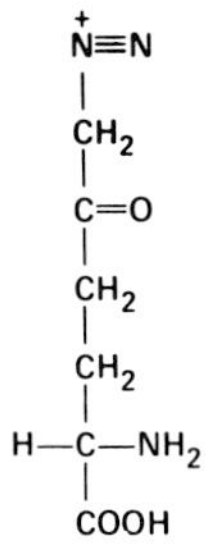

6-diazo-5-oxo-L-norleucine (DON)

(I)

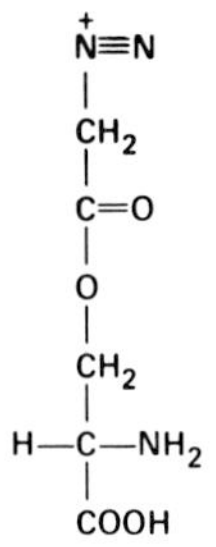

O-diazoacetyl-L-serine (azaserine)

(II)

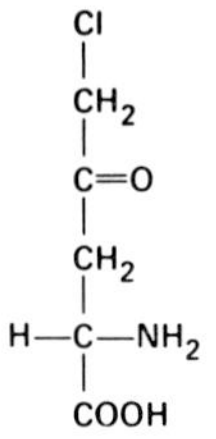

L-2-amino-4-oxo-5-chloro-pentanoic acid (chloroketone)

(III)

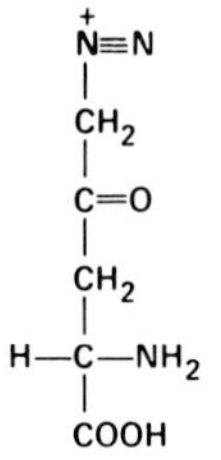

5-diazo-4-oxo-L-norvaline (DONV)

(IV)

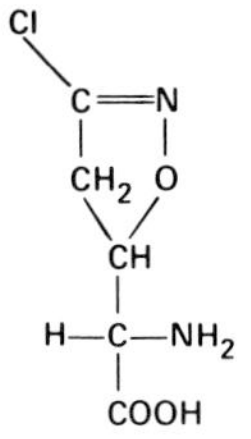

L-[αS,5S]-α-amino 3-chloro,4,5-dihydro-5-isoxazole acetic acid (isoxazole)

(V)

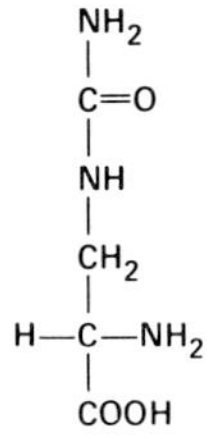

L-3-ureido-2-amino propionate (L-albizziin)

(VI)

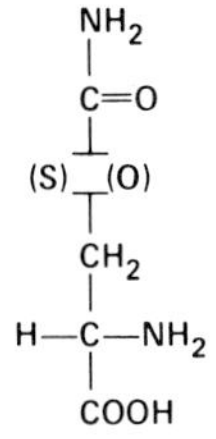

S-carbamyl cysteine
O-carbamyl serine

(VII)

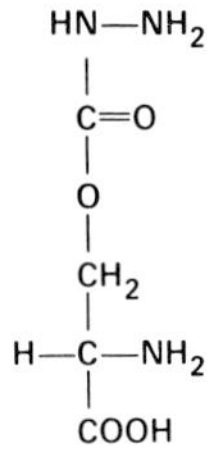

O-carbazyl L-serine

(VIII)

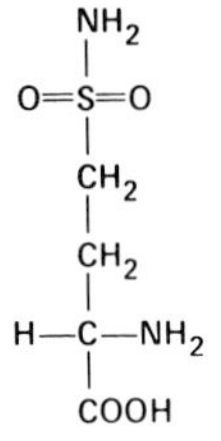

L-3-amino-3-carboxy-propane sulfonamide

(IX)

FIG. 1. Structures of glutamine analogs.

tumor activity,[2,3] and identified as analogs of glutamine which interfere with the biosynthesis of pyrimidines[4] and purines[5,6] through inhibition of amidotransferase reactions.[7,8] L-2-Amino-4-oxo-5-chloropentanoic acid (III) is another important analog. It was synthesized originally for studies

[2] H. W. Dion, S. A. Fusari, Z. L. Jakobowski, J. G. Zora, and Q. R. Bartz, *J. Am. Chem. Soc.* **78,** 3075 (1956).

[3] Q. R. Bartz, C. D. Elder, R. P. Fronhardt, S. A. Fusari, T. H. Haskell, D. W. Johannessen, and A. Ryder, *Nature (London)* **173,** 72 (1954).

[4] M. L. Eidinoff, J. E. Knoll, B. Marano, and L. Cheong, *Cancer Res.* **18,** 105 (1957).

[5] H. E. Skipper, L. L. Bennett, Jr., and F. M. Schabel, Jr., *Fed. Proc., Fed. Am. Soc. Exp. Biol.* **13,** 298 (1954).

[6] J. M. Buchanan, *in* "The Chemistry and Biology of Purines" (G. E. W. Wolstenholme and C. M. O'Conner, eds.), p. 233. Churchill, London, 1957.

[7] B. Levenberg, I. Melnick, and J. M. Buchanan, *J. Biol. Chem.* **225,** 163 (1957).

[8] S. C. Hartman, *J. Biol. Chem.* **238,** 3036 (1963).

with carbamyl phosphate synthetase II[9,10] but has since been found to inhibit the glutamine-dependent activity of other amidotransferases.[11]

6-Diazo-5-oxo-L-norleucine (DON). The analog may be prepared by the method of Weygand *et al.*[12] Radioactive [6-^{14}C]DON has been prepared by Hartman[8] starting with *N*-trifluoroacetyl-L-glutamic-γ-acid chloride-α-methyl ester and [^{14}C]diazomethane. DON may be recrystallized from water–acetone; m.p. 140°–150° (dec.). Its spectral properties are $\lambda_{max} = 274.7$ nm, $\epsilon_m = 11,800$, shoulder at 245 nm, $\epsilon_m = 5400$; $\lambda_{min} = 207$ nm. DON (free base) is quite soluble in water; solutions of up to 400 mM are easily prepared. At 25°, neutralized solutions slowly break down to form bubbles of nitrogen. Solutions of DON or the products resulting from its decomposition may exhibit significant yellow color and absorbance at 340 nm that may interfere in spectrophotometric studies if high concentrations of the inhibitor are present.

The lower homolog of DON, 5-diazo-4-oxo-L-norvaline (IV),[13] although similar in structure to the chloroketone (III), has not proved effective as an inhibitor of glutamine-utilizing enzymes. This compound functions as an irreversible inhibitor of L-asparaginase[14] and is therefore regarded as an asparagine analog.

O-Diazoacetyl-L-serine (Azaserine). The derivative may be prepared by selective diazotization of *O*-glycyl-L-serine as reported by Nicolaides *et al.*[15] It may be recrystallized from water-ethanol; m.p. 153°–155° (dec.); $\gamma_{max} = 249.5$ nm; $\epsilon_m = 20,000$. French *et al.*[16] prepared *O*-diazo-[2-^{14}C]acetyl L-serine by a modification of the above-cited procedure.

L-2-Amino-4-oxo-5-chloropentanoic Acid (5-Chloro-4-oxo-L-norvaline, CONV, or chloroketone). The chloroketone is best prepared by the method of Khedouri *et al.*[9] as modified.[10] The synthesis of [5-^{14}C]chloroketone in amounts of about 50 mg is described here.

N-Benzyloxycarbonyl-L-aspartic acid α-benzyl ester (108 mg, 0.3

[9] E. Khedouri, P. M. Anderson, and A. Meister, *Biochemistry* **5**, 3552 (1966).

[10] L. M. Pinkus and A. Meister, *J. Biol. Chem.* **247**, 6119 (1972).

[11] Also functioning as an apparent analog of aspartic semialdehyde, the chloroketone has been reported to specifically inactivate threonine-sensitive homoserine dehydrogenase [C. G. Hirth, M. Veron, C. Villar-Palasi, N. Hurion, and G. N. Cohen, *Eur. J. Biochem.* **50**, 425 (1975)].

[12] F. Weygand, H. J. Bestmann, and E. Klieger, *Chem. Ber.* **91**, 1037 (1958).

[13] Y. Lewschitz, R. D. Irsay, and A. I. Vincze, *J. Chem. Soc. London* **195**, 1308 (1959).

[14] See R. E. Handschumacher, this volume [47].

[15] E. D. Nicolaides, R. O. Westland, and E. L. Wittle, *J. Am. Chem. Soc.* **76**, 2887 (1954).

[16] T. C. French, I. B. Dawid, R. A. Day, and J. M. Buchanan, *J. Biol. Chem.* **238**, 2171 (1963).

mmole, Cyclo Chem Co.) is dissolved in 1 ml of freshly distilled thionyl chloride in a round-bottom flask equipped with a $CaCl_2$ drying tube. The flask contents are heated to 40° in a water bath for 30 min and the thionyl chloride is removed by flash evaporation under high vacuum at 35°. The remaining colorless oil is dissolved in 0.2 ml of thionyl chloride, and this solution is heated for 15 min at 40°. After removal of the thionyl chloride by evaporation under high vacuum, the colorless oily residue is dissolved in 1 ml of dry ethyl ether and used immediately in a reaction with [^{14}C]diazomethane. A flask containing the ethereal solution of acid chloride is cooled in a Dry Ice–acetone bath. To this solution 7 ml of a dry ethereal solution of [^{14}C]diazomethane [prepared[17] from 78 mg of [^{14}C]N-methyl-N-nitroso-p-toluene sulfonamide (specific activity, 4.63 Ci/mole) and 228 mg of unlabeled compound] are added rapidly and with swirling; a $CaCl_2$ drying tube is attached to the flask. The flask is allowed to stand in a bucket containing Dry Ice for 24 hr in a hood and gradually allowed to return to room temperature. After removal of the ether on a flash evaporator, the product, 5-diazo-4-oxo-N-benzoxycarbonyl L-[5-^{14}C]norvaline-α-benzyl ester, exhibits a sharp band at 2105 cm^{-1} ($-CH=N^+=N^-$), which is of about the same intensity as the carbonyl absorption at 1715 cm^{-1}. A band at 1789 cm^{-1} indicates formation of the side product, N-benzoxycarbonyl-L-aspartic acid-α-benzyl ester-γ-methyl ester. The diazoketone may be crystallized from dry ether.

The white crystalline diazoketone is dissolved in 10 ml of dry chloroform. Hydrogen chloride gas is bubbled slowly for 2 min into the solution that is cooled in ice. A spectrum of the product, 5-chloro-4-oxo-N-benzoxycarbonyl L-[5-^{14}C]norvaline-α-benzyl ester, should exhibit no diazo absorption. The chloroform is removed by flash evaporation and the residual material is suspended in 20 ml of 6 N HCl and heated at 70° for 10 hr. The supernatant solution is decanted and replaced with 20 ml of 6 N HCl, and the procedure is repeated until no insoluble material remains. The solutions are combined, and the products generated by removal of the protecting groups (benzyl alcohol, toluene) are removed by lyophilization; the remaining colorless, moist powder may be dried under reduced pressure over P_2O_5. The net yield of L-2-amino-4-oxo-5-chloropentanoic acid hydrochloride is about 80% (49 mg, 0.24 mmole; specific activity 1.16 Ci/mole). The chloroketone hydrochloride can be recrystallized from acetone–water to give the free base; m.p. 151°–152° (dec.).

On paper electrophoresis the chloroketone moves toward the cathode at pH 5.5 and reacts with ninhydrin to give a yellow color. It emerges

[17] C. H. Hirs, *J. Biol. Chem.* **227**, 611 (1956).

between proline and glycine on the amino acid analyzer.[18] Crystalline [^{14}C]chloroketone preparations were found to be stable at 4° for at least 1 year when stored in a vacuum desiccator over P_2O_5. No destruction of [^{14}C]chloroketone was observed on storage of frozen, dilute (0.1–10 mM) aqueous solutions (pH 4–5) at −20°. Acidic solutions of the chloroketone hydrochloride are stable at 37° for at least 2 hr. Neutralized concentrated solutions slowly decompose, turning brown on standing for several hours; the brown contaminant is removable with charcoal. Decomposition may be minimized by keeping the slightly acid chloroketone solution on ice and, if necessary, neutralizing it shortly before use. Aqueous solutions of the free base reach saturation at about 150 mM at 25°; the chloroketone hydrochloride is freely soluble.

L-[2S, 5S]-2-Amino-3-chloro-4,5-dihydro-5-isoxazoleacetic Acid (Isoxazole). The isoxazole is an antibiotic isolated from Streptomyces sviceus,[19] which has been shown to inhibit several amidotransferases.[20] It is known to irreversibly inactivate asparagine synthetase.[20,21]

L-2-Amino-3-ureidopropionic Acid (Albizzin). This compound was first isolated from seeds of the botanical species Albizzia julibrissin. Durazz by Gmelin et al.[22] L-[^{14}C]Albizzin can be synthesized by the method of Schroeder et al.[23] based on the original procedure of Kjaer and Oleson-Larson.[24] In the author's experience, albizzin functions, in most cases, as a competitive and reversible glutamine antagonist, but covalent attachment and irreversible inhibition have been reported with formylglycinamide ribonucleotide amidotransferase.[23]

S-Carbamyl-L-cysteine. The derivative may be synthesized by the method of Stark et al.[25,26] This compound is rather unstable and decomposes at alkaline pH to cysteine and cyanate.

Several other analogs might be considered as potential competitive inhibitors of glutamine utilization. These include O-carbamyl-L-serine,

[18] D. H. Spackmann, W. H. Stein, and S. Moore, Anal. Chem. 30, 1190 (1958).
[19] L. J. Hanka, D. G. Martin, and G. L. Neil, Cancer Chemother. Rep. 51, 141 (1973).
[20] H. N. Jayaram, D. A. Cooney, J. A. Ryan, G. Neil, R. L. Dion, and V. H. Bono, Cancer Chemother. Rep. 59, 481 (1975).
[21] The glutamine analogs discussed above are toxic substances and should be treated accordingly. Azaserine has been reported to be an active mutagen in bacterial systems [D. S. Longnecker, T. J. Curphey, S. T. James, D. S. Daniel, and N. J. Jacobs, Cancer Res. 34, 1658 (1974)].
[22] R. Gmelin, G. Strauss, and G. Hasenmaier, Z. Naturforsch Teil B 13, 252 (1958).
[23] D. Schroeder, J. Allison, and J. Buchanan, J. Biol. Chem. 244, 5856 (1969).
[24] A. Kjaer and P. Oleson-Larson, Acta Chem. Scand. 13, 1565 (1959).
[25] G. R. Stark, W. H. Stein, and S. Moore, J. Biol. Chem. 235, 3177 (1960).
[26] G. R. Stark, J. Biol. Chem. 239, 1411 (1964).

O-carbazyl-L-serine, L-3-amino-3-carboxyl propionic acid sulfonamide, γ-glutamyl hydrazide, and γ-glutamyl hydroxamate.[27,28]

Use of Glutamine Analogs

Of the analogs in Fig. 1, DON, azaserine, and the chloroketone have found the widest application as irreversible inhibitors of glutamine utilization. These compounds have been of particular value in studies on the mechanism of action of amidotransferases (as reviewed by Buchanan[29]). Enzymes known to be inhibited are listed in Table I. In general, amidotransferases can use either glutamine or ammonia as nitrogen donors. Some of these enzymes are formed from a single subunit containing the binding sites for both glutamine and ammonia. In others, containing nonidentical subunits, the glutamine and ammonia binding sites are found on different polypeptide chains. By choosing appropriate conditions the glutamine-dependent reaction can be inactivated in several of these enzymes without appreciably affecting the ammonia-dependent reaction. Inactivation of glutamine-dependent activity has been accomplished most simply by incubating the enzyme and inhibitor, usually at a concentration of 0.01 to 1 mM, in an appropriate buffer. Inhibitor concentration, pH, temperature, and substrates or effectors, which alter the structure and chemical reactivity of the protein or alter the mode of binding of the inhibitor, may affect the rate and specificity of the inactivation reaction. The rate of inactivation can be conveniently monitored by assaying aliquots of the incubation mixture under conditions wherein dilution of the inhibitor and a high glutamine concentration act to stop the inactivation reaction. If milligram quantities of a radioactively labeled enzyme are to be prepared, the inactivation reaction may be terminated by adding 100 mM glutamine followed by gel filtration or dialysis at 4° to remove free inhibitor.

One can be reasonably certain of specific inactivation if (1) dramatic inhibition of glutamine-dependent activity (or glutaminase activity) occurs without appreciable inhibition of other reactions catalyzed by the

---

[27] A. Meister "Biochemistry of the Amino Acids," 2nd ed., Vol. 1, p. 246. Academic Press, New York, 1965.

[28] Small quantities of DON (NSC-7365), azaserine (NSC-742), chloroketone (NSC-124412), and the isoxazole (NSC-163501) might be obtained from the Drug Development Branch, National Cancer Institute, Blair Building, 8300 Colesville Road, Silver Spring, Maryland 20910. Azaserine is available commercially from Calbiochem, La Jolla, California; albizzin and S-carbamylcysteine may be obtained from K&K Laboratories, Plainview, New York.

[29] J. M. Buchanan, *Advan. Enzymol.* **39**, 91 (1973).

TABLE I
Enzymes Inhibited by Glutamine Analogs

Enzymes	Effective analogs[a]	References
Anthranilate synthetase	I, III	b, c
Asparagine synthetase	I, III, V	d, e
Carbamyl phosphate synthetase	I, II, III	f, g
Cytidine triphosphate synthetase	I, II, III	h
Formylglycinamide ribonucleotide amidotransferase	I, II, III, VI	i, j
Fructose-6-phosphate amidotransferase	I	k–m
Glutamate synthase	I, III	n–p
Glutaminase (bacterial)	I	q, r
Glutaminase (mitochondrial)	I, III	s
γ-Glutamylcysteine synthetase	III	t
Guanosine-5-phosphate synthetase	I, II	u, v
NAD synthetase	I, II	w, x
Phosphoribosyl pyrophosphate amidotransferase	I, II	y

[a] Compounds reported to irreversibly inactivate glutamine dependent activity. Other analogs may also prove to be effective.

[b] Y. Goto, H. Zalkin, P. S. Keim, and R. L. Heinrikson, *J. Biol. Chem.* **251**, 941 (1976).

[c] H. Nagano, H. Zalkin, and E. J. Henderson, *J. Biol. Chem.* **245**, 3810 (1970).

[d] B. Horowitz and A. Meister, *J. Biol. Chem.* **247**, 6708 (1972).

[e] D. A. Cooney, H. N. Jayaram, J. A. Ryan, and V. H. Bono, *Cancer Chemother. Rep.* **58**, 793 (1974).

[f] E. Khedouri, P. M. Anderson, and A. Meister, *Biochemistry*, **5**, 3552 (1966).

[g] L. M. Pinkus and A. Meister, *J. Biol. Chem.* **247**, 6119 (1972).

[h] A. Levitski, W. B. Stallcup, and D. E. Koshland, Jr., *Biochemistry* **10**, 3371 (1971).

[i] B. Levenberg, I. Melnick, and J. M. Buchanan, *J. Biol. Chem.* **225**, 163 (1957).

[j] T. C. French, I. B. Dawid, R. A. Day, and J. M. Buchanan, *J. Biol. Chem.* **238**, 2171 (1963).

[k] S. Ghosh, H. J. Blumenthal, E. Davidson, and S. Roseman, *J. Biol. Chem.* **235**, 1265 (1960).

[l] C. J. Bates and R. E. Handschumacher, *Adv. Enzyme Regul.* **7**, 163 (1969).

[m] P. J. Winterburn and C. F. Phelps, *Biochem. J.* **121**, 721 (1971).

[n] P. P. Trotta, K. E. B. Platzer, R. H. Haschemeyer, and A. Meister, *Proc. Natl. Acad. Sci. U.S.A.* **71**, 460 (1975).

[o] P. Mäntsälä and H. Zalkin, *J. Biol. Chem.* **251**, 3294 (1976).

[p] D. W. Tempest, J. L. Meers, and C. M. Brown in "The Enzymes of Glutamine Metabolism" (S. Prusiner and E. R. Stadtman, eds.), p. 167. Academic Press, New York, 1973.

[q] S. C. Hartman and T. F. McGrath, *J. Biol. Chem.* **248**, 8506 (1973).

[r] J. Roberts, J. S. Holcenberg, and W. C. Dolowy, *J. Biol. Chem.* **247**, 84 (1972).

[s] L. Pinkus and H. G. Windmueller, *Arch. Biochem. Biophys.*, in press (1977).

[t] R. Sekura, *Fed. Proc., Fed. Am. Soc. Exp. Biol.* **35**, 1750 (1976).

[u] S. C. Hartman and S. Prusiner, in "The Enzymes of Glutamine Metabolism" (S. Prusiner and E. R. Stadtman, eds.), p. 409. Academic Press, New York, 1973.

[v] L. M. Iarovaia, S. R. Mardashev, and S. S. Debov, *Vopr. Med. Khim.* **13**, 176 (1967).

[w] J. Preiss and P. Handler, *J. Biol. Chem.* **233**, 493 (1958).

[x] K. Y. Cheng and L. S. Dietrich, *J. Biol. Chem.* **247**, 4794 (1972).

[y] S. C. Hartman, *J. Biol. Chem.* **238**, 3036 (1963).

enzyme, e.g., ammonia-dependent activity; (2) the binding of radioactive inhibitor (up to 1 mole/mole of active site) correlates with degree of inhibition; and (3) addition of glutamine protects against inactivation and prevents binding of the radioactive inhibitor.

Protein labeling that is not associated with inhibition of glutamine-dependent activity is "nonspecific." Inhibitor concentration should be as low as possible for inactivation, thereby minimizing the rate of any nonspecific binding reactions. If observed, nonspecific binding of gluta-mine analogs is most likely attributable to their potential for alkylating nucleophilic side chains, e.g., sulfhydryl groups. Some nonspecific binding of chloroketone was observed with carbamyl phosphate synthetase II.[10] It could be reduced as the pH was lowered from 9.0 to 7.0 and could be virtually eliminated by pretreatment with N-ethylmaleimide, which blocked several sulfhydryl groups but, fortuitously, did not react with the glutamine binding site.[30] Little or no "nonspecific" binding of gluta-mine analogs has been reported with several enzymes.[8,16,23,31]

The diazo group, present in azaserine and DON, is normally inert to nucleophilic attack, and azaserine does not react with free cysteine or activated papain at pH 6.5.[32] Buchanan[29] has suggested that the reaction of azaserine at the glutamine binding site of formylglycinamide ribonu-cleotide amidotransferase probably involves a diazonium salt (shown in Fig. 1) formed by protonation after the binding of azaserine to the active site. Loss of the diazo group as nitrogen gas presumably generates a carbonium ion that reacts to form a covalent bond. With the chloroke-tone, a nucleophilic displacement reaction might just as easily result in covalent attachment. The unique reactivity of an amino acid side chain in the glutamine binding site might be ascribed to increased nucleo-philicity and proper orientation, which could both facilitate its reaction with the inhibitor and allow its participation in the cleavage of the amide group during the enzymic reaction.[10,16]

Many amidotransferases exhibit a weak glutaminase activity, which can be greatly enhanced when all substrates are present, as is the case during the enzymic reaction. Although the presence of other substrates usually is not required for affinity labeling with glutamine analogs, changes at the glutamine site resulting from the binding of substrates other than glutamine may enhance or decrease the rate of inactivation. For example, with carbamyl phosphate synthetase II, ATP-Mg^{2+} and

[30] V. P. Wellner and A. Meister, *J. Biol. Chem.* **250**, 3261 (1975).
[31] P. P. Trotta, K. E. B. Platzer, R. H. Haschemeyer, and A. Meister, *Proc. Natl. Acad. Sci. U.S.A.* **71**, 460 (1975).
[32] I. B. Dawid, T. C. French, and J. M. Buchanan, *J. Biol. Chem.* **238**, 2178 (1973).

bicarbonate retarded the rate of inactivation by chloroketone[10]; evidence derived from the titration of sulfhydryl groups suggests that this combination of substrates induces a conformational change.[33] The inactivation of anthranilate synthetase component II by chloroketone was enhanced by chorismate in the presence of anthranilate synthetase component I.[34] In cytidine triphosphate synthetase[35] a "negative cooperativity" phenomenon has been reported in which only half of the glutamine binding sites react with DON. The effector, GTP, which stimulates glutaminase activity, also increases the rate of reaction with the inhibitor. A similar phenomenon has been reported in phosphoribosyl pyrophosphate amidotransferase,[8] in which 1 mole of DON reacts per tetramer of enzyme. PRPP and Mg^{2+} decreases the apparent K_i for DON by two orders of magnitude and increases the rate of inactivation without altering the amount of bound inhibitor.

In certain amidotransferases, the glutamine-dependent activity may be sensitive to inhibition, possibly because of oxidation during isolation procedures. Hydrogen peroxide, which can be produced by oxidation of mercaptans in the presence of metal ions, has been shown to irreversibly and specifically inactivate glutamine binding sites in at least two cases.[36,37] Whereas the use of mercaptans may be necessary to increase overall stability in sensitive enzymes, precautions to prevent peroxide formation, such as the use of metal chelating agents, a pH below 7, a nitrogen atmosphere, or catalase, should be considered. If activation of a partially oxidized and thereby inhibited enzyme is seen after reducing it with mercaptans, this should not be misinterpreted as reversal of the inhibition caused by a glutamine analog.[37] If mercaptans are observed to protect an enzyme against irreversible inhibition by the chloroketone (III), the possibility of a reaction between this analog and the mercaptan should also be considered.

As with other active site-directed irreversible inhibitors that form a reversible complex prior to covalent attachment, glutamine analogs normally exhibit saturation kinetics. The K_i for inactivation of the glutamine binding site may be calculated from a plot of inactivation half-time ($\tau_{1/2}$) against the reciprocal inhibitor concentration[38] or from competitive inhibition data.[8]

[33] P. M. Anderson and S. V. Marvin, *Biochemistry* 9, 171 (1970).
[34] Y. Goto, H. Zalkin, P. S. Keim, and R. L. Heinrikson, *J. Biol. Chem.* 251, 941 (1976).
[35] C. W. Long, A. Levitski, and D. E. Koshland, Jr., *J. Biol. Chem.* 245, 80 (1970).
[36] P. Trotta, L. M. Pinkus, and A. Meister, *J. Biol. Chem.* 249, 1915 (1974).
[37] P. Mäntsälä and H. Zalkin, *J. Biol. Chem.* 251, 3294 (1976).
[38] R. Kitz and J. B. Wilson, *J. Biol. Chem.* 237, 3245 (1962).

TABLE II
SEPARATION OF CARBOXYMETHYL AMINO ACIDS

	System			
Compound	I^a (cm)	II^b (cm)	III^c R_f	IV^d (R_f)
S-Carboxymethylcysteine	12	0–1	0.38	0.27
S-Carboxymethylcysteine sulfone	12	6	0.18	0.10
3-Carboxymethylhistidine	6.5	0–1	0.61	—
S-Carboxymethylhomocysteine	11	1.5	0.45	0.35
S-Carboxymethylhomocysteine sulfone	—	—	0.30	0.12
O-Carboxymethylserine	10	2	—	—
1,3-Dicarboxymethylhistidine	8.5	4	0.47	—

[a] Electrophoresis in 0.04 M sodium acetate-acetic acid (pH 5.5) on Whatman No. 3 MM paper, 10°, 50 volts/cm, 30 min.

[b] Electrophoresis in 0.58 M acetic acid adjusted with pyridine (pH 3.0); 25° 12.5 volts/cm, 120 min, Whatman No. 3 MM paper [C. S. Hexter and F. H. Westheimer, *J. Biol. Chem.* **246**, 3934 (1971)].

[c] Chromatography in 80% phenol, Whatman No. 3 MM paper.

[d] Chromatography in butanol–acetic acid–H$_2$O (4:1:5), Whatman No. 1 paper [P. Mäntsälä and H. Zalkin, *J. Biol. Chem.* **251**, 3294 (1976)].

Identification of Reactive Amino Acid Residues

Starting with enzyme in which the glutamine binding site has been blocked with a radioactively labeled moiety derived from a glutamine analog, two methods have been used to identify the reactive residue. Proteolytic cleavage has been followed by isolation of labeled peptides[32,39]; and performic acid oxidation has been followed by acid hydrolysis and isolation of labeled carboxymethyl amino acid.[10] The results indicate that a reactive cysteine is present in the glutamine binding site of each of the amidotransferases that have been examined.[40] This cysteine has also been specifically labeled in certain enzymes by [^{14}C]iodoacetate[34,41] or [^{14}C]iodoacetamide.[37] A [^{14}C]carboxymethylcysteine can then be identified after direct acid hydrolysis or a [^{14}C]carboxymethylcysteine-containing peptide can be isolated.

Table II presents a number of chromatographic methods for separating carboxymethyl amino acids. Although cysteine has been the only reactive

[39] T. C. French, I. B. Dawid, and J. B. Buchanan, *J. Biol. Chem.* **238**, 2186 (1963).

[40] In some cases a cysteine has been identified indirectly by loss of a titratable sulfhydryl group after reaction with the glutamine analog.

[41] S. Ohnoki, B. S. Hong, and J. M. Buchanan, *Fed. Proc., Fed. Am. Soc. Exp. Biol.* **35**, 1549 (1976).

amino acid identified in glutamine binding sites, certain other amino acids that might form carboxymethyl derivatives are included for comparison.

Examples

[6-^{14}C]DON. The enzyme phosphoribosyl pyrophosphate amidotransferase was inactivated in Tris chloride at pH 8.0 containing PRPP, MgCl$_2$, and 2-mercaptoethanol. The inactivation reaction was facilitated by the presence of substrates that greatly decrease the apparent K_i for binding of the inhibitor at the glutamine binding site. After removal of free radioactivity by gel filtration, ammonium sulfate precipitation, or TCA precipitation of protein, about 1 mole of inhibitor remained bound per mole of enzyme.[8] Glutaminase A[42] was irreversibly inactivated by DON in sodium acetate at pH 5.0. Prior to inactivation the enzyme also catalyzes an unusual cleavage reaction (monitored at 274 nm) in which nitrogen, methanol, and glutamic acid are formed from DON; about 70 moles of DON are destroyed for each mole that reacts irreversibly. It is of interest that asparaginase carries out a similar cleavage reaction with DONV; in that case, the inactivation reaction was promoted by treating with the inhibitor in 50% dimethyl sulfoxide.[43]

[2-^{14}C]Azaserine. This was used to identify a reactive cysteine residue in formylglycinamide ribonucleotide amidotransferase.[16,32,39] The labeled enzyme was digested with papain, and the resulting peptides were separated after gel filtration and ion-exchange chromatography with triethylammonium acetate or pyridinium acetate buffers. After paper electrophoresis at pH 4.5 or 6.0, most of the radioactivity remained at the origin. However, several procedures altered the electrophoretic mobility of the main radioactive products, which then moved toward the anode at a pH above 3.5. Such alteration was observed after incubation at pH 8.0 and 38° for 1 hr, after storage at pH 4.5 for several days, or after evaporation from triethylammonium acetate buffers. The change in electrophoretic behavior was explained by the isolation of a peptide, N-[2-L-2-aminocarboxyethylthioacetyl]-L-serine, which yielded S-[^{14}C]carboxymethylcysteine and serine after acid hydrolysis. It was proposed[32,39] that the inhibitor alkylates a sulfhydryl group and that the hypothetical protein bound S-[^{14}C]methylenecarbonyl serine moiety, derived from azaserine, rearranges during isolation from an O-acyl to an N-acyl linkage with the free amino group of serine. This behavior has been reported

[42] S. C. Hartman and T. F. McGrath, J. Biol. Chem. 298, 8506 (1973).
[43] R. C. Jackson and R. E. Handschumacher, Biochemistry 9, 3585 (1970).

for other O-acyl serine derivatives.[44] By including 0.5% thioglycerol in the buffers that are used, decomposition of cysteine-containing peptides resulting from the formation of sulfoxides was decreased.

A two-step method for selectively blocking the azaserine-reactive cysteine residue with iodoacetate was reported by Ohnoki et al.[41] In the presence of glutamine, two nonessential cysteine residues were allowed to react with iodoacetate. After removal of glutamine, the enzyme was incubated with [^{14}C]iodoacetate in the presence of formylglycinamide and Mg^{2+}, resulting in the incorporation of about one additional mole of the inhibitor per mole of enzyme. The specific and limited reaction with iodoacetate was promoted by incubating at an acid pH. A peptide with the sequence Gly·Val·[^{14}C]CM Cys·Asp·Asx·CM Cys·Glx was isolated from the glutamine binding site.

[5-^{14}C]Chloroketone. This compound has been used to identify a cysteine residue in the glutamine binding site of carbamyl phosphate synthetase II.[9,10] Carbamyl phosphate synthetase II consists of a light subunit that alone possesses glutaminase activity and of a heavy subunit that alone catalyzes ammonia-dependent synthetase activity.[45] The intersubunit association, which confers the ability to utilize glutamine upon the heavy subunit, also significantly affects the glutamine binding site; i.e., the light subunit glutaminase activity exhibits an extremely high K_m for glutamine, 130–180 mM, whereas the native enzyme exhibits a much lower K_m, about 1 mM.[46] Treatment with [^{14}C]chloroketone inhibited the glutamine-dependent reactions of the native enzyme, but not the ammonia-dependent activity. The heavy and light subunits of the labeled enzyme were separated by gel filtration in 1 M KSCN.[10,45] About 1.8 moles of [^{14}C]chloroketone were bound to the enzyme of which 1.3 moles were associated with the light subunit. In an attempt to eliminate "nonspecific" binding (above 1 mole of inhibitor per mole of enzyme), the native enzyme was treated with N-ethylmaleimide which reacted with several sulfhydryl groups[47] and inhibited all reactions catalyzed by the enzyme with the exception of glutaminase activity. The N-ethylmaleimide-treated enzyme catalyzed an activated glutaminase reaction[30] which was completely inhibited after the binding of only 1 mole of chloroketone. Thus in carbamyl phosphate synthetase II, the chloroketone reactive cysteine in the light subunit appears to be buried and unavailable for

[44] L. Benoiton and H. N. Rydon, J. Chem. Soc. 30, 3328 (1960).
[45] P. P. Trotta, M. E. Burt, R. H. Haschemeyer, and A. Meister, Proc. Natl. Acad. Sci. U.S.A. 68, 2599 (1971).
[46] P. P. Trotta, L. M. Pinkus, R. H. Haschemeyer, and A. Meister, J. Biol. Chem. 249, 1915 (1974).
[47] R. Foley, J. Poon, and P. M. Anderson, Biochemistry 10, 4562 (1971).

reaction with N-ethylmaleimide although it remains reactive with the glutamine analog.

The product of a reaction between [5-^{14}C]chloroketone and enzyme is a hypothetical enzyme bound 4-oxo [5-^{14}C]L-norvaline moiety. The susceptibility of such protein bound, aliphatic β-keto derivatives to rearrangement or degradation is poorly understood. Direct acid hydrolysis did not yield an identifiable derivative[48]; however, after the enzyme was treated with 0.2% hydrogen peroxide–88% formic acid at 0° for 3 hr and subjected to acid hydrolysis, a [^{14}C]carboxymethylcysteine derivative was isolated in good yield.[10] A Baeyer–Villiger rearrangement,[49,50] in which the adduct of a peracid and ketone rearranges to form an ester of a carboxymethylcysteine derivative, was proposed to explain the results.[10]

The chloroketone reactive sulfhydryl group in the glutamine binding site of carbamyl phosphate synthetase reacts also with cyanate.[51,52] However, the inactive cyanate–enzyme complex regains activity upon incubation at alkaline pH. An S-[^{14}C]carbamylcysteine derivative was isolated from the [^{14}C]cyanate–enzyme complex after digestion with Pronase. S-Carbamyl cysteine moves with $R_f = 0.17$ in butanol–acetic acid–water (12:3:5, v/v).

Glutamate synthase, of *Aerobacter aerogenes*[31] or *Escherichia coli* K12,[37] is another amidotransferase with nonidentical subunits; labeling

---

[48] Direct acid hydrolysis of carbamyl phosphate synthetase labeled with [5-^{14}C]-chloroketone or of asparaginase labeled with [5-^{14}C]DONV (L. Lachman, personal communication) results in several as yet unidentified radioactive breakdown products. Since both of these analogs yield the same hypothetical 4-oxonorvaline moiety when bound to protein, studies with model derivatives relevant to the inactivation of asparaginase [P. K. Chang, L. B. Lachman, and R. E. Handschumacher, presented at the 172nd meeting of the American Chemical Society, Medicinal Chemistry Section, abstract no. 84 (1976)] may also have relevance to studies on glutamine binding sites. It is known that the β-keto ethers resulting from the reaction between DONV and derivatives of threonine or serine are very labile but may be reduced to β-hydroxy ethers with NaBH$_4$. The resulting 5-substituted 4-hydroxy-L-norvalines, or their degradation products after acid hydrolysis, may be comparable to the derivatives obtained from protein binding sites after treatment with [5-^{14}C]chloroketone, reduction, and hydrolysis. The feasibility of this approach was illustrated in studies of the specific reaction of 5-chloro-4-oxo[3,5-^{3}H]pentanoic acid with pyruvate kinase. After reduction and acid hydrolysis a cysteine derivative 4-hydroxy[3,5-^{3}H]pentanoic acid alanine thioether, was obtained from the active site [R. A. Chalkley and D. P. Bloxham, *Biochem. J.* **159**, 213 (1976)].

[49] A. Baeyer and V. Villiger, *Chem. Ber.* **32**, 3625 (1899).

[50] C. H. Hassall, *Org. React.* **9**, 73 (1967).

[51] P. M. Anderson, J. D. Carlson, G. A. Rosenthal, and A. Meister, *Biochem. Biophys. Res. Commun.* **55**, 246 (1973).

[52] P. M. Anderson and J. D. Carlson, *Biochemistry* **14**, 3690 (1975).

with [^{14}C]chloroketone enabled localization of the glutamine binding site on the heavy subunit of the enzyme. After inactivation with about 1 mole of [^{14}C]chloroketone, dissociation of the subunits was accomplished by boiling the protein in a solution containing 1% SDS and 1% 2-mercaptoethanol at pH 8.5 followed by gel electrophoresis; after disintegration of the gel in 30% H_2O_2 substantial amounts of radioactivity were found associated only with the heavy subunit.[31] A cysteine that reacted with chloroketone or iodoacetamide was found in the active site.[37]

Comments

Studies with glutamine analogs have greatly increased our understanding of the action of glutamine-utilizing enzymes. Considerable evidence indicates that separate sites exist for glutamine and ammonia and that a γ-glutamyl thioester is a central intermediate in the cleavage of glutamine.[10,39,53] It has been suggested that amidotransferases may have developed by the association of a primordial glutaminase peptide, either in a separate subunit or covalently incorporated, with transferases specifically designed for different acceptors.[45,54] Comparison of peptides from glutamine binding sites of different enzymes is now possible and may lead to a better understanding of the evolution of this group of proteins. Inhibitors of glutamine utilization should also prove useful in studies of glutamine transport.

[53] H. Nagano, H. Zalkin, and E. J. Henderson, *J. Biol. Chem.* **245**, 3810 (1970).
[54] H. C. Li and J. M. Buchanan, *J. Biol. Chem.* **246**, 4713 (1971).

[46] Labeling of the Active Site of L-Aspartate β-Decarboxylase with β-Chloro-L-alanine

By NOEL M. RELYEA, SURESH S. TATE, and ALTON MEISTER

L-Aspartate-β-decarboxylase, a pyridoxal 5′-phosphate enzyme, catalyzes the β-decarboxylation of L-aspartate to L-alanine [Eq. (1)] as

$$\text{L-Aspartate} \rightarrow \text{L-alanine} + CO_2 \qquad (1)$$

well as a number of other reactions,[1] which include desulfination of L-cysteine sulfinate to L-alanine,[2] α-decarboxylation of aminomalonate to glycine,[3] decarboxylation of *threo-* and *erythro*-β-hydroxyl-L-aspartate

[1] S. S. Tate and A. Meister, *Adv. Enzymol.* **35**, 503 (1971).
[2] K. Soda, A. Novogrodsky, and A. Meister, *Biochemistry* **3**, 1450 (1964).
[3] A. G. Palekar, S. S. Tate, and A. Meister, *Biochemistry* **9**, 2310 (1970).

to L-serine,[4] various transamination reactions,[5-7] and α,β-elimination reactions such as that of β-chloro-L-alanine[8] [Eq. (2)].

$$
\begin{array}{c}
\text{CH}_2\text{Cl} \\
| \\
\text{CHNH}_3^+ \\
| \\
\text{COO}^-
\end{array}
+ \text{H}_2\text{O} \longrightarrow
\begin{array}{c}
\text{CH}_3 \\
| \\
\text{C}=\text{O} \\
| \\
\text{COO}^-
\end{array}
+ \text{NH}_4^+ + \text{H}^+ + \text{Cl}^-
\qquad (2)
$$

When the enzyme is incubated with β-chloro-L-alanine, the rate of the catalytic reaction declines rapidly as the enzyme becomes irreversibly inhibited. Such inactivation is associated with the binding of close to 1 mole of the 3-carbon chain of the substrate analog per mole of active site.[8] The enzyme, after inactivation by treatment with β-chloro-L-[14C]alanine, was treated with cyanogen bromide and a 14C-labeled peptide was then isolated. Treatment of the labeled peptide with mild alkali leads to release of β-hydroxy-[14C]pyruvate, suggesting an ester linkage. Ammonolysis of the labeled peptide leads to release of the label, and amino acid analyses of enzymic hydrolyzates of the peptide before and after ammonolysis show formation of an equivalent amount of glutamine[9] [Eq. (3)]. The findings indicate that the labeled derivative

$$
\begin{array}{c}
-\text{Glu}- \\
| \\
\text{O}=\text{C} \\
| \\
\text{O} \\
| \\
\text{CH}_2 \\
| \\
\text{C}=\text{O} \\
| \\
\text{COO}^-
\end{array}
\xrightarrow{\text{NH}_3}
\begin{array}{c}
-\text{Glu}- \\
| \\
\text{O}=\text{C}-\text{NH}_2 \\
\\
+ \\
\\
\text{CH}_2\text{OH} \\
| \\
\text{C}=\text{O} \\
| \\
\text{COO}^-
\end{array}
$$

(β-hydroxypyruvate) is bound by an ester linkage to a glutamate residue at the active center of the enzyme.

Preparation of β-Chloro-L-[14C]alanine

β-Chloro-L-alanine labeled with 14C may be prepared from any of the various preparations of [14C]serine that are commercially available

[4] E. Miles and A. Meister, *Biochemistry* **6**, 1734 (1967).
[5] S. S. Tate and A. Meister, *Biochemistry* **9**, 2626 (1970).
[6] A. Novogrodsky, J. S. Nishimura, and A. Meister, *J. Biol. Chem.* **238**, PC 1179 (1963).
[7] A. Novogrodsky and A. Meister, *J. Biol. Chem.* **239**, 879 (1964).
[8] S. S. Tate, N. Relyea, and A. Meister, *Biochemistry* **8**, 5016 (1969).
[9] N. M. Relyea, S. S. Tate, and A. Meister, *J. Biol. Chem.* **249**, 1519 (1974).

according to the procedure of Fischer and Raske[10] as modified,[11,12] which involves chlorination of serine methylester with phosphorus pentachloride. In this procedure it is of the utmost importance to maintain scrupulously dry conditions. L-[^{14}C]Serine-HCl (1.6 μmoles) and 75 mg (0.53 mmole) of unlabeled L-serine-HCl are dissolved in 3 ml of dry methanol. Dry hydrogen chloride gas (dried by bubbling through concentrated H_2SO_4) is bubbled through the methanolic solution of serine at 0° for 30 min. The solution is allowed to warm to room temperature and then refluxed for 30 min. After cooling, the solution is flash-evaporated to dryness and the residue is suspended in dry benzene; the benzene is removed by flash-evaporation. The benzene step is repeated. The crystalline residue is dissolved in 1 ml of dry methanol, and 50 ml of dry diethyl ether are added slowly at 26°. After chilling in ice for several hours, crystals of serine methylester hydrochloride are collected and washed with dry diethyl ether (m.p. 129°–131°).

Chloroform is passed through an alumina column (Woelm) and stored over a Linde 4A molecular sieve for not more than 2 hr before use. Phosphorus pentachloride (135 mg) is added to 2.5 ml of dry chloroform and dissolved by stirring at 26° for 30 min. The solution is then chilled on ice and 105 mg of the serine methyl ester hydrochloride (prepared as described above) is added in small portions over a 30-min period with stirring. The solution is stirred at 26° for 2 hr until it clarifies; crystals then begin to form. The mixture is placed at 0° for 18 hr; dry petroleum ether (b.p. 70°–90°) is added and the crystals of β-chloro-L-alanine methylester hydrochloride are collected and washed with dry petroleum ether. The crystals (92 mg) are suspended in 3 ml of 6 M HCl and refluxed for 1 hr. The solution is flash-evaporated, and the residue is dissolved in water and flash-evaporated again. This procedure is repeated once with water and then twice with benzene.

Labeling of the Enzyme

The inactivation and labeling of L-aspartate-β-decarboxylase probably involves nucleophilic attack by a group on the enzyme on the β-carbon atom of the α-aminoacrylate-Schiff base formed in the interaction of β-chloro-L-alanine and enzyme-bound pyridoxal 5′-phosphate.[8] Labeling of the enzyme thus requires conditions under which the catalytic reaction can occur. An excess of β-chloro-L-alanine is therefore required, since a substantial portion of the analog is converted to pyruvate before

[10] E. Fischer and E. Raske, *Ber. Deut. Chem. Ges.* **40**, 3717 (1907).
[11] C. T. Walsh, A. Schonbrunn, and R. H. Abeles, *J. Biol. Chem.* **246**, 6855 (1971).
[12] J. P. Greenstein and M. Winitz, "Chemistry of the Amino Acids," p. 2677. Wiley, New York, 1961.

the enzyme is completely inactivated. L-Aspartate-β-decarboxylase (1 mg/ml) is incubated in 0.2 M sodium acetate buffer (pH 5.5) containing 10 mM β-chloro-L-[^{14}C]alanine for 20 min at 37°.[9] Under these conditions, the enzyme is completely inactivated. The labeled enzyme is precipitated at 5° by adding solid ammonium sulfate to 65% of ammonium sulfate saturation. The precipitated enzyme is recovered by centrifugation and dissolved in a small volume of 1 M sodium acetate at pH 6 and then dialyzed exhaustively against water to remove salts; it is then lyophilized.

Cleavage of the Labeled Enzyme to Yield a Labeled Peptide

The labeled moiety introduced into the enzyme by treatment with β-chloro-L-[^{14}C]alanine is labile to alkali; the following procedures are therefore carried out at acid values of pH. The labeled enzyme (5 mg/ml) is suspended in 70% formic acid and a 5-fold (w/w) excess of cyanogen bromide is added. The solution is stirred with a magnetic stirrer in a tightly stoppered flask for 24 hr at 25°. The reaction is terminated by adding 10 volumes of cold water, and the solution is lyophilized.

The labeled peptide is purified by cation-exchange chromatography[13] as follows. A water-jacketed column (45 × 0.9 cm) containing Technicon type P cation exchange resin is used at 50°. The cyanogen bromide peptides are suspended in 2 ml of 0.2 M pyridine acetate at pH 3.1. After clarification by centrifugation, the solution is applied to the column and the peptides are eluted with a linear gradient established between 250 ml of 0.2 M pyridine acetate at pH 3.1, and 250 ml of 2 M pyridine acetate at pH 5.0, using a flow rate of 30 ml per hour; fractions of 2.5 ml are collected. Examination of the fractions for radioactivity and for amino acid content after hydrolysis (6 M HCl, 105°, 18 hr) leads to identification of the first peak obtained from the column (fractions 4–6) as the major labeled cyanogen bromide peptide. When this peptide is subjected to acid hydrolysis followed by amino acid analysis, all the radioactivity elutes in an early ninhydrin-negative peak. Complete enzymic hydrolysis of the labeled peptide can also be achieved by cleavage with α-chymotrypsin[14] followed by treatment with leucine aminopeptidase.[9] The amino acid composition found after complete enzymic hydrolysis is similar to but not the same as that found after acid hydrolysis. Thus, two aspartate residues are found after acid hydrolysis, and one aspartate and one asparagine residue are found after enzymic hydrolysis. On amino acid analysis after enzymic digestion, all the label appears in

[13] R. T. Jones, "Methods of Biochemical Analysis" (D. Glick, ed.), Vol. 18, pp. 205–258. Wiley (Interscience), New York, 1970.

[14] D. G. Smyth, this series, Vol. 11, p. 214 (1967).

a ninhydrin-negative peak that elutes early. The nature of the labeled material may be established by paper chromatography using appropriate standards[9]; thus, the labeled material moves with authentic β-hydroxypyruvate. It may also be identified as the corresponding 2,4-dinitrophenylhydrazone.[15]

Identification of the Protein Amino Acid Residue That Reacts with β-Chloro-L-alanine

The alkali lability of the labeled enzyme derivative suggests that an ester of either glutamic acid or aspartic acid is formed by interaction of the enzyme with β-chloro-L-alanine. Evidence derived from sequencing of the labeled peptide indicates that the label is not attached to the aspartate residue of the peptide, and thus suggests that the label is attached to a glutamate residue. Ammonolysis of esters to yield amides is a well known procedure,[16] which may be usefully applied to the study of β-chloro-L-alanine-labeled aspartate-β-decarboxylase. Thus, a sample of the labeled peptide (12 nmoles) is incubated in 0.5 ml of concentrated ammonium hydroxide (28% NH_3) for 16 hr at 26°. After flash-evaporation to dryness, the residue and a sample of the peptide not treated with ammonia are completely digested by successive treatment with chymotrypsin[14] and leucine aminopeptidase.[9] In studies in which this procedure was followed, amino acid analysis showed that glutamine is formed and that glutamate disappears in the sample subjected to ammonolysis. The amount of glutamine formed is in close quantitative agreement with the molar amount of labeled residue bound to the peptide and released during ammonolysis. No glutamine is found after enzymic digestion of the sample of peptide which was not subjected to ammonolysis. As a further control, a sample of unlabeled peptide obtained by cyanogen bromide treatment of the unlabeled enzyme may be subjected to ammonolysis; no glutamine is found in the enzymic digest of this material.[9]

Discussion

β-Chloro-L-alanine acts as a substrate and as an inhibitor of L-aspartate-β-decarboxylase. The relatively rapid catalytic reaction is accompanied by a slower reaction in which the enzyme becomes inactivated. Several experimental approaches support the conclusion that β-chloro-L-alanine is a true active site-directed inhibitor. (1) Only the L-isomer of β-chloroalanine is a substrate and an inhibitor. (2) Treatment of the

[15] A. Meister and P. A. Abendschein, *Anal. Chem.* **28**, 171 (1956).
[16] J. P. Greenstein and M. Winitz, "Chemistry of the Amino Acids," p. 1258. Wiley, New York, 1961.

enzyme with β-chloro-L-alanine leads to a spectral shift from a peak absorbance (native enzyme) at 355 nm to a peak at about 320 nm indicating that Schiff base formation is associated with inactivation. (3) When the apoenzyme or the 4'-deoxypyridoxine-5'-phosphate enzyme is treated with β-chloro-L-alanine, neither inactivation nor binding of the substrate analog occurs.

Cytoplasmic glutamate–aspartate transaminase also catalyzes α,β-elimination of β-chloro-L-alanine with concomitant inactivation of the enzyme.[17,18] In this inactivation, there is evidence that the substrate analog binds to the ϵ-amino group of the lysyl residue that is normally involved in binding pyridoxal 5'-phosphate. A number of reports have appeared on the esterification of specific protein carboxyl groups.[19] Specificity may be due to enhanced reactivity of a particular enzyme carboxyl group or to the ability of the substrate analog to bind specifically, or to both. Several methods have been used to identify ester linkages. Takahashi et al.[20] found that inactivation of ribonuclease T by iodoacetate is accompanied by the incorporation of one carboxymethyl group per molecule of enzyme. An ester linkage was suggested by the finding that glycolic acid was released when the labeled enzyme was treated with hydroxylamine. The ester bond was stable during enzymic hydrolysis of the protein, thus facilitating isolation of the labeled residue, which was shown to be identical to authentic γ-carboxymethyl ester of glutamic acid. Studies on triosephosphate isomerase, in which 3-halogenoacetol phosphate[21] and glycidol phosphate[22,23] were used as active site-directed inhibitors, showed that a specific enzyme glutamic acid residue is esterified.

[17] Y. Morino and M. Okamoto, *Biochem. Biophys. Res. Commun.* **47**, 498 (1972).
[18] Y. Morino and M. Okamoto, *Biochem. Biophys. Res. Commun.* **50**, 1061 (1973).
[19] P. E. Wilcox, this series, Vol. 25, p. 596 (1972).
[20] K. Takahashi, W. H. Stein, and S. Moore, *J. Biol. Chem.* **242**, 4682 (1967).
[21] F. C. Hartman, *Biochemistry* **10**, 146 (1971).
[22] J. C. Miller and S. G. Waley, *Biochem. J.* **123**, 163 (1971).
[23] S. G. Waley, J. C. Miller, I. A. Rose, and E. L. O'Connell, *Nature (London)* **227**, 181 (1970).

[47] Active Site of L-Asparaginase: Reaction with Diazo-4-oxonorvaline

By ROBERT E. HANDSCHUMACHER

L-Asparaginase[1] from *Escherichia coli* catalyzes the conversion of the diazo ketone analog of L-asparagine, diazo-4-oxo-L-norvaline (DONV),

[1] J. C. Wriston, Jr. and T. O. Yellin, *Adv. Enzymol.* **39**, 185 (1973).

to 5-hydroxy-4-oxo-L-norvaline and is also inactivated by covalent attachment of the analog to a region in the active site.[2] In aqueous buffers the decomposition of DONV is so rapid (6 μmoles/min per milligram of enzyme) in comparison to the inactivation reaction (3 nmoles of enzyme inactivated per minute) that labeling of the active site is not technically reasonable. If, however, L-asparaginase is permitted to react with DONV in the presence of 50% dimethyl sulfoxide (DMSO), there is virtually no catalytic decomposition of DONV, but a 400-fold increased rate of inactivation by DONV. The enzyme itself is stable in 50% DMSO for several hours at 25° in the absence of DONV, and at least 85% activity can be regained within a minute by dilution with aqueous buffers. The K_m for the inactivation reaction is similar in aqueous (70 μM) and DMSO buffers (90 μM). Under these inactivation conditions 4 moles of L-DONV bind to each tetrameric form of the enzyme. The specificity of the inactivation process can be demonstrated by the competitive inhibition caused by the natural substrate, L-asparagine, since in 50% DMSO the rate of hydrolysis of asparagine is less than 2% of that in aqueous buffers.

Synthesis of 5-[^{14}C]DONV

[^{14}C]Diazomethane is generated from N-methyl-[^{14}C]N-nitroso-p-toluenesulfonamide (diazald) according to an adaptation of the procedure reported in A. I. Vogel.[3] [^{14}C]Diazald (214 mg, 1 mCi/mmole, New England Nuclear) is placed in the distillation flask of a semimicro distillation apparatus and treated with an ice-cold solution of KOH (40 mg) in 95% ethanol (1 ml). After 5 min the ethereal diazomethane is distilled under nitrogen.

An ethereal solution of freshly prepared L-β-methoxycarbonyl-β-trifluoroacetamidopropionyl chloride[4] (110 mg in 5 ml of ether) is added dropwise at 0° to the ethereal solution of [^{14}C]diazomethane until only a pale yellow color remains with the reaction mixture, indicating a slight excess of diazomethane. After 20 min, the solvent and reagent are removed under reduced pressure, and the residual ester, a white solid, is dissolved in methanol (4 ml) and hydrolyzed with 0.2 N NaOH (8 ml) at —16° for 24 hr. The orange solution is adjusted to pH 6.2 with 0.1 N HCl at 4° and lyophilized. The dark red residue is dissolved in H$_2$O (1 ml), and

---

[2] R. C. Jackson and R. E. Handschumacher, *Biochemistry* **9**, 3585 (1970).

[3] A. I. Vogel, "Practical Chemistry," 3rd ed., p. 971. Longmans, Green, New York, 1957.

[4] Y. Liwschitz, R. D. Irsay, and A. I. Vincze, *J. Chem. Soc.* (*London*) **1959**, 1308.

the solution and washings (5×1 ml) are applied to a column (1×20 cm) of charcoal (Darco G-60, Atlas Chemical Industries) and Celite 535 (1:1 by weight). The column is eluted with 1% aqueous acetone at 4°. The fractions containing DONV are detected by the $A_{274 \text{ nm}}$ ($\epsilon^{1 \text{ cm}}_{274 \text{ nm}} = 11,000$) and lyophilized to remove acetone. The yield is consistently 35–50 mg of 5-[^{14}C]DONV (1 mCi/mmole).[5] The [^{14}C]DONV is diluted to give a specific activity of 0.34 mCi/mmole. Immediately before use, a 0.6 mM solution in sodium phosphate (0.05 M, pH 7.0) is filtered through an Amicon P-10 membrane to remove any polymeric material.

Reaction Conditions

L-Asparaginase (100 mg) dissolved in 45 ml of sodium phosphate (0.05 M, pH 7.0) was mixed with an equal volume of DMSO, and the temperature was maintained at 25° by cooling in ice. Fifty milliliters of the freshly prepared solution of [^{14}C]DONV (0.6 mM) were also diluted with an equal volume of DMSO. The [^{14}C]DONV and enzyme solutions were mixed rapidly at 25°, and appropriate samples were assayed for residual activity in the coupled enzyme method. If it is essential to minimize nonspecific labeling, a 10-fold excess of nonradioactive DONV (300 μmoles) in 20 ml of 50% DMSO:phosphate buffer can be added after 3 min when 30–50% loss of activity will have occurred. To assure maximal inactivation (90–95%) the reaction was allowed to proceed for 30 min at 25°. The reaction mixture was then dialyzed at 4° for 24 hr against the phosphate buffer (4 mM, pH 7.0) and lyophilized.

Under these conditions, the amount of [^{14}C]DONV bound to the enzyme is proportional to the degree of inactivation, and during the rapid inactivation phase 4 moles of DONV residue have been bound per mole of enzyme (135,000 daltons), a result consistent with the tetrameric struc-true of identical subunits. Continued exposure to DONV will result in more nonspecific labeling at a much slower rate. The DONV residue is stable to trichloroacetic acid precipitation but is released by 1 N HCl in 30 min at 110°. Using chymotryptic digestion, a decapeptide (Val-Gly-Ala-Met-Arg-Pro-Ser-Thr-Ser-Met) corresponding to residues 111 to 120 in the primary structure[6] has been isolated as the primary radioactive peptide.[7] The DONV residue is attached to either serine or threonine, presumably by a keto ether linkage.

[5] R. E. Handschumacher, C. J. Bates, P. K. Chang, A. T. Andrews, and G. A. Fischer, *Science* **161**, 62 (1968).

[6] T. Maita, K. Morokuma, and G. Matusuda, *J. Biochem.* **76**, 1351 (1974).

[7] R. G. Peterson, F. F. Richards, and R. E. Handschumacher, *J. Biol. Chem.*, in press (1977).

Comments

The buffer used in the 50% DMSO solution can change the balance between catalytic decomposition of DONV and enzyme inactivation. If instead of sodium phosphate, Tris or ammonium phosphate is used as buffer, sufficient catalytic activity is retained in 50% DMSO to render active site labeling difficult because of the rapid decomposition of the [^{14}C]DONV. The extreme of this effect may be seen with hydroxylamine (0.1 M), which, in aqueous buffer, accelerates the catalytic decomposition of DONV 3-fold and eliminates covalent inactivation in the DMSO buffer system. Inactivation by the D-isomer of DONV can be accomplished at 27% of the rate with L-DONV. Although the enzyme exhibits considerable stereospecificity, D-asparagine provides a useful means of occupying the active site since it has a K_m close to that of the L-isomer but is a poor substrate ($V_{max} = 2\%$ V_{max} for L-asparagine).[1]

Attempts to label the active site with 5-chloro or 5-bromo-4-oxo-L-norvaline have not been successful; these haloketones are also not substrates for the enzyme to a discernible degree. Although aspartic acid semialdehyde is a potent inhibitor of the enzyme,[s] attempts to bind it to asparaginase covalently by NaBH$_4$ reduction have not been successful. The specificity of the affinity label by DONV is seen in the lack of activity of the glutamine analog 6-diazo-5-oxo-L-norleucine (DON) as a covalent inactivator or substrate for $E.$ $coli$ L-asparaginase. This is of some interest, since glutamine is hydrolyzed at about 3% of the rate with asparagine and the glutaminase activity is lost coincident with asparaginase activity as inactivation proceeds. By contrast, the glutaminase of $E.$ $coli$ is covalently inactivated by DON. This analog also can serve as a very poor substrate for the enzyme but appears to yield glutamic acid and formaldehyde (see this volume [45]). Several enzymes that readily accept both asparaginase and glutamine as substrates have also been studied, and comments on affinity labeling of their active sites may also be found in this volume [45].

[s] J. O. Westerik and R. Wolfenden, $J.$ $Biol.$ $Chem.$ **249**, 6351 (1974).

[48] Labeling of Serum Prealbumin with N-Bromoacetyl-L-thyroxine

By Sheue-Yann Cheng

Human prealbumin is involved in the transport of thyroid hormones in plasma. N-Bromoacetyl-L-thyroxine [BrAcT$_4$; (II) in Scheme 2], an analog of L-thyroxine [T$_4$; (I) in Scheme 2], differs from T$_4$ only in the replacement of one amino hydrogen with —CO—CH$_2$Br, a group capable of reacting with nucleophiles in the active site of proteins. Its affinity constants ($K_1 = 1 \times 10^8$ M^{-1}, $K_2 = 1 \times 10^6$ M^{-1}), determined by fluorescence titration, are identical to those of T$_4$.[1]

When BrAcT$_4$ reacts with prealbumin, thyroxyl-N-carbonylmethyl-prealbumin (T$_4$CM-PA) is formed (Scheme 1). Modified amino acids

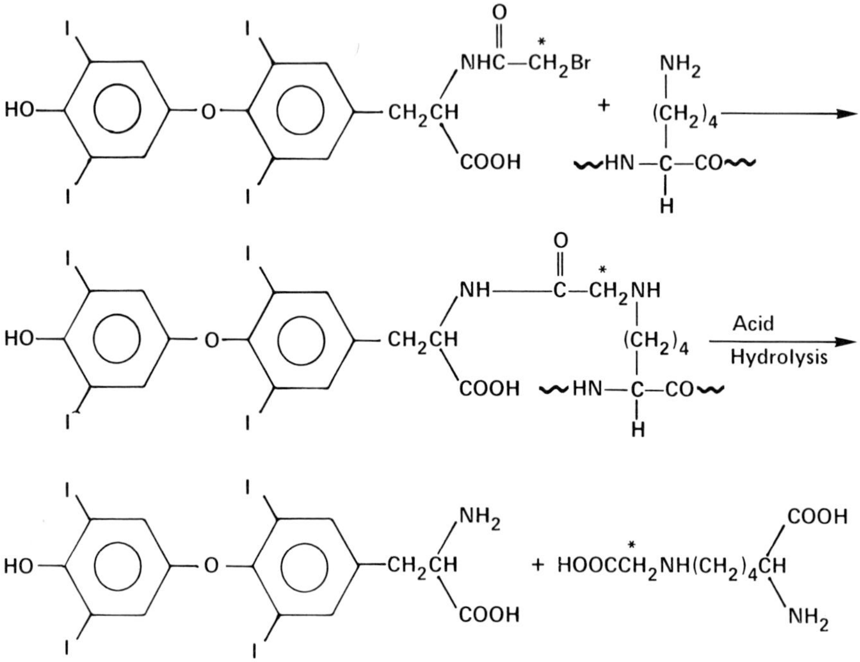

SCHEME 1. Formation and acid hydrolysis of thyroxyl-N-carbonylmethyl prealbumin; lysine is used as an example of nucleophile.

are obtained as carboxymethylamino acids after acid hydrolysis of T$_4$CM-PA (Scheme 1). These derivatives are easily identified by amino acid analysis or high-voltage paper electrophoresis.

Synthesis of N-Bromoacetyl-L-thyroxine

Method A (Scheme 2) is used in the small-scale synthesis of labeled Br-2-[^{14}C]AcT$_4$ for affinity labeling experiments. For the large-scale

---

[1] R. N. Ferguson, H. Edelhoch, H. A. Saroff, and J. Robbins, *Biochemistry* **14**, 282 (1975).

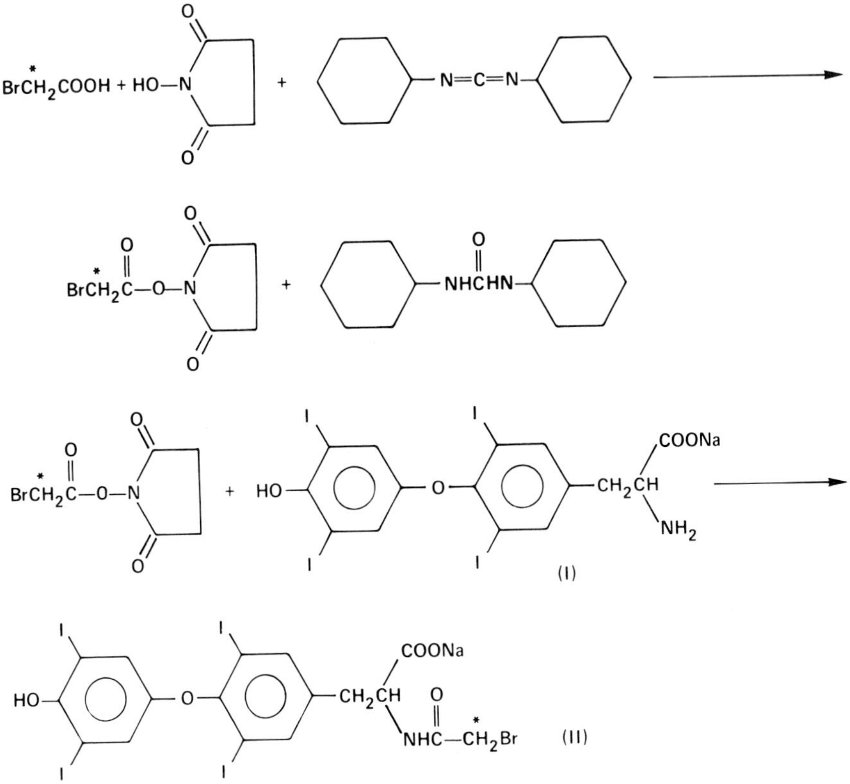

SCHEME 2. Synthesis of N-bromoacetyl-L-thyroxine (Method A and B).

synthesis of unlabeled BrAcT$_4$ to be used as a reference substance, Method B (Scheme 2) or Method C (Scheme 3) may be employed.

Method A. A solution of 1.1 mg (9.6 μmoles) of N-hydroxysuccinimide (purified by recrystallization from ethyl acetate) in 20 μl of redistilled dimethylformamide and a solution of 2.3 mg (11 μmoles) of N,N'-dicyclohexylcarbodiimide (purified by distillation) in 20 μl of the same solvent are added sequentially to a lyophilized powder of 2-[^{14}C]bromoacetic acid (1.3 mg, 9.3 μmoles; 27 Ci/mole). After mixing, the resultant clear solution is incubated at room temperature for 45 min. A precipitate of crystalline dicyclohexylurea forms in the course of the incubation. A micropipette, made by drawing out the narrow part of a disposable Pasteur pipette (9 inches long), is used to transfer the supernatant liquid, without disturbing the precipitate, to a 12-ml conical centrifuge tube containing 7.4 mg (8.3 μmoles) of T$_4$ (sodium salt pentahydrate). Two rinses, each with 20 μl of dimethylformamide, are transferred similarly.

The tube is shaken to dissolve the T_4, then kept at room temperature for 1.5 hr. The brownish solution is applied to a prewashed silica gel plate (Q1-F, 20×20 cm, Quantum Industries, Fairfield, New Jersey). This plate has a soft layer (250 μm) of silica gel containing a fluor. The prewashing is done with ethyl acetate–acetic acid (9:1), and a chromatogram is developed with the same solvent. The strong fluorescence-quenching band, with a R_f of approximately 0.4, corresponding to $BrAcT_4$, is scraped off, the powder is poured into a glass column fitted with a sintered-glass disk, and the $BrAcT_4$ is eluted with 4 ml of dioxane which had been freshly distilled over Drierite. (Handling of the powder is done in a hood.) For immediate use, an appropriate aliquot of the eluate is evaporated at $<30°$ in a rotating evaporator under reduced pressure, and the residue is taken up in a suitable amount of dioxane (see affinity labeling procedure). The eluate may be stored at $-25°$. The yield of Br-2-[^{14}C]AcT$_4$ is 15%.

For the large-scale preparation of labeled T_4CM-PA, it may be necessary to reduce the specific activity of the 2-[^{14}C]bromoacetic acid by addition of unlabeled bromoacetic acid. Quantitation of $BrAcT_4$ may be done spectrophotometrically using $\epsilon_{299 \text{ nm}}$ in ethyl acetate = 4248.

Method B. A solution of N,N'-dicyclohexylcarbodiimide (41 mg; 0.2 mmole) in 0.4 ml of dioxane is added to a glass-stoppered test tube containing a mixture of bromoacetic acid (27.8 mg, 0.2 mmole) and N-hydroxysuccinimide (23 mg, 0.2 mmole). (The reagents have been purified as in Method A.) The tube is shaken, causing dissolution of the solids, followed by precipitation of dicyclohexylurea. After incubation for 30 min at room temperature, the precipitate is removed by filtration and washed three times with 0.5-ml portions of redistilled dimethylformamide. Solid T_4 (sodium salt; 178 mg, 0.2 mmole) is added to the combined filtrate and washings. After mixing and incubation at room temperature for 20 min, the yellowish solution is poured into 20 ml of an ice-cold 20% (w/v) aqueous solution of citric acid. Two extractions with ethyl acetate ($\sim$20 ml each), followed by three washings of the combined extracts with 10-ml portions of a saturated aqueous solution of NaCl, drying over anhydrous $MgSO_4$, and concentration at $<30°$ to a small volume, yields slightly impure $BrAcT_4$. Final purification is accomplished by perparative thin-layer chromatography (Quantum Industries, PLQF-1000, 1-mm silica gel layer) as described in Method A.

Method C (Scheme 3). This method is based on that used by Matsuura and Cahnmann[2] for the synthesis of chloroacetyl-T_4. To a

---

[2] T. Matsuura and H. J. Cahnmann, *J. Am. Chem. Soc.* **81**, 871 (1959).

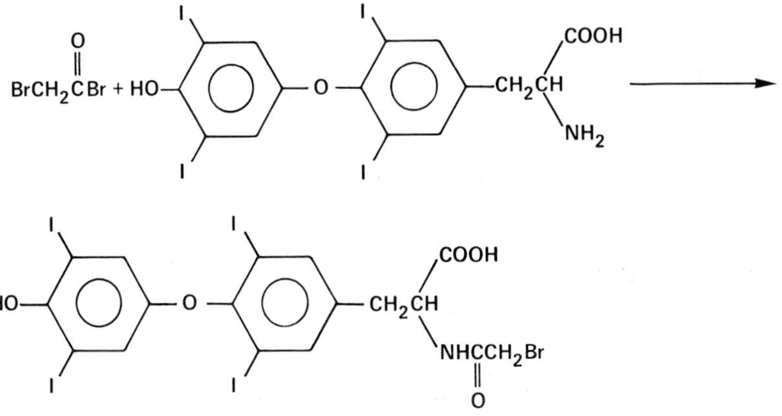

Scheme 3. Synthesis of N-bromoacetyl-L-thyroxine (Method C).

solution of 174 μl of bromoacetyl bromide (2 mmoles) in 10 ml of anhydrous ethyl acetate is added 155 mg (0.2 mmole) of solid T_4 (free acid).[3] The mixture is refluxed with a slow stream of dry nitrogen passing through it until a clear or almost clear solution is obtained (45–60 min). After concentration to a small volume at $<30°$, a small amount of precipitate is removed by filtration through a glass fiber filter (Reeve Angel 934 AH). The filtrate, containing $BrAcT_4$, is purified as described above by preparative thin-layer chromatography.[4]

Affinity Labeling of Prealbumin with N-Bromo-2-[14C]acetyl-L-thyroxine

Procedure. Prealbumin has both a strong and a weak binding site for $BrAcT_4$. In order to avoid or minimize nonspecific binding, a molar ratio of $BrAcT_4$ to PA of 2 is used.

---

[3] Bromoacetyl bromide is injurious to skin and mucous membranes on contact and should be handled with care. The free-acid form of T_4 may be prepared as described by H. J. Cahnmann *in* "The Thyroid" (J. E. Rall, ed.) [Vol. 1 of "Methods in Investigative and Diagnostic Endocrinology" (S. A. Berson, ed.)], p. 31. North-Holland Publ., Amsterdam, 1972.

[4] Repeated evaporation to dryness of solutions of $BrAcT_4$ in ethyl acetate should be avoided, since the dry residue thus obtained is no longer completely soluble in ethyl acetate. Overloading of the chromatographic plates (>50 mg per plate) results in poor separation. In that case, separation can be much improved by developing the chromatogram a second and third time, after brief drying of the plate.

To a solution of purified prealbumin[5] (2 mg, 37 nmoles) in 1.9 ml of 0.1 M NaHCO$_3$ at pH 8.6 is added Br-2-[^{14}C]AcT$_4$ (66 μg, 73 nmoles) in 0.1 ml of dioxane. The solution is mixed gently by swirling and then incubated at room temperature for 24 hr. Thereafter, it is dialyzed against 50 ml of 37 μM 3,3′,5,5′-tetraiodothyropropionic acid in 50 mM sodium phosphate–0.1 M NaCl at pH 8.1 for 24 hr with four changes to displace noncovalently bound labeling reagent. The salt and tetraiodothyropropionic acid are removed by further dialysis against two changes of deionized water for 2 hr at room temperature. The labeled protein (T$_4$CM-PA) is lyophilized.

Determination of the Molar Ratio of Covalently Linked Ligand (T$_4$CM) to PA. Owing to the presence of a small amount of tetraiodothyropropionic acid in T$_4$CM-PA obtained by the procedure described above, the amount of covalently linked ligand is best determined by amino acid analysis rather than spectrophotometrically. From the known amino acid composition[6] and the specific activity of BrAcT$_4$, a molar ratio of T$_4$CM to prealbumin of 1 is found.

Kinetic Study of the Labeling. The reaction of BrAcT$_4$ with nucleophiles in the protein is a slow process. Only 0.2 and 0.4 ml of T$_4$CM per mole of prealbumin are bound after 2 and 4 hr, respectively, under the described conditions. A plateau of 1 mole per mole of prealbumin is reached after 20 hr.

Identification of Modified Amino Acids. High-voltage paper electrophoresis[7] is used to identify the labeled amino acid residues. The dry residue obtained after acid hydrolysis[8] of 0.5 mg (9.1 nmoles) of T$_4$CM-PA is dissolved in 0.5 ml of deionized water and filtered through a Millipore filter (pore size, 0.22 μm), and the filtrate is evaporated under reduced pressure. The same process is repeated twice to remove all HCl. The dry residue is dissolved in 30 μl of pyridine–acetic acid–water (1:10:289, pH 3.6) and 20 μl of the solution is applied to Whatman No. 3 MM paper. Electrophoresis is carried out in the same solvent for 1 hr at 2500 volts. Two radioactive spots are identified as N^ε-carboxymethyllysine and iminodiacetic acid. The former moves slightly toward the cathode (1 cm from the origin), and the latter toward the anode (25 cm from the origin). The distribution of the radioactivity between N^ε-carboxymethyllysine and iminodiacetic acid is 7:3.

[5] S.-Y. Cheng, H. J. Cahnmann, M. Wilchek, and R. N. Ferguson, *Biochemistry* **19**, 4132 (1975).

[6] Y. Kanda, D. S. Goodman, R. E. Canfield, and F. J. Morgan, *J. Biol. Chem.* **249**, 6796 (1974).

[7] A. M. Katz, W. J. Dreyer, and C. B. Anfinsen, *J. Biol. Chem.* **234**, 2897 (1959).

[8] Constant-boiling HCl; 22 hr at 105° in evacuated and sealed tubes.

Results

Prealbumin is a tetramer with identical subunits. The subunits have the overall shape of elongated cylinders and are tetrahedrally arranged in such a manner that a central channel is formed which penetrates the entire molecule. The binding sites for thyroxine are located deep inside the central slot.[9]

By various enzymic degradation and fractionation techniques, Gly-1, Lys-9, and Lys-15 were shown to have been labeled in a molar ratio of 30:61:9. Gly-1, whose location has not been well defined by X-ray crystallography, is the N-terminal amino acid. Lys-15 is located within the central channel, whereas Lys-9 is near the channel entrance.

Comments

Dansyl chloride has also been used as a labeling reagent for prealbumin.[5] Although the structure of dansyl chloride resembles only remotely that of T_4, dansyl chloride also competes for the same two binding sites. By a methodology similar to that described in this chapter, 1.7 moles of the dansyl moiety were found to bind covalently when 2.1 moles of dansyl chloride per mole of prealbumin were used. In this case, virtually the entire label was found in Lys-15.

[9] C. C. F. Blake, M. J. Geisow, I. D. A. Swan, C. Rerat, and B. Rerat, *J. Mol. Biol.* **88**, 1 (1974).

[49] A Pyridoxamine Phosphate Derivative

By F. Riva, A. Giartosio, and C. Turano

Derivatives of pyridoxal-5'-phosphate or pyridoxamine-5'-phosphate modified in position 4' are able to bind with good affinity to most, if not all, B_6-dependent enzymes.[1,2] This property can be exploited to introduce in the 4' position substituents with a reactive function, thereby producing an affinity labeling reagent specific for the corresponding apoenzymes. Unfortunately, the choice of the reactive group to be introduced is severely restricted by some chemical properties of pyridoxal-5'-phosphate and/or of the complex formed between pyridoxal-5'-phosphate derivatives and the enzymes: many of the possible derivatives are too labile to be

[1] E. S. Severin, N. N. Gulyaev, E. N. Khurs, and R. M. Khomutov, *Biochem. Biophys. Res. Commun.* **35**, 318 (1969).
[2] C. Turano, C. Borri Voltattorni, A. Orlacchio, and F. Bossa *in* "Enzymes and Isoenzymes" (D. Shugar, ed.), p. 123. Academic Press, New York, 1970.

useful. However, the fluorodinitrophenyl moiety can be introduced at the 4' position of pyridoxamine-5'-phosphate; the resulting compound, 4'-N-(2,4-dinitro-5-fluorophenyl)pyridoxamine-5'-phosphate (FDNP-pyridox-amine-P), displays a remarkable affinity for B_6-dependent apoenzymes

(I)

and binds irreversibly to many of them (I). The closely related 4'-N-(2,4-dinitrophenyl)pyridoxamine-5'-phosphate (DNP-pyridoxamine-P), which lacks the reactive fluorine, can also be easily prepared and is useful in studying the first, reversible, phase of the affinity labeling reaction.

Procedures

Materials. The compounds used for the synthesis of FDNP-pyridox-amine-P and its analog are commercial products and can be used as such with the exception of pyridoxamine-5'-phosphate. Most commercial preparations of this compound contain significant amounts of contaminants and should be purified by chromatography on a carboxylic ion exchanger as described by Peterson and Sober.[3]

Preparation. 1,5-Difluoro-2,4-dinitrobenzene, 200 μmoles, in 0.5 ml of acetone is added to a solution of 40 μmoles of pyridoxamine-5'-phosphate hydrochloride in 5 ml of water, at pH 8.5 (adjusted with dilute sodium hydroxide). The reaction takes place, under continuous stirring, at room temperature in the dark and is practically completed within 3 hr. The pH should be maintained between 8.0 and 8.5 by adding dilute sodium hydroxide either manually or with a pH stat. It is important that the pH does not increase above 8.5 in order to avoid hydrolysis of the reagent and of the product. At the end of the reaction period, the yellow solution is extracted with ether, acidified with glacial acetic acid to a pH of about 3.5, and extracted again with ether. By this procedure the excess of reagent and its hydrolysis product are removed in the ethereal phase. The purification of the reaction product can be conveniently achieved by ion-exchange chromatography. The acid aqueous solution is applied to a SP-Sephadex C-25 column (35 $\times$ 3 cm) in H$^+$ form, equilibrated with 0.1 M acetic acid, and eluted with the same solvent. Two faster migrating fractions are eluted at about 30 and 100 ml, whereas FDNP-pyridox-

[3] E. A. Peterson and E. A. Sober, *J. Am. Chem. Soc.* **76**, 169 (1954).

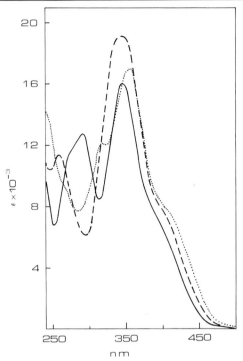

FIG. 1. Absorption spectra of 4'-N-(2,4-dinitro-5-fluorophenyl) pyridoxamine-5'-phosphate: ——, in 0.1 M HCl; – – –, in 0.1 M phosphate buffer, pH 7; · · · · · · ·, in 0.1 M NaOH.

amine-P is eluted between 400 and 500 ml. The last fraction is collected and lyophilized; the resulting product, obtained in about 40% yield, is an amorphous yellow powder.

DNP-pyridoxamine-P is prepared in the same way as FDNP-pyridoxamine-P, but with 2,4-dinitrofluorobenzene instead of 1,5-difluoro-2,4-dinitrobenzene.

The two derivatives of pyridoxamine-5'-phosphate have a limited stability, and preferably should be stored under dry conditions and in the dark at —20°. Aqueous solutions at pH below 7 may be kept for a few days at room temperature and for at least a month at —20°.

The spectra of FDNP-pyridoxamine-P are shown in Fig. 1. Absorbance contributions from the pyridoxamine moiety are recognizable particularly in acid (peak at 290 nm) and in alkali (shoulder at 310 nm). The peak in the 340–350 nm region and the shoulder at 410 nm are characteristic of the 2,4-dinitro-5-fluoroanilino compounds.[4]

[4] H. Zahn and J. Meienhofer, *Makromol. Chem.* **26**, 126 (1958).

A peak at 420 nm appears in strong alkali, due to the formation of a phenolate group following the hydrolysis of the reactive fluorine. A similar peak (430 nm) appears also upon substitution of the fluorine with an amino group (see Fig. 2).

FDNP- and DNP-pyridoxamine-P can be chromatographed on a cellulose thin layer developed in water/acetone/*tert*-amyl alcohol/acetic acid (20/35/40/5); their respective R_f values are 0.56–0.61 and 0.50–0.55; both compounds give yellow spots that turn blue on testing with dichloroquinone chlorimide.[5]

Reaction of FDNP-pyridoxamine-P with Apoenzymes

Pyridoxal-phosphate-dependent apoenzymes show great variability in their sensitivity to FDNP-pyridoxamine-P. In most cases the reaction takes place easily in the pH 7 to 8 range, with a reagent concentration from 10 to 500 μM. Under these conditions, and at room temperature, the irreversible binding to the apoenzymes takes place within a few minutes to about 3 hr, depending on the specific enzyme tested. The reaction should be performed in the dark because of the photosensitivity of FDNP-pyridoxamine-P.

The labeling of the active site by the reagent should be followed whenever possible by at least two procedures.

1. Reconstitution of the holoenzyme: the treated apoenzyme is incubated with an excess of pyridoxal-5'-phosphate and the resulting enzymic activity is measured. In order to assure the removal of all noncovalently bound reagent from the enzyme, the use of a high ionic strength and long incubation with the coenzyme, e.g., overnight, may be necessary. Binding of the reagent to the active site prevents reconstitution of the holoenzyme; the extent of labeling is therefore proportional to the decrease in activity measured after incubation with pyridoxal-5'-phosphate.

2. Spectral measurement of the enzyme-bound reagent: this method can be applied only when rather large (milligram) amounts of enzyme are available. The reagent–enzyme complex is precipitated with trichloroacetic acid or any other suitable denaturing agent; the precipitate should be collected, dissolved in urea, and analyzed spectrophotometrically for the presence of the DNP-pyridoxamine-P moiety. This test does not prove that the reagent is bound at the active site, but provides good evidence for covalent binding. Binding of the reagent at sites other than the active center can be detected by performing the reaction of FDNP-pyridoxamine-P on the holoenzyme, or by using a dephosphorylated re-

[5] V. W. Rodwell, B. E. Volcani, M. Ikawa, and E. E. Snell, *J. Biol. Chem.* **233**, 1548 (1958).

agent (FDNP-pyridoxamine). Because of the great importance of the phosphate group in the binding of pyridoxal-5'-phosphate to apoenzyme, the modified reagent has a very low affinity for the active site.[6]

The most likely candidates for the amino acid side chains of the protein which can react with FDNP-pyridoxamine-P are the ϵ-amino group of lysine, the terminal amino group, the phenolic hydroxyl of tyrosine, the sulfhydryl group of cysteine, and, possibly, the imidazole of histidine. Although conclusive identification requires hydrolysis of the protein and the isolation of the DNP-pyridoxamine-P-derivative, the spectral properties of the modified enzyme can give valid suggestions on the nature of the amino acid involved. Figure 2 shows the spectral variations that are to be expected upon reaction with an amino, a phenolic, and a sulfhydryl group. It may be useful to compare these spectra with those reported for a variety of 1,5-difluoro-2,4-dinitrobenzene derivatives by Zahn and Meienhofer[4] and by Marfey et al.[7]

Comments

FDNP-pyridoxamine-P has been tested with the following enzymes: supernatant[8] and mitochondrial[9] aspartate aminotransferases, tyrosine aminotransferase,[8] tryptophanase,[8] tyrosine decarboxylase,[8] cystathionase,[8] and glutamic decarboxylase.[9] The first five enzyme are irreversibly inhibited.

Favorable properties of FDNP-pyridoxamine-P are its water solubility and the satisfactory stability and spectral characteristics, which provide a convenient way of following reaction with an amino acid side chain. The yellow color is also an advantage when the binding of reagent to protein is studied by means of partial enzymic digestion and peptide separation; the presence of a phosphate group makes it possible to use the technique described by Strausbauch and Fischer[10] for rapid peptide isolation. Although the stability of the amino acid derivatives of FDNP-

---

[6] FDNP-pyridoxamine is easily prepared from its phosphorylated analog by the action of alkaline phosphatase: 1 μmole of FDNP-pyridoxamine-P is treated for 60 min at pH 8 with 5 units of alkaline phosphatase; the latter is then removed by means of a Sephadex G-25 column. The completeness of the hydrolysis is checked by thin-layer chromatography; under the conditions described, FDNP-pyridoxamine has an R_f of 0.92.

[7] P. S. Marfey, H. Nowak, M. Uziel, and D. A. Yphantis, J. Biol. Chem. 240, 3264 (1965).

[8] F. Riva, A. Giartosio, C. Borri Voltattorni, A. Orlacchio, and C. Turano, Biochem. Biophys. Res. Commun. 66, 863 (1975).

[9] A. Giartosio and F. Riva, unpublished data, 1976.

[10] P. H. Strausbauch and E. H. Fischer, Biochemistry 9, 233 (1970).

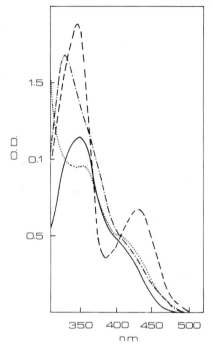

Fig. 2. Absorption spectra of 4'-*N*-(2,4-dinitro-5-fluorophenyl) pyridoxamine-5'-phosphate (FDNP-pyridoxamine-P) after reaction with lysine, cysteine, and tyrosine derivatives. All solutions are 0.5 *M* in $KHCO_3$. (1) ——, 64 μM FDNP-pyridoxamine-P; (2) ·–·–·–, same as (1) after 15 min incubation with 57 m*M* L-cysteine; (3) – – –, same as (1) after 24 hr incubation with 53 m*M* *N*-acetyl-L-lysine; (4) · · · · · · ·, same as (1) after 30 hr incubation with 9.3 m*M* *N*-acetyl-L-tyrosine ethyl ester. The blank values of added amino acid derivatives have been subtracted.

pyridoxamine-P is such that no problem is usually met during the purification of a reagent-bound peptide, the possibility should be kept in mind of a displacement of a thiol or phenol bound to DNP-pyridoxamine-P (as in a cysteine or a tyrosine derivative) by an amino group. Fortunately, such displacement can be detected easily by the appearance of a 430-nm absorption peak.

One drawback of the reagent is an unexplained reaction that takes place in the apoenzyme–reagent complex, one that leads to the formation of active coenzyme. This means that two competing reactions occur after the formation of the reversible apoenzyme–reagent complex: the irreversible labeling of the enzyme and the regeneration of the active holoenzyme. A high ratio of the rate constants of the first to the second reaction, respectively, is obviously required for satisfactory affinity labeling of the protein. In some cases the second reaction is sufficiently fast that a significant fraction of the enzyme remains uninhibited. Tyrosine aminotransferase,, for example, is irreversibly inhibited to only 70–80% by 50 μM reagent at pH 8. A proper choice of conditions, with particular regard to the type of buffer and the pH, can favor the affinity labeling reaction.

DNP-pyridoxamine-P, which cannot bind irreversibly, is very useful for assessing the unwanted reconstitution reaction and for a search of the experimental conditions that minimize it.

Acknowledgment

This work was supported by the Center of Molecular Biology of the C.N.R.

[50] Labeling of Steroid Systems

By James C. Warren and J. Robert Mueller

Unlike many other compounds of biologic importance, steroids are large, planar, rigid molecules that possess a variety of carbon atoms and functional groups capable of undergoing addition and substitution reactions. These characteristics make steroid derivatives ideal affinity-labeling reagents. Since synthetic procedures can be directed at any one of several sites on the steroid nucleus, it is possible to create a variety of derivatives that have their reagent-bearing "arms" attached to carbon atoms several Ångstrom units apart. By using sufficient numbers and types of such affinity-labeling derivatives, it is feasible to specifically label and identify several of the amino acid residues present at the steroid binding site of a protein molecule (Fig. 1). Such labeling techniques, when used in con-

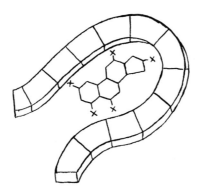

Fig. 1. Schematic representation of various affinity-labeling steroids at the catalytic site of an enzyme molecule. Each $\times$ represents the location of an affinity-labeling reagent group on a particular steroid derivative. In actual practice, an affinity-labeling steroid would possess only one $\times$ with the exception of a bifunction alkylating derivative.

junction with specific and nonspecific protein hydrolytic and amino acid sequencing procedures, can elucidate the topography of a macromolecular steroid binding site.

In an earlier volume of this series, we described the synthesis of 4-mercuriestradiol-17β, various bromoacetoxyprogesterones, iodoacetoxy and iodo derivatives of cortisone, and ring bromoprogesterone derivatives and their use as affinity-labeling reagents for study of the structure of steroid binding sites and in certain biological models where they were used to exclude natural steroids.[1] This current section summarizes more recent studies with affinity-labeling steroid derivatives and delineates both strategy and tactics of synthesis and application.

General Considerations

From our experiences in the affinity labeling of steroid interconverting enzymes we offer the following suggestions:

1. The protein under study should be used in homogeneous solution if possible. Fortunately, affinity chromatography[2] makes purification and electrophoretic diffusion[3] makes even crystallization of such enzymes possible.

2. When a new affinity-labeling steroid is synthesized, its stability under conditions of anticipated use (pH, buffer, temperature, etc.) should be evaluated. Studies should be done to determine which amino acid residues it will and will not alkylate.

3. We have generally done our studies under conditions that are as "physiologic" as possible, being concerned that wide deviation may induce drastic conformational changes.

4. Preliminary studies with a newly synthesized affinity-labeling steroid should ascertain whether it is a substrate or a competitive inhibitor of the enzyme. The substrate role proves that it undergoes the reversible binding step at the active site; the competitive inhibitor role strongly suggests that it does.

5. Rates of inactivation are actually studied by incubation of enzyme with affinity-labeling steroid. At the beginning and at given times thereafter, a small aliquot (50 μl) is withdrawn and assayed with a good substrate and cofactor at near saturating concentrations in 1–3 ml of assay solution. This minimizes interference by the affinity-labeling steroid.

6. We customarily assume that the K_m or K_i, obtained in (4), at least

[1] J. C. Warren, F. Arias, and F. Sweet, this series, Vol. 36, p. 374 (1975).

[2] P. Cuatrecasas and C. B. Anfinsen, this series, Vol. 22, p. 345 (1971); this series Vol. 34 (1975).

[3] C. C. Chin, J. B. Dence, and J. C. Warren, *J. Biol. Chem.* **251**, 3700 (1976).

approximates the K_s and then carry out inactivation over a 10-fold range of steroid concentration in that region. This is done with affinity-labeling steroid *always* in at least 10-fold molar excess of enzyme so that inactivation follows psuedo-first-order kinetics. If it does, calculation of k_{app} (apparent rate constant) is easy. A double reciprocal plot of k_{app} vs [S] should identify an approximate K_s for the affinity-labeling steroid if it undergoes the reversible binding step at the active site.

7. Now, the enzyme can be inactivated with concentrations of affinity-labeling steroid at, or only moderately in excess of, that saturating one-half of the sites, and the effects of nonaffinity-labeling substrates and cofactors can be evaluated.

8. Stoichiometry and identification of the amino acid residue alkylated will require radioactive labeling. The isotope must be present in some part of the steroid derivative that remains attached to the protein after hydrolysis, and consideration must be given to the specific activity required for successful identification.

9. Groman *et al.*[4] have recently pointed out that the ultimate proof of affinity labeling of the active site is display of catalytic competence by the enzyme molecule bearing the covalently bound steroid.

In the case of a nonenzyme protein, such as a steroid receptor, requirements of homogeneity, amounts of protein necessary, and absence of the "activity' handle pose additional problems. We have tried to affinity label partially purified progesterone receptor with O'Malley's 30% ammonium sulfate fraction from chick oviduct and bromoacetoxyprogesterones.[5] On subsequent sucrose density gradient analysis, we observed several radioactive peaks. Obviously, further purification is a prerequisite to intensive study.

Specific Considerations

Synthesis and Use of 16α-Bromoacetoxyestradiol 3-Methyl Ether
  [16α-Bromoacetoxy-1,3,5(10)-estratriene-3,17β-diol
  3-Methyl Ether][6]

Synthesis. To a stirred solution of 0.9 g (3 mmoles) of estriol 3-methyl ether in 110 ml of freshly distilled dry tetrahydrofuran (THF) at 25°, add in sequence (each in 5 ml of THF) 0.5 g (3.6 mmoles) of bromoacetic

[4] E. V. Groman, R. M. Schultz, and L. L. Engel, *J. Biol. Chem.* **250**, 5450 (1975). See also this volume [6].
[5] W. T. Schrader and B. W. O'Malley, *J. Biol. Chem.* **247**, 51 (1972).
[6] C. C. Chin and J. C. Warren, *J. Biol. Chem.* **250**, 7682 (1975).

Estriol 3-methyl ether

16α-Bromoacetoxyestradiol
3-methyl ether

acid and 1.24 g (6 mmoles) of dicyclohexylcarbodiimide. After 5 min of mixing, add 0.30 ml (0.36 mmole) of dry pyridine and stir for 2 hr at 25°. Remove the white precipitate of dicyclohexylurea that accumulates by suction filtration through a fritted Büchner funnel. Condense the filtrate and chromatograph it on preparative silica gel plates (0.25 mm) with benzene:ethanol (7:3) as developing solvent. Remove the forward migrating band, and extract the silica gel with 500 ml of diethyl ether. Take to dryness and crystallize the crude product from diethyl ether–petroleum ether. The physical characteristics of the purified product are: m.p. 106°–108°; $\lambda_{max}^{ethanol}$ 278 nm (ϵ 4690); infrared: λ_{max}^{KBr} 3500 cm^{-1} (C$_{17}$-OH); 1740 cm^{-1} (—C = O); mass spectra: M$^+$ 423 (100%); m/e 405 (—H$_2$O), (1.4%); m/e 284 (—BrCH$_2$COOH) (14%). It should be noted that if the bromoacetoxy moiety were on the 17 position of the steroid, the m/e 405 would be predominant. The synthesis of 16α-[2-³H]bromoacetoxy-1,3,5(10)-estratriene-3,17β-diol 3-methyl ether can be accomplished by first adding 134.6 mg of unlabeled bromoacetic acid to 18.3 mg (25 mCi) of [2-³H]bromoacetic acid to obtain a specific activity of 22.7 mCi/mmole. This mixture is allowed to react with 302 mg (1.0 mmole) of estriol 3-methyl ether, 412 mg of dicyclohexylcarbodiimide, and 0.1 ml of dry pyridine in 40 ml of dry THF. The purification procedure of the final product is as described above. Methylation of the phenolic hydroxy group, steric hindrance of the 17β-hydroxy group by the near-by angular methyl (C-18), and use of only a slight molar excess of bromoacetic acid yields the 16α-bromoacetoxy derivative almost exclusively.

The stability of 16α-bromoacetoxyestradiol 3-methyl ether is tested by incubating 127 μg (0.3 μmole) dissolved in 0.3 ml of ethanol with 4.7 ml of 0.05 M potassium phosphate at pH 7.0 and 25°. At various intervals, 0.5-ml aliquots are removed and extracted with 3 volumes of chloroform, and the extract is analyzed by thin-layer chromatography (TLC) on silica gel using acetone:chloroform (1:1) as the developing solvent. The R_f values of the aliquots should be identical to that of the starting material.

Use. Chin and Warren[6] demonstrated that L-cysteine, L-histidine,

L-methionine, L-lysine, and L-tryptophan are alkylated by 16α-bromo-acetoxyestradiol 3-methyl ether when the steroid (0.25 mM) is incubated with the respective amino acids (1.25 mM) in 0.05 M potassium phosphate (pH 7.0):methanol (50:50) at 25° for 24 hr. The steroid is a substrate for human placental estradiol 17β-dehydrogenase (EC 1.1.1.62). When estradiol 17β-dehydrogenase is treated with a 150-fold molar excess of 16α-bromoacetoxyestradiol 3-methyl ether, the inactivation of the enzyme follows pseudo-first-order kinetics over the first 3 hr of reaction (Fig. 2).

Estradiol-17β slows the inactivation and excess β-mercaptoethanol stops it. When tritiated 16α-bromoacetoxyestradiol 3-methyl ether is used, subsequent hydrolysis and amino acid analyses indicate dicarboxy-methylation of a histidyl residue at the site of enzyme catalysis. The histidyl residue labeled by this procedure must certainly be different

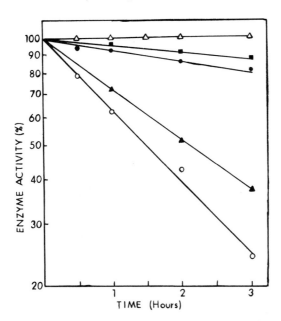

FIG. 2. Inactivation of estradiol 17β-dehydrogenase by 16α-bromoacetoxyestradiol 3-methyl ether and protection against enzyme inactivation by cofactor or substrate. Aliquots were assayed; each point is the mean of four determinations. Enzyme (1.4 nmoles) was preincubated with 16α-bromoacetoxyestradiol 3-methyl ether (0.21 μmole) in 1.5 ml of buffer A, 25°, pH 7.0 containing 20% ethanol, (○); 16α-bromoacetoxyestradiol 3-methyl ether (0.21 μmole) and NADH (0.42 μmole), (▲); 16α-bromoacetoxyestradiol 3-methyl ether (0.21 μmole) and NADPH (0.42 μmole) (●); 16α-bromoacetoxyestradiol 3-methyl ether (0.21 μmole) and estradiol (0.21 μmole) (■); and control or with bromoacetic acid (0.42 μmole) (△). From C. C. Chin and J. C. Warren, *J. Biol. Chem.* **250**, 7682 (1975).

from the one found by Pons *et al.*[7] and Boussioux *et al.*[8] since they employed an iodoacetoxy-3-estrone derivative. Furthermore, nonspecific labeling of cysteinyl residues is not observed when the 16α-bromoacetoxy-estradiol 3-methyl ether is used as the alkylating agent. Finally, a 300-fold molar excess of 2-[³H]bromoacetate does not inactivate the enzyme and does not label it. Thus, estradiol 17β-dehydrogenase and the previously studied 20β-hydroxysteroid dehydrogenase from *streptomyces hydrogenans*[9,10] both possess a histidyl residue at the catalytic region of the active site. While this may be pure coincidence, it is tempting to consider that such a residue is required for dehydrogenation of steroid substrates.

Synthesis and Use of 2,4-Bis(bromomethyl)estradiol-17β 3-Methyl Ether[11]

Synthesis. Synthesis of 2,4-bis(bromomethyl)estradiol-17β 3-methyl ether (BBE₂M) is accomplished by reducing 2,4-bis(bromomethyl)estrone

2,4-Bis(bromomethyl)-estrone   2,4-Bis(bromomethyl)estradiol-
3-methyl ether   17β 3-methyl ether

methyl ether (BBE₁M)[12] with sodium borohydride. To a one part solution of 2,4-bis(bromomethyl)estrone methyl ether in methanol at 4° is added 1.5 parts sodium borohydride. The progress of the reaction is monitored by TLC of the reaction mixture on silica gel strips developed with benzene:ethyl acetate (9:1). The disappearance of the nonpolar BBE₁M ($R_f = 0.66$) and the appearance of the polar BBE₂M ($R_f = 0.34$) is

[7] M. Pons, J. C. Nicolas, A. M. Boussioux, B. Descomps, and A. C. dePaulet, *FEBS Lett.* **36**, 23 (1973).
[8] A. M. Boussioux, M. Pons, J. C. Nicolas, B. Descomps, and A. C. dePaulet, *FEBS Lett.* **36**, 27 (1973).
[9] M. Ganguly and J. C. Warren, *J. Biol. Chem.* **246**, 3646 (1971).
[10] F. Sweet, F. Arias, and J. C. Warren, *J. Biol. Chem.* **247**, 3424 (1972).
[11] N. R. Kanamarlapudi and J. C. Warren, *J. Biol. Chem.* **250**, 6484 (1975).
[12] N. R. Kanamarlapudi, F. Sweet, and J. C. Warren, *Steroids* **24**, 63 (1974).

[50]

indicative of the completion of the reaction. The reaction mixture is neutralized with glacial acetic acid and evaporated to dryness under reduced pressure. The dried residue is partitioned between equal volumes of water and ethyl acetate, and the organic layer is purified by TLC on silica gel using benzene:ethyl acetate (9:1) as the developing solvent. Although the initial structural assignment of the product was accomplished by both nuclear magnetic resonance (NMR) and infrared (IR) spectrophotometry, a rapid "proof of conversion" of BBE_1M to BBE_2M can be effected by subjecting the two steroid derivatives to IR analysis. On conversion of BBE_1M to BBE_2M, the absorption band at 1740 cm^{-1} (17-keto group) will disappear while a new absorption band at 3440 cm^{-1} (17-hydroxy group) will be present.

The synthesis of $[17\alpha\text{-}^3H]BBE_2M$ is accomplished by adding 0.45 mg (8.4 Ci/mmole) of sodium borotritide to BBE_1M (4.0 mg) in 0.6 ml of methanol at 4°. After the reaction is complete, the mixture is neutralized and evaporated to dryness under a stream of nitrogen. Final purification of the product is effected by TLC as described above. Isotopic effects in this reaction reduce the specific activity of the product to about 2 Ci/mmole.

Use. Analysis of BBE_2M with model compounds reveals that it alkylates Ellman's anion, cysteine, tryptophan, and histidine. The stoichiometry of the reaction of BBE_2M with L-[2-^{14}C]cysteine was shown to be 2:1; thus, it is a bifunctional alkylating steroid. Before analysis of biological activity, it was shown to be free of estradiol-17β.[11] The effect of intraluminal administration of BBE_2M and analogous compounds on the activity of uterine glucose-6-phosphate dehydrogenase is shown in Table I. Further, the treatment of ovariectomized rats with BBE_2M resulted in hyperemia and water imbibition similar to that observed with estradiol-17β. Additional evidence clearly indicates that BBE_2M is a long-acting estrogen as compared to nonaffinity-labeling analogs.[11] When the *in vitro* actions of estradiol-17β and BBE_2M on calf endometrial nuclei (incubated with cytosol) were examined, both compounds displayed a "molecular estrogenic effect" as indicated by a significant increase in the incorporation of [^{3}H]UTP into RNA (Table II) using the assay of Mohla *et al.*[13] The long action *in vivo* and the effects with nuclei *in vitro* support a mechanism whereby BBE_2M delivers its reagent group to the estradiol binding site of the cytosol receptor where covalent reaction occurs with a cysteine, histidine, or tryptophan residue present at that site.

[13] S. Mohla, E. R. DeSombre, and E. V. Jensen, *Biochem. Biophys. Res. Commun.* **46**, 661 (1972).

TABLE I

ESTROGENIC ACTIVITY OF 2,4-BIS(BROMOMETHYL)ESTRADIOL-17β
3-METHYL ETHER (BBE$_2$M)[a]

Group	Treatment	Dose (ng/horn)	Glucose-6-phosphate dehydrogenase activity (milliunits/horn)
Experiment I			
1	Vehicle	0	111.91 ± 3.59
2	Estradiol-17β	5	150.99 ± 5.37[b]
3	BBE$_2$M	100	129.96 ± 8.27[c]
4	BBE$_2$M	200	135.55 ± 5.36[b]
5	E$_2$M	60	133.14 ± 7.13[c]
6	E$_2$M	120	149.54 ± 10.09[b]
Experiment II			
1	Vehicle		89.24 ± 2.80
2	m-Dibromoxylene	111	85.87 ± 2.00

[a] Animals (250–280 g) used in these experiments were ovariectomized 4 weeks prior to use. Steroids or vehicle were administered intraluminally, and animals were killed 24 hr later. Each value reported is mean of five animals per group. The molar equivalent of 100 ng of BBE$_2$M is 60 ng of E$_2$M. From N. R. Kanamarlapudi and J. C. Warren, *J. Biol. Chem.* **250**, 6484 (1975).
[b] $p = < 0.01$.
[c] $p = < 0.05$.

TABLE II

ACTION OF 2,4-BIS(BROMOMETHYL)ESTRADIOL-17β 3-METHYL ETHER (BBE$_2$M)
ON RNA SYNTHESIS IN CALF ENDOMETRIUM NUCLEI[a]

Nuclei incubated with	[^{3}H]UTP incorporated into RNA (mean ± SE) (cpm/100 μg DNA)
Cytosol	81.53 ± 7.19
Cytosol + estradiol-17β	124.25 ± 12.89[b]
Cytosol + BBE$_2$M	121.89 ± 18.10[c]

[a] Calf endometrium cytosols were prepared in 2.2 M sucrose, 3 mM MgCl$_2$. Preincubations of cytosol were carried out at 0° for 30 min with and without steroids (10 nM) and incubated with nuclei at 25° for 30 min. Nuclei were separated subsequently and assayed for RNA synthesis activity. From N. R. Kanamarlapudi and J. C. Warren, *J. Biol. Chem.* **250**, 6484 (1975).
[b] $p = < 0.01$.
[c] $p = < 0.05$.

Synthesis and Use of 4-Bromoacetamidoestrone Methyl Ether[14]

Synthesis. The synthesis of 4-bromoacetamidoestrone methyl ether involves four steps; the nitration of estrone to 4-nitroestrone, the con-

Estrone   4-Nitroestrone   4-Nitroestrone methyl ether

4-Bromoacetamidoestrone methyl ether   4-Aminoestrone methyl ether

version of 4-nitroestrone to 4-nitroestrone methyl ether, the reduction of 4-nitroestrone methyl ether to 4-aminoestrone methyl ether, and finally, the synthesis of 4-bromoacetamidoestrone methyl ether.

Step 1. Estrone (5 g) is dissolved in 150 ml of boiling glacial acetic acid. After cooling to 60°, 1.25 ml of concentrated nitric acid is added with stirring and the resulting solution is kept at room temperature for 24 hr. The yellow precipitate which forms is collected by filtration, washed with cold glacial acetic acid, and dried under reduced pressure. Additional precipitate can be obtained by condensing the filtrate to one-half its initial volume and cooling to 4° overnight. Recrystallization of the precipitate from 80% acetic acid yields the product, 4-nitroestrone; m.p. 270°.

Step 2. In a 50-ml round-bottom boiling flask fitted with a reflux condenser, 1.0 g (3.18 mmoles) of 4-nitroesterone is added to 0.15 g of sodium in 20 ml of ethanol. Then, 0.6 ml (7.7 mmoles) of methyl iodide is added dropwise through the condenser and the mixture allowed to reflux for 2 hr. During this period, light yellow crystals begin to precipitate from solution. Complete precipitation is effected by cooling the reaction mixture to room temperature. The crystals of 4-nitroestrone methyl ether are collected, washed with cold water, and dried under reduced pressure; m.p. 253°–255°.

Step 3. To a solution of 1.0 g (3 mmoles) of 4-nitroestrone methyl ether in 250 ml of acetone, add 50 ml of H_2O, 25 ml of 1 N NaOH, and

[14] Y. M. Bhatnagar, C. C. Chin, and J. C. Warren, unpublished observations, 1975.

5 g of sodium dithionite; reflux for 30 min and add 3.5 g of supplemental sodium dithionite. Keep the pH slightly alkaline by adding 1 N NaOH as required. After an additional 30-min reflux period, add another 3.5 g of sodium dithionite and readjust the pH as before. Reflux for an additional 10 min, cool to room temperature, and remove the acetone under reduced pressure. Neutralize the aqueous phase with 10% acetic acid and place in the cold (0°) for 2 hr. Collect the precipitate which forms by filtration, dry, and recrystallize from ethanol to yield pure 4-aminoestrone methyl ether. The physical characteristics of the purified product are: m.p. 162°–165°; $\lambda_{max}^{ethanol}$ 280 nm (ϵ 5052); IR: λ_{max}^{KBr} 3400 cm^{-1} and 3460 cm^{-1} (—NH$_2$), 1750 cm^{-1} (17—C=O).

Step 4. Dissolve 0.3 g (1 mmole) of 4-aminoestrone methyl ether in 4 ml of dry methylene chloride and add sequentially: 0.35 g (2.5 mmoles) of bromoacetic acid and 0.69 g (3 mmoles) of dicyclohexylcarbodiimide (each dissolved in 6 ml of methylene chloride). Allow the reaction to proceed for 1 hr at 0° and then for an additional hour at room temperature. The white precipitate of dicyclohexylurea that forms is removed by filtration. The filtrate is taken to dryness under reduced pressure. The residue is extracted with 10 ml of acetone, and the extract is filtered and concentrated to 2–3 ml. The concentrated extract is chromatographed on preparative silica gel plates (0.25 mm), developing with chloroform: acetone (95:5). Thereafter, the area corresponding to an R_f of 0.69 is removed and eluted with acetone. Evaporate the solvent and crystallize the residue from ethanol to yield pure 4-bromoacetamidoestrone methyl ether. The physical characteristics of the product are: m.p. 162°–164°; $\lambda_{max}^{ethanol}$ 280 nm (ϵ 5052); IR: λ_{max}^{KBr} 3250 cm^{-1} (—NH in —NHCOCH$_2$Br), 1750 cm^{-1} (17—C=O), 1670 cm^{-1} (—C=O in —NHCOCH$_2$Br).

The synthesis of radioactive 4-bromoacetamidoestrone methyl ether is accomplished by mixing 18.4 mg of unlabeled bromoacetic acid with 4.6 mg (5 mCi) of [2-^{3}H]bromoacetic acid to yield a final specific activity of 30 mCi/mmole. This bromoacetic acid is allowed to react with 45 mg (0.15 mmole) of 4-aminoestrone methyl ether and 100 mg (0.3 mmole) of dicyclohexylcarbodiimide in 8 ml of dry methylene chloride. The remainder of the procedure is the same as that described for the unlabeled compound.

Use. When the reactivity of 4-bromoacetamidoestrone methyl ether was examined, L-cysteine, L-methionine, L-histidine, L-lysine, L-tryptophan, L-tyrosine, and N-acetylhistidine were shown to be alkylated by the steroid derivative, but alanine was not. Assay revealed that 4-bromoacetamidoestrone methyl ether is a substrate for estradiol 17β-dehydrogenase and thus binds at the active site. It has a K_m of 0.14 mM. Radioactive 4-bromoacetamidoestrone methyl ether (0.2 mM) was used to inactivate estradiol 17β-dehydrogenase. Stoichiometry of the inactivation approximates 2.0. The inactivated enzyme was subjected to hydrolysis and amino acid analysis. Tritiated carboxymethyllysine and carboxymethyl cysteine were identified (3:1), suggesting that these amino acids proximate the lower A ring of the steroid on binding.[14]

In other work on this enzyme,[7,8] a histidyl and 2 cysteinyl residues have been labeled using iodoacetoxy-3-estrone as the affinity-labeling reagent. Since 4-bromoacetamidoestrone methyl ether does not affinity label the histidyl residue, it would seem that the latter is somewhere between the 2- and 3-carbons of the A ring. In as much as both affinity-labeling steroids (iodoacetoxy-3-estrone and 4-bromoacetamidoestrone methyl ether) give rise to labeled cysteinyl residues, it would seem that at least two such residues reside in that portion of the macromolecular binding site that surrounds the A ring of the steroid. In the presence of cofactor, neither steroid affinity labels a cysteine, suggesting that cofactor on binding induces a conformational change in the enzyme, perhaps by interposing a portion of itself between the steroids and cysteinyl residue in question. It is hoped that further affinity-labeling studies and X-ray diffraction of the now crystalline enzyme[3] will completely elucidate the topography of the active site.

One major advantage of compounds like 4-bromoacetamidoestrone methyl ether is that the amide linkage is strongly resistant to hydrolysis, and that therefore dicarboxymethylation of a residue via a transacylation mechanism will not occur.

Synthesis of 12β-Bromoacetoxy-19-nor-4-androstene-3,17-dione

Synthesis. The synthesis of 12β-bromoacetoxy-19-nor-4-androstene-3,17-dione is accomplished by specific hydroxylation of 19-nor-4-andro-

19-Nor-4-androstene-3,17-dione   12β-Hydroxy-19-nor-4-androstene-3,17-dione   12β-Bromoacetoxy-19-nor-4-androstene-3,17-dione

stene-3,17-dione at the 12 position and bromoacetylation to yield the 12β-bromoacetoxy derivative.

Step 1. The specific 12β-hydroxylation of 19-nor-4-androstene-3,17-

dione is accomplished using either *Diplodia natalensis* or *Colletotrichum gloeosporioides*.[15] We have found *C. gloeosporioides* to be most useful for this purpose particularly in respect to yield of product.[16] Surface growth cultures of *C. gloeosporioides* are maintained at 25° on a medium consisting of 10 g of glucose, 2.5 g of yeast extract, 1.0 g of K_2HPO_4, and 25 g of agar per liter of water. Growth of the organism in petri-dish cultures is allowed to proceed for 7–10 days prior to use. Then, inocula from the surface cultures are transferred to 500-ml Erlenmeyer flasks each containing 200 ml of a medium consisting of 10 g of dextrose, 6 g of corn steep liquor, 3 g of $NH_4H_2PO_4$, 2.5 g of yeast extract, and 2.5 g of $CaCO_3$ per liter of water. After 48–72 hr of incubation with continuous agitation (rotary or linear) at 25°, each of the 200-ml cultures is transferred to a 3-liter flask containing 1.5 liters of growth medium. Growth is allowed to proceed for an additional 24 hr with agitation at 25° as previously noted. Then, 1–2 g of 19-nor-4-androstene-3,17-dione, dissolved in 10 ml of *N,N*-dimethylformamide, is added to each flask. Incubation is allowed to proceed with continuous agitation for 10 days. The cells are collected using a fritted suction funnel and washed twice with 100-ml portions of hot water; the washings are added to the filtrate. The filtrate is extracted with an equal volume of chloroform, and the solvent is removed under reduced pressure. The residue so obtained is taken up in chloroform and subjected to TLC on preparative silica-gel plates (0.25 mm) with chloroform:methanol (95:5) used as the developing solvent. The R_f of 12β-hydroxy-19-nor-4-androstene-3,17-dione is 0.72, and the R_f of the starting material, 19-nor-4-androstene-3,17-dione, is 0.92 in this system. Elution of the steroid is accomplished with chloroform. The solvent is evaporated, and the product is crystallized with ethanol. Physical characteristics: m.p. 183°–185°; $[\alpha]_D = +121°$ (chloroform); $\lambda_{max}^{KBr} = 3448$ cm^{-1}. Current work indicates that Sephadex LH-20 may be useful in separating 12β-hydroxy-19-nor-4-androstene-3,17-dione from both the starting material and other organic impurities.

Step 2. The procedure for preparing 12β-bromoacetoxy-19-nor-4-androstene-3,17-dione from 12β-hydroxy-19-nor-4-androstene-3,17-dione is similar to that employed for the synthesis of 16α-bromoacetoxyprogesterone from 16α-hydroxyprogesterone.[10]

Use. As mentioned above, Chin and Warren[6] have shown that a histidyl residue is carboxymethylated when 16α-bromoacetoxyestradiol 3-methyl ether is incubated with human placental estradiol 17β-dehydrogenase. While it is reasonable to conclude that this histidyl residue is

---

[15] S. C. Pan, L. J. Lerner, P. A. Principe, and B. Junta, U.S. Patent 3324153 (E. R. Squibb and Sons, Inc.) (1967).

[16] J. R. Mueller, C. C. Chin, and J. C. Warren, unpublished observations, 1975.

indeed located at or near the catalytic region of the active site, we wish to ascertain whether it is in a position (in relation to the bound steroid) to actually serve as the catalytic residue. If one builds a model of 16α-bromoacetoxyestradiol 3-methyl ether it is clear that the leaving (bromo) group moves through a rotational hemisphere on an arm of some 4 Å, attached to the 16-carbon. Thus, the histidyl residue that it carboxymethylates may be spatially oriented immediately adjacent to the 17α-hydrogen that is removed by the catalytic dehydrogenation step or it may be located 5.4 Å away. A model of a hypothetical steroid bearing both 16α-bromoacetoxy and 12β-bromoacetoxy groups is shown in Fig. 3. It can be noted that the rotational hemispheres of the leaving groups overlap only in an area immediately adjacent to the 17α-hydrogen. It is our intention to use [2-³H]bromoacetic acid to synthesize tritiated 12β-bromoacetoxy-19-nor-4-androstene-3,17-dione. We already know that 19-nor-4-androstene-3,17-dione is a substrate for the human placental enzyme. If the affinity-labeling steroid alkylates a histidyl residue, specific enzymic hydrolysis, followed by fingerprinting and sequencing of

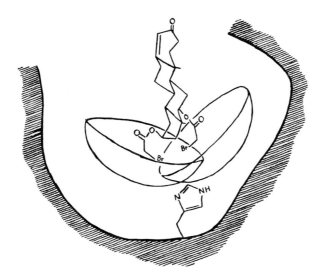

Fig. 3. Schematic representation of a hypothetical steroid bearing both 16α-bromoacetoxy and 12β-bromoacetoxy groups. The two overlapping hemispheres each represent regions of space within which alkylation of the imidazole ring of the histidine residue by the respective bromoacetoxy groups may occur. The volume within the overlap of the two hemispheres (which closely approximates the 17α-hydrogen on the steroid) represents the region in which both bromoacetoxy groups can alkylate the same residue. Note that the imidazole ring has been removed slightly from the region of the 17α-hydrogen to allow for construction of the hemispheres.

this peptide, will allow comparison with the peptide similarly obtained after affinity labeling with tritiated 16α-bromoacetoxyestradiol 3-methyl ether. If both compounds react and carboxymethylate the same histidyl residue, that residue must be in a very specific place, the site at which the catalytic dehydrogenation occurs.

Summary

We have followed a strategy of developing steroid analogs with affinity-labeling-reagent groups attached at various positions on the steroid nucleus. By carefully using several such compounds one may hope to affinity label sufficient numbers of amino acid residues in a given steroid binding site so as to elucidate its topography. The synthetic procedures reported here and in our previous summary[1] are clean, straightforward, and reproducible. Since many of these syntheses are carried out on a semimicro scale, care must be taken if satisfactory yields of product are to be obtained. This is particularly true in the synthesis of tritiated derivatives, since products with a high specific activity are desired.

The importance of proof of absolute structure and purity of the final product cannot be overemphasized. If one were to use a derivative to which the wrong structural assignment had been made or one that is contaminated with a second affinity-labeling derivative, confusing errors would be generated.

As is obvious, we have primarily relied on bromoacetoxy and bromoacetamido derivatives in our affinity-labeling studies. These compounds are reasonably stable, yet possess considerable reactivity with nucleophilic groups commonly found in amino acid side chains. The chloroacetoxy derivatives, on the other hand, are reasonably unreactive, and iodoacetoxy derivatives lack the stability of the bromoacetoxy compounds.

We have elected not to use photoaffinity-labeling derivatives in our steroid studies. Although we have used photoaffinity methods for the synthesis of affinity-labeling peptide derivatives because of a concern that bromoacetoxy groups on the serine residue would react with other amino acid residues in the LRF itself, this would not occur in the steroid series. Further, photoaffinity labeling may generate more than one derivative, even when reacting with a single amino acid residue, and this characteristic alone makes such technology less desirable.[17]

[17] D. T. Browne, S. S. Hixson, and F. H. Westheimer, *J. Biol. Chem.* **246**, 4477 (1971).

[51] Irreversible Inhibitors of Δ⁵-3-Ketosteroid Isomerase: Acetylenic and Allenic 3-Oxo-5,10-secosteroids

By F. H. Batzold, Ann M. Benson, Douglas F. Covey,
C. H. Robinson, and Paul Talalay

The Δ⁵-3-ketosteroid isomerase (EC 5.3.3.1) of *Pseudomonas testosteroni* catalyzes the conversion of a variety of unconjugated Δ⁵⁽⁶⁾- and Δ⁵⁽¹⁰⁾-3-ketosteroids into the corresponding Δ⁴-3-ketosteroids. Typical examples of the reaction are the conversion of Δ⁵-androstene-3,17-dione and Δ⁵-pregnene-3,20-dione to the respective Δ⁴-3-ketosteroids[1] (Scheme 1, A and B).

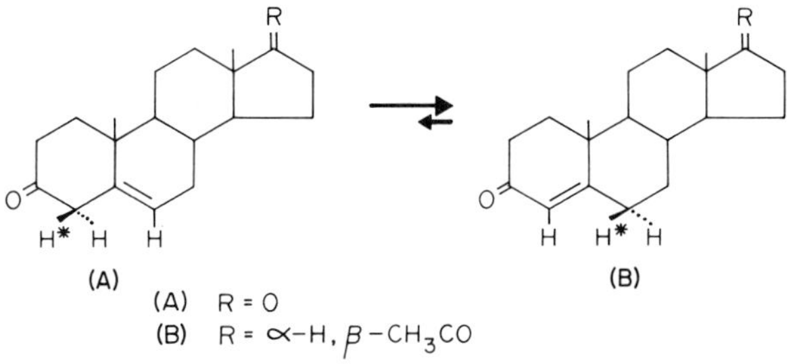

(A)   R = O
(B)   R = α-H, β-CH$_3$CO

SCHEME 1

Considerable information is available on the molecular properties of this enzyme, on the catalytic mechanism, and on the stereochemistry of the enzymic reaction. The isomerase is composed of identical associating subunits. The primary structure is known and comprises 125 residues (MW 13,394) including all the common amino acids except cysteine and tryptophan. The enzyme, which exhibits exceptionally high catalytic activity, has a molecular activity at saturating concentrations of Δ⁵-androstene-3,17-dione of 4.38×10^6 min⁻¹ per monomer at pH 7.0 and 25°. Mechanistic studies have disclosed that the isomerase catalyzes the reaction of Δ⁵⁽⁶⁾-3-ketosteroids by a direct, stereospecific, and intramolecular transfer of the 4β-proton to the 6β-position.[1,2] There is considerable evidence for the involvement of an enolic intermediate in the

[1] P. Talalay and A. M. Benson, *in* "The Enzymes" (P. D. Boyer, ed.), 3rd ed., Vol. 6, p. 591. Academic Press, New York, 1972.
[2] S. K. Malhotra and H. J. Ringold, *J. Am. Chem. Soc.* **87**, 3228 (1965).

reaction, and the participation in the catalytic process of both acidic (A) and basic (B) groups on the enzyme has been proposed (Scheme 2).

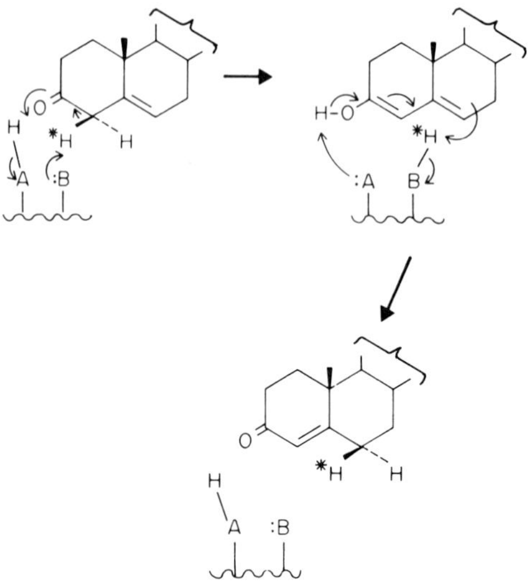

SCHEME 2

The present state of knowledge of the molecular properties and catalytic mechanism of this enzyme has been reviewed recently.[3]

Based on the proposed molecular mechanism of this reaction, a series of acetylenic 5,10-secosteroids has been prepared[4,5] in the belief that they might serve as substrates for Δ^5-3-ketosteroid isomerase. Abstraction of the proton at C-4 by the enzyme should then generate, via an enolic intermediate, the corresponding highly reactive conjugated allenic ketones, which might be expected to react covalently with a nucleophilic amino acid residue at the active site (Scheme 3). This proposal was based on expected conformational similarities between the acetylenic 5,10-seco-steroids and the normal $\Delta^{5(6)}$-3-ketosteroid substrates for the enzyme.

Synthesis of Acetylenic and Allenic 5,10-Seco-3-oxo-steroids

Details of the syntheses of acetylenic and allenic 5,10-secosteroid inhibitors of the isomerase have been presented elsewhere.[4-6] The essential

[3] F. H. Batzold, A. M. Benson, D. F. Covey, C. H. Robinson, and P. Talalay, Adv. Enzyme Regul. 14, 243 (1976).
[4] F. H. Batzold and C. H. Robinson, J. Am. Chem. Soc. 97, 2576 (1975).
[5] F. H. Batzold and C. H. Robinson, J. Org. Chem. 41, 313 (1976).
[6] D. F. Covey and C. H. Robinson, J. Am. Chem. Soc. 98, 5038 (1976).

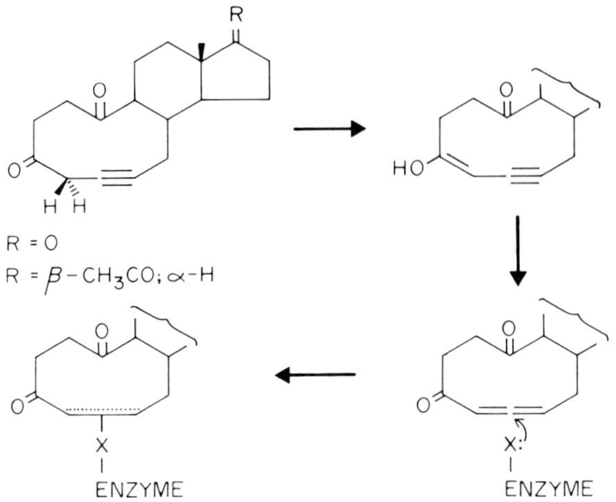

SCHEME 3

features of the synthetic routes leading to compounds carrying either a 17-oxo or a 17β-acetyl substituent are outlined in Scheme 4.

The nomenclature for the numbered allenic and acetylenic secosteroids in Scheme 4 is as follows. When R″ = O, compound **1** is 5,10-secoestr-5-yne-3,10,17-trione; compound **2** is (4R)-5,10-secoestra-4,5-diene-3,10,17-trione; and compound **3** is (4S)-5,10-secoestra-4,5-diene-3,10,17-trione. When R″ = α-H, β-CH₃CO, compound **1** is 5,10-seco-19-norpregn-5-yne-3,10,20-trione; compound **2** is (4R)-5,10-seco-19-norpregna-4,5-diene-3,10,20-trione; and compound **3** is (4S)-5,10-seco-19-norpregna-4,5-diene-3,10,20-trione.

Measurements of Enzyme Inhibition and Determination of Rate and Binding Constants

Measurement of Rates of Inhibition. The inhibition of Δ⁵-3-keto-steroid isomerase activity has been studied[4] in systems containing in a final volume of 500 μl: 1 mM potassium phosphate at pH 7.0,[7] 5–8 μM crystalline Δ⁵-3-ketosteroid isomerase of *P. testosteroni*,[8] and 10–200 μM allenic or acetylenic steroid (steroid added in 20 μl of 1,4-dioxane).[9] The reactions were run at 25° in small polypropylene tubes containing a

---

[7] At potassium phosphate concentrations above 1 mM, destruction of the acetylenic and allenic secosteroids occurs at a significant rate.

[8] The molar enzyme concentrations are based on a subunit weight of 13,394.

[9] The dioxane was freshly distilled from LiAlH₄.

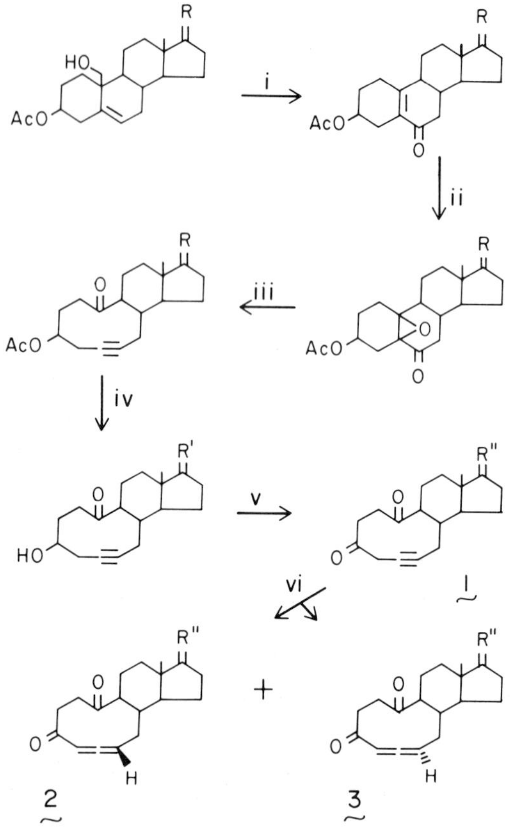

SCHEME 4. The substituents are as follows: $R = \alpha$-H, β-OCOCH$_3$; or $R = \alpha$-H, β-CH$_3$CHOCOCH$_3$; $R' = \alpha$-H, β-OH; or, $R' = \alpha$-H, β-CH$_3$CHOH; $R'' = O$; or, $R'' = \alpha$-H, β-CH$_3$CO. The reagents used to effect the reactions are (i) chromium trioxide–pyridine; (ii) m-chloroperbenzoic acid; (iii) p-toluenesulfonyl hydrazide; (iv) potassium hydroxide–methanol; (v) chromium trioxide–acetone; (vi) triethylamine–1,4-dioxane.

Teflon-coated magnetic stirrer. The enzyme activity in a parallel incubation system, containing all components except the steroid, was entirely stable for several hours. Small aliquots (1–20 μl) were removed from the reaction vessels at suitable time intervals (1–3 min) and diluted with 1.0% neutralized bovine serum albumin. The dilutions required to obtain measurable rates in the subsequent assays depended on the degree of inactivation of the enzyme, but were in the range of 2×10^3- to 2×10^6-fold. Aliquots of the diluted enzyme were then assayed for isomerase activity at 25° in systems containing in a final volume of 3.0 ml: 57.8 μM

Δ^5-androstene-3,17-dione (in 50 μl of methanol), 33 mM potassium phosphate at pH 7.0, and enzyme used to start the reaction. The rate of formation of Δ^4-androstene-3,17-dione ($a_m = 16,300$ M^{-1} cm^{-1}) was obtained by measuring the increase in absorbance at 248 nm.

Analysis of Kinetics of Inhibition. Under the above conditions, the rate of inactivation of the isomerase obeyed pseudo-first-order kinetics over several half-lives, for the range of inhibitor concentrations studied. The half-lives were determined[4,6] from linear regression plots of the logarithm of enzyme activity as a function of time.

Analysis of the kinetics of inactivation followed the procedure of Kitz and Wilson.[10] A slight modification of the original scheme[10] is necessary to accommodate the fact that in the present case the active alkylating species of inhibitor (I_A, allenic secosteroid) is generated from a precursor (I_0, the acetylenic secosteroid), which serves as a substrate for the enzyme. The scheme for the formation of the irreversibly inhibited enzyme species (EI_A') from either acetylenic or allenic secosteroid may be pictured as follows:

$$E + I_0 \underset{k_2}{\overset{k_1}{\rightleftharpoons}} E \cdot I_0 \overset{k'}{\rightleftharpoons} E \cdot I_A \overset{k_3}{\rightarrow} EI_A'$$
$$\updownarrow$$
$$E + I_A$$

The kinetic analysis is made on the basis of the following assumptions: the inhibitor concentration may be considered to remain constant during the initial phase of the inactivation process; $E \cdot I_0$ (the reversible enzyme–acetylene complex) is at all times in equilibrium with free enzyme (E) and free acetylene (I_0); and the rate of formation of the inhibitory species (I_A) far exceeds the rate of inactivation of the enzyme, i.e., $k' \gg k_3$. Kitz and Wilson[10] showed that the reciprocal of the pseudo-first-order rate constant of inactivation ($1/k_{app}$) and the reciprocal of the inhibitor concentration ($1/[I]$) obey the following linear relationship from which both the K_I and k_3 values can be obtained:

$$1/k_{app} = (1/k_3) + (K_I/k_3[I])$$

Double reciprocal plots of the pseudo-first-order rate constants of inactivation (k_{app}) with respect to inhibitor concentration are shown in Fig. 1 for two acetylenic steroids. The K_I and k_3 values were obtained from the slopes and intercepts. The half-lives of the enzyme in the presence

[10] R. Kitz and I. B. Wilson, *J. Biol. Chem.* **237**, 3245 (1962).

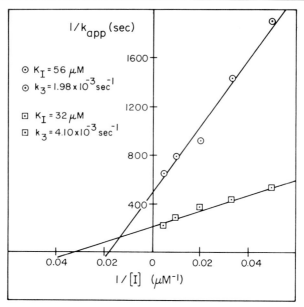

FIG. 1. Irreversible inactivation of Δ^5-3-ketosteroid isomerase by 5,10-secoestr-5-yne-3,10,17-trione ($\odot$) or 5,10-seco-19-norpregn-5-yne-3,10,20-trione ($\boxdot$). Double reciprocal plots are shown of the apparent pseudo-first-order rate constant of inactivation (k_{app}) with respect to inhibitor concentration. Values of K_I and k_3 were determined from slopes and intercepts. [F. H. Batzold and C. H. Robinson, *J. Am. Chem. Soc.* **97**, 2576 (1975)].

of two concentrations (20 and 200 μM) of each of the two acetylenic secosteroids, as well as the corresponding pairs of isomeric (4R and 4S) allenic secosteroids, are presented in the table.

Demonstration of the Enzymic Conversion of the Acetylenic Substrates into the Allenic Inhibitors

Addition of catalytic quantities of crystalline isomerase to solutions of either 5,10-secoestr-5-yne-3,10,17-trione or 5,10-seco-19-norpregn-5-yne-3,10,20-trione results in the rapid appearance of an absorption band at 227 nm of the intensity expected if the entire quantity of each acetylenic compound had been converted to the corresponding allenic compound.[11]

[11] The ultraviolet absorption maximum of (4 R)-5,10-secoestra-4,5-diene-3,10,17-trione is at 222 nm (CH₃CN) with $\alpha_m = 12,000 \ M^{-1} \ cm^{-1}$. The bathochromic shift of the maximum in aqueous solutions is a well recognized phenomenon (cf. L. F.

HALF-LIVES ($t_{1/2}$) OF THE INACTIVATION OF Δ^5-3-KETOSTEROID ISOMERASE BY
ACETYLENIC AND ALLENIC 3-OXO-5,10-SECOSTEROIDS[a]

| | $t_{1/2}$ at [I] of | |
Inhibitor	20 μM (sec)	200 μM (sec)
5,10-Secoestr-5-yne-3,10,17-trione	1644[b]	564[b]
(4R)-5,10-Secoestra-4,5-diene-3,10,17-trione	1520	540
(4S)-5,10-Secoestra-4,5-diene-3,10,17-trione	2916	660
5,10-Seco-19-norpregn-5-yne-3,10,20-trione	318[b]	152[b]
(4R)-5,10-Seco-19-norpregna-4,5-diene-3,10,20-trione	338	168
(4S)-5,10-Seco-19-norpregna-4,5-diene-3,10,20-trione	358	333[c]

[a] From D. F. Covey and C. H. Robinson, *J. Am. Chem. Soc.* **98**, 5038 (1976).
[b] These experimental values are slightly higher than those used to construct the plots shown in Fig. 1.
[c] The enzyme is apparently saturated by this inhibitor at 20 μM since the $t_{1/2}$ values are essentially the same at 20 and 200 μM. Inactivation with this inhibitor at 10 μM gave $t_{1/2} = 498$ sec.

Direct evidence for the enzymic conversion of the acetylenic seco-steroids to allenic secosteroids was obtained chromatographically.[6] The incubation mixture contained in a final volume of 500 μl: 1 mM potassium phosphate at pH 7.0, 200 μM acetylenic secosteroid (in 20 μl of 1,4-dioxane), and 4.8 μM isomerase. Two minutes after addition of the acetylenic substrate to the system, a 100-μl portion of the incubation mixture was removed by syringe and injected directly onto the column (μ-Bondapak/C_{18}; $\frac{1}{4}$ inch $\times$ 1 foot) of a Waters Associates High Pressure Liquid Chromatograph (HPLC) equipped with an OmniScribe recorder and a 254 nm fixed wavelength differential UV detector for the detection of the allenic secosteroids. The column was eluted with water/acetonitrile mixtures. After incubation for 2 min of 5,10-secoestr-5-yne-3,10,17-trione with isomerase, HPLC analysis (column eluted with 70% water/30% acetonitrile; flow rate 1.5 ml/min) showed the presence of a 2:8 (relative ratio) mixture of (4S)-5,10-secoestra-4,5-diene-3,10,17-trione (retention time, 13 min) and (4R)-5,10-secoestra-4,5-diene-3,10,17-trione (retention time, 15 min), respectively, which accounted for 94% of the acetylenic ketone present initially. In an analogous experiment, HPLC analysis (column eluted with 60% water/40% acetonitrile; flow rate 3 ml/min) showed that 90% of 5,10-seco-19-norpregn-5-yne-3,10,20-

Fieser and M. Fieser, "Steroids," p. 16. Van Nostrand-Reinhold, Princeton, New Jersey, 1959). The (4S) isomer has similar spectral characteristics.

trione had been converted to a mixture of (4S)-5,10-seco-19-norpregna-4,5-diene-3,10,20-trione (retention time, 6 min) and (4R)-5,10-seco-19-norpregna-4,5-diene-3,10,20-trione (retention time, 7 min) in a relative ratio of 1:9, respectively. Neither acetylenic secosteroid was converted into allenic ketones under the same conditions in the absence of isomerase.[6]

Nature of the Enzyme–Inhibitor Reaction Product

Although the precise nature of the linkage of the steroid to the protein in the enzyme–inhibitor complex remains to be determined, several lines of evidence support the premise that the highly reactive conjugated allenic oxosteroids enter into covalent linkage with the enzyme, and that this reaction is the basis for the loss of enzymic activity.

1. The enzymic conversion of the acetylenic secosteroids to a mixture of (4R)- and (4S)-allenic secosteroids is both rapid and complete. The half-lives of the enzyme in the presence of comparable concentrations of acetylenic secosteroid and the predominantly formed (4R)-allenic secosteroid are nearly the same (see the table). This suggests that the major reactive species is the (4R)-allenic secosteroid.

2. High degrees of dilution of partially inactivated enzyme preparations into 1% bovine serum albumin result in immediate arrest of the inactivation process. Such preparations retain constant activity for at least several days at 4°.

3. Enzyme inactivated by 5,10-seco-19-norpregn-5-yne-3,10,20-trione does not regain activity upon dialysis against 1 mM potassium phosphate at pH 7.0 for 24 hr at 4°.

4. The inactivated enzyme and the native isomerase are electrophoretically distinct, yielding R_f values of 0.51 and 0.54, respectively, upon electrophoresis on polyacrylamide gels at pH 9.

5. Isomerase inactivated by 5,10-secoestr-5-yne-3,10,17-trione is not reactivated to a significant extent by gel filtration on Sephadex G-25M. Restoration of 5% of the enzyme activity occurs during this procedure, possibly indicating that the linkage between steroid and enzyme is not entirely stable under the conditions used (25°, 10 mM potassium phosphate, pH 7.0).

Acknowledgments

These studies were supported by U.S. Public Health Service Research Grants AM 15918, AM 07422, CA 16418 and Training Grant GM 01183.

[52] Labeling of Δ⁵-3-Ketosteroid Isomerase by Photoexcited Steroid Ketones

By WILLIAM F. BENISEK

The Δ^5-3-ketosteroid isomerase of *Pseudomonas testosteroni* catalyzes the allylic isomerization of Δ^5- and Δ^{10}-3-ketosteroids to the Δ^4 isomers. The enzyme has been extensively studied structurally and mechanistically for many years. Much of what is known about the enzyme from these studies has been comprehensively reviewed by Talalay and Benson.[1] Several attempts to identify amino acid residues comprising the substrate binding site of the isomerase have been made using affinity labeling techniques. These include the use of 6β-bromotestosterone acetate as an active site-directed alkylating agent,[2] and the use[3] of 5,10-secoestr-5-yne-3,10,17-trione and 5,10-seco-19-norpregn-5-yne-3,10,20-trione as k_{cat} inhibitors.[4] Pursuing a different approach, Martyr and Benisek[5,6] investigated the use of Δ^4-3-keto steroids, products of the isomerase reaction, as photoexcitable affinity reagents. These workers found that several members of this class of steroids stimulated an ultraviolet light-dependent photoinactivation of the enzyme. The loss of enzymic activity could be correlated with destruction of a single residue of aspartic acid or asparagine.

Theory

In their electronic ground state, saturated and unsaturated ketones possess limited chemical reactivity toward protein functional groups, reactions being limited to the addition of nucleophiles across the carbon–oxygen double bond (as in Schiff's-base formation) and across a conjugated carbon–carbon double bond. The reactivity of the keto group can be enormously increased by the absorption of light. The $n \rightarrow \pi^*$ absorption band of ketones is centered in the range of 280–320 nm, the exact position depending upon the structure of the ketone and the solvent. Absorption of a photon in this wavelength range and at somewhat longer

---

[1] P. Talalay and A. M. Benson, "The Enzymes" (P. D. Boyer, ed.), 3rd ed., Vol. 6, p. 591. Academic Press, New York, 1972.

[2] K. G. Büki, C. H. Robinson, and P. Talalay, *Biochim. Biophys. Acta* **242**, 268 (1971).

[3] F. H. Batzold and C. H. Robinson, *J. Am. Chem. Soc.* **97**, 2576 (1975).

[4] R. R. Rando, *Science* **185**, 320 (1974). See also this volume [3] and [12].

[5] R. J. Martyr and W. F. Benisek, *Biochemistry* **12**, 2172 (1973).

[6] R. J. Martyr and W. F. Benisek, *J. Biol. Chem.* **250**, 1218 (1975).

wavelengths results in excitation of one of the nonbonding electrons of the oxygen to the antibonding π^* orbital. The initially produced excited state is a singlet. It is a very short-lived species (lifetime of about 10^{-9} sec), which is efficiently transformed to the triplet state. The triplet state is much more stable (lifetime = 10^{-3} to 10^{-2} sec at $77°K$), and it is from this state that much of the photochemistry of ketones proceeds.[7] Since the once paired nonbonding electrons of the ground state are in different molecular orbitals in the triplet excited state, this state exhibits a "diradical-like" character in its chemical properties. The unpaired electron on the oxygen imparts an alkoxy radical character to the ketone oxygen. Thus the first step in many of the photochemical reactions of ketones involves hydrogen atom abstraction from suitable donors by the oxygen atom generating a pair of free radicals. The radical pair can combine with each other, undergo additional hydrogen atom transfers, or other molecular rearrangements to yield the final products. If the hydrogen atom initially abstracted is part of a protein which binds the ketone, then photoaffinity labeling of the protein by the excited ketone is, in principle, possible.

A very large body of literature exists on the photochemical reactions of ketones.[7-10] Only a brief indication of reactions that might be of importance in affinity labeling of proteins by Δ^4-3-ketosteroids will be presented here.

Addition of Nucleophiles[11-13]

[7] See, for example, J. A. Barltrop and J. D. Coyle, "Excited States in Organic Chemistry." Wiley, New York, 1975.
[8] R. O. Kan, "Organic Photochemistry." McGraw-Hill, New York, 1966.
[9] O. L. Chapman and D. S. Weiss, *in* "Organic Photochemistry" (O. L. Chapman, ed.), Vol. 3, p. 197. Dekker, New York, 1973.

Addition of Hydrocarbon C—H Bonds[14-17]

[10] N. J. Turro, J. C. Dalton, K. Dawes, G. Farrington, R. Hautala, D. Morton, M. Niemczyk, and N. Schore, *Acc. Chem. Res.* **5**, 92 (1972).
[11] T. Matsuura and K. Ogura, *J. Am. Chem. Soc.* **88**, 2602 (1966).
[12] T. Matsuura and K. Ogura, *Bull. Chem. Soc. Jpn.* **40**, 945 (1967).
[13] W. G. Dauben, G. W. Shaffer, and N. D. Vietmeyer, *J. Org. Chem.* **33**, 4060 (1968).
[14] D. Bellus, D. R. Kearns, and K. Schaffner, *Helv. Chim. Acta* **52**, 971 (1969).
[15] N. C. Yang and D. D. H. Yang, *J. Am. Chem. Soc.* **80**, 2913 (1958).
[16] S. Wolff, W. L. Schreiber, A. B. Smith III, and W. C. Agosta, *J. Am. Chem. Soc.* **94**, 7797 (1972).
[17] R. E. Galardy, L. C. Craig, and M. P. Printz, *Nature (London), New Biol.* **242**, 127 (1973).

Photochemical Oxidation–Reduction[16,18]

Based on the foregoing reactions, some useful observations pertaining to the use of Δ⁴-3-ketosteroids as affinity reagents can be made. First, the chemical specificity of the photoexcited keto group is very broad since both nucleophiles and C—H groups are potential reactants. This would suggest that an enzyme's binding site need not contain reactive nucleophiles in order to be labeled. Second, reactions of the photoexcited steroid with a protein may or may not result in covalent attachment of the steroid to the protein. Third, the possibility of occurrence of nonspecific side reactions is a very real one, particularly in the case of a species that is as reactive as a ketone excited state.

Equipment

Light Source and Filters

Since the n → π* band of ketones is in the 280–320 nm range, a light source that emits radiation in this region is necessary. For many purposes

[18] R. Breslow, S. Baldwin, T. Flechtner, P. Kalicky, S. Liu, and W. Washburn, *J. Am. Chem. Soc.* **95**, 3251 (1973).

we have found a medium-pressure mercury arc lamp to be adequate. A suitable lamp is manufactured by Hanovia Lamp Division, Canrad Precision Industries, Inc., Newark, New Jersey[19] (Model 679A, rated at 450 watts). According to the manufacturer it produces an arc $4\frac{1}{2}$ inches long. The emission spectrum of the lamp is not continuous, but consists of a large number of discrete lines extending from the infrared through the ultraviolet. The most intense lines are located at 1014 nm, 578 nm, 546 nm, 436 nm, 404.5 nm, 366 nm, 334 nm, 313 nm, 302.5 nm, 297 nm, 280 nm, 265 nm, 254 nm, and 222 nm. The envelope of the lamp is made of quartz, which will pass all these wavelengths. It is therefore necessary to block those wavelengths likely to excite protein chromophores, a process known to result in the inactivation of many enzymes.[20] On the other hand, radiation effective in exciting the Δ^4-3-keto group $[\lambda_{max}(n \rightarrow \pi^*) \approx 310$ nm with a long wavelength tail extending to about 360 nm] must be transmitted to the sample. One useful filter that possesses suitable absorption characteristics is Pyrex glass (Corning 7740), which absorbs essentially all wavelengths below 280 nm and exhibits partial absorption of radiation between 280 and 350 nm. According to Calvert and Pitts,[21] 4 mm of Pyrex glass transmits 10% of incident 310 nm radiation, 30% at 319 nm, and 50% at 330 nm. The Hanovia lamp is cooled during use by means of a glass immersion well (dewar type) in which it operates.[19] This is a double-walled Pyrex or quartz dewar between the walls of which cooling water is circulated. When, as is usual, a Pyrex dewar is employed, 4 mm of Pyrex are traversed by the radiation. Additional filtration can be achieved by the insertion of glass filter sleeves[19] between the lamp and the innermost wall of the immersion well. Solution filters can also be employed, and several suitable for use in the ultraviolet are described by Calvert and Pitts.[21] Katzenellenbogen et al.[22] used a saturated solution of cupric sulfate to screen out wavelengths shorter than 315 nm.

Reaction Vessels

Pyrex nuclear magnetic resonance (NMR) sample tubes (0.4 mm wall thickness) are suitable for containing the sample to be irradiated. These can be purchased from Kontes Glass Co. in several different grades. The tubes can be stoppered with rubber septa during experiments requiring anaerobic conditions.

[19] Lamps, immersion wells, and filter sleeves are available from Ace Glass Company, Vineland, New Jersey 08360.
[20] R. Setlow and B. Doyle, Biochem. Biophys. Acta 24, 27 (1957).
[21] J. G. Calvert and J. N. Pitts, Jr., "Photochemistry." Wiley, New York, 1966.
[22] J. A. Katzenellenbogen, H. J. Johnson, Jr., K. E. Carlson, and H. N. Meyers, Biochemistry 13, 2986 (1974).

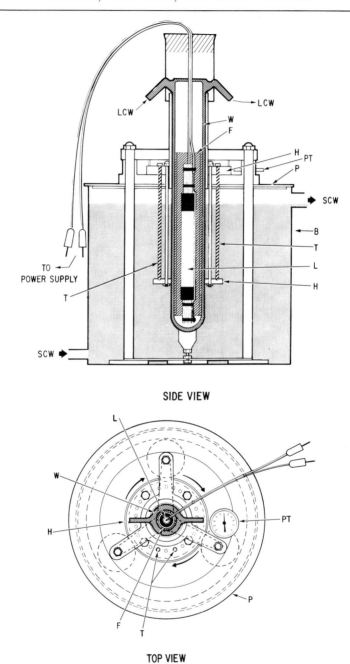

SIDE VIEW

TOP VIEW

Fig. 1. Apparatus used for photoaffinity labeling. Shown are schematic scale drawings of side and top views of the photochemical reactor in current use. The

Photolysis Apparatus

Early experiments[5] employed an apparatus in which the lamp and sample were contained in a light-tight wooden cabinet. The distance between the center of the mercury arc and the center of the sample tubes was 8.5–9.0 cm. Up to three tubes could be irradiated at the same time. The lamp was surrounded by a quartz cooling jacket, and the sample tubes were mounted in a Pyrex dewar flask that had a unsilvered window facing the lamp. Sample temperature was maintained by blowing cool dry nitrogen through the Pyrex dewar. This apparatus performed satisfactorily but suffered from several disadvantages. In order to protect personnel from exposure to dangerous radiation, the lamp had to be extinguished before the cabinet could be opened in order to withdraw aliquots from the samples. Only three samples could be irradiated simultaneously, and sample temperature control was rather poor ($\pm 5°C$).

This apparatus was superseded[6] by a "merry-go-round" type of device that does not have these disadvantages (Fig. 1). The lamp, L, is mounted vertically at the center and is surrounded by a filter sleeve, F, which, in turn, is surrounded by the Pyrex immersion well, W. Tap water flows between the walls of the immersion well in order to dissipate heat from the lamp. Surrounding the immersion well is a cylindrical "merry-go-round" tube holder, H, drilled with 20 holes (4.5 cm radial distance) for the NMR tubes, T. The holder is mounted on nylon ball bearings so that it can rotate about a vertical axis passing through the lamp. The holder is driven at 6 rpm by a small electric motor (not shown) coupled to the rim of the holder by a rubber rimmed wheel. Rotation of the holder about the lamp serves to average out any angular inhomogeneity in the light. Thus all tubes receive the same dose rate averaged over time. The entire apparatus below the mounting plate, P, is surrounded by a stainless steel bath, B, which is filled with distilled water. This water is circulated to a thermostatically controlled water baths, Forma Model 2095, and serves to control the sample temperature. The lamp is energized by an Ace Glass Co. mercury vapor lamp power supply (Cat. No. 6515–60).

General Procedures

Solutions to be irradiated are prepared in small test tubes from stock solutions of steroid (in ethanol), isomerase, and buffer and appropriate

---

meaning of the identifying symbols is as follows: L, mercury arc lamp; F, filter sleeve; W, immersion well; H, sample tube holder; T, NMR sample tube; B, water bath; PT, photographic thermometer; P, mounting plate; SCW, sample tube cooling water (direction of flow indicated by arrows); LCW, lamp cooling water (direction of flow indicated by arrows).

volumes (0.1–1 ml) transferred to NMR tubes. In the case of anaerobic samples the stock solutions are deoxygenated by bubbling nitrogen gas through them, and suitable aliquots are transferred directly to the NMR tubes under a nitrogen barrier. A nitrogen-purged glove box is convenient for this purpose. All tubes are securely stoppered with rubber septa. An identical set of tubes is prepared to serve as dark controls. The lamp cooling water and circulating bath water are turned on, followed by the lamp itself. After about 1 hr the lamp has reached a steady output, and the sample tubes are inserted into the holder and rotation begun. Zero time samples are withdrawn just prior to commencement of irradiation. Additional samples may be withdrawn at time intervals, care being taken to maintain the nitrogen atmosphere of the anaerobic samples.

Application to Affinity Labeling of Δ^5-3-Ketosteroid Isomerase

Kenetics of Inactivation

Tubes containing purified[23] isomerase either alone or in the presence of various steroid competitive inhibitors are irradiated with 4 mm Pyrex-filtered light. At various times aliquots are withdrawn from the tubes, diluted with neutral 1% bovine serum albumin, and assayed[5,23] for enzyme activity. The inactivation observed follows first-order kinetics. The first-order rate constants of inactivation are summarized in the table.

KINETICS OF PHOTOINACTIVATION[a]

Addition	$K_I{}^b$ (μM)	k_{app} (hr^{-1})
None	—	~0.001
Cyclohexenone	>5000	~0.006
Testosterone	48	0.023
19-Nortestosterone	13	0.06; 0.07
19-Nortestosterone acetate	7	0.10

[a] Reaction mixtures contained 5 μM isomerase, 3.3% ethanol, and 0.04 M sodium phosphate, pH 7.0, plus additions. The concentrations of the additions were cyclohexenone, 330 μM; testosterone, 23 μM; 19-nortestosterone, 24 μM; 19-nortestosterone acetate, 21 μM. Final volumes were 0.15 ml. Irradiations were performed using the "old" apparatus (see text).
[b] K_I values were determined under standard assay conditions by the method of M. Dixon [*Biochem. J.* **55**, 170 (1953)].

[23] R. Jarabak, M. Colvin, S. H. Moolgavkar, and P. Talalay, this series, Vol. 15, p. 642 (1969).

Three steroids that lack the Δ^4-3-keto chromophore [3β-hydroxy-5-androstene-17β-carboxylic acid, 3-methoxy-1,3,5(10)-estratrien-17β-ol, and 3β-hydroxy-5-pregnen-20-one] and are competitive inhibitors of the enzyme do not stimulate photoinactivation of the enzyme to a rate greater than that of a steroid-free control. The stimulation of photoinactivation required a steroid that bound at the catalytic site and possessed the Δ^4-3-keto chromophore.

Additional results were interpreted[5] to indicate that the inactivation was Δ^4-3-ketosteroid dependent and was probably active site-directed. Cyclohexeneone which possesses the same chromphore as Δ^4-3-ketosteroids but is not a competitive inhibitor of the enzyme, stimulated photoinactivation only to a small extent even when present at more than ten times the concentration of compounds that markedly stimulated photoinactivation. It was noted that the first-order rate constant for photoinactivation increases with increasing affinity of the enzyme for the particular steroid, a pattern to be expected if the photoinactivation is active site-directed. In addition, the rate of photoinactivation promoted by 19-nortestosterone acetate was decreased approximately 2-fold in the presence of 21 μM 3β-hydroxy-5-androstene-17β-carboxylic acid, a competitive inhibitor that does not support photoinactivation of the enzyme. This protective effect further strengthens the conclusion that photoinactivation is a catalytic site-directed reaction.

Partial Characterization of Photoinactivated Isomerase[6]

Chemical changes accompanying the 19-nortestosterone acetate-promoted photoinactivation were monitored by amino acid analysis. It was observed that both aspartic acid and histidine decreased with increasing extent of inactivation. In two kinetic studies comparing the rates of activity, aspartic acid and histidine losses, it was observed that the first-order rate constants for activity loss and loss of one residue of aspartate (asparagine) were nearly the same ($k_{\text{activity}} = -0.54$ hr^{-1}, -0.55 hr^{-1}; $k_{\text{asp}} = -0.56$ hr^{-1}, -0.52 hr^{-1}) whereas the rate of histidine loss was substantially slower ($k_{\text{his}} = -0.16$ hr^{-1}, -0.13 hr^{-1}), suggesting that aspartate destruction was responsible for the inactivation of the enzyme. Certain controls were run which indicated that aspartate destruction is a steroid binding site-specific process. Thus irradiation of isomerase in the presence of cyclohexenone, 3β-hydroxy-5-androstene-17β-carboxylic acid, or both together did not result in aspartate destruction (or in histidine destruction). Also, irradiation of performic acid oxidized bovine pancreatic ribonuclease A, to which steroids presumably do not bind, in the presence of 19-nortestosterone acetate induced no destruction of aspartic acid. Attempts to more fully characterize the nature of the aspartate-destroying reaction have been made recently. No significant differences are seen between peptide maps of the soluble tryptic peptides of native and photochemically inactivated isomerase. From chymotryptic digests of the insoluble material remaining after tryptic digestion, a peptide whose amino acid composition corresponds to residues 31–45[24] has been purified.[25] A similar peptide has been purified by the same procedure from the inactivated enzyme. The peptide from the photoinactivated enzyme contains 0.7 fewer residues of Asx and 0.7 more residues of alanine than the corresponding peptide from the native enzyme, suggesting that the aspartate destroying reaction includes a decarboxylation or decarboxamidation. Experiments to clarify the chemistry of the aspartate (asparagine) destroying reaction and its position in the polypeptide chain are in progress.

Effects of Oxygen[6]

The presence of oxygen was found to markedly influence the character of the photochemical events occurring during photoinactivation promoted by 19-nortestosterone acetate. Oxygen inhibited both the loss of activity and the destruction of aspartate whereas it stimulated destruction of histidine. Under anaerobic conditions, no significant loss of histidine occurred whereas 0.98 of 12 moles of aspartate (per mole of enzyme monomer, MW 13,394) was destroyed and 85% of the activity was lost.

Related Applications

Several reports describing photoaffinity labeling studies in which the photoactivation of ketones play a role have appeared recently. Glover et al.[26] have found that partially active α-chymotrypsin derivation containing phenacyl, 1-naphthacyl, and 2-naphthacyl moieties on the sulfur of methionine 192 (as sulfonium salts) undergo partial reactivation upon irradiation with λ >300 nm using an apparatus similar to the one described here. They found that about half of the protein-associated phenacyl group could be detached from the protein by irradiation, yield-

---

[24] This sequence is --- Ala Asp Asn Ala Thr Val Glu Asn Pro Val Gly Ser Glu Pro Arg ---.

[25] The peptide was purified by paper chromatography (butanol/acetic acid/H₂O 4:1:5 upper phase) followed by pH 6.5 electrophoresis at a right angle to the direction of chromatography, and then pH 1.9 electrophoresis.

[26] G. I. Glover, P. S. Mariano, T. J. Wilkonson, R. A. Hildreth, and T. W. Lowe, *Arch. Biochem. Biophys.* **162**, 73 (1974). See also this volume [71].

ing fully active α-chymotrypsin. They proposed that in addition to this reaction a concurrent photoprocess occurred yielding an inactive species. This process may involve an uncharacterized photochemical reaction between the excited keto group of the phenacyl moiety and a nearby amino acid side chain. Galardy *et al.*[17,27] reported that pentagastrin derivatives, 4-acetylbenzoyl pentagastrin and 4-benzoylbenzoyl pentagastrin, became covalently attached to bovine plasma albumin when irradiated with ultraviolet light. The photoattachment of the 4-acetylbenzoyl derivatives appeared to possess some specificity since oleate provided a substantial inhibition of the attachment process. Katzenellenbogen *et al.*[22] found that 6-oxoestradiol inhibited the estradiol binding activity of rat uterine cytosol during irradiation with $\lambda > 315$ nm and that the inhibition reaction was strongly suppressed by inclusion of estradiol in the photolysis reaction mixture.

[27] R. E. Galardy, L. C. Craig, J. D. Jamieson, and M. P. Printz, *J. Biol. Chem.* **249,** 3510 (1974).

[53] Affinity Labeling of Antibody Combining Sites as Illustrated by Anti-Dinitrophenyl Antibodies

By DAVID GIVOL and MEIR WILCHEK

The chemical structure of the combining site of antibodies was a central paradox in immunology: How could an apparently infinite range of combining specificities reside in what appeared to be a very monotonous group of proteins, closely related in their structure? Considered from this viewpoint, it is interesting that, relative to other proteins with combining sites, e.g., enzymes, antibodies are a late evolutionary acquisition, present only in vertebrates, and their synthesis is confined to only one type of cell: the lymphocyte. If we compare, on the one hand, the number of different combining sites that are being made by all cells of our body, and on the other, the number of different combining sites that lymphocytes can make in the form of antibodies, it is clear that the lymphocytes can produce a greater variety of proteins than all the other cells *in toto*. This is evident since the immune system can make more than one antibody against any specific foreign protein. It has therefore been a long-standing question whether the structure of the antibody site is based on principles that differ from those governing the structure of the active sites of other proteins.

All immunoglobulins have the same basic multichain subunit, a symmetrical molecule composed of two heavy chains (H) of molecular weight

50,000 and two light chains (L) of molecular weight of about 25,000.[1] The most unique feature of these polypeptide chains is that only the first 110–120 residues from the NH_2-terminus (one-half of L and one-fourth of H) have sequences that vary from one protein to another, whereas the remainder of the chain has an invariant sequence within a given class.

The second notable feature is that each polypeptide chain is composed of homology segments (domains) of approximately 110 residues,[2] which have the following characteristics: (a) they show sequence homology with other constant segments of both H and L chains, presumably as a result of gene duplication; (b) the homology segments contain one intrachain disulfide bond between two half-Cys, which are about 65 residues apart; (c) each homology segment folds, independently of other segments, as a globular domain; and (d) homologous globular domains from different chains interact noncovalently to yield a unit possessing a unique biological function, which may be retained even when such a fragment is split from the rest of the molecule. Thus, the antibody combining site resides in the first domains, V_L and V_H, and in one known mouse myeloma protein (protein 315, which binds dinitrophenyl ligands) has been isolated as a fragment, Fv, with a molecular weight of 25,000 and shown to contain all the properties of the combining site.[3,4] This fragment consists of only the first domains of both chains, i.e., the variable domains. The more conventional active fragment, Fab[5] (molecular weight 50,000), contains constant domains in addition to the variable domains, but from the demonstrable existence of Fv in one antibody and from the results of affinity labeling (to be described) it appears reasonable that the antibody combining site is entirely confined to the first 110–120 residues, i.e., the variable portion of both H and L peptide chains.

In addition to X-ray crystallography, two approaches have been used to locate the antibody combining site in immunoglobulins and to deduce the structural basis of antibody specificity. One method is based on the known dependence of protein conformation on amino acid sequence,[6] i.e., the specificity of antibodies must be a reflection of positions having specific amino acid residues. Hence, comparison of a set of homogeneous myeloma proteins would be expected to define regions of variable primary

---

[1] R. R. Porter, in "Essays in Biochemistry" (P. N. Campbell and G. D. Greville, eds.), Vol. 3, pp. 1–24. Academic Press, New York, 1967.

[2] G. M. Edelman, Biochemistry 9, 3197 (1970).

[3] D. Inbar, J. Hochman, and D. Givol, Proc. Natl. Acad. Sci. U.S.A. 69, 2659 (1972).

[4] J. Hochman, D. Inbar, and D. Givol, Biochemistry 12, 1130 (1973).

[5] R. R. Porter, Biochem. J. 73, 119 (1959).

[6] C. B. Anfinsen, Science 181, 223 (1973).

sequence. The most significant analysis of such a comparison of the variable regions of their L or H chains was made by Wu and Kabat.[7] They found that the sequences of three short segments of V_L and V_H are hypervariable as compared to the rest of the sequence. These segments comprise residues 24–34, 50–56, and 89–97 of L chain and 31–35, 50–65, and 95–102 of H chain.

A second and complementary method is the direct location of residues involved in contact with the ligand at the combining site. This can be achieved by *in situ* affinity labeling of the protein combining site with a reagent that is structurally similar to the ligand. Affinity labeling was first applied to antibodies by Wofsy *et al.*[8] in 1962; the method has been reviewed by Singer[9] and by Givol.[10] It has become clear that the results of these two approaches can provide clues to the chemical structure of the antibody combining site.

Affinity Labeling

Theory

This method aims at covalent binding of a ligand at the combining site of the antibody which binds that ligand. A labeling reagent for a particular active site is in fact an analog of the hapten. It is prepared so that (1) by reason of its steric complementarity to the combining site, the reagent first combines reversibly with the site, and (2) by reason of a small reactive group (x) on the reagent, it can combine with one or more amino acid residues (y) at the site to form irreversible covalent bonds. The formation of the initial reversible complex, C, increases the local concentration of the reagent at the site relative to its concentration in the solution. Hence, the reaction rate with residue y in the site will be greater than that with any similar residue y elsewhere on the protein. The specificity of the labeling reaction is due mainly to the increase in local concentration of the reagent at the site, and the labeled residue need not be an unusually reactive group. According to the elementary theory of the method,[8,9] the reaction may be written,

$$y + RX \underset{k_2}{\overset{k_1}{\rightleftharpoons}} C \overset{k_3}{\rightarrow} L \quad K_A = \frac{k_1}{k_2} \tag{1}$$

[7] T. T. Wu and E. A. Kabat, *J. Exp. Med.* **132**, 211 (1970).

[8] L. Wofsy, H. Metzger, and S. J. Singer, *Biochemistry* **1**, 1031 (1962).

[9] S. J. Singer, *Adv. Protein Chem.* **22**, 1 (1967).

[10] D. Givol, *in* "Essays in Biochemistry" (P. N. Campbell and G. D. Greville, eds.), Vol. 10, p. 73. Academic Press, New York, 1974.

in which y is a residue at the antibody site, RX the affinity labeling reagent, C the initial reversible complex, L the covalently labeled product, k the rate constant, and K_A the association constant. It is important to realize that, whereas the affinity labeling proceeds with first-order kinetics in the reversible complex C, the reagent may also react simultaneously in a bimolecular process with amino acid residues, y′, outside the site to give the undesirable product M [Eq. (2)].

$$y' + RX \xrightarrow{k_4} M \tag{2}$$

The ratio of the rates of formation of L and M ($\dot{L}/\dot{M}$) is the quantity that determines the specificity of affinity labeling and is termed the enhancement. It can be shown that

$$\frac{\dot{L}}{\dot{M}} = \frac{k_3 C}{k_4 y' RX} = \frac{k_3 K_A y_e}{k_4 y'} \tag{3}$$

where y_e is the equilibrium concentration of free y. It is obvious that enhancement increases with increasing K_A and that specificity depends on the ratio k_3/k_4, and not only on k_3. Under favorable conditions the initial rate of affinity labeling can occur with an enhancement of the order of magnitude of the association constant K_A.

Criteria for Affinity Labeling

The experimental criteria for affinity labeling may be summarized as follows: (a) Affinity labeling of an antibody should result in stoichiometric inactivation of the reversible binding activity; i.e., the number of moles of label found per mole of antibody should equal the fraction of binding sites inactivated. (b) Nonspecific immunoglobulin or unrelated antibody should not be labeled significantly. (c) Specific protection against labeling can be achieved by the specific hapten. It should be emphasized that this protection is a rate phenomenon, and affinity labeling will only be slowed, not eliminated, owing to the occupancy of a large portion of antibody sites by the hapten; given sufficient time, a protected sample may also be completely inactivated. (d) If possible, it is desirable to show that the covalently bound hapten can still be located at the binding site. For example, with anti-DNP[11] a red shift in the absorption

---

[11] The following abbreviations are used: MNBD, meta nitrobenzene diazonium; NAP-Lys, (4-azido-2-nitrophenyl)-L-lysine; BADE, N-bromoacetyl-N′-DNP-ethylenediamine; BADL, N^α-bromoacetyl-N-DNP-L-lysine; DNPN₃, 1-azida-2,4-dinitrobenzene; DNP-AD, dinitrophenyl alanyl diazoketone; BADB, N^α-bromoacetyl, N^γ-DNP-aminobutyric acid.

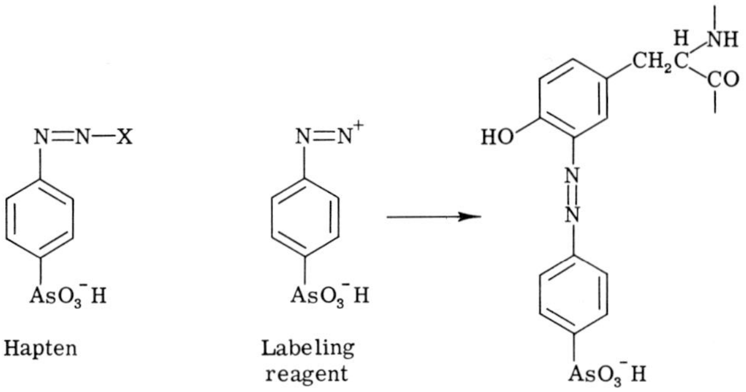

FIG. 1. Diazo reagent of the hapten arsonic acid and its reaction with a tyrosyl residue.

spectrum, characteristic of DNP-hapten bound to the anti-DNP combining site, may be used as a criterion.

Choosing the Labeling Reagent

Some obvious factors in the choice of affinity labeling reagents are noted: (a) The active group, x, should be small and should disturb neither the complementarity of the haptenic portion of the reagent nor the antibody combining site. (b) Stability, ease of synthesis, and availability of radiolabeled precursors are important factors in design of a reagent. (c) Ease of identification of the labeled residue, stability of the label to degradative methods used in analysis of protein sequence determination, including total hydrolysis or Edman degradation, are equally desirable features. (d) A wide range of chemical reactivity of the functional group is particularly helpful in affinity labeling of antibodies in which we do not expect the unusually reactive residues that are sometimes present at active sites of enzyme.

Singer and his colleagues introduced a diazonium labeling reagent (Fig. 1) that fulfills most of the above requirements except for its instability to acid hydrolysis or to Edman degradation, and its quite limited range of reactivity. Derivatives of diazonium compounds with either tyrosine, histidine, or lysine have absorption spectra in the visible range, and the reaction of covalent labeling can be followed spectrophotometrically. An attractive feature of the azo-linkage that is formed is the possibility of subsequently reducing it with dithionite to an amino group with the release of the ligand.[12] This was utilized by Hadler and Metzger[13]

[12] M. Gorecki, M. Wilchek, and A. Patchornik, *Biochim. Biophys. Acta* **229**, 590 (1971).

[13] N. Hadler and H. Metzger, *Proc. Natl. Acad. Sci. U.S.A.* **68**, 1421 (1971).

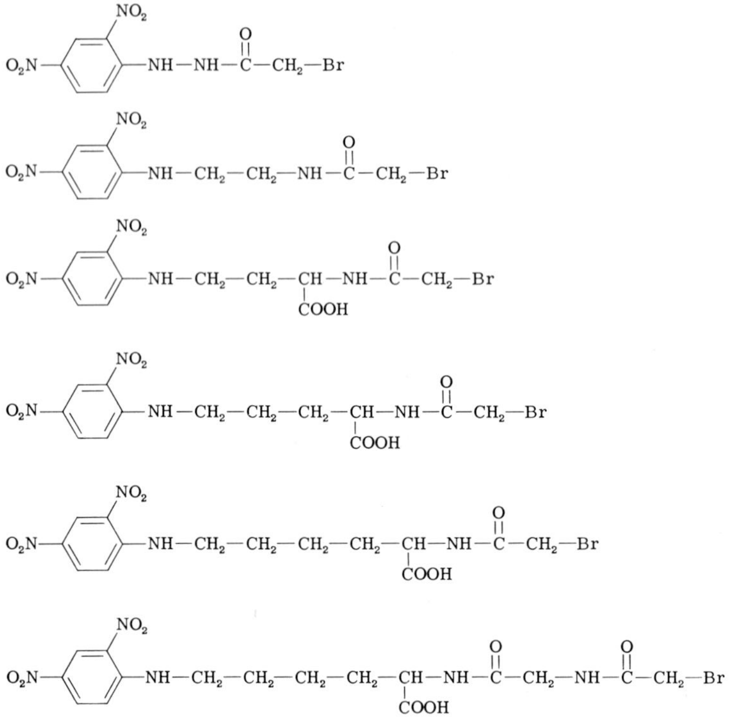

BADH

BADE

BADB

BADO

BADL

BADGL

FIG. 2. A homologous series of bromoacetyl derivatives of dinitrophenyl haptens [P. H. Strausbauch, Y. Weinstein, M. Wilchek, S. Shaltil, and D. Givol, *Biochemistry* **10**, 4342 (1971)].

to initiate cross-linking of the L and H chains via the aminotyrosine thus generated.

Another type of reagent, using the bromoacetyl group as the functional group X, was introduced by us.[14] Aside from the ease of preparation of the labeling reagents and of identification of the labeled residues, Such reagents allow the preparation of homologous series (Fig. 2), the synthesis of which is described here.

The bromoacetyl group was attached to the terminal amino group of the hapten at systematically increasing distances from the major haptenic moiety,[15] thereby permitting labeling residues at various distances in the combining site.

[14] Y. Weinstein, M. Wilchek, and D. Givol, *Biochem. Biophys. Res. Commun.* **35**, 694 (1969).

[15] P. H. Strausbauch, Y. Weinstein, M. Wilchek, S. Shaltiel, and D. Givol, *Biochemistry* **10**, 4342 (1971).

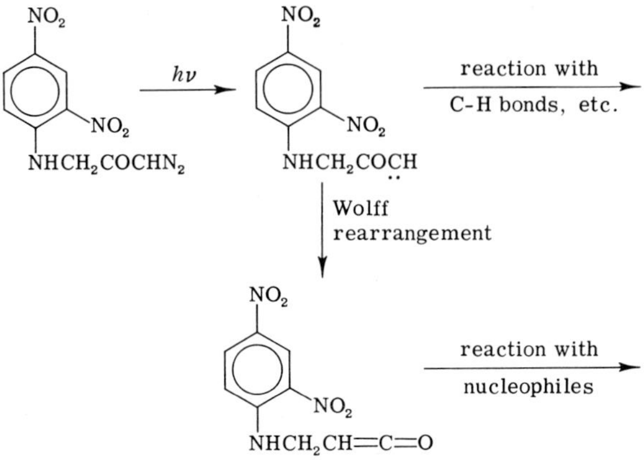

Fig. 3. The diazoketone reagent used for affinity labeling, attached to the hapten dinitrophenylglycine. The reactions as carbene and ketene are shown [C. A. Converse and F. F. Richards, *Biochemistry* **8**, 4431 (1969)].

Converse and Richards[16] have used a diazoketone group attached to the hapten (Fig. 3). The unique feature of this type of reagent is that it is unreactive until irradiated with ultraviolet light (300 nm), so that noncovalent and covalent binding can be separated into two steps, and excess reagent may be removed by gel filtration. Upon photolysis, diazoketones form ketocarbene derivatives (Fig. 3) which have the ability to insert into C—H bonds. However, it appears that the carbene rapidly rearranges by a Wolff type of rearrangement to the less reactive ketene. The ketene is capable of acylating nucleophiles, and its range of specificity is not as wide as is desirable. At least 50% of the labeling was due to reaction with the ketene. In addition, with high affinity antibodies the separation of excess reagent has no advantage since stoichiometric amounts of reagents can be used.

A major point of criticism of all these reagents is that the functional group X is an additional "tail" to the haptenic determinant and is not part of the determinant itself; thus, it may label only peripherally to the site. Another criticism deals with the limited number of amino acids with nucleophilic groups that are susceptible to attack by these reagents. It has been stressed that while we are constrained in selecting a reagent that fits an enzyme-active site, we can with antibodies select the protein to fit the reagent by immunizing animals with haptens so designed. To overcome the disadvantages encountered with previous reagents, Fleet *et*

[16] C. A. Converse and F. F. Richards, *Biochemistry* **8**, 4431 (1969).

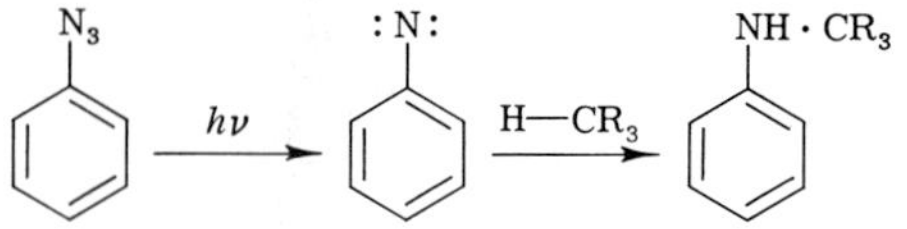

Nitrene reagent

FIG. 4. The aromatic azide reagent. On exposure to light the azide will decompose to give a nitrene that can insert into C–H bonds [G. W. J. Fleet, J. R. Knowles, and R. R. Porter, *Nature* (*London*) **224**, 511 (1969)]. See also this volume [8].

al.[17,18] introduced the aromatic azide (Fig. 4), which has the following advantages: (a) the azide is chemically inert and can be used as a hapten to prepare antibodies against it; (b) it can generate photochemically (by irradiation at 400 nm) the reactive nitrene at the combining site; (c) the reactive nitrene is very similar structurally to the azide used as hapten; and (d) the nitrene, unlike carbene, does not rearrange intramolecularly to a less reactive species but can insert into any C—H bond. Although this reagent seems to be ideal for labeling contact residues in protein, great difficulties were encountered in the identification of the labeled residue and in the isolation of modified peptides. The major features of the four types of affinity labeling reagents used to probe the antibody site are summarized in Table I.

TABLE I
COMPARISON OF AFFINITY LABELING REAGENTS

Property	Functional group			
	Diazonium	Bromoacetyl	Diazoketone	Aromatic azide
Functional group may be part of the antigenic determinant?	No	No	No	Yes
Specificity to residue	Tyr, His, Lys	Tyr, His, Lys, Cys, Met, Asp, Glu	C—H bonds or only nucleophiles	C—H bonds
Reaction rate	Fast	Slow	Slow	Very fast
Sites labeled	0.2–0.5	1.2	0.5–1.5	1–1.6
Identification of labeled residue	Spectral analysis	CM-amino acid	Peptide sequence	Peptide sequence
Possibility of preparing homologous series	No	Yes	Yes	No

[17] G. W. J. Fleet, J. R. Knowles, and R. R. Porter, *Nature* (*London*) **224**, 511 (1969).
[18] G. W. J. Fleet, J. R. Knowles, and R. R. Porter, *Biochem. J.* **128** (1972).

Affinity-Labeled Residues at the Combining Site of Antibodies

In view of the unique structure of antibodies (two dissimilar peptide chains and repeating domains) the questions to be answered by affinity labeling had to do with which chain bears the ligand, which residues are being tagged, and what are the positions of the labeled residues in the amino acid sequence. The answer to these questions will be illustrated mainly by the results obtained with anti-DNP antibodies and protein 315, a mouse myeloma protein possessing high affinity to nitrophenyl ligands.

The Labeled Chain

In an extensive series of studies, Singer and his colleagues analyzed many of the structural aspects of the antibody site by affinity labeling.[19,20] With a variety of antibodies of different hapten specificities and with anti-DNP antibodies from several species, affinity labeling with diazonium reagents labeled tyrosine residues on both L and H chains with a remarkable similarity in the distribution of label between the chains (H:L ratio of the label was between 2 and 4). This led to the hypothesis that all antibody sites are more or less uniform, having both H and L chains in close proximity at the combining site.

Similar studies were performed with myeloma proteins that have antibody activity. It appears that these homogeneous antibodies may be labeled exclusively on one of the two chains, depending on the type and size of reagent used. Table II illustrates the results of affinity labeling of protein 315 with different bromoacetyl reagents whose formulas are given in the table. BADE,[11] and to a great extent BADB,[11] labeled only Tyr on L chains, whereas BADL[11] labeled only Lys on H chains.[21,22] A difference of only 3 Å in length (between BADB and BADL) resulted in a marked change in specificity in both the residue modified and the chain labeled. These observations strongly support the view that labeling reactions are sterically directed in the reversible complex of protein and reagent. Furthermore, support for this contention is provided by the result with the optical isomers, D- and L-BADL. L-BADL labeled a heavy-chain lysine, but D-BADL was unreactive despite protein 315 having the same intrinsic affinity for each reagent.[22]

[19] A. H. Good, Z. Ovary, and S. J. Singer, *Biochemistry* **7**, 1304 (1968).

[20] S. J. Singer and R. F. Doolittle, *Science* **153**, 13 (1966).

[21] J. Haimovich, D. Givol, and H. N. Eisen, *Proc. Natl. Acad. Sci. U.S.A.* **67**, 1656 (1970).

[22] D. Givol, P. H. Strausbauch, E. Hurwitz, M. Wilchek, J. Haimovich, and H. N. Eisen, *Biochemistry* **10**, 3461 (1971).

TABLE II

LABELING OF PROTEIN 315 BY REAGENTS OF DIFFERENT LENGTH

		Labeled residue and chain	
Reagent		% Tyr (L)	% Lys (H)
BADE DNP—NH—CH$_2$—CH$_2$—NH—X[a]		96	4

BADB DNP—NH—CH$_2$—CH$_2$—CH—NH—X
         |
        COOH
		87	13

BADO DNP—NH—CH$_2$—CH$_2$—CH$_2$—CH—NH—X
           |
          COOH
		66	34

BADL DNP—NH—CH$_2$—CH$_2$—CH$_2$—CH$_2$—CH—NH—X
             |
            COOH
		5	95

[a] X = COCH$_2$ B$_r$.

Position of Affinity Labeled Residues

In addition to the conventional methods of peptide isolation, affinity chromatography may be used to isolate the labeled peptide. The method is based on the fact that the native antibody coupled to Sepharose can be used to bind specifically the labeled peptide by virtue of the ligand that is attached to the peptide. Thus peptides from affinity-labeled anti-DNP antibodies can be purified on anti-DNP Sepharose column.[23,24] By this means, the positions of affinity labeled residues in the heavy and light chains of protein 315 were established. These results together with those obtained by other reagents are presented in Table III.[25-31]

Synthesis of Reagents

N-Bromoacetyl-DNP Aniline. 2,4-Dinitroaniline, 5 g (13.5 mmoles), is suspended in 10 ml of benzene. To this suspension 4.2 g (20 mmoles) of

[23] D. Givol, Y. Weinstein, M. Gorecki, and M. Wilchek, *Biochem. Biophys. Res. Commun.* **38**, 825 (1970).

[24] M. Wilchek, V. Bocchini, M. Becker, and D. Givol, *Biochemistry* **10**, 2828 (1971).

[25] F. Franek, *Eur. J. Biochem.* **33**, 59 (1973).

[26] A. Ray and J. J. Cebra, *Biochemistry* **11**, 3647 (1972).

[27] N. O. Thorpe and S. J. Singer, *Biochemistry* **8**, 4523 (1969).

[28] C. E. Fisher and E. M. Press, *Biochem. J.* **139**, 135 (1974).

[29] E. J. Goetzl and H. Metzger, *Biochemistry* **9**, 3862 (1970).

[30] J. Haimovich, H. N. Eisen, E. Hurwitz, and D. Givol, *Biochemistry* **11**, 2389 (1972).

[31] C. L. Hew, J. Lifter, M. Yoshioka, F. F. Richards, and W. H. Konigsberg, *Biochemistry* **12**, 4685 (1973).

TABLE III
AFFINITY-LABELED RESIDUES AND THEIR LOCATION IN ANTIBODIES

Antibody protein	Reagent	Labeled residue	Chain	Reference[a]
Pig anti-DNP	MNBD	Tyr-33	H	25
Pig anti-DNP	MNBD	Tyr-33, Tyr-93	L	25
Guinea pig anti-DNP	MNBD	Tyr-33, Tyr-60, and Tyr-(99–119)	H	26
Mouse anti-DNP	MNDB	Tyr-86	L	27
Rabbit anti-DNP	MNDB	Tyr-90	H	27
Rabbit anti-DNP	NAP-Lys	(29–34), (95–114)(50–57)	H	28
Rabbit anti-NAP	NAP-Lys	(19–34), (95–114)(50–57)	H	28
Rabbit anti-NAP	NAP-Lys	Cys-92 Ala-93	H	18
315 anti-DNP	MNBD	Tyr-34	L	29
315 anti-DNP	BADE	Tyr-34	L	30
315 anti-DNP	BADL	Lys-54	H	30
460 anti-DNP	BADE	Lys-54	L	30
460 anti-DNP	DNPN$_3$	Lys-54	L	31
460 anti-DNP	DNP-AD	Lys-54	L	31

[a] Numbers refer to text footnotes.

bromoacetyl bromide are added dropwise over 2–3 min and refluxed for 10 hr. At the end of the reflux period, the solution is cooled and the benzene and excess bromoacetyl bromide are removed under reduced pressure. The residue is crystallized when washed and triturated with cyclohexane. The crude crystals are filtered and recrystallized from benzene–cyclohexane (1:1) as yellow crystals, m.p. 97°–100°. Yield, 3.5 g (80%).

N-DNP-N′-Z-ethylenediamine. A solution of 2.4 g (10 mmoles) of *N*-Z-ethylenediamine hydrochloride[32] in 15 ml of water is treated with 2.72 g of 2,4-fluorodinitrobenzene (15 mmoles) in 15 ml of ethanol in the presence of an excess of NaHCO$_3$. The reaction mixture is stirred for 2 hr at room temperature, and the precipitate formed is filtered and washed sequentially with water, 50% ethanol, and diethyl ether. The compound is recrystallized from ethyl acetate. Yield, 3 g (83%); m.p. 133°.

N-DNP-ethylenediamine Hydrobromide. A solution of 1.8 g (5 mmoles) of *N*-DNP-*N*′-Z-ethylenediamine in 10 ml of glacial acetic acid is mixed with 15 ml of HBr in CH$_3$COOH (45%). The reaction is allowed to proceed for 15 min and terminated by the addition of dry ether. The precipitate is washed with dry ether. Yield, 1.4 g (90%); m.p. over 250°.

N-Bromoacetyl-N′-DNP-ethylenediamine. *N*-DNP-ethylenediamine hydrobromide (0.77 g, 2.5 mmoles) is suspended in 10 ml of dioxane con-

[32] W. B. Lawson, M. D. Leafer, A. Tewes, and G. J. S. Rao, *Hoppe Seyler's Z. Physiol. Chem.* **349**, 251 (1968).

taining 2.5 ml of 1 N NaOH, and 5 ml of 1 N NaHCO$_3$ is added. This mixture is allowed to react with 0.6 g (2.5 mmoles) of N-hydroxysuccinimide ester of bromoacetic acid, which is dissolved in dioxane.[33] The reaction is allowed to proceed for 20 min after which the reaction mixture is acidified and the product is extracted with ethyl acetate. The extract is washed with water, dried over Na$_2$SO$_4$, and concentrated to dryness. Upon trituration with ether, the compound crystallizes and is recrystallized from ethyl acetate. Yield, 0.65 g (75%); m.p. 163°.

N-DNP-L-*ornithine Hydrochloride.* A sample of L-ornithine hydrochloride (1.68 g, 10 mmoles) is dissolved in 15 ml of boiling water, and 2 g of CuCO$_3$ are added with stirring over a period of 10 min. The excess of CuCO$_3$ is removed by filtration and, after washing with 5 ml of hot water, the filtrate is collected. To the blue filtrate is added an excess of NaHCO$_3$ and a solution of 3.6 g of 2,4-fluoradinitrobenzene in 20 ml of ethanol. The reaction mixture is stirred for 2 hr at room temperature. The green precipitate formed is filtered, washed successively with water, ethanol, and ether, and then dissolved in 10 ml of 3 N HCl. After about 10 min a yellow precipitate is formed, which is collected and washed sequentially with 5 ml of cold 3 N HCl, 5 ml of cold water, ethanol, and ether. Yield, 2.35 g (70%); m.p. 258°.

N^δ-DNP-N^α-*bromoacetyl*-L-*ornithine* (BADO). N^δ-DNP-L-ornithine hydrochloride (1.68 g, 5 mmoles) is dissolved in cold water containing 2 equivalents of NaOH. To this solution, 3 equivalents of BrCH$_2$COBr are added and the reaction is allowed to proceed at 0° for 20 min, during which the pH is maintained above 8 with NaHCO$_3$. The reaction is allowed to continue for another 10 min at room temperature, after which the mixture is acidified to pH 2 with HCl and the product is extracted into ethyl acetate. The extract is washed with water, dried over anhydrous Na$_2$SO$_4$, and concentrated to dryness. After recrystallization from ethyl acetate–petroleum ether the yield is 1.7 g (80%); m.p. 157°–159°.

N^γ-DNP-N^α-*bromoacetyl*-L-*diaminobutyric Acid* (BADB). This is prepared from N^γ-DNP-L-diaminobutyric acid hydrochloride and BrCH$_2$COBr as described for BADO. Yield, 75%; m.p. 81°.

N^ε-DNP-N^α-*bromoacetyl*-L-*lysine* (L-BADL) *and* N^ε-DNP-N^α-*bromoacetyl*-D-*lysine* (D-BADL). This is prepared by allowing ε-DNP-lysine hydrochloride (350 mg, 1 mmole) to react with bromoacetyl hydroxysuccinimide ester (590 mg, 2.5 mmoles) in 30 ml of 50% dioxane containing 1.5 g of NaHCO$_3$. After 1 min the solution is extracted with ethyl acetate and acidified by addition of HCl. The product is extracted into ethyl acetate, dried by Na$_2$SO$_4$, and evaporated. It is crystallized from acetone–water. Yield, 80%; m.p. 128°.

---

[33] P. Cuatrecasas, M. Wilchek, and C. B. Anfinsen, *J. Biol. Chem.* **244**, 4316 (1969).

Radioactivity Labeled Reagents. The labeled reagents are prepared by a scaled-down modification. Solutions of 1.2 mg of N-hydroxysuccinimide (13 μmoles) in 0.1 ml of ethyl acetate and 2.0 mg of dicyclohexylcarbodiimide (14 mmoles) in 0.1 ml of dioxane are mixed with 0.1 ml of a stock solution of [^{14}C]BrCH$_2$COOH (10 μmoles, 4.1 μCi/μmole in dioxane, Amersham) and allowed to react for 3 hr at 25°. The precipitate of dicyclohexylurea is removed by filtration through cotton wool in a Pasteur pipette, and the filtrate is added immediately to a solution of the appropriate DNP compound (10 μmoles) in 1 ml of 50% aqueous dioxane containing 10 mg of NaHCO$_3$. After 10 min at 25° the solution is acidified, the [^{14}C]bromoacetyl-DNP reagent is extracted into ethyl acetate, and the extract is evaporated to dryness. The reagent is further purified by thin-layer chromatography on silica gel (Riedel-DE Haen AG) developed with chloroform–*tert*-amyl alcohol–acetic acid (70:30:3 v/v). Radioactive reagents are located by markers of the purified non-radioactive reagents and then eluted with dioxane. The R_f values of the several reagents follow: BADB, 0.58; BADE, 0.72; BADO, 0.72; D- and L-BADL, 0.80.

Antibodies

Rabbits are immunized with DNP-bovine serum albumin and goats with DNP-keyhole limpet hemocyanine. All animals are immunized by two injections, 3 weeks apart. Each injection consists of 1 mg of the antigen in 1 ml of a buffer composed of 0.15 M NaCl and 0.01 M sodium phosphate (pH 7.4) emulsified with 1 ml of complete Freund's adjuvant (Difco). Injections are given at several intradermal sites. Animals are bled weekly after the second injection and, when antibody titers decrease to below 0.5 gm/ml, are boosted by reinjection in the same manner. Rabbit antisera are pooled whereas the antiserum of each bleeding from each goat is separately maintained. Anti-DNP antibodies are isolated on a DNP rabbit serum albumin–Sepharose immunoadsorbent.[23] The adsorbed antibodies are eluted by incubation (1 hr) with 0.1 M acetic acid at 37° and dialyzed against a buffer composed of 0.01 M sodium phosphate and 0.15 M NaCl (pH 7.4). The myeloma protein 315 is isolated from the serum of tumor-bearing mice by a described procedure.[21]

Affinity Labeling of Antibodies

Affinity labeling of immunoglobulins is carried out in a reaction mixture composed of antibody (1 μM), the ^{14}C affinity-labeling reagent (4 μM), and 0.1 M NaHCO$_3$ (pH 9.0). This represents a ratio of labeling reagent to binding sites of two to one, since each mole of antibody contains 2 moles of hapten binding sites. The reaction is allowed to proceed at 37°, and the extent of covalent bonding is determined periodically by precipitation of the protein with 25% CCl_3COOH, filtration on selection filters,[14] and assessment of radioactivity. Upon acid hydrolysis (6 N HCl, 108°, 24 hr), the labeled amino acids yield the corresponding carboxymethyl derivatives. The [14C]carboxymethyl amino acids are identified by paper electrophoresis (4000 V, pH 3.5).[21] Nonradioactive O-carboxymethyltyrosine and N^{ε}-carboxymethyllysine are added to the acid hydrolyzates as internal markers. The electrophorogram is stained with ninhydrin, and the spots containing the above markers are traced. After bleaching the paper with acidic acetone (1 ml of 1 N HCl in 40 ml of acetone), the traced spots are cut and counted. The accuracy of these determinations is confirmed by analyses with an amino acid analyzer to which a scintillation flow cell is attached.[14] In the experiments reported here, only tyrosine and lysine were found to be labeled.

Concluding Remarks

The results of affinity labeling of antibodies as illustrated with anti-DNP can be summarized as follows: (1) Labeling by different reagents demonstrates the stereospecificity of the affinity labeling reaction. The chemical reaction of the bound affinity labeling reagent depends on exact positioning of the reactive bromoacetyl group in the vicinity of the residue to be labeled. Therefore, use of such a homologous series of reagents would facilitate mapping the structure of the site. This is certainly due to the rigidity of the antibody combining site, which restricts the mobility of the hapten in the site. (2) Both L and H chains participate in the combining site and contribute residues that are complementary to the hapten and can be labeled by affinity-labeling reagents. (3) The affinity-labeled residues in all antibodies analyzed fall within the hypervariable segments of the variable region, implying that these are the complementarity-determining segments in the antibody combining site. These results have been verified by X-ray analysis of crystals of complexes of Fab fragments and haptens.

[54] p-Azobenzenearsonate Antibody

By MANUEL J. RICARDO and JOHN J. CEBRA

Affinity labels provide a useful experimental tool for analyzing the specificity of antibodies. With the information acquired from affinity labeling experiments, immunochemists have gained considerable insight into the location and topography of the antigen-binding site.[1-5] In this method, chemically reactive ligands become covalently attached specifically to amino acid residues in or very near the binding site of antibody molecules such that subsequent noncovalent antibody–hapten interactions are irreversibly blocked. Further progress in comparing antigen-binding sites of different myeloma proteins or of different preparations of antibodies has been made possible through the use of pairs or a series of homologous affinity labels.[6,7]

Principle

We describe here the preparation of two affinity labels, bromoacetyl-mono-(p-azobenzenearsonic acid)-L-tyrosine (BAAT) and bromoacetyl arsanilic acid (BAA), which have the same immunospecific group (benzene arsonate) but differ in that BAAT has a tyrosine spacer between the reactive bromoacetyl moiety and the haptenic determinant making the molecule 8 Å longer than BAA and allowing greater rotational freedom (Fig. 1). The procedures outlined for the synthesis of these two affinity reagents and their acetyl analogs are rather simple and give good reproducible results both in the quality and quantitative recovery of the reaction products. The synthesis of another bromoacetyl reagent having arsonate specificity, 4-(p-bromoacetyl phenylazoamide)phenylarsonic acid, is adequately described elsewhere.[8] When used in a complementary way, this simple series of bromoacetyl labels should permit a more comprehensive assessment of the residues that might be involved in conferring antiarsonate binding specificity in a population of antibodies prior to primary structural analysis. Advantages of bromoacetyl reagents over diazonium and aromatic azide derivatives include reaction properties favoring site-directed labeling and the ease of identifying the modified

[1] S. J. Singer and R. F. Doolittle, *Science* **153**, 13 (1966).
[2] D. Givol, Y. Weinstein, M. Gorecki, and M. Wilchek, *Biochem. Biophys. Res. Commun.* **38**, 825 (1970). See also this volume [11], [53], and [55].
[3] J. J. Cebra, P. Koo, and A. Ray, *Science* **186**, 263 (1974).
[4] J. Haimovich, D. Givol, and N. H. Eisen, *Proc. Natl. Acad. Sci. U.S.A.* **67**, 1656 (1970).
[5] J. J. Cebra, A. Ray, D. Benjamin, and B. Birshtein, *in* "Progress in Immunology" (B. Amos, ed.), p. 269. Academic Press, New York, 1971.
[6] P. H. Strausbaugh, Y. Weinstein, M. Wilchek, S. Shaltiel, and D. Givol, *Biochemistry* **10**, 4342 (1971).
[7] M. J. Ricardo, Jr. and J. J. Cebra, *Biochemistry*, in press (1977).
[8] M. J. Becker, D. Givol, and M. Wilchek, *Immunol. Commun.* **2**, 383 (1974).

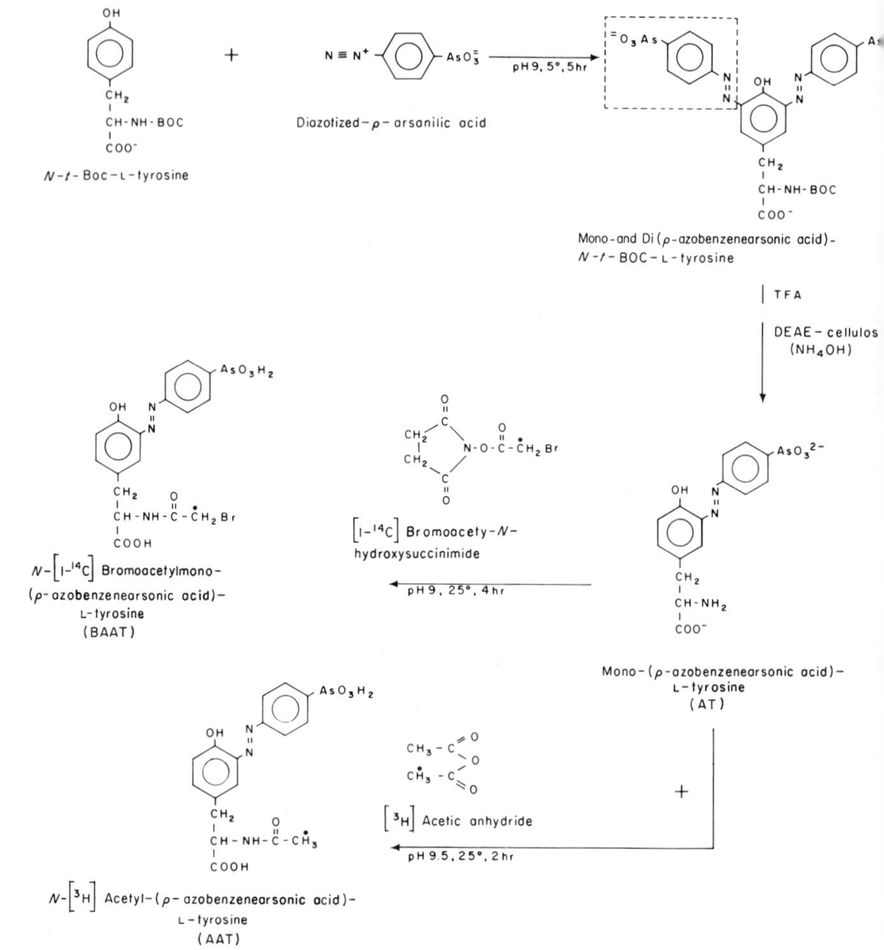

FIG. 1. Reaction scheme for preparing radioactive (*) bromoacetyl and acetyl (ρ-azobenzenearsonic acid)-L-tyrosine ligands. t-Boc, tert-butyloxycarbonyl.

residues as carboxymethylamino acids. The use of such pairs of bromo-acetyl affinity labels may also be valuable in analyzing the degree of clonal restriction in a population of antiarsonate antibodies, as well as for detecting differences in the quality of the antigen-binding sites of antibodies produced by animals whose immune responsiveness has been manipulated by various immunization regimens.[7,9,10]

[9] D. A. Hart, A. L. Wang, L. L. Pawlak, and A. Nisonoff, *J. Exp. Med.* **135,** 1293 (1972).
[10] S. H. Liu, P. Koo, and J. J. Cebra, *J. Immunol.* **113,** 677 (1974).

Preparations

Synthesis of Mono-(p-azobenzenearsonic acid)-L-tyrosine[11,12]

Mono-(*p*-azobenzenearsonic acid) *N-t*-Boc-L-tyrosine is prepared by allowing, at a rate of 1 ml/min with constant stirring, 7 mmoles of diazotized arsanilic acid (1.7 g) dissolved in 130 ml of 0.17 *N* HCl to react with 7 mmoles of *N-t*-Boc-L-tyrosine (2 g) in 100 ml of 0.02 *M* sodium borate at pH 9. The reaction is performed at 5°, maintaining the pH at 9 with 4 *N* NaOH. After the final addition of diazotized arsanilic acid, the mixture is incubated for an additional 2 hr and is then titrated to pH 2 with 6 *N* HCl. The dark brown precipitate that develops is filtered, dried under reduced pressure, dissolved in absolute ethanol, and crystallized from the ethanol solution by adding petroleum ether, b.p. 30°–60° (permanganate purified). The yellow-brown precipitate is filtered, dried under nitrogen, and dissolved in trifluoroacetic acid (10 ml) for 5 min to remove the α-NH_2 protective group. The mono-(*p*-azobenzenearsonic acid)-L-tyrosine (AT) thus obtained is precipitated from the trifluoroacetic acid (TFA) by adding 35 ml of anhydrous ether. The brown AT precipitate is collected by centrifugation (2000 rpm for 10 min), dried under nitrogen, dissolved in 15 ml of 0.2 *N* NH_4OH, and applied to a column of DEAE-cellulose (2.4 $\times$ 35 cm) equilibrated in, and eluted with, 0.05 *M* NaCl in 0.1 *N* NH_4OH. The first 100 ml of orange-red component to be eluted contain most of the monosubstituted tyrosine-*p*-azobenzenearsonic acid. This fraction is collected and freeze-dried. The disubstituted components and most of the impurities are retained by the column as a purple fraction. AT is further purified by recrystallizing twice from hot water at pH 2. The purity of the product is assessed by thin-layer chromatography on silica gel with ethanol–glacial acetic acid–water solvent system (5:2:3 v/v). Only one yellow, ninhydrin-positive component is detected. The yield of AT is approximately 30% (0.6 g). The absorption spectrum and E_{max} for AT is $E_{325} = 20{,}500 \ M^{-1}$ at pH 7.3.

Synthesis of N-[1-^{14}C]Bromoacetylmono-(p-azobenzenearsonic acid)-L-tyrosine[12]

Radioactive bromoacetylmono-(*p*-azobenzenearsonic acid)-L-tyrosine (BAAT) is prepared by acylating mono-(*p*-azobenzenearsonic acid)-L-tyrosine with [1-^{14}C]bromoacetyl-*N*-hydroxysuccinimide ester. The ester

---

[11] M. Tabachnick and H. Sobotka, *J. Biol. Chem.* **234**, 1726 (1959).
[12] P. Koo and J. J. Cebra, *Biochemistry* **13**, 184 (1974).

is synthesized by allowing 50 μmoles of [1-^{14}C]bromoacetic acid (7 mg, 0.5 mCi) to react with a mixture of 65 μmoles of N-hydroxysuccinimide (7 mg) and 70 μmoles of dicyclohexylcarbodiimide (14 mg) in 2 ml of dioxane.[13] The reaction is performed at ambient temperature for 3 hr with frequent stirring with a spatula to promote good precipitation of dicyclohexylurea. The reaction mixture is then filtered to remove the urea and, to the radioactive ester, is added 50 μmoles of mono-(p-azobenzenearsonic acid)-L-tyrosine (21 mg) dissolved in 3 ml of dioxane–water (23:7 v/v) that has been adjusted to pH 9 with triethylamine. The mixture is incubated at room temperature for 1 hr, after which two additions of 50 μmoles of nonradioactive bromoacetyl-N-hydroxysuccinimide ester (2 $\times$ 12 mg) are added at 1-hr intervals while keeping the pH of the reaction at 9 with small additions of triethylamine. After an additional hour of incubation, the reaction mixture is brought to pH 2 with glacial acetic acid and freeze-dried. The orange-red residue is dissolved in 5 ml of absolute ethanol to which petroleum ether (b.p. 30°–60°) is added until a red oil (BAAT) completely settles out. The supernatant fraction is discarded, and the red oil fraction is dried under nitrogen for 10 min. BAAT is subsequently dissolved in 5 ml of ethyl acetate–methanol–2 N aqueous acetic acid (100:25:10 v/v) and applied to a column of Sephadex LH-20 (2.4 $\times$ 100 cm) equilibrated in, and eluted with, the same solvent system at room temperature. [1-^{14}C]BAAT is the only major eluted fraction (deep yellow color) with an absorption maximum at 325 nm. Most of the unreactive mono-(p-azobenzenearsonic acid)-L-tyrosine is retarded by the column. [1-^{14}C]BAAT is judged pure by thin-layer chromatography in the solvent system of absolute ethanol–glacial acetic acid–water (5:2:3 v/v) developed at room temperature. The label migrates as a single, yellow, ninhydrin-negative spot and accounts for 90% of all the radioactivity applied to the chromatographic sheet. The R_f value of [1-^{14}C]-BAAT relative to AT is 0.42. When 20 μmoles of purified [1^{14}C]BAAT (1 mg) are allowed to react with 40 nmoles of poly-L-lysine (1 mg) in 0.1 M sodium bicarbonate (pH 9) at 37° for 24 hr, more than 85% of the radioactivity is covalently incorporated into the polymer. [1-^{14}C]BAAT is stored at −20° in the ethyl acetate–methanol–acetic acid solvent mixture. A portion of this solution is dried under nitrogen and suspended in 0.1 M sodium bicarbonate at pH 9 before use in affinity labeling experiments. The specific radioactivity of [1-^{14}C]BAAT is 8500 cpm/nmole; the amount of product recovered is approximately 9 mg and represents a 45% yield.

[13] Y. Weinstein, M. Wilchek, and D. Givol, *Biochem. Biophys. Res. Commun.* **35**, 694 (1969).

Synthesis of N-[³H]Acetyl-(p-azobenzenearsonic acid)-L-tyrosine (AAT)

The radiolabeled ligand is prepared by allowing 50 μmoles of mono-(p-azobenzenearsonic acid)-L-tyrosine (21 mg) dissolved in 5 ml of alkaline water (pH 10) to react with 50 μmoles of [³H]acetic anhydride (5 mg, 2.5 mCi) for 2 hr at ambient temperature while stirring slowly and maintaining the pH at 9.5 with concentrated triethylamine. Then the reaction mixture is brought to pH 3 with 0.1 N HCl, and the precipitate which forms is recovered by centrifuging at 2000 rpm for 10 min; the supernatant liquid is discarded. The orange-red product is washed once with 0.1 N HCl, centrifuged, dried under reduced pressure, and stored at −20°. [³H]AAT is at least 95% pure as judged by thin-layer chromatography, using the same solvent system as in the BAAT analysis. The derivative migrates as a single, yellow, ninhydrin-negative spote with an R_f value of 0.50 relative to AT. The specific radioactivity of [³H]AAT is 22,000 cpm/nmole. The amount of product recovered is approximately 15 mg, corresponding to a 70% yield.

Synthesis of p-[-1-¹⁴C]Bromoacetylarsanilic Acid

Radioactive BAA is prepared by acylating the monosodium salt of *p*-arsanilic acid with [1-¹⁴C]bromoacetic anhydride.[7,8,14] The sodium salt of *p*-arsanilic acid is prepared by dissolving 10 mmoles of *p*-arsanilic acid (2.1 g) in 45 ml of deionized water at pH 8 with 5 ml of 4 N NaOH. Crystals of sodium *p*-arsanilic acid are obtained by adding 45 ml of absolute ethanol slowly to the solution. The product is filtered, and the crystals are dried under reduced pressure.

Bromoacetic acid (reagent grade) is recrystallized twice from 2,2,4-trimethylpentane (spectrograde) prior to use in preparing the anhydride. Approximately 50 mmoles of bromoacetic acid (7 g) are dissolved in 40 ml of 2,2,4-trimethylpentane (isooctane) with heating to 90°. The upper transparent layer of the solution is recovered immediately and allowed to cool slowly at room temperature. The bromoacetic acid crystals that form are filtered and dried under reduced pressure for 2 hr.

Immediately before the synthesis of [1-¹⁴C]bromoacetic anhydride, 0.25 mmole of radioactive bromoacetic acid (35 mg, 1 mCi) is dissolved in 1 ml of dichloromethane (methylene chloride, spectrograde) at 0° with 1.7 mmoles of recrystallized bromoacetic acid (235 mg). To this solution is added 1 mmole of *N,N′*-dicyclohexylcarbodiimide (206 mg), also dissolved in 1 ml of dichloromethane. The reaction mixture is incu-

[14] P. Ehrlich and A. Bertheim, *Ber. Deut. Chem. Ges.* **40**, 3292 (1907).

bated at 0° for 15 min and stirred frequently to promote urea formation. The mixture is filtered through a Pasteur pipette that contains glass wool, and the effluent is collected in a round-bottom flask. The volume of dichloromethane is kept small in order to minimize solubilizing the urea. To the bromoacetic anhydride solution, 0.33 mmole of p-arsanilic acid monosodium salt (79 mg) is added and the reaction mixture is immediately flash evaporated to dryness at room temperature. [1-^{14}C]BAA is precipitated with cold 1 N HCl, filtered, and sequentially washed with 15 ml each of cold 1 N HCl, cold 95% ethanol, and cold anhydrous ether. The derivative is purified further by crystallizing twice from hot water carefully brought to boiling and slowly allowed to cool for 1 hr at room temperature and subsequently for 3 hr at 0°. The [1-^{14}C]BAA crystals are filtered, rinsed with cold ethanol, anhydrous ether, dried under reduced pressure, and stored at −20°. The label is judged pure by silica gel thin-layer chromatography in the solvent system of butanol–glacial acetic acid–water (4:1:2 v/v). [1-^{14}C]BAA migrated as a single component with an R_f value of 0.66 compared to 0.54 for p-arsanilic acid. The specific radioactivity of [1-^{14}C]BAA is 2300 cpm/nmole. The absorption maximum at $E_{260\ nm}$ for [1-^{14}C]BAA, dissolved in dioxane–water (1:1 v/v), is 19,600 M^{-1}. Approximately 34 mg of [1-^{14}C]BAA are recovered, corresponding to a 43% yield.

Synthesis of p-[³H]Acetylarsanilic Acid

[³H]Acetylarsanilic acid (AA) is prepared by mixing 0.1 mmole of [³H]acetic anhydride (50 mg, 25 mCi) with 4.9 mmoles of reagent grade acetic anhydride (460 mg) in 10 ml of dichloromethane and then allowing this solution to react with 1.7 mmoles of monosodium p-arsanilic acid (400 mg). The reaction mixture is immediately flash evaporated at room temperature until the solvent is completely removed and a white residue remains. The [³H]AA is precipitated with cold 1 N HCl and filtered; the precipitate is washed twice sequentially with cold 95% ethanol, anhydrous ether, dried under reduced pressure, and stored at −20°. Thin-layer chromatography analysis of [³H]AA in butanol–acetic acid–water solvent system detected only a single radioactive fluorescent spot with an R_f value of 0.60. The specific radioactivity of [³H]AA is 12,000 cpm/nmole. Approximately 46% (185 mg) of [³H]AA was recovered.

Synthesis of Radiolabeled Nε-Carboxymethyl-L-lysine (CM-Lysine) and O-Carboxymethyl-L-tyrosine (CM-Tyrosine)

Carboxymethylpolylysine is prepared by radioalkylating 40 nmoles of poly-L-lysine (1 mg) with 3.7 μmoles of [^{14}C]iodoacetic acid (0.7 mg,

0.05 mCi) at 37° for 24 hr in 0.05 M sodium borate at pH 9. The radio-alkylated polymer is dialyzed against deionized water and subsequently hydrolyzed for 20 hr at 108°. The hydrolysis product, $[^{14}C]N^{\varepsilon}$-CM-lysine, is purified by paper electrophoresis (4 kV for 1 hr) on Whatman No. 3 MM with pyridine–glacial acetic acid–water as the solvent system (1:10:89 v/v), pH 3.6. The electrophoretically neutral, ninhydrin-positive $[^{14}C]N^{\varepsilon}$-CM-lysine is eluted from the paper with 0.1 N acetic acid, flash evaporated, and stored at −20°. The specific radioactivity of $[^{14}C]N^{\varepsilon}$-CM-lysine is 3000 cpm/nmole.

[^{14}C]O-CM-tyrosine is synthesized by allowing 3.2 μmoles of N-CBZ-L-tyrosine (1 mg) to react with 10 μmoles of $[1\text{-}^{14}C]$bromoacetic acid (1.4 mg 0.05 mCi) for 20 hr at ambient temperature. The reaction mixture is then acidified to pH 3 with 2 N HCl. The derivative, $[^{14}C]N$-CBZ-O-CM-L-tyrosine, is extracted into ethyl acetate (10 ml), dried over anhydrous Na_2SO_4, and flash evaporated to dryness. The product is dissolved in 5 ml of 4 N HBr that is in glacial acetic acid, and is incubated for 15 min to remove the N-CBZ (N-carbobenzoxy) groups. $[^{14}C]O$-CM-L-tyrosine is precipitated from the reaction mixture by adding anhydrous ether (15 ml). The precipitate is collected by centrifugation (2000 rpm, 10 min), dried under N_2, dissolved in pyridine–acetate at pH 3.6, and purified further by paper electrophoresis (4 kV for 45 min) in the same solvent system. $[^{14}C]O$-CM-L-tyrosine gives a brownish-yellow color when stained with a tyrosine stain and contains most of the radioactivity applied to the paper whereas the L-tyrosine spot gives a rose color and is not radioactive.[15] The derivative is eluted with 0.1 N acetic acid, flash-evaporated to dryness, and stored at −20°. The specific radioactivity attained for $[^{14}C]O$-CM-L-tyrosine is 2500 cpm/nmole. The amount of CM-lysine and CM-tyrosine recovered is approximately 10%. Both of these products are then used as standards on paper electrophoresis followed by autoradiography to assess whether lysyl or trosyl residue(s) are modified by the ligands $[1\text{-}^{14}C]$BAAT or $[1\text{-}^{14}C]$BAA.

Comments

The applicability of using pairs of homologous ligands to detect variations in the combining sites of specific antibodies within a restricted population is supported by the data in Table I. $[1\text{-}^{14}C]$BAAT probably resembles more closely the major haptenic determinant on the immunogen (arsanilic acid coupled to hemocyanin) than $[1\text{-}^{14}C]$BAA and apparently is modifying the binding sites of higher affinity. The slow rate of

---

[15] C. W. Easley, *Biochim. Biophys. Acta* **107**, 386 (1965).

TABLE I

DETERMINATION OF AFFINITY AND KINETIC PARAMETERS FOR ANTIARSONATE
ANTIBODIES USING PAIRS OF HOMOLOGOUS LIGANDS[a]

Ligand	Average affinity (1/mole)	Heterogeneity index	Percent of sites labeled[b]	Half-time of reaction (hr)[c]
[³H]AAT	5×10^5	0.58	—	—
[³H]AA	6×10^4	0.51	—	—
[1-¹⁴C]BAAT	—	—	50	11
[1-¹⁴C]BAA	—	—	30	19

[a] The same population of antiarsonate antibodies elicited in inbred guinea pigs are used in all the determinations.
[b] Represents the moles of ligand bound per mole of antibody at 72 hr, beyond which little to no significant labeling occurs.
[c] The time required to permanently label half of the sites which finally are modified at 72 hr.

reaction shown by both affinity reagents demonstrate their potential for use in attempting to label arsonate receptors on the surfaces of lymphoid cells under the proper conditions.

The majority of the residues chemically substituted by both affinity labels occur in Hv2 (Table II).[7] This may reflect a similar orientation of the ligands within the various binding sites. However, the subpopulations of antiarsonate antibodies modified by [1-¹⁴C]BAA probably contain a greater mixture of binding sites, since three specificity related residues are labeled and the degree of nonsite modifications exceeds that

TABLE II

DISTRIBUTION OF AFFINITY LABELS ON THE HEAVY CHAIN OF
ANTIARSONATE ANTIBODIES[a]

Ligand	Percent of label bound to				Residues modified
	Hv1	Hv2[b]	Hv3	Nonsite[c]	
[¹⁴C]BAAT[d]	6	67	7	7	Lys-59
[¹⁴C]BAA[d]	4	58	6	16	Tyr-57 > Lys-59 > Tyr-50

[a] The same population of guinea pig antiarsonate antibodies is affinity labeled with either [1-¹⁴C]BAAT or [1-¹⁴C]BAA.
[b] Hv represents hypervariable region and Hv2 spans residues from N-48 to N-59.
[c] Percent of label residing outside the variable region (N1–N120) of the heavy chain.
[d] Approximately 13% of the [1-¹⁴C]BAAT label is found associated with the light chain, and that of [1-¹⁴C]BAA 16%.

of [1-^{14}C]BAAT which predominately substitutes one residue. Both reagents are site-directed and react negligibly with nonspecific antibodies. Further, the rate and extent of the reaction is not influenced by increasing the concentration of either affinity label. Acetyl analogs of the affinity reagents effectively protect the specific antibodies from modification, but unrelated haptens do not.

[55] Affinity Cross-Linking of Heavy and Light Chains

By MEIR WILCHEK and DAVID GIVOL

In multichain proteins in which more than one chain forms a combining site, it is possible to design bifunctional affinity labeling reagents that will cross-link the two chains participating in the site. This will be illustrated by the cross-linking of the heavy and light chains of protein 315, a mouse myeloma protein that binds nitrophenyl ligands.

Two bromoacetyl derivatives of DNP ligands of different length (BADB and BADL) were shown to specifically label either light or heavy chains.[1,2] BADB (α-N-bromoacetyl-γ-N-DNP-$\alpha\gamma$-diamino-L-butyric acid) labels Tyr-34 on the light chain of protein 315, whereas BADL (α-N-bromoacetyl-ϵ-N-DNP-L-lysine) labels Lys-54 on the heavy chain of protein 315.[1,2]

If specificity of the chemical reaction is indeed a function of the positioning of the bromoacetyl group, the results would indicate that the distance between the labeled lysine on the H chain and the labeled tyrosine on the L chain is approximately equal to the difference in the length of BADB and BADL. Hence, a bifunctional reagent with two bromoacetyl groups, separated by a distance equal to the difference in length between BADB and BADL, would react simultaneously with both lysine and tyrosine and thereby would cross-link the heavy and light chains.[3] The data are entirely consistent with the bifunctional reagent, DNPHN-(CH$_2$)$_2$CH-(-HNCOCH$_2$Br)CO(NH)$_2$COCH$_2$Br, γ-DNP-α-bromoacetyl-L-diaminobutyric acid bromoacetyl hydrazide (DIBAB), synthesized according to Scheme 1 and covalently cross-linking the H and L chains to yield a molecule with a molecular weight of 72,000. The H-L band contains labeled Tyr and Lys in a ratio of 1:1, indicating that

[1] J. Haimovich, H. N. Eisen, E. Hurwitz, and D. Givol, *Biochemistry* **11**, 2389 (1972).

[2] D. Givol and M. Wilchek, this volume [53].

[3] D. Givol, P. H. Strausbauch, E. Hurwitz, M. Wilchek, J. Haimovich, and H. N. Eisen, *Biochemistry* **10**, 3461 (1971).

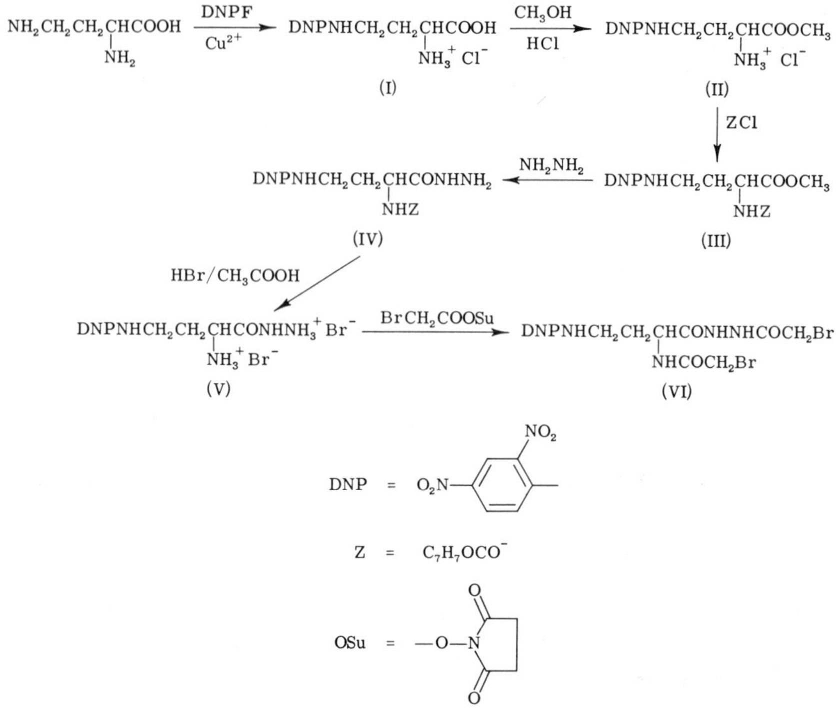

$$\text{NH}_2\text{CH}_2\text{CH}_2\underset{\underset{\text{NH}_2}{|}}{\text{CH}}\text{COOH} \xrightarrow[\text{Cu}^{2+}]{\text{DNPF}} \text{DNPNHCH}_2\text{CH}_2\underset{\underset{\text{NH}_3^+ \text{Cl}^-}{|}}{\text{CH}}\text{COOH} \xrightarrow[\text{HCl}]{\text{CH}_3\text{OH}} \text{DNPNHCH}_2\text{CH}_2\underset{\underset{\text{NH}_3^+ \text{Cl}^-}{|}}{\text{CH}}\text{COOCH}_3$$

(I)   (II)

$$\Big\downarrow \text{ZCl}$$

$$\text{DNPNHCH}_2\text{CH}_2\underset{\underset{\text{NHZ}}{|}}{\text{CH}}\text{CONHNH}_2 \xleftarrow{\text{NH}_2\text{NH}_2} \text{DNPNHCH}_2\text{CH}_2\underset{\underset{\text{NHZ}}{|}}{\text{CH}}\text{COOCH}_3$$

(IV)   (III)

$$\text{HBr}/\text{CH}_3\text{COOH} \Big\swarrow$$

$$\text{DNPNHCH}_2\text{CH}_2\underset{\underset{\text{NH}_3^+ \text{Br}^-}{|}}{\text{CH}}\text{CONHNH}_3^+ \text{Br}^- \xrightarrow{\text{BrCH}_2\text{COOSu}} \text{DNPNHCH}_2\text{CH}_2\underset{\underset{\text{NHCOCH}_2\text{Br}}{|}}{\text{CH}}\text{CONHNHCOCH}_2\text{Br}$$

(V)   (VI)

DNP = O$_2$N— (2,4-dinitrophenyl ring with NO$_2$)

Z = C$_7$H$_7$OCO$^-$

OSu = —O—N (succinimidyl)

SCHEME 1

Tyr on the light chain and Lys on the heavy chain are involved in the cross-linking.

Synthesis of the Reagent

γ-DNP-L-*diaminobutyric Acid Hydrochloride* (*I*). 1,3-Diaminobutyric acid (1.9 g) is dissolved in 15 ml of boiling water, and 2 g of CuCo₃ are added with stirring over a period of 10 min. The excess CuCO₃ is removed by filtration and washed with 5 ml of hot water. NaHCO₃ (2 g) and fluorodinitrobenzene (3.6 g in 20 ml of ethanol) are added to the blue filtrate and the reaction mixture is stirred for 2 hr at room temperature. The green product is filtered, washed successively with water, ethanol, and ether, and then dissolved in 10 ml of 3 N HCl. After 10 min, a yellow precipitate is formed which is collected and washed successively with 5 ml of cold 3 N HCl, 5 ml of cold water, ethanol, and ether. The yield is 2.1 g (65%), m.p. 254°.

γ-DNP-L-*diaminobutyric Acid Methyl Ester Hydrochloride* (*II*). Compound (I) (1.5 g) is added to a chilled solution of thionyl chloride (1 ml) in methanol (10 ml), and the mixture is shaken for a few minutes until the solid dissolves. The reaction mixture is left for 5 hr at room temperature, after which 50 ml of dry ether are added. The precipitate is filtered and washed with ether. The yield is 1.69 g (95%), m.p. 125°.

*γ-DNP-α-Cbz-*L-*diaminobutyric Acid Methyl Ester (III).* Compound II (1.69 g) is treated with 1 equivalent of carbobenzoxy chloride (1.25 g) in a mixture of 5% NaHCO₃ solution and chloroform. After shaking for 20 min at 0° the reaction is allowed to proceed for an additional 20 min at room temperature. The chloroform layer is washed with a 5% solution of pyridine, dried over Na₂SO₄, and concentrated to dryness. This semi-crystalline compound is used for the next step.

*γ-DNP-α-Cbz-*L-*diaminobutyric Acid Hydrazide (IV).* Compound (III) (1.1 g) is dissolved in 15 ml of anhydrous ethanol, and hydrazine hydrate (0.5 ml) is added. The mixture is allowed to stand overnight at room temperature. The crystalline is filtered, and washed with cold ethanol and ether. The yield is 0.9 g (80%), m.p. 104°–106°.

*γ-DNP-*L-*diaminobutyric Acid Hydrazide Dihydrobromide (V).* Compound (IV) (0.86 g) is dissolved in glacial acetic acid (3 ml) and allowed to react with a 5-ml solution of HBr in acetic acid (45%). After 15 min at room temperature, 50 ml of dry ether are added. The precipitate is collected and washed with dry ether. The yield is 0.79 g (85%), m.p. 235°.

*γ-DNP-α-bromoacetyl-*L-*diaminobutyric Acid Bromoacetyl Hydrazide (VI).* Compound (V) (0.72 g) is dissolved in dimethylformamide (5 ml), to which is added triethylamine (0.56 ml), and treated with 5 equivalents (1.3 g) of bromoacetic anhydride. After 1 hr at room temperature, 0.3M sodium bicarbonate is added to adjust the pH to 7. After another 20 min, water is added and the resultant precipitate is collected and washed with water. The yield is 75%, m.p. 135°–138°.

The ¹⁴C-labeled compound is prepared as follows: Triethylamine (1 mmole) is added to a solution of (V) (0.5 mmole) in 3 ml of dimethyl-formamide. This is mixed with a solution of [¹⁴C]bromoacetyl-*N*-hydroxysuccinimide ester[4] (1.5 mmoles) in dioxane (3 ml). After 30 min at room temperature, 5 ml of 1 *N* NaHCO₃ are added and the resultant precipitate is collected. This is purified on a thin layer of silica gel developed with chloroform–benzyl alcohol–acetic acid (70:30:3, v/v; R, 0.82).

Cross-Linking of Protein 315

Protein 315 (1 μM), isolated from sera of mice bearing tumor MOPC 315,[5] was incubated with the bromoacetyl-DNP derivative (VI) (2 μM)

[4] P. Cuatrecasas, M. Wilchek, and C. B. Anfinsen, *J. Biol. Chem.* **244**, 4316 (1969).
[5] J. Haimovich, D. Givol, and H. N. Eisen, *Proc. Natl. Acad. Sci. U.S.A.* **67**, 1056 (1970).

in 0.10 M NaHCO$_3$ (pH 9.0) at 37°. Samples were assayed for covalent binding at various times by filtration of trichloroacetic acid-precipitated protein on selectron filters BA 85/0 (Schleicher & Schuell). Acid hydrolyzates of ^{14}C-labeled protein were analyzed for carboxymethyl-amino acid derivatives by high-voltage electrophoresis at pH 3.5. The electrophoresis runs included internal markers of carboxymethyl-lysine and carboxymethyl-tyrosine. Analytical separation of heavy and light chains by polyacrylamide gel electrophoresis was carried out on 5% gels in 0.1% sodium dodecyl sulfate–0.14 M β-mercaptoethanol.[6] The 5% cross-linked gels were stained with Coomassie brilliant blue and treated by the following procedures: Gels are analyzed for protein-containing bands by scanning at 570 nm in a spectrophotometric gel scanner.[7] Radioactivity in the gel is analyzed by slicing the gel into 50 thin slices, which are solubilized with Soluene (Packard Instrument Co.) and counted in toluene-containing scintillation fluid. Protein-containing bands are analyzed for carboxymethyl-amino acid derivatives by hydrolysis in 6 N HCl, performed directly upon thin slices of gel containing this protein; after 24 hr of acid hydrolysis, the hydrolysis tubes were opened, cooled, and centrifuged to remove the residual polyacrylic acid. The hydrolyzates are analyzed by high-voltage electrophoresis at pH 3.5, as mentioned above. By these techniques, gels prepared with standard samples of labeled protein showed quantitative release of amino acids.

Comments

In a slightly different manner, Hadler and Metzger[8] have also succeeded in cross-linking H and L chains on the basis of affinity labeling. They labeled tyrosine 34 on the L chain of protein 315 with m-nitrobenzene diazonium salt followed by reduction of the diazo bond with dithionite. The generated NH$_2$-tyrosine was then used, due to the low pK of its amino group, as the "first hook" for the bifunctional reagent, 1,5-difluoro-2,4-dinitrobenzene. Subsequently, the pH of the solution was raised and the second fluorine of the reagent was allowed to react with a neighboring residue on the heavy chain. This site-directed cross-linking is consistent with a distance of 3–4 Å between Tyr-34 (the affinity-labeled residue) on the L chain and an as yet unidentified residue on the H chain.

[6] A. L. Shapiro, E. Viñuela, and J. V. Maizel, *Biochem. Biophys. Res. Commun.* **28**, 815 (1967).
[7] J. B. Gressel and J. Wolowelsky, *Anal. Biochem.* **24**, 157 (1968).
[8] N. Hadler and H. Metzger, *Proc. Natl. Acad. Sci. U.S.A.* **68**, 1421 (1971).

[56] Bivalent Affinity Labeling Haptens in the Formation of Model Immune Complexes

By Paul H. Plotz

Because immune complexes appear to be important in many diseases, efforts have been made to understand their pathogenic role in detail. Factors such as the size, solubility, and stoichiometry of aggregates and the classes of the antibodies appear to affect the interaction between complexes, complement, and cell receptors for immunoglobulins and complement. Such interactions may thereby influence the biological effects of the complexes. Since both naturally occurring complexes and complexes formed experimentally by mixing antigen and antibody are noncovalent aggregates, their physical state is unstable to many laboratory manipulations. It would, therefore, be useful to have antigen–antibody complexes with the spatial structure of natural complexes but held together by covalent bonds.

In an attempt to provide such model complexes, I have synthesized a family of symmetric, bivalent affinity labeling haptens. The compounds (Fig. 1) are bis-p-nitrophenyl (PNP) esters of dicarboxylic acids. The PNP group serves as a hapten for anti-PNP antibodies and also serves to activate the ester bond so that the dicarboxylic acid can form a covalent bond in the vicinity of the antigen-binding site. These molecules incorporate useful features of compounds previously used to study antigen–antibody interactions. Antigen-binding sites have been intensively studied by affinity labeling[1]; the PNP group has previously been used as activating group for an affinity labeling hapten[2]; and bishaptens have been used successfully to study antibody structure in several laboratories.[3-5] The reagents depicted in Fig. 1 have been shown to produce covalently bonded polymers of anti-PNP antibodies, and the specificity of polymer formation has been established.[6]

[1] D. Givol, in "Essays in Biochemistry" (P. N. Campbell and G. D. Greville, eds.), Vol. 10, p. 73. Academic Press, New York, 1974.

[2] P. H. Plotz, in "Cell Interactions and Receptor Antibodies in Immune Responses" (O. Mäkelä, A. Cross, and T. U. Kosunen, eds.), pp. 171–179. Academic Press, New York, 1971.

[3] R. C. Valentine and N. M. Green, J. Mol. Biol. 27, 615 (1967).

[4] R. L. Wilder, G. Green, and V. N. Schumaker, Immunochemistry 12, 39 and 49 (1975).

[5] D. Carson and H. Metzger, Immunochemistry 11, 355 (1974).

[6] P. Plotz, submitted for publication.

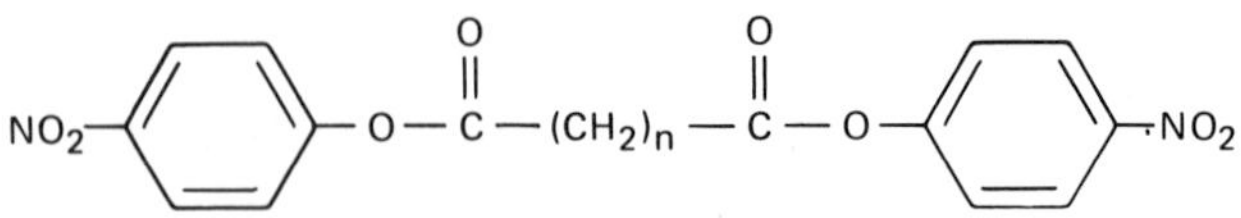

Fig. 1. Bis-*p*-nitrophenyl esters of dicarboxylic acids.

Synthesis of Bis-PNP-dicarboxylic Esters

Synthesis is performed by carbodiimide condensation, and the details of synthesis are similar for the three compounds produced. Pimelic acid (heptanedioic acid, Eastman), adipic acid (hexanedioic acid, Sigma), suberic acid (octanedioic acid, Sigma), and dicyclohexylcarbodiimide (Eastman) may be used without further purification. *p*-Nitrophenol (Eastman) is twice recrystallized from dilute hydrochloric acid.

To produce bis-PNP-pimelic ester, 0.025 mole of pimelic acid and 0.05 mole of *p*-nitrophenol are dissolved in 25 ml of K_2CO_3-dried ethyl acetate, and 0.05 mole of dicyclohexylcarbodiimide is added. An insoluble precipate begins to form within minutes, but the reaction mixture is allowed to remain in the dark overnight at room temperature. Dicyclohexylurea is removed by filtration, and the ethyl acetate is removed by rotary evaporation under vacuum. The residual yellow oil is dissolved in a small amount of chloroform, and absolute methanol is added until the solution becomes cloudy. Crystals appeared after cooling in ice. Two further crystallizations from chloroform–methanol are performed. The resulting off-white crystals are washed with methanol and air-dried. The crystals are stable at room temperature and do not appear to be light sensitive. Yield, 36%; m.p. 81°.

Thin-layer chromatography on prewashed silica gel plates developed with chloroform–hexane (3:1) yields a single band visible only with ultraviolet light, clearly separated from marker *p*-nitrophenol. The compounds made with adipic and suberic acids are also off-white crystals of similar stability. When hydrolyzed in 0.05 *M* NaOH, each of the compounds yields about 1.9 moles of *p*-nitrophenol per mole of starting compound as measured by absorbance at 403 nm. When radioactive compounds are synthesized, the whole reaction mixture is centrifuged and the supernatant liquid is applied directly to prewashed silica gel plates and developed with chloroform–hexane. It is beneficial to repeat the chromatographic procedure if the compound is not fully separated from material closer to the origin. Radioactive compounds are eluted from the gel with chloroform and stored in solution. The bis-2,4-dinitrophenyl ester of pimelic acid can be synthesized in an analogous way and has similar properties.

Polymer Formation

Anti-PNP antibodies are produced by rabbits immunized with PNP_{20}-bovine γ-globulin and are purified from a 50% ammonium sulfate precipitate of serum by affinity chromatography on ϵ-2,4 dinitrophenyllysine agarose with elution by 2,4-dinitrophenylglycine.[7] The preparation is dialyzed against 0.2 M sodium borate–0.15 M NaCl, pH 8.0, to remove hapten.

To produce polymers, a solution of anti-PNP antibodies, 30–120 μM in borate–saline, pH 8.0, is mixed with a 5-fold molar excess of reagent dissolved in 0.05 volume of dimethylformamide. The reaction is allowed to continue for 2.5 hr at room temperature in the dark and is terminated by the addition of 0.2 volume of 0.02 M L-lysine in water. Incubation for 1 hr results in less polymer formation; overnight incubation does not increase polymer formation. At a lower reagent to polymer ratio (1:1), polymer formation is less; a higher ratio (20:1) does not increase the yield. Polymer formation with the reagents having one less methylene bridge (adipic) or one more (suberic) is less than with the bis-PNP-pimelic ester. Temperature and pH optima have not been studied in detail. Increasing antibody concentration increases polymer formation, but, at high concentrations, nonspecific aggregation can be demonstrated.[6]

Polymer formation can be shown by gel filtration on both agarose A 1.5 M (Bio-Rad) in neutral buffer, which partially resolves polymers larger than dimers, and on Sephadex G-200 in either neutral buffers or 1 M acetic acid to dissociate noncovalent complexes. The resolution of oligomers on Sephadex G-200 is not as good as on agarose A 1.5 M. Electrophoresis with gradient gels of polyacrylamide in a sodium dodecyl sulfate buffer has been shown by Segal and Hurwitz to resolve clearly monomer, dimer, trimer, and even higher polymers produced by the reagents described here and by another bivalent affinity labeling compound, bis-(α-bromacetyl-ϵ-DNP-Lys-Pro)ethylenediamine, which they have synthesized and studied.[8] Segal and Hurwitz have also been able to resolve dimers and trimers by successive gel filtration on Sephadex G-200 and Ultragel AcA22 (LKB).[8]

Because the bis-PNP-dicarboxylic esters are also nonspecific alkylating reagents, some nonspecific bond formation might be expected. In experiments with bis-PNP-[14C]pimelic ester, as many as 3 moles of pimelate were bound per mole of antibody. However, the binding was usually below 2 moles per mole or in the range of 0.5 to 1.5 as predicted for a dimer if binding was limited to antigen-binding sites or if only

[7] E. J. Goetzl and H. Metzger, *Biochemistry* **9**, 1267 (1970).
[8] D. Segal and E. Hurwitz, *Biochemistry* **15**, 5253 (1976).

1 molecule was bound per antigen-binding site. The factors that influence the extent of covalent bond formation are not well understood. Whether the polymers are open chain or rings remains unknown, and a method of separating rings from chains is unavailable: affinity chromatography appears to be unsuitable for detecting unfilled antigen-binding sites since the hapten, *p*-nitrophenol, is released in the binding reaction, allowing the reacted antibody to bind to an affinity column.

The biological properties of the complexes produced by these compounds are under active investigation.

[57] DNP-Based Diazoketones and Azides

By FRANK F. RICHARDS and WILLIAM H. KONIGSBERG

Affinity labels have been used in an attempt to identify amino acid residues in and around the binding sites for protein ligands, substrates, and inhibitors. They have also been used to identify proteins with particular binding specificities when the protein is part of a more complicated organelle, such as a membrane or ribosome.

An ideal affinity reagent for proteins should have the following properties. It should ligate to the protein in exactly the same manner as the "natural" ligand upon which it is based. It should have the additional property of forming a covalent bond with a conveniently placed adjacent reactive amino acid residue without altering the position of the bound ligand. The resulting ligand–amino acid residue complex should be sufficiently stable to isolate. The affinity of the ligand for the protein should be high, so that the only high concentrations of ligand are the combining site and finally, it is helpful to have a reactive moiety on the reagent whose chemical half-life, after activation in the presence of the protein, is very short.

Since, therefore, the design of a perfect affinity reagent requires considerable knowledge of the geometry of the binding site, it is not surprising that most affinity labels represent only crude approximations to the ideal. Affinity reagents with the potential for a two-stage labeling process are potentially useful. With this type of binding reagent, binding can occur at the active site, excess reagent can be removed, and the reagent or the protein activated by a change in pH or by photolysis. The active species should then form covalent bonds.

Photoreactive affinity reagents do not form covalent bonds with proteins until they are photolyzed, thus a two-stage labeling procedure is possible. The reactant can be placed in the active site, and if the energy of interaction is sufficiently high, excess reagent can be removed rapidly with little dissociation of the reagent–protein complex. The complex is then irradiated and covalent bonds are formed. Since effective irradiation is a process which may take minutes or hours, some dissociation of the complex may take place, and a reactive reagent species may be formed outside the site. Site-specific labeling will, in general, be favored by reducing the period of irradiation and by employing a reactive label with a short chemical half-life in aqueous media[1] or in the presence of small "scavenger" molecules that react with the activated reagent.[2] The reactivity of photoaffinity labels is perhaps in practice a more important feature. Generation *in situ* of highly reactive reagent species in relatively high dilution tends to reduce unwanted side reactions and allows one to employ relatively unstable intermediates capable of reacting with many amino acid residues in the vicinity of the binding site.

Principles

[^{3}H]2,4-Dinitrophenylalanyldiazoketone ([^{3}H]DNP-AD) reacts according to the following scheme. DNP-AD undergoes photolysis to a carbene, which may insert into heteroatomic (containing N, O, or S) side chains of amino acid residues.[3] Analogous carbenes with ester rather than ketene linkages also insert into tyrosine and histidine[4] and insertion by carbenes into C-H bonds has also been reported.[5] The carbene undergoes

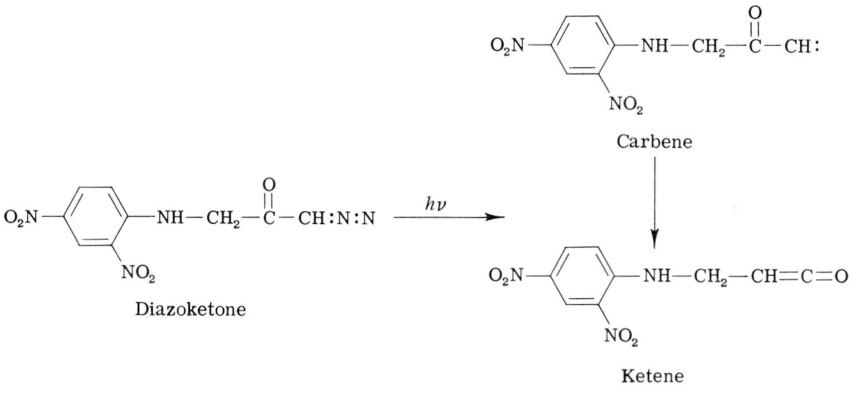

Wolff rearrangement to a ketene that may acylate nucleophilic carboxyl, hydroxyl, amino, and thiol functions. Lysine, tyrosine, cysteine, and

---

[1] J. R. Knowles, *Acc. Chem. Res.* **5**, 155 (1972). See also this volume [8].

[2] A. E. Ruoho, H. Kiefer, P. E. Roeder, and S. J. Singer, *Proc. Natl. Acad. Sci. U.S.A.* **70**, 2567 (1973).

[3] W. Kirmse, "Carbene Chemistry." Academic Press, New York, 1964.

[4] J. Shafer, P. Baronowsky, R. Laursen, F. Finn, and W. H. Westheimer, *J. Biol. Chem.* **241**, 412 (1966).

[5] W. von E. Doering and L. H. Know, *J. Am. Chem. Soc.* **78**, 4947 (1956).

perhaps also histidine and aspartic and glutamic acids may be acylated by ketenes.[3]

[³H]2,4-Dinitrophenyl-1-azide (DNP-N$_3$) generates a nitrene on photolysis, which also has the potential of inserting into heteroatomic linkages such as C=O, C—S—, C—N, C—H, or N—H.

Protein heteroatomic insertion reactions

Azide — Nitrene — Dinitrophenylaniline

4-Nitrobenzene-1,2-furazan n-oxide

The reactive intermediate[6] may hydrolyze to dinitrophenylalanine or cyclize to an n-oxyfurazan derivative. There is no information on the reactivity of this derivative, but the relatively high incorporation of the reagent into protein suggests that it is not a rapidly formed major intermediate.

Synthesis of Reagents

DNP-AD and [³H]DNP-AD

Synthesis of DNP-Alanine. 1-Fluoro-2,4-dinitrobenzene (20 ml) in 30 ml absolute dry ethanol is added over a period of 20 min to 11 g of L-alanine and 42 g of Na$_2$CO$_3$ in 250 ml of water and stirred at 20° in the dark for 2 hr. The yellow precipitate is filtered, washed with ether,

[6] G. W. Fleet, J. R. Knowles, and R. R. Porter, *Nature* (*London*) **224,** 511 (1969).

and dissolved in a minimal volume of water; the resulting solution is acidified to pH 1.0 with HCl, and the reprecipitated DNP-alanine is filtered and dried in the dark. [³H]DNP-L-alanine may be prepared by similar methods from [3,5,6-³H]1-fluoro-2,4-dinitrobenzene on a 1–2 mg scale except that centrifugation in 1 ml vials replaces the filtration steps. Pure DNP-L-alanine should give a single component on polyamide thin-layer chromatography in benzene–acetic acid (80:20 v/v; solvent I); or in *t*-amyl alcohol saturated with pH 6.0 0.2 *N* sodium phthalate buffer (solvent II); or by electrophoresis at pH 1.8 and pH 6.0.

Synthesis of DNP-Alanyl Acid Chloride. DNP-Alanine, 250 mg, is suspended in 300 ml of thoroughly dried benzene and 1.0 ml of redistilled SO₂Cl₂[7] in 20 ml of dry benzene is added slowly over 20 min; the mixture is refluxed for 2 hr and stored at −20° overnight, moisture being carefully excluded at all stages. The course of synthesis is followed by the rosaniline reaction.[8]

Synthesis of DNP-Alanyl Diazoketone. DNP-Alanyl acid chloride (40 mg) is made up to 100 ml with anhydrous diethyl ether at 0° and added slowly to 10 ml of dry ether containing 0.01 mole of diazomethane at 0°. (For preparation of ethereal diazomethane, see deBoer and Backer.[9]) The resulting mixture is stored in a moisture-proof container in the dark at 4° for 12 hr. During large-scale preparations, the DNP-AD is precipitated from the ether solution and washed with ether at 4°. For small-scale synthesis of [³H]DNP-AD, the product is purified on a basic alumina (Brockman activity I) column, 1 × 15 cm, which is eluted with dry dioxane. Ultraviolet absorption λ_{max} was at 255 and 342 nm; IR spectra gave a 2115 cm⁻¹ diazo stretch band. Mass spectroscopy gave no molecular ion, but two major peaks at m/e 210 and 69 were consistent with the two halves (DNP-NH CH CH₃)⁺ and (O=CCH=N=N)⁺.

Synthesis of DNP-N₃ and [³H]DNP-N₃. DNP-N₃ and [³H]DNP-N₃ are synthesized by direct displacement of fluorine in 1-fluoro-2,4-dinitrobenzene or the 3, 5, 6-³H analogs.[10] Fluorodinitrobenzene (14.7 mg) is dissolved in 0.25 ml of redistilled dimethylformamide. Solid sodium azide 10 mg, is added, and the mixture is stirred in the dark at 20° for 2 hr, then dried under reduced pressure. The dried film is dissolved in a minimal volume of dioxane and streaked onto thin-layer chromatography (TLC) polyamide plates (Chen-Ching Trading Co., Ltd., Taiwan). The

[7] A. I. Vogel, "A Textbook of Practical Organic Chemistry," 3rd ed., p. 189. Wiley, New York, 1956.

[8] C. A. Converse and F. F. Richards, *Biochemistry* **8**, 4431 (1969).

[9] T. J. deBoer and H. J. Backer, *Org. Synth. Coll. Vol.* **IV**, 250 (1963).

[10] P. A. Grieco and J. P. Mason, *J. Chem. Eng. Data* **12**, 623 (1967).

polyamide plates are chromatographed with chloroform and dried prior to sample application. The plate is exposed to ammonia gas for 5 sec after sample application, but prior to chromatography in order to convert unreacted fluorodinitrobenzene to dinitroaniline. The TLC plate is developed with chloroform. DNP-N_3 runs at the front; dinitroaniline, the major contaminant, has an R_f of 0.5. DNP-N_3 is eluted from the TLC plate with dry ethanol and may be stored dry or as an ethanolic solution in the dark at $-10°$. The melting point of DNP-N_3 is 65° (uncorrected). High-resolution mass spectrometry gave a high-intensity M-N_2 ion peak at m/e 181.01252 corresponding to a composition $C_6H_3N_3O_4$ (calcd. 181.01236). The IR spectrum gave an N_3 stretching band at 2145 cm^{-1} indicating that the azide rather than the corresponding 4-nitrobenzo-1, 2-furazan N-oxide was present. The NMR spectrum in CDCl$_3$ showed two doublets at 7.55 and 8.47 ppm and a singlet at 8.97 ppm.

Reaction Conditions

Modification of Protein 460 with Photoaffinity Reagents

The choice of optimal reaction condition will depend both on the nature of the photoaffinity label and the immunoglobulin chain that is to be labeled. Thus, DNP reacts predominantly with the light chain of protein 460 and requires a larger concentration of DNP-AD and a longer irradiation time than the reaction of DNP-N_2, which reacts predominantly with the heavy chain. The considerations for a choice of reaction conditions are complex and will differ for each protein. The reader is referred to references[11-13] for the conditions employed with protein 460.

Controls

It is common practice to check that the label is "site-directed" by inhibiting the covalent attachment of the affinity label (for instance, a DNP-based reagent) by an excess of a nonreactive analog. i.e., DNP-lysine. It should be stressed that, as evidence for attachment of the label to a single binding site, this control is by itself insufficient and is based on circular reasoning. Few, if any, affinity reagents are universally reac-

[11] M. Yoshioka, J. Lifter, C.-L. Hew, C. A. Converse, M. Y. K. Armstrong, W. H. Konigsberg, and F. F. Richards, *Biochemistry* 12, 4679 (1973).
[12] C.-L. Hew, J. Lifter, M. Yoshioka, F. F. Richards, and W. H. Konigsberg, *Biochemistry* 12, 4685 (1973).
[13] J. Lifter, C.-L. Hew, M. Yoshioka, F. F. Richards, and W. H. Konigsberg, *Biochemistry* 13, 3567 (1974).

tive, and, if labeling outside the most strongly binding ligand site occurs, it is almost certainly also based on weaker interactions between protein and ligand at specific loci. A high concentration of analog bearing the same binding group, i.e., DNP, will inhibit both in the strongly binding site and the more weakly binding sites. Small changes in the specific radioactivity of the protein after addition of low concentrations of analog, i.e., of the order of concentration needed to half-saturate the high affinity site, are much better evidence of reaction with that avid site than is the abolition of all labeling by 10 mM ligand or its analog. Control experiments are also needed to answer the following questions: Are photoaffinity labels catalyzed to reactive intermediates by the protein in the absence of light? Are the number of affinity label molecules that are attached matched by the number of ligating sites lost? Do all the affinity reagent–protein complexes involve covalent bonds? Is radioactivity released by high concentrations of nonradioactive analog under dissociating conditions for the protein–ligand complex?

Summary of Results

After reaction of protein 460 with [³H]DNP-AD, 85% of the total radioactivity was found on the light chain and 20% on the heavy chain. The light chain had 70% of the total radioactivity in a single peptide spanning residues 25–85. All the radioactivity of this peptide was in a single modified residue, Lys-54, of the L chain. The remaining 30% of radioactivity outside this peptide appeared to be distributed among several different sites, but there was no evidence of a second single residue containing a substantial proportion of radioactivity. The heavy chain contained 15% of [³H]DNP-AD radioactivity. The amount of radioactivity in any one peptide of the heavy chain was insufficient to make isolation of the labeled residue(s) a rewarding procedure. Most of the 15% radioactivity was scattered over 3 peptides spanning residues 29–58, 62–77, and 78–108. In contrast to DNP-AD, DNP-N₃ reacted mainly with two tyrosine residues, 33 and 88, in the heavy chain of protein 460. This was shown by isolation of a CNBr fragment containing 60% of the total radioactivity in the H chain and by further degradation to show the location of the radioactivity. Although no other major locus of radioactivity was identified, sequential automated Edman degradation released some radioactivity into the chlorobutane washes during the first three cycles of degradation, suggesting that there may have been present other acid-unstable photoaffinity reagent–protein adducts. Our photoaffinity labeling experiments are consistent with the following conclusions.

[³H]DNP-AD labels preferentially the light chain reacting by nucleophilic attack on Lys-54. It seems likely therefore that DNP-AD underwent Wolff rearrangement to a ketene prior to reaction. A smaller quantity of radioactivity was, however, widely spread over the heavy chain. [^{3}H]DNP-N$_3$ reacted principally with two tyrosines in position 33 and 88,[14] respectively. The distance from the center of the ring to the reactive moiety of the photolyzed intermediate derived for DNP-AD is 5.6 Å and 3.6 Å for DNP-N$_3$. The position of residues analogues to those labeled is known on X-ray crystallographic models of protein NEW, a human IgG myeloma protein. Unless the overall arrangement of the combining region of protein NEW differs substantially from that of protein 460, it seems unlikely that all the residue modifications observed could have occurred with the DNP moiety bound to a single locus on the protein. Even when allowance is made for the 2 Å difference in the ring center to reactive moieties distance in the two reagents, and for the uncertainty that the provisional assignation of Tyr-88 is correct, the distances separating the modified residues are too large. The data on which these conclusions are based have been published in more detail.[11–13,15]

Comments

It is probably incorrect to interpret affinity labeling patterns only in terms of the quantitatively major residues that are modified, without considering the totality of the modified residues. Whether or not formation of covalent bonds occurs between affinity reagents and amino acid residues while the haptenic portion of the reagent is bound to protein probably depends on the proximity of a reactive amino acid residue and also on the reaction rate of the activated reagent with that residue. Thus, it is not possible to forecast whether covalent bond formation occurs while the affinity reagent (or photoaffinity reagent) is bound by the protein or whether it occurs at a highly reactive residue near, but not directly at, the binding site. Multiple low-affinity binding sites for a

[14] Dr. Eduardo Padlan and Dr. Elvin Kabat have drawn our attention to the possibility that another assignment to the residue we tentatively identify as Tyr-88 can be made. The primary amino acid sequence of this region of protein 460 is not known. If, however, the cysteine residue normally found at or near position 96 were displaced or if there were unexpected deletions present, it would be possible that our isolated peptide could be homologous with residues 49–60 of a related myeloma protein derived from MOPC 315. In this event, the modified tyrosine residue would be homologous to Tyr-53 of protein 315. When the amino acid sequence of protein 460 is known, we shall be able to choose between these two possibilities.

[15] F. F. Richards, J. Lifter, C.-L. Hew, M. Yoshioka, and W. H. Konigsberg, *Biochemistry* **13**, 3572 (1974).

single ligand have been demonstrated by X-ray crystallography to occur in an immunoglobulin combining region.[16] If, in addition to a high-affinity binding site, a combining region contained additional low-affinity binding sites, this might be difficult to detect by conventional hapten binding techniques, such as equilibrium dialysis or fluorescence quenching. Since covalent bond formation with an affinity label depends both on geometric considerations and reactivity, it does not follow necessarily that the most avidly binding site will be the most extensively modified. Ruoho et al.[2] expected that covalent bond formation between an azide affinity reagent and acetylcholinesterase would occur during the lifetime of the reagent–protein complex, but demonstrated with the use of a scavenger molecule (p-aminobenzoic acid) that this probably does not occur. In another instance, these workers used 4-avido-2-nitophenyl-L-lysine as a photoaffinity reagent for anti-DNP antibodies and concluded that in this system, reaction had occurred prior to hapten–antibody complex dissociation. It is likely that few predictions about rates or mechanisms with either conventional affinity or photoaffinity reagents can be made without detailed knowledge of the geometry of the interaction and the lifetimes of the intermediates. Kinetic measurements of the association of DNP-based ligands and another mouse IgA myeloma immunoglobulin, protein 315, suggest that there are multiple binding sites in this immunoglobulin combining region. The data support the existence of an initial "encounter complex" followed by the formation of a "final" complex in which a distinct subsite for the DNP ring and three other ligand-binding subsites may be identified.[17] Lancet and Pecht have more recently examined the kinetics of binding of protein 460 with a DNP-based ligand by chemical relaxation methods.[18] Their results support a mechanism in which two interconvertible conformational states of the protein bind the hapten with different association constants. Hapten binding shifts the equilibrium to an energetically more favored state.[18] Hapten ligation of antibodies is therefore probably not a simple process involving a single second-order interaction between one group of spatially juxtaposed residues forming a single binding site and one region of the hapten. A more likely interpretation of available data suggests that, in addition to high-affinity binding sites, there may be other lower-affinity binding sites present. On evolutionary grounds it would also be expected that a ligating protein would tend to retain secondary mutational changes

[16] R. J. Poljak, L. M. Amzel, H. P. Avey, B. L. Chen, R. P. Phizackerley, and F. Saul, Proc. Natl. Acad. Sci. U.S.A. **70**, 3305 (1973); L. M. Amzel, R. J. Poljak, J. M. Varga, and F. F. Richards, Proc. Natl. Acad. Sci. U.S.A. **71**, 1427 (1974).
[17] D. Haselkorn, S. Friedman, D. Givol, and I. Pecht, Biochemistry **13**, 2210 (1974).
[18] D. Lancet and I. Pecht, Proc. Natl. Acad. Sci. U.S.A. **73**, 3549 (1976).

in amino acid sequence that produce low-affinity binding sites for the ligand. This might increase the local concentration of ligand and increase the efficiency of binding. Antigens appear to have several complicated equilibria with the immunoglobulin molecule. Affinity and photoaffinity reagents on wandering over the molecular surface may leave footsteps of their journey in the form of those modified amino acid residues that are sufficiently reactive to form covalent bonds with the affinity reagents.

Acknowledgments

The studies summarized here are the work of Drs. Martine Armstrong, Carolyn Converse Cooper, Choy Hew, John Lifter, and Masanori Yoshioka.

[58] Labeling of Antilactose Antibody

By P.V. GOPALAKRISHNAN, U. J. ZIMMERMAN, and FRED KARUSH

Affinity labeling studies with anticarbohydrate antibodies have been very limited. In earlier studies, diazoniumphenyl glycosides were employed as affinity labeling reagents for rabbit and equine anti-p-azophenyl-β-lactoside and p-azophenyl-β-galactoside antibodies.[1,2] Although these antibodies were heterogeneous, it was possible to identify the labeled residues in the heavy or light chains since the modified residues had characteristic absorption spectra. With the discovery that bacterial cell walls of *Streptococcus* groups A and C induced antipolysaccharide antibodies of restricted heterogeneity,[3] such antibodies have been used for affinity labeling studies.[4-7]

Described herein is the production of antilactose antibodies, synthesis of affinity labeling reagents containing lactose, the labeling conditions, and the distribution of label in the polypeptide chains of the antibodies.

Antilactose Antibodies

These antibodies were induced by immunization with an antigen prepared by conjugation of p-azophenyl-β-lactoside to a protein.[8] Immuni-

[1] L. Wofsy, J. Kimura, D. H. Bing, and D. C. Parker, *Biochemistry* **6**, 1981 (1967).
[2] L. Wofsy, N. R. Klinman, and F. Karush, *Biochemistry* **6**, 1988 (1967).
[3] R. M. Krause, *Fed. Proc., Fed. Am. Soc. Exp. Biol.* **29**, 50 (1970).
[4] D. C. Parker, R. M. Krause, and L. Wofsy, *Immunochemistry* **10**, 727 (1973).
[5] L. Wofsy, D. C. Parker, I. Corneil, and B. Burr, "Developmental Aspects of Antibody Formation and Structure," p. 425. Academia, Prague, 1973.
[6] P. V. Gopalakrishnan and F. Karush, *Immunochemistry* **12**, 449 (1975).
[7] P. V. Gopalakrishnan, FASEB Meetings, Anaheim, Abstract No. 372 (1976).
[8] N. R. Klinman, J. H. Rockey, G. Frauenberger, and F. Karush, *J. Immunol.* **96**, 587 (1966).

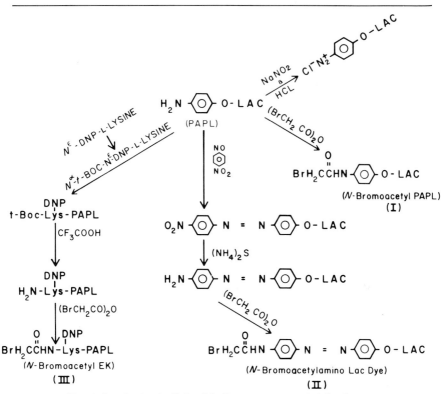

FIG. 1. Synthesis of affinity labeling reagents containing lactose.

zation of rabbits and a horse with *Streptococcus faecalis* (strain N) was also used to produce IgG and IgM antilactose antibodies of restricted heterogeneity. Purification was carried out by affinity chromatography.[9,10] Rabbits produced anti-*p*-aminophenyl-*β*-lactoside antibodies of restricted heterogeneity (anti-PAPL)[11] when injected with (PAPL)$_n$-*Pneumococcus*-R36A.[12]

Synthesis of Affinity Labeling Reagents

Three types of affinity labeling reagents were synthesized by the steps shown in Figs. 1 and 2 to label the antilactose antibodies: (a) diazonium-

[9] A. Ghosh and F. Karush, *Biochemistry* **12**, 2437 (1973).

[10] Y. Kim and F. Karush, *Immunochemistry* **10**, 365 (1973).

[11] The following abbreviations are used: PAPL, *p*-aminophenyl-*β*-lactoside; amino-Lac dye, *p*-(*p*-aminophenylazo)phenyl-*β*-lactoside; EK compound, *N*-(*N*$^\epsilon$-DNP-L-lysyl)-*p*-aminophenyl-*β*-lactoside; PBS-EDTA, 0.01 *M* phosphate, 0.15 *M* NaCl pH 7.4, containing 1 m*M* ethylenediaminetetraacetate.

[12] P. V. Gopalakrishnan and F. Karush, *J. Immunol.* **113**, 769 (1974).

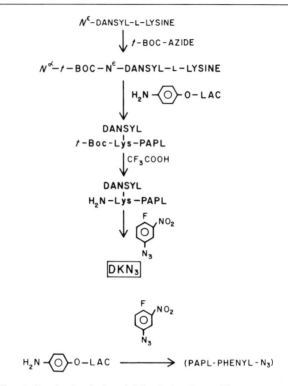

FIG. 2. Synthesis of photolabile derivatives of lactose.

phenyl β-glycosides,[1,2] (b) bromoacetyl-containing lactose derivatives,[6,7] and (c) photolabile derivatives of lactose.[13]

The starting material for the preparation of these reagents is either *o*- or *p*-aminophenyl-β-galactoside or β-lactoside, obtained by condensing the respective acetobromosugar with *o*- or *p*-nitrophenol,[14] deacetylating with sodium methylate in methanol,[15] and catalytic reduction in aqueous methanol.[16]

a. Diazonium Reagents.[1,2]

Aminophenyl-β-glycoside (10 μmoles) is dissolved in 0.1 M HCl (0.45 ml), and the temperature is maintained between 0° and 5°. NaNO$_2$ (12

[13] U. J. Zimmerman and F. Karush, unpublished results.
[14] H. G. Latham, E. F. May, and E. Mossetig, *J. Org. Chem.* **15**, 884 (1950).
[15] A. Thompson, M. L. Wolfrom, and E. Pascu, "Methods in Carbohydrate Chemistry" (R. L. Whistler and M. L. Wolfrom, eds.), Vol. 2, p. 215. Academic Press, New York, 1963.
[16] F. H. Babers and W. F. Goebel, *J. Biol. Chem.* **105**, 473 (1934).

μmoles) in water (50 μl) is added. After 10 min, ice water (4–5 ml) is added to make a stock solution of 2 mM in reagent.

b. Bromoacetyl-Containing Lactose Derivatives.

Three bromoacetyl derivatives containing tritium label have been synthesized.

i. BrAcPAPL[11,12] [*N-Bromoacetyl-p-aminophenyl-β-lactoside*]. A solution of p-aminophenyl-β-lactoside (Cyclochemical Corporation, about 10 μmoles) in dimethyformamide (0.1 ml) is treated with tritiated bromoacetic acid (Amersham, 26 μmoles, 5 mCi), and dicyclohexylcarbodiimide (38 μmoles) is added. After 3 hr at room temperature, the mixture is filtered to remove dicyclohexylurea, and ether (10 volumes) is added. The precipitated product is dissolved in n-butanol:water:glacial acetic acid (4:5:1) and fractionated in a countercurrent distribution apparatus for 50 transfers in the same solvent system. The contents of tubes 8 through 20 are pooled and flash evaporated to dryness; the residue is triturated with ether, dried, and stored as a solid.

ii. BrAcNH Lac Dye[6] [*p-(N-bromoacetyl-p-aminophenylazo)phenyl-β-lactoside*]. Amino Lac dye[17] (6.5 μmoles) in dimethylformamide (0.2 ml) is added to a vial containing tritiated bromoacetic acid (26 μmoles, 5 mCi, Amersham), followed by dicyclohexylcarbodiimide (26 μmoles). After 3 hr at room temperature, the precipitated dicyclohexyl urea is removed by filtration and washed with dimethylformamide. The filtrates are mixed with anhydrous ether (20 ml), and the precipitated material is further purified by countercurrent distribution as described.[6]

iii. BrAcEK [*N-(N$^\alpha$-Bromoacetyl-N$^\varepsilon$-DNP-L-lysyl)-p-aminophenyl-β-lactoside*]. To a solution of bromoacetic acid (1 mmole) in p-dioxane (3 ml) are added N-hydroxysuccinimide (1.1 mmoles) and dicyclohexylcarbodiimide (1.5 mmoles). After reaction at room temperature for 90 min, the precipitated dicyclohexyl urea is removed by filtration and washed with dimethylformamide; the combined filtrates are added to a solution of EK compound[11,18] (0.25 mmole) in dimethylformamide (2 ml). After 2 hr at room temperature, the solution is flash-evaporated to dryness, dissolved in the upper phase of n-butanol:water:glacial acetic acid (4:5:1, 10 ml), and separated for 50 transfers by countercurrent distribution. The upper phase from tubes 31 through 45 and lower phase from 34 through 42 are combined and flash-evaporated to dryness. The residue is triturated with ether and dried. The tritiated form of BrAcEK is obtained by using

[17] P. V. Gopalakrishnan, W. S. Hughes, Y. Kim, and F. Karush, *Immunochemistry* **10**, 191 (1973).
[18] P. V. Gopalakrishnan and F. Karush, *J. Immunol.* **114**, 1359 (1975).

tritiated bromoacetic acid (26 moles, 5 mCi, Amersham) and lower quantities of the other reactants.

c. Photolabile Lactose Reagents

i. DKN$_3$ {N-[N$^\alpha$(2-Nitro, 4-azidophenyl), N$^\varepsilon$-Dansyllysyl]-p-amino-phenyl-β-lactoside}. The trifluoroacetate salt of N-(N^ε-dansyllysyl)p-aminophenyl-β-lactoside[13] (0.1 mmole) is dissolved in dimethylformamide (1 ml) and neutralized with triethylamine (0.4 mmole). After the addition of 4-fluoro-3-nitrophenylazide (FNPA, 0.12 mmole), the mixture is stirred in the dark at 60° for 5 hr. Progress of the reaction is monitored by measurement of absorption at 460 nm, since neither starting material absorbs at this wavelength. The reaction mixture is precipitated with ether (10 ml) and centrifuged. The orange-colored solid is dissolved in n-butanol:water:glacial acetic acid (4:5:1, 10 ml) and fractionated with 60 transfers. The product can be identified by its orange color under visible light and by green fluorescence under UV. The contents of tubes 45 through 56 are pooled and flash evaporated. TLC with upper layer of n-butanol:water:glacial acetic acid (4:5:1) shows a single spot with R_f 0.74. The IR spectrum in a KBr pellet shows a distinct N$_3$ band at 2000 cm^{-1}.

ii. PAPL-phenyl-N$_3$[11] [N-(4-Azido-2-nitrophenyl)-p-aminophenyl-β-lactoside]. (1) *PAPL*[11] (H*): The lactoside, p-aminophenyl-β-lactoside (18 μmoles), is dissolved in water (1 ml) and galactose oxidase (0.2 ml, 1 mg/ml, Worthington, 30 units/mg) is added, followed by catalase (1 mg, Worthington, 250 units/ml). The mixture is incubated at room temperature for 19 hr in the dark. The reaction is stopped by placing the mixture in a 60° waterbath for 15 min. After centrifugation, the pH of the supernatant liquid is adjusted to 8.0 with 1 N NaOH and is followed by the addition of tritiated sodium borohydride (0.1 mmole, 25 mCi, New England Nuclear). The mixture is allowed to react for 1 hr, after which it is brought to pH 4.0 with 1 N HCl and flash evaporated. The solid is dissolved in a small amount of methanol, applied to a silica gel column (17 cm $\times$ 1 cm, 100–200 mesh), and eluted with ethyl acetate–methanol (1:1) and 3-ml fractions are collected. Fractions 2 and 3, showing absorption at 290 nm and having redioactivity, are pooled and flash evaporated. A specific activity of 2.7×10^6 cpm/μmole has been obtained for PAPL. TLC showed a single spot with R_f 0.15 in n-butanol:water:glacial acetic acid (4:5:1, upper phase) and a positive sugar test. (2) *Coupling of FNPA to PAPL* (H*): To PAPL (H*) (0.1 mmole) in dimethylformamide (1.5 ml) are added triethylamine (0.2 mmole) and FNPA (0.2 mmole). The mixture is incubated at 50° for 26 hr. The coupling reaction may be monitoried by the appearance of a new band at 460 nm. The reaction product is purified by 60 transfers in n-butanol:water:glacial acetic acid (4:5:1) employing absorption at 290 nm and 460 nm to identify the desired fractions. A dark red solid is obtained after flash evaporation. This shows a single spot on TLC with the upper phase of the above solvent; $R_f = 0.75$.

Affinity Labeling Conditions

With diazonium reagents, a 22 μM solution is allowed to react with 18 μM anti-p-azophenyl-β-glycoside antibodies in sodium borate at pH 8.0. After 90 min at 5°, the reaction is terminated by the addition of 75% cold EtOH or 50% cold SAS. p-Nitrophenyl-β-lactoside, 1 mM, is used as inhibitor to demonstrate the specificity of the reaction.[1,2]

With bromacetyl-containing lactose derivatives as affinity labeling reagents, an extensive study for maximum labeling has been done.[6] 10 μM antilactose antibody was treated with 18–20 μM affinity labeling reagent in 0.1 M NaHCO$_3$ (pH 9.0). After 2 hr at 37°, the labeled antibody was either precipitated with two volumes of 25% TCA[6] or dialyzed extensively against PBS-EDTA.[7,11]

Photolysis of DKN$_3$ (and PAPL-phenyl-N$_3$)[11] can be achieved by irradiating the compounds within either the broad weak 460 nm band or the main UV band; the latter is photochemically more efficient. With an Osram 200-watt mercury or 500-watt Xenon arc lamp plus appropriate optics and a mirror behind the cells to increase the light intensity, the duration of irradiation is 30–45 min. The damage to the protein is minimal when Pyrex cells are used. A cell containing antibody solution (about 10 μM) and DKN$_3$ (about 0.1 mM) in PBS-EDTA and another identical cell, containing the same solution plus 0.1 M lactose as an inhibitor, are simultaneously irradiated. The extent of photolysis can be estimated from the disappearance of the 460-nm band. The photolyzed sample is precipitated with either 25% TCA or ethanol, repeatedly washed with dimethylformamide, and dissolved in either 0.1 M phosphate buffer containing 0.25% SDS and 0.5 M urea, or in 0.1 M NaOH.

In the case of DKN$_3$, the number of moles of label bound per mole of protein can be estimated from the fluorescence emission at 545 nm when the solution is excited at 335 nm. The protein concentration is obtained from the absorbance at 280 nm of a control which contains no added hapten or inhibitor and has undergone the same photochemical and subsequent treatment as the sample. The conditions for maximum specific labeling are being worked out.

In the other cases, the molar extinction coefficients of the affinity

MOLAR EXTINCTION COEFFICIENTS OF AFFINITY LABELING REAGENTS

Affinity labeling reagent	Molar extinction coefficient	Wavelength (nm)	Reference
BrAcPAPL	1.0×10^4	250	11
BrAcNHLac Dye	1.0×10^4	311	6
BrAcEK	1.52×10^4	365	17
OD Lac and OD Gal	1.50×10^4	370 (pH 9.0)	1
PD Lac	1.10×10^4	370	1

labeling reagents, shown in the table, were employed, to calculate the number of moles of affinity labeling reagent bound per mole of antibody, or radioactive assays were used.[6]

Distribution of Label in the Polypeptide Chains of the Antibodies

Distribution of label in the chains can be calculated by separating the chains on a Sephadex G-100 column in 1 M propionic acid by the method of Fleischman et al.[19] or on 5% SDS gels.[6] The residue labeled has been identified in some instances.[1,2]

Comments

All three bromoacetyl compounds have been found to be suitable affinity labeling reagents for the antilactose antibodies. Under optimal conditions, the extent of specific labeling can be as high as 1.5 moles of label per mole of IgG. However, there were significant differences in the behavior of the reagents with respect to inhibition of labeling in the presence of lactose. Thus, with BrAcPAPL and rabbit anti-PAPL antibody, 100% inhibition by 10 mM lactose was observed. With bromoacetylamino-Lac dye, the inhibition was only 75%, and the extent of labeling was 33% less. One of the major problems in using BrAcNHLac dye as an affinity labeling reagent is the nonspecific labeling, probably due to enhanced hydrophobic interaction of the azophenyl group. BrAcPAPL and BrAcNH-Lac dye are also somewhat unstable in solution. After a few days' storage in solution, the extent of labeling was reduced from 1.05 to 0.66. This instability might be due to the aryl character of the substituted amine. For maximum labeling, the affinity labeling reagent must be stored free from moisture and fresh solutions must be prepared on the day of the experiment. In spite of this disadvantage, bromoacetylamino-Lac dye proved to be a promising reagent for labeling different chains, depending on

[19] J. B. Fleischman, R. R. Porter, and E. M. Press, Biochem. J. 88, 220 (1963).

whether the reaction is carried out in dark or light. This has been attributed to the photoisomerization phenomenon.[20]

Preliminary studies showed that BrAcPAPL could not label equine IgM antilactose antibodies. Whether this is due to affinity, to the geometry of the reagent, or to other factors is not clear. With bromoacetyl EK, nonspecific labeling was considerably reduced and the specificity of the reaction has been demonstrated.

[20] P. V. Gopalakrishnan, *Immunol. Commun.* **4**, 499 (1975).

[59] The Ouabain-Binding Site on $(Na^+ + K^+)$ Adenosinetriphosphatase

By ARNOLD RUOHO and JACK KYTE

$(Na^+ + K^+)$ Adenosinetriphosphatase is the membrane-bound enzyme that transports sodium and potassium in opposite directions across the cell membrane to create the ion gradients that are utilized to perform physiological functions. The cardiac glycosides and aglycons, such as digoxin, ouabain, and strophanthidin, are natural products that have been used therapeutically for centuries in the treatment of heart disease. Their only known biochemical effect is to inhibit specifically $(Na^+ + K^+)$ adenosinetriphosphatase and consequently decrease or completely stop active cation flux.

The purified enzyme is a specific complex of two polypeptide chains.[1] In order to determine which polypeptide contains the cardiac glycoside-binding site, a radioactive affinity label can be used.[2] Haloacetyl derivatives of the cardiac glycosides strophanthidin and hellebrigenin[3,4] were synthesized for affinity labeling of the cardiac glycoside-binding site of $(Na^+ + K^+)$ adenosinetriphosphatase. They did not, however, react covalently and specifically with the binding site.[5] One disadvantage of such electrophilic reagents is the requirement for a suitably positioned nucleophile in the active site. On the other hand, a photoaffinity label can be converted by photolysis into an exceedingly reactive intermediate, and under appropriate circumstances, even insertion into a C-H bond is possible.[6] Compounds that generate carbenes or nitrenes upon photolysis

[1] J. Kyte, *J. Biol. Chem.* **247**, 7642 (1972).

[2] S. J. Singer, *Adv. Protein Chem.* **22**, 1 (1967).

[3] L. E. Hokin, M. Mokotoff, and S. M. Kupchan, *Proc. Natl. Acad. Sci. U.S.A.* **55**, 797 (1966).

[4] A. E. Ruoho, L. E. Hokin, R. J. Hemingway, and S. M. Kupchan, *Science* **159**, 1354 (1968).

[5] A. E. Ruoho, R. Blaiklock, and L. E. Hokin, unpublished observations.

[6] J. R. Knowles, *Acc. Chem. Res.* **5**, 155 (1972). See also this volume [8].

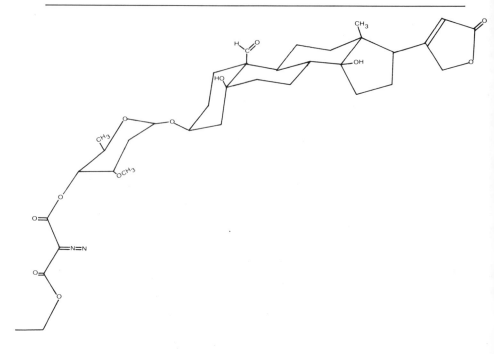

FIG. 1. Structure of 4'-(ethyldiazomalonyl)cymarin (DAMN-cymarin).

have been used for the photoaffinity labeling of enzyme active sites,[7] antibody ligand sites,[8] and membrane-bound adenosine 3':5'-cyclic monophosphate binding sites.[9] We have synthesized and used an ethyl diazomalonyl derivative of the cardiac glycoside, cymarin (DAMN-cymarin[10]; Fig. 1), for photoaffinity labeling of this specific binding site in Na$^+$ + K$^+$) adenosinetriphosphatase.

Synthesis of Photolabels

Method A. Acylation of cymarin at the 4'-hydroxyl of the cymarose was performed with redistilled ethyl diazomalonyl chloride[11] by a modi-

[7] Y. Stefanovsky and F. H. Westheimer, *Proc. Natl. Acad. Sci. U.S.A.* **70**, 1132 (1973).

[8] G. W. J. Fleet, J. R. Knowles, and R. R. Porter, *Biochem. J.* **128**, 499 (1972).

[9] C. E. Guthrow, H. Rasmussen, D. J. Brunswick, and B. S. Cooperman, *Proc. Natl. Acad. Sci. U.S.A.* **70**, 3344 (1973).

[10] Abbreviations: DAMN-cymarin, 4'-(ethyldiazomalonly)cymarin; CM-cymarin, 4'-(ethylchloromalonyl)cymarin; AMPPNP, adenylyl imidodiphosphate.

[11] F. Weygand, H. J. Bestmann, and H. Fritsche, *Chem. Ber.* **93**, 2340 (1960).

fication of the method of Jacobs.[12] Cymarin (Sigma) (200 mg, 0.365 mmole) is dissolved in pyridine, previously dried by distillation from phthalic anhydride. Diazomalonyl chloride (0.476 ml, 3.65 mmoles) is added dropwise to a swirling solution of the steroid in pyridine. A precipitate, probably the pyridinium adduct of the acyl halide, forms immediately. The reaction is allowed to proceed for 1 hr at room temperature in the dark. Thin-layer chromatography on silica gel plates containing 4% inorganic fluorescent indicator in 7% methanol/chloroform can be used to follow the quantitative conversion of cymarin to the acylated derivative. Pyridine is removed under vacuum overnight, and the yellow-brown residue is dissolved in chloroform.

The chloroform solution is extracted exhaustively with water and dried with anhydrous magnesium sulfate. Upon concentration, the chloroform solution yields a yellow oil. The 4'-diazomalonyl cymarin (DAMN-cymarin) is crystallized from methanol–ether.

The product has been characterized by the diazo absorption at 2100 cm^{-1} in the infrared spectrum and nuclear magnetic resource.

Method B. In order to perform the reaction on the small scale required for the synthesis of the radioactive derivative, an alternative method of preparing DAMN-cymarin may be used. Ethyl diazomalonyl chloride is synthesized by adding 64 μl (0.6 mmole) of ethyl diazoacetate (Aldrich) to [^{14}C]phosgene (New England Nuclear) (0.2 mmole, 5 mCi/mmole) in 0.1 ml of dry benzene. The benzene solution of phosgene is placed in an ice bath initially while the ethyl diazoacetate is being added and is then gently warmed at room temperature until no further bubbling can be detected. This is essentially the method of Vaughan and Westheimer.[13] Approximately a one-fifth molar equivalent of cymarin (20 mg, 0.0365 mmole) is added in 0.1 ml of dry pyridine, and the reaction is allowed to proceed overnight in the dark with stirring. The reaction is quantitative.

The reaction mixture is taken to dryness under pressure, dissolved in acetone, and chromatographed on 20 cm $\times$ 20 cm silica gel G plates with fluorescent indicator in benzene:acetonitrile (1:1). The products are routinely found to be 80% DAMN-cymarin ($R_f = 0.45$) and 20% another cymarin derivative ($R_f = 0.57$). The later compared has been tentatively identified as 4'-(ethylchloromalonyl)cymarin (CM-cymarin).[14] DAMN-cymarin can be visualized on the plate by its ultra-

---

[12] W. A. Jacobs and A. Hoffmann, *J. Biol. Chem.* **67**, 609 (1926).

[13] R. Vaughan and F. H. Westheimer, *Anal. Biochem.* **29**, 305 (1969).

[14] This tentative characterization is based on the following observations: (a) the 220-MH$_z$ nuclear magnetic resonance spectrum of CM-cymarin is identical to that of DAMN-cymarin except for an additional singlet (1 proton) at $\delta = 4.89$; (b)

violet absorbance or by autoradiography. [^{14}C]CM-cymarin can be localized by autoradiography. The radioactive compounds cochromatograph with the pure nonradioactive derivatives in four separate solvent systems on silica gel G thin-layer plates: (a) 7% methanol/chloroform, (b) 50% benzene acetonitrile, (c) 10% ethanol/benezene, and (d) 30% tetrahydrofuran/chloroform.

Both compounds were reversible inhibitors of $(Na^+ + K^+)$ adenosine-triphosphatase when assayed in dim light. The K_I's were 1.3 μM for DAMN-cymarin and 0.9 μM for CM-cymarin.

Photolysis

In order to perform the photolysis experiments, radioactive photoaffinity label (in 50% benzene–ethanol) was dried in a cuvette of 2-mm path length; it was dissolved in 20 μl of dimethyl sulfoxide, and enzyme[15] and buffer were added, so that the final concentrations were 50 mM phosphoenolpyruvate, 150 mM Na^+, 0.5 mM ATP, 1.5 mM EDTA, 7 mM 2-mercaptoethanol, 2.5 mM $MgCl_2$, 0.2 mg/ml of pyruvate kinase, and 30 mM imidazolium·Cl, pH 7.0, in a final volume of 0.1 ml. The mixture was flushed with N_2 for 5 min and photolyzed at 0° under N_2, 1 cm from the filament of a H85A medium pressure Hg lamp. The solution was diluted to 2.0 ml with 1 mM cymarin, incubated at 37° for 15 min, and centrifuged at 100,000 g for 20 min.

Preparation of Ethylene Diacrylate Cross-Linked Polyacrylamide Gel[16]

The pellets were dissolved in 0.1 ml of 4% sodium dodecyl sulfate (SDS), 1% mercaptoethanol at 100°, and subjected to electrophoresis on ethylene diacrylate (EDA) cross-linked 5.7% polyacrylamide gels containing 0.1% SDS. The gels were composed as follows: 40% w/v acrylamide + 2.85% w/v EDA, 1.4 ml; 20% w/v SDS, 0.05 ml; concentrated

---

the largest ion in its mass spectrum is at m/e = 678, the molecular weight of CM-cymarin·H_2O; (c) both nuclear magnetic resonance and mass spectra show that the strophanthidin core is present in an unaltered state; (d) there is no diazo absorption band at 2100 cm^{-1} in the infrared spectrum; otherwise CM-cymarin and DAMN-cymarin display very similar infrared spectra; (e) it is radioactive when the synthesis is carried out with [^{14}C]phosgene; (f) CM-cymarin gives a positive Beilstein test; and (g) upon saponification, the radioactivity is released from [^{14}C]CM-cymarin as [^{14}C]chloromalonic acid (determined by thin-layer chromatography on polyamide sheets).

[15] J. Kyte, *J. Biol. Chem.* **246**, 4157 (1971).

[16] G. L. Choules and B. H. Zimm, *Anal. Biochem.* **13**, 336 (1965).

buffer (0.4 M Tris chloride 0.2 M sodium acetate, 0.03 M Na_2 EDTA, pH 7.4), 1.0 ml; 1.5% v/v $N,N,N'N'$-tetramethylethylenediamine (TEMED), 0.5 ml; 1.5% w/v ammonium persulfate, 0.5 ml; water 6.55 ml.

This volume of gel solution was sufficient to prepare 4–6 gel tubes 8.5 cm long. Fresh solutions of ammonium persulfate and TEMED were prepared each time gels were made. The gels were overlayered with a solution of 0.075% TEMED, 0.075% ammonium persulfate, 0.1% SDS. The electrophoresis buffer consisted of a 10-fold dilution of the concentrated buffer described above containing 0.1% SDS. Electrophoresis was performed at a constant voltage of 50 volts.

After electrophoresis, gels were scanned at 280 nm and sliced into 2-mm disks. Gel slices were dissolved by overnight incubation at 37°C in 15 ml of a standard PPO-POPOP toluene scintillation solution containing 10% Nuclear Chicago Solubilizer (NCS). This method allowed for gel solution without decarboxylation of the photolabel and loss of radioactivity as CO_2. Samples were counted for 20 min each.

Conditions for Photoaffinity Labeling of the Ouabain-Binding Site of $(Na^+ + K^+)$ Adenosinetriphosphatase with DAMN-Cymarin and CM-Cymarin

Cardiac glycosides bind most tightly to $(Na^+ + K^+)$ adenosine triphosphatase when it is in the phosphorylated form,[17] which requires the presence of Na^+ and MgATP[18] or Mg^{2+} and P_i.[19] Since there is a low level of contaminating, nonspecific adenosinetriphosphatase present in the enzyme prepartaion,[15] it was necessary to add an ATP-generating system, i.e., phosphoenolpyruvate and pyruvate kinase, when Na^+ and ATP were used to phosphorylate the enzyme. This avoided using a high concentration of MgATP, which would absorb the light used to photolyze the enzyme inhibitor complex. In the presence of Na^+, MgATP, and the ATP-generating system, [14C]DAMN-cymarin or [14C]CM-cymarin (2 nmoles) and $(Na^+ + K^+)$ adenosinetriphosphatase (0.5 mg, 2 nmoles of cardiac glycoside-binding sites[20]) were mixed together in a 2-mm pathlength quartz cuvette in a final volume of 0.1 ml. Under these circumstances, both enzyme and inhibitor are present at concentrations in excess of the dissociation constant for the enzyme–inhibitor complex, and

[17] H. Matsui and A. Schwartz, *Biochim. Biophys. Acta* **151**, 655 (1968).
[18] R. W. Albers, S. Fahn, and G. J. Koval, *Proc. Natl. Acad. Sci. U.S.A.* **50**, 474 (1963).
[19] G. J. Siegel, G. J. Koval, and R. W. Albers, *J. Biol. Chem.* **244**, 3264 (1969).
[20] J. Kyte, *J. Biol. Chem.* **247**, 7634 (1972).

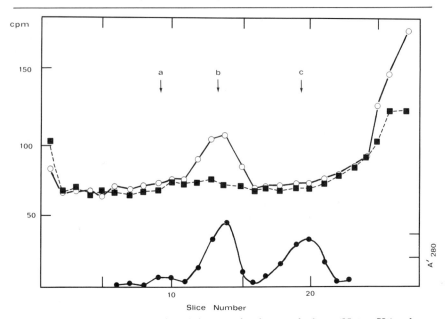

FIG. 2. Distribution of radioactivity covalently attached to $(Na^+ + K^+)$ adenosine triphosphatase that has been photolyzed in the presence of $[^{14}C]CM$-cymarin. Samples of enzyme that had been photolyzed either in the presence of the complete MgATP phosphorylating system (○) or the complete system plus a 25-fold excess of cymarin as protector (■) were run on sodium dodecyl sulfate gels which were scanned, sliced, and counted. The A_{280} trace from the gel with the unprotected sample was divided into segments exactly as the gel had been sliced. The mean A'_{280} of each of these segments was calculated, and the values were plotted (●). The units of A'_{280} are arbitrary since the scanner was uncalibrated. The three protein components are: (a) cross-linked $\alpha\beta$ dimer; (b) large chain; (c) small chain.

most of the inhibitor is bound to sites on the enzyme. After a short preincubation under nitrogen, the solution was exposed to strong ultraviolet light for 5 min. The reaction product was analyzed on SDS–polyacrylamide gels.[21] The distribution of ^{14}C counts and protein that resulted are shown for $[^{14}C]DAMN$-cymarin in Fig. 2. The results were analogous for $[^{14}C]CM$-cymarin. There are three protein components resolved by the electrophoresis; the two polypeptides of $(Na^+ + K^+)$ adenosinetriphosphatase and a covalently cross-linked dimer of the two chains, which is formed during photolysis. It can be seen that the large chain of $(Na^+ + K^+)$ adenosinetriphosphatase is radioactively labeled.

Several controls were performed to demonstrate that the labeling ob-

[21] A. L. Shapiro, E. Viñuela, and J. V. Maizel, *Biochem. Biophys. Res. Commun.* **28**, 815 (1967).

RADIOACTIVITY COVALENTLY ATTACHED TO THE LARGE CHAIN OF (Na+ + K+)
ADENOSINETRIPHOSPHATASE THAT HAS BEEN PHOTOLYZED IN THE
PRESENCE OF [14C]CM-CYMARIN[a]

Incubation mixture	Total cpm	Relative cpm/A_{280}
Complete ATP system	227 (9)	1.00
+ Cymarin	8 (4)	0.06
− Mg^{2+}	11	0.06
− ATP + AMPPNP	59	0.28
− Pyruvate kinase	27	0.11
− Na^+ + K^+	67	0.12
Dark control	17	0.09
Mg^{2+} + P_i	274 (3)	1.80
+ Cymarin	65 (2)	0.35
− Mg^{2+} + Ca^{2+}	38	0.30
− Mg^{2+}	15 (2)	0.10

[a] Enzyme was photolyzed under the conditions noted and run on gels that were scanned, sliced, and counted. The position of the large chain was determined from the scan, and the total number of cpm in all the slices from that region was calculated after background was subtracted (determined from those controls that contained the least amount of radioactivity). These data are presented as total cpm. The total cpm were then divided by the area of the A'_{280} peak of the large chain, calculated from the scan of each gel, and each of these quotients was normalized to that of the complete system to obtain relative cpm/A_{280}. The number of times that each experiment was performed is in parentheses.

served with the photolabels was occurring at the cardiac glycoside site. They are summarized in the table for [14C]CM-cymarin. Much less labeling was observed when AMPPNP was substituted for ATP or when photolysis was omitted. The presence of the ATP-generating system is essential to obtain labeling, since none occurs when pyruvate kinase is omitted. When Na^+ is replaced by K^+, no labeling is observed. It has been shown that K^+ partially competes with the binding of cardiac glycoside.[22]

Specific labeling was observed when the phosphorylated enzyme was formed in the presence of Mg^{2+} and P_i. Although the incorporation of label in the presence of Mg^{2+} and P_i occurs only at the cardiac glycoside site, the specific activity (cpm/A_{280}) of the large chain labeled under these conditions is twice that of large chain labeled in the presence of Na^+ and ATP. This probably reflects the fact that the absorbance of the former reaction mixture is much lower and more extensive photolysis is possible. However, there is also an increase in the amount of photolytic

[22] E. T. Dunham and I. M. Glynn, *J. Physiol.* (*London*) **156**, 274 (1961).

cross-linking so that less unaggregated large chain remains in the sample after photolysis.

The radioactivity present at the bottom of the gels when ($Na^+ +$ K^+) adenosinetriphosphatase is photolyzed in the presence of either DAMN-cymarin or CM-cymarin (Fig. 2) is not due to the covalent labeling of any protein component. It is present in the dark control samples, and it can be removed from the gels, before slicing and counting, by soaking them in isopropanol–acetic acid overnight, a procedure that does not remove any of the radioactivity from the large chain or any protein from the gels. This material probably is label that was still noncovalently bound to the enzyme when it was dissolved in SDS.

It is concluded that the large chain of ($Na^+ + K^+$) adenosinetriphosphatase is specifically photoaffinity labeled with these derivatives of cymarin and that it, therefore, contains a portion or all of the cardiac glycoside-binding site.

Problems Arising in These Photolysis Experiments

1. The incorporation of radioactivity specifically into the large polypeptide of the enzyme with [^{14}C]CM-cymarin, which lacks the diazo group, raised the possibility that photolysis generated a reactive species in the glycoside itself, which proceeded to insert into some active site residue(s) in the protein. To test this hypothesis [^3H]ouabain (17 Ci/ mmole) was photolyzed, under the conditions previously described for [^{14}C]CM-cymarin. No specific labeling of the polypeptides occurred upon analysis on SDS gels. This argues against a photoactivation of the steroid moiety per se and also eliminates the possibility of "reverse photoaffinity" labeling, i.e., photoactivation of a residue in the protein which inserts into the ligand.

2. It was not possible to photolyze the enzyme inhibitor complex for extended periods of time owing to photolytic cross-linking that acts as a competing reaction. Cross-linked material appeared at a position on the SDS gels consistent with a large polypeptide–small polypeptide dimer. With prolonged photolysis (30–60 min) almost all the enzyme was cross-linked to a very large molecular weight species that would not penetrate 5.6% polyacrylamide gels. The rate of disappearance of the large polypeptide occurred faster when ultraviolet-absorbing compounds were removed from the photolysis medium. The fastest rate occurred in phosphate buffer at ph 7.5. Whether the inactivation of enzyme activity with photolysis was correlated with cross-linking was not determined. It is possible that the level of photoincorporation into the large polypeptide was limited because of this photolytic cross-linking.

3. Because of the relatively low specific radioactivity of these photo-labels and the low efficiency of photoincorporation (less than 2% of total added photolabel), the possibility is not ruled out that the small polypeptide of this enzyme participates in the binding site.[23] If the amount of label incorporated into the small chain was less than 5–10% of the level of incorporation into the large chain, it would not have been detectable with these photolabels.

Acknowledgments

A. R. was a Helen Hay Whitney Postdoctrol Fellow, and J. K., a Damon Runyon Fund Postdoctoral Fellow. We thank S. J. Singer, in whose laboratory this work was performed, for his advice and support. This research was supported by U.S. Public Health Service Grant No. GM1597 (to S. J. Singer). We also thank Dr. Charles Perrin for helpful discussions.

[23] Although both the heavy and light chains of IgG immunoglobulin together form the binding site, it has been observed that the ratio of covalent incorporation of an affinity label into the two chains of different anti-DNP immunoglobulins varies from 0.1 to 10 (heavy/light). Where two polypeptides contribute to an active site, the ratio of affinity label attached to residues in the two chains can vary over a wide range as a result of relatively small differences in the free energy of activation for reaction with groups on the two chains.

[60] Penicillin Isocyanates for β-Lactamase

By HIROSHI OGAWARA

β-Lactamase (penicillin and cephalosporin amido-β-lactam hydrolase) catalyzes the hydrolysis of the β-lactam ring of penicillins and cephalosporins. The amino acid sequences of some β-lactamases[1-3] from gram-positive bacteria and a preliminary crystallographic analysis[4] of a β-lactamase from *Escherichia coli* were reported. Histidine[5-7] and tyro-

[1] R. P. Ambler and R. J. Meadway, *Nature (London)* **222,** 24 (1969).

[2] D. R. Thatcher, *Biochem. J.* **147,** 313 (1975).

[3] R. P. Ambler, *Biochem. J.* **151,** 197 (1975).

[4] J. R. Knox, P. E. Zorsky, and N. S. Murthy, *J. Mol. Biol.* **79,** 597 (1973).

[5] R. H. Dupue, A. G. Moat, and A. Bondi, *Arch. Biochem. Biophys.* **107,** 374 (1964).

[6] H. Ogawara and H. Umezawa, *J. Antibiot.* **27,** 567 (1974).

[7] H. Ogawara and H. Umezawa, *in* "Microbial Drug Resistance" (S. Mitsuhashi and H. Hashimoto, eds.), p. 375. Univ. of Tokyo Press, Tokyo, 1975.

sine[8,9] residues and amino[6,7] and carboxyl[10,11] groups have been implicated in the binding and/or catalytic sites of the enzymes.

In exploring the structural features of the active site, derivatives of penicillins which are substrates or competitive inhibitors of all β-lactamases should be useful for affinity labeling and active-site analysis. The method of preparation of penicillin isocyanates[12] and the inactivation of an *E. coli* β-lactamase, β-lactamase$_{75}$, and a β-lactamase from *Bacillus cereus* are described herein.

Preparation of Affinity Labels

Benzylpenicillin Isocyanate. A suspension of 435 mg (1.0 mmole) of benzylpenicillin triethylammonium salt in 3 ml of methylene chloride is cooled to $-10°$ with ice and NaCl, and 100 μl (1.0 mmole) of ethylchloroformate are added in one portion. The resulting mixture is stirred at $-10°$ for 1.5 hr during which the solution becomes clear. Thereafter, a solution of 65 mg (1.0 mmole) of NaN$_3$ in 1 ml of water is added in 50-μl portions over a 30-min period. Stirring is continued for a further 10 min at $-5°$ to $-10°$, then the reaction is stopped by diluting with 2.5 ml of ice-cold water and extracting with three portions of 2.5 ml of ice-cold methylene chloride. The extracts are pooled, washed with 2.5 ml of ice-cold 1% NaHCO$_3$ and three portions of 2.5 ml of ice-cold water, dried over MgSO$_4$, and evaporated to dryness at room temperature under reduced pressure. The product, 260 mg of an oily substance (73%), shows an azide band at 2145 cm^{-1}, a β-lactan band at 1785 cm^{-1}, and aromatic ring bands at 770 and 695 cm^{-1}. When the benzylpenicillin azide thus obtained has been dried under reduced pressure for 3 days at room temperature, it is quantitatively converted to the corresponding isocyanate, which is also characterized by its infrared spectrum: isocyanate, 2270 cm^{-1}; β-lactam, 1785 cm^{-1}; and aromatic ring, 770 and 700 cm^{-1}.

6-Phthalimidopenicillanic Acid Isocyanate. This substance was prepared according to the method of Perron *et al.*[13] The infrared spectrum showed an intense isocyanate band at 2260 cm^{-1}; phthalimide at 1803 and 1723 cm^{-1}; β-lactam at 1780 cm^{-1}; and aromatic ring at 790 and 720 cm^{-1}.

Isocyanates of other penicillins can be similarly prepared.

[8] R. J. Meadway, *Biochem. J.* **115**, 12p (1969).

[9] V. Csányi, I. Ferencz, and I. Mile, *Biochim. Biophys. Acta* **236**, 619 (1971).

[10] G. V. Patil and R. A. Day, *Biochim. Biophys. Acta* **293**, 490 (1973).

[11] S. G. Waley, *Biochem. J.* **149**, 547 (1975).

[12] H. Ogawara and H. Umezawa, *Biochim. Biophys. Acta* **327**, 481 (1973).

[13] Y. G. Perron, L. B. Crast, J. M. Essery, R. R. Fraser, J. C. Godfrey, C. T. Holdrege, W. F. Minor, M. E. Neubert, R. A. Partyka, and L. C. Cheney, *J. Med. Chem.* **7**, 483 (1964).

β-Lactamases

β-Lactamase$_{75}$ is prepared as described previously[14] whereas β-lactamase of *B. cereus* is obtained commercially (Calbiochem) and used after dialysis against the appropriate buffer. This enzyme is thought to belong to the β-lactamase I group of Kuwabara.[15]

Assay Methods

Method A. β-Lactamase activity is assayed as follows: The reaction mixture in a total volume of 1.0 ml contained 6 mM benzylpenicillin and 0.1% bovine serum albumin in 0.1 M sodium phosphate buffer at pH 7.0. Serum albumin is added to stabilize the β-lactamase enzyme in the course of the activity determination, although it is not essential. A suitable aliquot of enzyme in the above solution is incubated for 30 min at 30°, heated for 1 min in a boiling water bath, and cooled in an ice bath. Thereafter, a 5-ml portion of 50 mM iodine solution is added, and the remaining iodine is determined by measurement of the absorbance at 520 nm.

Method B. For the determination of K_m values, the microiodometric assay of Novick[16] is used. The reaction is carried out at 30° in 0.1 M sodium phosphate buffer at pH 7.4.

Affinity Labeling

Procedure. Affinity labeling of β-lactamase$_{75}$ is carried out at 30° in a reaction mixture composed of the enzyme (approximately 150 nM) and benzylpenicillin isocyanate, which is added in 5 μl of acetone to a total volume of 0.5 ml of 0.1 M sodium phosphate containing 1 mM dithiothreitol at either pH 7.4 or 5.9. The use of dithiothreitol is not essential for the stabilization of the enzyme. In a control, 5 μl of acetone without the isocyanate were added to the incubation mixture. After appropriate incubation, the remaining enzymic activity was determined by method A. Table I summarizes the effect of various isocyanates on the activity of β-lactamase$_{75}$: carbenicillin isocyanate shows the strongest effect in reducing the activity, particularly when allowed to react under acidic conditions. Affinity labeling of a β-lactamase of *B. cereus* is carried out in a similar way.

[14] H. Ogawara, K. Maeda, and H. Umezawa, *Biochim. Biophys. Acta* **289**, 203 (1972).
[15] S. Kuwabara, *Biochem. J.* **118**, 457 (1970).
[16] R. P. Novick, *Biochem. J.* **83**, 236 (1962).

TABLE I
EFFECT OF VARIOUS ISOCYANATES ON THE ACTIVITY OF β-LACTAMASE$_{75}$

Reagents	Concentration (μM)	pH	Duration of reaction (min)	Residual activity (%)
Benzylpenicillin isocyanate	20	7.4	15	56
	10	5.9	30	47
6-Phthalimidopenicillanic acid isocyanate	20	7.4	15	55
	10	5.9	30	74
Carbenicillin isocyanate	20	7.4	15	45
	10	5.9	30	9.5
n-Butylisocyanate	200	7.4	15	92
	10	5.9	30	81
Phenylisocyanate	100	7.4	15	87
Control	—	7.4	15	93
	—	5.9	30	89

Effects of pH. Mixtures of β-lactamase$_{75}$ (approximately 150 nM) and benzylpenicillin isocyanate (10 μM), β-lactamase$_{75}$ (approximately 750 nM) and carbenicillin isocyanate (CPI, 200 μM), or a β-lactamase of *B. cereus* (approximately 200 nM) and carbenicillin isocyanate (200 μM) were incubated at 30° in 1 ml of 0.1 M sodium phosphate at various pH levels. The incubation time for the mixture of β-lactamase$_{75}$ and benzylpenicillin isocyanate was 30 min, and that for the other combinations was 1 hr. The results illustrated in Fig. 1 show that both benzyl-

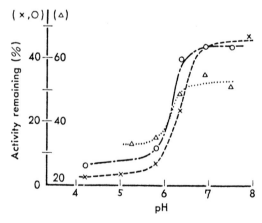

FIG. 1. Effect of pH on the inactivation of β-lactamases by benzylpenicillin isocyanate (BPI) and carbenicillin isocyanate (CPI). △---△, BPI and β-lactamase$_{75}$; ✕---✕, CPI and β-lactamase$_{75}$; ○·-·-○, CPI and a *Bacillus cereus* β-lactamase.

penicillin isocyanate and carbenicillin isocyanate inactivate the enzymic activity more strongly in an acidic condition than in a neutral or an alkaline condition, and that the pK that may be estimated from the data is about 6 to 6.5. This is also the case with β-lactamase$_{75}$ and 6-phthalimidopenicillanic acid isocyanate and with β-lactamase of $B. cereus$ and carbenicillin isocyanate. From the pK of the inactivation reaction in combination with other results, it is suggested that the isocyanates react first with a histidyl residue in its protonated form and then proceed to react with a neighboring group, probably an amino group to form a stable but inactivated protein.

Kinetics. Incubation of β-lactamase$_{75}$ with various concentrations of isocyanates produced a rapid inactivation of the enzyme which followed pseudo-first-order kinetics. The double-reciprocal plot of the rate of inactivation against concentration of benzylpenicillin isocyanate (BPI) showed that the inactivation proceeded by the formation of a reversible enzyme–reagent complex (E·BPI) with K and k_3 of 40 μM and 9×10^{-2} min^{-1}, respectively; K is equal to $(k_{-1} + k_2 + k_3)/k_1$.

$$E + BPI \underset{k_{-1}}{\overset{k_1}{\rightleftharpoons}} E \cdot BPI \overset{k_3}{\rightarrow} E - BPI$$
$$\downarrow k_2$$
$$E + BPI'$$

If the inactivation is due to a chemical reaction on a group at the active site of the enzyme, the inactivation rate should be decreased in the presence of a substrate to an extent depending on the affinity of a substrate for the enzyme. Actually, the presence of benzylpenicillin in the reaction mixture reduced the inactivation rate of the enzyme. In addition, methicillin, but not the alkaline degradation product of methicillin, reduced the inactivation rate of β-lactamase$_{75}$ by either carbenicillin isocyanate (Table II) or 6-phthalimidopenicillanic acid isocyanate (Table III).

TABLE II

EFFECT OF METHICILLIN AND ITS DEGRADATION PRODUCT ON THE
INACTIVATION OF β-LACTAMASE$_{75}$ BY CARBENICILLIN ISOCYANATE (CPI)

Reagents	Concentration (μM)	Residual activity (%)
CPI	0.1	11.8
CPI	0.1	63.2
Methicillin	1.0	
CPI	0.1	15.2
Methicillin degradation product	1.0	
Control	—	84.0

TABLE III
EFFECT OF METHICILLIN AND ITS DEGRADATION PRODUCT ON THE
INACTIVATION OF β-LACTAMASE$_{75}$ BY 6-PHTHALIMIDOPENICILLANIC
ACID ISOCYANATE (PPI)

Reagents	Concentration (μM)	Residual activity $(\%)$
PPI	0.1	15.9
PPI	0.1	54.7
Methicillin	1.0	
PPI	0.1	12.1
Methicillin degradation product	1.0	
Control	—	75.4

Other Affinity Labels

Patil and Day[10] described the irreversible inactivation of β-lactamase from *B. cereus* 569/H with diazotization products of 6-aminopenicillanic acid and ampicillin. Preliminary experiments and β-lactamase$_{75}$ and the diazotization products from many penicillins and cephalosporins showed no inactivation.

Other reagents, such as derivatives of oxacillin and methicillin with a varied length of "methylene arm" attached to the carboxyl group and a reactive group in the terminal region of the arm, did not inactivate β-lactamase$_{75}$.

$n = 1, 2, 3,$ and 4

Recently, Virden *et al.*[17] reported that the substrate quinacillin bound covalently to the active site of β-lactamase from *Staphylococcus aureus* PC1.

Comments

Although a large-scale affinity labeling experiment has not been performed because of a shortage of the β-lactamase enzyme, it is possible

[17] R. Virden, A. F. Bristow, and R. H. Pain, *Biochem. J.* **149**, 397 (1975).

to do a large-scale experiment using penicillin isocyanates and to isolate a peptide containing a labeled amino acid residue.

Benzylpenicillin isocyanate has a β-lactam ring that could be hydrolyzed by the enzyme. When benzylpenicillin isocyanate binds to the catalytic site and hydrolysis and inactivation proceed simultaneously, these reactions should take place through the same reversible complex (E·BPI) as represented above. Although the rate was over 100 times less than that with benzylpenicillin, the hydrolysis of benzylpenicillin isocyanate did occur with concomitant inactivation of the enzyme. The apparent Michaelis constant, $(k_{-1} + k_2 + k_3)/k_1$, was coincident with K (400 μM) obtained in the inactivation reaction described. Other penicillin isocyanates may be hydrolyzed in a similar manner by β-lactamase.

[61] Active Site-Directed Addition of a Small Group to an Enzyme: The Ethylation of Luciferin

By EMIL H. WHITE and BRUCE R. BRANCHINI

In most cases of derivation of enzymes, a major part of the reactive substrate becomes attached to the protein.[1] The appending of a large group to the active site probably accounts for the observation that most modified enzymes prepared in this way are inactive.

Recently, the use of labeling agents that deliver a small group to the enzyme has been reported.[2-4] Nakagawa and Bender have shown that methyl 4-nitrobenzenesulfonate (I) inhibits α-chymotrypsin by the specific methylation of His-57. Hydrolysis produced 2-amino-3-(1-methyl-4-imidazolyl)propanoic acid, indicating that ". . . the hydrogen atom at position 3 of this histidine (No. 57) is essential to the mechanism of hydrolysis of substrates by this enzyme."[2]

[1] (a) B. R. Baker, "Design of Active Site Directed Irreversible Enzyme Inhibitors." Wiley, New York, 1967. (b) E. Shaw, *in* "The Enzymes" (P. D. Boyer, ed.), 3rd ed., Vol. 1, pp. 91–146, Academic Press, New York, 1970. (c) L. A. Cohen, *ibid.,* pp. 147–211. (d) E. Shaw, *Physiol. Rev.* **50**, 244 (1970). (e) General chemical modification of enzymes is discussed in this series, Vol. 25, p. 387 (1972). (f) E. Shaw, this series, Vol. 25, p. 655 (1972). (g) F. C. Hartman, this series, Vol. 25, p. 661 (1972).

[2] (a) Y. Nakagawa and M. L. Bender, *J. Am. Chem. Soc.* **91**, 1566 (1969); (b) *Biochemistry* **9**, 259 (1970).

[3] J. D. Rawn and G. E. Lienhard, *Biochem. Biophys. Res. Commun.* **56**, 654 (1975).

[4] E. H. White and B. R. Branchini, *J. Am. Chem. Soc.* **97**, 1243 (1975).

RSO$_3$R'

(I) R = 4-nitrophenyl   ; R' = CH$_3$

(II$_a$) R = 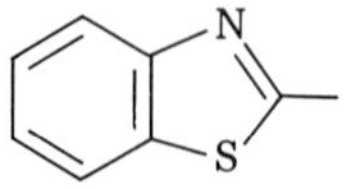   ; R' = C$_2$H$_5$

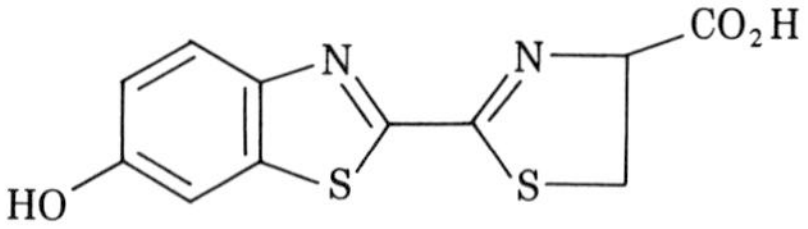

(III)

(II$_b$) R = 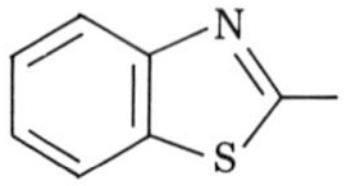   ; R' = (CH$_3$)$_2$CH

(II$_c$) R = 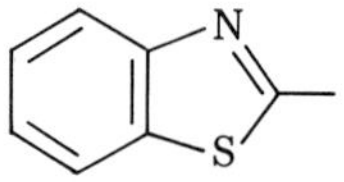   ; R' = CH$_3$

In the present case, a small alkyl group (ethyl) is transferred to firefly luciferase. The resulting modified enzyme is still active, but in a modified way.[4]

$$\text{Firefly luciferase} + \text{compound (IIa)} \rightarrow \text{"ethyl-luciferase"} \qquad (1)$$

The enzyme still catalyzes the oxidation of luciferin, but red light is produced rather than the normal yellow-green. The small size of the ethyl group presumably accounts for the fact that alkylation does not totally inhibit the enzyme, although it is also possible that the enzyme has two active sites and one of them has been fully inhibited. Our interpretation of the color shift was that the enzyme utilized two basic groups, one to initiate the reaction and one to switch the light emission from red to yellow-green, and that our inhibitor preferentially "ethylated" the second group.

In the modification of α-chymotrypsin and firefly luciferase, the agents used to transfer small alkyl groups were sulfonate esters. In the sulfonate esters, unlike their carboxylic acid counterparts, the O-alkyl carbon atom is susceptible to nucleophilic attack [Eq. (2)].[5]

$$X - \overset{H}{\underset{R}{\text{C}}} - O - SO_2 - Ar \longrightarrow X - \overset{H}{\underset{R}{\text{C}}} - H + ArSO_3^- \qquad (2)$$

For this reason, sulfonate esters are effective alkylating agents by the simple SN$_2$ displacement mechanism. Thus, any enzyme that contains a

[5] (a) C. M. Suter, "The Organic Chemistry of Sulfur." Wiley, New York, 1944. (b) J. Hine, "Physical Organic Chemistry," 2nd ed. McGraw-Hill, New York, 1974.

nucleophilic amino acid residue essential for its normal functioning is, in theory, susceptible to modification by sulfonate esters. In order to select for important active site residues, one would design the sulfonic acid portion of the molecule to interact favorably with the substrate binding area of the enzyme. In this manner, ethyl 2-benzothiazolesulfonate (IIa), an analog of the natural substrate firefly luciferin (III), was used to direct a small group (C_2H_5) onto the active site of firefly luciferase, leaving it active, but modifying its chemical properties.

Synthesis and Stability. Sulfonate esters are conveniently prepared from precursor sulfonyl halides and the appropriate alcohol or alkoxide [Eq. (3)]; this synthetic pathway is particularly suitable for the introduction of a ^{14}C label into the alkylating moiety since radioactive primary alcohols can be readily synthesized.[6]

$$Ar-SO_3H \xrightarrow{PCl_5} Ar-SO_2Cl \xrightarrow[NaOR]{R-OH} Ar-SO_3R \tag{3}$$

A alternative approach to the preparation of the sulfonyl halide was used in the preparation of 2-benzothiazolesulfonyl chloride [Eq. (4)].[7]

$$\tag{4}$$

The sulfonate esters decompose in aqueous solutions to the corresponding sulfonic acids and alcohols. The rate of decomposition of the sulfonate esters of primary alcohols depends on the stability of the sulfonate ion formed upon hydrolysis.[8] For example, in sodium phosphate at pH 7.9 and 25°, ethyl 2-benzothiazolesulfonate (IIa) and methyl 4-nitrobenzenesulfonate (I) have half-lives of about 12 and 60 min, respectively. Isopropyl 2-benzothiazolesulfonate (IIb) proved to be very reactive to

(IV$_c$)

water, presumably because of the availability of an ionization type decomposition (Sn 1 pathway). Attempts to prepare methyl 2-benzothia-

[6] R. B. Wagner and H. D. Zook, "Synthetic Organic Chemistry." Wiley, New York, 1953.
[7] R. O. Roblin, Jr. and J. W. Clapp, *J. Am. Chem. Soc.* **72**, 4890 (1950).
[8] M. S. Morgan and L. H. Cretcher, *J. Am. Chem. Soc.* **70**, 375 (1948).

zolesulfonate (IIc) led to the thiazolium salt (IVc) presumably because of the occurrence of a chain displacement reaction. The ethyl ester (IIa) also rearranges (to IVa), but at a slower rate (weeks at 25°, or about 10 hr at 55°).

Preparations

2-Mercaptobenzothiazole. Commercial mercaptan was purified prior to use by dissolving in 60% aqueous ammonium hydroxide, filtering insoluble material, and precipitating the mercaptan by addition of 50% sulfuric acid. The desired product was dried in a vacuum oven; m.p. 177°–178° (lit.[9] m.p. 179°).

2-Benzothiazolesulfonyl Chloride. The crude material was prepared by oxidative chlorination of 2-mercaptobenzothiazole according to the method of Roblin and Clapp.[7] The desired sulfonyl chloride was extracted into boiling petroleum ether. However, nearly 90% of the crude reaction mixture consisted of an uncharacterized impurity. Attempts to minimize impurity formation by varying reaction time and using vigorous stirring were futile. The solid residue obtained by evaporating the petroleum ether extract to dryness under reduced pressure was recrystallized three times from petroleum ether to yield yellow needles in 6% yield; m.p. 107°–110° (lit.[7] 108°–110°); IR (CS_2) 1395, 1320, 1190, 1180, and 1080 cm^{-1}.

Ethyl 2-Benzothiazolesulfonate (IIa). A solution of 0.75 M sodium ethoxide in absolute ethanol (6.1 ml, 4.6 mmoles) was pipetted into a round-bottom flask containing 2-benzothiazolesulfonyl chloride (1.05 g, 4.5 mmoles) in 7 ml of absolute ether at 0°. The mixture was stirred under nitrogen for 2 min, and the solvents were removed by vacuum distillation at 0° (using a receiver cooled by Dry Ice or liquid nitrogen). Ethyl acetate (5 ml) and hexane (30 ml) were added to the solid residue, and the resulting suspension was filtered. The solvents were evaporated under reduced pressure at 0°, and the white residue obtained was recrystallized from hexane to give white needles (500 mg, 2.06 mmoles, 46%): m.p. 73.5°–74.5°; IR (CS_2) 1380, 1185, 945, and 725 cm^{-1}; UV (CH_2Cl_2) λ_{max} (log ϵ) 274 nm (3.98), 238 (3.95), 229 (3.93), and 298 (sh) (3.54); NMR ($CDCl_3$) δ 1.44 (t, 3H), 4.72 (q, 2H), and 7.44–8.52 (m, 4H). The material was stored at −25° to prevent thermal rearrangement.

---

[9] "Handbook of Chemistry and Physics" (R. C. Weast, ed.). Chem. Rubber Publ. Co., Cleveland, Ohio, 1966.

Inhibition of Firefly Luciferase with Ethyl 2-Benzothiazolesulfonate

Luciferase activity was assayed by measuring the height of the initial flash of light emitted upon rapid injection of excess ATP into a solution of luciferase[10] and excess luciferin.[11] The assay system consisted of 2.0 ml of 0.025 M glyclglycine at pH 7.9, 0.1 ml of 0.1 M magnesium sulfate, 0.1 ml of 1.2 mM luciferin, 15 μl of 15 μM luciferase (or modified luciferase), and 0.2 ml of 0.02 M ATP. The modified enzyme was prepared by adding 1.5 μl of a 0.3 M solution of ethyl 2-benzothiazole sulfonate in tetrahydrofuran to 0.5 ml of 15 μM luciferase in 0.05 M sodium phosphate at pH 7.9. A second aliquot on inhibitor solution is added about 30 min later for a total reaction time of 1–2 hr. The reaction solutions are initially cloudy, clarity usually returning within minutes except when large excesses of inhibitor are used.

Light emission intensities are determined with a photomultiplier (RCA IP 21) powered by a fluke 4128 dc power supply. The signal from the photomultiplier is fed into a Keithley Instruments Model 610C electrometer and recorded with a Technirite high-speed recorder.

During the reaction of ethyl 2-benzothiazolesulfonate with luciferase, the color of light emitted during the assay changed from the normal yellow-green to red. Experimental details have been presented elsewhere.[4]

[10] A. A. Green and W. D. McElroy, *Biochim. Biophys. Acta* **20**, 170 (1956).
[11] W. D. McElroy and H. H. Seliger, *in* "Light and Life" (W. D. McElroy and B. Glass, eds.), p. 219. Johns Hopkins Press, Baltimore, Maryland, 1961.

[62] Mandelate Racemase

By George L. Kenyon and George D. Hegeman

Mandelate racemase (EC 5.1.2.2) of *Pseudomonas putida* catalyzes the reversible interconversion of the D and L enantiomers of mandelate:

$$\overset{\ominus}{}O_2C \blacktriangleright \overset{\displaystyle OH}{\underset{\displaystyle C_6H_5}{C}} \blacktriangleleft H \rightleftharpoons H \blacktriangleright \overset{\displaystyle OH}{\underset{\displaystyle C_6H_5}{C}} \blacktriangleleft CO_2^{\ominus}$$

The enzyme has been isolated in a high state of purity[1-3] and has been shown to be a tetramer composed of identical subunits, each with a molecular weight of 69,500.[3] The enzyme acts without flavin or pyridine nu-

[1] G. D. Hegeman, this series, Vol. 17, p. 670.
[2] G. D. Hegeman, E. Y. Rosenberg, and G. L. Kenyon, *Biochemistry* **9**, 4029 (1970).
[3] J. A. Fee, G. D. Hegeman, and G. L. Kenyon, *Biochemistry* **13**, 2528 (1974).

cleotide cofactor, but has an absolute divalent metal ion requirement (e.g., Mg^{2+}) for activity.[3] Mechanistic studies[4,5] have indicated that the hydrogen alpha to the carboxylate is removed as a proton in the course of the enzyme-catalyzed racemization, generating a carbanion intermediate; the intermediate can be represented by the resonance structures shown in Eq. (1).

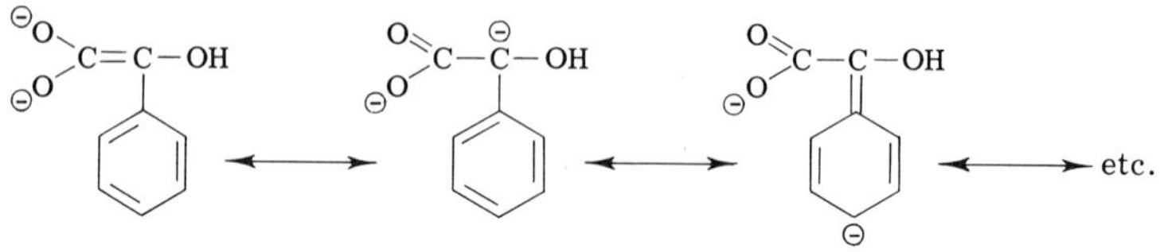

$$\tag{1}$$

It followed that a basic group on the enzyme itself was participating in the generation of the carbanion intermediate by abstracting the α-hydrogen from mandelate. In an attempt to identify such a basic group on the enzyme, structural analogs of mandelate were sought that could also serve as alkylating agents. These were screened as potential affinity labels for the enzyme.[6,7]

Design of the Affinity Labeling Reagent

A variety of structural analogs of mandelate had earlier been tested as potential substrates or reversible inhibitors of the racemase.[2,5,6] Consequently, much was known about the minimal structural requirements for binding to the enzyme. It was clear that both an aromatic ring and a carboxylate group (or other similar anionic group) generally aided in this binding. Therefore, initial candidates for the affinity label all included both a phenyl and carboxylate substituent.

Chloro- or bromoethylketone derivatives were not considered because of the anticipated incompatibility of these alkylating groups with the carboxylate anion. The idea of incorporating an epoxide group into the affinity label was triggered by the discovery of the Merck group[8-10] that

[4] G. L. Kenyon and G. D. Hegeman, *Biochemistry* 9, 4036 (1970).
[5] E. T. Maggio, G. L. Kenyon, A. S. Mildvan, and G. D. Hegeman, *Biochemistry* 19, 1131 (1975).
[6] Judy A. Fee, Ph.D. Dissertation, University of California, Berkeley, California, 1973.
[7] J. A. Fee, G. D. Hegeman, and G. L. Kenyon, *Biochemistry* 13, 2533 (1974).
[8] D. Hendlin, E. O. Stapley, M. Jackson, H. Wallick, A. K. Miller, F. J. Wolf, T. W. Miller, L. Chaiet, F. M. Kahan, E. L. Foltz, H. B. Woodruff, J. M. Mata, S. Hernandez, and S. Mochales, *Science* 166, 122 (1969).

the antibiotic fosfomycin (previously "phosphonomycin") was acting as an affinity label for the enzyme, phosphoenolpyruvate-uridine-5'-diphospho-N-acetyl-2-amino-2-deoxyglucose 3-0-pyruvyltransferase. The epoxide group seemed ideally suited for our purposes. It is compatible with

$$H_3C-\overset{H}{\underset{}{C}}\overset{O}{\diagup\diagdown}\underset{}{C}\overset{H}{\underset{PO_3^{2-}}{}}$$

Fosfomycin dianion

anions, is reasonably stable chemically at a basic pH, and it has been used successfully as an alkylating group incorporated into affinity labels for enzymes.

The following series of epoxides was therefore synthesized[6]:

$$C_6H_5-\overset{H}{\underset{}{C}}\overset{O}{\diagup\diagdown}\underset{H}{\overset{CO_2^{\ominus}}{C}}$$

trans-β-Phenyl-glycidate

$$C_6H_5-\overset{\ominus OOC}{\underset{}{C}}\overset{O}{\diagup\diagdown}\underset{H}{\overset{H}{C}}$$

α-Phenyl-glycidate

$$C_6H_5-\overset{H}{\underset{}{C}}\overset{O}{\diagup\diagdown}\underset{H}{\overset{H}{C}}$$

Styrene oxide

$$H-\overset{H}{\underset{H}{C}}\overset{O}{\diagup\diagdown}\underset{H}{\overset{CO_2^{\ominus}}{C}}$$

Glycidate

Both trans-β-phenylglycidate and α-phenylglycidate were separately examined as potential irreversible inhibitors by incubating 3 mM of each with a small amount of racemase (0.5 mM) at 30° and pH 7.0.[6] After 1 hr, about 25% of the initial enzymic activity remained after incubation with trans-β-phenylglycidate. After only 30 min, enzyme incubated with the α-phenylglycidate was entirely inactive. After exhaustive dialysis to remove the α-phenylglycidate, racemase activity did not return. Preliminary studies also showed that the presence of mandelate itself slowed the inactivation process.

Under similar conditions, both styrene oxide and glycidate were also examined as potential irreversible inhibitors of the racemase. Styrene oxide did show partial inhibition, but glycidate was not inhibitory.

---

[9] B. G. Christensen, W. J. Leanza, T. R. Beattie, A. A. Patchett, B. H. Arison, R. E. Ormond, F. A. Kuehl, Jr., G. Albers-Schonberg, and O. Jardetzky, *Science* **166**, 123 (1969).

[10] P. J. Cassidy and F. M. Kahan, *Biochemistry* **12**, 1364 (1973).

All efforts were therefore concentrated on examining the potential of α-phenylglycidate as an affinity label. Structurally it is very similar to mandelate. Moreover, the-CH_2-group of the epoxide ring seemed strategically placed to intercept the key basic group postulated to be involved in the α-proton extraction process associated with enzymic catalysis:

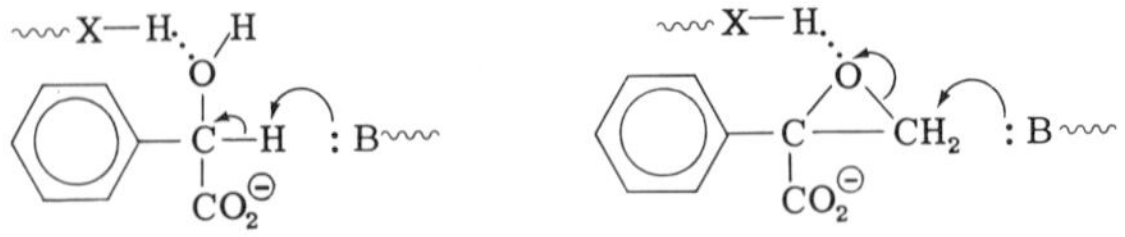

In both cases a specific hydrogen bond to the oxygen attached to the α-carbon could be assisting attack by the hypothetical base ~B. α-Phenylglycidate had another very attractive structural feature: an asymmetric center at the same relative position as the substrate.

Synthesis of Radioactivity Labeled Affinity Label

A convenient synthesis of the inhibitor, adaptable to a small scale and by which an appropriate radioactively labeled precursor could be introduced, was selected[6,7]:

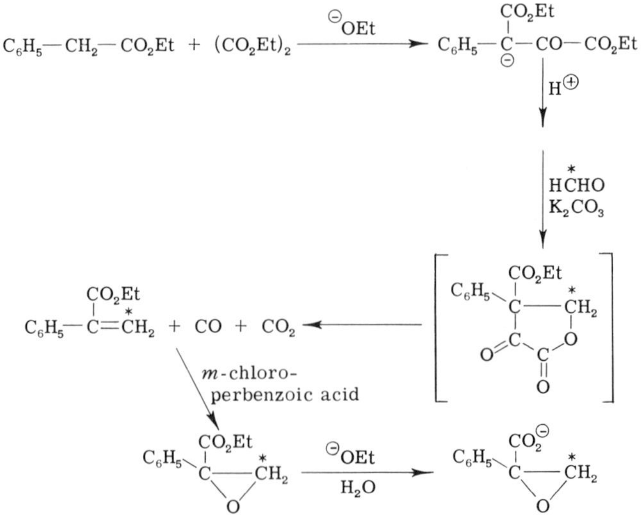

This scheme has the following advantages: radioactively labeled formaldehyde of high specific activity is available commercially at a reasonable cost; only two intermediates containing this radioactive label need be isolated; all the steps give reasonably good yields.

Synthesis of Sodium α-Phenylglycidate

The synthesis of ethyl atropate was adapted from the procedure of Ames and Davey.[11] Freshly distilled diethyl oxalate (Eastman, 109.5 g) and ethyl phenylacetate (Eastman, 162.9 g) are successively added to 51.5 g of sodium ethoxide (freshly prepared from Na metal and ethanol) in 300 ml of benzene. After standing overnight, the sodium salt of diethyl phenyloxaloacetate separates from the solvent. This product is washed with ether and acidified with concentrated H_2SO_4 to yield diethyl phenyloxaloacetate. At this stage, 90 ml of 38% aqueous formaldehyde and 300 ml of H_2O are added at 15°. Then, 81 g of K_2CO_3 in 150 ml of water are added over a period of 30 min. After standing for about 4 hr, the reaction mixture is extracted with ether to yield 116 g of ethyl atropate, b.p. 76°–77° (1–2 mm).

Following the procedure of Singh and Kagan,[12] ethyl atropate is converted to ethyl α-phenylglycidate by treatment with a slight excess of 85% m-chloroperoxybenzoic acid (Aldrich) in $CHCl_3$ at reflux for 20 hr. The solution is cooled, extracted with 5% aqueous sodium bicarbonate, and dried. The crude product, obtained in about 90% yield, is purified by silica gel chromatography using petroleum ether as eluent. Ethyl phenylacetate emerges just ahead of ethyl α-phenylglycidate.

Saponification of the ethyl α-phenylglycidate is performed by treating with one equivalent of sodium ethoxide in water, again according to Singh and Kagan.[12] After standing overnight, the solid product is filtered, washed with ether, and dried. Sodium α-phenylglycidate is obtained in 96% yield. The product may be recrystallized from ethanol giving analytically pure needles, m.p. > 300° (sintering at 260°).[12]

The synthesis of the radioactively labeled sodium α-phenylglycidate is similar and is presented elsewhere.[7]

Determination of the Stoichiometry of Binding of the Affinity Label to the Racemase

The stoichiometry of binding of sodium [³H]α-phenylglycidate to the enzyme was determined to be 0.98 mmole of α-phenylglycidate per millimole of subunit of racemase.[7] During storage, radioactivity associated with the enzyme decreased, either by loss of the α-phenylglycidate label or by exchange with the medium. To distinguish between these two possibilities, racemase was labeled with sodium [¹⁴C]α-phenylglycidate

[11] G. R. Ames and W. Davey, J. Chem. Soc. London 1958, 1794 (1958).
[12] S. P. Singh and J. Kagan, J. Org. Chem. 35, 2203 (1970).

synthesized in a completely analogous manner to that of the tritiated affinity label. The results showed that 1.09 moles of α-phenylglycidate were bound per mole of subunit of racemase.[7] Unlike the case of the [³H]α-phenylglycidate, no loss of radioactivity from the blocked enzyme was observed even after prolonged storage. It was, therefore, concluded that the tritium of the [³H]α-phenylglycidate-blocked enzyme was slowly exchanging with the medium without loss of the affinity label. This interpretation was also consistent with the fact that no enzymic activity returned upon prolonged storage of the blocked racemase.

For subsequent studies aimed at determining which amino acid residue (or residues) is alkylated by the affinity label, [¹⁴C]α-phenylglycidate was used exclusively.[7]

Kinetic Studies of the Irreversible Inhibition

α-Phenylglycidate apparently forms a complex with the enzyme since a plot of $t_{1/2}$ (determined from the initial portion of semilogarithmic plot of activity vs time) vs the reciprocal of various inhibitor concentrations[13] had a y intercept at a maximal inactivation half-time of 1.34 min, corresponding to a first-order rate constant, k, of 8.61×10^{-3} sec^{-1}.[3,7] The inactivation process follows first-order saturation kinetics.

In a separate series of kinetic studies, the inhibition of racemase by α-phenylglycidate was studied in the presence of various concentrations of substrate; DL-mandelate showed strict competitive inhibition of the labeling process.

These kinetic studies on mandelate racemase were interpreted by assuming a simple Michaelis–Menten kinetic scheme for this "one-substrate" enzymic reaction. This assumption has recently received some experimental verification.[5]

Effect of Divalent Metal Ion on the Affinity Labeling Process

Mandelate racemase, free of metals, is completely devoid of catalytic activity, but full activity can be restored to this inactive enzyme by the addition of divalent metal ion (e.g., Mg^{2+}).[3] The affinity labeling process by α-phenylglycidate was shown to have a parallel divalent metal ion requirement. Without added divalent metal ion, little or no inactivation occurs in the presence of α-phenylglycidate even after prolonged incubation.[7] In both the normal enzyme-catalyzed reaction and the affinity labeling process, the divalent metal ion presumably is situated close to

[13] H. P. Meloche, *Biochemistry* 6, 2273 (1967).

the substrate or inhibitor and promotes attack by the key basic group of the enzyme.[5]

Attempts to Identify the Residue of the Racemase Alkylated by the Affinity Label

A number of affinity labels have been found to alkylate either glutamate or aspartate residues of enzymes.[14–18] The resulting carboxylate ester linkages are generally base-labile. Such carboxylate ester groups can generally be cleaved even more rapidly in the presence of basic NH_2OH solution. To examine the possibility that affinity labeling of mandelate racemase by α-phenylglycidate involved formation of such an ester linkage, racemase (0.5 mg/ml) labeled with [^{14}C]α-phenylglycidate was incubated at 33° for 38 hr in 7 M urea and 2M NH_2OH at pH 9.[6] During this period, 20% of the label was lost. In a separate experiment, labeled enzyme was incubated under identical conditions except that the NH_2OH was deleted; only 10% of the label was lost. At least a portion of the radioactivity released from enzyme cocrystallized three times with an authentic sample of the expected diol, α-phenylglyceric acid. This result suggests that at least a portion of the α-phenyglycidate alkylates either

$$\begin{array}{c} \text{OH} \\ | \\ C_6H_5\text{---}C\text{---}CH_2OH \\ | \\ COOH \end{array}$$

α-Phenylglyceric acid

an aspartate or glutamate residue of the enzyme.

Compared to similar cases,[17,19–22] however, NH_2OH-induced release of label from mandelate racemase appears to be rather slow. This may mean either that the active site of blocked racemase is relatively inaccessible to small molecules such as NH_2OH or that the α-phenyglycidate alkylates more than one type of amino acid residue.[6]

[14] G. Legler and S. N. Hasnain, *Hoppe-Seyler's Z. Physiol. Chem.* **351**, 25 (1970).

[15] J. A. Hartsuck and J. Tang, *J. Biol. Chem.* **247**, 5248 (1972).

[16] J. Moult, Y. Eshdat, and N. Sharon, *J. Mol. Biol.* **75**, 1 (1973).

[17] K. J. Schray, E. L. O'Connell, and I. A. Rose, *J. Biol. Chem.* **248**, 2214 (1973).

[18] E. L. O'Connell and I. A. Rose, *J. Biol. Chem.* **248**, 2225 (1973).

[19] G. M. Hass, M. A. Govier, D. T. Grahn, and H. Neurath, *Biochemistry* **11**, 3787 (1972).

[20] T. H. Plummer, Jr., *J. Biol. Chem.* **246**, 2930 (1971).

[21] G. M. Hass and H. Neurath, *Biochemistry* **10**, 3541 (1971).

[22] I. A. Rose and E. L. O'Connell, *J. Biol. Chem.* **294**, 6548 (1969).

Caution must be exercised in interpreting loss of label induced by NH$_2$OH or base as being indicative of modification of an aspartate or glutamate residue.[6] In the affinity labeling of pepsin by 1,2-epoxy-3-(p-nitrophenoxy)propane, for example, two residues of the enzyme were alkylated.[23] Since all label could be removed by prolonged incubation in base, these two residues were initially identified as carboxylate groups.[23] Subsequently, it was shown that the two residues were actually methionine and aspartate.[24] It is also possible that in basic solutions NH$_2$OH can cleave peptide linkages; the bond between asparagine and glycine is especially susceptible to this treatment.[25] In the presence of base and oxygen, NH$_2$OH can decompose to a number of products, including peroxonitrite and H$_2$O$_2$,[26] which could complicate a desired specific attack on a given carboxylate ester linkage.

[23] J. Tang, *J. Biol. Chem.* **246**, 4510 (1971).
[24] K. C. S. Chen and J. Tang, *J. Biol. Chem.* **247**, 2566 (1972).
[25] P. Borstein, *Biochemistry* **9**, 2408 (1970).
[26] M. N. Hughes and H. G. Nicklin, *J. Chem. Soc. A* **1971**, 164 (1971).

[63] Dimethylpyrazole Carboxamidine and Related Derivatives

By FRANK DAVIDOFF and PARLANE REID

The frequency and great biological importance of divalent metal binding sites on proteins makes them of particular interest as the object of affinity labeling studies. Organic cations of many types interact with metal ion sites in biological materials[1]; the binding of amidine and guanidinium compounds at these sites has been extensively studied.[2] In 1972, a systematic exploration was undertaken[3] of the interaction of phenethylbiguanide and related compounds with the glycolytic enzyme pyruvate kinase (ATP:pyruvate phosphotransferase, EC 2.7.1.40). This enzyme is activated by K$^+$ and similar monovalent cations, and is dependent for catalytic activity upon Mn^{2+} or Mg^{2+} bound at the active site. Even though phenethylbiguanide is monovalent at physiologic pH, this compound demonstrates strictly competitive kinetics with Mn^{2+} or Mg^{2+}, but is noncompetitive with K$^+$. Structure–activity studies indi-

[1] R. L. Marois and C. Edwards, *Fed. Proc., Fed. Am. Soc. Exp. Biol.* **28**, 669 (1968); B. C. Pressman and J. J. Park, *Biochem. Biophys. Res. Commun.* **11**, 182 (1963); M. Otsuka and M. Endo, *J. Pharmacol. Exp. Ther.* **128**, 273 (1960).
[2] F. Davidoff, *N. Engl. J. Med.* **289**, 141 (1973); F. N. Fastier, *Pharmacol. Rev.* **14**, 37 (1962).
[3] F. Davidoff and S. Carr, *Proc. Natl. Acad. Sci. U.S.A.* **69**, 1957 (1972).

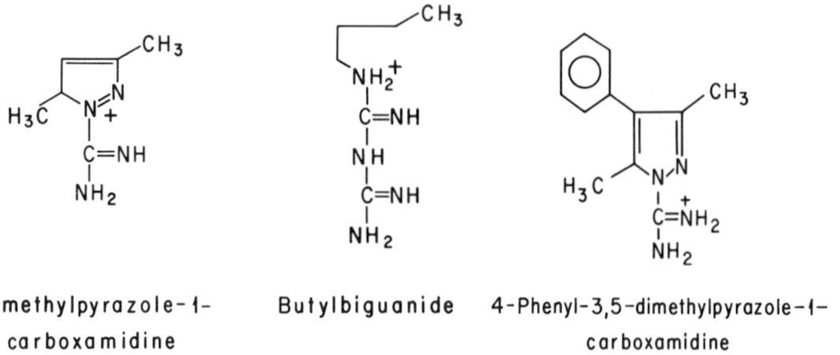

3,5-Dimethylpyrazole-1-   Butylbiguanide   4-Phenyl-3,5-dimethylpyrazole-1-
carboxamidine   carboxamidine

FIG. 1. Structures of 3,5- and 4-phenyl-3,5-dimethylpyrazole-1-carboxamidine and of butylbiguanide.

cate that the largest proportion of binding energy is furnished by the hydrophobic residue of the phenethylbiguanide molecule.[3]

In 1960, 3,5-dimethylpyrazole-carboxamidine (DMPA) was introduced as a guanidinating reagent for proteins.[4] In comparison with O-methylisourea and S-methylpseudourea, which had previously been employed for this purpose, the carboxamidine provides several advantages: greater stability in water, and selective reaction with α- and ε-amino nitrogens of proteins at lower pH values, i.e., below 10. Inspection of the structures of 3,5-dimethylpyrazole-carboxamidine and butylbiguanide (Fig. 1) reveals considerable similarity, suggesting that the carboxamidine compound should also bind competitively at the divalent metal ion site of proteins such as pyruvate kinase. Whereas the guanidinium residue of butylguanidine is unreactive, the amidine carbon of the carboxamidine derivative is rendered labile by its attachment to the stable pyrazole ring. It seems reasonable, therefore, that after reversible and selective binding of the carboxamidine compound to the divalent ion site in pyruvate kinase in a rapid step, the amidine carbon should be subject to slower covalent attack by ε-aminolysine groups in the vicinity, thus providing an affinity labeling reagent. The experimental data indicate that both of these steps occur[5]; we have also extended the principles involved in this affinity labeling technique to mitochondria by using a more hydrophobic derivative, 4-phenyl-3,5-dimethylpyrazole-1-carboxyamidine.

Preparation of Reagents

3,5-Dimethylpyrazole-1-carboxamidine is available commercially as the nitrate and is easily recrystallized from ethanol. This reagent may also be synthesized without difficulty by condensation of aminoguanidine with

[4] A. F. S. A. Habeeb, Can. J. Biochem. Physiol. 38, 493 (1960).

[5] F. Davidoff, S. Carr, M. Lanner, and J. Leffler, Biochemistry 12, 3017 (1973).

pentane-2,4-dione.[6] The hydrophobic analog, 4-phenyl-3,5-dimethylpyra-
zole-1-carboxamidine (Fig. 1), is synthesized in similar fashion by condensation of aminoguanidine with 3-phenylpentane-2,4-dione in 50%
ethanol. The final product melts at 171° and shows an absorption maximum at 246 nm as compared with the peak of 234 nm for DMPA (sodium
borate at pH 9.5). NMR spectrum indicates that the structure is as in
Fig. 1.

The synthesis of these pyrazole derivatives labeled in the amidine
carbon with ^{14}C has been accomplished by allowing [^{14}C]cyanamide to
react with hydrazine in water.[7] After crystallization, the labeled aminoguanidine so produced is then condensed with either of the pentanediones
to produce the labeled pyrazolecarboxamidines.

Preparation of Affinity-Labeled Pyruvate Kinase

Enzyme Preparation. This enzyme from rabbit muscle is available
commercially (Boehringer Manheim Corp.) or may be prepared by the
method of Tietz and Ochoa.[8] Further purification may be achieved by
the technique of Cottam *et al.*[9] Pyruvate kinase activity is conveniently
assayed by coupling the reduction of pyruvate, produced from phosphoenolpyruvate, to lactate, catalyzed by lactate dehydrogenase. The
NADH consumed in this reaction is assayed spectrophotometrically.[3]
Studies of reversible inhibition are carried out directly in the reaction
cuvette, with appropriate controls to eliminate the possibility that inhibition is due to inactivation by DMPA of the lactate dehydrogenase or
other assay reagents.

Affinity Labeling. Affinity labeling of pyruvate kinase, which results
in irreversible inactivation, is carried out by incubating the enzyme in a
relatively small volume of buffer with the chosen concentration of
DMPA. To assay for residual enzyme activity, small aliquots of pyruvate kinase are withdrawn at increasing time intervals and diluted successively into 0.05 M Tris buffer, then into the final assay cuvette containing 1 mM Mn^{2+} and 75 mM K^{+} but no inhibitor. An overall enzyme
dilution of between 1500- and 6000-fold is easily achieved, thereby bringing the final concentration of DMPA in the assay mixture well below
that which inhibits reversibly.

Reversible Inhibition. We have found that DMPA inhibits the enzyme
competitively with Mn^{2+} or Mg^{2+}; with 0.16 mM Mn^{2+} as the activating

[6] F. L. Scott and J. Reilly, *J. Am. Chem. Soc.* **74,** 4562 (1952).

[7] J. T. Thurston, *Inorg. Synth.* **3,** 45 (1950).

[8] A. Tietz and S. Ochoa, *Arch. Biochem. Biophys.* **78,** 477 (1958).

[9] G. L. Cottam, P. F. Hollenberg, and M. J. Coon, *J. Biol. Chem.* **244,** 1481 (1969).

metal, 50% inhibition of enzyme activity is observed at a DMPA concentration of 36 mM. Using the kinetic techniques described by Levenberg et al.,[10] the K_I for DMPA is found to be 18 mM.

Factors Affecting Affinity Labeling. Pyruvate kinase incubated with DMPA for more prolonged periods of time is irreversibly inactivated. The characteristics of the inactivation process are as follows:

1. Activity loss is linear with time on a semilog plot; incubation of enzyme with 160 mM DMPA at pH 7.5 and 20° leads to 50% inactivation in 1.2 hr. At 0°, 50% inactivation occurs in about 2 hr.

2. Inclusion of 5 mM Mn^{2+} markedly prolongs the 50% inactivation time as does Ca^{2+}, which is known to bind competitively at the Mn^{2+} site.[11] This protective action of specific ligands is one of the major criteria for the recognition of an affinity labeling process.[12]

3. High concentrations of the monovalent cations K$^+$ and NH$_4^+$ also protect the enzyme from irreversible inactivation. The exact mechanism of this effect is not understood, but it is probably related to the partial protective effect of monovalent cations against reversible Ca^{2+} inhibition of the enzyme.[13] Since activating monovalent cations result in a "V" rather than a "K" effect kinetically,[14] and since such cations actually reduce the binding affinities of the enzyme for divalent metals,[15] the protection is probably due to a change in protein conformation that ultimately slows the rate of attack on the bound carboxamidine, e.g., by increasing the pK_a of specific lysine ϵ-amino groups in the vicinity, or by increasing their distance from the site of bound reagents.

4. The rate of irreversible inactivation increases as a function of DMPA concentration, but the change in rate is not a linear function of concentration. The ratio of inactivation rates at concentrations equal to 2K_I and K_I is about 1.20 ± 0.15 ($n = 4$), which does not differ significantly from the value of 1.3 associated with an affinity labeling process.[16]

5. Irreversible inactivation is a linear function of time at all DMPA concentrations, even as low as 8 mM, when the concentration of pyruvate kinase in the incubation mixture is kept below 0.3 mg/ml. At higher protein concentrations, a significant lag period in irreversible inactivation occurs, possibly because protein–protein interactions protect the enzyme.

[10] B. Levenberg, I. Melnick, and J. M. Buchanan, *J. Biol. Chem.* **225**, 163 (1957).
[11] A. S. Mildvan and M. Cohn, *J. Biol. Chem.* **240**, 238 (1965).
[12] B. R. Baker, "Design of Active-Site-Directed Irreversible Enzyme Inhibitors," p. 126. Wiley, New York, 1967.
[13] J. F. Kachmar and P. D. Boyer, *J. Biol. Chem.* **200**, 669 (1953).
[14] L. J. DeAsua, E. Rozengurt, and H. Carminatti, *J. Biol. Chem.* **245**, 3901 (1970).
[15] J. Reuben and M. Cohn, *J. Biol. Chem.* **245**, 6539 (1970).
[16] B. R. Baker, *J. Biol. Chem.* **245**, 166 (1970).

6. The guanidination rate of model compounds and of α- and ϵ-amino nitrogens in proteins by DMPA increases with increasing pH,[4] indicating that the reaction probably involves attack by unprotonated amino nitrogen on the N-C bond between the pyrazole ring and the amidine carbon. The rate of irreversible inactivation of pyruvate kinase by DMPA is strongly pH-dependent, the $t_{1/2}$ for inactivation decreasing about 1000-fold from $\sim$100 to 0.1 hr over the range of pH 6.5–9.5. Inactivation at pH values above 9.5 is difficult to accomplish since the enzyme begins to precipitate visibly. Since the pK_a of ϵ-aminonitrogens in proteins is 9.6–9.8,[17] while that of DMPA is 8.3,[5] these data indicate that the non-protonated form of ϵ-aminonitrogen, the reagent, or both are involved in irreversible enzyme inactivation and support the relationship between affinity labeling and the inactivation process.

7. The hydrophobic 4-phenyl-DMPA derivative (PDMPA) also irreversibly inactivates pyruvate kinase. Indeed, the half-time of inactivation is significantly more rapid than with DMPA itself ($t_{1/2}$ = 20 min for PDMPA, 75 min for DMPA at 33 mM, 20°, pH 8).

Characteristics of the Affinity-Labeled Enzyme. Kinetic studies of the inactivated enzyme indicate that with loss of activity V_{max} decreases, but the apparent K_m for Mn^{2+} remains unchanged. The modified enzyme, therefore, does not lose activity because of partial reduction in affinity for divalent metals; rather, divalent metal affinity is either lost altogether, or, alternatively, divalent metal affinity does not change, but catalytic activity is destroyed because of alterations in active site function unrelated to divalent metals.

For analysis of homoarginine formation, the enzyme is separated from DMPA either by precipitation and washing of the protein,[18] or by passing the incubation mixture over a column of Sephadex G-25 equilibrated with 10 mM NH_4HCO_3 at pH 7.4. The protein content of the eluate is determined spectrophotometrically at 280 nm, with correction for the anomalously low absorbance of the protein at this wavelength.[19] The protein is then lyophilized and hydrolyzed in 6 N HCl at 110° under reduced pressure, the HCl is removed, and aliquots are analyzed on a Spinco amino acid analyzer, Model 121. Arginine and homoarginine may conveniently be quantitated by cutting out and weighing peaks using standards of authentic amino acids.

Native pyruvate kinase contains no detectable homoarginine. During the irreversible inactivation associated with prolonged exposure to DMPA, increasing quantities of homoarginine are found. The number

[17] B. L. Vallee and W. E. C. Wacker, "The Proteins" (H. Neurath, ed.), 2nd ed., Vol. 5, p. 61. Academic Press, New York, 1970.
[18] W. J. Ray, Jr. and D. E. Koshland, Jr., *J. Biol. Chem.* **237**, 2493 (1962).
[19] J. Bücher and G. Pfleiderer, this series, Vol. 1, p. 435 (1955).

of lysine residues converted to homoarginine during the loss of the initial 75% of enzyme activity is a very small fraction of the total. Thus, with native arginine as an internal standard and assuming the mole content of arginine to be 113,[9] a 50% loss of enzyme activity is accompanied by the accumulation of a total of 2 ± 1 mole of homoarginine/237,000 g of enzyme.

This estimate of homoarginine includes not only lysines guanidinated in the affinity labeling process, but also nonspecific conversions occurring elsewhere on the molecule, the latter process apparently occurring much more slowly but reaching significant proportions. To refine the estimate of specific active site-related homoarginine formation, a differential protection technique may be employed. In this procedure, two samples of enzyme are incubated in parallel, both containing DMPA, but only one containing a high concentration of a specific competitive protector ligand. When this is done with pyruvate kinase, the differential, hence specific, homoarginine formation corresponding to complete, irreversible loss of enzyme activity does not differ significantly from 2 moles of homoarginine/237,000 g of protein.

These latter data are of particular interest because although pyruvate kinase contains 4 Mn^{2+} binding sites, it has only two substrate binding sites per MW 237,000 tetramer. The data are thus compatible with the existence of only two catalytic sites per enzyme tetramer. Furthermore, the data are consistent with the hypothesis that each catalytic site contains one lysine that is intimately involved with the catalytic process. This lysine can be converted selectively to homoarginine in an affinity labeling reaction with DMPA bound to the Mn^{2+} region of the catalytic site, and when so converted, eliminates catalytic activity.

In contrast to this sensitive and highly selective response of pyruvate kinase to affinity labeling of a small number of lysine residues, lactate dehydrogenase loses activity exactly in proportion to the fraction of total lysines in the molecule converted to homoarginine with the nonspecific reagent O-methylisourea.[20] As would also be expected of an affinity labeling process, the rate of reaction of DMPA with the specific lysines in the catalytic sites of pyruvate kinase is very fast compared with the rate of reaction for a nonaffinity process, the former being of the order of 2500 times faster than the latter.[4]

Characterization of 4-Phenyl-3,5-Dimethylpyrazole-1-carboxamidine (PDMPA) as a Potential Affinity Label for Mitochondria

DMPA does not significantly inhibit mitochondria at concentrations up to 5 mM. However, since increasing the hydrophobicity of guani-

[20] P. C. Yang and G. W. Schwert, *J. Biol. Chem.* **245**, 4886 (1970).

dinium derivatives generally augments their effectiveness as mitochondrial inhibitors,[21] we explored the effects of PDMPA. This latter compound was studied with guinea pig liver mitochondria isolated[22] and incubated in a medium containing 0.25 M sucrose, 10 mM HEPES at pH 7.4, and substrate. When O_2 uptake is monitored with an oxygen electrode, PDMPA is revealed as an effective inhibitor of NAD-linked, but not of succinate supported, respiration; i.e., it is specific for site 1, 50% inhibition being observed at 0.3 mM. Energized Ca^{2+} uptake is also inhibited by PDMPA, as is observed with other guanidine derivatives.[22] In contrast to the slow onset and reversal, the energy dependence and the sensitivity to K^+ that characterize mitochondrial inhibition by phenethylbiguanide,[23] PDMPA inhibition is virtually instantaneous in onset, is rapidly reversed by washing the mitochondria, and is not affected by dinitrophenol or by K^+.

Binding Studies with [^{14}C]PDMPA. The filtration techniques applied to the study of labeled phenethylbiguanide binding to mitochondria[23] can not be applied to PDMPA, since the filter blanks contain relatively large amounts of retained radioactivity and since labeled PDMPA is easily removed from the mitochondria during the wash procedure. However, determination of radioactivity in aliquots of the supernatant from rapidly centrifuged samples give consistent data, and such studies, employing [^{14}C]PDMPA, fully parallel the functional data described above. In addition, such studies reveal the presence of at least one class of 20 binding sites per milligram of mitochondrial protein, with an affinity constant of 5×10^4. At least one class of a somewhat larger number of lower affinity sites also appears to be present. Several unlabeled guanidinium compounds, including butylbiguanide, phenethylbiguanide, and phenethylguanidine, have been tested as potential binding competitors, hence protective ligands, but none are effective, nor is Ca^{2+} competitive. Conversely, unlabeled PDMPA does not compete for the nonenergized binding of $^{45}Ca^{2+}$.[22]

[21] B. C. Pressman, *J. Biol. Chem.* **238**, 401 (1963).
[22] F. Davidoff, *J. Biol. Chem.* **249**, 6406 (1974).
[23] F. Davidoff, *J. Biol. Chem.* **246**, 4017 (1971).

[64] Labeling of Catechol-*O*-methyltransferase with *N*-Haloacetyl Derivatives

By RONALD T. BORCHARDT and DHIREN R. THAKKER

Catechol-*O*-methyltransferase (COMT; EC 2.1.1.6) plays an important role in the extraneuronal inactivation of catecholamines and in the

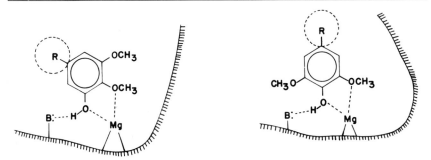

Fig. 1. Proposed enzymic binding of 3,4-dimethoxy-5-hydroxyphenylethylamine and 3,5-dimethoxy-4-hydroxyphenylethylamine. Dashed circles denote areas on the enzyme surface to which the side chain (R) would be exposed.

detoxification of xenobiotic catechols.[1] In an effort to elucidate the relationship between the chemical structure and catalytic function of this enzyme, our laboratory has developed several classes of affinity labeling reagents for the enzyme.[2-7] One approach to affinity labeling this enzyme has involved the preparation of chemically reactive derivatives of 3,4-dimethoxy-5-hydroxyphenylethylamine and 3,5-dimethoxy-4-hydroxyphenylethylamine; both amines are known reversible inhibitors of the transferase.[8] Since 3,4-dimethoxy-5-hydroxyphenylethylamine resembles a *para*-methoxylated product and 3,5-dimethoxy-4-hydroxyphenylethylamine resembles a *meta*-methoxylated product, both compounds would be expected to bind to the active site through the *ortho*-methoxyphenol functionality as depicted in Fig. 1. These binding modes would dictate that the respective ethylamine side chains (R) of these molecules will be exposed to slightly different environments on the enzyme surface, thereby serving as useful reagents to map the topography surrounding the catalytic site.

The chemically reactive derivatives of 3,4-dimethoxy-5-hydroxy-

[1] H. C. Guldberg and C. A. Marsden, *Pharmacol. Rev.* **27**, 135 (1975).
[2] R. T. Borchardt and D. Thakker, *Biochem. Biophys. Res. Commun.* **54**, 1233 (1973).
[3] R. T. Borchardt and D. R. Thakker, *J. Med. Chem.* **18**, 152 (1975).
[4] R. T. Borchardt and D. R. Thakker, *Biochemistry* **14**, 4543 (1975).
[5] R. T. Borchardt, *Mol. Pharmacol.* **11**, 436 (1975).
[6] R. T. Borchardt, E. E. Smissman, D. Nerland, and J. R. Reid, *J. Med. Chem.* **19**, 30 (1976).
[7] R. T. Borchardt, *in* "Chemical Tools in Catecholamine Research" (G. Jonsson, T. Malmfors, and C. Sachs, eds.), Vol. 1, p. 33. North-Holland Publ., Amsterdam, 1975.
[8] B. Nikodejevic, S. Senoh, J. W. Daly, and C. R. Creveling, *J. Pharmacol. Exp. Ther.* **174**, 83 (1970).

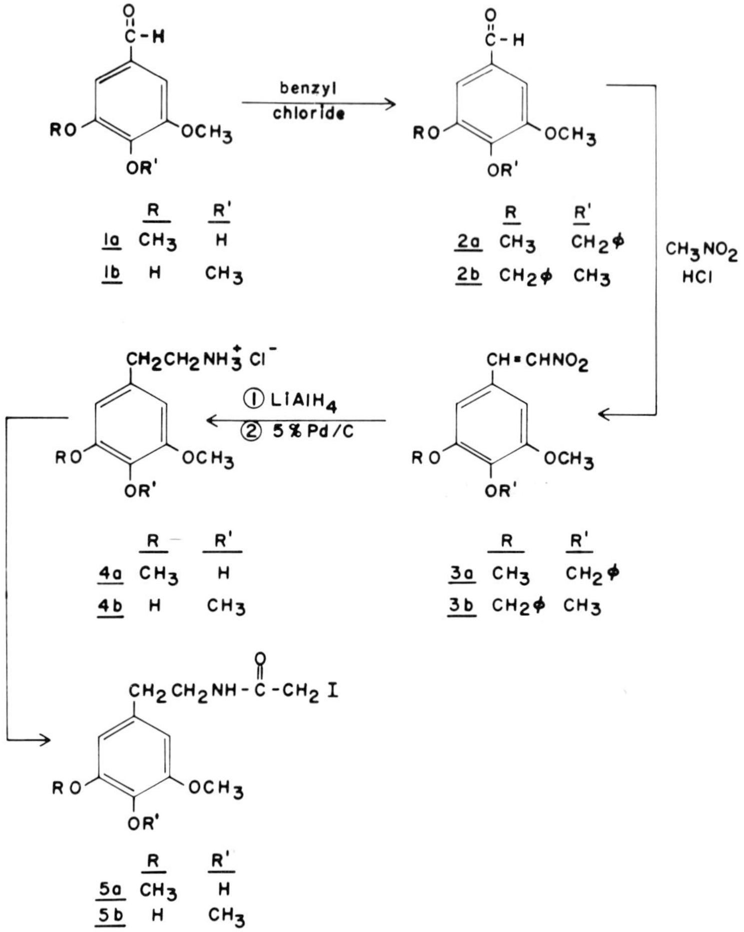

FIG. 2. Steps used for the synthesis of affinity labeling reagents 5a and 5b.

phenylamine and 3,5-dimethoxy-4-hydroxyphenylethylamine, which were synthesized and subsequently shown to be useful affinity labeling reagents, were the N-haloacetyl derivatives 5a and 5b.[2-4] Figure 2 summarizes the steps employed in the synthesis of these affinity labeling reagents. The detailed synthetic procedures are given below.

Synthesis of Affinity Labeling Reagents

3,5-Dimethoxy-4-benzyloxybenzaldehyde (2a). To 3,5-dimethoxy-4-hydroxybenzaldehyde[9] (1a, 5.0 g, 27.5 mmoles) in 350 ml of methanol are

[9] Aldrich Chemical Company.

added 4.93 g (35.8 mmoles) of anhydrous potassium carbonate and 4.5 g (34.8 mmoles) of benzyl chloride. The reaction mixture is refluxed for 72 hr, after which the solvent is removed under reduced pressure. The residue is dissolved in benzene, and the benzene solution is washed sequentially with water, 5% sodium hydroxide solution, and saturated sodium chloride solution and is dried over anhydrous magnesium sulfate. The drying agent is removed by filtration, and the filtrate is concentrated under reduced pressure to yield a yellow oil. The product can be crystallized from ethanol. Yield, 4.94 g (66%); m.p. 61°–62°.

3,4-Dimethoxy-5-benzyloxybenzaldehyde (2b). 3,4-Dimethoxy-5-hydroxybenzaldehyde[10] (*1b*, 5.0 g, 27.5 mmoles) is treated with benzyl chloride using conditions identical to those described above for the preparation of *2a* except that the product is recrystallized with ethyl acetate and hexane. Yield, 5.25 g (70%); m.p. 67°–68°.

3,5-Dimethoxy-4-benzyloxy-β-nitrostyrene (3a). Aldehyde *2a* (4.08 g, 15.0 mmoles) is dissolved in 150 ml of methanol to which is added 1.0 g (16.5 mmoles) of nitromethane. The reaction mixture is cooled to 10°–15°, after which a solution of 0.66 g (16.5 mmoles) of sodium hydroxide in 4 ml of water is added dropwise at such a rate that the temperature does not rise above 15°. When addition is complete, the reaction mixture is slowly poured into 15 ml of 6 N HCl. The resulting precipitate is collected by filtration and washed with cold methanol. The solid product is recrystallized using ethyl acetate and methanol. Yield, 4.14 g (88%); m.p. 131°–132°.

3,4-Dimethoxy-5-benzyloxy-β-nitrostyrene (3b). Aldehyde *2b* (2.0 g, 7.35 mmoles) is converted to the β-nitrostyrene derivative *3b* using a procedure identical to that described above for the preparation of *3a*. The product can be recrystallized using ethyl acetate and methanol. Yield, 2.1 g (91%); m.p. 102°–103°.

3,5-Dimethoxy-4-hydroxyphenylethylamine Hydrochloride (4a). Phenylethylamine *4a* can be prepared using a general procedure for the reduction of β-nitrostyrenes to the corresponding phenylethylamines using lithium aluminum hydride (LiAlH₄).[11] To 0.938 g (25 mmoles) of LiAlH₄ in 100 ml of dry tetrahydrofuran[12] is added dropwise a solution of 3,5-dimethoxy-4-benzyloxy-β-nitrostyrene (*3a*, 1.5 g, 4.76 mmoles) dissolved in 100 ml of dry tetrahydrofuran. The addition takes approximately 1 hr, after which the reaction mixture is refluxed for 3 hr and then cooled in an ice bath. Excess LiAlH₄ is decomposed by dropwise addition of ether that is saturated with water, followed by dropwise addition of

[10] F. Mauthner, *Justus Liebigs Ann. Chem.* **499**, 102 (1926).
[11] F. A. Ramirez and A. Burger, *J. Am. Chem. Soc.* **72**, 2781 (1950).
[12] Freshly distilled from lithium aluminum hydride.

water. The resulting precipitate is removed by filtration and extracted with chloroform to remove any occluded product. The ethereal filtrate and the chloroform extract are combined and concentrated under reduced pressure to yield a yellow oil. The oil is dissolved in 50 ml of ether, and the ethereal solution is dried over anhydrous magnesium sulfate. The drying agent is removed by filtration. Dry HCl (gas) is bubbled into the cooled ethereal filtrate, and the resulting precipitate is collected by filtration to afford the intermediate, 3,5-dimethoxy-4-benzyloxyphenylethylamine hydrochloride. This hydrochloride salt can be recrystallized from methanol and ether to yield 1.24 g (47%); m.p. 154°–155°.

The 3,5-dimethoxy-4-benzyloxyphenylethylamine hydrochloride (0.5 g, 1.55 mmoles) is dissolved in 50 ml of methanol to which is added 0.05 g of 5% palladium on carbon. The reaction mixture is hydrogenated at 25° under 2 atm of hydrogen for 2 hr. The catalyst is removed by filtration, and the filtrate is concentrated under reduced pressure. The residue is crystallized from methanol and ether. Yield, 0.34 g (95%); m.p. 254°–255°.

3,4-Dimethoxy-5-hydroxyphenylethylamine   Hydrochloride   (4b). Using a procedure identical to that described above for the preparation of *4a*, 3,4-dimethoxy-5-benzyloxy-β-nitrostyrene (*3b*, 2.6 g, 8.25 mmoles) can be reduced with LiAlH₄ to the intermediate, 3,4-dimethoxy-5-benzyloxyphenylethylamine hydrochloride, in 47% yield (1.25 g); m.p. 154°–155°.

The   3,4-dimethoxy-5-benzyloxyphenylethylamine   hydrochloride (1.01 g, 3.12 mmoles) can then be converted by catalytic reduction to 3,4-dimethoxy-5-hydroxyphenylethylamine hydrochloride (*4b*). Yield, 0.66 g (91%); m.p. 186°–188°.

N-Iodoacetyl-3,5-dimethoxy-4-hydroxyphenylethylamine (5a). To a solution of 3,5-dimethoxy-4-hydroxyphenylethylamine (*4a*, 0.245 g, 1.24 mmoles) and *N,N'*-dicyclohexylcarbodiimide (DCC) (0.256 g, 1.24 mmoles) in 150 ml of dry acetonitrile is added 0.231 g (1.24 mmoles) of iodoacetic acid. The reaction mixture is stirred for 72 hr at ambient temperature, after which the solution is filtered and the solvent is removed under reduced pressure. The residue is applied to a silica gel thick-layer chromatography plate [1000 μm; solvent system: ethanol/chloroform (1:9)]. The desired product is extracted from the developed silica gel plate with ethyl acetate. The ethyl acetate solution is filtered and the filtrate is concentrated under reduced pressure. The residual oil can be crystallized from chloroform and hexane. Yield, 0.19 g (42%); m.p. 122°–123.5°.

N-Iodoacetyl-3,4-dimethoxy-5-hydroxyphenylethylamine (5b). 3,4-Dimethoxy-5-hydroxyphenylethylamine (*4b*, 0.32 g, 1.63 mmoles) is treated with iodoacetic acid and DCC in dry acetonitrile using conditions similar to those described above for the synthesis of *5a*. The product can be crystallized from chloroform and hexane. Yield 0.22 g (37%); m.p. 109°–111°.

COMT Isolation and Assay

The enzyme used in these affinity labeling experiments can be purified from rat liver (male, Sprague-Dawley, 80–200 g) according to methods previously described.[13] Purification is generally carried through the affinity chromatography step, resulting in a preparation having a specific activity of approximately 500 nmoles of product per milligram of protein per minute. The enzyme activity is determined using S-adenosylmethionine–[14]CH$_3$ and 3,4-dihydroxybenzoic acid[9] as substrates according to a previously described radiochemical assay.[2-7]

Procedures for Affinity Labeling the Enzyme

a. *With N-Iodoacetyl-3,5-dimethoxy-4-hydroxyphenylethylamine (5a)*. A typical inactivation experiment using affinity labeling reagent *5a* is carried out in the following manner: A preincubation mixture is prepared consisting of the following components: water to a final volume of 3.20 ml; magnesium chloride (4.8 μmoles); potassium phosphate at pH 7.60 (400 μmoles); affinity labeling reagent and 500 μg of the purified transferase. The preincubation step is started by the addition of enzyme. After an appropriate period at 37°, an aliquot (0.2 ml) of the mixture is removed and assayed for residual enzyme activity.[2-7]

b. *With N-Iodoacetyl-3,4-dimethoxy-5-hydroxyphenylethylamine (5b)*. Since affinity labeling reagent *5b* is a very potent reversible inhibitor of the transferase, a special procedure is necessary to accurately measure the residual enzyme activity. For example, the preincubation samples have to be diluted prior to assaying for residual enzyme activity in order to minimize the reversible inhibition produced by *5b*. Affinity labeling experiments with *5b* are generally carried out according to the following procedure. A preincubation mixture is prepared consisting of the following components: water to a final volume of 0.1 ml; magnesium chloride (0.30 μmole); potassium phosphate at pH 7.60 (10 μmoles); affinity labeling reagent; and 40 μg of the purified enzyme preparation. The preincubation step is started by addition of enzyme and after the

[13] R. T. Borchardt, C. F. Cheng, and D. R. Thakker, *Biochem. Biophys. Res. Commun.* **63**, 69 (1975).

appropriate incubation period at 37°, the reaction is stopped by diluting the reaction mixture with 0.9 ml of 50 mM potassium phosphate at pH 7.6 (1°–4°). Aliquots of 0.2 ml of the reaction mixtures are assayed for residual enzyme activity.[2–7]

Properties of the Affinity Labeling Reagents

Both N-iodoacetyl-3,5-dimethoxy-4-hydroxyphenylethylamine (5a) and N-iodoacetyl-3,4-dimethoxy-5-hydroxyphenylethylamine (5b) satisfy the criteria established for affinity labeling reagents.[2–4] The evidence in support of N-iodoacetyl-3,5-dimethoxy-4-hydroxyphenylethylamine (5a) as an affinity labeling reagent includes its kinetic behavior; this reagent inactivates by a unimolecular reaction within a dissociable complex rather than by a nonspecific bimolecular reaction.[2] The kinetic order of the reaction can be shown to be one, suggesting that one molecule of the inhibitor (5a) is bound to one molecule of transferase when inactivation occurs.[4] By incorporation studies it can be shown that only one molecule of radiolabeled 5a is incorporated into the enzyme.[4] The amino acid residue being modified in this inactivation can be shown to have a pK_a = 8.5, consistent with the idea that the modified group is a sulfhydryl residue. Protection experiments have shown that the enzyme can be partially protected from inactivation by 5a if the catechol substrate is included in the preincubation mixture. However, inclusion of both the catechol substrate and the methyl donor (S-adenosylmethionine) in the preincubation mixture affords complete protection from inactivation. In contrast, the inclusion of S-adenosylmethionine alone in the preincubation mixture produces enhanced rates of inactivation by 5a. This increased rate of inactivation apparently results because the methyl donor enhances the binding of the affinity labeling reagent 5a.[4]

The results obtained for N-iodoacetyl-3,4-dimethoxy-5-hydroxyphenylethylamine (5b) indicate that this compound is also an affinity labeling reagent for catechol O-methyltransferase, although it appears to modify an amino acid residue different from that being modified by 5a. A kinetic order of two for the reaction of 5b with transferase suggests a complex protein–ligand interaction. A sigmoidal relationship is observed when plots of the pseudo-first-order rate constants of inactivation vs inhibitor concentration are constructed.[4] Such data would suggest positive cooperativity during this inactivation process. The kinetic data can be fit to a model system in which two molecules of the inhibitor are reversibly bound to the enzyme at the time of inactivation, with only one eventually becoming covalently bound. Incorporation studies show that only one molecule of 5b is being incorporated per molecule of enzyme. The amino acid residue modified in this inactivation process has a pK_a greater than 9.0. Dramatic differences are observed in the substrate protection profile for 5b vs 5a. Of particular interest is the observation that the enzyme can be partially protected from inactivation by 5b when S-adenosylmethionine is included in the preincubation mixture. This result is in sharp contrast to the rate enhancement effect that S-adenosylmethionine has on enzyme inactivation by 5a. The S-adenosylmethionine protection by 5b suggests that when this ligand binds to the enzyme there is either a physical protection of the nucleophile or a conformational change of the enzyme that reduces accessibility of the protein nucleophile to the affinity labeling reagent.

These basic differences in the properties of the affinity labeling reagents 5a and 5b would suggest that the two classes of affinity labeling reagents are modifying different nucleophilic residues on the enzyme. The different modes of binding of these reagents (5a and 5b) to catechol-O-methyltransferase, as depicted in Fig. 1, are probably responsible for the interaction of these ligands with different protein nucleophiles.

Acknowledgments

This investigation was supported by a Research Grant from the National Institute of Neurological Diseases and Stroke (NS-10918). RTB gratefully acknowledges support by the American Heart Association for an Established Investigatorship.

[65] Affinity Labeling of Binding Sites in Proteins by Sensitized Photooxidation

By JOHNNY BRANDT

Affinity labeling of ligand-binding sites in proteins is an important tool in the study of their structure–function relationship. Precise localization of amino acid residues near or in the active site can in this way be obtained by degrading the protein and identifying the labeled fragments. Several review articles have been devoted to this subject, and a large number of different active-site-directed reagents have been described.[1-3] A new approach to such affinity labeling of binding sites in proteins is to use photooxidative coupling reactions for the attachment of

[1] S. J. Singer, Adv. Protein Chem. **22,** 1 (1967).

[2] E. Shaw, in "The Enzymes" (P. Boyer, ed.), 3rd ed., Vol. 1, p. 91. Academic Press, New York, 1970.

[3] J. R. Knowles, Acc. Chem. Res. **5,** 155 (1972). See also this volume [8].

the ligand. This technique seems to offer considerable promise.[4] This report presents the principles involved, the general procedure, and some results obtained in studies with a model system.

Principles

Dye-sensitized photooxidation[5,6] is a well-known method which has been used to modify certain amino acid residues in proteins. In this technique, a mixture of a dye and the protein under study is irradiated in the presence of oxygen at wavelengths at which only the dye can absorb light. In this way one can, under special conditions, accomplish very specific chemical modification of one or several amino acid residues in proteins. By using a ligand as the sensitizer, modification can be obtained preferentially at the ligand-binding site of the protein.[7]

The amino acid modifications previously described using sensitized photooxidation are, however, relatively minor and include such systems as the oxidation of methionine to methionine sulfoxide or of tryptophan to kynurenine. Such modification can be difficult to detect and result in difficulties in subsequent separation required to locate the modified residues. However, we have found that by using a high concentration of dye and a rather high light intensity one obtains not only some modification of amino acid residues, but also covalent coupling of the dye to the protein. This observation suggests a new, simple method for affinity labeling of protein ligand-binding sites. The method can easily be applied to the detection of dye-binding sites in proteins, but can also be extended to other ligands by using dye–ligand conjugates as affinity-labeling reagents. In Fig. 1 are shown the structural formulas of some of the dyes that have been successfully coupled to biomolecules using sensitized photooxidation. I stress that the coupling reaction is not restricted to proteins but can occur as well with ribonucleic acid and carbohydrate. The presence of oxygen is not necessary for the coupling of dye but strongly enhances the reaction.

Although detailed information about the mechanism of the dye–protein coupling reaction is unavailable, a general outline can be given. When irradiated by light the dye undergoes electronic excitation followed by radical formation. The radicals formed can react and couple to the protein, probably as the result of radical–radical reactions. By using a

---

[4] J. Brandt, M. Fredriksson, and L.-O. Andersson, *Biochemistry* **13**, 4758 (1974).

[5] W. J. Ray, Jr., this series, Vol. 11, p. 490 (1967).

[6] E. W. Westhead, this series, Vol. 25, p. 401 (1972).

[7] G. Gennari, G. Jori, G. Galiazzo, and E. Scoffone, *J. Am. Chem. Soc.* **92**, 4140 (1970).

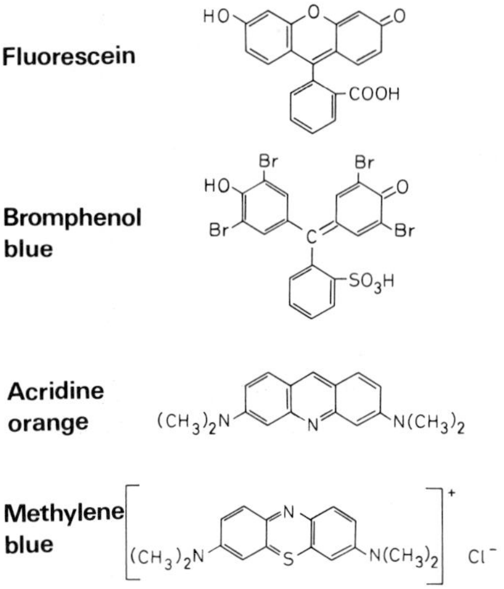

Fluorescein

Bromphenol blue

Acridine orange $(CH_3)_2N$ $N(CH_3)_2$

Methylene blue $\left[(CH_3)_2N \right]^+$ $N(CH_3)_2$ Cl^-

FIG. 1. Structural formulas of some dyes effective in photooxidative coupling reactions.

dye or a dye–ligand conjugate as a sensitizer with affinity for a protein binding site, coupling to amino acid residues in the binding site should be favored, because of the localization and high concentration of the dye or dye–ligand conjugate in the binding site. Specific labeling of binding sites for a dye or a dye–ligand conjugate in proteins should thus be accomplished.

Methods

Photooxidative coupling is fairly easy to perform. The only special equipment required is a suitable irradiation source that gives a high intensity at those wavelengths that are readily absorbed by the dye. For the model experiment presented here, a high-pressure mercury lamp (Gates 420-U2, 250 watts) was used, but a tungsten lamp would have been satisfactory. A typical experimental arrangement is illustrated in Fig. 2. The sample solution is contained in an open, flat-bottom beaker ($d = 15$ cm) at a distance of 20 cm from the irradiation source. A filter is inserted between the lamp and the sample solution to absorb the infrared irradiation and the wavelengths absorbed by the protein; a flat-bottomed beaker of glass containing 0.5 M sodium nitrite is suitable (lower wavelength cutoff is about 400 nm). The temperature of the sample is controlled by a water bath connected to a thermostat, or more simply, by keeping the sample in an ice bath. A stirring motor is placed under the water bath for a gentle driving of a long magnetic stirrer that

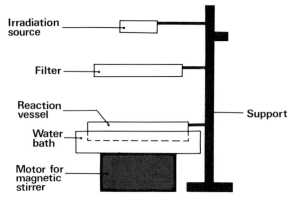

FIG. 2. Schematic illustration of simple experimental arrangements suitable for the photooxidative coupling method.

is in the sample solution. When the volume of sample is 50 to 100 ml, the degree of coupling will be about 1 mg of dye per gram of protein after irradiation for 1 hr. Longer irradiation periods should be avoided because the protein denaturation that often accompanies modification by photooxidation processes increases the risk of unspecific labeling.

The degree of coupling of dye can be determined after removal of excess dye from a known amount of protein by gel filtration on a Sephadex G-25 column. Usually dyes are strongly adsorbed to the gel and a rapid and efficient separation of free from protein-bound dye is easily achieved. The concentration of protein-bound dye is then determined spectrophotometrically in the visible region. It is worth noting that the strong adsorption of dyes to dense Sephadex gels greatly facilitates the isolation of the labeled peptides after fragmentation of the protein. Small labeled peptides are easily fractionated and separated from unlabeled peptides by Sephadex G-25 because of the strong adsorption of the dye moiety to the gel.

Model System

For our pilot studies of the specificity of the dye-sensitized photo-oxidative coupling reaction we have, for several reasons, chosen the coupling of fluorescein to bovine serum albumin. The first advantage of this system is that the photochemistry of fluorescein[8-11] is well known.

[8] L. Lindqvist, *Ark. Kemi* **16**, 79 (1960).
[9] L. Lindqvist, *J. Phys. Chem.* **67**, 1701 (1963).
[10] V. Kasche and L. Lindqvist, *J. Phys. Chem.* **68**, 817 (1964).
[11] V. Kasche, *Photochem. Photobiol.* **6**, 643 (1967).

However, the main reason for our choice was that the binding of fluorescein to bovine seurm albumin[12] has been extensively studied in our laboratory and we have considerable experience in the isolation of peptides from digests of bovine serum albumin.

Application

Preparation of Reagents

Fluorescein (G. T. Gurr, London) is purified by precipitation from 0.01 M NaOH solution with acetic acid. The procedure is repeated three times. A stock solution of 6 mM fluorescein is then prepared at pH 11. Bovine serum albumin (Statens Bakteriologiska Laboratorium, Stockholm), about 100 mg/ml in 0.1 M Tris chloride at pH 8.0, is extensively dialyzed against water. A 5% stock solution is prepared.

Affinity Labeling

A 1% albumin solution containing 1 mM fluorescein is prepared by mixing 10 ml of 6 mM fluorescein and 12 ml of 5% albumin with 38 ml of 30 mM sodium phosphate at pH 7.7 to give a final pH of 8.0. The mixture is illuminated for 60 min at 20° using the apparatus described above. Excess fluorescein is removed by gel filtration on a Sephadex G-25 column (3.2 × 30 cm) in 0.1 M ammonia–ammonium acetate at pH 9.2. The green color of the free dye changes to red upon coupling. The deep red color of the protein fraction eluting at the void volume of the column thus serves as a visible check on successful labeling. The degree of coupling, based upon spectrophotometric measurement, is about 0.20 mole of dye per mole of protein. The labeled protein fraction is concentrated by ultrafiltration after adjustment to pH 8 with neutral 1 M sodium phosphate.

Isolation of Labeled Peptides

The labeled albumin fraction (600 mg in 8.5 ml) is extensively digested with trypsin (36 mg) at pH 7.7 to 7.9. The tryptic digest is subjected to gel filtration on a Sephadex G-50 column (3.2 × 90 cm) in 0.1 M ammonia–ammonium acetate at pH 9.2. The main part of the labeled peptides is eluted at the total volume of the column. This peptide fraction is further fractionated by ion-exchange chromatography on a SE-

---

[12] L.-O. Andersson, A. Rehnström, and D. L. Eaker, Eur. J. Biochem. **20**, 371 (1971).

Sephadex column (1.6 × 40 cm) using pH gradient elution. The gel is equilibrated with 0.1 M acetic acid–0.1 M formic acid buffer, pH 3.0, and the peptides are eluted by increasing the pH of the eluent with 0.4 M ammonia. One main peptide labeled with fluorescein is eluted at pH 4.3. A complete purification of the labeled peptide is performed by gel filtration on a Sephadex G-25 column (1.5 × 40 cm) in 0.1 M ammonia–ammonium acetate at pH 9.2.

Characterization of Labeled Peptide

Amino acid analysis after acid hydrolysis of the pure peptide yields leucine and tyrosine in a 1:1 molar ratio. The peptide material (100 nmoles) is digested with carboxypeptidase A (10 μg) at pH 8.3 and 37° for 2 hr. The digestion mixture is then passed through a Sephadex G-25 column (1.5 × 30 cm) in 0.1 M ammonia–ammonium acetate at pH 9.2. A complete separation of the peptide fragments is obtained. On the basis of N-terminal and amino acid analysis data on the peptide fragments, and from the known sequence data of bovine serum albumin,[13] it has been concluded[4] that the labeled peptide is the Tyr-Leu-Tyr sequence of residues 137–139 with the fluorescein molecule attached to Tyr-137.

Control Experiment

The specificity of the coupling reaction can be checked in a control experiment by labeling the protein in denaturing medium. The conditions were the same as in the model experiment described above except that the solution was 6 M in urea during irradiation. The degree of coupling of fluorescein to albumin was about the same as that observed in the absence of urea. The labeled protein was digested with trypsin and gel filtered on Sephadex G-150, the low molecular weight peptides were separated by ion-exchange chromatography on SE-Sephadex as described above. The elution curve obtained showed that labeling is much less specific under denaturing conditions and that no major labeled peptide was found.

Conclusions

The studies on the model system described show that the photooxidative coupling reaction results in a covalent bond between fluorescein and albumin. The labeled protein has a spectral absorption maximum at 505

[13] J. R. Brown, Fed. Proc., Fed. Am. Soc. Exp. Biol. 34, 591 (1975).

nm compared to 499 nm for the fluorescein adduct obtained after degradation. This spectral shift suggests that the fluorescein molecule is in a hydrophobic region within the intact albumin molecule. We also know from earlier studies[12] that bovine serum albumin has hydrophobic binding sites for fluorescein that cause a red shift in the absorption maximum upon binding. Moreover, the separation of peptides after degradation of labeled protein shows selective labeling of a single peptide. The specificity for labeling of this peptide is lost if coupling is performed under denaturing conditions. It thus seems probable that the main labeled peptide obtained derives from a fluorescein-binding site in the albumin molecule.

Comments

Although the pilot studies described in this report indicate that the principle is valid, affinity labeling of binding sites in proteins using sensitized photooxidation cannot yet be considered a standard method. The approach may form a valuable complement to existing affinity labeling methods, although more information is needed. Further developmental work is now in progress with protein-dye systems for which detailed knowledge of the active-site structure of the protein is available.

[66] Bromocolchicine as a Label for Tubulin

By Daphne Atlas and Henri Schmitt

Microtubules are found in all eukaryotic cells, wherein they participate in a wide variety of functions including mitosis, cell shaping, secretion, motility, axonal growth, and transport.[1,2] Microtubules are polymers of a protein called tubulin, which is itself a dimer composed of two similar but not identical subunits (α, β), each of molecular weight 55,000.[3] Drugs, such as colchicine, that inhibit mitosis and axonal function through disruption of microtubules have been shown to interact *in vitro* with tubulin isolated from several sources.[4] Knowledge of the precise localization and properties of these drug receptors would contribute to the better understanding of the mechanism of tubulin assembly to form microtubules.

[1] J. B. Olmsted and G. G. Borisy, *Annu. Rev. Biochem.* **42**, 507 (1973).

[2] M. L. Shelanski and H. Feit, *in* "The Structure and Function of Nervous Tissue" (G. H. Bourne, ed.), Vol. 6, pp. 47–80. Academic Press, New York, 1972.

[3] R. E. Stephens, *in* "Subunits in Biological Systems" (S. N. Timasheff and G. D. Fasman, eds.), Part A, pp. 355–391. Dekker, New York, 1971.

[4] L. Wilson and J. Bryan, *Adv. Cell Mol. Biol.* **3**, 21 (1974).

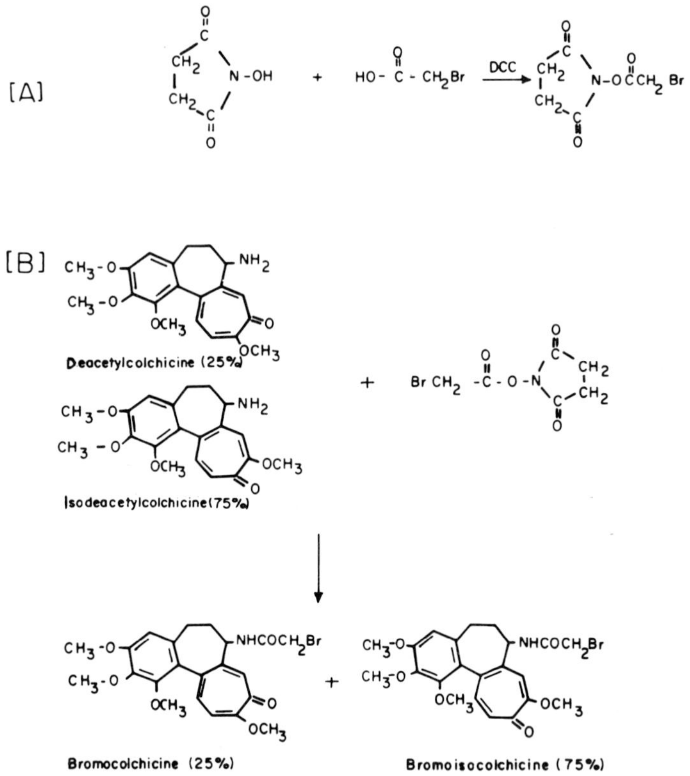

FIG. 1. Outline of synthesis of bromocolchicine.

We have synthesized an affinity label which appears to mimic the effects of colchicine.[5] It binds covalently to tubulin by specifically alkylating its α and β subunits. The active function is the bromoacetyl group: the modified ligand has one bromine which replaces one of the hydrogens of the acetyl group of colchicine.

Outline of Synthesis

Synthesis of Bromoacetyl-N-hydroxysuccinimide (See Fig. 1). Bromoacetic acid (0.5 mmole, 66 mg) is dissolved in 0.6 ml of dry dioxane and 0.2 ml of dry ethyl acetate. N-Hydroxysuccinimide, 0.6 mmole, is added, and the mixture is cooled to 4° in an ice-water bath. Dicyclohexylcarbodiimide, 0.5 mmole, dissolved in 0.5 ml of dry dioxane at 0°, is then added to the mixture. The reaction is stirred for 1 hr in an ice-water

[5] H. Schmitt and D. Atlas, *J. Mol. Biol.* **102**, 743 (1975).

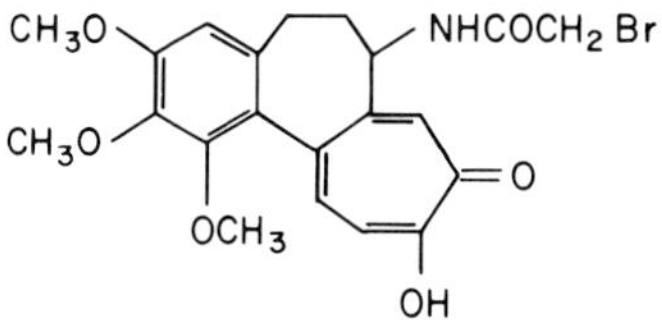

Fig. 2. The structure of bromocolchicinic acid.

bath. After 12 hr at 4°, the dicyclohexylurea is removed by filtration and washed with a cold mixture of dioxane–ethyl acetate (3:1). The filtrate is evaporated to dryness, and the compound is recrystallized from 2-propanol.

Radioactive N-hydroxysuccinimide ester of bromoacetic acid is prepared by the procedure described above, but with 30 mCi (33 mg) of the radioactive compound diluted with an equal amount of unlabeled bromoacetic acid.

Synthesis of a Mixture of Deacetylcolchicine and Isodeacetylcolchicine. A mixture of deacetylcolchicine and isodeacetylcolchicine (25:75) is prepared according to the method of Raffauf et al.[6] and Wilson and Friedkin.[7]

Synthesis of Bromocolchicine [See Fig. 1(B)]. The mixture of deacetylcolchicine and isodeacetylcolchicine, 0.147 mmole, is dissolved in 0.5 ml of dry dioxane. Bromoacetylsuccinimide ester (0.212 mmole) is then added, and after 1 hr of stirring the reaction mixture is allowed to stand overnight at room temperature. After evaporation to dryness under reduced pressure, the residue obtained is suspended in water (4 ml) and stirred for 30 min at room temperature. The suspension is filtered and dried in a desiccator over P_2O_5. The absorption spectrum of this compound is identical to that of colchicine, indicating that the delicate ring structure of the alkaloid remains intact during the coupling procedure. Purity of the nonradioactive compounds may be demonstrated by thin-layer chromatography on silica gel plates (Merck), using two solvent systems, the organic phase of butanol/acetic acid/water 4:1:4; or absolute methanol. Radioactive bromocolchicine is prepared in an identical manner.

Synthesis of Bromotrimethylcolchicinic Acid (See Fig. 2). This compound is prepared by allowing trimethylcolchicinic acid to react with bromoacetylsuccinimide ester as described above for the preparation of

---

[6] R. F. Raffauf, A. L. Farren, and B. E. Ullyot, *J. Am. Chem. Soc.* **75**, 5292 (1953).

[7] L. Wilson and M. Friedkin, *Biochemistry* **5**, 2463 (1966).

bromocolchicine. The trimethylcolchicinic acid is obtained by acid hydrolysis according to Wilson and Friedkin.[7]

Analysis and Quantitation of the Proteins Labeled with [³H]Bromocolchicine

The binding of colchicine to tubulin is time and temperature dependent.[4] The binding capacity decays rapidly in the absence of a stabilizing agent such as GTP (0.5 mM)[8] or of a high concentration of either sucrose (0.8 M) or glycerol (4 M).[9] An estimate of the number of specific colchicine binding sites can be obtained by measuring the [³H]colchicine bound to tubulin.[4,5]

To measure the amount of [³H]bromocolchicine covalently bound, 70-μl portions of the incubation mixture are applied to filter paper disks (2.5 cm, Whatman No. 3 MM) that are then placed in cold 7% trichloroacetic acid. The disks are either washed twice for 10 min in cold trichloroacetic acid and sequentially rinsed with ethanol and diethyl ether, or heated for 30 min at 95° and washed twice for 10 min in trichloroacetic acid, ethanol, and diethyl ether. Radioactivity is measured in a toluene-based scintillant.

As will be discussed below, bromocolchicine is very reactive and can eventually bind covalently in a nonspecific way to a number of proteins other than tubulin. To assess the degree of specific labeling of tubulin in a complex protein mixture, analysis on sodium dodecyl sulfate (SDS)–polyacrylamide slab gels in one[10] or two dimensions[11] is a convenient method.

Quantitation of the radioactivity present in the tubulin band can be achieved by fluorography[12,13] or by direct counting of gel slices, treated with a solubilizer such as NCS (Amersham, Searle) in a toluene-based scintillant.

Binding of [³H]Bromocolchicine to Brain Extracts and Tubulin Purified by Self-Assembly

Our work[5] has centered on adult brain, a tissue that contains a high proportion of tubulin (20%), which allows its purification by self-

[8] R. C. Weisenberg, G. G. Borisy, and E. W. Taylor, *Biochemistry* **7**, 4466 (1968).

[9] F. Solomon, D. Monard, and M. Rentsch, *J. Mol. Biol.* **78**, 569 (1973).

[10] I. Gozes, H. Schmitt, and U. Z. Littauer, *Proc. Natl. Acad. Sci. U.S.A.* **72**, 701 (1975).

[11] P. H. O'Farrel, *J. Biol. Chem.* **250**, 4007 (1975).

[12] W. F. Bonner and R. A. Laskey, *Eur. J. Biochem.* **46**, 83 (1974).

[13] R. A. Laskey and A. D. Mills, *Eur. J. Biochem.* **56**, 335 (1975).

assembly[14] without loss of its biological and pharmacological properties.

A K_i of 23 μM was found for bromocolchicine. The affinity label, synthesized from a 1:3 mixture of deacetylcolchicine and isodeacetylcolchicine, is therefore composed of a mixture of these two isomers. Isocolchicine is 20 times less potent as an antimitotic agent than colchicine[15] and is at least 40 times less efficient *in vitro* in its capacity to inhibit [³H]-colchicine binding.[16] Thus the active isomer of bromocolchicine has a real K_i value of 5.7 μM. We obtained a much higher value ($K_i = 0.5$ mM) for bromoacetyl trimethylcolchicine acids, emphasizing the important contribution of the methoxy group in the C ring of colchicine for the binding to tubulin.

A large excess of affinity label is bound to both purified tubulin and soluble brain extract as compared to the amount of [³H]colchicine retained.[5] This could be explained by the existence of two classes of binding sites for colchicine. In one, the alkaloid is irreversibly bound and can also be detected by the [³H]colchicine binding assay. In the other, colchicine is reversibly bound and can be measured only using the covalent binding of [³H]bromocolchicine.

Autoradiography of soluble brain proteins separated on SDS–polyacrylamide gels showed that four proteins in addition to tubulin were labeled by [³H]bromocolchicine. This result is expected since bromocolchicine contains a very reactive bromine in an allylic position to a carboxyl group, which can alkylate exposed amino acids of any protein in the extract. Most of the nonspecific binding to soluble brain proteins can be eliminated by pretreatment of the extract with bromoacetic acid, yielding about 70% of the protein-bound [³H]bromocolchicine localized in the tubulin band. Bromoacetic acid had little effect on the labeling of tubulin itself, a good indication that specific sites on the protein were involved in the binding. When tubulin purified by polymerization *in vitro* was used in the labeling procedure, with subsequent separation of the subunits on SDS/urea/polyacrylamide slab gels, about 30% of the label was attached to the α-subunit and 70% to the β-subunit. By varying the concentration of [³H]bromocolchicine in the presence of cold colchicine, which competitively inhibits the binding of the affinity label, it was shown that the irreversible binding site of colchicine is localized on the α-subunit of tubulin. Other sites on both subunits react irreversibly with the affinity label whereas they bind colchicine reversibly.

[14] M. L. Shelanski, F. Gaskin, and C. R. Cantor, *Proc. Natl. Acad. Sci. U.S.A.* **70**, 765 (1973).
[15] E. W. Taylor, *J. Cell Biol.* **25**, 145 (1965).
[16] M. H. Zweig and C. F. Chignell, *Biochem. Pharmacol.* **22**, 2141 (1973).

[67] Affinity Labeling of Receptors

By Nava Zisapel and Mordechai Sokolovsky

The term receptor in the context used currently in drug and hormonal studies is defined operationally as those molecules that specifically recognize and bind the drug or hormone and, as a consequence of this recognition, can lead to other changes, which ultimately result in biological response. Most drug and hormone receptors that are membrane bound are in the category of "integral" membrane proteins; i.e., they are not readily solubilized from the membrane and require detergents or other hydrophobic bond-breaking agents to release them.[1,2] After release of the receptor from the membrane, the physiological functions by which the receptor activity is recognized are usually lost, and even the capacity of the receptor to bind its specific drug or hormone may be destroyed or radically modified.

Positive identification of receptors in preparations of subcellular fractions is feasible only on conditions that the receptor protein has been labeled before the disruption of the membrane. The labeling techniques commonly employed for the identification of receptor proteins are inadequate, however, since the latter constitute only a minor component of the total cell proteins. In principle, affinity labeling[3,4] provides the high specificity needed for this purpose. By this method, the reagent interacts reversibly and specifically with the active site of the receptor and then forms a stable covalent bond with amino acid residues located at or near the active site. The attached moiety can be detected by virtue of a tailored chromophore that is radioactive, fluorescent, or spin label. In this connection it is of interest that affinity labeling of the norepinephrine α-receptor was reported as early as 1945.[5,6]

The application of affinity labeling to receptor sites may have one of several objectives.

[1] S. J. Singer, *in* "Structure and Function of Biological Membranes" (L. I. Rothfield, ed.), p. 145. Academic Press, New York, 1971.

[2] S. J. Singer and G. L. Nicolson, *Science* **175**, 720 (1972).

[3] L. Wofsy, H. Metzger, and S. J. Singer, *Biochemistry* **1**, 1031 (1962).

[4] S. J. Singer, *Adv. Protein Chem.* **22**, 1 (1967).

[5] M. Nickerson and L. S. Goodman, *Fed. Proc., Fed. Am. Soc. Exp. Biol.* **2**, 109 (1945).

[6] M. Nickerson and W. S. Gump, *J. Pharmacol. Exp. Ther.* **97**, 25 (1949).

Identification and Quantitation

Specific labeling that may be achieved by affinity labels is often needed in order to follow changes in the quantity or distribution of certain receptors. For example, in developing or differentiating systems, the correlation between the appearance of a functional activity and a particular receptor population or its ultrastructural distribution may be significant. As mentioned above, general labeling of membranes by nonspecific nonpermeant reagents is not satisfactory, since the receptor proteins in question often comprise only a very small fraction of the total protein of the membrane preparation. The application of the method to the identification and quantitation of receptors has been reviewed.[7-10] A recent illustration for the objectives mentioned above can be found in the identification of the acetylcholine receptor in muscle cell cultures that was demonstrated by specific binding of radioactive neurotoxin from elapid snake venom.[11,12] Similarly, the appearance and disappearance of the acetylcholine receptor during differentiation in chick skeletal muscle was followed *in vitro* by autoradiography using $[^{125}I]\alpha$-bungarotoxin.[13]

Purification

After disrupting the membrane or extracting the receptor protein, the physiological functions by which its activity is recognized are usually lost. For the detection of receptor proteins during isolation, labels that are firmly attached and easy to recognize should be used. Disaggregation or denaturation of the receptor in the course of such manipulations will not then affect the ability of detection of the receptor constituents. Moreover, in such cases, it should be demonstrated that the labeled receptors have preserved their biochemical characteristics. This is of major importance in order to ensure that the isolated protein represents not only one conformational type of the receptor, but the average population of

[7] J. F. Moran and D. J. Triggle, *in* "Fundamental Concepts in Drug–Receptor Interaction" (J. F. Danieli, J. F. Moran, and D. J. Triggle, eds.), p. 133. Academic Press, New York, 1970.

[8] S. J. Singer, *in* "Molecular Properties of Drug Receptors" (R. Porter and M. O'Conner, eds.), p. 229. Churchill, London, 1970.

[9] S. I. Chavin, *FEBS Lett.* 14, 269 (1971).

[10] K. L. Carraway, *Biochim. Biophys. Acta* 415, 379 (1975).

[11] J. Patrick, S. F. Heinemann, J. Lindstrom, and D. Schubert, *Proc. Natl. Acad. Sci. U.S.A.* 69, 2762 (1972).

[12] Z. Vogel, A. J. Sytkowski, and M. W. Nirenberg, *Proc. Natl. Acad. Sci. U.S.A.* 69, 3180 (1972).

[13] J. Prives, I. Silman, and A. Amsterdam, *Cell* 7, 543 (1976).

receptor molecules as demonstrated by Sugiyama and Changeux.[14] Furthermore, if the receptor is an aggregated protein complex, only the chains that contain the labeled binding site would be detectable during purification, and, since the state of aggregation of the receptor may depend on the binding of the drug, an isolation procedure based on the monitoring of an affinity label may result in the loss of some structural components of receptor constituents. It must be emphasized that attempts to isolate receptor material by using the affinity labeling method (see Moran and Triggle[7] and references cited therein) have generally not yet resulted in success.

Pharmacological Studies

Covalent binding of a drug to its receptor either *in vivo* or *in vitro* produces long-lasting inactivation or stimulation of the receptor system. This type of perturbation can be utilized to follow biochemical, physiological, and behavioral molecular events[15] that follow receptor stimulation. Affinity labeling can also be used to evaluate stoichiometric relationships between the state of occupancy of receptors and the response. The spare receptor theory, which states that maximum response does not require total occupation of the receptors by an agonist, and theories related to activity of partial agonists[16-21] may thereby be examined. Some of the experiments performed on the adrenergic and the cholinergic muscarinic receptors in order to determine the stoichiometry of the physiological response have been reviewed by Moran and Triggle.[7]

Biochemical Structure–Function Studies

Many receptors exhibit multiple modes of binding with agonists and antagonists. Often, cooperativity (either negative or positive) is observed.[22-24] The effects of covalent binding of a drug or drug analog to

[14] H. Sugiyama and J.-P. Changeux, *Eur. J. Biochem.* **55**, 505 (1975).
[15] B. R. Baker, "Design of Active Site Directed Irreversible Enzyme Inhibitors." Wiley, New York, 1967.
[16] E. J. Ariens, *Arch. Int. Pharmacodyn.* **99**, 32 (1954).
[17] E. J. Ariens, *Adv. Drug Res.* **3**, 235 (1966).
[18] R. F. Furchgott, *Adv. Drug Res.* **3**, 21 (1966).
[19] J. M. Van Rossum, *Adv. Drug Res.* **3**, 49 (1966).
[20] D. R. Waud, *Pharmacol. Rev.* **20**, 49 (1968).
[21] R. P. Stephanson, *Br. J. Pharmacol.* **11**, 379 (1956).
[22] J.-P. Changeux, *Mol. Pharmacol.* **2**, 369 (1966).
[23] D. J. Triggle, "Neurotransmitter–Receptor Interactions," p. 424. Academic Press, New York, 1971.
[24] A. Levitzki, *J. Theor. Biol.* **44**, 367 (1974).

its receptor on the kinetic parameters of reversible interactions with other drugs can be used to study the characteristics of binding processes as well as such other phenomena as desensitization of receptors and the existence of low- and high-affinity binding sites. Such investigations, carried out with intact cells, membrane fragments, or purified receptors, can supply valuable information on the basis of which the characterization of differences in type and site of binding could be attained.

Clearly, affinity labeling can be used to study the primary structure of the receptor binding site and to identify the amino acid residues in its close vicinity. It can also serve as a model for a "turned on" receptor to be compared structurally with a "turned off," ligand-free, receptor. A special variant of affinity labeling will be the result of attempts to obtain "superactivated" receptors by chemical modification. This approach was recently demonstrated with enzymes (see Blumberg and Vallee[25] and references therein).

Successful application of the affinity labeling method depends on the degree to which the receptor molecule has been exclusively labeled at the specificity site. A number of criteria should be applied in the design and evaluation of such experiments.

1. Specific protection against inactivation. If a specific binding is inactivated by reaction with an affinity label the inactivation should be slower when the reaction is carried out in the presence of a specific protector (reversible inhibitor, agonist, or antagonist) of the active site in question. However, inactivation may result also from reaction conditions or from nonspecific modification of the receptor protein; the effect on the receptor by a nonspecific analog that possesses the same reactive group as the affinity label should therefore be examined in control experiments. This criterion is not altogether adequate, since a particular receptor will probably account for less than 1% of the total cellular protein and specific labeling of a single protein, when this protein is a minor component in a mixture, has to be demonstrated. Therefore, the following criterion must be met.

2. Exclusivity.[8] The estimation of nonspecific labeling of proteins other than the receptor by the affinity label may be achieved by combination with the differential labeling technique. A preparation to be labeled is first treated with a nonradioactive form of the affinity labeling reagent in the presence of a reversible protective agent (inhibitor, agonist, antagonist). The protective agent is then removed, e.g., by dialysis or centrifugation, and the sample is exposed to the radioactive affinity labeling reagent. Comparison of the results of this experiment

[25] S. Blumberg and B. L. Vallee, *Biochemistry* **14**, 2410 (1975).

with those of direct labeling of another portion of the sample with a radioactive agent, allows an estimate of the exclusivity of the procedure. Furthermore, the exclusivity measurements should be performed with suitable controls, since exposure of new nonspecific sites might occur during labeling of the receptor with the nonradioactive form of the reagent. Thus, the sample that was modified in the presence of the protector is divided into two portions, to one of which the protector is added back. Both portions are then treated with the radioactive form of the reagent.

Other criteria that are used for affinity labeling of soluble enzymes, e.g., saturability and stoichiometric inactivation, are not applicable to receptors because of the high degree of nonspecific labeling that usually occurs. A theoretical kinetic description of affinity labeling process was carried out by Metzger et al.[26] This includes the formation of a reversible complex (L---R) and then, while in the complexed form, it undergoes an accelerated covalent-bound-forming reaction with one or more appropriately located residues (L—R)

$$L + R \underset{k_2}{\overset{k_1}{\rightleftharpoons}} L\text{-}\text{-}\text{-}R \overset{k_3}{\to} L\text{—}R \tag{1}$$

where L = affinity label reagent and R = receptor.

The affinity labeling reagent may react simultaneously, although usually at a slower rate, in a bimolecular process with amino acid residues outside the active site. Reaction with sites of lower specificity would yield a series of undesired products (L–A_i).

$$L + A_i \overset{k_{4i}}{\to} L\text{–}A_i \tag{2}$$

The ratio between the specific L—R and the sum of the nonspecific products, L–A_i, obtained after a reaction time, t, will markedly influence the choice of a labeling reagent. Let us assume for simplicity that all nonspecific sites are equally reactive, i.e., that the individual rate constants can be approximated by a mean rate constant, $\bar{k}_4$ ($k_{4i} = \bar{k}_4$). Assume also that the total sum of nonspecific sites available for modification, A, is equal to ΣA_i, and that the total amount of nonspecific reagent bound, L–A, is equal to ΣL–A. Then, the parameters that measure the specificity, $X_{(t)}$, of the reaction can be reduced to

$$X_{(t)} = \frac{(LR)}{(L\text{–}A)} = \frac{k_3 \cdot (k_1/k_2)(R)}{\bar{k}_4(A)} \tag{3}$$

Equation (3) shows that specificity is directly proportional to the equilibrium constant, K, for the formation of the complex R---L ($K = k_1/k_2$) [Eq. (1)]. Hence, the greater the association constant of the reagent

[26] H. Metzger, L. Wofsy, and S. J. Singer, *Biochemistry* **2**, 979 (1963).

with the receptor, the greater the specificity of labeling. Specificity also depends on the ratio, $k_3/\bar{k}_4$, i.e., on the relative reactivity of the labeling groups in and out of the binding site. This implies that the absolute reactivity of the reactive group of the reagent is not very important. However, its selectivity may be crucial, since a group of low selectivity will have more sites to react with. An increase in A (ΣA_i) would thus lead to a corresponding decrease in specificity. Nevertheless, reactivity considerations are important because the reactivity of the labeling group and the reaction conditions can determine which residues are labeled. Thus, bromoacetyl groups have little selectivity since they will react with amino, phenolic, sulfhydryl, and carboxyl residues. However, performing the reaction at a pH lower than 7 might increase selectivity. It is also obvious that, since specificity (X) [Eq. (3)] is proportional to the free receptor concentration (R), it will be markedly reduced when the receptor comprises only a minor fraction of the total groups exposed.

In choosing a reagent suitable for affinity labeling of a receptor, two main factors have to be considered: (a) the affinity constant of the reagent toward the target receptor; and (b) the nature of the chemically reactive moiety. Some examples of reactive groups that have been employed for affinity labeling of a group of receptors are listed in the table.

In this context it is of interest that the use of photoactive labels introduced by Westheimer et al.[27-29] has great potentialities for labeling receptors owing to the instantaneous reaction of the photolyzed group. In this method (photoaffinity labeling) a reagent is used that can combine specifically and reversibly with the binding site and contains a photolyzable group, P. P is unreactive in the dark, but when photolyzed, is converted to an extremely reactive intermediate, P*; if certain conditions are satisfied, P* may then react to form a covalent bond, with residue(s) in the binding site of the receptor before the reagent dissociates from the site [Eq. (4)].

$$L + R_p \underset{k_2}{\overset{k_1}{\rightleftharpoons}} L\text{---}R_p \overset{h\nu}{\rightarrow} \underset{\substack{k_1*\uparrow\downarrow k_2* \\ L + R_p*}}{L\text{---}R_p*} \overset{k_3*}{\rightarrow} L\text{---}R \tag{4}$$

Under such circumstances $k_3* > k_2*$ and the formation of a covalent L–R will occur before the affinity label can dissociate from the binding site.

[27] A. Singh, E. R. Thorn, and F. H. Westheimer, J. Biol. Chem. 237, PC3006 (1962). See also this volume [8].

[28] H. R. Chaimovich, R. J. Vaughan, and F. H. Westheimer, J. Am. Chem. Soc. 90, 4088 (1968).

[29] R. J. Vaughan and F. H. Westheimer, J. Am. Chem. Soc. 91, 217 (1969).

REAGENTS FOR AFFINITY LABELING OF RECEPTORS

Reactive moiety	Reagent	Receptor	Tissue	Reaction conditions used	Specificity	References
Aromatic diazonium ion	p-Nitrophenyl diazonium fluoroborate	Acetylcholine (muscarinic antagonist)	Guinea pig ileum	10 μM reagent, pH 7.4, 37°, 30 min	—	32, 33
Aziridinum ion produced by cyclization of N-haloalkyl amines	N-2-Chloroethyl-N-propyl-2-benzilylethylamine	Acetylcholine (muscarinic antagonist)	Guinea pig ileum	2 μM reagent, pH 7.4, 30°, 10 min	—	36, 34
	N-2-Chloroethyl-N-methyl-2-acetoxyethylamine	Acetylcholine (muscarinic agonist)	Guinea pig ileum	50 μM reagent, pH 7.4, 30° in the presence of 1 μM physostigmine, 20 min	—	34, 35
Bromoacetyl	N(2-hydroxy-3-naphthoxy-propyl)-N-bromoacetyl-ethylenediamine	β-Adrenergic	Turkey erythrocyte ghosts	25 μM reagent, pH 7.4, 25°, 15 min	—	37
	Bromoacetylcholine (bromide)	Acetylcholine (antagonist)	Electroplax-exitable membrane; frog neuromuscular junction	(1) Reduction of sample with dithiothreitol; (2) 1–5 μM reagent, pH 8.0, room temp, 20 min	Reacts with the anionic site of the receptor	38–40
Maleimide	4-(N-Maleimido)-benzyl trimethylammonium iodide	Acetylcholine	Purified and membranous receptor from electric tissue of Electrophorus and electroplax	(1) Reduction of sample with dithiothreitol; (2) 100 nM reagent, pH 7.0, 25°, 2 min	One major protein, MW 40,000, is labeled; labeling is blocked by reversible cholinergic ligands or snake neurotoxins	41–46

Trialkyl oxonium	Trimethyl oxonium tetrafluoroborate	Acetylcholine	Purified and membranous receptor from *Torpedo californica*	0.1 mM reagent, pH 7.0, 4°, 30 min	Acetylcholine but not α-bungarotoxin binding affected by the modification; low degree of specificity	47
Carbene generators	2,2-Hydroxy-3-isopropyl amino propoxyiodobenzene	β-Adrenergic	Taenia isolated from guinea pig cecum	10 μM reagent, pH 8.0, 38°, UV irradiation (270–350 nm), 2 hr	—	48
	1-Isoprenaline	β-Adrenergic	Taenia isolated from guinea pig cecum	10 μM reagent, pH 8.0, 38°, UV irradiation (270–350 nm), 2 hr	—	48
	N^6-(Ethyl 2-diazomalonyl)adenosine 3':5'-cyclic monophosphate	Adenosine 3':5'-cyclic-monophosphate	Human erythrocyte ghosts	1 nM reagent, pH 5.5, 4°, in the presence of 40 mM dithiothreitol, UV irradiation (253.7 nm), 12 min	One protein was labeled in the intact membranes	49

REAGENTS FOR AFFINITY LABELING OF RECEPTORS (Continued)

Reactive moiety	Reagent	Receptor	Tissue	Reaction conditions used	Specificity	References
Nitrene generators	4-Azido-2-nitro-benzyl trimethyl-ammonium fluoroborate	Acetylcholine (nicotinic antagonist)	*Torpedo californica*-excitable membrane fragments	0.5 mM reagent, pH 7.4, room temp., UV irradiation, 10 min	Four proteins were labeled; neurotoxin from *Naja Naja siamensis* protects from labeling only 1 out of the 4 proteins. Agonists and antagonists containing quaternary ammonium groups protect all 4 chains against labeling	50
	3-Azidohexestrol	Estrogen	Rat uterus	—	—	51
	2,4-Dinitrophenyl 1-azide	Antibody (immunoglobulin that binds 2,4-dinitrophenyl groups)	Mouse IgA myeloma immunoglobulin protein (460)	300 μM reagent, pH 8.0, 4°, UV irradiation, 60 min	The number of moles of reagent incorporated was equivalent to the loss of binding sites for ε-2,4-dinitrophenyl-lysine	52, 53
	ε-(4-Azido-2-nitro-phenyl)-L-lysine	Antibody (against 4-azido-2-nitro-phenyl determinants)	Rabbit antibodies	0.1 μM hapten, pH 7.4, 0°, UV irradiation, 24 hr	Labeling of the binding site is considerably exclusive	54

As noted by Kiefer *et al.*,[30] these conditions could lead to a very high degree of specificity of labeling at active sites; any reagent molecules that underwent photolysis in free solution would be expected to react with the solvent or with an added scavenger, and nonspecific reactions with the membrane proteins might be greatly reduced. The photolyzable groups commonly employed are diazoacyl reagents and arylazides, which upon irradiation are converted to carbene and nitrene radicals, respectively. However, these reagents exhibit a high degree of nonspecificity either because k_3^* [Eq. (4)] is not sufficiently high or because of the low selectivity of the carbene or nitrene moieties formed in the photolytic process, which react by a variety of insertions at any chemical bonds, including C–H bonds, in their vicinity.[31] Examples of photoaffinity labeling of receptors are also presented in the table.[32-54]

[30] H. Kiefer, J. Lindstrom, E. S. Lennox, and S. J. Singer, *Proc. Natl. Acad. Sci. U.S.A.* **67**, 1688 (1970).

[31] S. J. Singer, A. Ruoho, H. Kiefer, J. Lindstrom, and E. S. Lennox, in "Drug Receptors" (H. P. Rang, ed.), p. 183. Univ. Park Press, Baltimore, Maryland, 1973.

[32] C. Lebbin, A. Hofmann, and P. G. Waser, *Naunyn-Schmiedeberg's Arch. Pharmacol.* **289**, 237 (1975).

[33] H. G. Mautner and E. Burtels, *Proc. Natl. Acad. Sci. U.S.A.* **67**, 74 (1970).

[34] D. A. Robinson, J. G. Taylor, and J. M. Young, *Br. J. Pharmacol.* **53**, 363 (1975).

[35] J. G. Clement and E. W. Colhoun, *Can. J. Physiol. Pharmacol.* **53**, 264 (1975).

[36] A. S. V. Burgen, C. R. Hiley, and J. M. Young, *Br. J. Pharmacol.* **51**, 279 (1974).

[37] D. Atlas and A. Levitzki, *Biochem. Biophys. Res. Commun.* **69**, 397 (1976).

[38] N. Kalderon and I. Silman, *Isr. J. Chem.* **9**, 12BC (1971).

[39] D. Ben-Haim, E. M. Landau, and I. Silman, *J. Physiol. (London)* **234**, 305 (1973).

[40] C. Y. Chiou, *Eur. J. Pharmacol.* **26**, 268 (1974).

[41] M. J. Reiter, D. A. Cowburn, J. M. Prives, and A. Karlin, *Proc. Natl. Acad. Sci. U.S.A.* **69**, 1168 (1972).

[42] A. Karlin, *Fed. Proc., Fed. Am. Soc. Exp. Biol.* **32**, 1847 (1973). See also this volume [68].

[43] A. Karlin and M. Winnik, *Proc. Natl. Acad. Sci. U.S.A.* **60**, 668 (1968).

[44] A. Karlin, J. Prives, W. Deal, and M. Winnik, *Mol. Prop. Drug Recep. Ciba Found. Symp. 1970*, p. 247.

[45] C. L. Weill, M. G. McNamee, and A. Karlin, *Biochem. Biophys. Res. Commun.* **61**, 997 (1974).

[46] A. Karlin and D. Cowburn, *Proc. Natl. Acad. Sci. U.S.A.* **70**, 3636 (1973).

[47] Y. Chao, R. L. Vandlen, and M. A. Raftery, *Biochem. Biophys. Res. Commun.* **63**, 300 (1975).

[48] I. Takayanagi, M. Yoshioka, K. Takagi, and T. Tamura, *Eur. J. Pharmacol.* **35**, 121 (1976).

[49] C. E. Guthrow, H. Rasmussen, D. J. Brunswick, and B. S. Cooperman, *Proc. Natl. Acad. Sci. U.S.A.* **70**, 3344 (1973).

[50] F. Hucho, P. Layer, H. R. Kiefer, and G. Bandini, *Proc. Natl. Acad. Sci. U.S.A.* **73**, 2624 (1976).

[51] J. A. Katzenellengogen, J. J. Johnson, Jr., and H. W. Myers, *Biochemistry* **12**, 4085 (1973).

Although affinity labeling has important contributions to make in solving the complex problems of receptor structure–function relationship, the technique must be used with caution since the perturbation caused by the presence of a covalently bound ligand is capable of altering the system under study in ways that are not as yet fully understood.

[52] M. Yoshioka, J. Lifter, L. L. Hew, C. A. Converse, M. Y. K. Armstrong, W. H. Konigsberg, and F. F. Richards, *Biochemistry* **12**, 4679 (1973).

[53] F. F. Richards, J. Lifter, C. L. Hew, M. Yoshioka, and W. H. Konigsberg, *Biochemistry* **13**, 3572 (1974). See also this volume [57].

[54] R. A. G. Smith and J. R. Knowles, *Biochem. J.* **141**, 51 (1974).

[68] Nicotinic Acetylcholine Receptors[1]

By Arthur Karlin

Nicotinic acetylcholine receptors are integral proteins of the subsynaptic membranes of skeletal muscle and of homologous electrocytes of electric fish. These receptors translate the binding of acetylcholine into an increase in the permeability of the membrane to sodium, potassium, and calcium ions. The acetylcholine binding site has been affinity labeled by a two-step procedure, the first step of which is the reduction of a disulfide group close to the acetylcholine binding site and the second, the affinity alkylation of one of the sulfhydryl groups formed by reduction. Two types of affinity alkylating agents have been used. One type, exemplified by 4-(N-maleimido)benzyltrimetylammonium iodide (MBTA), fixes the receptor in an inactive state.[2,3] A second type, exemplified by bromoacetylcholine bromide, fixes the receptor in an active state.[4] In all cases, these affinity labels act as completely reversible activators or competitive inhibitors of the *unreduced* receptor. Radioactively tagged MBTA has been used to quantitate receptor in intact cells,[5] in membrane fragments,[6] and in solution,[7,8] and to identify the receptor subunit bear-

[1] This research was supported by research grants from the U.S. Public Health Service (NS07065) and from the National Science Foundation (BMS75-03026).
[2] A. Karlin and M. Winnik, *Proc. Natl. Acad. Sci. U.S.A.* **60**, 668 (1968).

[3] A. Karlin, *J. Gen. Physiol.* **54**, 245s (1969).

[4] I. Silman and A. Karlin, *Science* **164**, 1420 (1969).

[5] A. Karlin, J. Prives, W. Deal, and M. Winnik, *J. Mol. Biol.* **61**, 175 (1971).

[6] A. Karlin and D. A. Cowburn, *in* "Neurochemistry of Cholinergic Receptors" (E. DeRobertis and J. Schacht, eds.), p. 37. Raven, New York, 1974.

[7] A. Karlin and D. A. Cowburn, *Proc. Natl. Acad. Sci. U.S.A.* **70**, 3636 (1973).

[8] A. Karlin, M. G. McNamee, and D. A. Cowburn, *Anal. Biochem.* **76**, 442 (1976).

ing the acetylcholine binding site.[7,9,10] Other potential affinity labels for receptor have been described, such as p-(trimethylammonium)benzenediazonium fluoroborate,[11,12] the photolabel, 4-azido-2-nitrobenzyltrimethylammonium fluoroborate,[13,14] and trimethyloxonium fluoroborate,[15] which react directly with unreduced receptor but apparently lack specificity for the acetylcholine binding site.

Synthesis of 4-(N-Maleimido)benzyltri-[³H]methylammonium Iodide[5]

p-Nitro-α-dimethylaminotoluene (I). Twenty-five grams of p-nitrobenzylbromide[16] (0.12 mole) in 200 ml of benzene and 0.24 mole of dimethylamine (25–40% aqueous solution) are stirred together at room temperature for 2 hr. The mixture is washed five times with 200 ml of water. The organic phase is dried with anhydrous $MgSO_4$, and the benzene is removed on a flash evaporator. About 20 g (95% yield) of a light yellow oil results.

p-Amino-α-dimethylaminotoluene (II). To 9 g of I (0.05 mole) and 15 g of 30-mesh tin in a round-bottom flask are added 36 ml of concentrated HCl in portions over 15 min while mixing. After spontaneous refluxing ceases, the mixture is heated at 100° for 30 min. It is then cooled and poured into 50 ml of ice-cold 10 N KOH. This mixture is extracted 5 times with 50 ml of ethyl ether. The extract is dried with $MgSO_4$ and filtered, giving a final volume of about 200 ml.

4-(N-Maleamido)benzyldimethylamine (III). The 200 ml of ether solution of (II) are added at about 2 ml/min to a vigorously stirred solution of 15 g (0.15 mole) of maleic anhydride in 600 ml of ethyl ether. The stirring is continued for an additional hour. The mixture is chilled and filtered, and the precipitate is washed with ethyl ether and dried.

[9] C. L. Weill, M. G. McNamee, and A. Karlin, *Biochem. Biophys. Res. Commun.* **61**, 997 (1974).

[10] M. J. Reiter, D. A. Cowburn, J. M. Prives, and A. Karlin, *Proc. Natl. Acad. Sci. U.S.A.* **69**, 1168 (1972).

[11] J.-P. Changeux, T. R. Podleski, and L. Wofsy, *Proc. Natl. Acad. Sci. U.S.A.* **58**, 2063 (1967).

[12] H. G. Mautner and E. Bartels, *Proc. Natl. Acad. Sci. U.S.A.* **67**, 74 (1970).

[13] A. E. Ruoho, H. Kiefer, P. Roeder, and S. J. Singer, *Proc. Natl. Acad. Sci. U.S.A.* **70**, 2567 (1973).

[14] F. Hucho, P. Layer, H. Kiefer, and G. Bandini, *Proc. Natl. Acad. Sci. U.S.A.* **73**, 2624 (1976).

[15] Y. Chao, R. L. Vandlen, and M. A. Raftery, *Biochem. Biophys. Res. Commun.* **63**, 300 (1975).

[16] Available from Aldrich Chemical Co.

The crude yield is 8.7 g (35 mmoles). The product is recrystallized twice from 80% ethanol to give about 5 g of white platelets of m.p. 208°–209°.

4-(N-Maleimido)benzyldimethylamine (IV). Cyclization[17] of (III) to (IV) is as follows: To 0.5 g (2 mmoles) of (III) and 90 mg of anhydrous sodium acetate is added 1 ml of acetic anhydride, and the pastry mixture is stirred at 100° for about 5 min until a deep yellow solution forms. This is added to 60 ml of cold water. The mixture is extracted twice with 25 ml of methylene chloride, and the extracts are discarded. The mixture is adjusted to pH 7.0–7.5 with about 30 ml of 1 M NaHCO$_3$ and extracted three times with 25 ml of methylene chloride. These extracts are combined, dried over MgSO$_4$, and filtered, and the solvent is removed. The product is recrystallized from carbon tetrachloride and hexane to give bright yellow crystals of m.p. 86°–87° in about 80% yield. Compound (IV) polymerizes readily; however, the polymer is insoluble in carbon tetrachloride and can be removed by filtration.

[³H]MBTA. Freshly recrystallized (IV) (130 mg, 0.56 mmole) is dissolved in 0.5 ml of methylene chloride in a tube with a male 14/35 standard taper joint. The tube is connected through a female taper joint and a stopcock to a vacuum line, and the solution is degassed by two freeze–thaw cycles. A break-seal vial fitted with a male 14/35 standard taper joint and containing 13 mg (0.09 mmole) of tritiated methyl iodide (about 2 Ci/mmole) is similarly connected to the vacuum line. Above the seal is placed an iron rod sealed in glass and 0.5 ml of ethyl ether. The vacuum line is evacuated and closed off from the pump. The methylene chloride solution and the methyl iodide are frozen, and the breakseal is broken using a magnet to raise and drop the iron rod.

The stopcocks connecting the tubes to the vacuum line are opened, and the methyl iodide–ethyl ether solution is allowed to come to room temperature, while the methylene chloride solution is kept frozen. After all the liquid is gone from the break-seal vial, that vial and the vacuum line are heated with a hot-air gun to force the last traces of volatile material into the methylene chloride solution. This solution is then sealed in its tube and stored in the dark at room temperature for 3 to 4 days. The tube is opened, 5 ml of ethyl ether are added, the tube is centrifuged, and the supernatant is removed. The precipitate is washed with ethyl ether three more times. The precipitate is dissolved in 1 ml of acetonitrile and added dropwise to 10 ml of stirred ethyl ether. The precipitate is collected by centrifugation and dried under reduced pressure. The yield is approximately 100% (33 mg, 0.09 mmole).

The NMR and IR spectral characteristics of MBTA are in Karlin

---

[17] M. P. Cava, A. A. Deana, K. Muth, and M. J. Mitchell, *in* "Organic Synthesis," p. 93. Wiley, New York, 1961.

APPROXIMATE R_f'S OF 4-(N-MALEIMIDO)BENZYLTRIMETHYLAMMONIUM IODIDE (MBTA), PRECURSORS, AND PRODUCTS

Compound	BuAcEt[a]/ cellulose[b]	Acetonitrile/ silica[c]	Acetone/ silica
MBTA	0.70	0.75	0.65
Maleamidobenzyl-trimethylammonium iodide[d]	0.45	1.0	0
(IV)	0.8	0.5	—
(III)	0.40	0	—

[a] 35 ml of n-butanol–10 ml of 1 N acetic acid–10 ml of ethanol.
[b] Eastman Chromagram Sheet 13254 cellulose.
[c] Eastman Chromagram Sheet 13181 silica gel.
[d] Hydrolysis product of MBTA.

et al.[5] MBTA has an m.p. of 204°–205°. The synthesis may be conveniently monitored by TLC (see the table).

The [³H]MBTA is stored as a 1 mM solution in acetonitrile in a liquid N₂ refrigerator. It is stable for years under these conditions. The acetonitrile solution may be stored for a couple of months at −20° without change if water is excluded.

Preparation of [³H]MBTA Solutions

MBTA is readily hydrolyzed. The following hydrolysis rates are to be taken into account in storing and diluting the label: at pH 4, 0.004 per day at −20°, 0.01 per hour at 0°, 0.08 per hour at 25°, and at pH 7.1, 0.016 per minute at 25°. Because of the rapidity of hydrolysis at pH 7.0, the label is kept at a low pH until it is added to receptor. [³H]MBTA is first prepared for use as a solution in 0.1 mM HCl, in which it can be stored for several days at −20°. A small aliquot (0.2 ml) of [³H]MBTA in acetonitrile is dried under reduced pressure, and the residue is redissolved in 4 ml of 0.1 mM HCl to give an approximate concentration of 30 μM. The exact concentration of [³H]MBTA is determined spectrophotometrically.

Present in the solution are [³H]MBTA and its hydrolysis product, 4-(N-maleamido)benzyltrimethylammonium iodide. The concentrations in moles per liter of [³H]MBTA (a) and its hydrolysis product (b) are determined from their molar extinction coefficients[5] and absorbances at 224, 236.7 (isosbestic point), 260, and 290 nm. The absorbances at any two wavelengths suffice to determine a and b; somewhat arbitrarily the following equations have been used:

$$a = (4930\, A_{224} - 20{,}700\, A_{290}) \div 1.51 \times 10^8 \tag{1}$$
$$b = (32{,}500\, A_{290} - 451\, A_{224}) \div 1.51 \times 10^8 \tag{2}$$
$$a = (8990\, A_{224} - 20{,}700\, A_{260}) \div 2.73 \times 10^8 \tag{3}$$
$$b = (32{,}500\, A_{260} - 942\, A_{224}) \div 2.73 \times 10^8 \tag{4}$$
$$a + b = (A_{236.7}) \div 15{,}900 \tag{5}$$

The values of a and b obtained from the first pair of equations, (1) and (2), and the second pair, (3) and (4), are averaged. Also, the quantity, $a + b$, the total concentration of labeled species, is taken as the average of the values obtained from the first pair, the second pair, and Eq. (5).

Reduction of Receptor

Complete reduction of the disulfide at the acetylcholine binding site is obtained with the receptors of *Electrophorus* and *Torpedo* electric tissues by reaction with 0.2 mM dithiothreitol at pH 8.0 for 20 min at 25°. At the conclusion of the reduction, the pH is brought to 7.0 with NaH_2-PO_4 solution. Significant spontaneous reoxidation of reduced receptor does not occur within 30 min.

Kinetics of the Reactions of MBTA

In general, three types of sulfhydryls are available for reaction with MBTA: the sulfhydryls formed by reduction at the acetylcholine binding site (RSH),[18] sulfhydryls[19] either preexisting or formed by reduction elsewhere on the receptor or on other proteins (PSH), and the sulfhydryls of dithiothreitol (QSH). The rate constant, k_R, for reaction with RSH is 6×10^5 M^{-1} sec^{-1}; the average rate constant, k_P, for reaction with PSH is roughly 600 M^{-1} sec^{-1}; and that, k_Q, for reaction with QSH is 5×10^3 M^{-1} sec^{-1}.[7,8] The kinetic equations may be written

$$(1/k_R)(dr/dt) = (1/k_P)(dp/dt) = (1/k_Q)(dq/dt) = -m$$

[18] It is assumed that only one sulfhydryl per site is affinity labeled.
[19] Other nucleophiles, such as amino groups, which in general react with maleimides much more slowly than sulfhydryls, may be included in PSH.

where $r = [\text{RSH}]$, $p = [\text{PSH}]$, $q = [\text{QSH}]$, and $m = [\text{MBTA}]$. Integration yields

$$(r/r_0)^{1/k_R} = (p/p_0)^{1/k_P} = (q/q_0)^{1/k_Q}$$

where r_0, p_0, and q_0 are initial concentrations. These relationships determine the extent and the specificity of the alkylation.

The specificity is defined as $R'/(R' + P')$ where R' is the adduct of RSH with MBTA and P' is the adduct of PSH with MBTA. Operationally, $(R' + P')$ is the total labeling of protein and P' is the labeling in the presence of an agent that protects the binding site, such as carbamylcholine, dithiobischoline, or one of the α-neurotoxins.[5-10] Theoretically, when 95% of RSH is alkylated (i.e., $r/r_0 = 0.05$), 0.3% of PSH is alkylated (i.e., $p/p_0 = 0.997$), and 2.5% of QSH is alkylated (i.e., $q/q_0 = 0.975$). Purified receptor contains 1 mole of RSH per 250,000 g[9] and 1 mole of PSH per ~15,000 g.[8] Consequently, when $r' = 0.95r_0$, $p' = (0.003)$ $(250,000/15,000)r_0$, and the specificity $(r'/r' + p') = 0.95/(0.95 + 0.05) = 0.95$.

Alkylation in the Absence of Dithiothreitol

In this procedure dithiothreitol is removed from the protein before [³H]MBTA is added. It is applicable in general to cells, membrane fragments, and solubilized receptor. The extent of labeling can be controlled by varying the ratio of $m_0:r_0$. The procedure for labeling 300 μg of purified receptor is as follows: To 300 μl of a solution containing 1 mg of receptor protein per milliliter of TNP50[20] are added 100 μl of TNP50 and 50 μl of 2 mM dithiothreitol in 200 mM Trischloride (pH 8.3). (The Tris brings the final pH to 8.0.) The mixture is incubated for 20 min at 25°, and then 75 μl of 0.56 M NaPO$_4$ (pH 6.7) are added, bringing the pH to 7.0. The mixture is cooled to 4° and layered on a 24 × 0.9 cm column of Bio-Gel P-6, 100–200 mesh (Bio-Rad Laboratories) preequilibrated with TNP150 and eluted with TNP150 at 0.4 ml per min at 4°. About 90% of the initial protein is recovered at the void volume in 2.5 ml, completely separated from the dithiothreitol. It is not necessary to determine the protein concentration at this stage since the elution properties of the column are nearly constant and may be checked before use with cytochrome c. To the 2.5 ml of reduced receptor (about 1 nmole)

[20] TNPxx stands for a buffer containing 0.2% Triton X-100, xx mM NaCl, 10 mM NaPO$_4$, 1 mM EDTA, 3 mM NaN$_3$ (pH 7.0).

at 25° are added 50 μl of 25 μM [³H]MBTA (1.25 nmoles). After 2 min,[21] the reaction is quenched with 25 μl of 0.1 M β-mercaptoethanol. The mixture is cooled to 4° and chromatographed on the P-6 column as before. The eluted fractions are sampled for radioactivity (10 μl into 5 ml of scintillant), and fractions containing about 90% of the radioactive protein are pooled. There is complete separation from small labeled products. The pooled fractions of labeled receptor contain about 200 μg of protein in 2.5 ml. In this procedure, at least 95% of the available sites are labeled.

With receptor of a specific activity of 2 nmoles of MBTA reacting sites per milligram of protein,[22] or greater, there is 10% or less non-specific reaction of [³H]MBTA under the above conditions. The extent of nonspecific reaction of [³H]MBTA can be determined by adding α-neurotoxin[23] at a final concentration of 1 μg/ml to an aliquot of the reduced receptor and incubating for about 15 min at 25° before adding [³H]MBTA.

Alkylation in the Presence of Dithiothreitol

It is possible to avoid the removal of the dithiothreitol before the addition of [³H]MBTA even though the concentration of dithiothreitol sulfhydryls (q_0) is much greater than m_0. As shown above, when 95% of RSH is alkylated, 2.5% of QSH is alkylated; hence, if the concentration of ³H-MBTA (m_0) is about 2.5% of q_0, the reaction with receptor will be 95% complete by the time the bulk of the [³H]MBTA has reacted with dithiothreitol. This approach is suitable for the assay of small quantities (1 μg) of receptor[7,8] as well as for the labeling of larger quantities.

The assay procedure is as follows: In 15-ml, screw-top, plastic tubes (Becton–Dickinson), quadruplicate 50-μl samples containing 1–50 μg of

---

[21] At 0.5 μM [³H]MBTA, the half-time for reaction with receptor is 2.3 sec.

[22] The number of molecules of ³H-MBTA that react specifically with receptor is exactly one-half the number of molecules of α-neurotoxin that are bound. The specific activities of presumably pure receptor from *Electrophorus* or *Torpedo* are 4 nmoles of "MBTA-sites" per milligram of protein and 8 nmoles of "toxin-sites" per milligram of protein; see A. Karlin, C. L. Weill, M. G. NcNamee, and R. Valderrama, *Cold Spring Harbor Symp. Quant. Biol.* **40,** 203 (1975).

[23] The principal α-neurotoxin of the venom of *Naja naja siamensis*, purified according to E. Karlsson, H. Arnberg, and D. Eaker, *Eur. J. Biochem.* **21,** 1 (1971), is conveniently used. Venoms and purified toxins are available from Miami Serpentarium Laboratories.

protein in TNP50 are mixed with 50 μl of 0.4 mM dithiothreitol in 0.2%
Triton X-100, 150 mM NaCl, 20 mM Trischloride, 1 mM EDTA, and
3 mM NaN$_3$ (pH 8.3) to give a final buffer composition of 0.2% Triton
X-100, 100 mM NaCl, 5 mM NaPO$_4$, 10 mM Trischloride, 1 mM EDTA,
3 mM NaN$_3$ (pH 8.0). (The concentrations of the components of the
medium may be varied somewhat, but the final pH should be 8.0.) All
incubations are at 25°. After 20 min, 10 μl of 0.53 M NaPO$_4$ (pH 6.7)
are added to bring the pH to 7.0. To two of the samples (set A) 400 μl
of TNP150 are added, and to the other two samples (set B), 400 μl of
TNP150 containing 3 μg of *Naja naja siamensis* α-neurotoxin per milliliter. After 15–20 min, 500 μl of 2 μM [³H]MBTA in 0.2% Triton X-
100–150 mM NaCl (unbuffered) are added to each sample. After a fixed
interval, which can be from 0.5 to 20 min, 25 μl of approximately 0.1 M
β-mercaptoethanol[24] are added. Finally, about 100 μg of reduced, carboxamidomethylated, succinylated lysozyme[25] are added as carrier, and
the tubes are capped and placed in ice. The tubes may be put in a freezer
for later processing, or the procedure may be carried forward immediately. It is possible to combine the 0.53 M NaPO$_4$ and TNP150 into one
solution, and also the β-mercaptoethanol and carrier, in order to reduce

---

[24] The addition of β-mercaptoethanol has little effect on the extent of labeling when
the samples are precipitated and filtered immediately, but it prevents an increase
in nonspecific labeling that results when samples are stored overnight or longer
at −20° prior to filtering. Since the primary reactions of label with dithiothreitol
and with protein are complete within 30 sec, the slow reaction leading to the increase in nonspecific labeling is probably due to the formation of mixed disulfides
between proteins and the dithiothreitol-[³H]MBTA adduct. This reaction is suppressed by the presence of a 2500-fold higher concentration of β-mercaptoethanol.
[25] Reduced, carboxamidomethylated, succinylated lysozyme (CSL) is required for
good recovery of small quantities of receptor during precipitation and filtration.[8]
In the presence of 100 μg of CSL, more than 90% of 1 μg of labeled receptor is
recovered in filtration under assay conditions. Substitution of lysozyme, bovine
serum albumin, bovine γ-globulin, or gelatin for CSL results in poor recoveries.
CSL is prepared as follows: Hen egg-white lysozyme, 100 mg, is dissolved in 10 ml
of saturated guanidine hydrochloride solution. The pH is adjusted to 8.5 with
concentrated NaOH. The solution is continuously stirred on a magnetic stirrer,
and 46 mg of dithiothreitol are added. After 20 min at room temperature, 555 mg
of iodoacetamide are added over a few minutes. The pH is adjusted back to 8.5
with 3–5 M NaOH and maintained at 8.5 while adding 400 mg of succinic
anhydride over about 10 min. After the pH no longer changes, the solution is
dialyzed against about 7 volumes of water, changed every 2 hr, for 8 hr total. A
precipitate may form in the bag, which eventually redissolves. The outside solution is changed to 10 mM Na$_2$CO$_3$ for 2 hr and then changed to water again for
2 additional 2-hr periods. Any precipitate remaining is removed by centrifugation.
The protein solution is directly lyophilized, or the protein is precipitated with
9 volumes of acetone, washed with acetone, and dried in a vacuum desiccator.

the number of separate additions. Forty tubes can be conveniently carried through all the steps by one person making the appropriate additions at 30-sec intervals.

In the assay procedure, labeled protein is isolated and separated from small labeled products by filtration. To each tube is added 14 ml of saturated ammonium sulfate solution, the top is replaced, and the tube is inverted. The contents are filtered under suction through a 24-mm glass fiber filter (Whatman GF/A) held in a polypropylene filter holder (Gelman). The tubes, including the top, are washed twice with 15 ml of 50% saturated ammonium sulfate solution, and the washings are filtered. Each filter is washed three times with about 25 ml of 50% saturated ammonium sulfate solution, twice with 25 ml of ice-cold 1 N HCl, and twice with 15 ml of acetone; the vacuum source is turned off while the funnel is being filled at each of these washing steps. The extensive wash procedure is required to obtain low blanks of the order of 500 cpm (0.05 to 0.1% of the total radioactivity in the sample filtered). At the conclusion of the washes, the filter is dry and free of ammonium sulfate and HCl.

The dry filters are placed in counting vials and treated with 0.1 ml of water and 0.6 ml of NCS solubilizer (Amersham-Searle) for 1 hr at 50°. Finally, 10 ml of a toluene-base scintillant (5 g PPO and 0.3 g of dimethyl POPOP per liter of toluene) are added. Typically, 1000–10,000 cpm is the tritium activity obtained per sample. The extent of site-specific labeling is the difference between the means of set A (fully labeled) and set B (protected with toxin). The factor converting from cpm to moles of label is obtained by dividing the total moles of labeled species $(a + b)$ by cpm in a counting standard; i.e., it is assumed that the counting efficiencies of [^{3}H]MBTA and of its hydrolysis product are the same. A standard is prepared by spotting an aliquot of diluted label onto a glass filter in a vial. Such standards in duplicate are digested and counted identically to the assay samples.

The assay is linear with quantity of receptor up to about 10 μg, with a slight but increasing deviation from linearity at higher quantities. The optimal range is 0.5 to 5 μg of receptor or 2 to 20 pmoles of sites. Even 50 μg of receptor in a final volume of 1 ml, however, will be labeled to at least 90% of completion under assay conditions. Proportionately larger reaction volumes may be used to label larger quantities of receptor. If these conditions are used to label receptor for other purposes than for assay, the labeled receptor can be separated from small labeled products by gel filtration as in the previous section.

The specificity of the reaction, $(A - B) \div A$ or $R'/(R' + P')$, varies with the purity of the receptor sample from 12% for 1% pure receptor to 95% for pure AChR.

[69] β-Adrenergic Receptors

By DAPHNE ATLAS

Recently, successful experiments on the probing of β-adrenergic receptors using radioactive *reversible* β-antagonists possessing high affinity such as propranolol,[1-4] alprenolol,[5] chloropractolol,[6] and [^{125}I]hydroxybenzylpindolol[7,8] have been reported. The binding studies have made it possible to calculate a turnover number for the hormone-stimulated adenylate cyclase.[3] However, additional understanding as to the molecular organization of the hormone–receptor–enzyme complex in the membrane can be tremendously advanced by developing a ligand that can bind *covalently* at the receptor site. A covalent affinity label for the β-receptor can become a powerful technique for the identification, characterization, and, it is hoped, the isolation of the β-receptor. Solubilization of the β-receptor adenylate cyclase complex causes the loss of the responsiveness of the enzyme. Thus the ability to monitor the receptor by activity measurements in its solubilized state is lost. A radioactive affinity label for the β-receptor can provide one with a specific marker for the receptor even subsequent to solubilization.

Synthetic Procedures

The derivatives that we have prepared are based on the structure of norepinephrine, practolol, and propranolol. All compounds synthesized were tested as β-blockers by their capacity to inhibit the turkey erythrocyte[9] β-receptor-dependent adenylate cyclase activity. Adenylate cyclase

[1] A. Levitzki, D. Atlas, and M. L. Steer, *Proc. Natl. Acad. Sci. U.S.A.* **71**, 2773 (1974).
[2] D. Atlas, M. L. Steer, and A. Levitzki, *Proc. Natl. Acad. Sci. U.S.A.* **71**, 4246 (1974).
[3] A. Levitzki, N. Sevilla, D. Atlas, and M. L. Steer, *J. Mol. Biol.* **97**, 35 (1975).
[4] S. R. Nahorski, *Nature (London)* **259**, 488 (1976).
[5] R. J. Lefkowitz, C. Mukherjee, M. Coverstone, and M. H. Caron, *Biochem. Biophys. Res. Commun.* **60**, 703 (1974).
[6] M. Erez, M. Weinstock, S. Cohen, and G. Shtacher, *Nature (London)* **255**, 635 (1975).
[7] G. D. Aurbach, S. A. Fedak, C. J. Woodard, J. S. Pamer, D. Hauser, and F. Troxler, *Science* **186**, 1223 (1974).
[8] M. E. Maguire, R. A. Wiklund, H. J. Anderson, and A. G. Gilman, *J. Biol. Chem.* **251**, 1221 (1976).
[9] M. L. Steer and A. Levitzki, *J. Biol. Chem.* **250**, 2080 (1975).

activity was assayed as previously described.[10] All protein determinations were performed according to Lowry et al.[11]

A. Epinephrine Derivative

Outline of Synthesis of 1,2-Epoxy-3-norepinephrine Propane

Procedure. Norepinephrine (bitartarate salt, 0.2 mmole, 67.2 mg) is partially dissolved in 0.2 ml of dioxane and neutralized with triethylamine (0.4 mmole, 0.056 ml); the amine formed is added to epichlorohydrin (excess). The reaction mixture is stirred at room temperature for 12 hr. The white crystalline triethylammonium bitartarate is filtered, and the filtrate is concentrated under reduced pressure. The residue is precipitated with ether. UV spectra of the compound shows that the catechol ring remains intact.

B. Outline of Synthesis of Bromopractolol
[Compound (I) in Tables I and II]

(*i*) *p*-Nitrophenol (2.8 g; 20 mmoles) is partially dissolved in 2 ml aqueous solution of NaOH (20 mmoles). The reaction is stirred for 15 min and epichlorohydrin (20 mmoles) is added slowly. The reaction is refluxed for 1 hr at 50°, allowed to stand overnight at room temperature, and extracted into chloroform three times. All the chloroform fractions are combined and evaporated under reduced pressure. Crystallization from petroleum ether yields needlelike crystals (m.p. 60°).

(*ii*) 1,2-Epoxy-3-*p*-nitrophenoxypropane (4.18 mmoles, 620 mg) is dissolved in 1 ml of absolute dioxane, and isopropylamine (3.2 mmoles; 186 mg) is added. The reaction is heated (70°) overnight in a sealed vessel. The oily residue is acidified with concentrated hydrochloric acid. A white precipitate is obtained upon the addition of ether (m.p. 140°–142°).

[10] Y. Salomon, C. Londos, and M. Rodbell, *Anal. Biochem.* **58**, 541 (1974).
[11] O. H. Lowry, N. J. Rosebrough, A. L. Farr, and R. J. Randall, *J. Biol. Chem.* **193**, 265 (1951).

TABLE I
New Reversible β-Blockers

Compound no.	Structure of free amine analog	K displacement of [³H]propranolol binding (M)	K activity for adenylate cyclase (M)
(I)	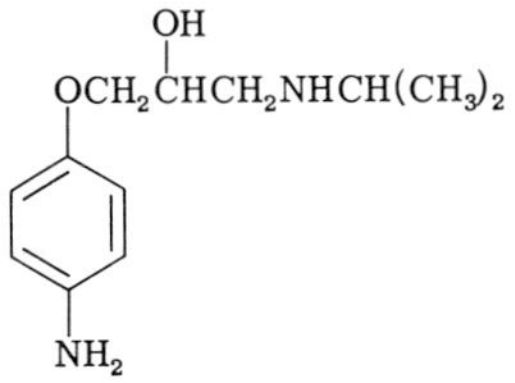 OH OCH₂CHCH₂NHCH(CH₃)₂ (para-NH₂ benzene)	—	(8 ± 1) × 10⁻⁶
(II)	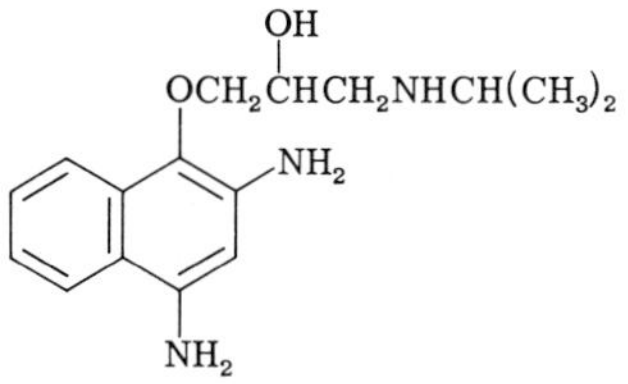 OH OCH₂CHCH₂NHCH(CH₃)₂ (naphthalene, 2-NH₂, 4-NH₂)	(7.5 ± 1) × 10⁻⁶	(1.5 ± 1) × 10⁻⁶
(III)	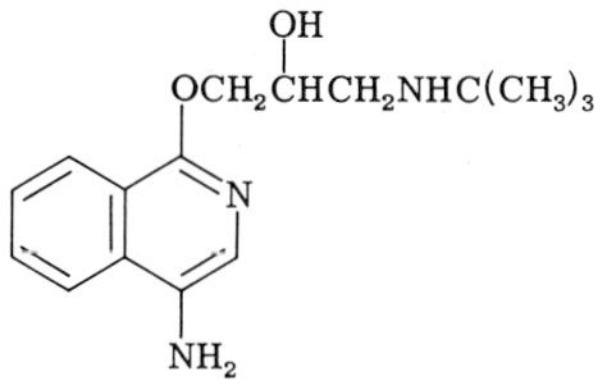 OH OCH₂CHCH₂NHC(CH₃)₃ (isoquinoline, NH₂)	(1 ± 1) × 10⁻⁴	—
(IV)	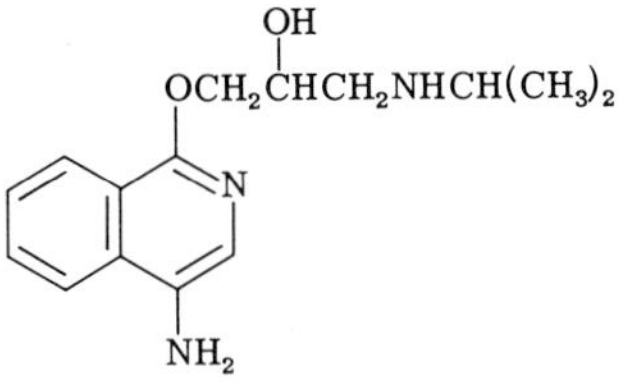 OH OCH₂CHCH₂NHCH(CH₃)₂ (isoquinoline, NH₂)	(5 ± 1) × 10⁻⁵	—
(V)	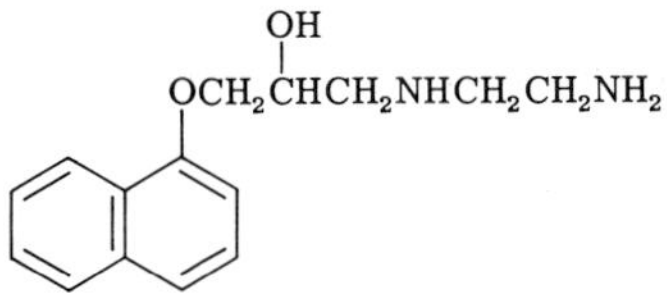 OH OCH₂CHCH₂NHCH₂CH₂NH₂ (naphthalene)	(2 ± 1) × 10⁻⁷	(2.5 ± 1) × 10⁻⁷

The structures are depicted below:

(I)
$$\text{OH} \atop \text{OCH}_2\text{CHCH}_2\text{NHCH(CH}_3)_2$$
para-aminophenoxy, with NH₂

(II)
$$\text{OH} \atop \text{OCH}_2\text{CHCH}_2\text{NHCH(CH}_3)_2$$
naphthalene with NH₂ and NH₂

(III)
$$\text{OH} \atop \text{OCH}_2\text{CHCH}_2\text{NHC(CH}_3)_3$$
isoquinoline with NH₂

(IV)
$$\text{OH} \atop \text{OCH}_2\text{CHCH}_2\text{NHCH(CH}_3)_2$$
isoquinoline with NH₂

(V)
$$\text{OH} \atop \text{OCH}_2\text{CHCH}_2\text{NHCH}_2\text{CH}_2\text{NH}_2$$
naphthalene

TABLE II

CONCENTRATIONS OF BROMOACETYL DERIVATIVES REQUIRED FOR 50% EFFECT[a]

Bromoacetyl derivative	$D_{0.5}$ (M) for [³H]propranolol displacement	$D_{0.5}$ (M) for l-epinephrine activity
(I)	—	$(1.5 \pm 1) \times 10^{-4}$
(II)	$(3 \pm 1) \times 10^{-4}$	$(2 \pm 2) \times 10^{-4}$
(III)	$(1.70 \pm 0.5) \times 10^{-4}$	—
(IV)	$(2 \pm 1) \times 10^{-4}$	—
(V)	$(2.5 \pm 1) \times 10^{-5}$	$(1.0 \pm 1.0) \pm 10^{-5}$

[a] The concentration of [³H]propranolol in the binding assay was 1.0×10^{-8} M (or 4 times its dissociation constant) and the concentration of l-epinephrine in the adenylate cyclase assay was 2.5×10^{-4} M (or 25 times its dissociation constant).

(iii) The normal procedure for catalytic hydrogenation is performed [see $C_I(iii)$ below]. The anilinium chloride derivative obtained separates as an oil.

(iv) The anilinium hydrochloride (0.68 mmole; 200 mg) obtained is partially dissolved in N,N'-dimethylformamide (0.1 ml). The equivalents of triethylamine are added to neutralize the amines, and the triethyl ammonium hydrochloride is removed by filtration. The filtrate is added to bromoacetic anhydride (0.68 mmole), and the reaction mixture is then stirred for 4 hr at room temperature. The compound is extracted with ethyl acetate (20 ml).

C. Propranolol Analogs

C_I. *Synthesis of 2,4-Bromoacetamidopropranolol [Compound (II) in Tables I and II]*

(i) Preparation of the 1-naphthoxy-2,3-epoxypropane of 2,4-dinitro-naphthol is carried out by a procedure similar to that used for the preparation of 1-naphthoxy-2,3-epoxypropane (C_{IV}).

(ii) 1,(2,4-Dinitro)-naphthoxy-2,3-epoxypropane (0.65 mmole; 190 mg) is partially dissolved in 2 ml of dioxane, and isopropylamine is added in excess. The reaction mixture is stirred and heated slightly (50°) in a sealed vessel. After 24 hr the reaction mixture is evaporated almost to dryness. The amine is acidified with HCl to yield the hydrochloride salt.

(*iii*) The hydrochloride salt (0.382 mole, 150 mg) is dissolved in methanol and submitted to catalytic hydrogenation at atmospheric pressure overnight. The catalyst (Pd 10% on C) is removed by filtration, and the residue is evaporated to dryness. The reddish residue is precipitated with ether. The compound is dried under reduced pressure.

(*iv*) The hydrochloride salt (see *iii*) is partially dissolved in *N,N'*-dimethylformamide. Triethylamine (3 equivalents) is added to neutralization. The amine is added to an excess of an insoluble ortho-hydroxy-polystyrene polymer[12] loaded with bromoacetic acid. The reaction mixture is stirred very mildly overnight. The polymer is then removed by filtration and washed with *N,N'*-dimethylformamide. The filtrate is evaporated almost to dryness, and the residue is precipitated with water, filtered, and dried over H_2SO_4 and $CaCl_2$.

C_{II}. *Synthesis of 1-[(1-Oxy-2-hydroxy-3-tertbutylamino)propane]-3-bromoacetamidoisoquinoline [Compound (III) in Tables I and II]*

(*i*) See preparation of 1,2-epoxy 3-(*p*-nitroisoquinoline) propane ($C_{III}i$).

(*ii*) The epoxide (see above) (3.33 mmoles) is dissolved in *tert*-butyl-amine (4 mmoles, excess) and stirred with heating at 70° for 12 hr. The solution is subsequently evaporated under reduced pressure in order to remove free *tert*-butylamine. The residue is acidified with hydrochloric acid. Then the solution is evaporated, and the residue is precipitated with ether (m.p. 210°).

(*iii*) Hydrogenation of the nitro group. See procedure for catalytic hydrogenation of compound (I) (step *iii*).

(*iv*) The free aromatic amine of the isoquinoline derivative (*iii*) (0.309 mmole; 100 g) is suspended in dioxane (0.5 ml), and bromoacetyl succinimide ester (0.34 mmole; 81 mg) is added in solid form. The reaction mixture is stirred for 12 hr and then evaporated to dryness. The residue is treated with ether to yield a yellowish precipitate, which is collected on a sintered-glass funnel and dried over H_2SO_4 and $CaCl_2$.

C_{III}. *Synthesis of 1-(Oxy-2-hydroxy-3-isopropylamine)propane-3-bromo-acetamidoisoquinoline [Compound (IV) in Tables I and II]*

(*i*) *4*-Nitroisocarbostyryl (5 mmoles; 950 mg) is partially dissolved in 2 ml of dioxane and 0.5 ml of water. Potassium hydroxide (5 mmoles) is added in a solid form, and the yellow color of the 4-nitroisocarbostyryl turns brown. After 15 min, epichlorohydrin (5 mmoles, 280 mg) is added

[12] D. Atlas, R. Kalir, and A. Patchornik, *FEBS Lett.* **58**, 179 (1975).

dropwise to the stirred naphthoxy solution. Stirring is continued for 12 hr and evaporated to dryness. The dried residue is extracted three times with ethyl actate (45 ml), after which the solvent is evaporated to dryness. The compound is dried over KOH and H_2SO_4.

(*ii*) The opening of the epoxide ring of compound (*i*) is carried out according to a procedure discussed above ($C_I ii$).

(*iii*) Hydrogenation of the 4-nitro group was carried out according to the procedure described in previous sections ($C_I iii$).

(*iv*) The amine obtained after hydrogenation in step (*iii*) (0.2 mmole; 70 mg) is suspended in chloroform (2 ml) and traces of dimethylformamide. Triethylamine (0.4 mmole; 0.056 ml) is then added. To the solution, an active polymer, 4-hydroxy-3-nitro benzylated polystyrene, loaded with bromoacetic acid, is added in excess.[12] After 4 hr the insoluble polymer is removed by filtration, chloroform is evaporated, and the residue is precipitated with water. The precipitate is dried under reduced pressure.

C_{IV}. *Outline of Synthesis of N-(2-Hydroxy-3-naphthyloxypropyl)ethylenediamine NHNP-E [Fig. 1, Compound (V) in Table I]*

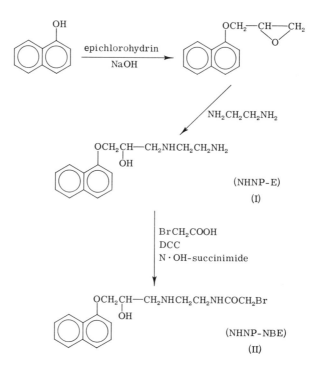

Fig. 1. Outline of synthesis of NHNP-NBE.

α-Naphthol (1 equivalent) is allowed to react with epichlorohydrin in the presence of one equivalent of sodium hydroxide to form the α-naphthyloxy-1-propane-2,3-oxide. The α-naphthyloxy-1-propane-2,3-oxide, without further purification, and ethylenediamine in equimolar quantities are heated overnight on a water bath at 70°; no solvent is needed. The crude hydrochloride salt[13] (300 mg) is purified on a Sephadex G-10 column (2 × 56 cm) and separated. The product is identified by its UV absorption at 295 nm, and by ninhydrin staining. The hydrochloride salt of the amine contains 1.5 molecules of water.

Cv. *Outline of synthesis of N-(2-Hydroxy-3-napthyloxypropyl)-N'-bromoacetylethylenediamine (NHNP–NBE) [Fig. 1, Compound (V) in Table II]*

The affinity label [compound (II), Fig. 1] is obtained by allowing the amine analog [see Compound (I)] (α-naphthyloxypropanolethylenediamine) to react with the N-hydroxysuccinimide ester of bromoacetic acid.

The amine derivative [NHNP-E, compound (I)] (100 mg, 0.3 mmole) is partially dissolved in 2 ml of ether–dioxane mixture (1:1). Triethylamine (0.08 ml, 0.57 mmole) is added, and the mixture is heated to 50° to complete neutralization. Then N-hydroxysuccinimide ester of bromoacetic acid, dissolved in 0.2 ml of dioxane, is added (0.3 mmole; 10% excess), and the reaction is carried out at room temperature with occasional stirring. After 1 hr the reaction mixture is evaporated and the residue is washed with ether and water at pH 5.5. The white precipitate is dried over H_2SO_4. Overall yield 40%.

Cvi. *Synthesis of Bromomethyl-(2-hydroxy-3-naphthoxypropylaminomethyl)ketone (BHNPK)*

Figure 2 outlines the synthesis of compound (III) (BHNPK), which was not effective as an affinity label for the β-adrenergic receptor. 1-Naphthoxy-2,3-epoxypropane (Fig. 1, 0.5 mmole; 100 mg) is added to benzylamine (0.8 mmole; 85.6 g) and the mixture is refluxed for 36 hr at 45°. The product separates as white, rodlike crystals and is washed with ether (yield, 81%, m.p. 110°). The catalytic hydrogenation of the benzyl group to yield toluene and the appropriate free amine is conducted according to Liwschitz et al.[14] After reduction the separated hydrochlo-

[13] H. R. Ing and W. E. Ormerad. *J. Pharmacol.* 9, 21 (1952).
[14] Y. Liwschitz, Y. Rabinsohn, and D. Perera, *J. Chem. Soc. London* **1962**, 1116 (1962).

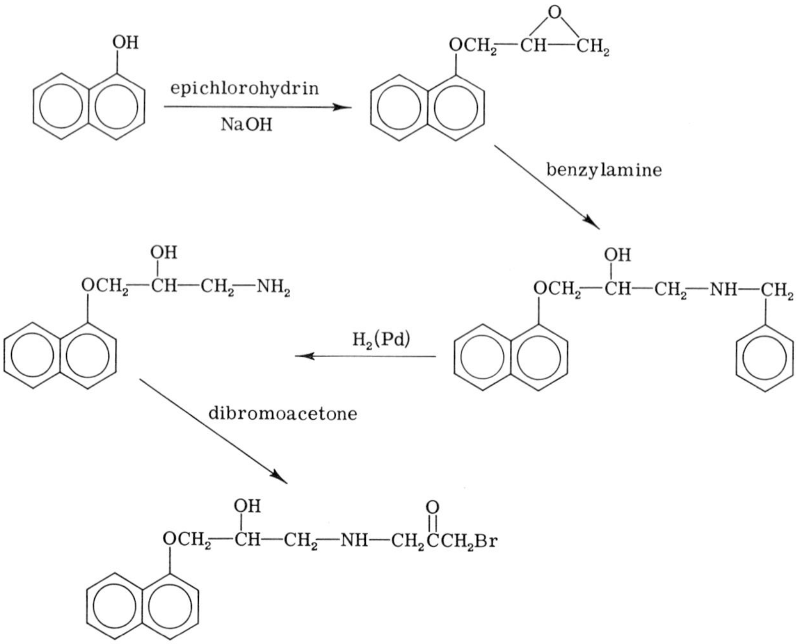

FIG. 2. Outline of synthesis of BHNPK.

ride salt of the naphthoxy derivative (0.4 mmole, 100 mg) is suspended in dioxane, and two equivalents of N,N'-diisopropylethylamine are added (0.8 mmole, 103 mg). The reaction mixture is added to dibromoacetone (0.44 mmole, 95 mg) dissolved in traces of dioxane. After 4 hr, the bromide and chloride salts are removed by filtration, and the filtrate is evaporated under reduced pressure. The compound is obtained by precipitation from ether.

Conclusions

Reversible β-Blockers. A number of reversible β-blockers have been synthesized (Table I).[15,16] The effectiveness of these compounds in inhibiting *l*-catecholamine-dependent adenylate cyclase activity and their ability to displace specifically bound *dl*[^{3}H]propranolol was tested.[15] Using both techniques, the compounds listed in Table I were demonstrated to be competitive inhibitors. The dissociation constants for the β-

[15] D. Atlas and A. Levitzki, *Biochem. Biophys. Res. Commun.* **69**, 397 (1976).
[16] D. Atlas, M. L. Steer, and A. Levitzki, *Proc. Natl. Acad. Sci. U.S.A.* **73**, 1921 (1976).

blockers to the β-receptors were calculated from the kinetic measurements (Table I) and from the direct-binding experiments as previously described.[1-3]

Irreversible β-Blockers. The bromoacetyl derivatives of the compounds shown in Table I were found to be β-blockers for the β-adrenergic receptors (Table II). Their effects were studied as described above. All compounds tested were shown to be devoid of any effect on the fluoride activation of adenylate cyclase. However, BHNPK is without inhibitory effect, either on the binding of [³H]propranolol or on the hormone-dependent adenylate cyclase activity. The most effective affinity label found was N-(2-hydroxy-3-napthoxypropyl)-N'-bromoacetylethylenediamine (Table II), which will be discussed in detail in the following sections.

Irreversible β-Agonists. Bromoacetyl compounds such as N-bromoacetylnorepinephrine have the same effect as bromoacetic acid (unpublished). The very low affinity of the compound toward the receptor demands the use of high concentrations of these ligands, thereby resulting in nonspecific alkylation of the protein membrane. 1,2-Epoxy-3-norepinephrine-propane (see synthetic procedure A) retains its properties as a full agonist but fails to form a covalent bond at or near the receptor site.

Inhibition of Epinephrine-Dependent Activity and [³H]Propranolol Binding by NHNP-E

The compound NHNP-E, which is the parent compound for the bromoacetyl analog, NHNP-NBE, was found to be a competitive inhibitor of the β-receptor. From the concentration of NHNP-E needed to displace 50% of the bond [³H]propranolol, one can calculate a dissociation constant of $(2.0 \pm 10) \times 10^{-7}$ M to the β-receptor. From the concentration of NHNP-E required to inhibit 50% of the l-epinephrine-dependent activity, one can calculate a dissociation constant of $(2.5 \pm 1.0) \times 10^{-7}$ M. Treatment of the membranes with up to 5×10^{-4} M bromoacetic acid, or other bromoacetyl compounds such as BHNPK, has no effect on the [³H]propranolol binding capacity.

The Rate of Inactivation

The routine experimental conditions of irreversible blockage of either epinephrine-dependent activity or irreversible inhibition of [³H]propranolol binding included incubation of the membranes with NHNP-NBE for 15 min. However, it was found that even after less than 1 min incubation, essentially the same values were obtained, thereby indicating that the covalent labeling step is extremely fast.[17]

The Irreversible Adenylate Cyclase Inhibition and [³H]Propranolol Binding by NHNP-NBE

NHNP-NBE inactivates irreversibly the hormone-dependent adenylate cyclase activity, whereas the fluoride-stimulated activity remains intact. Bromoacetic acid as well as bromoacetamide causes enzyme inactivation at concentrations two orders of magnitude higher than those obtained with NHNP-NBE. Other bromoacetyl derivatives of propranolol, e.g., BHNPK (Fig. 2),[15,16] which do not react specifically with the β-receptor, were similar in behavior to bromoacetic acid (see below) and bromoacetamide. Specific binding of [³H]propranolol to the β-receptor was inhibited irreversibly by the compound NHNP-NBE.[17]

Protection of the β-Receptor against Affinity Label by dl-Propranolol

The reversible β-blocker, propranolol, was found to protect the β-receptor against irreversible loss of l-epinephrine-dependent activity as well as against irreversible loss of [³H-]propranolol binding. The results are summarized in Table III. It may be seen that the protection offered by the propranolol is only partial.

TABLE III

PROTECTION OF THE β-RECEPTOR IN RED BLOOD CELL MEMBRANES AGAINST
THE IRREVERSIBLE β-BLOCKER NHNP-NBE BY PROPRANOLOL[a]

Affinity label concentration (M)	Percent inhibition of epinephrine-dependent activity		Percent inhibition of [³H]propranolol binding	
	No addition	With propranolol[b]	No addition	With propranolol[c]
5 × 10⁻⁶	27 ± 3	0.0	21 ± 5	0.0
5.0 × 10⁻⁵	—	—	65 ± 10	20 ± 5

[a] Experimental details are given in the text. 100% epinephrine-dependent activity is 100 pmoles of cAMP per milligram per minute; 100% binding of [³H]propranolol is 1.8 ± 0.3 pmole per milligram protein.
[b] dl-Propranolol, 2 × 10⁻⁵ M.
[c] dl-Propranolol, 1.0 × 10⁻⁶ M.

[17] F. C. Hartman, Biochemistry 10, 146 (1971).

*Irreversible Inactivation of Epinephrine-Dependent cAMP Formation in
Intact Red Blood Cells by NHNP-NBE*

NHNP–NBE was found to block irreversibly the epinephrine-dependent cAMP formation in intact erythrocytes. As in the case of the erythrocyte membranes, propranolol protects the intact cell against the irreversible β-blocker.[16] Similarly, *l*-epinephrine protects against the irreversible inhibition by NHNP–NBE. Propranolol and *l*-epinephrine offer only partial protection in both the intact erythrocytes and membranes prepared from them. This behavior is probably due to the extremely fast covalent labeling step.

Acknowledgments

D.A. is supported by a fellowship from the Lady Davis Fellowship Trust and by a research grant from the Bat-Sheva de Rothschild Fund for the Advancement of Science and Technology.

[70] Opiate Receptors

By Brian M. Cox

The potential contribution that affinity labeling of opiate receptors appears to offer in the isolation of the receptors and their chemical analysis and in studies of their distribution and function has not as yet been achieved. However, the results of preliminary work offer some guidelines for future experimental approaches to this problem. Useful opiate receptor labels must have high reactivity with the receptor while eschewing interactions with other molecules, and hence selected compounds are likely to be closely related to ligands with high affinity. So far, however, only relatively simple opiates, with low or moderate potency, have been derivatized in the search for possible affinity labels. Reactive groupings through which covalent attachment is achieved need to be located on those parts of the molecule that lie in close proximity to the receptor surface. Most potent opioid drugs have structures related to morphine, and critical features for receptor interactions appear to be the phenolic hydroxyl group in position 3, oriented in a particular relationship to a nitrogen atom protonated at physiological pH. Up to the present time, reactive groups employed in opiate affinity labels have been coupled through this nitrogen atom, where a number of different substituents are compatible with high receptor affinity.

Other potential attachment sites include the 6-position, where relatively bulky substituents can be accommodated without substantial loss of activity.[1] In the 6,14-endoethenotetrahydrothebaine series, considerable increases in potency are associated with the insertion of the 6,14-endoetheno bridge and substitution of the 7-position[2]; reactive groups might usefully be attached to the 7-position substituents. The marked loss in activity associated with such limited modification of the 3-position substituent as replacement of OH by OCH_3 (cf. morphine and codeine) suggests that insertion of reactive groups in this position might be associated with considerable reduction in receptor affinity.

Consideration must also be given to the lipophilicity of the affinity label; accumulation by cell lipid results in high levels of nonspecific labeling. It is necessary to differentiate between persistent high-affinity reversible binding, which is often observed with very lipophilic compounds, and the specific covalent attachment of the label to the receptor. These types of binding may be distinguished by washing the tissue or receptor preparation with denaturing solvents, which will also extract small lipophilic compounds unless these are attached to tissue macromolecules. Washing media that have been used include trichloracetic acid and diethyl ether. Conventional histological tissue preparation procedures (fixation in formalin–saline, dehydration in ethanol, and clearing in xylene) are also very effective in removing reversibly bound lipophilic drugs.

In designing a photoaffinity label for opiate receptors, additional consideration must be given to the photosensitive group. There is some evidence that opiate receptors in brain homogenate are destroyed by intense UV irradiation over periods of several minutes.[3] It is therefore desirable to select precursor groups that are activated at an absorption wavelength where receptor damage due to photolysis is less probable.

The demonstration of the covalent attachment of the label to the receptor requires the use of *in vitro* opiate assay systems since persistent changes in *in vivo* opiate sensitivity that might be induced by potential affinity labels cannot be interpreted unambiguously. Two *in vitro* systems have been used to study opiate drug–receptor interactions in some detail.

High-affinity, saturable, stereospecific binding of tritium-labeled

---

[1] E. J. Simon, W. P. Dole, and J. M. Hiller, *Proc. Natl. Acad. Sci. U.S.A.* **69**, 1835 (1972).

[2] K. W. Bentley and J. W. Lewis, *in* "Agonist and Antagonist Actions of Narcotic Analgesic Drugs" (H. W. Kosterlitz, H. O. J. Collier, and J. E. Villarreal, eds.), p. 7. Macmillan, New York, 1973.

[3] B. M. Cox, K. E., Opheim, and A. Goldstein, *in* "Tissue Responses to Addictive Drugs" (D. H. Ford and D. H. Clouet, eds.), p. 373. Spectrum, New York, 1976.

opiate drugs has been demonstrated in homogenates of brain tissue.[4-7] Opiates that have been used (at a final concentration 0.1 to 10 nM) include [³H]etorphine and [³H]dihydromorphine (both agonists) and [³H]naloxone and [³H]naltrexone (both antagonists).[8] Separation of free from bound drug can be achieved by centrifugation or by rapid filtration through glass fiber filters (Whatman GF/B or GF/C).

The myenteric plexus–longitudinal muscle preparation from guinea pig ileum[9,10] offers the advantage that both receptor binding[3,11] and functional effects of receptor occupation can be observed in the same tissue. The preparation is set up in an organ bath in oxygenated Krebs–bicarbonate buffer at 37° and stimulated through platinum electrodes with single pulses at 10-sec intervals. Opiate drugs produce a reversible inhibition of the elicited contractions by reducing the output of acetylcholine from neurons of the myenteric plexus.[9] This effect is specifically blocked by the opiate antagonist naloxone (10 to 100 nM). In both these systems there is a close correlation between potency *in vitro* and analgesic potency in animals and man.[7,11] Other *in vitro* preparations in which specific opiate effects have been demonstrated, and which may be useful in future receptor labeling studies, include the electrically stimulated mouse vas deferens preparation[12] and nuroblastoma–glioma hybrid cells in culture.[13]

Alkylating Agents as Opiate Receptor-Affinity Labels

There are only two studies of the *in vivo* effects of alkylating derivatives of opioid drugs, and all the compounds employed had relatively low potency. May *et al.*[14] showed that two *N*-2-bromoalkyl-substituted benzomorphans produced prolonged analgesic, hypothermic, and depres-

[4] A. Goldstein, L. I. Lowney, and B. K. Pal, *Proc. Natl. Acad. Sci. U.S.A.* **68**, 1742 (1971).

[5] L. Terenius, *Acta Pharmacol. Toxicol.* **32**, 317 (1973).

[6] E. J. Simon, J. M. Hiller, and I. Edelman, *Proc. Natl. Acad. Sci. U.S.A.* **70**, 1947 (1973).

[7] C. B. Pert and S. H. Snyder, *Science* **179**, 1011 (1973).

[8] Tritium-labeled opiate drugs with high specific activity are available from Amersham/Searle Corp. and New England Nuclear Corp.

[9] W. D. M. Paton, *Br. J. Pharmacol.* **12**, 119 (1957).

[10] R. Schulz and A. Goldstein, *J. Pharmacol. Exp. Ther.* **183**, 404 (1972).

[11] I. Creese and S. H. Snyder, *J. Pharmacol. Exp. Ther.* **194**, 205 (1975).

[12] J. Hughes, H. W. Kosterlitz, and F. M. Leslie, *Br. J. Pharmacol.* **53**, 371 (1975).

[13] S. K. Sharma, M. Nirenberg, and W. A. Klee, *Proc. Natl. Acad. Sci. U.S.A.* **72**, 590 (1975).

[14] M. May, L. Czoncha, D. R. Garrison, and D. J. Triggle, *J. Pharm. Sci.* **57**, 884 (1968).

sant effects after systemic injections in mice. However, their experiments do not clearly demonstrate the opioid nature of the observed effects, or an irreversible interaction with the receptor. Portoghese et al.,[15] in a study of a series of N-acyl-substituted anileridines, showed that the fumarido ethyl ester produced analgesia in mice with about one fourth the potency of morphine. Unlike other analgesically active compounds in this series, the fumarido ethyl ester reduced the effects of a subsequently administered dose of morphine. The initial analgesic effects of the compound were prevented if the specific opioid antagonist naloxone was given previously. These results could be interpreted to imply that the compound produces initial receptor activation, followed by irreversible blockade, but further experiments are needed to confirm this.

Irreversible interactions between these alkylating analogs of opioid drugs and opiate receptors in vitro have not been demonstrated. Other groups have shown that the stereospecific binding of labeled opiates to brain homogenates can be reduced by prior exposure to alkylating agents, such as phenoxybenzamine,[16] or a 2-chloroethyl-substituted local anaesthetic.[17] Compounds with high reactivity toward sulfhydryl groups (N-ethylmaleimide, 0.1 mM; iodoacetamide, 5 mM) will also reduce opiate receptor binding.[6] Opiate agonist binding is reduced more than antagonist binding,[18] a discrimination which implies that the effect is more probably related to an action on the opiate receptor than to non-specific changes in the membrane. Attempts to introduce groups with high sulfhydryl reactivity into structures with high opiate receptor affinity to yield potential affinity labels have not been reported.

Photoaffinity Labeling of Opiate Receptors

Winter and Goldstein have described the synthesis of a potential opiate receptor photoaffinity label, N-(2-p-azidophenylethyl-1,1-^{3}H$_2$)-norlevorphanol ([^{3}H]APL),[19] and a modified synthetic procedure for both the tritium and carbon-14 labeled compounds has recently been published (Fig. 1).[20] The quaternary analog of APL [N-methyl-N-(2-p-azidophenylethyl-1,1-^{3}H$_2$)norlevorphanol; [^{3}H]MAPL] has also been

---

[15] P. S. Portoghese, V. G. Telang, A. E. Takemori, and G. Hayashi, J. Med. Chem. 14, 144 (1971).

[16] T. J. Cicero, C. E. Wilcox, and E. R. Meyer, Biochem. Pharmacol. 23, 2349 (1974).

[17] J. M. Musacchio and G. L. Craviso, Fed. Proc., Fed. Am. Soc. Exp. Biol. 35, 263 (1976).

[18] H. A. Wilson, G. W. Pasternak, and S. H. Snyder, Nature (London) 253, 448 (1975).

[19] B. A. Winter and A. Goldstein, Mol. Pharmacol. 8, 601 (1972).

[20] J. I. DeGraw and J. S. Engstrom, J. Labelled Compd. 11, 233 (1975).

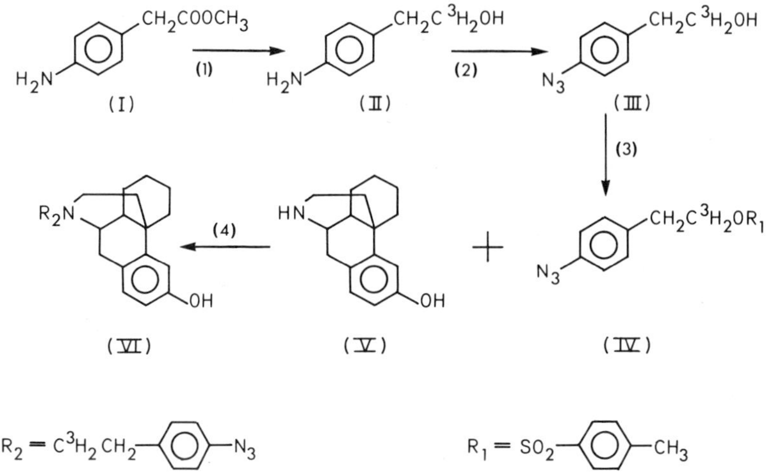

Fig. 1. Synthesis of N-(2-p-azidophenylethyl-1,1-^{3}H$_2$)norlevorphanol (APL) [J. I. DeGraw and J. S. Engstrom, *J. Labelled Compd.* **11**, 233 (1975)]. Compounds: (I) methyl-p-aminophenylacetate; (II) 2-(p-aminophenyl)ethanol-(1,1-^{3}H$_2$); (III) 2-(p-azidophenyl)ethanol-(1,1-^{3}H$_2$); (IV) 2-(p-azidophenyl)ethanol-(1,1-^{3}H$_2$)-p-toluene-sulfonate; (V) norlevorphanol; (VI) N-(2-p-azidophenylethyl-1,1-^{3}H$_2$)norlevorphanol, [^{3}H]APL.

(1) Dry bis(2-methoxyethyl)ether (4 ml) is mixed with 7.5 mmoles of AlCl$_3$ at 0°, followed by the addition of 5 mmoles of I [for synthesis of (I) see references cited in text footnotes 20 or 21]. After 10 min, a slurry of 20.5 mmoles of NaBH$_4$ in 15 ml of bis(2-methoxyethyl)ether is added dropwise, and the mixture is stirred for 15 hr. Cold water, 5 ml, is added and the mixture is extracted three times with 10-ml portions of dichloromethane. Combined extracts are dried over MgSO$_4$ and then evaporated to yield (II) (64% yield). [To obtain tritiated (II), NaBH$_4$ is replaced by NaB^{3}H$_4$ in the same molar proportions.]

(2) To 1.04 mmoles (II) in 0.71 ml of 4 N HCl, cooled to 0°, are added 1.04 mmoles of NaNO$_2$ in 0.36 ml water. After 20 min, 1.04 mmoles of NaN$_3$ in 0.36 ml of water are added, and the mixture is stirred at 0° for 1 hr. After extraction with three 6-ml portions of dichloromethane, the combined extracts are washed with 1 ml of cold water, dried over MgSO$_4$, and evaporated to yield (III) (85%).

(3) To 0.88 mmole of (III) in 2.4 ml of pyridine are added 1.78 mmoles of p-tolu-ene-sulfonyl chloride, and the mixture is maintained at 0° for 3 hr. Water, 13, ml, is added, and the mixture is extracted with 10 ml of dichloromethane. The extract is dried over MgSO$_4$ and evaporated to give (IV) (75%).

(4) Powdered anhydrous K$_2$CO$_3$ 0.76 mmole, norlevorphanol 0.778 mmole (Hoff-mann-La Roche Inc.), 0.655 mmole of (IV), and 1.1 ml of methanol are stirred at 41°–43° in a sealed flask protected from light. After 60 hr, 4.5 ml of water and 10 ml of a 2:1 benzene–ether mixture are added, and the stirring is continued for 30 min. The organic layer is removed and an additional extraction with 10 ml of the benzene–ether mixture is performed. The combined extracts are washed with 3.0 ml of water and dried over MgSO$_4$. The hydrochloride salt is precipitated by addition of ethereal hydrogen chloride. The precipitate is dissolved in 4.5 ml of water, and the solution adjusted to pH 8 with 1.5 N NH$_4$OH. Two 1.5-ml portions of chloro-form are used to extract the precipitated oil. The extracts are dried over MgSO$_4$ and evaporated. Compound (VI) is further purified by thick-plate TLC (SiO$_3$:methanol: chloroform, 15:85, R_f = 0.45, 30% yield).

synthesized by methylation of APL with excess methyl iodide in the presence of sodium bicarbonate, followed by purification by chromatography on Dowex 1-X8 anion exchange resin.[3]

Both APL and MAPL should be handled under subdued light until photolytic activation of the azide group is required. Exposure to UV light results in the generation of the very reactive nitrene, capable of insertion into carbon–hydrogen bonds, hydrogen abstraction, or addition to double bonds. Incubates containing tissues such as brain homogenates or strips of guinea pig ileum and [³H]APL or [³H]MAPL are placed in Kimax or Pyrex test tubes (which filter out low wavelength radiation and hence reduce receptor damage) and exposed to light from a Hanovia medium-pressure mercury arc lamp with a Kimax sleeve placed inside a quartz immersion well. Both lamp and sample tubes are immersed in a cooled water bath. In clear solution, conversion of azide to nitrene is complete within 5 min, but the presence of protein, tissue, or fragments of tissue may exert a filtering effect necessitating an increase in irradiation time.

APL has been shown to produce a naloxone-reversible analgesia in mice.[19] In the dark both APL and MAPL exert typical opiate effects in *in vitro* assays, with dissociation constants of about 1 nM and 50 nM, respectively.[3] UV irradiation of APL with bovine serum albumin results in the formation of a covalent complex,[19] but attachment to opiate receptors remains to be demonstrated. The quaternary analog, MAPL, was synthesized in an attempt to reduce the accumulation of label by cell lipid, which complicates analysis of APL binding. Attachment of [³H]MAPL to opiate receptors in myenteric plexus of guinea pig ileum following UV irradiation has been demonstrated by autoradiography,[3] and Schultz and Goldstein[21] have shown that irradiation of MAPL with the guinea pig ileum preparation leads to a permanent opioid effect, i.e., inhibition of acetycholine release. The specificity of the MAPL attachment was demonstrated by the protective effect exerted by naloxone when present during the irradiation, and its irreversible nature was confirmed by the failure of naloxone to antagonize the reduction in acetylcholine release when administered after the irradiation.

Acknowledgments

Some of the work reported here was supported by National Institutes of Health Grants DA-972 and DA-1199. The author is grateful to Dr. Avram Goldstein for helpful discussion.

[21] R. Schulz and A. Goldstein, *Life Sci.* 16, 1843 (1975).

[71] Amino Acid Transport Proteins

By G. I. GLOVER

Rationale for the Approach

Historically, it has been useful to approach the study of a complex system by breaking it down into its components to simplify the investigation and then reconstructing it. The study of active transport by this general methodology has taken a variety of directions. Membrane vesicles have allowed the study of active transport apart from possible interactions with cytoplasmic and genetic material and have given insight into the coupling of energy to transport.[1] Other techniques have been applied to the identification and purification of the molecular components (proteins) of transport systems. Osmotic shock has led to the release of proteins capable of binding a variety of transport substrates.[2] In the event the transport system is inducible, differential labeling techniques appear to be useful in introducing radioactive markers into transport proteins to guide their isolation.[3,4] Utilizing equilibrium dialysis to assay for binding, proline binding proteins have been solubilized by means of detergents from the membranes of *Eschericia coli.*[5]

These approaches have all depended on one or more characteristics of individual transport systems that are not common to all systems: the transport proteins must be removable from the membrane by osmotic shock, or be inducible, or retain binding activity upon solubilization from the membrane or removal by osmotic shock. An alternative that depends on a transport system having specificity for a particular substrate, or class of substrate, and high affinity for the substrate(s) is affinity labeling. A radioactive affinity label could, in principle, be used to specifically label the substrate binding proteins of a transport system and the label could be used as a marker to follow the purification of the protein. The experience gained could then be applied to isolation of the potentially active, unlabeled protein.

Few reports of affinity labeling of transport sites have appeared.

[1] H. R. Kaback, *CRC Critical Reviews in Microbiology,* p. 333 (1973).

[2] D. L. Oxender, *Annu. Rev. Biochem.* **41,** 777 (1972).

[3] C. F. Fox and E. P. Kennedy, *Proc. Natl. Acad. Sci. U.S.A.* **54,** 891 (1965).

[4] A. R. Kolber and W. D. Stein, *Curr. Mol. Biol.* **1,** 244 (1967).

[5] A. S. Gordon, F. J. Lombardi, and H. R. Kaback, *Proc. Natl. Acad. Sci. U.S.A.* **69,** 358 (1972).

Becker *et al.*[6] specifically inactivated the biotin transport system of *E. coli* using biotin *p*-nitrophenyl ester. The lactose transport protein of *E. coli* was labeled by *N*-bromoacetyl-β-D-galactopyranosylamine.[7] In the only reported attempt to isolate an affinity labeled protein,[8] glucose 6-isothiocyanate, which is an affinity label for the glucose transport system in human erythrocytes, gave enough nonspecific labeling of other membrane proteins to render identification of the transport protein difficult.

It is not unexpected that electrophilic affinity labeling reagents, by virtue of their inherent chemical reactivity, might react randomly with nucleophilic groups found on the surfaces of proteins in addition to their site-specific reactions. This criticism[9] is valid, but affinity labeling reagents with reactive alkylating groups on them have shown great selectivity and given stoichiometric alkylation of purified proteins.[10]

Design of an Affinity Labeling Reagent

The approach to selecting a suitable site for introducing a reactive grouping into the substrate is the same whether one is seeking to label a purified enzyme or one that is membrane bound. We investigated the substrate specificity of the general amino acid transport system in *Neurospora crassa*[11] and the tyrosine/phenylalanine transport system in *Bacillus subtilis*[12] by testing amino acids and amino acid derivatives and analogs as competitive inhibitors. In *Neurospora* we found that all common amino acids, with exception to the acidic amino acids and their amides, had high affinities for the general system. *N*-Acyl amino acids were poor inhibitors while those amino acids modified in the carboxylate group (esters or amides) were reasonable inhibitors with K_i values in the millimolar range. The tyrosine/phenylalanine transport system is specific for these amino acids, and it was found that carboxyl modification led to better inhibitors than did alteration of the amino groups.

[6] J. M. Becker, M. Wilcheck, and E. Katchalski, *Proc. Natl. Acad. Sci. U.S.A.* **68**, 2604 (1971).

[7] J. Yariv, A. J. Kolb, and M. Yariv, *FEBS Lett.* **27**, 27 (1972).

[8] R. D. Taverna and R. G. Langdon, *Biochem. Biophys. Res. Commun.* **54**, 593 (1973). See also this volume [13].

[9] S. I. Chavin, *FEBS Lett.* **14**, 269 (1971).

[10] E. Shaw, *Physiol. Rev.* **50**, 244 (1970).

[11] C. W. Magill, S. O. Nelson, S. M. D'Ambrosio, and G. I. Glover, *J. Bacteriol.* **113**, 1320 (1973).

[12] S. M. D'Ambrosio, G. I. Glover, S. O. Nelson, and R. A. Jensen, *J. Bacteriol.* **115**, 673 (1973).

Accordingly, the logical site for modification of the substrate molecule in either case was the carboxyl group. We chose to utilize the chloromethyl ketones of leucine (LUCK) and lysine (LCK) for *Neurospora*, and those of phenylalanine (PCK) and tyrosine (TCK) for *Bacillus*.

$$
\begin{array}{c}
\overset{+}{N}H_3 \\
/ \\
R-CH_2-\overset{}{C}H \\
\backslash \\
\underset{O}{\overset{\parallel}{C}}-CH_2-Cl
\end{array}
$$

Synthesis of Amino Acid Chloromethyl Ketones

Syntheses of TCK,[12] PCK,[13] LUCK,[14] and LCK[15] have been reported. With the exception of TCK, the procedures are relatively successful. TCK is obtained in low yield and is sufficiently unstable and difficult to purify that we did not use it in our extensive studies. We have developed a general synthetic procedure involving a minimum of handling and purification. Although the published procedures use crystallization for purification, we found it to be inadequate, particularly with regard to the radioactive reagents.[16] TCK,[12] PCK, and LUCK can be purified by ion-exchange chromatography on SE-Sephadex, but impurities cochromatographed with tritiated PCK. A new method for adsorption chromatography of water-soluble organic compounds on Sephadex G-10[17,18] has allowed the purification of tritiated LUCK and PCK to homogeneity. The detailed procedure presented below is applicable to any amino acid with an unfunctionalized side chain and to lysine in which the amino groups can be protected with the *t*-butyloxycarbonyl group. The procedure generally follows that reported for synthesis of LUCK[14] except that intermediates are not isolated at any point and the products are purified by adsorption chromatography on Sephadex G-10. We believe that this is a superior method, particularly for radioactive syntheses.

It should be noted that taking melting points of these compounds is not a reliable method of identification and that nuclear magnetic resonance (NMR) spectroscopy is the only sure method of confirming the

[13] E. Shaw and J. Ruscica, *Arch. Biochem. Biophys.* **145**, 484 (1971).

[14] P. L. Birch, H. A. El-Obeid, and M. Akhtar, *Arch. Biochem. Biophys.* **148**, 447 (1972).

[15] E. Shaw and G. Glover, *Arch. Biochem. Biophys.* **139**, 298 (1970).

[16] S. M. D'Ambrosio, G. I. Glover, and R. A. Jensen, *Arch. Biochem. Biophys.* **167**, 754 (1975).

[17] G. I. Glover, P. S. Mariano, and T. J. Wilkinson, *Sep. Sci.* **10**, 795 (1975).

[18] G. I. Glover, P. S. Mariano, and S. Cheowtirakul, *Sep. Sci.* **11**, 147 (1976).

structures of the products. In this regard, the chloromethyl ketone methylene proton singlet (—CO—CH$_2$—Cl) appears reliably at —4.1 ppm relative to internal tetramethylsilane in trifluoroacetic acid solvent.[15]

General Synthetic Procedure[19]

The N-t-butyloxycarbonyl amino acids are prepared according to the published procedure[20] and used without crystallization after drying the crude product at reduced pressure for several days in the presence of powdered phosphorus pentoxide. The BOC-amino acid (10 mmoles) is stirred at —15° in 100 ml of anhydrous ether[21] and 1.4 ml (1.0 g, 10 mmoles) of triethylamine,[22] and 0.95 ml (1.08 g, 10 mmoles) of ethyl chloroformate[23] is added. The suspension, which contains a white precipitate of triethylamine hydrochloride, is stirred at 0° for 10 min and added during a few minutes to a solution of about 30 mmoles of diazomethane in ether.[24] A vigorous evolution of nitrogen occurs, and the intensity of the yellow color of the diazomethane should decrease, but not disappear. The suspension is stirred for 30 min at 25°, then is freed of excess diazomethane by a vigorous stream of nitrogen; the yellow color fades. The solution is extracted sequentially with four 50-ml portions of deionized water to remove triethylamine hydrochloride and with 50 ml of a 5% sodium bicarbonate solution to remove unreacted BOC-amino acid. The solution is dried by stirring over several grams of powdered anhydrous sodium sulfate and filtered. The diazoketone is converted to the chloromethyl ketone by cooling the solution to 0° and bubbling anhydrous hydrogen chloride gas through it. If a precipitate forms within 5 min, interrupt the reaction, remove the triethylamine hydrochloride by filtration, and continue the hydrogen chloride treatment for 30 min. Store the hydrogen chloride-saturated solution at 0° for at least 2 hr, during which time crystals of the amino acid chloromethyl ketone hydrochloride form. The yield of crystals obtained by filtration is about 50%. If no crystals appear, evaporate the solvent *in vacuo* and proceed with purification of the residue.

[19] NOTE: All procedures must be carried out in a fume hood.
[20] A. Ali, F. Fahrenholz, and B. Weinstein, *Angew. Chem. Int. Ed. Engl.* **11,** 289 (1972).
[21] Anhydrous ether can be purchased in 1-pint cans and used freshly opened.
[22] Dried by storage of reagent grade over potassium hydroxide pellets.
[23] Used as purchased from Aldrich Chemical.
[24] Diazomethane is prepared from Diazald using the diazomethane kit and instructions from Aldrich. Request literature from Aldrich prior to carrying out the reaction. The standard procedure yields a yellow solution of about 3 g (**70** mmoles) of diazomethane in about **250** ml of ether.

Purification

The crude product is first chromatographed on SE-Sephadex as described for TCK,[12] removing impurities not removed by absorption chromatography on Sephadex G-10.

A column, 2.5 × 200 cm, of Sephadex G-10 is poured and equilibrated with 1 mM hydrochloric acid. Do not substitute another brand or cross-linkage type, since this is not gel filtration, but adsorption chromatography.[17,18] The crystalline amino acid chloromethyl ketone obtained above is dissolved in 2–5 ml of 1 mM HCl and applied and eluted from the column with the same solvent. The purified amino acid chloromethyl ketone elutes as a broad peak at an elution volume of several hundred milliliters. The peaks can be detected by the UV absorption of the ketone grouping at 280 nm or by spotting samples on filter paper and using ninhydrin to detect the amino group. The tubes containing the product are combined and lyophilized to yield the slightly off-white crystals of the chloromethyl ketone hydrochlorides.

Affinity Labeling Studies

We found that LUCK irreversibly inhibited the neutral and general transport systems of *Neurospora* whereas LCK had little effect on these systems.[25] Leucine is a substrate for both systems inactivated, whereas lysine is a substrate for only the general system. Both TCK and PCK inactivated the specific tyrosine/phenylalanine transport system of *Bacillus* as well the transport system(s) for neutral, aliphatic amino acids.

The advantage of using affinity labeling reagents to "tag" proteins in complex mixtures is that inactivation via an enzyme–inhibitor complex can be distinguished kinetically from simple bimolecular alkylation. There are two criteria that should be met by a site-specific reagent: (1) The rate of inactivation should be retarded by substrate; (2) an appropriate kinetic analysis should reveal the intermediacy of the enzyme–inhibitor complex in the inactivation process. In whole-cell systems additional controls are needed: (1) Incubation of the cells with the reagent must have no effect on cell viability as determined by dilution and replicate plating. (2) The inactivation could be due to a general effect on transport, such as inactivation of the system coupling energy to active transport. This must be tested by assaying the effect of the affinity

---

[25] S. O. Nelson, G. I. Glover, and C. W. Magill, *Arch. Biochem. Biophys.* **168**, 483 (1975).

labeling on related and unrelated transport systems found in the same cells.

Neurospora[25]

The rate of inactivation of the neutral and general transport systems by LUCK could not be retarded by three substrates (phenylalanine, leucine, or histidine), a finding indicative that this reagent was not functioning as an affinity label. This is interesting, since LUCK is a competitive inhibitor of the general transport system and, therefore, binds to the site of interest. We have shown that the mode of inactivation is the alkylation of sulfhydryl groups that are essential for transport. This attempted affinity labeling study is an example of what one can learn from a chemical approach even though original expectations are not realized. For example, LCK, which should be as good an alkylating agent as LUCK, does not inactivate the transport systems for amino acids or glucose. Furthermore, N-ethylmaleimide and iodoacetamide irreversibly inhibit amino acid and glucose transport. Thus, it is clear that LUCK is remarkably specific for sulfhydryl groups essential to the neutral and general transport systems even though there are sulfhydryl groups in the membrane that are essential to other transport systems and are accessible to other alkylating agents.

The selectivity observed suggests that the use of radioactive LUCK to inactivate the neutral and general transport systems could result in the alkylation of no more than two sulfhydryl group-containing proteins that are involved in transport. Although these proteins may not be the substrate binding proteins, they may have some activity or characteristic that would suggest their role in the transport process. We are presently working with a mutant that lacks the neutral transport system in an effort to selectively alkylate the protein involved in the general amino acid transport system.

Bacillus

TCK and PCK appeared to give the same results, and we elected to continue our work with PCK since TCK was more difficult to prepare and purify.[12] PCK is a competitive inhibitor of the transport of tyrosine and phenylalanine, and the rate of inactivation of the tyrosine/phenylalanine transport system can be effectively retarded by either natural ligand. On the other hand, the rate of inactivation of the transport system(s) for the neutral, aliphatic amino acids is unaffected by any of the substrates or by phenylalanine and tyrosine. Clearly, PCK is an affinity label for the tyrosine/phenylalanine transport system and inactivates and the transport of neutral, alphatic amino acids by simple bimolecular alkylation.

We determined the apparent first-order rate constant for the inactivation of the tyrosine/phenylalanine transport system by several concentrations of PCK. By plotting these data in double-reciprocal form we were able to demonstrate the intermediacy of an enzyme–PCK complex in the inactivation reaction and determine K_i values of 194 and 177 μM for PCK inhibition of tyrosine and phenylalanine transport, respectively. In addition, this analysis gives the actual first-order rate constants for the inactivation of tyrosine and phenylalanine transport by the enzyme–PCK complex of 0.016 and 0.012 M^{-1}, respectively. Given the level of imprecision in the assays using whole-cell systems, these numbers are in agreement. The rate of loss of leucine transport activity is comparable to that for tyrosine and phenylalanine transport activity.

The transport systems for adenine and basic and acidic amino acids are unaffected by PCK. Cells treated with PCK and then diluted and plated showed no decrease in viability compared to untreated controls. The effects of PCK are, therefore, limited to those observed.

[72] The Biotin Transport System

By EDWARD A. BAYER and MEIR WILCHEK

In the following account we used a well characterized active transport system as a model for affinity labeling studies on intact cells. The biotin transport system in yeasts has been characterized as a high-affinity, carrier-mediated, energy-requiring process.[1,2] Since the carrier recognizes the ureido ring of the biotin molecule, a broad range of modifications of the valeric acid side chain are possible without affecting the inherent affinity. Thus, several potential candidates for affinity labeling studies were synthesized, and biotinyl-*p*-nitrophenyl ester (pBNP) was found to be a potent inhibitor of biotin transport.[3] Evidence, which supports the contention that this compound acts as an affinity label, includes (a) time and concentration dependence of pBNP inactivation at relatively low concentrations; (b) protection of the transport system from pBNP inactivation by high concentrations of free biotin; (c) inability of

[1] T. O. Rogers and H. C. Lichstein, *J. Bacteriol.* **100**, 557 (1969).
[2] T. O. Rogers and H. C. Lichstein, *J. Bacteriol.* **100**, 564 (1969).
[3] J. M. Becker, M. Wilchek, and E. Katchalski, *Proc. Natl. Acad. Sci. U.S.A.* **68**, 2604 (1971).

model compounds, lacking the biotin moiety or chemically reactive group, to inhibit biotin uptake; and (d) the neutrality of pBNP with respect to other transport systems.

The major disadvantage in the use of affinity labels is the inherent loss of activity. Subsequent isolation of such affinity-labeled components from a heterogeneous mixture is valueless, since there is no direct method of proving their association with the original system; no additional mechanistic information can be obtained from an inactivated component, the identity of which is in question. Therefore, a salient feature of this system is the discovery that the pBNP-affinity-labeled biotin transport components can be chemically reactivated with thiols.[4] This discovery, the first of its kind in whole cells, is of multifold significance, since it provides (a) chemical information as to the nature of the bond between the affinity label and the transport protein; (b) a rapid method for the localization, quantification, and identification of transport components during various stages of purification; (c) a potential method for the ultimate regaining of binding activity in the purified transport component(s); and (d) a method for switching on and off the biological transport system of whole cells.

In an expansion of the above approach, a homologous series of affinity labeling reagents was synthesized in which the chemically reactive p-nitrophenyl ester group is located at increasing distances from the biologically active ureido moiety.[5] The affinity labels were selected in order to encompass chain lengths corresponding to those of biotin and biocytin, both of which occurs in nature. The inhibitory capacity of each derivative was examined and compared in order to map the topography of the biotin receptor. The results suggest the existence of at least two nucleophilic amino acid residues adjacent to the binding site of the biotin transport system in yeast. From studies of this type, structural information may be obtained for heterogeneous systems in which X-ray crystallographic analysis is inconceivable.

Preparation of Biotinyl-p-nitrophenyl Ester (BNP)

p-Nitrophenol (175 mg, 1.3 mmoles) and dicyclohexylcarbodiimide (206 mg, 1 mmole) are added to a suspension of biotin (244 mg, 1.0 mmole) in 3 ml of methylene chloride. The above mixture is stirred for 24 hr at 25°, then the reaction mixture is taken to dryness under reduced pressure, and the yellow, gummy residue is washed several times with

---

[4] T. Viswanatha, E. Bayer, and M. Wilchek, *Biochim. Biophys. Acta* **401**, 152 (1974).
[5] E. A. Bayer, T. Viswanatha, and M. Wilchek, *FEBS Lett.* **60**, 309 (1975).

absolute ether. The residue is taken up in isopropanol. Following filtration, the solution is reduced to minimum volume and allowed to crystallize overnight. The crystals of pBNP are collected by filtration and washed with anhydrous ether. Yield: 120 mg (33%), m.p. 156°–158°.

The synthesis of norbiotinyl- and homobiotinyl-p-nitrophenyl esters is accomplished in a similar fashion to that of the biotinyl derivative by substituting the appropriate analog.

Preparation of Model Affinity Labels

Members of the homologous series of biotin-containing affinity labels (see the table) are synthesized by reacting biotinyl-N-hydroxysuccini-

HOMOLOGOUS SERIES OF BIOTIN-CONTAINING AFFINITY LABELS

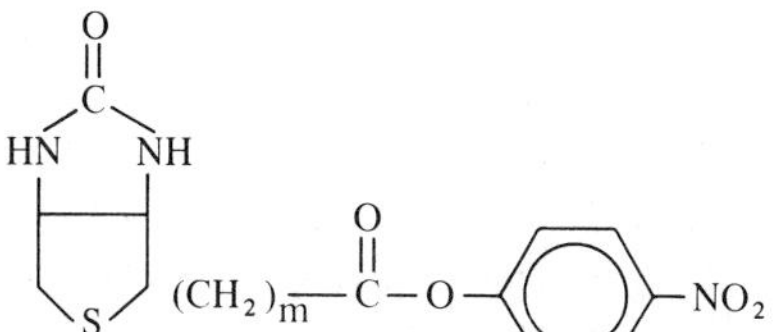

(I)

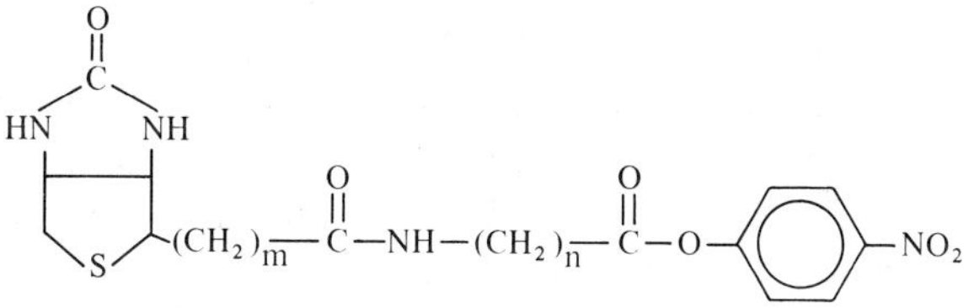

(II)

p-Nitrophenyl ester	Abbreviation	m	n	Δ Chain length[a]
Norbiotinyl	−1BNP	3	—	−1
Biotinyl	BNP	4	—	0
Homobiotinyl	+1BNP	5	—	1
Norbiotinyl-glycyl	−1B3NP	3	1	2
Biotinyl-glycyl	B3NP	4	1	3
Biotinyl-β-alanyl	B4NP	4	2	4
Biotinyl-γ-aminobutyryl	B5NP	4	3	5
Biotinyl-δ-aminovaleryl	B6NP	4	4	6
Biotinyl-ε-aminocaproyl	B7NP	4	5	7

[a] Deviation of side arm from reference affinity label (BNP) in number of backbone atoms.

mide ester (341 mg, 1 mmole),[6] dissolved in a minimum of dimethylform-amide, with the corresponding ω-amino carboxylic acid (1.2 mmoles) dissolved in 0.1 M sodium bicarbonate. The reaction is allowed to proceed for 4 hr at room temperature. The solution is acidified with 1 N HCl, and the resulting precipitate is filtered, washed well with deionized water, and dried in a desiccator over calcium chloride. The purity of the above intermediates is checked by thin-layer chromatography (TLC) (chloro-form–methanol–acetic acid, 9:1:1). The biotinyl-amino carboxylic acid derivative is allowed to react further with p-nitrophenol (20% excess) as described above for the synthesis of pBNP. Purity is established by TLC (chloroform–methanol 9:1) and by the stoichiometric release (A_{400}) of p-nitrophenol from a weighed sample hydrolyzed in 0.2 M sodium hydroxide.

Biotin Uptake Assay

Saccharomyces cerevisiae, Fleischman strain 139 (ATCC 9896) is grown at 30° in a "biotin-sufficient"[1] Vogel's medium,[7] with the following additions: inositol (36 μg/ml), calcium pantothenate (2 μg/ml), pyri-doxine hydrochloride and thiamine hydrochloride (4 μg/ml each), biotin (0.25 ng/ml), and glucose (1%) instead of sucrose. In a typical experi-ment, growth cultures (200 ml) are harvested, washed three times, and resuspended to 10^8 cells/ml with 50 mM potassium phosphate at pH 4.0. Cell samples (9 ml) are treated for 30 min at 30° with an ethanolic solution (1 ml) containing one of the affinity label homologs (to a final concentration of 25 μM); ethanol (1 ml) is added to control cells. Cells treated in this manner are washed three times with distilled water and resuspended to 10 ml. A sample (0.5 ml) is added to 50 mM potassium phosphate at pH 4.0 containing glucose (1% final concentration; 5.0 ml final volume). After incubation for 20 min at 30°, 50 μl of [¹⁴C]biotin (5 μg/ml) are added. Aliquots (1 ml) are taken at intervals, and the cells are collected and washed on glass-fiber filters (Tamar, Israel), after which radioactive samples are counted.[8]

The remainder of the inhibited cells is centrifuged and treated with a solution (5.0 ml) of 0.2 M mercaptoethanol (pH 4.0). After 30 min at 30°, the cells are washed three times and resuspended to the original volume (9.5 ml). A sample (0.5 ml) is assayed as above for biotin uptake.

[6] E. Bayer and M. Wilchek, this series, Vol. 34, pp. 265–267.
[7] H. J. Vogel, *Microbiol Gen. Bull.* 13, 42 (1956).
[8] G. A. Bray, *Anal. Biochem.* 1, 279 (1960).

Comments

The above affinity labeling reagents were found to inhibit specifically the uptake of biotin, but not other nutrients, into yeast cells. From the reactivation studies, one can conclude that an essential cysteine or histidine residue was modified in the transport system. The biotinylated protein can be specifically adsorbed to avidin columns, and subsequent elution with thiols should yield biotin transport component(s) in an active form.

Section III

Specific Procedures for Nucleic Acids and Ribosomal Systems

[73] Affinity Labeling of Ribosomal Functional Sites

By ADA ZAMIR

Ribosomes perform multiple functions in the course of protein synthesis; recognition and binding of the initiation region of mRNA, decoding of mRNA by binding of cognate aminoacyl-tRNAs, catalysis of peptide bond formation, and the translocation of nascent peptidyl-tRNA and mRNA. In addition, ribosomes can specifically bind a number of antibiotic compounds that interfere with one or another facet of protein biosynthesis. This diversity of function is matched by the complexity of ribosomal structure, an intricate arrangement of protein and RNA molecules, whose highly cooperative interaction is responsible for the formation of ribosomal functional sites.[1]

This unique feature of the ribosome greatly obstructs the identification of ribosomal components *directly* located at specific active centers, and it renders difficult the distinction between directly and indirectly involved components. One approach that in principle circumvents this difficulty is based on the method of affinity labeling. Employing to date several classes of chemically or photoreactive site-specific ligands, this method has been applied to identify ribosomal components located within close range of several functional sites on the *Escherichia coli* ribosome.

This chapter reviews methods applied in affinity labeling studies of ribosomes and describes the sites under study, major types of ligand used, criteria for establishing the specificity of labeling, and the characterization of modified ribosomal components. It does not include procedures for the synthesis of affinity labeling probes or a detailed discussion of the results obtained in the different studies. These can be found in the cited publications or elsewhere in this volume. For a general consideration of particular problems related to affinity labeling of ribosomes, see Cantor *et al.*[2] and this volume [15]. Early experiments are reviewed by Pongs *et al.*[3]

mRNA Analogs

Site Studied. In the normal functioning of the ribosome, mRNA initially binds to the ribosome in a manner allowing translation to start

[1] H. G. Wittmann, *Eur. J. Biochem.* **61**, 1 (1976).
[2] C. R. Cantor, M. Pellegrini, and H. Oen, *in* "Ribosomes" (M. Nomura, A. Tissières, and P. Lengyel, eds.), p. 573. Cold Spring Harbor Laboratory, Cold Spring Harbor, New York, 1974. See also this volume [15].
[3] O. Pongs, K. H. Nierhaus, V. A. Erdmann, and H. G. Wittmann, *FEBS Lett.* **40** (suppl.), 28 (1974).

at the correct initiation codon. Initiation and subsequent codons will then be decoded by selection and binding to the ribosome of the respective cognate aminoacyl-tRNAs. This process is actively affected by the ribosome.[4]

Affinity labeling studies of the decoding site (defined as the ribosomal site in contact with the codon to be decoded) are based on the existence of relatively simple model compounds for mRNA. These include short oligonucleotides regarded as single isolated codons, or homopolynucleotides such as poly(U). Comparable to codons in natural mRNA, such analogs can direct binding of cognate aminoacyl-tRNAs to the ribosome and can, therefore, be considered to bind at the ribosomal decoding site.

Two such sites may exist on the ribosome corresponding to the donor and acceptor positions of tRNA derivatives. While the initiation triplet is likely to direct binding to the donor site, all other triplets are decoded at the acceptor site.

Reagents. Table I specifies all mRNA analogs reported to date as affinity labeling probes for ribosomal decoding sites.[5-16] The table lists a group of oligonucleotides of defined length and sequence where chemically reactive groups have been introduced as substituents of the phosphate (a–c), pyrimidine ring (d–f), or ribose (g) groupings. Whereas the haloacetyl group in compounds a–f will make likely a reaction with nucleophilic groups on ribosomal proteins, reagent g was designed to interact with G residues in the rRNA. In addition, the list includes poly(U) and poly(4-thio-U), both of which are photoreactable.

Oligonucleotides. Tests of Function

The binding and coding properties of modified oligonucleotides were examined in order to establish their adequacy as affinity labeling probes.

[4] L. Gorini, *Nature (London), New Biol.* **234**, 261 (1971).
[5] O. Pongs and E. Lanka, *Proc. Natl. Acad. Sci. U.S.A.* **72**, 1505 (1975).
[6] O. Pongs, G. Stöffler, and E. Lanka, *J. Mol. Biol.* **99**, 301 (1975).
[7] O. Pongs and E. Rossner, *Nucleic Acids Res.* **3**, 1625 (1976).
[8] J. Petre and D. Elson (1976). Submitted for publication.
[9] R. Lührmann, U. Schwarz, and H. G. Gassen, *FEBS Lett.* **32**, 55 (1973).
[10] R. Lührmann, H. G. Gassen, and G. Stöffler, *Eur. J. Biochem.* **66**, 1 (1976).
[11] O. Pongs, G. Stöffler, and R. W. Bald, *Nucleic Acids Res.* **3**, 1635 (1976).
[12] R. Wagner and H. G. Gassen, *Biochem. Biophys. Res. Commun.* **65**, 519 (1975).
[13] I. Fiser, K. H. Scheit, G. Stöffler, and E. Kuechler, *Biochem. Biophys. Res. Commun.* **60**, 1112 (1974).
[14] E. Kuechler, A. Barta, I. Fiser, R. Hauptmann, P. Margaritella, W. Maurer, and G. Stöffler, *Abstr. Int. Congr. Biochem. 10th,* Hamburg, 03-3-153 (1976).
[15] M. L. Schenkman, D. C. Ward, and P. B. Moore, *Biochim. Biophys. Acta* **353**, 503 (1974).
[16] I. Fiser, P. Margaritella, and E. Kuechler, *FEBS Lett.* **52**, 281 (1975).

TABLE I
mRNA ANALOGS

Reagent[a]	Modified group	References
(a) p*ApUpG		5, 6
(b) p*UpGpA		7
(c) p*ApApA		8
(d) UpUpUpU*		9
(e) GpUpUpU*		10
(f) ApUpGpU*		11
(g) UpUpUpU**		12
(h) Poly(4-thio-U)		13, 14
(i) Poly(U)		15, 16

[a] Asterisks indicate modified component.

Although these tests are best performed in the absence of covalent binding, in most cases no care has been taken to avoid irreversible binding during the assay. Thus, for example, GpUpUpU* has been shown to direct the binding of Val-tRNA to an even greater extent than the non-modified oligonucleotide,[10] possibly reflecting a stabilization due to the irreversible binding of template. However, other oligonucleotide probes did not exhibit such enhanced activity.[e.g. 11]

Another criterion for functionality of short oligonucleotide templates is the stimulation of their binding to ribosomes by cognate aminoacyl-tRNAs.[8]

Labeling Reaction. Labeling media were buffered at pH 7.2 to 7.4 and usually contained: 6–20 mM Mg^{2+}, 80–150 mM NH$_4$Cl, and radioactively labeled affinity reagent at 10- to 100-fold excess over ribosomes added in the 70 S form or as isolated 30 S subunits (sometimes preactivated[17]). To enhance the specificity and stability of the reversible complex of oligonucleotide and ribosome, some reaction mixtures were supplemented with the cognate aminoacyl-tRNAs.[8,10,12] Labeling mixtures were usually incubated for 1–2 hr at 0°–37°.

Maximal labeling of 30 S subunits or 70 S ribosomes by the various reagents ranged between 0.1 and 0.7 mole of reagent per mole of ribosome. However, the routine determination of reagent uptake by Millipore filtration does not appear to be reliable, since the values measured by this method by far exceed the recovery of label after partial fractionation.[6]

The efficiency and pattern of labeling are markedly affected by the state of the ribosomes. Differences in reaction pattern were noted with 70 S ribosomes as compared to 30 S subunits,[6] with freshly prepared and stored 70 S ribosomes,[6] and with crude and salt-washed 70 S ribosomes or 30 S subunits.[11] Particularly striking are the changes observed in the labeling of 30 S subunits with p*ApUpG, where freezing and thawing of the reaction mixture greatly enhanced uptake by making available new labeling sites on the ribosome.[6]

Specificity of Labeling. Several tests have been performed to determine whether affinity labeling probes were attached at functionally significant sites on the ribosome.

The most crucial examination of the specificity of labeling is based on the ability of the modified ribosomes to bind the tRNA specified by the covalently bound oligonucleotide in the absence of any free template. Such tests were performed, for example, with p*ApUpG-modified 30 S subunits where IF-2 stimulated binding of fMet-tRNA$_f^{Met}$ was demon-

---

[17] A. Zamir, R. Miskin, and D. Elson, *J. Mol. Biol.* **60**, 347 (1971).

strated,[6] and with GpUpUpU*-modified 70 S ribosomes that were shown to be capable of binding Val-tRNA.[10] These results indicate that at least part of the modification in these cases has occurred at genuine decoding sites.

Another criterion for specificity is based on the stimulation of covalent binding by cognate aminoacyl-tRNAs.[8,12] Reduced covalent reaction to ribosomes precomplexed with poly(U) and Phe-tRNA[10] and competition by nonmodified oligonucleotides[5,8] have also been taken as indications for the specificity of labeling.

As indicated above, it is possible that codons bind to ribosomes at two positions corresponding to donor and acceptor sites. The nature of the site occupied by attached oligonucleotides has only been tentatively determined by testing the reactivity in the puromycin reaction of the aminoacyl-tRNA bound to the modified ribosomes.[6,10] In another case, the ability of ApUpGpU*-modified 70 S ribosomes to bind Met-tRNA$_m^{Met}$ in the presence of EF-Tu has been considered indicative of acceptor site location of the bound codon.[11]

Polynucleotides

Very little use has been made so far of polynucleotides as affinity labeling probes. The only reported cases are poly(U)[15,16] and its analog poly(4-thio-U).[13,14] Both can be made to react with ribosomes by irradiation at 253 nm and 335 nm, respectively. The adequacy of poly(4-thio-U) to serve as probe was indicated by its ability to direct Phe-polymerization.[18] In both cases irradiated mixtures contained Phe-tRNA in addition to 70 S ribosomes and polynucleotide. As cross-linking of mRNA to the ribosome will probably arrest translocation, loss of polymerizing activity has been taken to indicate the specificity of binding.[13] It is worth noting that the average chain length of the polynucleotide probe is likely to affect the course of the reaction, as the use of long poly(U) chains has been reported to result in the formation of large cross-linked ribonucleoprotein aggregates.[16]

Analysis of Modified Ribosomal Components. As a general rule in affinity labeling of ribosomes, excess of reagent was removed from the labeled ribosomes before any fractionation was attempted, to avoid any reaction from taking place after the structure of the ribosome had been disrupted. In cases in which 70 S ribosomes were labeled, the first step was normally the dissociation into subunits and determination of label

---

[18] W. Bähr, P. Faerber, and K. H. Scheit, *Eur. J. Biochem.* **33,** 535 (1973).

distribution in the two subunits, and subsequently in separated protein and RNA fractions. All mRNA analogs listed in Table I modified specifically the 30 S subunit, and all but compound g, which modified 16 S RNA, reacted with proteins.

Modification with oligonucleotides introduces some difficulty in the identification of the labeled proteins. The attachment of a highly charged oligonucleotide probe or polynucleotide fragment alters considerably the net charge of the labeled protein and hence its electrophoretic mobility. This makes unreliable the identification of proteins by their position on the map obtained by 2-dimensional gel electrophoresis.[6,8] This problem has been considered in the design of compound c, where, after the labeling reaction, the bulk of the oligonucleotide can be removed by a mild acid hydrolysis of the phosphoamide linkage.[8] In the other cases, partial splitting of proteins from 30 S subunits by cesium chloride gradient centrifugation[6] and application of various one-dimensional gel electrophoretic techniques (quoted in references cited in footnotes 6, 10, 11) allowed a characterization of modified proteins.

The most reliable and sensitive techniques for identifying the oligonucleotide-labeled proteins employ antibodies raised against purified individual ribosomal proteins.[19] Techniques used were radioimmunodiffusion,[6,10] and sucrose gradient centrifugation of soluble antigen–antibody complexes.[13]

Peptidyl-tRNA and tRNA Derivatives

Sites Studied. Interactions of tRNA with the ribosome can be broadly classified into two types. (a) Interactions with the peptidyltransferase center of the 50 S ribosomal subunit involving the 3′ terminus of peptidyl- or aminoacyl-tRNA occupying, respectively, the donor (peptidyl) or acceptor (aminoacyl) site of the catalytic center. (b) Interactions likely to involve internal parts of the tRNA molecule and designed to align and stabilize the binding of cognate tRNAs to the ribosome.

Affinity labeling probes for mapping of ribosomal components involved in interactions of the first type were constructed by introducing suitable substituents at the free terminal amino group of aminoacyl or peptidyl tRNA. To study the second class interactions, sites within the tRNA molecule must be rendered capable of reacting with potential binding sites on the ribosome. Most studies to date have centered on the elucidation of ribosomal components involved in peptidyl transfer. Attempts at unraveling components involved in the second class of interactions are still very few.

Mapping of Components of the Peptidyltransferase Center

Reagents. Compounds used for affinity labeling are listed in Table II.[20-37] Probes are for the most part derivatives of Phe-tRNA and in some cases of Met-tRNA$_f^{Met}$. For charging with phenylalanine, unfractionated or purified tRNA from *E. coli* or yeast was used. A systematic comparison of affinity probes made with the different tRNAs[22] did not reveal differences in labeling pattern. Analogs of the initiator tRNA were invariably prepared starting from purified tRNA$_f^{Met}$.

Modifying groups fall into two general categories: (1) a–g are chemically reactive (haloacetyl, *p*-nitrophenylcarbamyl, chlorambucyl); (2) h–n are photosensitive (ethyldiazomalonyl, aryl azide, aryl keto). Chemically reactive and photosensitive probes differ mainly in two respects. Compounds of the first type can react only with nucleophilic groups on the ribosome, whereas the extensive chemical reactivity of the photolysis

[19] G. Stöffler *in* "Ribosomes" (M. Nomura, A. Tissières, and P. Lengyel, eds.), p. 615. Cold Spring Harbor Laboratory, Cold Spring Harbor, New York, 1974.

[20] M. Pellegrini, H. Oen, and C. R. Cantor, *Proc. Natl. Acad. Sci. U.S.A.* **69**, 837 (1972).

[21] H. Oen, M. Pellegrini, D. Eilat, and C. R. Cantor, *Proc. Natl. Acad. Sci. U.S.A.* **70**, 2799 (1973).

[22] M. Pellegrini, H. Oen, D. Eilat, and C. R. Cantor, *J. Mol. Biol.* **88**, 809 (1974).

[23] J. B. Breitmeyer and H. F. Noller, *J. Mol. Biol.* **101**, 297 (1976).

[24] M. Sopori, M. Pellegrini, P. Lengyel, and C. R. Cantor, *Biochemistry* **13**, 5432 (1974).

[25] D. Eilat, M. Pellegrini, H. Oen, Y. Lapidot, and C. R. Cantor, *J. Mol. Biol.* **88**, 831 (1974).

[26] M. Yukioka, T. Hatayama, and S. Morisawa, *Biochim. Biophys. Acta* **390**, 192 (1975).

[27] A. P. Czernilofsky and E. Kuechler, *Biochim. Biophys. Acta* **272**, 667 (1972).

[28] A. P. Czernilofsky, E. E. Collatz, G. Stöffler, and E. Kuechler, *Proc. Natl. Acad. Sci. U.S.A.* **71**, 230 (1974).

[29] R. Hauptmann, A. P. Czernilofsky, H. O. Voorma, G. Stöffler, and E. Kuechler, *Biochem. Biophys. Res. Commun.* **56**, 331 (1974).

[30] E. S. Bochkareva, V. G. Budker, A. S. Girshovich, D. G. Knorre, and N. M. Teplova, *FEBS Lett.* **19**, 121 (1971).

[31] L. Bispink and H. Matthaei, *FEBS Lett.* **37**, 291 (1973).

[32] A. S. Girshovich, E. S. Bochkareva, U. M. Kramarov, and Y. A. Ovchinnikov, *FEBS Lett.* **45**, 213 (1974).

[33] N. Hsiung, S. A. Reines, and C. R. Cantor, *J. Mol. Biol.* **88**, 841 (1974).

[34] N. Hsiung and C. R. Cantor, *Nucleic Acids Res.* **1**, 1753 (1974).

[35] A. Barta, E. Kuechler, C. Branlant, J. Sriwidada, A. Krol, and J. P. Ebel, *FEBS Lett.* **56**, 170 (1975).

[36] N. Sonenberg, M. Wilchek, and A. Zamir, *Proc. Natl. Acad. Sci. U.S.A.* **72**, 4332 (1975).

[37] N. Sonenberg, M. Wilchek, and A. Zamir, *Abstr. Int. Congr. Biochem. 10th,* Hamburg, 03-3-160; and *Biochem. Biophys. Res. Commun.* **72**, 1534 (1976).

TABLE II
Peptidyl-tRNA Analogs

Aminoacyl-tRNA modified	N-blocking group	References
(a) Phe-tRNA	$BrCH_2\overset{O}{\overset{\|}{C}}-$	20–23
(b) Met-tRNA$_f^{Met}$	$BrCH_2\overset{O}{\overset{\|}{C}}-$	24
(c) Phe-tRNA	$BrCH_2\overset{O}{\overset{\|}{C}}(Gly)_n-$	25
(d) Phe-tRNA	$ICH_2\overset{O}{\overset{\|}{C}}-$	26
(e) Phe-tRNA	$O_2N-\langle\text{benzene}\rangle-O-\overset{O}{\overset{\|}{C}}-$	27, 28
(f) Met-tRNA$_f^{Met}$	$O_2N-\langle\text{benzene}\rangle-O-\overset{O}{\overset{\|}{C}}-$	29
(g) Phe-tRNA	$(ClCH_2CH_2)_2-N-\langle\text{benzene}\rangle-(CH_2)_3-\overset{O}{\overset{\|}{C}}-$	30
(h) Phe-tRNA	$N_2-\underset{\underset{O}{\overset{\|}{C}-OC_2H_5}}{\overset{\overset{O}{\overset{\|}{C}-}}{C}}$	31
(i) Phe-tRNA	$N_3-\langle\text{benzene, }NO_2\rangle-\overset{O}{\overset{\|}{C}}-$	32
(j) Phe-tRNA	$N_3-\langle\text{benzene, }NO_2\rangle-O-\langle\text{benzene}\rangle-CH_2-\overset{O}{\overset{\|}{C}}-$	33

TABLE II—*Continued*

Aminoacyl-tRNA modified	N-blocking group	References
(k) Phe-tRNA	N_3—⟨benzene ring, with NO_2⟩—NH—CH_2—$\overset{O}{\overset{\|}{C}}$—	34
(l) Phe-tRNA	⟨benzene ring⟩—$\overset{O}{\overset{\|}{C}}$—⟨benzene ring⟩—$(CH_2)_3$—$\overset{O}{\overset{\|}{C}}$—	35
(m) Phe-tRNA	N_3—⟨benzene ring⟩—CH_2—$\overset{HN\text{-}tBoc}{\underset{}{CH}}$—$\overset{}{\underset{O}{\overset{\|}{C}}}$—	36
(n) Met-tRNA$_f^{Met}$	N_3—⟨benzene ring⟩—CH_2—$\overset{HN\text{-}tBoc}{\underset{}{CH}}$—$\overset{}{\underset{O}{\overset{\|}{C}}}$—	37

products of probes of the second type eliminates such restrictive requirements. Another advantage of the photosensitive probes lies in the ability to accurately control the timing and duration of the labeling reaction. Nonspecific side reactions, which usually afflict photoaffinity labeling experiments, are not very likely in this particular case, since the high affinity of tRNA for the ribosome eliminates the necessity of having an excess of reagent free in solution.

Tests of Function. To be considered as proper functional analogs, affinity labeling probes were examined by a number of criteria: formation of binding complexes with ribosomes; dependence of binding on mRNA and with derivatives of Met-tRNA$_f^{Met}$; dependence on initiation factors. Additional tests were designed to establish the mode of binding by determining the ability of the probe to serve as donor substrate in forming a peptide bond with aminoacyl-tRNA or puromycin as acceptor substrates.

Binding of Phe-tRNA derivatives was usually performed in media buffered at pH 7.2 to 7.5 and containing, Mg^{2+} at 10–30 mM, K^+ or NH_4^+ at 0.03–0.15 M, and poly(U). With initiator tRNA analogs, the mRNAs employed were R17-RNA,[29] f2-RNA,[24] or T4-mRNA,[37]; Mg^{2+} concentration was reduced to about 5 mM, and the mixtures were also supplemented with initiation factors and GTP. Concentration ratios of affinity labeling probes to ribosomes (70 S) varied from about 1.0 to large excesses of ribosomes. With chemically reactive probes, incubation conditions usually allowed both reversible and irreversible binding to take place.[e.g. 20] However, a clear distinction between reversible and irreversible binding is possible with the photosensitive probes.

Whenever tested, binding of labels to the ribosome has been shown to be stimulated severalfold by the appropriate mRNA[20,23,24,32,33,36,37] or by initiation factors.[29,37] Based on reactivity with puromycin, all peptidyl-tRNA probes tested were considered to occupy predominantly the ribosomal donor site. This conclusion was supported by the observed ability of several Phe-tRNA derived probes to form a peptide bond with added Phe-tRNA as acceptor.[22,33,37] Similarly, modified Met-tRNA$_f^{Met}$ bound with f2-RNA has been shown to interact with Ala-tRNA to form the Met-Ala initiation sequence typical of the phage coat protein.[24]

Another characterization of the binding of peptidyl-tRNA analogs to the ribosome is based on the response to antibiotic inhibitors of peptidyl-transferase activity. Although Oen et al.[38] observed no inhibition of reversible binding of N-bromoacetyl Phe-tRNA by chloramphenicol, lincomycin, or streptogramin A, the same antibiotics as well as erythromycin significantly inhibited the binding of p-azido-N-tBoc-Phe-Phe-tRNA.[36] It should be noted, however, that in both cases the antibiotics markedly reduced the extent of the covalent reaction of the affinity probes with the ribosome.

Labeling Reaction. As indicated above, the same media were usually employed for reversible and irreversible binding reactions. One notable exception is the reaction with N-iodoacetyl-Phe-tRNA, which exhibited maximal specificity at pH 5.0.[26] With chemically reactive ligands, incubations were usually at 37° and for periods varying from 30 min to 18 hr. Irradiation of ribosomes with the different photosensitive probes was carried under a variety of conditions; at wavelengths above or below 300 nm, with different light intensities, in the presence or the absence of oxygen, at temperatures from 0° to room temperature, and for different periods of time. It is noteworthy that, whenever tested, irradiation did not cause inactivation of the ribosomes, as they retained activity for both binding of, and peptide bond formation with, aminoacyl-tRNA added after irradiation.[33,37]

In estimating the extent of covalent binding to the ribosome, assays must be employed that can clearly distinguish between reversibly and irreversibly bound material. Toward this end, the following methods were used: (a) treatment with buffers of low magnesium concentration that would completely dissociate reversible binding complexes,[20] and (b)

[38] H. Oen, M. Pellegrini, and C. R. Cantor, *FEBS Lett.* **45**, 218 (1974).

mild digestion with RNase in the presence of EDTA to cleave preferentially the unreacted probe.[31]

Specificity of Labeling. The specificity of modification by affinity labels was judged by several criteria. The most widely used test examines the dependence of the covalent reaction on former formation of a reversible binding complex. Thus, labeling has been shown to depend on mRNA (in all experiments cited), to be inhibited by deacylated tRNA,[22] and by puromycin and other antibiotics.[28,36,38]

The most critical test for the functionality of the covalently bound peptidyl-tRNA is based in its ability to serve as donor substrate for peptide bond formation. This test was performed usually by first allowing a covalent reaction to take place with added nonradioactive affinity probe, and subsequently adding a radioactive acceptor substrate, aminoacyl-tRNA[22,33,37] or puromycin.[22] Covalent attachment of radioactivity to the ribosome now indicates that the attached nonradioactive probe served as donor substrate in peptide bond formation. For this test to be meaningful it is absolutely necessary that affinity labeling does not continue after addition of the acceptor substrate. This condition is easily fulfilled with photosensitive probes, but only to a lesser extent with chemically reactive substrates.

A similar technique has been used to scan the pattern of labeling with peptidyl-tRNA bound at the acceptor site.[22,33,37] In this case peptide bond formation is allowed to occur prior to the covalent reaction, resulting in the transfer of the reactive peptidyl moiety to the acceptor site. This indirect method of labeling also ensures that the reactive species is bound to the ribosome in a functional form. Another method of directing the affinity probe to the acceptor site has been by increasing the magnesium concentration and adding deacylated tRNA to the binding mixtures.[39]

It is evident from Table II that reactive groups in different probes are placed at different distances from the 3′ end of the tRNA. It is therefore possible that a comparison of labeling data will provide information on the relative arrangement of components at the binding site. In two instances, this problem has been dealt with more systematically by using homologous series of peptidyl-tRNAs, where a varying number of Gly residues were attached to Phe-tRNA, and which bore a chemically reactive[25] or photosensitive[37] group at the N-terminal position of the peptide.

Analysis of Modified Ribosomal Components. The nature of affinity labeling probes discussed in this section poses a special problem in analy-

---

[39] D. Eilat, M. Pellegrini, H. Oen, N. de Groot, Y. Lapidot, and C. R. Cantor, *Nature (London)* **250**, 514 (1974).

TABLE III
tRNA DERIVATIVES

tRNA	Modified base	Modifying group	References
(a) tRNA^{Val}		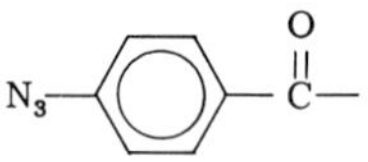	40–42
(b) tRNA^{Phe}	4-thio-U	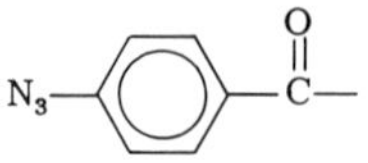	42
(c) tRNA$_f^{Met}$		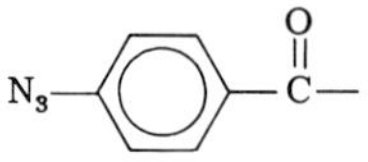	43
(d) tRNA^{Phe}	3-(3-Amino-3-carboxypropyl)U	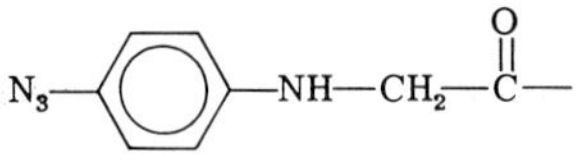	42

sis of the modified components. This is due to the presence of the bulky, highly charged tRNA molecule as part of the affinity probes. It has been indicated that conditions of isolation and analysis of individual ribosomal components are sufficient to hydrolyze the ester bond linking the peptide to tRNA. In other instances, modified ribosomes were treated so as to cleave this bond prior to analysis.[23,26,31,36]

Modification by peptidyl moieties does not alter drastically the electrophoretic mobility of the ribosomal proteins, and despite small shifts in position[22,28] modified proteins can be identified by two-dimensional gel electrophoresis. Immunoprecipitation has also been applied in the analysis of the modified proteins.[28,29]

In a considerable number of cases, peptidyl-tRNA affinity probes were shown to modify primarily not the ribosomal proteins, but the 23 S rRNA.[23,26,31,32,35-37] Further characterization of modified sites was obtained by splitting of the 23 S rRNA into 13 S and 18 S fragments including the 5' third and the 3' two-thirds of the molecule, respectively.[35,36]

Mapping of tRNA Binding Sites. Identification of ribosomal components interacting with parts of tRNA other than the 3' end has been attempted by preparing tRNAs derivatized in odd bases with photosensitive residues. Affinity probes are listed in Table III.[40-43] Assays for func-

---

[40] I. Schwartz and J. Ofengand, *Proc. Natl. Acad. Sci. U.S.A.* **71**, 3951 (1974).
[41] I. Schwartz, E. Gordon, and J. Ofengand, *Biochemistry* **14**, 2907 (1975). See also this volume [82].

tionality of derivatized tRNAs included formation of aminoacyl-tRNA, binding to ribosomes, and participation in peptide bond formation.[38] Different labeling patterns depending on the nature of the probe and mode of binding to the ribosomes have been reported.[38-41]

GTP Analogs

GTP is involved in the interaction of initiation, elongation, and to a certain extent also of termination factors with the ribosome. GTP hydrolysis usually accompanies these interactions.[44] In addition to taking part in the GTPase reaction, ribosomes can also stably bind GDP in the presence of EF-G and fusidic acid.

The only reported attempt at identifying the ribosomal components directly interacting with GTP is based on the formation of the GDP–ribosome complex. The GDP derivative, 1-(4-azidophenyl)-2-(5′-guanyl) pyrophosphate, served as a photoaffinity probe.[45] The analogy of the probe with unmodified GDP was demonstrated by the inhibition of GDP binding to the ribosome by the analog. Irradiation of the photoaffinity label with the ribosome resulted in a wide distribution of label in 50 S subunit proteins. Specific and nonspecific labeling were partly resolved by comparing the labeling pattern in the presence and in the absence of fusidic acid.

Antibiotics

Sites Studied. Numerous antibiotic compounds exert their antibiotic function by interfering with ribosome activity.[46,47] The mechanism of interference involves binding of the drug molecules to specific sites on the ribosome, resulting in most cases in inhibition of specific ribosomal functions. Sites involved in antibiotic binding and in the normal functioning of the ribosome may, or may not, overlap. Mapping of antibiotic binding sites may therefore reveal ribosomal components directly located at functional sites, or others that may have regulatory functions.

[42] G. Chinali, I. Schwartz, E. Gordon, R. Tejwani, and J. Ofengand, *Abstr. Int. Congr. Biochem. 10th,* Hamburg, 03-3-155 (1976).

[43] I. Schwartz, R. Tejwani, and J. Ofengand, *Abstr. Int. Congr. Biochem. 10th,* Hamburg, 03-3-154 (1976).

[44] W. Möller, *in* "Ribosomes" (M. Nomura, A. Tissières, and P. Lengyel, eds.), p. 711. Cold Spring Harbor Laboratory, Cold Spring Harbor, New York, 1974.

[45] J. A. Maassen and W. Möller, *Proc. Natl. Acad. Sci. U.S.A.* **71,** 1277 (1974).

[46] S. Pestka, *Annu. Rev. Microbiol.* **25,** 487 (1971).

[47] G. Stöffler and G. W. Tischendorf, *in* "Topics in Infectious Diseases" (J. Drews and F. E. Hahn, eds.), Vol. 1, p. 117. Springer-Verlag, Berlin and New York, 1975.

Antibiotics can be grouped into several classes based on the ribosomal subunit location of the binding site and the specific ribosomal function affected. One group of compounds interferes with peptide bond formation by binding to 50 S ribosomal subunits at, or close to, the peptidyl-transferase center. Chloramphenicol, employed in several affinity labeling studies, belongs to this group. Although puromycin falls into the same category, its activity as acceptor in peptide bond formation makes it a special case. Antibiotics typically affecting the 30 S subunit are the aminonucleosides, of which streptomycin has been the most widely studied.

Reagents. Modified and unmodified antibiotics employed in reported affinity labeling studies are listed in Table IV.[48-58] The synthesis of additional derivatives of chloramphenicol of potential use in photoaffinity labeling has been described.[59,60] Interestingly, the list includes two unmodified antibiotics; chloramphenicol and puromycin. These two as well as reagents e and i react with the ribosome photolytically.

Tests of Function. Modified antibiotics were compared to the natural compounds in reversible binding to ribosomes and in inhibitory or other effects on ribosome function. In one case, the examination also included the *in vivo* effects of an antibiotic analog.[50] In *in vitro* experiments, the interference by irreversible binding was prevented or minimized by performing the tests under conditions unfavorable for covalent reaction, e.g., low temperatures[49] or suboptimal pH.[51] Reversible binding studies were carried out with the chloramphenicol and streptomycin derivatives and included the subunit localization of binding sites,[58] estimation of the

[48] R. Bald, V. A. Erdmann, and O. Pongs, *FEBS Lett.* **28**, 149 (1972).

[49] O. Pongs, R. Bald, and V. A. Erdmann, *Proc. Natl. Acad. Sci. U.S.A.* **70**, 2229 (1973).

[50] O. Pongs and W. Messer, *J. Mol. Biol.* **101**, 171 (1976). See also this volume [79] and [80].

[51] N. Sonenberg, M. Wilchek, and A. Zamir, *Proc. Natl. Acad. Sci. U.S.A.* **70**, 1423 (1973).

[52] N. Sonenberg, A. Zamir, and M. Wilchek, *Biochem. Biophys. Res. Commun.* **59**, 693 (1974).

[53] O. Pongs, R. Bald, T. Wagner, and V. A. Erdmann, *FEBS Lett.* **35**, 137 (1973).

[54] B. S. Cooperman, E. N. Jaynes, D. J. Brunswick, and M. A. Luddy, *Proc. Natl. Acad. Sci. U.S.A.* **72**, 2972 (1975).

[55] R. J. Harris, P. Greenwell, and R. H. Symons, *Biochem. Biophys. Res. Commun.* **55**, 117 (1973).

[56] P. Greenwell, R. J. Harris, and R. H. Symons, *Eur. J. Biochem.* **49**, 539 (1974).

[57] O. Pongs and V. A. Erdmann, *FEBS Lett.* **37**, 47 (1973).

[58] A. S. Girshovich, E. S. Bochkareva, and Y. A. Ovchinikov, *Mol. Gen. Genet.* **144**, 205 (1976). See also this volume [77] and [78].

[59] P. E. Nielsen, V. Leick, and O. Buchardt, *Acta Chem. Scand.* **B29**, 662 (1975).

[60] N. Sonenberg, A. Zamir, and M. Wilchek, this volume [83] and [84].

TABLE IV
ANTIBIOTIC ANALOGS

Antibiotic	Modification	References
Chloramphenicol	$-\overset{O}{\overset{\|}{C}}-CHCl_2$ replaced by	
(a)	$-\overset{O}{\overset{\|}{C}}-CH_2I$	48–50
(b)	$-\overset{O}{\overset{\|}{C}}-CH_2Br$	48, 51
(c)	None	52
Puromycin	$-NH_2$ group substituted with	
(d)	$-\overset{O}{\overset{\|}{C}}-CH_2I$	53
(e)	$-\overset{O}{\overset{\|}{C}}\diagdown_{\begin{smallmatrix}C-N_2\\C_2H_5O-\overset{\|}{\underset{O}{C}}\end{smallmatrix}}$	
	5′ OH substituted with	
(f)	$-\overset{O}{\overset{\|}{P}}_{\underset{OH}{\|}}-O-\!\!\bigcirc\!\!-NH-\overset{O}{\overset{\|}{C}}-CH_2Br$	55, 56
(g)	None	54
Streptomycin	Formyl group derivatized with	
(h)	$ICH_2-\overset{O}{\overset{\|}{C}}-HN-\!\!\bigcirc\!\!-NHNH_2$	57
(i)	$N_3-\!\!\bigcirc\!\!\underset{NO_2}{\,}-\overset{O}{\overset{\|}{C}}-NHNH_2$	58
	Hydrazone reduced	

number of binding sites,[50] binding affinity,[50,51] competition with the parent, unmodified compound,[51] or with other antibiotics.[50] Derivatives of antibiotics were also tested for their effect on poly(U)directed polymerization of phenylalanine[48,49,53,54,57,58] and on the fragment reaction.[51,56] In addition, puromycin derivative f was tested for acceptor activity in the fragment reaction.[56] The blockage of the free amino group in the other two puromycin derivatives eliminates acceptor function.

Labeling Reaction. Media for the reaction were usually similar to those used for the reversible binding of antibiotics with some exceptions, as, for example, the use of a higher pH to increase the rate of the reaction[51] or of high salt concentration to suppress nonspecific binding.[58] Reactions with chemically reactive derivatives were usually carried out at 37° for various periods of time, and mixtures with photosensitive reagents were irradiated in the cold or at higher temperatures with light of wavelengths below[54] or above[52] 300 nm.

Specificity of Labeling. Contrary to reagents described in sections on mRNA analogs and peptidyl-tRNA and tRNA derivatives, no positive function of the bound antibiotic could be demonstrated except for reagent f.[56] Thus, specificity of labeling relied mostly on tests of the effect of different factors on the labeling reaction and on the irreversible inactivation of the ribosomes.

It is normally to be expected that the inclusion of unmodified antibiotic in the labeling mixture should result in reduction in labeling. Competition of this nature was indicated in a few cases,[53] but not in others.[51] Other antibiotics known to interfere with the reversible binding of the probe were also tested for inhibitory effects on the labeling reaction.[52,56] Other criteria applied to support specificity were based on the dependence of the extent of labeling on the concentration of the affinity probe[54] and on the comparison of labeling patterns with affinity probes and with chemically similar but nonspecific reagents.[51] Correlating the extents of labeling reaction and inactivation also served to distinguish between specific and nonspecific modification.[52,56]

Analysis of Modified Ribosomal Components. All reagents, except for reagent f, modified ribosomal proteins. While some derivatives reacted at a limited number of sites (48–50), the others reacted with a large number of components. Modified proteins were identified by gel electrophoresis techniques.

Concluding Remarks

It is evident from the studies summarized in this chapter that the method of affinity labeling is applicable to the study of ribosomal functional sites. Labeling patterns with a variety of site-specific reagents form the basis for the identification of protein and RNA molecules located at, or in the vicinity of, various ribosomal sites. Functional roles have been assigned to individual ribosomal components using a score of other approaches,[61] but a detailed evaluation of the results obtained is beyond the scope of this chapter.

The results of the affinity labeling studies bring to light several advantages and shortcomings of the method as applied to ribosomes. To only mention a few, specific labeling reactions have been detected not only in proteins, but also in rRNA, and may prove instrumental in the elucidation of the role of rRNA in ribosome function. Results of labeling experiments are often variable, a fact perhaps attributable to the inhomogeneity and variability of ribosome preparations. This, and the lack of clear-cut criteria for distinguishing between specific and nonspecific modifications, may account for the large number of ribosomal components often implicated in a single ribosomal functional site.

Solution of these problems as well as a score of others is to be expected in future studies.

Acknowledgments

Work from the author's laboratory was supported by a grant from the United States—Israel Binational Science Foundation (BSF), Jerusalem, Israel. The help of Dr. N. Sonenberg in the preparation of this manuscript is gratefully acknowledged.

[61] R. R. Traut, R. C. Heimark, T. T. Sun, J. W. B. Hershey, and A. Bollen, in "Ribosomes" (M. Nomura, A. Tissières, and P. Lengyel, eds.), p. 271. Cold Spring Harbor Laboratory, Cold Spring Harbor, New York, 1974.

[74] Photoaffinity Labeling of 23 S RNA in Ribosomes

By Ludwig Bispink and Heinrich Matthaei

Photoaffinity labeling—a well known technique for studying active sites in enzymes—has attracted considerable attention as a tool to ascertain structure–function relationships in more complex biological structures, such as ribosomal binding sites[1-7] or membrane receptor sites.[8,9]

[1] L. Bispink and H. Matthaei, *FEBS Lett.* **37**, 291 (1973).

[2] N. Hsiung, S. A. Reines, and C. R. Cantor, *J. Mol. Biol.* **88**, 841 (1974).

[3] A. S. Girshovich, E. S. Bochkareva, V. M. Kramarov, and Y. A. Ovchinnikov, *FEBS Lett.* **45**, 213 (1974). See also this volume [77] and [78].

[4] B. S. Cooperman, E. N. Jaynes, D. J. Brunswick, and M. A. Luddy, *Proc. Natl. Acad. Sci. U.S.A.* **72**, 2974 (1975).

[5] A. Barta, E. Kuechler, C. Branlant, J. Sriwidada, A. Kvol, and J. P. Ebel, *FEBS Lett.* **56**, 170 (1975).

In order to localize the peptidyltransferase region of ribosomes, substrate analogs bearing a reactive group at the α-amino group of the amino acid moiety of aminoacyl-tRNA have been used.[10] These might react with structures in the neighborhood of this active center.

The method described here involves a photoreactive diazoacyl group coupled to the terminal amino group of Phe-tRNA[Phe]. Photolysis of diazoketones and diazoesters yields highly reactive carbenes, capable of insertion even into aliphatic side chains. The nearly universal reactivity of such compounds in photoaffinity labeling makes them specially suitable for labeling the physically most adjacent residue(s); labeling may thus not depend on the existence of a nucleophilic group in or near the active site. The ultraviolet-reactive peptidyl-tRNA analog, N-(ethyl-2-diazomalonyl)Phe-tRNA, has been shown to label specifically 23 S RNA rather than ribosomal proteins in the peptidyltransferase region of *Escherichia coli* ribosomes.[1]

Preparation of N-(Ethyl-2-diazomalonyl)Phe-tRNA[Phe]

N-(Ethyl-2-diazomalonyl)Phe-tRNA[Phe] is obtained by acylation of Phe-tRNA[Phe] with ethyl-2-diazomalonyl-N-hydroxysuccinimide ester.[11] The praparation of the reagent and its reaction with tRNA are schematically represented in Fig. 1.

Procedure. Caution: the following operations should be done with suitable precautions (light protection) in a well-ventilated hood.

a. Synthesis of Ethyl-2-diazomalonyl Chloride.[12] A solution of 0.15 mole (17 g) diazoacetic acid in 50 ml toluene is added dropwise under stirring to an ice-cooled solution of 0.05 mole of phosgene in 25 ml of toluene (20% solution, Fluka). After 2 hr at 4°, the reaction mixture is stirred at room temperature until nitrogen escapes. The solvent, with an

[6] N. Sonenberg, M. Wilchek, and A. Zamir, *Proc. Natl. Acad. Sci. U.S.A.* **72**, 4332 (1975).

[7] I. Schwartz and J. Ofengand, *Proc. Natl. Acad. Sci. U.S.A.* **71**, 3951 (1974).

[8] H. Kiefer, J. Lindstrom, E. S. Lennox, and S. J. Singer, *Proc. Natl. Acad. Sci. U.S.A.* **67**, 1688 (1970).

[9] C. E. Guthrow, H. Rasmussen, D. J. Brunswick, and B. S. Cooperman, *Proc. Natl. Acad. Sci. U.S.A.* **70**, 3344 (1973).

[10] C. R. Cantor, M. Pellegrini, and H. Oen, *in* "Ribosomes" (M. Nomura, A. Tissières, and P. Lengyel, eds.), p. 573. Cold Spring Harbor Laboratory, Cold Spring Harbor, New York, 1974.

[11] Y. Lapidot, N. de Groot, S. Rappoport, and A. D. Hamburger, *Biochim. Biophys. Acta* **149**, 532 (1967).

[12] R. J. Vaughan and F. H. Westheimer, *Anal. Biochem.* **29**, 305 (1969).

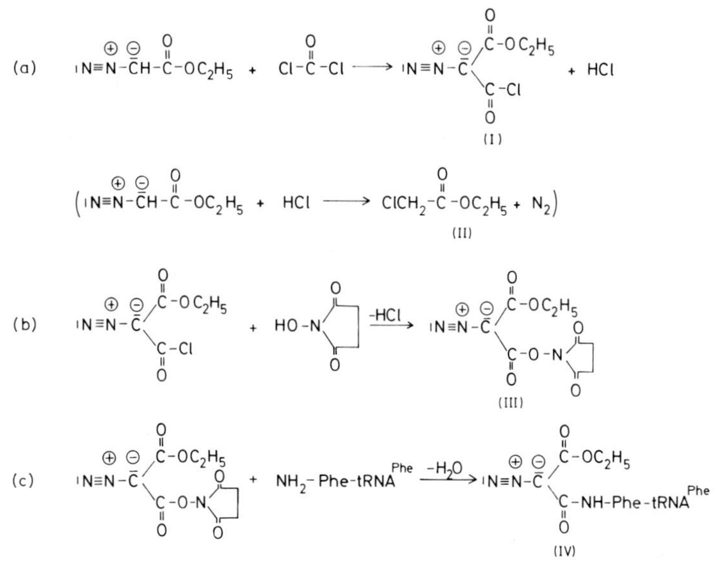

(a) $\overset{\oplus\ \ominus}{N\equiv N}-\overset{}{C}H-\overset{O}{\overset{\|}{C}}-OC_2H_5$ + $Cl-\overset{O}{\overset{\|}{C}}-Cl$ $\longrightarrow$ $\overset{\oplus\ \ominus}{N\equiv N}-\overset{}{C}\overset{\overset{O}{\overset{\|}{C}}-OC_2H_5}{\underset{\underset{O}{\overset{\|}{C}}-Cl}{}}$ + HCl

(I)

$\left(\overset{\oplus\ \ominus}{N\equiv N}-\overset{}{C}H-\overset{O}{\overset{\|}{C}}-OC_2H_5 + HCl \longrightarrow ClCH_2-\overset{O}{\overset{\|}{C}}-OC_2H_5 + N_2 \right)$

(II)

(b) $\overset{\oplus\ \ominus}{N\equiv N}-\overset{}{C}\overset{\overset{O}{\overset{\|}{C}}-OC_2H_5}{\underset{\underset{O}{\overset{\|}{C}}-Cl}{}}$ + $HO-N$ [succinimide] $\xrightarrow{-HCl}$ $\overset{\oplus\ \ominus}{N\equiv N}-\overset{}{C}\overset{\overset{O}{\overset{\|}{C}}-OC_2H_5}{\underset{\underset{O}{\overset{\|}{C}}-O-N}{}}$ [succinimide]

(III)

(c) $\overset{\oplus\ \ominus}{N\equiv N}-\overset{}{C}\overset{\overset{O}{\overset{\|}{C}}-OC_2H_5}{\underset{\underset{O}{\overset{\|}{C}}-O-N}{}}$ [succinimide] + $NH_2-Phe\text{-}tRNA^{Phe}$ $\xrightarrow{-H_2O}$ $\overset{\oplus\ \ominus}{N\equiv N}-\overset{}{C}\overset{\overset{O}{\overset{\|}{C}}-OC_2H_5}{\underset{\underset{O}{\overset{\|}{C}}-NH\text{-}Phe\text{-}tRNA^{Phe}}{}}$

(IV)

Fig. 1. Reaction sequence for synthesis of N-(ethyl-2-diazomalonyl)Phe-tRNA.

excess of phosgene and some chloroacetic acid ester (II) produced in the reaction, is evaporated under reduced pressure (10 torr) in a rotating evaporator at 30°–40°. The liquid residue containing the crude product (I) is used without purification for further synthesis.

b. *Synthesis of Ethyl-2-diazomalonyl-N-hydroxysuccinimide Ester.* The crude ethyl-2-diazomalonyl chloride (I) is dissolved in 20 ml of tetrahydrofuran, and 10 g of N-hydroxysuccinimide are added. After cooling the suspension, 4 ml pyridine in 30 ml tetrahydrofuran are added slowly at 4° while stirring. The reaction mixture is further stirred overnight at room temperature (15°–20°). Excess of N-hydroxysuccinimide is filtered and the filter is washed with tetrahydrofuran; the solvent is removed by evaporation under reduced pressure. The solid residue is dissolved in isopropanol by heating; the hot solution is filtered, and petroleum ether (40°–60°) is added until precipitation occurs. The solution is heated again and then allowed to stand at 4° overnight. The product crystallizes in fine needles. Recrystallization from dichloro-methane/petroleum ether yields 6.5 g (51%) of the pure diazo compound (III): m.p. = 114°; IR (KBr) 2118 cm⁻¹ (diazo stretching band); λ_{max} (methanol) = 254 nm (disappears under ultraviolet irradiation).

c. *Synthesis of N-(Ethyl-2-diazomalonyl)-[³H]Phe-tRNA^{Phe}.* Transfer RNA^{Phe} (yeast) (Boehringer) is charged with [³H]phenylalanine (770 Ci/mole, Amersham) with a purified enzyme fraction obtained from baker's yeast, as described by von der Haar.[13] A solution of 10 mg ethyl-2-diazomalonyl-N-hydroxysuccinimide ester (III) dissolved in 1.5 ml of dimethyl sulfoxide is added slowly with shaking to 2 mg [³H]Phe-tRNA^{Phe} (yeast) (54.5 nmoles of [³H]Phe/40 A_{260} units tRNA) in 1.2 ml 150 mM potassium acetate at pH 6.5 containing 10 mM MgCl$_2$. The reaction mixture is shaken in the dark for 7 hr at 4°. The tRNA is precipitated by addition of 6 ml of precooled ethanol and kept at −20° for 30 min; it is isolated by centrifugation (10 min 20,000 g). The pellet is dried under reduced pressure, dissolved in 0.2 ml of 1 mM MgCl$_2$–10 mM KCl, and applied to a column of Sephadex G-25 (fine) (1 × 20 cm) that is equilibrated and eluted with 1 mM MgCl$_2$–10 mM KCl. The elution peak contains about 90% of the basic quantity of tRNA subjected to the procedure.

Product Analysis by Thin-Layer Chromatography

Of the Phe-tRNA derivative solution, 5 μl are subjected to hydrolysis by addition of 3 μl of 1 N NaOH (40 min at 37°). The hydrolyzate is applied to a silica gel thin-layer plate (Merck); the chromatogram is developed in 2-butanol–90% HCOOH–water (75:15:10, v/v); R_f(Phe) = 0.41, R_f (N-acyl-Phe) = 0.85. The yield in the aminoacylation step varies between 80 and 100%, as calculated by the ratio between N-acyl-[³H]Phe and [³H]Phe.

Irradiation

Diazoketones have two absorption bands in the ultraviolet region, one intense (λ_{max} = 254 nm, ϵ_{max} ≈ 7000) and the other rather weak (λ_{max} = 350 nm, ϵ_{max} ≈ 10). Carbenes may be generated by irradiation at either wavelength. Prolonged irradiation at the shorter wavelength (λ = 253.7 nm) is found to inactivate ribosomes, possibly owing to cross-linking of proteins and ribosomal RNA.[14] Therefore, photolysis is performed at longer wavelengths (λ > 300 nm) in a simple device using Pyrex glass tubes, which absorb the short wavelength radiation: ten Pyrex tubes (16 × 100 mm) are placed in a tube rack, made from two 100 mm diameter × 5 mm PVC disks kept 70 mm apart from each other by three PVC rods. The centers of ten 17-mm diameter holes for taking up the tubes are spaced evenly around the upper plastic disk, 38 mm outside of the center of a 20 mm diameter hole in the middle of

[13] F. von der Haar, *Eur. J. Biochem.* 34, 84 (1973).
[14] L. Gorelic, *Biochemistry* 14, 4627 (1975).

the disk. A rod-shaped high-pressure mercury lamp (type Q 81, 70 W, Quarzlampen GmbH, Hanau) is inserted through the central hole, placing the burner opposite to the samples. The tube rack stands in a 2-liter beaker, which contains, near its inner surface, a cylindrical glass coil through which cooling water is pumped by a refrigerated constant-temperature circulator to control the desired temperature (2° or 13°) in the reaction vessels. Thermal contact is maintained by water in the beaker, which is stirred by a magnetic spinning bar.

Cross-Linking Reaction

In general, 1-ml reaction mixtures contain 500 pmoles of *E. coli* ribosomes,[15] up to 1.4 nmoles i.e., a saturating amount of N-(ethyl-2-diazomalonyl)-[^{3}H]Phe-tRNA^{Phe}, 100 μg of poly(U), 10 mM MgCl$_2$, 100 mM NH$_4$Cl, 1 mM dithiothreitol, and 10 mM Tris chloride at pH 7.4 or 10 mM potassium acetate at pH 6.5. The mixture is placed in Pyrex tubes and incubated at 37° in the absence of light to form a reversible complex. Both in kinetics and extent of poly(U)-stimulated complex formation with ribosomes, N-(ethyl-2-diazomalonyl)Phe-tRNA behaves like N-acetyl-Phe-tRNA.[1] After 10 min at 37°, the tubes are cooled in ice and then inserted into the tube rack. For quantitative determination of photoinduced irreversible binding of N-acyl-Phe-tRNA, 100-μl aliquots are withdrawn from the reaction mixtures. Then 0.1 μg of RNase A, 5 μl of 0.5 M potassium EDTA at ph 7.0, and 5 μl of 1 M Tris chloride at pH 7.5 are added, and the samples are incubated at 37° for 30 min to release the tRNA from the complex. Aliquots of 75 μl are plated on GF/A filters that are placed in 10% trichloroacetic acid at 4° for precipitation of macromolecular material. The filters are washed with 5% trichloroacetic acid, ethanol/ether (1:1), and ether, dried, and counted in 2 ml of toluene containing 0.4% diphenyloxazole. After filtration of unirradiated samples or irradiated reaction mixtures without ribosomes, no radioactivity is found on the filter. Photolysis of the ribosomal complexes, however, results in a high-molecular-weight product. The maximum yield of covalently bound material—about 4% of the complexed N-(ethyl-2-diazomalony)Phe-tRNA—is reached within 2.5-3 hr.[1] This incorporation is poly(U)-dependent and thus presumably takes place within the specific complex N-(ethyl-2-diazomalonyl)Phe-tRNA 70 S ribosomes poly(U). When reaction mixtures are irradiated in quartz rather than glass tubes, the kinetics of the reaction is eight to ten times faster. However, under these conditions, photolysis leads to simultaneous

---

[15] M. Noll, B. Hapke, and H. Noll, *J. Mol. Biol.* **80**, 519 (1973).

TABLE I

EFFECT OF pH ON THE EXTENT OF COVALENT ATTACHMENT OF
N-(ETHYL-2-DIAZOMALONYL)-[³H]PHE-tRNA

| | Bound compound[a] | | | |
	pH 5.0	pH 6.0	pH 6.5	pH 7.4
With poly(U)	4.1	3.85	3.5	2.4
Without poly(U)	2.3	0.9	0.6	0.5

[a] Picomoles of N-(ethyl-2-diazomalonyl)-[³H]Phe-tRNA^{Phe} covalently bound per 100 pmoles of ribosomes, 10 min at 37°, 3 hr photolysis at 2°.

inactivation of the ribosomal complex within an irradiation period of 5–10 min. As shown in Table I, a more acidic pH increases the extent of covalent attachment of the photoaffinity label (up to 7%); however, nonspecific labeling is markedly enhanced at the lower pH region (pH 5.0–6.0). Hence, a pH of about 6.5 should be optimal for the photoreaction.

Analysis of Reaction Products

In order to analyze the irradiated complexes for reaction products among proteins and ribosomal RNA, proteins are extracted with acetic acid following the method of Hardy et al.[16] From a reaction mixture irradiated at 13° for 3 hr and digested with RNase as described above, 200-μl aliquots are shaken at 4° for 40 min after addition of 400 μl of acetic acid and 20 μl of 1 M MgCl$_2$. After centrifugation (10 min, 20,000 g), the supernatant liquid is plated on GF/A filters. Macromolecular material is precipitated with 10% trichloroacetic acid at 4°. The rRNA pellet is dissolved in 100 μl of 150 mM NaCl, 50 mM EDTA pH 7.0, 0.5% SDS, and precipitated on GF/A filters with 10% trichloroacetic acid as described above. The results (Table II) show that N-acyl-Phe-tRNA is covalently bound to rRNA rather than to ribosomal proteins. When RNase digestion is carried out as described above but with a 10-fold increased quantity of RNase A ($= 1$ μg/100 μl), no radioactivity is found in the rRNA fraction, indicating hydrolysis of the reaction product.

To further examine the site of the reaction, ribosomes are incubated at 37° for 10 min with N-(ethyl-2-diazomalonyl)-[³H]Phe-tRNA and then photolyzed at 13° for 3 hr. The complex is recovered by centrifugation and resuspended in standard buffer. Total rRNA is isolated by

[16] S. J. Hardy, C. G. Kurland, P. Voynow, and G. Mora, Biochemistry 8, 2897 (1969).

TABLE II

EXTENT OF COVALENT ATTACHMENT OF N-(ETHYL-2-DIAZOMALONYL)-[³H]-
PHE-tRNA^Phe TO RIBOSOMAL PROTEINS AND rRNA[a]

	Bound compound[b]	
	Ribosomal proteins	rRNA
With poly(U)	0.5	9.4
Without poly(U)	0.2	2.5

[a] Data from L. Bispink and H. Matthaei, *FEBS Lett.* **37**, 291 (1973).

[b] Picomoles of N-(ethyl-2-diazomalonyl)-[³H]Phe-tRNA^Phe covalently bound per 500 pmoles of ribosomes, 1 ml, 10 min at 37°, 3 hr photolysis at 13°.

phenol/SDS extraction as described by Traub *et al.*[17] and analyzed by sucrose gradient centrifugation. In control experiments, poly(U) is omitted from the reaction mixture (b), or N-(ethyl-2-diazomalonyl)-[³H]-Phe-tRNA is replaced by N-acetyl-[³H]Phe-tRNA (c). The results show that the total amount of N-(ethyl-2-diazomalonyl)-[³H]Phe-tRNA irreversibly reacted with ribosomes has modified 23 S RNA.[1]

Photoreaction with N-(ethyl-2-diazomalonyl)Phe-tRNA does not inactivate peptidyltransferase. This observation agrees with results of affinity labeling experiments with other reactive peptidyl-tRNA analog; these also suggest that the fixed N-acyl-aminoacyl-tRNA is able to participate in at least one peptidyl transfer.[2,18] Incubation of the modified complex with Phe-tRNA yields an irreversible fixation of one to two additional phenylalanyl-groups to 23 S RNA.

Photolyzed complexes of N-(ethyl-2-diazomalonyl)-[³H]Phe-tRNA · 70 S ribosomes·poly(U) are incubated with a saturating amount of [¹⁴C] Phe-tRNA at 37° for 10 min, recovered by centrifugation, and subjected to sedimentation in a sucrose gradient. The 50 S subunits are precipitated by addition of two volumes of cold ethanol to the pooled fractions. To characterize the reaction product(s), labeled 50 S subunits are submitted to limited hydrolysis with pancreatic ribonuclease according to the method of Allet and Spahr.[19] Under defined conditions, 23 S RNA in the 50 S subunit is cleaved into two unequal fragments. Separation of the degradation products by polyacrylamide gel electrophoresis shows that the peptide is covalently bound to the larger, 3′-terminal fragment of 23 S RNA.[18]

[17] P. Traub, S. Mizushima, C. V. Loury, and M. Nomura, this series, Vol. 20, p. 391 (1971).

[18] B. Allet and P. F. Spahr, *Eur. J. Biochem.* **19**, 250 (1971).

[19] M. Pellegrini, H. Oen, D. Eilat, and C. R. Cantor, *J. Mol. Biol.* **88**, 809 (1974).

Remarks

The yield of the labeling reaction does not exceed about 4–7% of complex-bound N-(ethyl-2-diazomalonyl)Phe-tRNA; the low yield seems to reflect a general problem in photoaffinity labeling. Since the highly nonspecific reactivity of the carbene generated from the diazo compound favors reaction with the solvent, irradiation of the reaction mixture largely results in hydrolytic inactivation of the photogenerated carbene. Moreover, photolyzed N-(ethyl-2-diazomalonyl)Phe-tRNA, inactivated within the binding site, is poorly exchanged by fresh reagent owing to high stability of the complex.

Another point of importance is the pH-dependent formation of a 1,2,3-triazole, by the Dimroth rearrangement,[20] that is characteristic of α-diazoamides. Since the triazole fails to efficiently photolyze and yield a carbene, the influence of pH on the reaction was examined (Table I). Although more acidic pH favors covalent bond formation, the extent of nonspecific labeling is increased simultaneously.

[20] O. Dimroth, *Justus Liebigs Ann. Chem.* **373**, 131 (1910).

[75] Photoaffinity Labels for Nucleic Acids

By WILLIAM E. WHITE, JR., and K. LEMONE YIELDING

Photoaffinity labeling is a valuable approach for studying nucleic acids *in vitro* and *in vivo*. In contrast to most protein binding, the specific ligands that bind nucleic acids do so with high affinity (K C$_a$ 10^{-6} to 10^{-7} are not unusual). The reagents of interest are of sufficient size and complexity to offer many opportunities for attachment of reactive groups, and many are strong chromophores in the visible and near-UV range where cells are sufficiently transparent to permit photolysis of drug nucleic acid complexes *in situ*. Examples of biological processes that lend themselves to study by this approach include: mutagenesis; carcinogenesis, regulation of transcription, translation, and replication; and nuclear-cytoplasmic interactions. Today, the evidence is overwhelming that carcinogens are mutagens and that most are covalently attached to DNA at some stage. The path from mutagenesis, a rapid process following a DNA insult, to carcinogenesis, a slow process that may not be observed for years, is unknown. Many carcinogens must be modified into species capable of reacting covalently with a biopolymer before they are active as carcinogens or mutagenic in bacteria. External activation is necessary for bacteria because they lack some of the activating enzymes found in animal tissues. Metabolic activation is also rather inefficient, since it must occur at some distance from the target site.

The photoaffinity labeling technique affords the opportunity to bypass the metabolic activation process for both mutagenesis and carcinogenesis by using photosensitive moieties to attach the drug to the biopolymer after initial binding has occurred in the dark. Nitrenes may be generated photolytically from azides, and carbenes from diazocompounds and ketenes.

Generally, carbenes are higher energy than nitrenes and therefore react more randomly, i.e., with their nearest neighbor. This is an advantage for labeling specific sites on enzymes because the carbene need not search for a secondary or tertiary carbon or other more reactive moiety. Carbenes have the disadvantage of undergoing the Wolff rearrangement to isocyanates, which may then react chemically with nucleophiles. Frequently, carbene precursors absorb at short wavelengths, so that irradiation with nucleic acids might produce pyrimidine dimers and other photoproducts. In our work we have concentrated on nitrene labeling. Since the carcinogens are attached to DNA by electrophilic addition, we are not hampered severely by the less reactive nitrenes seeking a more reactive site. Nitrenes generated by photolysis of aryl azides undergo few rearrangements. The aryl azide is relatively small, consisting of an additional two nitrogen atoms if substitution has been for an amine. The azide is conjugated with the aryl ring(s), which absorb at long wavelength, depending on substitution, so that irradiation and nitrene generation can be effected without exposing the DNA, cell, or animal to short-wavelength UV light. The reaction of the unbound nitrene with water during *in vivo* experiments may present problems. Many of the aromatic amines are metabolized to hydroxylamine and subsequently to various esters, which are believed to be ultimate mutagens or carcinogens. Thus, in some cells the covalent labeling could occur either by the photo-labeling method or by metabolism of the hydroxyl amines formed from the reaction of the nitrene with water.

Although phenyl azide can be prepared from phenyl hydrazine hydrochloride by addition of $NaNO_2$ and HCl, we have used two other methods exclusively. If the ring contains a deactivating heteroatom, a direct nucleophilic substitution of a halogen by N_3 in a suitable solvent is usually successful. If the ring is homoannular, then generation of the diazonium salt from nitrous acid, followed by substitution with N_3, produces the azide. For example, 9-azidoacridine can be prepared from 9-chloroacridine, and 3-azidoacridine is prepared from 3-aminoacridine.[1]

---

[1] A. C. Mair and M. F. G. Stevens, *J. Chem. Soc. Perkin Trans.* 1, 161 (1972).

We were never able to prepare a diazonium salt from a heterocycle, and our observations are supported in the literature.

Separation and purification of reaction products is extremely important in using the complex reagents that interact with DNA since multiple reaction products may be possible. Examples of synthesis of drug derivatives of interest to the nucleic acid investigator follow.

Preparation of 2-Azidofluorene

2-Aminofluorene (Aldrich) (500 mg, 2.75 mmoles) is dissolved in 25 ml of dimethyl sulfoxide (DMSO) and cooled to 5°; 1 ml of concentrated HCl is added. Sodium nitrite (120 mg, 2.9 mmoles) is dissolved in 2 ml of water and added slowly, followed by approximately 5 ml of water. After 2 min, NaN_3 (190 mg, 2.9 mmoles), dissolved in approximately 2 ml of water, is added to the yellow solution. A white solid forms as N_2 evolves. After 10 min, 50–60 ml of water are added and the precipitate is filtered and recrystallized from MeOH–water. Yield: 77%.

Preparation of 4-Azido-7-chloroquinoline

4,7-Dichloroquinoline (Aldrich; recrystallized from absolute ethanol) (2.73 g, 0.013 mole) and sodium azide (0.91 g, 0.014 mole) are dissolved in 50 ml of absolute ethanol and 10 ml of dimethylformamide (DMF) and refluxed for 26 hr. The precipitate is filtered, washed, and recrystallized from EtOH–H$_2$O, giving 1.73 g (62%); m.p. = 114°–115°. The quinoline nucleus is not as reactive as the acridine. No reaction occurred when the reagents were refluxed in methanol (b.p. 65°), thereby necessitating a higher boiling solvent. Substitution at C-7 was neither observed nor expected.[2]

3-Azidoacridine

3-Aminoacridine (3.0 g) in acetic acid, 30 ml, reacts with NaNO$_2$ (1.2 g in 5 ml of water) at 0°. Sodium azide, 2.0 g, is added slowly and stirred for 3 hr. The solution is poured into aqueous ammonia, and the red precipitate is filtered and recrystallized from petroleum ether (b.p. 60°–80°). Yield: 1.5 g; m.p. 113°–115°.[1]

9-Azidoacridine

9-Chloroacridine (recrystallized from absolute ethanol) (1.08 g, 5.1 mmoles) and NaN$_3$ (350 mg, 5.4 mmoles) are dissolved in methanol and warmed to 60° for 20 min. NaCl forms after a few minutes. The methanol is evaporated, and 50 ml of water are added to dissolve NaCl and to hydrolyze the remaining 9-chloroacridine to acridone. Recrystallization is from MeOH–H$_2$O (0.85 g, 77%). The compound has been prepared by refluxing 9-chloroacridine and sodium azide in aqueous acetone and also by oxidizing the hydrazo compound with HNO$_2$ in 2 N HCl.[1]

Ethidium Azides

Ethidium, the phenanthridine derivative usually purchased as the bromide salt, has two amino groups that can be diazotized.

[2] L. A. Paquette, "Principles of Modern Heterocyclic Chemistry," p. 292. Benjamin, New York, 1968.

Bastos has prepared the diazido derivative by using a large excess of $NaNO_2$ in 2 N HCl.[3] Originally, we reported the preparation of the monoazido derivative.[4] Subsequently, Graves and Yielding in this laboratory found that the product contained some diazido isomer, and modified the procedure to obtain the monoazido derivative in crystalline purity.[5] The best conditions for production of the monoazide are diazotization of at least a 3-fold excess of ethidium at pH 1.6 followed by addition of nitrite and elevation of the pH to 2.6. After a reaction time of 15 min, the product is precipitated by raising the pH to 11. Separation of the diazide from monoazide is effected by adsorption onto a carboxymethyl cellulose column and elution with a pH gradient from 3.5 to approximately 2. The diazide elutes first, followed by the monoazide. The details of the preparation and characterization of the mono- and diazide products will appear elsewhere,[5] as do reports of photolytically induced mutations in yeast mitochondria[4] and *Salmonella*,[6] and light-provoked repair synthesis in human lymphocyltes[7] and rat lens.[8]

Propidium Azide

Propidium is supplied by Calbiochem as the diiodide salt. It is necessary to remove the iodide prior to diazotization because nitrous acid oxidizes iodide to iodine. Removal can be accomplished by dissolving the propidium in water and applying it to a carboxymethyl cellulose column. The column is washed with water until I$^-$ no longer elutes. Iodide can be detected by adding nitrous acid to the beaker collecting the eluent and noting the formation of dark yellow or tan iodine. The propidium is eluted with dilute HCl at pH 1.5 to 2.0. The HCl then serves as the

[3] R. N. Bastos, *J. Biol. Chem.* **350**, 7739 (1975).
[4] S. C. Hixon, W. E. White, Jr., and K. L. Yielding, *J. Mol. Biol.* **92**, 319 (1975).
[5] D. Graves and L. W. Yielding, "Synthesis and Separation of the Monoazide Analog of Ethidium Bromide" (to be submitted).
[6] L. W. Yielding, W. E. White, Jr., and K. L. Yielding, *Mutat. Res.* **34**, 351 (1976).
[7] C. E. Cantrell and K. L. Yielding, *Photochem. Photobiol.* **25** (2), 191 (1977).
[8] J. Jose, "Repair Synthesis of DNA in the Lens Epithelium: Relevance for Cataractogenesis." Dissertation, University of Alabama, Birmingham, 1976.

acid for converting nitrite to nitrous acid. The remainder of the procedure is essentially the same as that for azido ethidium.

Stability and Photolysis Conditions

Photolysis conditions and light stability depend on the spectral properties of the compounds prepared. In all synthesis, storage, and subsequent experiments, the extent of illumination must be rigorously controlled to prevent spurious losses in reagent or production of irregular results. Even a few minutes in ordinary laboratory light will be seriously detrimental to many of the azides, which absorb strongly in the visible range. The colorless azidofluorenes and azidoquinoline require photolysis with light at about 350 nm. Ethidium azide has been used under a 35-watt Thomas super safelight for as long as 30 min without significant loss in potency or change in UV or visible spectra.

Comments

Photosensitive analogs may be prepared from a variety of reagents that interact with nucleic acids by noncovalent forces. These agents may be used, therefore, for photoaffinity labeling and provide potential solutions to special problems in nucleic acid research.

[76] Identification of GDP Binding Sites with 4-Azidophenyl-GDP

By J. A. MAASSEN and W. MÖLLER

To identify the proteins located at or near the GDP binding site in a multicomponent system like ribosomes, a GDP derivative (APh-GDP) with the photoreactive 4-azidophenyl group (APh) bound to the β-phosphate was synthesized.[1]

Properties

APh-GDP has a UV-absorption maximum at 252 nm with an absorption tail that stretches down to 400 nm. This property makes it possible to photolyze the compound at wavelengths above 320 nm thereby reducing the danger of UV damage to biological material. Concentra-

---

[1] J. A. Maassen and W. Möller, *Proc. Natl. Acad. Sci. U.S.A.* **71**, 1277 (1974).

tions of APh-GDP in solution are determined by measuring the A_{260} value. The molar extinction coefficient is assumed to be 2.2×10^4, i.e., the sum of the molar extinction coefficients of GMP and 4-azidophenylphosphoric acid.

The initial photolysis of the compound is rapid, but the process results in formation of dark products that absorb significant amounts of light, making prolonged irradiation times required. This drawback can be avoided by using high-intensity UV lamps. The compound has considerable thermal stability. In aqueous solution it may be kept in the dark for 3 hr at room temperature without measurable decomposition.

Methods

Chemicals

Dimethylformamide was distilled under reduced pressure and stored over 3A molecular sieves. Methanol was similarly stored. Ether, pentane, and 1,2-dimethoxyethane were held over sodium wire; tri-n-butylamine (TBA), pyridine, and dioxane over calcium hydride. Diphenylphosphoric acid chloride was distilled under reduced pressure prior to use, and phosphorus oxytrichloride at atmospheric pressure. All these chemicals (Merck-Darmstadt) were of analytical grade and dry.

The tri-n-butylammonium salt of guanosine 5'-monophosphate (TBA-GMP) was prepared by passing a solution of 1 g of the disodium salt of GMP in 40 ml of 50% aqueous methanol through a Dowex 50W-X8 column in the pyridinium form. The solvent was evaporated under reduced pressure, and the residue was dissolved in 40 ml of 50% aqueous methanol and TBA (0.5 ml). The solvent was evaporated again, and excess TBA was removed by shaking the residue with pentane. The TBA-GMP was dried by dissolution in dry pyridine–dimethylformamide (1:1, v/v), followed by evaporation of the solvent under reduced pressure. This procedure was repeated three times. To obtain [³H]APh-GDP, [³H]GMP (Amersham) was added to TBA-GMP before drying.

Detection of compounds on silica gel thin-layer plates (Merck-Darmstadt, F254) was by UV light. All derivatives containing an azido group become dark when the thin-layer plates are held for several minutes under a UV lamp.

Synthesis

The procedure for preparing APh-GDP is shown schematically in Fig. 1. Intense light should be excluded as far as possible. For the prepara-

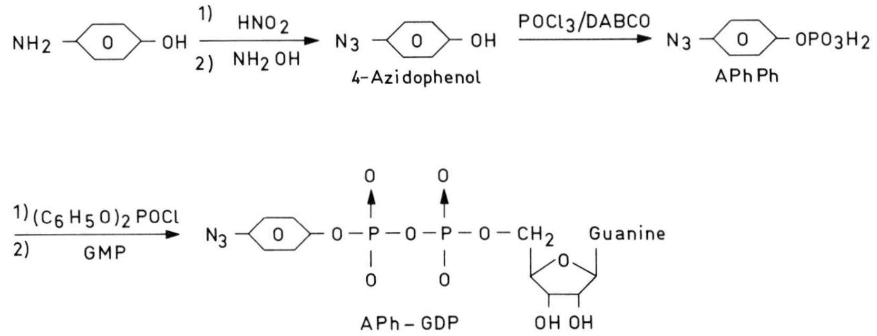

Fig. 1. Outline of the synthesis of 4-azidophenyl-GDP. DABCO, 1,4-diazabicyclo-[2.2.2]octane.

tion of azidophenyl phosphoric acid (APh-P) and APh-GDP, it is important that the glassware, reagents, and solvents be completely dry.

4-Azidophenol. A solution of 4-aminophenol, 10 g (92 mmoles), in 10% hydrochloric acid (100 ml) is cooled to —5° in an ice–salt bath. A solution of $NaNO_2$ (6.25 g) in water (12.5 ml) is added dropwise under vigorous stirring. It is essential that the temperature remains, below 5° during this procedure. A solution of hydroxylamine hydrochloride (6.25 g) in water (12.5 ml) is then added at once. After mixing, the mixture is poured into an ice-cold suspension of $Na_2CO_3 \cdot 10H_2O$ (125 g) in water (375 ml). This solution is stirred in the dark for 18 hr at 0°. The resulting dark-brown solution is acidified with concentrated hydrochloric acid (100 ml), the color becoming reddish. The solution is extracted three times with 125 ml of ether, and the combined ether layers are dried on anhydrous K_2CO_3. After removal of the drying agent by filtration, the ether is removed by rotary evaporation, leaving a dark oily residue, which is extracted six times with 65 ml each of boiling pentane (boiling point 35°). This extraction is performed by vigorously shaking the residue with pentane in a flask that is occasionally immersed in a warm-water bath until the pentane begins to boil. After each extraction, the light-yellow pentane layer is decanted. The combined pentane layers are kept at —30° overnight, resulting in the crystallization of 4-azidophenol as yellow or brown crystals. The pentane is decanted, and the crystals are dried under reduced pressure. Yield, 3.5 g (26 mmoles, 31%) of 4-azidophenol. Analytical data: NMR ($CDCl_3$) $\delta 6.84$ (singlet, 4H), $\delta 5.70$ (broad singlet, 1H); IR (KBr) 2100 cm^{-1} (—N$_3$), 3400 cm^{-1} (—OH).

NOTE: 4-Azidophenol is an unstable substance, easily oxidized by air, and producing black crystals. However, this material can be used without difficulty in the next synthetic step. Storage of larger amounts should be avoided because of the intrinsic explosive properties of azido compounds.[2]

4-Azidophenylphosphoric Acid. To a solution of 1,4-diazabicyclo-[2.2.2.]octane (1.5 g, 13 mmoles) in dry 1,2-dimethoxyethane (25 ml), freshly distilled $POCl_3$ (0.4 ml, 4.2 mmoles) is added, after which the mixture is stirred in a closed flask for 1 hr at room temperature. A white precipitate forms. Under stirring, a solution of 4-azidophenol (135 mg, 1 mmole) in dry dimethoxyethane (5 ml) is added, and the mixture is stirred in a closed flask in the dark for 16 hr at room temperature. The solvent is removed by rotary evaporation at room temperature, and ice-cold water is added to the residue. Insoluble material is removed by filtration. The filtrate is charged onto column (16 × 2.5 cm) of Dowex 50W-X8, 30–80 mesh, equilibrated with 10 mM hydrochloric acid.

The column is washed with 10 mM hydrochloric acid (300 ml). Fractions of 5.5 ml are collected and the absorbance at 300 nm is monitored. APh-P elutes as a broad peak between fractions 5 and 30. The purity of the compound in these fractions may be checked by thin-layer chromatography. About 2 μl of these fractions are spotted on silica gel thin-layer plates, which are developed with methanol. Fractions giving one spot with an R_f value of 0.79 are pooled. Removal of the solvent by lyophilization provided 195 mg (0.9 mmole; 90%) of brown crystals of APh-P. Analytical data: NMR(D_2O) δ6.95 and 6.72 (A_2B_2 pattern); IR (KBr) 1210 cm^{-1}(PO_3H_2), 2120 cm^{-1}($-N_3$), 2600–3100 cm^{-1}(PO_3H_2).

Azidophenyl-GDP (APh-GDP). To a solution of APh-P (215 mg, 1 mmole) in dry methanol (2 ml), 0.5 ml of TBA is added. Methanol is removed by rotary evaporation, and excess TBA by shaking the residue with dry pentane (10 ml); the pentane is decanted. The residue is made anhydrous by dissolving it in dry pyridine (1 ml), followed by evaporation of the pyridine under reduced pressure. This procedure is repeated three times. The resulting TBA salt of APh-P is dissolved in dry dioxane (2 ml). Diphenylphosphoric acid chloride (0.3 ml, 1.5 mmoles) and dry TBA (0.48 ml) are added under vigorous stirring. After 2 hr of stirring at room temperature, the solvent is evaporated under reduced pressure. The remaining oily residue is shaken vigorously with a mixture of dry ether–pentane (1:1, v/v; 20 ml) and allowed to settle for 1 hr at 4°, after which the ether–pentane is decanted. Residual ether–pentane is removed by vacuum evaporation, followed immediately by dissolving the residue in dry dimethylformamide–pyridine (1:1, v/v; 10 ml) and addition of this solution to dry TBA-GMP (350 mg, 0.7 mmole). The resultant mixture is stirred for 2 hr at room temperature. It is important that mois-

---

[2] M. E. C. Biffin, J. Miller, and D. B. Paul, *in* "The Chemistry of the Azido Group" (S. Patai, ed.), p. 61. Wiley (Interscience), New York, 1971.

ture be rigorously excluded throughout these steps. Evaporation of the solvent under reduced pressure gives an oily residue, which is extracted with 0.1 M triethylammonium bicarbonate (TEAB) at pH 7.6 (10 ml). The aqueous extract is chromatographed on DEAE-Sephadex A-25 (2.5 × 15 cm) equilibrated with 0.1 M TEAB buffer at pH 7.6. The column is developed with a linear gradient of the pH 7.6 buffer (0.1–0.5 M; total volume 800 ml). Fractions of 3.5 ml are collected, and the absorbance at 295 nm is monitored. Three large fractions are eluted from the column (at 0.17, 0.26, and 0.34 M TEAB). The middle fraction is pooled, and the solvent and TEAB are removed by repeated lyophilization. The residue is dissolved in ethanol–1 M TEAB at pH 7.6 (8:2 v/v; 4 ml) and chromatographed with the same solvent on three preparative silica gel plates (Merck-Darmstadt, 60F254). Three bands are obtained: one is near the origin (GMP), one near the front (unknown), and one with an R_f of 0.8. The last of these is scraped out and extracted with aqueous methanol (10%). Silica gel is removed by centrifugation. The supernatant liquid is lyophilized, providing 0.3 mmole of APh-GDP.

The purity of the synthesized APh-GDP can be checked by two methods: (1) Silica gel thin-layer chromatography; solvent ethanol–0.1 M TEAB at pH 7.6 (8:2, v/v). R_f of APh-GDP = 0.75; R_f of APh-P = 0.46; R_f of GMP = 0.30. (2) High-voltage paper electrophoresis (Whatman No. 3 MM) with 0.05 M TEAB at pH 7.6 R_f of GMP = 1.00; R_f of GDP = 1.11; R_f of APh-P = 1.30; R_f of APh-GDP = 0.98.

Furthermore, incubation of APh-GDP with alkaline phosphatase (intestinal) and phosphodiesterase (snake venom) liberates 4-azidophenol. This leads to an increase in absorption at 303 nm. Incubation with phosphodiesterase alone produces APh-P, and incubation with alkaline phosphatase alone is without effect.

Irradiation Procedure

Equipment

Photolysis has been conducted with a super-high-pressure mercury lamp (Philips SP500W, 500 watts) equipped with a glass window cutting off light below 350 nm. Solutions are irradiated in a glass tube. For cooling purposes the tube is placed in a double-walled glass vessel filled with water, and the temperature is maintained by circulating alcohol at a constant temperature through a jacket (Fig. 2). The distance between the lamp and the vessel is about 3 cm. Vigorous stirring of the irradiation mixture is imperative, since otherwise local heating may occur

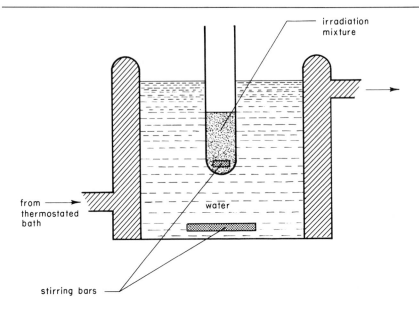

FIG. 2. Glass vessel used for the irradiations.

owing to light absorption by the dark photodecomposition products of APh-GDP.

Photolysis

A detailed description of the identification of proteins involved in GDP binding in ribosomes and a 5 S RNA–protein complex has been given elsewhere,[1,3] and only a general outline is presented here. Two irradiations are carried out simultaneously. In one tube, the biological material is irradiated with a 10- to 100-fold molar excess of [³H]APh-GDP in the absence of GDP. In the other tube, the same irradiation is performed in the presence of 10-fold mloar excess of GDP relative to the amount of APh-GDP. It is advisable to use organic buffers, e.g., Tris or HEPES, in order to suppress pseudo-photoaffinity labeling[4] (see under Discussion).

Photolysis of APh-GDP is complete in about 30 min. After the irradiation, excess radioactivity is removed by gel filtration or, if possible, by removing the biological material by centrifugation. A difference in

³ J. A. Maassen and W. Möller, *Biochem. Biophys. Res. Commun.* **64**, 1175 (1975).
⁴ A. E. Ruoho, H. Kiefer, P. E. Roeder, and S. J. Singer, *Proc. Natl. Acad. Sci. U.S.A.* **70**, 2567 (1973). See also this volume [59].

the incorporation of radioactivity between the two samples indicates the occurrence of active site labeling, which can be analyzed further by the appropriate techniques.

Discussion

Labeling Efficiency

The main condition for APh-GDP labeling of the GDP binding site is that this compound be a competitive inhibitor for GDP binding. In a rough approximation, it can be estimated that APh-GDP is suitable as a labeling agent if it inhibits GDP binding with a k_I less than 1 mM; this assumes a rate constant for the formation of the APh-GDP active site complex of between 10^6 and 10^8 M^{-1} sec^{-1}.[5-7]

For labeling of the GDP binding site in ribosomes, APh-GDP is likely to have a K_I of 1 mM (as determined for the similar compound phenyl-GDP[8]). This is the upper limit for photoaffinity labeling. However, addition of an antibiotic, fusidic acid, slows down the dissociation of the ribosome APh-GDP complex,[9] thus making it possible to use the compound. In the presence of the antibiotic, labeling occurs in ribosomal proteins with an efficiency of 2-5%. Omission of fusidic acid lowers the incorporation by a factor of 6.[1,4] The percentage of labeling can be increased by repeating the irradiation procedure several times. This may be useful when little biological material is available, e.g., after electrophoresis. Because the reaction mixture becomes brown after several irradiations, it is necessary to use high-intensity lamps.

Pitfalls

Ruoho et al.[4] have shown that the phenomenon of pseudophotoaffinity labeling can represent a complication. In the identification of the proteins involved in GDP binding in a 5 S RNA protein complex or in ribosomes, pseudophotoaffinity labeling was suppressed by the use of Tris buffers, which scavenge the freely migrating reactive species.

Sometimes, labeling of a ribosomal protein was observed which was due to a noncovalent sticking between photodecomposition products from

[5] M. Eigen and G. G. Hammes, Adv. Enzymol. 25, 1 (1963).
[6] R. A. G. Smith and J. R. Knowles, Biochem. J. 141, 51 (1974).
[7] J. R. Knowles, Acc. Chem. Res. 5, 155 (1972).
[8] P. Eckstein, W. Bruns, and A. Parmeggiani, Biochemistry 14, 5225 (1975).
[9] J. H. Highland, L. Lin, and J. W. Bodley, Biochemistry 10, 4404 (1971).

APh–GDP and the protein. This sticking even persisted during electrophoresis in urea–polyacrylamide gels. Repeated washing of the ribosomes diminished this effect, which can easily be distinguished from real photoaffinity labeling because of its insensitivity to the addition of GDP.

[77] Photoactivated GTP Analogs

By ALEXANDER S. GIRSHOVICH, VALERY A. POZDNYAKOV, and YURI A. OVCHINNIKOV

To localize the GTP-binding center at the site of interaction of GTP with ribosomes and elongation factor G, we applied the affinity labeling method using two types of GTP analogs: a photoactivated group is attached to the ribose or to the γ-phosphate of the GTP molecule.[1-3] As a photoactivated group, an aromatic azide was used to generate with visible light the nitrene radical, which is capable of attacking any sterically close chemical bond up to the C—H bond.[4] The high reactivity of the radical generated from azide allows an assessment of the components and topography of any studied center.

The methods of preparing the proposed GTP analogs in a radioactive form and some of their properties are described here. All procedures were carried out in the dark or in dim daylight.

Synthesis of Ribose Photoanalog of GTP (2-Nitro-4-azidobenzoyl Hydrazone of Periodate Oxidized [14C]GTP)

The photoactivated component, 2-nitro-4-azidobenzoyl hydrazide was obtained by hydrazinolysis of the N-hydroxysuccinimide ester of 2-nitro-4-azidobenzoic acid. The latter can be easily prepared, as shown in Scheme 1, from 2-nitro-4-aminobenzoic acid by the usual methods of arylazide preparation (see, e.g., Galardy *et al.*[5]). The GTP photoanalog was obtained in analogy to the synthesis of thiosemicarbazone of nucleoside 5'-monophosphates described by Dulbecco and Smith.[6]

[1] A. S. Girshovich, V. A. Pozdnyakov, and Y. A. Ovchinnikov, *Dokl. Akad. Nauk SSSR* **219**, 481 (1974).

[2] A. S. Girshovich, E. S. Bochkareva, and V. A. Pozdnyakov, *Acta Biol. Med. Ger.* **33**, 639 (1974).

[3] A. S. Girshovich, V. A. Pozdnyakov, and Y. A. Ovchinnikov, *Eur. J. Biochem.* **69**, 321 (1976).

[4] J. R. Knowles, *Acc. Chem. Res.* **5**, 155 (1972). See also this volume [8].

[5] R. E. Galardy, L. C. Craig, J. D. Jamieson, and M. P. Printz, *J. Biol. Chem.* **249**, 3510 (1974).

[6] R. Dulbecco and L. D. Smith, *Biochim. Biophys. Acta* **39**, 361 (1960).

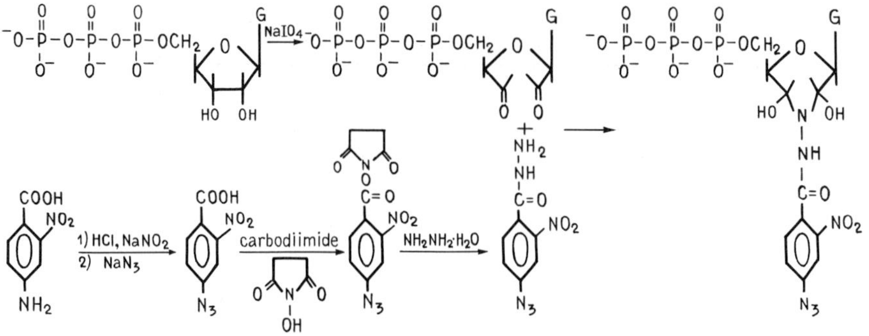

SCHEME 1

Preparation of 2-Nitro-4-azidobenzoylhydrazide

A solution of 145 μl (3 mmoles) of freshly distilled hydrazine hydrate in 15 ml of dioxane is added to a solution of 915 mg (3 mmoles) of N-hydroxysuccinimide ester of 2-nitro-4-azidobenzoic acid in 100 ml of dioxane, and the mixture is stirred for 1 hr at 0°. It is necessary to add hydrazine hydrate in the same molar quantity as the ester, since the azide group would be decomposed by an excess of hydrazine (the 2100 cm^{-1} band disappears in the infrared spectrum). The mixture is evaporated to an oil under reduced pressure, and the oil is crystallized by addition of light petroleum ether (40°–70°) and repeated evaporation. Recrystallization is from water or aqueous ethanol. Yield, 420 mg (58%); m.p., 163°–165° (decomposition); IR (KBr), cm^{-1}: 3300–3400 (amide), 2100 (—N$_3$), 1530 (—NO$_2$); UV spectrum (H$_2$O): λ_{max} 250 nm ($\epsilon = 21,000$), λ_{min} 227 nm ($\epsilon = 11,000$). The product is chromatographically and electrophoretically homogeneous (see the table). The compounds are detected in UV light and darken under irradiation by UV or visible light.

Preparation of the Ribose Photoanalog of [^{14}C]GTP

A mixture of 2.5 nmoles of [^{14}C]GTP (Amersham, England, specific activity 500 Ci/mole) and 1.4 μmoles of NaIO$_4$ in 25 μl of water is incubated for 30 min at 0°. To separate the GTP dialdehyde formed from excess periodate, the mixture is applied as a band to FN-15 paper (Filtrak, GDR) with unlabeled GTP as a marker, and subjected to electrophoresis at 4° (20 mM citrate at pH 5.0; 15 volts/cm). After 1 hr electrophoresis is stopped and a solution of 8 μmoles of 2-nitro-4-azidobenzoylhydrazide in about 0.1–0.2 ml of 50% aqueous ethanol is added to the band of GTP dialdehyde. After incubation in the cold for 1 hr,

CHROMATOGRAPHIC AND ELECTROPHORETIC MOBILITIES
OF THE SYNTHESIZED COMPOUNDS

| | R_f | | | Electrophoretic mobility |
Compound	A[a]	B[a]	C[b]	(M)[c]
GTP	—	—	0.03	1
N-Hydroxysuccinimide ester of 2-nitro-4-azidobenzoic acid	0.54	—	—	—
2-Nitro-4-azido benzoylhydrazide	0.1	0.33	0.8	−0.1
Ribose photoanalog of GTP	—	—	0.07	0.75
4-Azidobenzylamine	—	0.1	0.8	—
γ-Phosphate photoanalog of GTP	—	—	0.2	0.9

[a] Thin-layer chromatography on Silufol UV$_{254}$ (Kavalier, Czechoslovakia) in the solvent system (v/v): (A) chloroform:ethanol (95:5); (B) ethanol.

[b] Paper chromatography on FN-15 (Filtrak, GDR) in the solvent system ethanol:1 M ammonium bicarbonate, pH 8.0 (7:3).

[c] Paper electrophoresis on FN-15 (Filtrak, GDR) in 0.02 M sodium citrate at pH 5.0 and 15 volts/cm.

electrophoresis is continued for 2.5–3 hr in the same conditions. The band containing the product is detected by radioactivity scanning and is cut out and washed with ethanol to remove the buffer. The product is eluted with 5 ml of water.

The analog may be stored at −70° or in a dry state after lyophilization. Its mobility upon paper chromatography and electrophoresis is given in the table.

Synthesis of γ-Phosphate Photoanalog of GTP, γ-(4-Azidobenzyl)amide of [^{14}C]GTP

The photoactivated component, 4-azidobenzylamine, is obtained from 4-nitrobenzylbromide by the method described by Smith and Hall[7] for

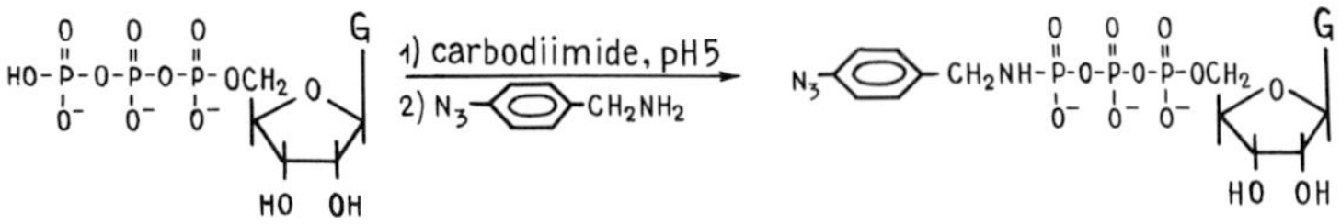

SCHEME 2

4-azidoaniline. 4-Azidobenzylamine is an oil that is soluble in ethanol, ether, dimethylformamide, and dioxane. Its chromatographic mobility is

[7] P. A. S. Smith and J. H. Hall, J. Am. Chem. Soc. 84, 480 (1962).

given in the table. The substance moves as a single band visible in UV light; it stains with ninhydrin and darkens upon irradiation.

Synthesis of the γ-phosphate photoanalog of GTP was by a modification of the method of Babkina *et al.*,[8] developed to obtain the γ-amide of ATP.

N-Cyclohexyl-*N'*-β-(4-methylmorpholinium)ethylcarbodiimide *p*-toluenesulfonate, 7 μmoles, is added to 20 nmoles of [^{14}C]GTP in 50 μl of 0.1 *M* 2-(*N*-morpholino)ethanesulfonate at pH 5.0, and the mixture is incubated for 30 min at room temperature. A solution of 100 μmoles of 4-azidobenzylamine in 20 μl of dimethylformamide is added, and the mixture is incubated for an additional 2–3 hr. The product is separated by ascending paper chromatography in system C (see the table). The band containing the analog is cut out and eluted with 5 ml of water. After several evaporation steps under reduced pressure with addition of ethanol to remove the buffer, the product is dissolved in water and stored at −70°.

The γ-phosphate analog of GTP is thus obtained homogeneous upon paper chromatography and electrophoresis (see the table). Treatment with 0.1 *M* HCl at 37° for 1 hr quantitatively converts it into the initial nucleotide by hydrolysis of the γ-phosphoamide bond.

Functional Properties of the Photoanalogs of GTP[1–3]

The functional activity of both GTP analogs was tested by their ability to participate in the formation of a ternary complex with ribosomes and elongation factor G from *Escherichia coli* MRE 600. The criteria of specificity were the following: (1) Binding of the analogs only in the presence of both the elongation factor G and ribosomes; i.e., the absence of binding by ribosomes alone or by the elongation factor G alone. (2) Inhibition of binding by an excess of the native nucleotide (GTP). (3) Stimulation of the antibiotic fusidic acid. (4) The ability of the analogs to be hydrolyzed by the ribosome- and factor G-dependent GTPase.

By the above criteria, both analogs form a specific ternary complex with the ribosomes and elongation factor G that is close in efficiency to the native GTP complex. In particular, the ribose photoanalog of GTP is a substrate of the GTPase reaction; to test this activity it was synthesized from [γ-^{32}P]GTP. Some characteristics of the γ-phosphate analog of GTP (the absence of hydrolysis by GTPase, noneffectivity of

[8] G. T. Babkina, V. F. Zarytova, and D. G. Knorre, *Bioorg. Khim.* (*USSR*) **1**, 611 (1975).

fusidic acid) are similar to those of the well known GTP analogs GMPPCP or GMPPNP.

Upon irradiation of the ternary complex by visible light (incandescent lamp, 400 W) both analogs form a covalent bond only with elongation factor G, the labeling of which depends on irradiation and is inhibited by the excess of the native nucleotide. Labeling of ribosomes is insignificant, i.e., less than 15%, and nonspecific. Thus, both analogs of GTP are true photoaffinity reagents, whose site of covalent binding in the specific ternary complex is localized on elongation factor G.

The preparation of these photoanalogs of GTP is simple and can be used practically without modification for the radioactive microsynthesis of other photoactivated or chemically specific affinity analogs of GTP. It cannot be concluded, however, that the several groups attached to the nucleotide may disturb its function. In our case, for example, a GTP derivative similar in structure to the γ-(4-azidobenzyl)amide of GTP, but containing an aromatic amine, the γ-(4-azido)anilide of GTP, is almost without inhibitory ability for the cell-free poly(U)-dependent synthesis of polyphenylalanine and does not form a ternary complex with ribosomes and elongation factor G.

[78] A Photoactivated Analog of Streptomycin

By ALEXANDER S. GIRSHOVICH, ELENA S. BOCHKAREVA, and YURI A. OVCHINNIKOV

Streptomycin operates at the level of translation: in the ribosomal protein-synthesizing system, it inhibits polypeptide synthesis and induces miscoding. To identify the components of the streptomycin-binding center of *Escherichia coli* MRE 600 ribosomes, we applied the method of affinity labeling using the photoactivated analog of streptomycin[1] synthesized as described below. The method is based on the possibility of modifying the antibiotic aldehyde group without affecting its functional specificity. As a photoactivated component, we used 2-nitro-4-azidobenzoylhydrazide interacting with the aldehyde group of streptomycin to form the corresponding hydrazone. Treatment of the latter by NaB^3H_4 reduces the hydrazone double bond and permits introduction of a radioactive label into the analog molecule.

[1] A. S. Girshovich, E. S. Bochkareva, and Y. A. Ovchinnikov, *Mol. Gen. Genet.* **144**, 205 (1976).

Method of Synthesis of the Photoactivated Analog of Streptomycin

Synthesis of the photoactivated component, 2-nitro–4-azidobenzoyl-hydrazide, is described elsewhere.[2] All the procedures are carried out in the dark or in dim daylight.

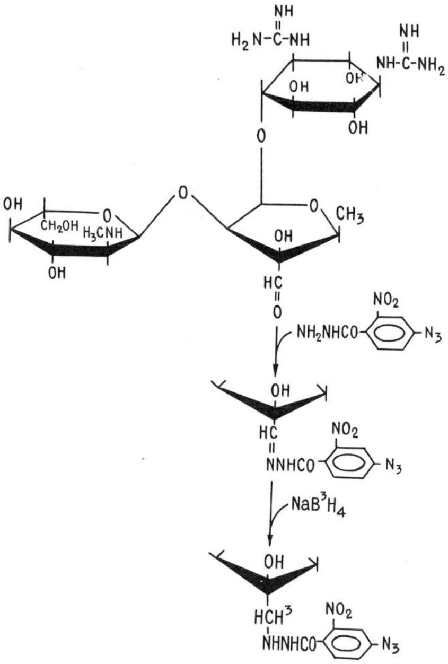

SCHEME 1

2-Nitro-4-azidobenzoylhydrazide, 20 μmoles, in 0.4 ml of freshly distilled dioxane is added to a solution of 10 μmoles of streptomycin in 1.3 ml of 50 mM potassium phosphate at pH 6.7, and the mixture is incubated for 3 hr at room temperature. After cooling in an ice bath, 100 μmoles of NaB³H₄ (Amersham, England, specific activity about 400 Ci/mole) were added. After 1 hr, the product is precipitated and washed with ethanol, dissolved in 10 mM potassium acetate at pH 5.0, and applied to a carboxymethyl cellulose column (CM 32, Whatman), 1×5 cm. The column is washed with the same buffer to remove excess radioactivity, and the analog of streptomycin is eluted in a minimal volume with 0.5 M KCl and stored at $-20°$.

The purity of the product may be determined by thin-layer chromatography on Silufol (Kavalier, Czechoslovakia) in ethanol (the R_f

[2] A. S. Girshovich, V. A. Pozdnyakov, and Y. A. Ovchinnikov, this volume [77].

of the analog was zero; of 2-nitro-4-azidobenzoylhydrazide, 0.5; of NaB^3H_4 and the products of its hydrolysis, 0.9–1.0). By paper electrophoresis on FN-15 (Filtrak, GDR) in 0.03 M Tris chloride at pH 7.5 (900 V, 2 hr), the analog migrated to the cathode with a relative R_e value of −1 as compared to picric acid. In both systems, the radioactivity was observed as a single spot, which coincided with that stained according to Sakaguchi (the guanidine group of streptomycin) and with that visible in UV light. Upon chromatography on phosphocellulose according to Hardy et al.,[3] the analog is eluted at 0.28 M NaCl as a single symmetrical peak and is not precipitated by 5% TCA.

The concentration of the streptomycin analog in solution was determined according to a modification of the Sakaguchi method (Tomlinson and Viswanatha[4]) using native streptomycin to obtain a calibration curve. The specific activity of the analog obtained was about 50 Ci/mole.

Functional Properties of the Streptomycin Photoanalog[1]

The photoactivated analog retains the functional activity of streptomycin by two criteria: (1) it binds only to the 30 S ribosomal subparticle, and (2) it inhibits the factor-free ("nonenzymic") p-chloromercuribenzoate-stimulated poly(U)-dependent ribosomal translation system.[5]

Upon irradiation (incandescent lamp, 400 W) of the reaction mixtures of the streptomycin analog with 30 S or 50 S ribosomal subparticles taken separately, the analog covalently binds predominantly to the 30 S subparticle. There is no covalent binding without irradiation. Therefore, we have concluded that the compound is an active photoaffinity analog of streptomycin. An analysis of components labeled with this analog permitted identification of the proteins of the 30 S ribosomal subparticle located in or near the streptomycin-binding site, i.e., proteins S 7 and S 14.

[3] S. J. S. Hardy, C. G. Kurland, P. Voynow, and G. Mora, Biochemistry 8, 2897 (1969).
[4] G. Tomlinson and T. Viswanatha, Anal. Biochem. 60, 15 (1974).
[5] A. S. Spirin, O. E. Kostiashkina, and J. Jonák, J. Mol. Biol. 101, 553 (1976).

[79] Haloacylated Streptomycin and Puromycin Analogs

By OLAF PONGS and ERWIN REINWALD

The binding sites of streptomycin and of puromycin on bacterial ribosomes have been probed with chemically reactive derivatives of both

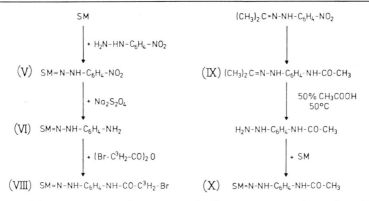

FIG. 1. Scheme of the reaction sequence leading to streptomycin analogs (VIII) and (X), respectively. For further details see Methods.

antibiotics.[1-5] In both cases, bromoacetic acid or iodoacetic acid were attached to the antibiotics so that they did not lose their general antibiotic properties but could irreversibly react at their respective binding sites.

Synthesis of Chemically Reactive Streptomycin Analogs

The formyl residue in position 3 of the streptose moiety of streptomycin can be chemically modified without changing significantly its antibiotic action.[6] Accordingly, streptomycin was allowed to react with 4-aminobenzoylhydrazine and subsequently bromoacetylated.[2] However, the product of this reaction pathway could not unambigously be identified. The product could have been the 4-bromoacetamidobenzoylhydrazone of streptomycin [compound (IV), see Methods] or, alternatively, an azomethine. Therefore, another similar streptomycin analog, the 4-bromoacetamidophenylhydrazone of streptomycin, was synthesized by an unambiguous route, that which is illustrated in Fig. 1. A critical step in the synthesis of compound (VIII) was the selective reduction of the nitro group. Reduction with hydrogen in the presence of palladium

[1] O. Pongs, R. Bald, T. Wagner, and V. A. Erdmann, *FEBS Lett.* **35**, 137 (1973).
[2] O. Pongs and V. A. Erdmann, *FEBS Lett.* **37**, 47 (1973).
[3] O. Pongs, R. Bald, V. A. Erdmann, and E. Reinwald, *in* "Drug Receptor Interactions in Antimicrobial Chemotherapy" (J. Drews and F. E. Hahn, eds.), pp. 179–190. Springer-Verlag, Berlin and New York, 1974.
[4] R. Bald, O. Pongs, and V. A. Erdmann, *in* "Ribosomes and RNA Metabolism," pp. 161–182. Slov. Acad. Sci., CSSR, 1974.
[5] O. Pongs, *Acta Biol. Med. Ger.* **33**, 629 (1974).
[6] H. Heding, G. J. Fredericks, and O. Lützen, *Acta Chem. Scand.* **26**, 3251 (1972).

catalyst led to the destruction of the hydrazone. The selective reduction of the nitro group by sodium dithionite[7] was chosen instead.

The streptomycin analogs (IV) and (VIII) differ by one carbonyl group. The similarities of both analogs in electrophoretic and chromatographic, as well as spectroscopic, properties indicate that both derivatives [not only compound (VIII)] are the corresponding hydrazones of streptomycin. The synthesis of 4-acetamidophenylhydrazone of streptomycin can be achieved by still another reaction pathway, (IX) to (X), as illustrated in Fig. 1. 4-Acetamidophenylhydrazine was prepared by acid treatment of the 4-acetamidophenylhydrazone of acetone. The resulting hydrazine was then allowed to react with streptomycin. Attempts to prepare 4-bromoacetamidophenylhydrazine by a similar reaction pathway were unsuccessful. However, compound (X) served as an important standard in all analytical procedures.

Purification of Bromoacetylated Streptomycin Analogs

Efforts to isolate bromoacetylated streptomycin analogs by preparative paper chromatography were hampered by the instability of the analogs. Elution from paper resulted in the appearance of a new, unidentified product, which had a very slow electrophoretic mobility. This indicated that the charged amino and/or guanidinium groups of streptomycin were altered as a result of the isolation procedure. Therefore, bromoacetylated streptomycin analogs were separated from the reactants by simple anion exchange chromatography. This method did not lead to decomposition and yielded compounds of 90% purity.

Methods

4-Aminobenzoylhydrazone of Streptomycin (I). Streptomycin sulfate (Boehringer, Germany) and 4-aminobenzoylhydrazine (Fluka-AG, Switzerland) are mixed in water in a molar ratio of 1:2. After 2 hr at room temperature, the hydrazone is precipitated by the addition of 10 volumes of ethanol and is maintained for at least 2 hr at $-20°$. The precipitate is collected by centrifugation and is then dissolved in water. It is reprecipitated twice with cold ethanol. Final purification is carried out by descending paper chromatography with butanol(1)/ethanol/water (3:2:5) as solvent system. R_f relative to dihydrostreptomycin, 1.77; absorption spectrum (pH 6), λ_{max} 298 nm, λ_{min} 255 nm, A_{298}/A_{255} 2.36. The reaction is quantitative.

---

[7] H. Franzen and P. Steinführer, *Chem. Ber.* **54**, 862 (1921).

4-Acetamidobenzoylhydrazone of Streptomycin (II). Compound (I), 100 μmoles, dissolved in 4 ml of 30% aqueous dioxane and 2 mmoles of acetic anhydride are added together with 2 mmoles of 2,6-lutidine. The reaction mixture is allowed to stand for 3 hr at room temperature. The streptomycin analog is precipitated by the addition of 10 volumes of cold ethanol and kept overnight at $-20°$. The precipitate is collected by centrifugation, dissolved in 1 ml of water, and reprecipitated with ethanol. The reaction product is chromatographically pure as shown by paper chromatography in the above solvent system (A): R_f relative to dihydrostreptomycin, 5.0; and by high-voltage paper electrophoresis in 0.05 M sodium citrate pH 3.5 buffer: R_e relative to dihydrostreptomycin, 0.74. Absorption spectrum: λ_{max} 280 nm, λ_{min} 238 nm; A_{280}/A_{238} 2.15.

4-(Bromo)acetamidobenzoylhydrazone of Streptomycin (III). Compound (I), 20 μmoles, is dissolved in 2 ml of water. After addition of 10 μl of 2,6-lutidine (86 μmoles), a solution of 86 μmoles of bromoacetic anhydride in 1 ml of dioxane is added. The reaction mixture is allowed to stand on ice for 14 hr. The streptomycin analog is precipitated by addition of 20 ml of ethanol. After 2 hr at $-20°$, the precipitate is collected by centrifugation and dissolved in 1 ml of water. Purity may be checked by paper chromatography and by paper electrophoresis as above. R_e is 0.73 and R_f is 4.5 relative to dihydrostreptomycin. Absorption spectrum (pH 6): λ_{max} 281 nm, λ_{min} 240 nm; A_{281}/A_{240} 1.94.

4-Bromo[2-³H]acetamidobenzoylhydrazone of Streptomycin (IV). Compound (IV) is prepared similarly to compound (III) but with the use of α,α'-dibromo[2-³H]acetic anhydride (300 Ci/mole). The ethanol precipitate is dissolved in water and then applied to an anionic exchange column (1 $\times$ 20 cm; Dowex 1 $\times$ 2, Cl⁻ form) in order to remove coprecipitated bromoacetate, which migrates on paper chromatograms in solvent system A at the same rate as compound (IV) itself. The aqueous effiuent of the column, which contains the streptomycin analog, is collected, concentrated by lyophilization, and stored in 250-μl portions in liquid nitrogen. Each portion is used only once for affinity labeling experiments.

4-Nitrophenylhydrazone of Streptomycin (V). Streptomycin (1 mmole) and 4-nitrophenylhydrazine hydrochloride (2 mmoles) (Merck, Germany) are dissolved in water and allowed to stand at room temperature for 12 hr. The hydrazone is precipitated by addition of 10 volumes of ethanol and kept at $-20°$ for 1 hr. The precipitate is dissolved in water and reprecipitated three times. The final aqueous solution is lyophilized. R_e is 0.8 and R_f is 4.83 relative to dihydrostreptomycin. Absorption spectrum (pH 6): $\lambda_{max,1}$ 390 nm, $\lambda_{max,2}$ 250 nm, λ_{min} 300 nm; A_{390}/A_{300} 10.7, A_{250}/A_{300} 4.3.

4-Aminophenylhydrazone of Streptomycin (VI). Compound (V), 1 μmole, is dissolved in 50 μl of 0.5 M potassium phosphate (pH 7.5), and 20 μl of freshly prepared 1 M sodium dithionite are added. The reaction is completed within a few seconds at room temperature and is indicated by decolorization of the reaction mixture. The reduction product is directly used in the subsequent acylation steps without further isolation.

4-Acetamidophenylhydrazone of Streptomycin (VII). Compound (VI), 1 μmole, in 0.25 M potassium phosphate at pH 7.5 is mixed with 5 μmole of acetic anhydride; dioxane is added to a final concentration of 30%. The reaction mixture is allowed to stand at room temperature for 5 hr and is then passed through an anionic exchange column (1 $\times$ 20 cm, Dowex 1 $\times$ 2, Cl⁻ form). The water eluate, containing compound (VII), is concentrated under reduced pressure and used for descending paper chromatography in solvent system A. R_e is 0.75 and R_f is 4.9 relative to dihydrostreptomycin. Absorption spectrum (pH 6): λ_{max} 283 nm, λ_{min} 235 nm; A_{283}/A_{235} 2.7.

4-Bromo[2-³H]acetamidophenylhydrazone of Streptomycin (VIII). Compound (V), 10 μmoles, in 0.25 M potassium phosphate at pH 7.5 is mixed with 50 μmoles of α,α'-dibromo[2-³H]acetic anhydride (300 Ci/ mole), and allowed to react for 14 hr in the dark at 0°. Workup of the reaction mixture is the same as for compound (III). R_e is 0.71 and R_f is 4.4 relative to dihydrostreptomycin. Absorption spectrum (pH 6): λ_{max} 285 nm, λ_{min} 235 nm; A_{285}/A_{235} 2.13.

4-Acetamidophenylhydrazone of Acetone (IX). The 4-nitrophenylhydrazone of acetone, 10 μmoles, is dissolved in 50 μl of anhydrous dioxane and, after addition of 1 μl of acetic anhydride (10.6 μmoles), is hydrogenated under normal pressure in the presence of palladium catalyst. Within 10 min the calculated hydrogen uptake necessary for the reduction of the nitro to an amino group is reached. The reaction is terminated by removal of the catalyst by filtration under nitrogen. The filtrate contains the desired 4-acetamidophenylhydrazone of acetone, which is purified and isolated by SiO_2 column chromatography with CH_2Cl_2/CH_3OH (99/1) as solvent system. R_f value on SiO_2/TLC sheets (Riedel de Haen, Germany) in CH_2Cl_2/CH_3OH (98/2) is 0.17.

4-Acetamidophenylhydrazone of Streptomycin (X). Compound (IX), 10 μmoles, is evaporated to dryness and dissolved in 1 ml of 50% aqueous acetic acid. The solution is heated for 2 hr at 50° and then again evaporated to dryness. The solid residue is dissolved in 500 μl of water (pH 3), and a solution of 20 μmoles of streptomycin in 500 μl water is added. The reaction mixture is allowed to stand at room temperature for 15 hr. Ethanol, 10 volumes, is added and, after 2 hr at −20°, the precipitate is collected by centrifugation and dissolved in a small vol-

REACTION OF STREPTOMYCIN ANALOG (IV) AND OF IODOACETAMIDE WITH
Escherichia coli 70 S RIBOSOMES[a]

Compound	Label/ribosome (mole/mole)
(IV)	1.30
(IV) + streptomycin	0.70
Iodoacetamide	0.10

[a] 70 S ribosomes (0.75 μM) were incubated at 37° in 300 μl of 6 mM MgCl$_2$, 80 mM NH$_4$Cl, 50 mM Tris chloride at pH 7.8 with compound (IV) (3 μM) or with iodoacetamide (10 μM). The streptomycin concentration was 2.3 mM. After 2 hr, 30-μl samples of the incubation mixtures were diluted with 3 ml of the cold incubation buffer and subjected to Millipore filtration. Filters were washed three times with 2 ml of cold buffer, and radioactivity on the filters was determined.

ume of water. Paper chromatography and paper electrophoresis of the reaction product are identical to those of compound (VII).

Bromoacetamidophenylhydrazones of Streptomycin as Affinity Probes

The derivatization of the formyl residue of streptomycin with bromo-acetamidophenylhydrazine does not influence the antibiotic's effect on the function and activity of *Escherichia coli* ribosomes as determined in various assays.[2,3] The specificity and, hence, the usefulness of chemically reactive streptomycin analogs can be tested by comparison of the reaction of the affinity probes with streptomycin-sensitive and resistant ribosomes.[7a] Furthermore, the specificity of the labeling reactions can be investigated by comparison with the reaction of iodoacetamide and ribosomes and by inhibition studies. This is illustrated in the table. For example, 1.3 moles of compound (IV) become cross-linked per mole of ribosome at 37° after 2 hr of incubation. This compares to 0.10 mole of iodoacetamide that react per mole of ribosome under similar reaction conditions. Addition of streptomycin to the incubation mixtures inhibits the reaction of the streptomycin analog with ribosomes. However, the inhibition is not complete, even at 1000-fold higher concentrations of streptomycin. This suggests that the streptomycin analog has more than one binding site on the ribosome, only one of which corresponds to the streptomycin binding site. This notion finds further support from two other observations. Streptomycin-resistant ribosomes also react with the streptomycin analogs. The dependence of streptomycin analog bind-

[7a] O. Pongs, E. Reinwald, E. Lanka, and G. Stöffler, manuscript in preparation.

ing on ribosome concentration shows at least two independent binding sites (Pongs et al.[7a]). This indicates that the specificity of the streptomycin labeling reaction is not exclusively confined to a reaction in the streptomycin binding site. This is somewhat surprising, considering the fact that the apparent binding constant of streptomycin analog (IV) to 70 S ribosomes is more than one order of magnitude higher than that of chemically reactive oligonucleotides (see this volume [80]).

Synthesis of N-Iodoacetylpuromycin

The synthesis of this chemically reactive puromycin analog is relatively simple. Since iodoacylated puromycin is not very stable, a simple and rapid isolation procedure by DEAE-column chromatography was devised.

[2-^{14}C]Iodoacetic acid (20 μmoles; 51 Ci/mole) and 20 μmoles of DCC are allowed to react in 200 μl of anhydrous dioxane in the dark for 2 hr at room temperature. The precipitate is removed by centrifugation, and the supernatant liquid, which contains iodoacetic anhydride, is used directly in the acetylation reaction. It is added to a solution of 10 μmoles of puromycin in 400 μl of potassium acetate (pH 5.0). The reaction mixture is kept for 24 hr at 4° and is then passed through a DEAE-cellulose column (1.5 × 5 cm), which is equilibrated with the acetate buffer. The column is washed with 50% v/v methanol. The eluate, containing N-iodoacetylpuromycin, is collected and carefully evaporated under reduced pressure. The residue is taken up in methanol and used as such in labeling experiments. The puromycin derivative is about 90% pure as judged by thin-layer chromatography on TLC Al$_2$O$_3$ plates.[1] R_f is 0.84 in chloroform/methanol (9/1) as solvent system.

N-Iodoacetylpuromycin as Affinity Probe

Puromycin functions as an analog of the 3′ terminus of aminoacyl-tRNA in the peptidyltransferase center of the ribosome.[8] Accordingly, N-iodoacetylpuromycin would mimic the 3′-terminal end of acylated aminoacyl-tRNA, i.e., of peptidyl-tRNA, for the peptidyltransferase center. This was investigated by measuring the binding of the terminal tRNA fragments C-A-C-C-A-Leu and C-A-C-C-A-Leu-N-ac to 50 S ribosomal subunits, which had been allowed to react with N-iodoacetylpuromycin.[9] The results of such experiments indicate that the binding of fragment C-A-C-C-A-Leu to N-iodoacetylpuromycin-labeled ribosomes is not in-

[8] D. Nathans and F. Lipmann, Proc. Natl. Acad. Sci. U.S.A. 47, 491 (1961).
[9] G. Wischnath, Ph.D. thesis, Freie Universität Berlin, W.-Berlin, 1976.

hibited, whereas that of fragment C-A-C-C-A-Leu-N-ac is. This suggests that N-iodoacetylpuromycin has a preferential affinity to the P site of the peptidyltransferase center of the ribosome. Accordingly, puromycin only weakly inhibits the labeling reaction of N-iodoacetylpuromycin with ribosomes. On the contrary, preincubation of ribosomes with puromycin stimulates the labeling reaction.[9,10]

Puromycin binds weakly and relatively unspecifically to ribosomes. Its use as an affinity probe is limited.[11] Not unexpectedly, reactions of N-iodoacetylpuromycin with ribosomes exhibit similar characteristics. On the other hand, a high binding constant does not necessarily guarantee a good affinity probe as exemplified by the streptomycin affinity labels. Vice versa, a compound with a weak binding constant might still be a superb affinity probe, since the outcome of experiments with haloacetyl compounds depends so much on the presence of a properly oriented reactive amino acid side chain in the substrate binding site. This, of course, is a matter of chance.[12]

[10] J. Stahl, K. Dressler, and H. Bielka, *FEBS Lett.* **47**, 167 (1974).
[11] B. S. Cooperman, E. N. Jagnes, D. J. Brunswick, and M. A. Luddy, *Proc. Natl. Acad. Sci. U.S.A.* **72**, 2974 (1975).
[12] O. Pongs and E. Lanka, this volume [80].

[80] Chemically Reactive Oligonucleotides

By Olaf Pongs and Erich Lanka

Bromoacetamidophenyl derivatives of thymidylic acid have been used to elucidate the structure of the substrate-binding site of *Staphylococcus aureus* nuclease.[1] Similarly, bromoacetamidophenyl derivatives of the initiation codon, A-U-G, and the termination codon, U-G-A, have been employed as affinity probes of the ribosomal codon binding sites.[2-5] The synthesis of these chemically reactive oligonucleotides combines chemical and enzymic methods as outlined for the synthesis of 5'-[4-(bromo-[2-^{14}C]acetamido)phenylphospho]adenylyl-(3'-5')-uridiylyl-(3'-5') guanosine (*A-U-G) in Fig. 1.

[1] P. Cuatrecasas, M. Wilchek, and C. B. Anfinsen, *J. Biol. Chem.* **244**, 4316 (1969). See also this volume [38].
[2] O. Pongs and E. Lanka, *Hoppe Seyler's Z. Physiol. Chem.* **356**, 449 (1975).
[3] O. Pongs and E. Lanka, *Proc. Natl. Acad. Sci. U.S.A.* **72**, 1505 (1975).
[4] O. Pongs and E. Rössner, *Hoppe Seyler's Z. Physiol. Chem.* **356**, 1297 (1975).
[5] O. Pongs, G. Stöffler, and E. Lanka, *J. Mol. Biol.* **99**, 301 (1975).

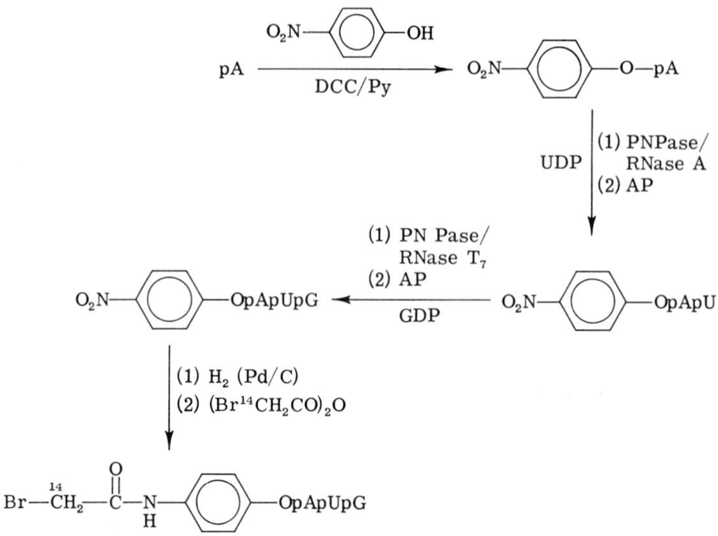

$O_2N-\langle\bigcirc\rangle-OH$

$pA \xrightarrow{\text{DCC/Py}} O_2N-\langle\bigcirc\rangle-O-pA$

UDP $\begin{array}{l}\text{(1) PNPase/}\\ \quad\text{RNase A}\\ \text{(2) AP}\end{array}$

$O_2N-\langle\bigcirc\rangle-OpApUpG \xleftarrow[\text{GDP}]{\begin{array}{l}\text{(1) PN Pase/}\\ \quad\text{RNase T}_7\\ \text{(2) AP}\end{array}} O_2N-\langle\bigcirc\rangle-OpApU$

$\begin{array}{l}\text{(1) H}_2\ \text{(Pd/C)}\\ \text{(2) (Br}^{14}\text{CH}_2\text{CO)}_2\text{O}\end{array}$

$Br-\overset{14}{C}H_2-\overset{\overset{\textstyle O}{\|}}{C}-\underset{H}{N}-\langle\bigcirc\rangle-OpApUpG$

FIG. 1. Synthesis of a 5′-modified, chemically reactive A-U-G analog. Abbreviations are DCC, *N,N*′-dicyclohexylcarbodiimide; PNPase, polynucleotide phosphorylase (EC 2.7.7.8); AP, alkaline phosphatase (EC 3.1.3.1).

Synthesis of 5′-Modified Oligonucleotides

p-Nitrophenyladenosine 5′-phosphate is prepared similarly to the synthesis of *p*-nitrophenyldeoxythymidine 5′-phosphate.[6] 5′-Adenylic acid is converted to the tri-*n*-hexylammonium salt, dissolved in anhydrous pyridine, and then condensed with *p*-nitrophenol in the presence of *N,N*′-dicyclohexylcarbodiimide (DCC). *p*-Nitrophenol and DCC are used in large excess, i.e., about 20-fold. This minimizes undesirable side reactions, such as pyrophosphate formation and cyclization to 3′,5′-phosphates. The nitrophenyl ester is isolated by preparative paper chromatography in an acidic solvent system. This rather tedious procedure is preferable to cellulose column chromatography. Attempts to isolate *p*-nitrophenyladenosine 5′-phosphate by DEAE-cellulose column chromatography with triethylammonium bicarbonate (pH 7.5) as the buffer system were unsuccessful. The phosphodiester bond of *p*-nitrophenyladenosine 5′-phosphate is unstable in mild alkali. Thus, the removal of triethylammonium bicarbonate by repeated evaporation under reduced pressure is harmful to the compound.

p-Nitrophenyluridine 5′-phosphate has been synthesized by condensation of a 2′,3′-protected nucleoside with di-4-nitrophenyl hydrogen phos-

[6] W. E. Razzell and H. G. Khorana, *J. Biol. Chem.* **234**, 2105 (1959).

phate and di-4-tolylcarbodiimide in anhydrous dioxane.[7] Subsequently, a synthesis of p-nitrophenyl-deoxythymidine 5'-phosphate was described, which used 3'-protected deoxythymidine and phosphoric acid-bis-(4-nitrophenyl)ester chloride.[8] Both syntheses require protected nucleosides and also give rise to considerable side reactions. The above-outlined synthesis has the advantage that it uses simple, commercially available starting materials, that it gives almost no side reactions, and that it is applicable to all four nucleosides alike (including guanosine).

Reaction conditions have been described for the synthesis of oligonucleotides by the use of polynucleotide phosphorylase,[9] which uses a dinucleoside (3'-5')-monophosphate as primer and nucleoside 5'-phosphates. The enzyme also used as primers "unnatural" phosphodiesters, such as nitrophenylated nucleoside 5'-phosphates. The yields of these reactions, however, are considerably lower as compared to reactions with natural phosphodiesters as primer. The yields of the reaction were found to be relatively insensitive to variations in salt concentration or in pH of the incubation mixtures. Optimal salt conditions, i.e., mM $MgCl_2$ and 0.4–1.0 M$NaCl$, are similar to those that are used in oligonucleotide synthesis with normal primers. However, it is advisable that incubation mixtures not exceed volumes of 1 ml. In our experience, it is better to split larger volumes into smaller ones, if large quantities of oligonucleotide are to be synthesized.

It is generally advantageous to have combinations in the oligonucleotide sequence that allow the simultaneous use of polynucleotide phosphorylase and ribonucleases. One example is illustrated in Fig. 1. p-Nitrophenyladenosine 5'-phosphate can be easily extended to 5'-(4-nitrophenylphospho)adenylyl-(3'-5')-uridine by a simultaneous incubation of primer, uridine 5'-phosphate, polynucleotide phosphorylase, and pancreatic ribonuclease A. This yields only the desired product besides the starting materials. Oligonucleotides that contain more than one uridine residue are split by ribonuclease A to the desired ApU derivative. Similarly, 5'-(4-nitrophenylphospho)uridylyl-(3'-5')guanosine can be synthesized by an incubation of p-nitrophenyluridine 5'-phosphate with guanosine 5'-diphosphate, polynucleotide phosphorylase, and ribonuclease T_1.[5] Once a nitrophenylated dinucleotide has been synthesized, a further chain extension is carried out with the help of polynucleotide phosphorylase in the usual manner.[9]

[7] J. G. Moffatt and H. G. Khorana, *J. Am. Chem. Soc.* **79**, 3741 (1957).

[8] R. P. Glinski, A. B. Ash, C. L. Stevens, M. B. Sporn, and H. M. Lazarus, *J. Org. Chem.* **36**, 245 (1971).

[9] R. E. Thach, "Procedures in Nucleic Acid Research" (G. L. Cantoni and D. R. Davies, eds.), pp. 520–534, Harper & Row, New York, 1966.

Reduction of nitrophenylated oligonucleotides prior to acetylation increases the alkaline lability of the nitrophenyl phosphoester bond. Aminophenyl derivatives should, therefore, be used in acetylation reactions without unnecessary delay. It has been reported that aminophenylated thymidylic acid can be bromoacetylated with the N-hydroxysuccinimide ester of bromoacetic acid.[1] This method was devised for acetylations of primary aliphatic amino groups.[10] It is not feasible for acetylation of aminophenylated oligonucelotides. Instead, the anhydride of bromoacetic acid is employed in the acetylation reactions. Nucleosides are not attacked under the conditions of the acetylation reactions employed.

Purification of Modified Oligonucleotides

Compounds are isolated by preparative paper chromatography on Whatman No. 3 MM paper. Chromatograms are developed in freshly made n-butanol/water/glacial acetic acid (5:3:2) as solvent system for 2–3 days. Although this system is very slow, a more convenient system was not found because of the alkaline lability of the compounds. Purity of isolated products can be checked by paper chromatography in 1 M ammonium acetate (pH 7.4)/ethanol (1:1) and by paper electrophoresis in 0.05 M potassium phosphate at pH 7.0. Analysis of oligonucleotides is best carried out by enzymic degradation with snake venom phosphodiesterase or ribonuclease T_2.[2,5]

Methods

4-Nitrophenyladenosine 5'-Phosphate. 5'-Adenylic acid (4.2 g; 12 mmoles) is converted to the pyridinium salt. Tri-n-hexylamine (3.23 g; 12 mmoles) is added, and the mixture is evaporated in anhydrous pyridine about three times. Then, 5.84 g of 4-nitrophenol (42 mmoles) are added, and evaporation with anhydrous pyridine is repeated twice more. The final residue is taken up in a minimum amount of anhydrous pyridine in order to keep it in solution, and 8.65 g of N,N'-dicyclohexylcarbodiimide are added. The reaction mixture is kept in the dark for 8 days at room temperature, after which it is evaporated under reduced pressure. The residual material is resuspended in the cold in 200 ml of methanol. Dicyclohexylurea is removed by filtration. The clear filtrate is dropped slowly into 2 liters of dry ether. The resulting precipitate is collected by centrifugation. It is dissolved in water and purified by pre-

[10] N. de Groot, Y. Lapidot, A. Panet, and Y. Wolman, *Biochem. Biophys. Res. Commun.* **25**, 17 (1966).

parative paper chromatography. Yield: 3.04 g (52%). The mobilities relative to uridine 5'-phosphate in paper chromatography and paper electrophoresis are: 0.55 (isopropanol/water/conc. ammonia, 7:2:1; solvent system A); 1.57 (n-butanol/water/glacial acetic acid, 5:3:2; solvent system B); 0.72 (1 M ammonium acetate/ethanol, 1:1; solvent system C); 0.35 (high-voltage paper electrophoresis, 60 V/cm, in 0.05 M potassium phosphate at pH 7.0).

5'-(4-Nitrophenylphospho)adenylyl-(3'-5')-uridine. The ammonium salt of 4-nitrophenyl adenosine 5'-phosphate (422 μmoles), 442 μmoles of UDP, and 280 μg of pancreatic ribonuclease A are incubated in 10 tubes of 1 ml each containing 3.2 mM MgCl$_2$, 0.3 mM EDTA, 0.12 M Trischloride at pH 9.0 with 4 mg polynucleotide phosphorylase for 16 hr at 37°. The reaction is terminated by boiling for 2 min, then 10 mg of alkaline phosphatase and 10 μmoles of ZnCl$_2$ are added to the pooled and cooled incubation mixtures. Incubation at 37° is carried out for additional 24 hr. The whole reaction mixture is directly applied to Whatman No. 3 MM paper for preparative paper chromatography in solvent system B. Mobility relative to 5'-uridylic acid is 1.12; in solvent system C, 0.5. Yields vary between 10 and 20%.

5'-(4-Nitrophenylphospho)adenylyl-(3'-5')-uridylyl-(3'-5')-guanosine. Nitrophenylated ApU (57.4 μmoles), 45 μmoles of GDP, and 350 μg ribonuclease T1 are incubated for 3 hr at 37° with 1 mg of polynucleotide phosphorylase in 4 ml of 0.4 M NaCl, 10 mM MgCl$_2$, 0.3 mM EDTA, 0.1 M Trischloride at pH 9.0. The reaction is terminated by boiling for 2 min. Treatment of the reaction mixture with alkaline phosphatase and purification of the nitrophenylated ApUpG are carried out as above. Yields vary between 60 and 75%. Mobility relative to 5'-uridylic acid is 0.55 in solvent system B; 0.29 in solvent system C.

Reduction of Nitrophenyl Derivatives to Aminophenyl Derivatives. Nitrophenylated compounds are dissolved in a minimum amount of 50% aqueous methanol and hydrogenated under normal pressure in the presence of Adams palladium catalyst. The methanolic solutions are cooled to −20° in this procedure when the oxygen atmosphere is exchanged for a hydrogen atmosphere, in order to avoid evaporation of the methanol. After hydrogenation is completed, the catalyst is removed by filtration under reduced pressure. All subsequent steps are carried out under nitrogen because of the high sensitivity toward oxygen of aminophenyl compounds. The methanolic solution is evaporated. The residue is taken up in 1 M potassium phosphate at pH 6.0 to provide a final concentration of 25 mM. The solution is directly used in acetylation reactions.

Bromacetylation of Aminophenylated Oligonucleotides. A 25 mM solution of the aminophenylated oligonucleotide in 1 M phosphate at pH 6.0, which is obtained as described above, is mixed with a solution of α,α'-dibromoacetic anhydride in dioxane. The final concentration of anhydride is 50 mM. The ratio of phosphate buffer to dioxane should not exceed 5:1. The reaction mixture is maintained in the dark for 10 hr at room temperature. It is directly applied afterward to a paper chromatogram, which is developed in solvent system B for 2.5 days. For affinity labeling purposes, the anhydride is usually radioactive. A simple method for the preparation of α,α'-bromoacetic anhydride is described below. Yields are between 50 and 60% with a nitrophenyl derivative as starting material.

Synthesis of α,α'-Dibromo-[2-^{14}C]acetic Anhydride. Bromo-[2-^{14}C] acetic acid (5 μmoles, 22.6 Ci/mole) is dissolved in 50 μl of dry dioxane. N,N'-dicyclohexyl carbodiimide, 2.5 μmoles, dissolved in 50 μl of dioxane, is added. After 1 hr in the dark at room temperature, the precipitated dicyclohexyl urea is removed by centrifugation. The supernatant liquid contains the anhydride and can be used directly in acetylation reactions. The dioxane used in these reactions should be absolutely free of peroxides.

Bromoacetylated Oligonucleotides as Affinity Probes

Affinity labeling experiments with bromoacetyl compounds are biased by two important limitations, which often make them inferior to comparable photoaffinity labels. The number of properly oriented amino acid functional groups that can undergo a nucleophilic displacement reaction in the active site of a protein is limited to histidine, lysine, tyrosine, cysteine, and glutamic acid. Reactions are strongly influenced by the intrinsic pK of the respective amino acid residue and by the pH of the incubation mixture. It should be noted that bromoacetyl compounds can also react with RNA[11]. The other limitation is that the time point for the affinity labeling reaction to occur cannot be freely chosen. One can only incubate the reactants and let them react for a given time. Reactions are usually quite slow and take considerable time for completion, which can vary between 1 and 20 hr.[1,2]

Maximum velocities for the reaction of bromoacetyl derivatives of A-U-G and of U-G-A with *Escherichia coli* 70 S ribosomes have been determined to be 5×10^{-9} mole-sec^{-1}.[12] The K_m values of the A-U-G and of the U-G-A label, on the other hand, differ by 20-fold. This difference in apparent affinity seems to be without influence on the maximum veloc-

[11] J. B. Breitmeyer and H. F. Noller, *J. Mol. Biol.* **101**, 297 (1976).
[12] O. Pongs and E. Rössner, *Nucleic Acids Res.* in press (1976).

ity of the reaction. The specificity of the reactions can be markedly improved if the reaction between components A and B is made dependent on the addition of a component, C. The K_m value of bromoacetylated A-U-G binding to ribosomes is 50 μM, but it is 0.5 μM in the presence of initiator tRNA and initiation factors. This means that the apparent affinity of the A-U-G label to ribosomes is increased by the addition of tRNA and factors such that conditions can be chosen, wherein the label reacts only with ribosomes in a ternary ribosome–*A-U-G–tRNA complex and does not react by itself with ribosomes.[13] Again, this increase in apparent affinity does not accelerate the velocity of the chemical reaction.

Affinity labeling experiments with ribosomes give rise to a number of potential pitfalls, which one has to try to circumvent. In most cases, this problem is handled indirectly by competition and/or inhibition experiments. This usually gives only negative answers that are not completely satisfactory. However, affinity labeling experiments with chemically reactive oligonucleotides have an important advantage. As discussed above, experiments can be carried out such that the reaction is made dependent on the formation of a ternary complex between ribosomes, label, and tRNA. Ribosomes can also be directly incubated with the affinity label. After the oligonucleotide analog has reacted at its ribosomal binding site, ribosomes have to be programmed for the coding objective of the oligonucleotide sequence. Thus, it is possible to provide direct and positive evidence that the labeling experiment was indeed an affinity labeling experiment, not merely a chemical modification reaction.[5] Both types of labeling experiments have been carried out with A-U-G derivatives and *E. coli* ribosomes. This included studies on ribosomal conformations, on the structure of ribosomal active sites, and on the mechanism of mRNA–tRNA interaction on the ribosome.[2-5,12,13]

Identification of Oligonucleotide-Labeled Proteins

Cross-linking of oligonucleotides to proteins changes their charge considerably. Therefore, protein identification is not meaningful in electrophoretic systems in which separation is mainly based on charge differences.[5] The identification of oligonucleotide-labeled ribosomal proteins by polyacrylamide gel electrophoresis is generally inadequate. The method of choice for analysis is autoradioimmunodiffusion. In this case, labeled proteins are precipitated on double-diffusion Ouchterlony plates by their specific antisera. The immunoprecipitates are then investigated for their

[13] O. Pongs, E. Lanka, and G. Stöffler, manuscript in preparation.

content of radioactive oligonucleotide by autoradiography, which results in an unambiguous identification of the protein that was labeled.[5]

Acknowledgment

This work was supported by a grant of the Deutsche Forschungsgemeinschaft, which is gratefully acknowledged.

[81] Aromatic Ketone Derivatives of Aminoacyl-tRNA as Photoaffinity Labels for Ribosomes

By Ernst Kuechler and Andrea Barta

Several derivatives of aminoacyl-tRNA have been studied as affinity labels for the ribosomal peptidyltransferase center. Attachment of the reactive substituent can be obtained most easily by allowing the amino group of the aminoacyl moiety to react with N-hydroxysuccinimide esters according to the method of Rappaport and Lapidot.[1] For this reason most analogs studied until now are acyl derivatives of aminoacyl-tRNA.[2] One of the main obstacles faced in these studies are the low yields of the affinity labeling reaction usually attained. This problem is particularly serious in the case of photoaffinity labels.

Described herein is the synthesis of aromatic ketone derivatives of aminoacyl-tRNA. The method has been employed for the synthesis of 3-benzoylpropionyl- and 3-(4-benzoylphenyl)propionyl-Phe-tRNA (Fig. 1), but can be applied to other systems as well. According to Turro,[3] the carbonyl group of aromatic ketones can be photoactivated with UV light of 320 nm. This causes an $n \rightarrow \pi^*$ transition of the free electron pair on the oxygen. The activation is followed by a rapid spin inversion; the resulting triplet state has a triplet energy of about 70 kcal. The activated molecule is capable of reacting with covalent bonds of similar energy, as shown in Fig. 2. Since the bond energy of the O—H bond in the water molecule amounts to 125 kcal, an aromatic ketone in its activated state does not react with water. On the other hand, the photoreaction between benzophenone and N-acetylglycine methyl ester occurs in high yield as demonstrated by Galardy *et al.*,[4] resulting in a photoaddi-

[1] S. Rappaport and Y. Lapidot, this series, Vol. 29, p. 685.
[2] C. R. Cantor, M. Pellegrini, and H. Oen, *in* "Ribosomes" (M. Nomura, A. Tissières, and P. Lengyel, eds.), p. 573. Cold Spring Harbor Laboratory, Cold Spring Harbor, New York, 1974.
[3] N. J. Turro, "Molecular Photochemistry," p. 44, Benjamin, New York, 1967.
[4] R. E. Galardy, L. C. Craig, and M. P. Printz, *Nature (London), New Biol.* **242**, 127 (1973).

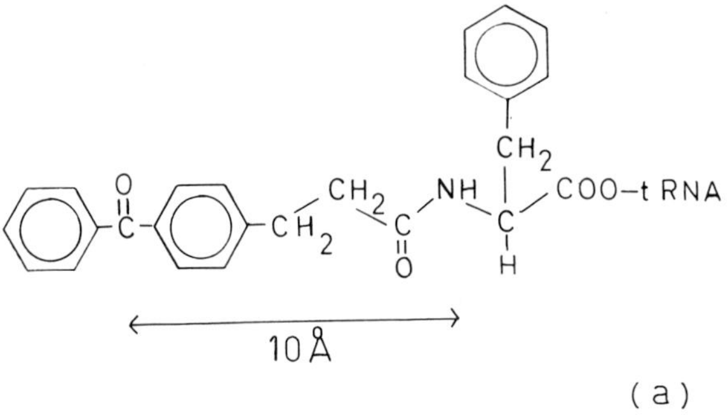

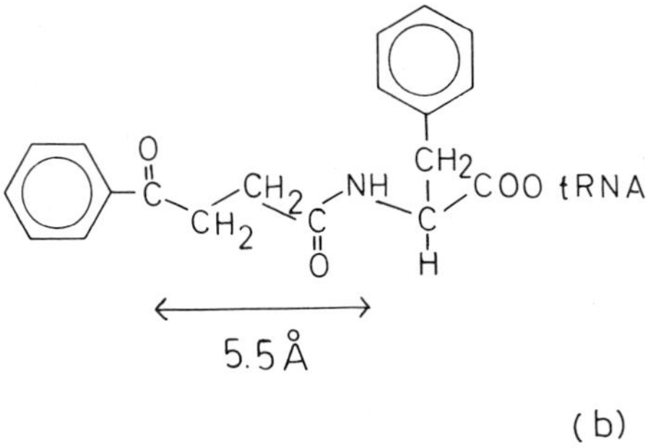

(a)

(b)

FIG. 1. Structure of 3-(4-benzoylphenyl)propionyl-Phe-tRNA (a) and 3-benzoyl-propionyl-Phe-tRNA (b).

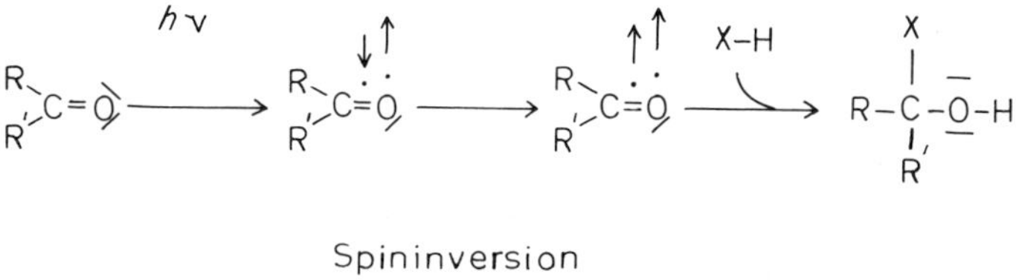

Spininversion

FIG. 2. Mechanism of reaction of the photoactivated aromatic carbonyl group.

tion to the α—C of the glycine molecule. A photoreaction between benzo-phenone and individual nucleotides has also been observed.[5] Since there is no reaction with water the photoreaction can be carried out in aqueous

[5] A. Barta and E. Kuechler, unpublished results.

medium without reduction in yield. This property makes aromatic ketone derivatives extremely valuable tools for photoaffinity labeling in biological systems.

Preparation of Aromatic Ketone Derivatives

Preparation of 3-(4-Benzoylphenyl)propionic Acid and 3-Benzoylpropionic Acid. 3-(4-Benzoylphenyl)propionic acid is prepared by a modification of the procedure of Borsche and Sinn.[6] Anhydrous aluminum chloride (6.6 g, 50 mmoles) is suspended in 25 ml of carbon disulfide or dichloromethane in a 100-ml three-neck, round-bottom flask equipped with a stirrer, a dropping funnel, and a reflux condenser fitted with a calcium chloride drying tube. The flask is cooled in ice, and freshly distilled benzoylchloride (3.5 g, 25 mmoles) is added dropwise. It is allowed to warm to room temperature, and hydrocinnamic acid ethyl ester (4.7 g, 25 mmoles; Fluka, Basel) is added slowly. After standing overnight at room temperature, the mixture is boiled for a few hours and subsequently poured onto ice. A precipitate of aluminum hydroxide forms, which is dissolved by addition of concentrated HCl. The organic phase is separated, and the aqueous phase is extracted with 25 ml of carbon disulfide or dichloromethane. The organic phases are combined and washed successively with water, 2% sodium hydroxide, and again with water until the solution is neutral. The solvent and unreacted materials are removed by distillation (benzoylchloride, b.p.$_{20}$ 75°; hydrocinnamic acid ethyl ester, b.p.$_{20}$ 130°). The remaining oil is subjected to distillation in high vacuum (use of a Kugelrohr is recommended). At 150°–170° (0.1–0.2 mmHg pressure) the 3-(4-benzoylphenyl)propionic acid ethyl ester is obtained as a slightly yellow oil (n_{D21}:1.5628). It is dissolved in 3 M methanolic KOH (20 ml) and refluxed on a boiling water bath for 5 hr. After removal of methanol by distillation, the residue is dissolved in water (30 ml) and transferred to a separatory funnel. Ether (30 ml) is added and the mixture is acidified by addition of concentrated HCl in small portions; the two phases are mixed by shaking vigorously after each addition. After the phases separate, the aqueous phase is discarded and the ether layer is washed with water and subsequently dried over anhydrous Na$_2$SO$_4$. After evaporation of the ether, a glassy mass remains. The product is crystallized from methanol. Yield, 1.9 g (30%); m.p. 98.5°–99°. Proton magnetic resonance at 60 MHz (chloroform-d), δ, multiplicity, number of protons, assignment: 8.85 (broad signal, 1, COO*H*), 7.6 (multiplet, 9, aromatic *H*), 2.9 (A,A',

[6] W. Borsche and F. Sinn, *Justus Liebigs Ann. Chem.* **553**, 260 (1942).

B,B' system, 4, X-CH_2CH_2-Y) ppm. 3-Benzoylpropionic acid is synthesized as described.[7]

Preparation of N-Hydroxysuccinimide Esters. For the synthesis of the *N*-hydroxysuccinimide esters of 3-(4-benzoylphenyl)propionic acid and 3-benzoylpropionic acid, the method of Rappaport and Lapidot[1] has been modified. 3-(4-Benzoylphenyl)propionic acid (765 mg, 3 mmoles) or 3-benzoylpropionic acid (534 mg, 3 mmoles) and *N*-hydroxysuccinimide (345 mg, 3 mmoles) are suspended in dry ethyl acetate (6 ml) and cooled in ice. Dicyclohexylcarbodiimide (618 mg, 3 mmoles), dissolved in ethyl acetate (2 ml), is added under stirring. After 15 min in ice, the mixture is allowed to warm to room temperature and stirring is continued for 2 hr. Dicyclohexylurea is removed by filtration with suction and washed with ethyl acetate (20 ml). The combined filtrate is evaporated to dryness under reduced pressure (under 40°). The residue is crystallized from methanol. Yield: 3-(4-benzoylphenyl)propionic acid *N*-hydroxysuccinimide ester 810 mg (72%), m.p. 114°–115°; 3-benzoylpropionic acid *N*-hydroxysuccinimide ester 540 mg (65%), m.p. 132°–134°.

Preparation of Aminoacyl-tRNA Derivatives

Preparation of 3-(4-Benzoylphenyl)propionyl-[^3H]Phe-tRNA^{Phe}. tRNA^{Phe} from yeast (Boehringer, Mannheim) is charged with [^{3}H]Phe (11 Ci/mmole, Amersham) according to Bartmann *et al.*[8]; a 75–90% charging of the Phe-tRNA is attained. [^{3}H]Phe-tRNA^{Phe} (5 nmoles) is dissolved in 0.2 ml of 0.05 M HEPES (*N*-2-hydroxyethylpiperazine-*N*'-2-ethanesulfonic acid) sodium salt, pH 6.8, and 3-(4-benzoylphenyl)-propionic acid *N*-hydroxysuccinimide ester (30 mg) in freshly distilled dimethyl sulfoxide (0.9 ml) is added. After the mixture is incubated for 3 hr at 37°,[1] potassium acetate at pH 5 is added to 2% and the tRNA is precipitated with 2 volumes of ethanol for 1 hr at −20°. The precipitate is washed twice with ethanol and stored dry at −20°.

The extent of the reaction is determined after hydrolysis in 0.35 M triethylamine for 1 hr at 37°. The hydrolyzate is spotted on a 4 × 20 cm silica–thin-layer plate (Polygram SIL G, Macherey & Nagel, Düren), and the chromatogram is developed with benzene–pyridine–glacial acetic acid (70/30/3, v/v). Strips of 1 cm are cut, and the radioactivity is determined with a toluene-based scintillator. R_f of 3-(4-benzoylphenyl)-

[7] L. F. Somerville and C. F. H. Allen, *in* "Organic Synthesis" (A. H. Blatt, ed.), Coll. Vol. II, p. 81. Wiley, New York, 1943.
[8] P. Bartmann, T. Hanke, R. Hammer-Raber, and E. Holler, *Biochemistry* **13**, 4171 (1974).

propionylphenylalanine, 0.4; phenylalanine does not migrate in this system.

Alternatively, an aliquot is dissolved in water (1 ml) and divided into two equal parts. One part is immediately treated with 5% trichloroacetic acid, and the precipitate is filtered onto a Millipore filter (HAWP, 0.45 μm, 2.5 cm). The filter is washed three times with 5% trichloroacetic acid and twice with 70% (v/v) ethanol, dried, and counted in a toluene-based scintillator. To the remaining portion, 0.4 M sodium acetate at pH 5.5 (0.5 ml), containing 0.02 M CuSO$_4$, is added and the mixture is incubated for 20 min at 37°.[9] The tRNA is then precipitated with trichloroacetic acid and filtered onto a Millipore filter; the radioactivity is determined as above. Cu^{2+} ions catalyze the hydrolysis of the ester bond between the amino acid and the terminal adenosine in aminoacyl-tRNA, whereas an acylated aminoacyl-tRNA is completely resistant to this treatment.[10] The degree of acylation can therefore be calculated from the percentage of radioactivity precipitable after treatment with CuSO$_4$. About 90–95% of the phenylalanine residues were found to be acylated in several preparations.

Preparation of 3-Benzoylpropionyl-[^{3}H]Phe-tRNA^{Phe}. [^{3}H]Phe-tRNA^{Phe} (5 nmoles) is dissolved in 0.05 M potassium phosphate at pH 6.8 (0.1 ml) at 0°. 3-Benzoylpropionic acid N-hydroxysuccinimide ester, 1 mg, dissolved in acetonitrile (0.05 ml) is then added, and the mixture is incubated for 5 hr at 37°. The tRNA derivative is precipitated with ethanol and stored dry at −20°. The degree of acylation is determined by thin-layer chromatography (R_f of 3-benzoylpropionylphenylalanine is 0.5) or by treatment with CuSO$_4$ as described. The degree of acylation was usually about 95%.

Photoaffinity Labeling of *Escherichia coli* Ribosomes

Apparatus for Irradiation

Light source: Super-high-pressure mercury lamp, Philips Type SP 500 watt, Cat. No. 57 300 ZB/51, with reflector and metal lampholder, cooled with tap water, operated with a Philips transformer, Cat. No. 59 300 BE/00.

Filter: WG 320, thickness 2 mm, diameter 50 mm (Schott & Gen., Mainz). The filter cuts off light below 300 nm.

Focusing lens: Infrasil I, biconvex, diameter 50 mm, f 100 mm (Heraeus, Hanau).

---

[9] E. Kuechler and A. Rich, *Nature (London)* **225**, 920 (1970).
[10] P. Schofield and P. C. Zamecnik, *Biochim. Biophys. Acta* **155**, 410 (1968).

The distance between the lamp and the focusing lens was 15 cm. The filter was placed close to the focusing lens. Irradiation of the sample was carried out in a quartz tube (10 mm diameter) at a distance of 5 cm from the lens. Caution: During irradiation high-quality safety goggles ("welder's goggles") must be worn.

Photoaffinity Reaction. Ribosomes, 70 S ("vacant couples," which do not dissociate at 6 mM magnesium ion concentration) are obtained from *E. coli,* strain MRE 600, according to the procedure of Noll *et al.*[11,12] Incubation and irradiation are carried out in volumes of 0.5–1 ml. In a total volume of 1 ml are contained the following: *E. coli* ribosomes (1.0 mg, 0.38 nmole) ; 3-(4-benzoylphenyl)propionyl-[^{3}H]Phe-tRNA^{Phe} (0.01 mg, 0.4 nmole) or 3-benzoylpropionyl-[^{3}H]Phe-tRNA^{Phe} (0.01 mg, 0.4 nmole) ; 0.2 mg of polyuridylic acid (Boehringer, Mannheim) dissolved in 50 mM NH$_4$Cl; 50 mM HEPES (N-2-hydroxyethylpiperazine-N'-2-ethanesulfonic acid) sodium salt (pH 7.4) ; 10.5 mM magnesium acetate; 0.5 mM EDTA; and 6 mM mercaptoethanol. Samples are mixed at 0° and incubated for 15 min at 37°. When sensitivity to puromycin is tested, the drug is added to a concentration of 1 mM and the incubation is continued for an additional 15 min at 37°.

In order to remove oxygen, the samples are flushed with argon. This is done best by attaching a three-way stopcock through a rubber stopper to the quartz tube. The tube is first evacuated and then filled with argon, and the procedure is repeated twice. Finally, a small balloon filled with argon is attached to the tube. During irradiation the tube is cooled by running tap water (8°). Samples containing 3-(4-benzoylphenyl)propionyl-[^{3}H]Phe-tRNA are irradiated for 45 min; those containing 3-benzoylpropionyl-[^{3}H]Phe-tRNA are irradiated for 2 hr. Ribosomal RNA is prepared by phenol extraction and centrifugation on sucrose-SDS gradients.[13] Radioactivity is found to be incorporated almost exclusively into 23 S RNA. The specific activity of the 23 S RNA labeled with 3-(4-benzoylphenyl)propionyl-[^{3}H]Phe-tRNA varied between 7.2 $\times 10^5$ and 1.2 $\times 10^6$ cpm/mg. From this information it was calculated that 12–20% of the 23 S RNA molecules were labeled. When 3-benzoylpropionyl-[^{3}H]Phe-tRNA was used for the photoaffinity labeling, the specific activity was about 1.5 $\times 10^5$ cpm/mg, indicating that 2.5% of the 23 S RNA molecules became labeled in the reaction.

Evidence for Specificity and the Site of the Reaction. The specificity of the photoaffinity labeling reaction was checked by the control experi-

[11] M. Noll, Ph.D. Thesis, Northwestern University, Evanston, Illinois, 1972.

[12] M. Noll, B. Hapke, M. X. Schreier, and H. Noll, *J. Mol. Biol.* **75,** 281 (1973).

[13] E. Wagner, L. Katz, and S. Penman, *Biochem. Biophys. Res. Commun.* **28,** 152 (1967).

SPECIFICITY OF THE REACTION BETWEEN 3-(4-BENZOYLPHENYL)-
PROPIONYL-[³H]PHE-tRNA AND 23 S RNA

Addition to reaction mixture		Radioactivity incorporated (³H cpm/10 pmoles of 23 S RNA)	Yield (% of 23 S RNA labeled)
Poly(U)	Puromycin		
+	−	10,300	15.6
−	−	300	0.45
+	+	350	0.53

ments presented in the table. Omission of poly(U) from the incubation mixture results in a large reduction of the incorporation of [³H]phenylalanine into 23 S RNA, indicating that messenger RNA is required for the photoaffinity reaction with the 23 S RNA. Addition of puromycin to the complete mixture before irradiation reduces the incorporation dramatically. This demonstrates that the aromatic ring system attached to the aminoacyl moiety does not interfere with the binding at the correct ribosomal site. Since the puromycin reaction can occur only with aminoacyl-tRNA attached to the donor site, the puromycin control provides primary evidence that the 23 S RNA becomes labeled from the donor site.

More direct evidence for this conclusion comes from experiments in which the capability of the covalently attached Phe-tRNA derivative to carry out peptide-bond formation is tested.[14] For this purpose a complex of ribosomes, poly(U), and nonradioactive 3-(4-benzoylphenyl)propionyl-Phe-tRNA is formed and irradiated as described. [³H]Phe-tRNA is then added and the incubation is continued. Radioactivity is again incorporated into 23 S RNA by means of peptide bond formation with the covalently attached Phe-residue.[5] Together with the strong inhibition observed in the presence of puromycin (see the table), the experiment provides evidence that the photoaffinity labeling occurs from the ribosomal donor site.

Comments

The distance between the amino group of the amino acid and the photoreactive carbonyl group in 3-(4-benzoylphenyl)propionyl-Phe-tRNA is 10 Å; with 3-benzoylpropionyl-Phe-tRNA it is 5.5 Å. The fact that reagents of different length react with 23 S RNA proves that part of the 23 S RNA is a major component of the ribosomal donor site. By specific cleavage of the 23 S RNA, it can be demonstrated that the sites of reaction of both photoaffinity labels are in the "18 S" fragment, which

[14] M. Pellegrini, H. Oen, D. Eilat, and C. R. Cantor, *J. Mol. Biol.* **88**, 809 (1974).

stems from the 3′ part of the 23 S RNA.[5,15] Similar results have been obtained by Sonenberg et al.[16] using the dipeptidyl-tRNA derivative p-azido-N-tBoc-Phe-Phe-tRNA as photoaffinity label.

The higher efficiency of the 3-(4-benzoylphenyl)propionyl derivative as compared to that of the 3-benzoylpropionyl derivative is primarily due to the higher absorbance coefficient of the $n \rightarrow \pi^*$ transition in the former compound. The yield of the photoaffinity labeling can be increased further by reisolation of the labeled ribosomes on Sepharose 6B columns[17] and by repeated irradiation with the photoreactive tRNA derivative. Such a procedure is of potential interest for the determination of the RNA sequence at the site of the reaction.

[15] A. Barta, E. Kuechler, C. Branlant, J. Sri Widada, A. Krol, and J. P. Ebel, *FEBS Lett.* **56**, 170 (1975).

[16] N. Sonenberg, M. Wilchek, and A. Zamir, *Proc. Natl. Acad. Sci. U.S.A.* **72**, 4332 (1975). See also this volume [83] and [84].

[17] J. Modolell, B. Cabrer, and D. Vazquez, *Proc. Natl. Acad. Sci. U.S.A.* **70**, 3561 (1973).

[82] Photoaffinity-Probe-Modified tRNA for the Analysis of Ribosomal Binding Sites

By James Ofengand, Ira Schwartz, Gianni Chinali, Stephen S. Hixson, and Susan H. Hixson

I. Principle

The use of tRNA suitably modified with a chemically reactive group in order to probe the nature of the tRNA binding site in macromolecular complexes such as ribosomes, aminoacyl-tRNA (AA-tRNA) synthetases, and elongation factors is a logical extension of the use of chemically reactive substrate analogs to study the interaction of these small-molecular-weight ligands with proteins. In the same way that changing the location of the chemically reactive group on small molecule substrates alters the probability for covalent linking and thus provides a crude topographical analysis of the binding site, varying the site of the affinity probe on the tRNA molecule can also be expected to affect the extent of covalent binding and thus provide some insight into the contact areas between a tRNA molecule and its binding site. The main difference is one of scale. Whereas small molecule studies are concerned with ligands of the order of 10 Å in length, the tRNA molecule is 73 Å in its largest dimension. Consequently, consideration of possible changes in orientation or flexibility of the reactive functional group are of less concern here because of the more macroscopic nature of the studies.

Most previous work with tRNA modified with an affinity probe has placed the probe at the 3' end of AA-tRNA because of the reactivity of the amino group of the amino acid.[1-3] Such probes are limited to analysis of one area of the binding site, and moreover, because these tRNAs are N-acylamino-acyl-tRNAs, they cannot be used to analyze elongation factor Tu complexes or tRNA bound at the ribosomal A site except under special conditions. Complexes with synthetase have been studied with such derivatives,[4-7] but, since N-acylaminoacyl-tRNA is functionally inactive,[8] the relationship between the tagged region of the enzyme and the active site for the aminoacyl end of tRNA is unclear.

By contrast, the chemical reactivity of many of the rare bases found in tRNA allows one to place affinity probes at strategic sites along the tRNA molecule and to use such modified tRNA to probe any of the binding sites without limitation as to function if the probes themselves do not interfere. In this article, attachment of photoaffinity probes to two rare bases found in many tRNAs is described. The procedures do not inactivate the tRNA, and the bases involved, 4-thiouridine (^{4}S) and 3-(3-amino-3-carboxypropyl)uridine (nbt^{3}U), are located on opposite faces of the central region in the three-dimensional structure of tRNA.[9,10] Thus, the covalent linking results should be approximately complementary and provide useful topographical information about tRNA binding sites. Moreover, since the bases occur but once in a given tRNA molecule, and at a known site which is the same for all tRNAs containing the rare nucleoside, the location of the affinity probe in tRNA is defined simply by the chemistry of the system.

The probes described here (Figs. 1–3) are photoaffinity probes of the aromatic azide class.[11] p-Azidophenacyl bromide (APA-Br) and its analog, the p-azidophenacyl ester of bromoacetic acid (APAA-Br), were used

[1] M. Pellegrini and C. R. Cantor, in "Molecular Mechanisms of Protein Biosynthesis" (H. Weissbach and S. Pestka, eds.), pp. 203–245. Academic Press, New York, 1977.
[2] A. Zamir, this volume [73].
[2a] N. Sonenberg, M. Wilchek, and A. Zamir, this volume [84].
[3] L. Bispink and H. Matthaei, this volume [74].
[4] C. J. Bruton and B. S. Hartley, J. Mol. Biol. 52, 165 (1970).
[5] D. V. Santi and S. O. Cunnion, Biochemistry 13, 481 (1974).
[6] P. Bartmann, T. Hanke, B. Hammer-Raber, and E. Holler, Biochem. Biophys. Res. Commun. 60, 743 (1974).
[7] O. I. Lavrik and L. F. Khutoryanskaya, FEBS Lett. 39, 287 (1974).
[8] F. Chapeville, in "The Mechanism of Protein Synthesis and Its Regulation" (L. Bosch, ed.), p. 5. North-Holland Publ., Amsterdam, 1972.
[9] G. J. Quigley, A. H. J. Wang, N. C. Seeman, F. L. Suddath, A. Rich, J. L. Sussman, and S. H. Kim, Proc. Natl. Acad. Sci. U.S.A. 72, 4866 (1975).
[10] J. E. Ladner, A. Jack, J. D. Robertus, R. S. Brown, D. Rhodes, B. F. C. Clark, and A. Klug, Proc. Natl. Acad. Sci. U.S.A. 72, 4414 (1975).
[11] H. Bayley and J. R. Knowles, this volume [8].

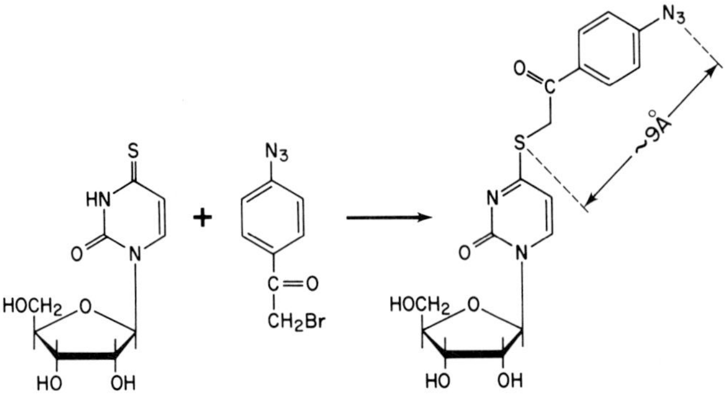

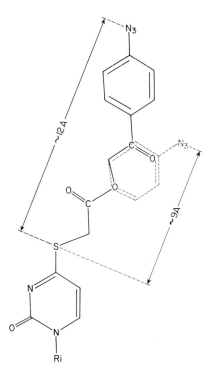

Fig. 1. Reaction of *p*-azidophenacyl bromide with 4-thiouridine. The distance from the sulfur atom to the azido group is approximately 9 Å when maximally extended.

Fig. 2. Structural comparison of the adducts of 4-thiouridine with APA-Br and APAA-Br. The APAA adduct is shown by solid lines, and the APA adduct by dashed lines.

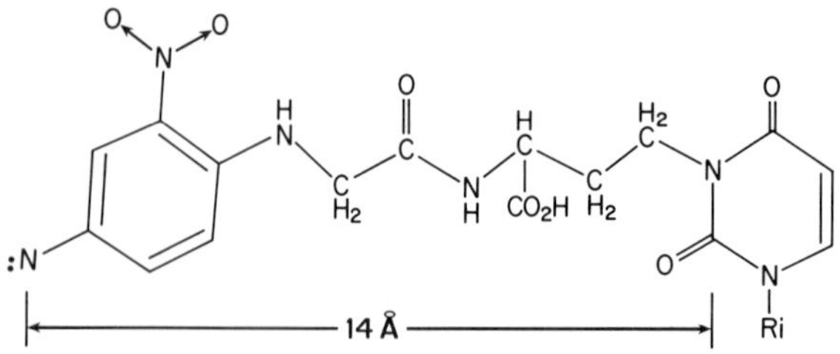

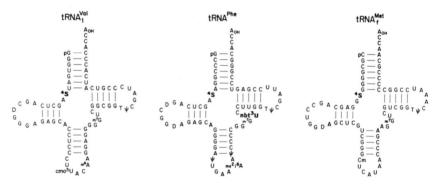

FIG. 3. Structure of the adduct of 3-(3-amino-3-carboxypropyl)uridine (nbt³U) with NAG. The distance from the uridine ring to the azido group is 14 Å in the maximally extended form.

to modify the ⁴S residue (Figs. 1 and 2) while the N-hydroxysuccinimide ester of N-(4-azido-2-nitrophenyl)glycine (NAG-NOS) was used to derivatize the nbt³U residue which has a free α-NH₂ group (Fig. 3). The location of these residues in the primary sequence of the three tRNAs studied, tRNA^Val, tRNA^Phe, and tRNA_f^Met, is shown in Fig. 4.[12]

FIG. 4. Primary sequence of tRNA^Val, tRNA^Phe, and tRNA_f^Met. The nucleotides that have been affinity labeled are shown in boldface type.

II. Synthesis of Affinity Probes

A. [carbonyl-¹⁴C]p-Azidophenacyl Bromide ([carbonyl-¹⁴C]APA-Br)

The synthesis of unlabeled APA-Br has been described,[13] and the compound is available commercially.[14] We present here the synthesis of [carbonyl-¹⁴C]APA-Br.

[12] B. G. Barrell and B. F. C. Clark, "Handbook of Nucleic Acid Sequences." Joynson-Bruvvers, Eynshaw, Oxford, England, 1974.
[13] S. H. Hixson and S. S. Hixson, *Biochemistry* 14, 4251 (1975).
[14] Pierce Chemical Company, Rockford, Illinois, Cat. No. 20106.

A suspension of 16.3 mg (0.119 mmole) of [*carboxyl*-¹⁴C]*p*-amino-benzoic acid (ICN Pharmaceuticals, Inc., 8.4 mCi/mmole) and 13.7 mg (0.100 mmole) of unlabeled *p*-aminobenzoic acid in 5 ml of 11% sulfuric acid is cooled in an ice bath. A solution of 47.7 mg (0.692 mmole) of sodium nitrite in 0.3 ml of water, cooled to 0°, is added dropwise over 1 min to the stirred suspension. After the mixture has been stirred for 0.5 hr in the ice bath, a 5-ml portion of ether is added. A solution of 73.0 mg (1.12 mmoles) of sodium azide in 0.3 ml of water is added dropwise quickly to the reaction mixture. The reaction is stirred for 15 min at 0°, and then for 0.5 hr at room temperature. The ether layer is decanted. The remaining water layer is stirred with 5 ml of ether, and the ether is decanted; the procedure is repeated with 3 ml of ether and decanted. The combined ether layers are stirred with 3–4 ml of saturated aqueous NaCl, decanted, dried ($MgSO_4$), and filtered into a 5-ml round-bottom flask. Removal of the ether with a stream of nitrogen yields 32 mg (0.197 mmole, 90%) of [*carboxyl*-¹⁴C]*p*-azidobenzoic acid.

A 1.0-ml portion of thionyl chloride is added to the [¹⁴C]*p*-azido-benzoic acid in the 5-ml round-bottom flask containing a stirring bar. The solution is refluxed 0.5 hr and then cooled to room temperature. A micro distillation head is substituted for the condenser and the thionyl chloride is removed by distillation under reduced pressure (about 20 mm). A 0.5-ml portion of dry benzene is added and distilled; this process is repeated twice to ensure thionyl chloride removal. The solid *p*-azidobenzoyl chloride is kept under reduced pressure until used (~2 hr).

Dimethylcadmium is then prepared. A solution of 3 ml of ether and 0.75 ml (2.25 mmoles) of methylmagnesium bromide solution (Ventron, 3 *M* in ether) in a 3-necked round-bottom flask equipped with a condenser and a mechanical stirrer is maintained under nitrogen and cooled in an ice bath. Then 207 mg (1.13 mmoles) of cadmium chloride (dried at 110° and stored in a desiccator over Drierite) are added to the methyl-magnesium bromide solution in one portion. The mixture is refluxed with stirring for 20 min in an oil bath and then cooled to room tempera-ture. A Gilman[15] test should be negative. A distilling head is added to the flask, and most of the ether is distilled while stirring is continued. A 7-ml portion of dry benzene is then added to the reaction mixture, and distillation is continued until the temperature of the distilling vapors is about 70°. The distillation head is removed and the reaction mixture containing dimethylcadmium (under N_2) is cooled in an ice bath.

The acid chloride prepared above is dissolved in 0.3 ml of dry benzene

---

[15] L. F. Fieser, "Experiments in Organic Chemistry," 3rd ed. revised, p. 269. Heath, Boston, Massachusetts, 1957.

and transferred quickly via syringe to the stirred dimethylcadmium solution. The acid chloride flask is rinsed twice with 0.3-ml portions of benzene which are added to the flask containing dimethylcadmium. After 0.5 hr of stirring at room temperature, several pieces of ice are added to the reaction mixture followed by the sequential addition of 2 ml of 2 N HCl, 3 ml of ether, and additional water. The ether–benzene layer is decanted. The water layer is stirred twice with 3-ml portions of ether, which are decanted, and the combined organic layers are stirred with about 5 ml of saturated aqueous NaHCO$_3$, dried (MgSO$_4$), filtered, and blown dry under nitrogen. The crude product is chromatographed on a 1 × 30 cm column of deactivated (10% H$_2$O) silica gel with 1:10 ether: petroleum ether (b.p. 61°–70°) as eluent to give 18.5 mg of [^{14}C]p-azidoacetophenone. The 18.5 mg (0.115 mmole) [carbonyl-^{14}C]p-azidoacetophenone is dissolved in 0.5 ml of anhydrous ether plus 0.25 ml of dioxane (sodium-dried), placed in a 5-ml, 3-necked round-bottom flask equipped with a stirring bar and serum cap, and placed in a water bath at about 10°. Bromine, 5.9 μl (0.115 mmole), is added to the solution by syringe. After stirring for 15 min, the reaction mixture is diluted with 1 ml of water and 1 ml of ether and the ether layer is separated. The water layer is stirred twice with additional 1-ml portions of ether. The combined ether layers are washed sequentially with 1-ml portions of water, saturated aqueous NaHCO$_3$, and saturated aqueous NaCl, then dried (MgSO$_4$), filtered, and blown dry under a nitrogen stream.

The crude product is chromatographed on a 1 × 30 cm column of deactivated (10% H$_2$O) silica gel with 1:5 ether:petroleum ether (b.p. 61°–70°) as eluent. Removal of the solvent under reduced pressure yields 18.5 mg (0.077 mmole) of [carbonyl-^{14}C]APA-Br (35.2% overall yield based on p-aminobenzoic acid) with a specific activity of 4.57 mCi/mmole. The product is stored as a solid in a vial over Drierite at −20°.

B. p-Azidophenacyl Bromoacetate (APAA-Br) and p-Azidophenacyl Iodoacetate (APAA-I)

A solution of 4.76 g (19.9 mmoles) of APA-Br in 15 ml of acetone is added to a solution of 8.26 g (59.4 mmoles) of bromoacetic acid and 6.9 ml (49.5 mmoles) of triethylamine (distilled from calcium oxide) in 60 ml of acetone. The reaction mixture is stirred 50 min at room temperature in the dark, poured into 50 ml of water, and extracted with ether (twice with 50 ml each). The ether extract is washed sequentially with 5% NaHCO$_3$, 0.1 N HCl, and saturated aqueous NaCl, dried (MgSO$_4$), filtered, and concentrated under reduced pressure. The resulting product is crystallized from 95% ethanol to give a first crop of 4.09 g (13.7

mmoles, 69%) of APAA-Br: m.p. 76°–77.5° (after recrystallization);
NMR (CDCl$_3$) δ4.02 (s,2,-CH$_2$Br), 5.43 (s,2,-CH$_2$O-), and 7.07–8.00
(m,4, arom) ppm; UV (methanol) λ$_{max}$ 287 nm (ε = 2.16 × 10^4), ε$_{280}$/
ε$_{300}$ = 1.29, ε$_{300}$/ε$_{270}$ = 1.12.

To a solution of 500 mg (1.68 mmoles) of APAA-Br in 3.0 ml of acetone is added a solution of 277 mg (1.85 mmoles) of sodium iodide in 3.0
ml of acetone. A precipitate forms immediately; after 30 min, the precipitate is filtered and rinsed with an additional 5 ml of acetone. The
combined filtrates are evaporated to dryness under reduced pressure to
give a residue that is dissolved in 10 ml of ether. The ether solution is
washed sequentially with water, twice with 5% aqueous Na$_2$S$_2$O$_3$, water,
and saturated aqueous NaCl, dried (MgSO$_4$), filtered, and then combined
with an equal volume of petroleum ether (b.p. 61°–70°) and stored at
0°. Crystalline APAA-I, 311 mg (0.903 mmole, 54%), is isolated by filtration: m.p. 54.5°–56° (after crystallization from ether–petroleum ether);
NMR (CDCl$_3$) δ4.00 (s,2,-CH$_2$I), 5.47 (s,2,-CH$_2$O-) and 7.15–8.10 (m,4,
arom); UV (methanol) λ$_{max}$ 287 nm (ε = 2.35 × 10^4), ε$_{280}$/ε$_{300}$ = 1.28,
ε$_{300}$/ε$_{270}$ = 1.11.

C. [carbonyl-^{14}C]p-Azidophenacyl Bromoacetate ([carbonyl-^{14}C]APAA-Br) and [carbonyl-^{14}C]p-Azidophenacyl Iodoacetate ([carbonyl-^{14}C]APAA-I)

A solution of 10.4 mg (0.043 mmole) of [carbonyl-^{14}C]APA-Br (4.57
mCi/mmole) in 0.1 ml of acetone is added with stirring to a solution of
38.7 mg (0.279 mmole) of bromoacetic acid and 30 μl (0.216 mmole) of
triethylamine in 0.12 ml of acetone. The vial containing the labeled
ketone is rinsed with acetone (3 × 0.1 ml) and the rinses are added to
the reaction mixture which is stirred at room temperature in the dark.
No APA-Br remains after 40 min according to TLC analysis [Merck
Silica Gel G-coated microscope slide; ether–petroleum ether (b.p. 61°–
70°); R$_f$: APA-Br, 0.67; APAA-Br, 0.33]. After an additional 10 min,
0.5 ml of ether and 1.0 ml of water are added to the reaction. The ether
layer is decanted and the aqueous layer is washed twice more with 0.5
ml portions of ether. The combined ether layers are washed sequentially
with 0.5-ml portions of 5% NaNCO$_3$ (2×), 0.1 N HCl, water, and saturated aqueous NaCl, dried (MgSO$_4$), filtered. The ether is evaporated
under a stream of nitrogen, and the residue is dissolved in a few drops
of ether and applied as a band 10-cm wide to a 20-cm silica gel preparative TLC plate (Merck Silica Gel GF-254 for TLC according to Stahl;
the plates are previously dried for 0.5 hr at 100° and then allowed
to equilibrate with the atmosphere at room temperature).

The plate is developed with 1:1 ether (anhydrous)–petroleum ether (b.p. 35°–60°). R_f: APA-Br, 0.63. APAA-Br has a lower R_f value than APA-Br, and the two are readily separated. APAA-Br is eluted from the silica gel with ether, and the ether solution is concentrated to a few milliliters under reduced pressure, dried (MgSO$_4$), and filtered. The residual ether is evaporated under a nitrogen stream to give 6.9 mg (0.023 mmole, 53%) of [*carbonyl*-^{14}C]APAA-Br. The product is stored as a solution in anhydrous ether at $-20°$.

A solution of 5.6 mg (0.037 mmole) of sodium iodide in 0.1 ml of acetone is added to a solution of 4.7 mg (0.016 mmole) of [*carbonyl*-^{14}C]-APAA-Br in 0.1 ml of acetone. A precipitate forms immediately, and the reaction mixture turns yellow. The mixture is swirled occasionally over a 10-min period and filtered. The precipitate is rinsed several times with 0.1-ml portions of acetone, and the combined acetone solutions are evaporated under a nitrogen stream. The residue is treated with 1.0 ml each of ether and water. The water layer is decanted, and the ether layer is washed successively with 5% aqueous Na$_2$S$_2$O$_3$ ($2\times$), water, and saturated aqueous NaCl; it is dried (MgSO$_4$), filtered, and evaporated under a nitrogen stream. The residue is dissolved in a few drops of ether and purified by preparative TLC, exactly as was done for [^{14}C]APAA-Br (see above). The R_f value for APAA-I is greater than that for APAA-Br. A yield of 1.8 mg (0.0052 mmole, 32%) of [*carbonyl*-^{14}C]APAA-I is obtained. The product is stored as a solution in anydrous ether at $-20°$.

D. *p-Azidophenacyl [2-^{14}C]bromoacetate ([acetate-^{14}C]APAA-Br) and p-Azidophenacyl [2-^{14}C]iodoaceate ([acetate-^{14}C]APAA-I)*

[2-^{14}C]Bromoacetic acid, 1.2 mg (0.0086 mmole, specific activity 58.2 mCi/mmole), is obtained commercially (ICN Pharmaceuticals, Inc.) as an ether solution. After ether is removed with a stream of air, a solution of 4.2 mg (0.0175 mmole) of APA-Br plus 1.2 µl (0.87 mg, 0.0086 mmole) of triethylamine (distilled from CaO) in 0.5 ml of acetone are added directly to the vial. The reaction mixture is stirred for 2 hr at room temperature and then applied directly to a 20 × 20 cm silica gel preparative TLC plate (see the procedure for [*carbonyl*-^{14}C]APAA-Br) which is developed with 1:2 ether–petroleum ether (b.p. 61°–70°). The [*acetate*-^{14}C]APAA-Br is recovered from the plate as described for [*carbonyl*-^{14}C]APAA-Br. The product is stored as a solution in anhydrous ether at $-20°$.

The [*acetate*-^{14}C]APAA-Br is dissolved in 0.3 ml of acetone, and 1.5

mg (0.01 mmole) of sodium iodide is added. After stirring for 15 min, 0.5 ml of 5% aqueous $Na_2S_2O_3$ is added. The mixture is extracted with ether $(3 \times 1$ ml$)$, and the ether extracts applied directly to a silica gel preparative TLC plate as before. The [acetate-^{14}C]APAA-I, 1.97 mg (0.00572 mmole, 67% overall yield based on starting [^{14}C]bromoacetic acid), is recovered from the plate as described for [carbonyl-^{14}C]APAA-Br. The product is stored as a solution in anhydrous ether at $-20°$.

E. N-Hydroxysuccinimide Ester of N-(4-Azido-2-nitrophenyl)glycine (NAG-NOS)

The following is a modification of the procedure originally described by Fleet et al.[16] for the preparation of N-(4-azido-2-nitrophenyl)glycine. The synthesis of NAG-NOS has also been described by Hsiung and Cantor.[17]

Glycine (5.0 mg, 0.066 mmole), 36.1 mg (0.198 mmole) of 4-fluoro-3-nitrophenylazide, and 19.8 mg (0.198 mmole) of Na_2CO_3 are dissolved in 4 ml of 4:1 dimethyl sulfoxide–water and stirred for 24 hr at 35°. The reaction mixture is cooled, diluted with 4 ml of water, checked to ensure that the pH is between 8 and 9, and then extracted with ether $(2 \times 100$ ml$)$. The aqueous layer is acidified (pH 3–5) with dilute HCl and again extracted with ether $(2 \times 50$ ml$)$. These latter ether extracts are combined, dried ($MgSO_4$), filtered, and concentrated under reduced pressure to yield 15.0 mg (95%) of N-(4-azido-2-nitrophenyl)glycine, m.p. 170°– 172° (lit.[16] 173°–174°).

To a solution of 15.0 mg (0.063 mmole) of N-(4-azido-2-nitrophenyl)-glycine in freshly distilled (CaH_2) dimethoxyethane are added 9.6 mg (0.063 mmole) of N-hydroxysuccinimide. This solution is stirred and cooled to 4°, and 16.6 mg (0.063 mmole) of dicyclohexycarbodiimide are added. The reaction is stirred at 4°. Progress of the reaction is monitored by silica gel TLC using 1:3 ether–petroleum ether (b.p. 61°–70°) as eluent [R_f: 4-fluoro-3-nitrophenyl azide, 0.47; N-(4-azido-2-nitrophenyl)glycine, 0.23]. After 100 min, the mixture is warmed to room temperature, filtered to remove the dicyclohexylurea, and concentrated under reduced pressure. The resulting product is dried in a vacuum desicator over $CaCl_2$ to give 20.6 mg (0.061 mmole, 96%) of the N-hydroxysuccinimide ester of N-(4-azido-2-nitrophenyl)glycine (NAG-NOS): m.p. 164°–166°; ir (KBr) 3400, 2140, 1820, 1780, and 1740 cm^{-1}.

---

[16] G. W. J. Fleet, J. R. Knowles, and R. R. Porter, *Biochem. J.* **128**, 499 (1972).

[17] N. Hsiung and C. R. Cantor, *Nucleic Acids Res.* **1**, 1753 (1974).

F. N-Hydroxysuccinimide Ester of N-(4-Azido-2-nitrophenyl) [2-³H]-glycine ([³H]NAG-NOS)

[2-³H]Glycine hydrochloride, 0.59 mg (0.0053 mmole, specific activity 9.39 Ci/mmole), in 70% ethanol is combined with 9.4 mg (0.125 mmole) unlabeled glycine. The solvent is removed under reduced pressure, and the resulting solid is dissolved in 10 ml of 4:1 dimethyl sulfoxide–water. To this solution is added 72 mg (0.395 mmole) of 4-fluoro-3-nitrophenyl azide and sufficient solid Na_2CO_3 to ensure that the reaction mixture is slightly basic (pH 8–9). The mixture is stirred 60 hr at 35°, cooled, diluted with 10 ml of water, and extracted twice with 100 ml of ether. The aqueous layer is acidified (pH 3–5) with dilute HCl and then extracted twice with 50 ml of ether. The last ether extracts are dried $(MgSO_4)$ and filtered, and the ether is evaporated under nitrogen. The solid residue is crystallized from 5 ml of 100% ethanol to give, in two crops, 28.4 mg (0.119 mmole, 92%) of N-(4-azido-2-nitrophenyl)[2-³H]-glycine.

The N-(4-azido-2-nitrophenyl) [2-³H]glycine, 22 mg (0.092 mmole), is immediately dissolved in 3 ml of freshly distilled (CaH_2) dimethoxyethane, stirred, cooled to 4°, and 10.8 mg (0.093 mmole) of N-hydroxysuccinimide plus 21 mg (0.101 mmole) of dicyclohexylcarbodiimide are added. The mixture is stirred at 4°. Progress of the reaction is monitored by silica gel TLC as above (Section II,E). After 90 min the reaction mixture is warmed to room temperature and filtered to remove the dicyclohexylurea. The addition of 6–7 volumes of cold water to the filtrate precipitates the desired product, which is extracted twice with 25 ml of chloroform. The resulting chloroform solution is dried, filtered, and stored in a sealed vial at −20°. The total yield of the N-hydroxysuccinimide ester of N-(4-azido-2-nitrophenyl) [2-³H]glycine is 20.7 mg (0.062 mmole, 34 mCi, 68% yield, specific activity 381 mCi/mmole). This radioactive ester should be stored in solution in the cold. Maintaining the product as a solid leads to noticeable decomposition.

III. Coupling of Affinity Probes to tRNA

A. p-Azidophenacyl Bromide (APA-Br)

1. Preparation and Characterization of the Adduct with 4-Thiouridine. S-(p-azidophenacyl)-4-thiouridine (APA-⁴S) and its nonphotolyzable analog, S-(phenacyl)-4-thiouridine (PA-⁴S), are prepared by incubation of a 60% methanol solution containing 40 mM potassium phosphate (pH 7.4), 15 mM 4-thiouridine (Sigma), and 20 mM APA-Br or PA-Br at 30° for 20 min. The reaction mixture is streaked out on Merck 2 mm-thick silica gel TLC plates (Cat. No. 5766, E. Merck, Darmstadt), 1 ml per plate, and chromatographed in CHCl$_3$–MeOH (85:15). R_f: APA-^{4}S, 0.49; PA-^{4}S, 0.47; ^{4}S, 0.25; APA-Br, 0.91; PA-Br, 0.89. Products are eluted by chromatography twice in MeOH at right angles to the bands to concentrate them to a small area. The gel is placed in a tube and serially eluted with 3 ml of dioxane until the absorbance at 300 nm is sufficiently low. The dioxane eluates are concentrated in a stream of air at room temperature. All operations with APA-Br are carried out under yellow safelights (Wratten 00 filters, Eastman Kodak).

Evidence for the structure shown in Fig. 1 was provided by the presence of ir bands in the isolated product for azide (2120 cm^{-1}), acetophenone ketone (1690 cm^{-1}), and ribose hydroxyl groups (3300–3400 cm^{-1})[18]; by the absence of a thioketone functional group as defined by the iodine–azide spot test of Feigl, which is diagnostic for thiols and thioketones[19]; and by the characteristic shift in uv spectrum of ^{4}S upon reaction.[19,20] In fact, with PA-Br, it was possible to follow the kinetics of the reaction spectroscopically. In 40 mM potassium phosphate, pH 7.4, at 23°, the pseudo-first-order rate constant was 133 min^{-1} M^{-1}.[21]

The products in both cases lack the 330 nm absorption band characteristic of the nonionized form of unsubstituted or of N-substituted 4-thiouridine. PA-^{4}S possesses two bands, one at 250 nm due to the phenacyl group and a second at 301 nm due to the S-substituted thiouridine. The 301:250 ratio is 0.86 and the spectrum is constant over the pH range of 2 to 9.8. Similarly, the APA-^{4}S spectrum is a single 300-nm band consisting of an overlap of the azidophenacyl band and the S-substituted thiouridine band, and this spectrum is also invariant with pH from 2 to 10.2, although a slow decomposition was noted at pH 10.2. Slow decomposition of PA-^{4}S probably also occurs at pH 10, although this has not been directly examined.

2. Preparation of the Adduct with Pseudouridine. The *p*-azidophenacyl derivative of pseudouridine (Ψ) is prepared by incubation of a 65% MeOH solution containing 0.09 M NaHCO$_3$ (pH 9), 14 mM Ψ_c (the natural β-furanose isomer), and 32 mM APA-Br at 37° for 16 hr in the dark. By this time, most of the Ψ has reacted. The reaction mixture is streaked out on Merck 2 mm-thick silica gel TLC plates, 1 ml per plate,

[18] I. Schwartz and J. Ofengand, *Proc. Natl. Acad. Sci. U.S.A.* **71**, 3951 (1974).

[19] J. Ofengand, *J. Biol. Chem.* **242**, 5034 (1967).

[20] S. A. Kumar, M. Krauskopf, and J. Ofengand, *J. Biochem. (Tokyo)* **74**, 341 (1973).

[21] I. Schwartz, E. Gordon, and J. Ofengand, *Biochemistry* **14**, 2907 (1975).

R_f Values of S-(p-Azidophenacyl)-4-thiouridine and p-Azidophenacyl-
pseudouridine in Various Solvent Systems[a]

	R_f value	
Solvent	APA-4S	APA-Ψ
n-BuOH:HOAc:H$_2$O (4:1:5) upper phase	0.69	0.61
BuOH:H$_2$O (86:14)	0.66	0.57
CHCl$_2$:MeOH (90:10)	0.79	0.51
CHCl$_3$:MeOH (93:7)	0.64	0.36
CHCl$_3$:MeOH (95:5)	0.47	0.21
CHCl$_3$:MeOH:HOAc (85:10:5)	0.80	0.60

[a] Chromatography was performed on silica gel TLC sheets (Cat. No. 6060, Eastman Kodak) containing fluorescent indicator, and visualization was done under
UV lights. In all cases only a single spot was observed.

and chromatographed in the upper phase of n-BuOH:HOAc:H$_2$O (4:
1:5). R_f: APA-Ψ, 0.72; Ψ, 0.43; APA-Br, 0.93. The product is eluted as
described above except that MeOH is used in place of dioxane. R_f values
for this compound and APA-4S in several solvents are listed in the table.

The rate of reaction under these conditions is slow, 1 or 2.5 hr incubation yielding considerably less product than 18.5 hr; pH 9 is better than
pH 10, possibly owing to a competing alkaline hydrolysis reaction. The
structure of the product has not been clarified although, by analogy with
other studies,[22] the azidophenacyl group is expected to be attached to
the N$_1$ of Ψ.

In an experiment designed to compare the relative reactivities of 4S
and Ψ toward APA-Br under the same conditions, 5 mM Ψ or 6.2 mM
4S were incubated in 60% MeOH, 20 mM potassium phosphate (pH 7.4)
with 20 mM APA-Br for 5 min at room temperature. After addition of a
6-fold excess of mercaptoethanol to react with excess APA-Br, the
samples were chromatographed. By visual inspection, all the 4S had been
converted to APA-4S, but less than 2% of Ψ had reacted.

3. *Reaction with tRNA$_1^{Val}$.* The above conditions were modified to
keep both APA-Br and tRNA in solution by replacing the 60% MeOH
with 90% DMSO, 10 mM KPO$_4$ (pH 7.4). Under these conditions, at
least 20 μM tRNA was soluble and the reaction with 500 μM APA-Br
or PA-Br was complete in less than 10 min at room temperature. Mercaptoethanol was added to consume the excess acyl bromide and the
tRNA was dialyzed against water to remove most of the DMSO before

[22] C.-H. Yang and D. Söll, *Biochemistry* **13**, 3615 (1974).

precipitation by the addition of 0.1 volume of 20% potassium acetate (pH 5) and 2 volumes of ethanol. If desired, the dialysis step may be omitted. After 30 min at $-20°$, the precipitate was redissolved in 2% potassium acetate (pH 5), reprecipitated, and finally dissolved in water.

When using unlabeled acyl bromide, the extent of reaction with ^4S can be followed by the loss of photochemical cross-linking to C_{13},[23] which is measured by the fluorescence of the reduced ^4S_s-C_{13} binucleotide[24]; 98–100% reaction is usually obtained. When [^{14}C]APA-Br is used, the incorporation can be directly measured and is approximately stoichiometric with the acceptor activity.[21]

A useful procedure for this assay is as follows: Up to 40 μl of a reaction mixture is absorbed on a 2.4 cm circle of Whatman No. 3 MM paper and washed sequentially in 90% ethanol–10% TCA, 67% ethanol–33% CHCl$_3$, twice with 50% ethanol–50% ether and ether. The dried disk is then counted under 5 ml of a toluene-based scintillation fluid. This procedure avoids the high background obtained when solutions containing [^{14}C]APA-Br are treated with TCA and filtered through Millipore membranes.

Two types of controls show that only ^4S reacts under these conditions. First, cross-linking of ^4S_s-C_{13} before derivatization blocked the incorporation of [^{14}C]APA-Br (1.3 nm bound/A_{260} before cross-linking, <0.05 nm bound/A_{260} after cross-linking).[25] Second, nucleoside analysis of tRNA^{Val} derivatized with [^{14}C]APA-Br (2.2 mM APA-Br, 23 μM tRNA, 10 min) showed only one radioactive spot corresponding to authentic APA-^4S synthesized as described above (Section III,A,1). In particular, there was no radioactivity associated with the marker of APA-Ψ.[21]

4. Reaction with tRNA^{Phe} and tRNA$_f^{Met}$. The same conditions described above were used for these two tRNAs also. The specificity of the reaction for 4-thiouridine was monitored as follows. In each case, reaction with APA-Br blocked the ability to form the ^4S_s-S_{13} photoinduced cross-link showing that ^4S had indeed reacted. Conversely, formation of the cross-link before reaction with [^{14}C]APA-Br blocked its uptake, which in the absence of the cross-link was approximately stoichiometric with the acceptor activity.[25] This shows that under these conditions reaction with Ψ, nbt^{3}U, or other bases did not occur.

5. Comments. A marked anomalous additional incorporation has been observed at lower DMSO concentrations (60–70%), especially at pH 8.5 and above. The site(s) of this reaction are unknown but, based on

[23] F. A. Favre, M. Yaniv, and A. M. Michelson, *Biochem. Biophys. Res. Commun.* **37,** 266 (1969).
[24] J. Ofengand, P. Delaney, and J. Bierbaum, this series Vol. 29, p. 673 (1974).
[25] J. Ofengand and G. Chinali, unpublished results.

model studies, are probably accessible guanine residues.[25] This side reaction can be prevented by carrying out the reaction at 90–95% DMSO (pH 7.4 to 8.5). For example, no anomalous incorporation could be detected with either tRNA^{Val} or tRNA^{Phe} when tested with both native and cross-linked species after 2 hr incubation at 23° in 90% DMSO with 0.3 mM APA-Br at pH 8.5.

Because of the speed of the reaction at neutral pH, it is possible to derivatize already acylated tRNA without reaction of the α-amino group or loss of amino acid by hydrolysis. In particular, N-acylaminoacyl-tRNA, useful for P- or I-site binding studies, can be readily derivatized without loss of the amino acid label.

B. p-Azidophenacyl Ester of Bromoacetic Acid (APAA-Br)

Reaction with tRNA$_f^{Met}$. The same general conditions were used for the reaction of APAA-Br with tRNA$_{f1}^{Met}$ and tRNA$_{f3}^{Met}$. By the ^4S_s-C_{13} cross-linking assay, 95% reaction had occurred. In view of the almost identical nature of APAA-Br and APA-Br, no differences in reactivity are expected, and it is likely that the same reactions described for APA-Br will occur with APAA-Br for all three tRNAs.

C. N-(4-Azido-2-nitrophenyl)glycine (NAG)

This derivative, activated at the carboxyl group by formation of the N-hydroxysuccinimide ester (NOS), reacts with free α-NH$_2$ groups, such as are at the aminoacyl end of tRNA or in the nbt^{3}U residue of tRNA^{Phe}. Reaction with the nbt^{3}U residue of tRNA^{Phe} was carried out at room temperature for 2 hr at a final concentration of 72% DMSO, 20 mM triethanolamine (pH 8.0), 50 μM tRNA^{Phe}, and 8 mM [^{3}H]NAG-NOS (160-fold excess added serially). Forty percent of the NAG-NOS was added at 0 and 30 min, and the remaining 20% at 60 min. The product was isolated by an initial precipitation with 5% TCA at zero degrees for 1 hr and subsequently freed of excess reagent by cyclic extraction into 2% potassium acetate (pH 5) and reprecipitation at 67% EtOH to a constant specific activity (cpm/A$_{260}$). Usually, three cycles of extraction sufficed. The extent of derivatization was consistent with the acceptor activity for the tRNA. All operations with NAG derivatives were performed under safelights (Wratten ML2 or 1A filters, Eastman Kodak). Less NAG-NOS gives incomplete derivatization, 2.5 and 5 mM NAG-NOS corresponding to 60 and 80% reaction, respectively.

In the absence of radioactive NAG-NOS, the extent of derivatization can be measured by the absorbance of the product at 470 and 260 nm. The λ$_{max}$ of NAG-NOS in 20 mM KPO$_4$ (pH 7.4) is 460 nm and the extinction coefficient is 4300. In ethanol, ϵ is 4970 in good agreement with the value of 4800 reported by Fleet et al.[16] for NAG. In tRNA, the λ_{max} is shifted to 470 nm. Assuming no change in ϵ, the picomoles of NAG bound/A_{260} of tRNA can be calculated from the 470:260 ratio of absorbance.

The chemistry of the reaction makes it very unlikely that addition takes place at sites other than the nbt³U residue, but this could be readily checked by use of tRNA^{Val} or tRNA$_f^{Met}$ in place of tRNA^{Phe} since the latter do not contain nbt³U. Similarly, ⁴S$_s$-C$_{13}$ cross-linked tRNA^{Phe}, in which the ⁴S is blocked to further reaction, may be used.

IV. Functional Activity of Affinity Probe-Modified tRNAs

Generally speaking all functional activities of tRNA^{Val}, tRNA^{Phe}, and tRNA$_f^{Met}$ are preserved when modified with any of the probes described here. Although in some cases the *rates* of reaction may be slower with the modified tRNAs, the *extent* of reaction in assays in which the limiting component is the tRNA is not affected. Full aminoacylation takes place with PA-, APA-, and APAA-tRNA^{Val}, PA- and APA-tRNA^{Phe}, APA- and APAA-tRNA$_f^{Met}$, and NAG-tRNA^{Phe}. Other combinations have not been tested. The rate of acylation is slower for all the modified tRNA^{Val} species, appears not to be appreciably altered in the tRNA$_f^{Met}$ cases, and APA-tRNA^{Phe} is acylated at only about one-half the rate of PA-tRNA^{Phe}. Ternary complex formation with Val-tRNA is not affected by the PA modification.[18] Although other modified species have not been tested directly, those that have been examined (PA-, APA-, APAA-tRNA^{Val}; PA-, APA-, NAG-tRNA^{Phe}) are fully able to bind to the ribosomal A site in a T-factor-dependent reaction. Ribosomal P-site binding was tested with PA-, APA-, and APAA-tRNA^{Val}, APA- and NAG-tRNA^{Phe}, and APA-tRNA$_f^{Met}$ and was normal in extent in all cases. Peptidyltransferase activity was normal both in rate and extent as judged by polypeptide synthesis using PA-tRNA^{Val}.[18] The extent of the reaction with puromycin was normal when tested with APA-tRNA^{Phe} and APA-tRNA$_f^{Met}$. Also, formylation and initiation factor-dependent binding to 70 S ribosomes were normal for both APA- and APAA-tRNA$_f^{Met}$.

V. Affinity Labeling of Ribosome Binding Sites

A. Binding of tRNA to A, P, and I (Initiation) Sites

It is beyond the scope of this chapter to describe in detail the conditions used to define the binding of tRNA to the various ribosomal sites, since they must be optimized for each tRNA species and mRNA used, particularly when synthetic polynucleotides are employed. The conditions used for binding tRNA^{Val} to both P and A sites have been reported.[18,21] Operationally, we have adopted the following general criteria. I-site binding is binding of fMet-tRNA to 70 S ribosomes dependent on the presence of initiation factors, and releasable by reaction with puromycin. There is also a dependency on mRNA, but in our hands it is never as great as the initiation factor dependence. A site binding is defined as EF-Tu factor-dependent binding performed in the presence of a 5- to 10-fold excess of unacylated tRNA in order to largely block the P site. Tetracycline should inhibit, and the AA-tRNA so bound should not react with puromycin. P-site binding is defined as binding in the absence of elongation or initiation factors that is not affected by the addition of tetracycline, but is released by treatment with puromycin. Additional specificity for P-site binding is achieved by the use of N-acetyl-AA-tRNA, which does not interact with EF-Tu and is therefore effectively excluded from the ribosomal A site at Mg^{2+} concentrations that still allow P-site binding.

B. Irradiation Conditions

After complex formation has occurred under safelights, the mixtures are irradiated. APA and APAA derivatives are irradiated at 0° in Pyrex vessels in a Rayonet RPR-100 reactor (Southern New England Ultraviolet Company, Middletown, Connecticut, 06457) equipped with sixteen 350-nm lamps (total of 24 watts). Under these conditions, less than 0.05% of the light energy is transmitted below 310 nm. The mixtures are continually stirred by means of a magnetic flea, and when multiple samples are irradiated simultaneously a "merry-go-round" holder is used[24] to ensure an even distribution of radiant energy. Based on preliminary kinetic experiments, 4 hr of irradiation gives maximal yield, but this should be checked for each case. The lengthy irradiation time is probably due to the poor overlap between the absorption band of the photolabile groups and the output spectrum of the lamps as well as to competing light absorption by the ribosomes or buffer. Since the complexes are quite stable at 0°, use of these lamps ensures the absence of short uv radiation.

The NAG derivatives are irradiated at 0° in Pyrex vessels, with a 650-watt incandescent movie lamp source 10 cm from the samples. The mixtures are stirred and a merry-go-round is used for multiple samples. Judged by spectral changes in NAG-NOS after irradiation, 3–5 min is sufficient under these conditions. However, the kinetics should be checked for each experimental system.

C. Controls

1. Nonspecific Linking to the Ribosome. A frequent problem with affinity labeling experiments, dealt with in depth elsewhere in this volume, is to show that true site labeling has occurred. Fortunately, binding of tRNA to the ribosome is readily controlled since functionally significant binding is highly dependent on at least one additional component. P-site binding requires mRNA, A-site binding requires EF-Tu·GTP as well as mRNA, and I-site binding is dependent on initiation factors and, to a lesser extent, on mRNA. Consequently, by relying on this third factor dependency, the extent of nonspecific covalent linking can be readily assessed.

2. Affinity Labeling by Other Mechanisms. Two types of controls are useful to establish that covalent linking takes place from the photo-affinity-labeled base of tRNA and not from another residue. First, there should be no linking in the absence of photolysis. While this may appear to be a trivial control, we have in fact observed such a light-independent covalent reaction with NAG-tRNA^{Phe} at the ribosomal A site, which otherwise had all the characteristics of a specific reaction in that labeling required EF-Tu and poly(U), and no labeling was observed from the ribosomal P site.

A second control should be either unmodified tRNA or tRNA modified with a suitable analog lacking the azido group and thus nonphoto-lyzable in the expected way. In the above example of NAG-tRNA^{Phe}, the use of unmodified tRNA showed that the NAG-nbt^{3}U residue was not involved and, therefore, that labeling must be due to some hitherto unsuspected reaction. In other instances some caution must be exercised in the choice of control. For example, PA-tRNA has been used as a nonphoto-lyzable analog of APA-tRNA since it lacks the azido group. However, the PA group is actually an acetophenone which is potentially light-activated to a reactive triplet state.[26] Unmodified tRNA may also be unsuitable when the residue to be modified is ^{4}S, since ^{4}S itself can be converted into a reactive species by irradiation.[27,28] For these cases, the iodoacetamide derivative of ^{4}S[20] may be the better control since S-substitution appears to largely prevent activation of ^{4}S by irradiation.[29]

[26] R. E. Galardy, L. C. Craig, J. D. Jamieson, and M. P. Printz, *J. Biol. Chem.* **249**, 3510 (1974).

[27] A. M. Frischauf and K. H. Scheit, *Biochem. Biophys. Res. Commun.* **53**, 1227 (1973).

[28] I. Friser, K. H. Scheit, G. Stöffler, and E. Kuechler, *Biochem. Biophys. Res. Commun.* **60**, 1112 (1974).

[29] V. G. Budker, D. G. Knorre, V. V. Kravchenko, O. I. Lavrik, G. A. Nevinsky, and N. M. Teplova, *FEBS Lett.* **49**, 159 (1974).

Examples from our studies illustrate the importance of these controls. In all cases of thiouridine-modified tRNAs that have been studied, irradiation was required to effect covalent linking to any of the ribosomal sites. In addition, the linking of APA-tRNA^{Val} and APA-tRNA^{Phe} to the A site was dependent on the azido group as substitution of APA- by PA- drastically reduced the extent of linking.[21] In this case, it is clear that not only did linking involve the modified ^{4}S, but also that attachment by the aromatic ketone moiety did not occur since PA-tRNA, also an aceto- phenone derivative, was inactive. However, linking of APA-tRNA^{Val} to the P site was not dependent on the presence of the APA group. PA-, unmodified, or idoacetamide-treated tRNA^{Val} are all equally linked to 30 S subunits by irradiation, despite the absence of any other suitable chromophores in the system, and we have concluded that the extensive covalent linking to 16 S RNA which we observed[18] is not by the APA-^{4}S residue but instead occurs at a still unknown site on the tRNA.

An alternative way to generate a nonphotolyzable analog at the same site is by photolysis of the photolabile group prior to complex formation. This procedure, termed prephotolysis, is an important and quite general test, since it should block any true photoaffinity labeling reaction. How- ever, suitable controls must be included in order to check for inhibitory photolytic side reactions occurring elsewhere in the system.

3. Effect of Irradiation on the Receptor. These probes were designed to obviate light-induced changes in either the tRNA or the ribosome, and a direct test of ribosome function after irradiation confirmed this.[18] A better way to look for structural changes due to irradiation is to examine the covalent labeling pattern after different extents of irradiation, i.e., after one-third and full linking has occurred. In this way, changes that affect the subject of interest, namely the covalent linking distribution, can be monitored directly.

4. Independence of Nature of the Probe. If the results are really a reflection of the topography of the active site, they should be independent of the nature of the probe, provided that the active group is functionally similar and the probe is of equivalent length. This type of control has not been employed up to now due to the limited selection of available reagents.

D. Results

A brief summary of the results we have obtained are included here to illustrate the type of conclusions obtainable by this approach.[18,21] APA- tRNA^{Val} at the A site links to both 30 S and 50 S subunits in a 3:2 ratio. 30 S linking is mostly to 16 S RNA (70%) but also to some 30 S proteins (30%), while 50 S linking is only to proteins and not to 5 S or 23 S RNA. On the other hand, APA-tRNA[Phe] at the A site links only to 30 S subunits. Neither tRNA is covalently bound by the APA-[4]S group at the P site. NAG-tRNA[Phe] does not link to either the P or A site. An unknown residue on tRNA[Val] is closely associated with 16 S RNA and is efficiently linked after irradiation at the P site, but no reaction occurs at the A site. This reaction does not occur with either tRNA[Phe] or tRNA$_f$[Met] at the P site. Covalent linking of tRNA$_f$[Met] to the initiator binding site can be tripled by use of a 12A long probe (APAA) instead of one 9A long (APA), but binding to the P site is virtually absent.

These results have allowed us to conclude (1) A and P sites are structurally distinct, (2) the major contacting region in both P and A sites is RNA, (3) tRNA[Val] and tRNA[Phe] bind in the A site in a similar but not identical manner, (4) the [4]S region of tRNA in the A site probably straddles both subunits, (5) the opposite face of tRNA (bearing the nbt[3]U residue) appears to be greater than 14A from the ribosomal surface when bound at either the P or A site, and (6) P and I sites are distinguished by fMet-tRNA at least in so far as the [4]S region of the tRNA is concerned.

VI. Additional Possibilities

Reaction of the APA probe with Ψ has been described above (Section III,A,2). It should be possible to carry out this reaction on intact tRNA, and by suitable choice of a tRNA containing only the Ψ in loop IV, introduce a specific affinity probe into that loop. Reaction with [4]S can be obviated by choice of a tRNA lacking [4]S, by conversion of [4]S to U,[30] or by chemical or photochemical blocking, i.e., by reaction with iodoacetamide[20] or by [4]S$_8$–C$_{13}$ cross-linking.[24] Other possible residues which should react with APA and similar compounds are the 2-thiouridine derivatives found in the anticodons of a number of tRNAs,[31] 2-thiocytidine found in the anticodon loop,[31] which can be introduced at the acceptor end of tRNA,[32] and 2-thiothymidine located in loop IV of some thermophilic bacteria.[33] It is essential that attachment of the APA or APAA groups at these sites does not impair the function of tRNA. This will have to be tested for each individual case.

[30] K. L. Wong and D. R. Kearns, *Biochim. Biophys. Acta* **395**, 381 (1975).
[31] J. Ofengand, *in* "Molecular Mechanisms of Protein Biosynthesis" (H. Weissbach and S. Pestka, eds.), pp. 7–79. Academic Press, New York, 1977.
[32] M. Sprintzl, K. H. Scheit, and F. Cramer, *Eur. J. Biochem.* **34**, 306 (1973).
[33] K. Watanabe, T. Oshima, M. Saneyoshi, and S. Nishimura, *FEBS Lett.* **43**, 59 (1974).

In addition to varying the site of attachment, the length of the probe can be varied as was done in the case of APA and APAA (Fig. 2). In general, an increased extent of reaction is to be expected when the probe is longer (as was found experimentally with fMet-tRNA at the initiation site), but the selectivity may be reduced as the volume of the potentially reactive region is increased.

The nitrene derived from APA is relatively long-lived and presumably less reactive than the more substituted analogs with absorption bands at higher wavelength. α-Haloketones of these compounds should be even more useful than the present ones, as they may result in a higher covalent yield and could be activated at wavelengths even further removed from the ultraviolet range.

Although not emphasized in this chapter, it should be clear that exactly the same approach can be applied to any other tRNA binding site such as exists in AA-tRNA synthetases, in elongation factors, and in eukaryotic ribosomes. Some studies on Phe-tRNA synthetase of *Escherichia coli*[29,34] are illustrative of this approach.

The APAA derivative has another virtue. Since it is an oxygen ester, it can be readily cleaved by alkali after the insertion reaction has taken place. Thus, after covalent linking, isolation, and characterization of the product, one macromolecule can be cleaved from the other, leaving only part of the probe with each. As each part can be separately labeled in the synthesis of APAA (Sections II,C, and II,D), it should be possible to label the contact site on each macromolecule with minimal structural disturbance. This feature may be important both for the test of regeneration of functional activity as well as for analysis of the site of attachment in certain cases.

Finally, it should be noted that the reagents described here are also reactive with proteins at both -SH and -NH$_2$ groups and could find potential wide utility in the conversion of these groups in proteins into specific photoaffinity probes.[13] In this regard, the iodo derivatives APA-I and APAA-I described in Section II may be of particular value.

[34] I. I. Gorshkova, D. G. Knorre, O. I. Lavrik, and G. A. Nevinsky, *Nucleic Acids Res.* 3, 1577 (1976).

[83] Analogs of Chloramphenicol and Their Application to Labeling Ribosomes

By N. Sonenberg, A. Zamir, and M. Wilchek

The presence of specific binding sites for various antibiotics on the ribosome offers the possibility of using suitable analogs of such compounds as affinity labeling probes. Specifically labeled ribosomal components can be regarded as being involved in antibiotic binding sites as well as related to the ribosomal site whose function is affected by the antibiotic under study. Chloramphenicol was selected for affinity labeling studies in view of its specific interaction with the 50 S ribosomal subunit at, or in the vicinity of, the peptidyltransferase center.[1,2]

Analogs of chloramphenicol suitable for affinity or photoaffinity labeling were synthesized with N-bromoacetyl or N-diazoacetyl replacing the naturally occurring N-dichloroacetyl group (Scheme 1). In another analog, the nitro group was replaced by an azido group (Scheme 2). While the bromoacetyl derivative is chemically reactive, the other two analogs are photoreactable.

The biological activity of the synthetic analogs was assessed by measuring their ability to inhibit peptidyltransferase activity, or to compete with chloramphenicol in reversible binding to 50 S subunits. The latter test was performed only with bromoamphenicol. There are indications that all three analogs bind covalently to the 50 S ribosomal subunit under appropriate conditions and cause irreversible inactivation of peptidyltransferase activity (unpublished data; see also Sonenberg et al.[3]).

Experimental Procedures

Synthesis of Chloramphenicol Analogs

[14C]Bromamphenicol[4]

[14C]Bromoacetic acid (3.7 mg, 50 Ci/mole, Radiochemical Centre) is mixed with 6.5 mg of unlabeled bromoacetic acid in 100 μl of

---

[1] R. E. Monro and D. Vazquez, J. Mol. Biol. **28**, 161 (1967).

[2] S. Pestka, Annu. Rev. Microbiol. **25**, 487 (1971).

[3] N. Sonenberg, M. Wilchek, and A. Zamir, Proc. Natl. Acad. Sci. U.S.A. **70**, 1423 (1973).

[4] Names of compounds: (a) bromamphenicol = D-threo(1R,2R)-1-p-nitrophenyl-2-bromoacetamido-1,3-propanediol; (b) chloramphenicol base = D-threo(1R,2R)-1-p-nitrophenyl-2-amino-1,3-propanediol; (c) glycylamphenicol = D-threo(1R,2R)-1-p-nitrophenyl-2-glycylamido-1,3-propanediol; (d) diazoacetylamphenicol = D-threo-(1R,2R)-1-p-nitrophenyl-2-diazoacetamido-1,3-propanediol; (e) p-aminochloramphenicol = D-threo(1R,2R)-1-aminophenyl-2-dichloroacetamido-1,3-propanediol; (f) p-azidochloramphenicol = D-threo(1R,2R)-p-azidophenyl-2-dichloroacetamido-1,3-propanediol; (g) iodamphenicol = D-threo(1R,2R)-1-p-nitrophenyl-2-iodo-acetamido-1,3-propanediol; (h) tBoc = tert-butyloxycarbonyl.

dioxane. Dicyclohexylcarbodiimide, 12 mg, and 6 mg of N-hydroxysuc-cinimide are added, and the mixture is left for 1 hr at room temperature. Chloramphenicol base (Sigma), 12 mg, is added and incubation is continued for 1 hr. After addition of 2 ml of ethyl acetate, dicyclohexylurea is removed by centrifugation. The clear solution is extracted twice with water and evaporated to dryness. The residue, dissolved in methanol, is purified by thin-layer chromatography on silica gel developed with chloroform–methanol 3:1 ($R_f = 0.76$).

Unlabeled bromamphenicol is prepared from bromoacetylbromide and chloramphenicol base[5]; m.p. $= 132°$.

Diazoacetylamphenicol (Scheme 1)

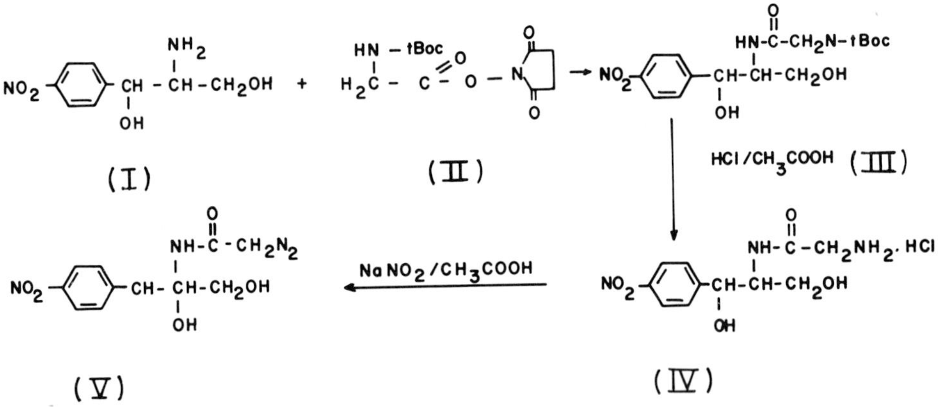

SCHEME 1

1. Glycylamphenicol Hydrochloride. To a solution of chloramphenicol base (545 mg, 2.2 mmoles) in 5 ml of chloroform and 0.3 ml of triethylamine is added the N-hydroxysuccinimide ester of N-tBoc-glycine (600 mg, 2.1 mmoles) in 5 ml of chloroform. After 2 hr at room temperature the solution is washed successively with 1 M sodium bicarbonate, water, 10% citric acid, and water, and dried over sodium sulfate. The solution is evaporated to a small volume and petroleum ether is added to precipitate the product. The tBoc group is removed by treatment with 1 N hydrochloric acid in acetic acid for 15 min at room temperature. The product is precipitated with diethyl ether. The yield is 510 mg. Thin-layer chromatography on silica gel with two solvent mixtures, n-butanol:

[5] M. C. Rebstock, J. Am. Chem. Soc. **72**, 4800 (1950).

acetic acid:water (4:1:4) or (5:3:2), reveals one spot when sprayed with ninhydrin.

2. *Diazoacetylamphenicol.* Glycylamphenicol hydrochloride (400 mg, 1.3 mmoles) is dissolved in 1.4 ml of 1 M sodium acetate at 0°. To this solution is added sodium nitrite (80 mg, 0.9 mmole) followed by 80 μl of acetic acid. The mixture is left for 2.5 hr at 0°, at which time a yellow precipitate is observed. The precipitate is dissolved in ethyl acetate, dried with sodium sulfate, and precipitated with petroleum ether. The yield is 95 mg.

p-Azidochloramphenicol (Scheme 2)

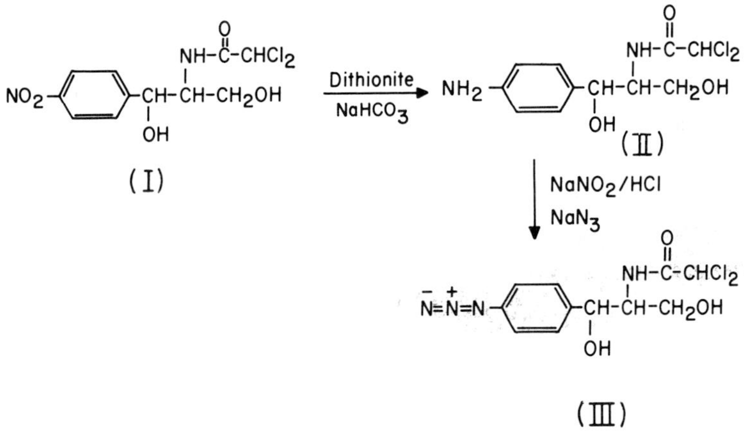

SCHEME 2

1. *p-Aminochloramphenicol Hydrochloride.* Sodium dithionite (1.6 g, 9 mmoles) is added to a solution of chloramphenicol (500 mg, 1.6 mmoles) in 50 ml of 0.1 M sodium bicarbonate. The pH is maintained at about 7.0 by the addition of sodium bicarbonate. After 10 min at room temperature the product is extracted with ethyl acetate, and the solution is acidified by the addition of 1 ml of 4 N hydrochloric acid in dioxane. On addition of diethyl ether the product precipitates. The yield is 95 mg; m.p. 135°–138°. The product gives a single spot on thin layer chromatography on silica gel developed with chloroform:methanol (3:1) ($R_f = 0.66$).

2. *p-Azidochloramphenicol.* To a solution of p-aminochloramphenicol (70 mg, 0.21 mmole) in 100 μl of 2 N hydrochloric acid at 0° is added sodium nitrite (14 mg, 0.20 mmole). The excess of nitrous acid formed is removed under reduced pressure, and sodium azide (13 mg, 0.20

mmole) is added slowly. After stirring for 15 min at 0°, the product is extracted with ethyl acetate, dried over sodium sulfate, and evaporated to dryness. The yield is 35 mg; m.p. 89°–91°. $R_f = 0.85$ on thin-layer chromatography as in (1) above. The same compound has also been synthesized by other methods.[6,7]

Covalent Binding of Chloramphenicol Analogs to 50 S Subunits

Bromamphenicol

The 50 S subunits (25 mg/ml) are incubated in a buffer containing 0.1 M ammonium chloride, 1 mM magnesium acetate, and 0.02 M Tris chloride (pH 8.6) for 2 hr at 37° with [^{14}C]bromamphenicol (0.3 mM). The mixture is dialyzed overnight against 0.1 M ammonium chloride, 0.01 M magnesium acetate, and 0.02 M Tris chloride (pH 7.3) to remove unreacted reagent. Covalent binding is determined as described.[3]

Diazoacetylamphenicol, p-Azidochloramphenicol

The 50 S subunits (18.8 mg/ml) in a solution containing 0.1 M ammonium chloride, 1 mM magnesium acetate, and 0.02 M Tris chloride (pH 7.4) are incubated at 0° for 15 min with each of the photoaffinity analogs of chloramphenicol to allow for the formation of reversible binding complexes, and then placed in a water bath at room temperature and irradiated for 30 min at an average distance of 8 cm from a mercury high pressure lamp (450 watts, Hanovia). Light of wavelengths below 300 nm is eliminated by the use of Pyrex glassware and a Pyrex jacket around the lamp. After irradiation, the mixtures are dialyzed against 0.1 M ammonium chloride, 0.01 M magnesium acetate, and 0.02 M Tris chloride (pH 7.4) as above.

Comments

The three compounds, the synthesis of which is described here, are functional analogs of chloramphenicol according to at least one criterion. When reversibly bound to ribosomes they all inhibit peptidyltransferase activity. Bromamphenicol has also been shown to compete with chloramphenicol in reversible binding to what appears to be the same binding site on the ribosome.[3] The results of labeling experiments indicated that bromamphenicol binds covalently to the 50 S ribosomal subunit and

[6] F. Seela and F. Cramer, *Hoppe-Seyler's Z. Physiol. Chem.* **356**, 1185 (1975).
[7] P. E. Nielsen, V. Leick, and O. Buchardt, *Acta Chem. Scand.* **B29**, 662 (1975).

causes irreversible inactivation of peptidyltransferase activity by selectively alkylating cysteine-SH groups in proteins L2 and L27 as identified by amino acid analysis, and gel electrophoretic separation of the ribosomal proteins.[3] Iodoamphenicol prepared from iodoacetyl-N-hydroxysuccinimide ester and chloramphenicol base was reported to label protein L16.[8]

In unpublished experiments it has been shown that irradiation with each of the two photoaffinity labeling reagents resulted in the irreversible inactivation of peptidyltransferase. The presence of erythromycin in the irradiation mixture with p-azidochloramphenicol protected against inactivation. It thus appears that these analogs react with the ribosome at functionally significant sites. The modified ribosomal components in these two cases have not been identified, however.

Finally, a photoinduced reaction of unmodified chloramphenicol with the ribosome has been observed.[9] The reaction is expressed in the labeling of several ribosomal proteins and in a pronounced shift in the electrophoretic mobility of protein L19.[10]

Acknowledgments

We are indebted to Mr. D. Haik for ribosome preparations. This work was supported by a grant from the United States–Israel Binational Science Foundation (BSF), Jerusalem, Israel.

[8] O. Pongs, R. Bald, and V. A. Erdmann, *Proc. Natl. Acad. Sci. U.S.A.* **70**, 2229 (1973).

[9] N. Sonenberg, A. Zamir, and M. Wilchek, *Biochem. Biophys. Res. Commun.* **59**, 693 (1974).

[10] M. Israel, N. Sonenberg, M. Wilchek, and A. Zamir, in preparation.

[84] A Homologous Series of Photoreactive Peptidyl-tRNAs for Probing the Ribosomal Peptidyltransferase Center

By N. Sonenberg, M. Wilchek, and A. Zamir

Aminoacyl- and peptidyl-tRNA are the natural substrates of ribosomal peptidyltransferase. Derivatives of these substrates modified chemically or with photoreactive groups can therefore be used as affinity labeling reagents for the location of ribosomal components at, or close to, the peptidyltransferase active center.

Here we describe the synthesis of photoreactable derivatives of Phe-tRNA of the general structure: (AP)(Gly)$_n$-Phe-tRNA ($n = 0, 2, 4$).[1]

[1] Abbreviations: (a) (AP) = p-azido-N-tBoc-Phe; (b) tBoc = *tert*-butyloxycarbonyl.

These peptidyl-tRNA analogs bind reversibly, in the presence of poly(U), to 70 S ribosomes and attach covalently upon irradiation to a specific site(s) on the 23 S RNA of the 50 S subunit.

Experimental Procedures

Synthesis of Reagents

p-Azido-N-tBoc-Phe-[³H]Phe-tRNA

(AP)Phe-tRNA is synthesized according to Scheme 1 by coupling[2] the *N*-hydroxysuccinimide ester of *p*-azido-*N*-*t*Boc-phenylalanine[3] with

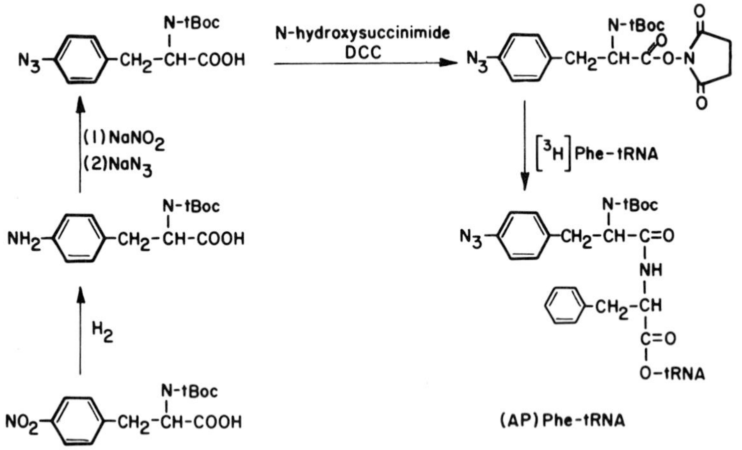

SCHEME 1

(³H)Phe-tRNA (unfractionated). In a standard preparation, 3.5 nmoles of (³H)Phe-tRNA are dissolved in 3.0 ml of 0.2 M triethanolamine-HCl (pH 8.0) or 0.2 M *N*-2-hydroxyethylpiperazine-*N'*-2-ethanesulfonic acid (HEPES) (pH 8.0), and 600 mg of the *N*-hydroxysuccinimide ester of *p*-azido-*N*-*t*Boc-phenylalanine in 30 ml of freshly distilled dimethyl sulfoxide are added. The mixture is incubated for 2 hr at 30°, chilled to 0°, and 3.7 ml of 50% dichloroacetic acid are added. After 2 hr at 0°, the mixture is centrifuged and the pellet is washed once with dimethyl-formamide and twice with ethanol. The dried pellet is finally dissolved

---

[2] Y. Lapidot, D. Eilat, S. Rappoport, and N. deGroot, *Biochem. Biophys. Res. Commun.* **190**, 558 (1970).

[3] A. Schwyzer and M. Caviezel, *Helv. Chim. Acta* **54**, 1395 (1971).

in water and titrated to pH 6.0 with Tris chloride. The preparation contained, at most, 5% of unreacted Phe-tRNA as determined by paper electrophoresis following alkaline hydrolysis.[4]

p-Azido-N-tBoc-Phe-(Gly)$_n$[^{3}H]Phe-tRNA (n = 2, 4)

1. p-Nitro-N-tBoc-Phe-(Gly)$_n$COOH (n = 2, 4). To a solution of glycylglycine (1.60 g, 12 mmoles) in 50 ml of 0.45 M sodium bicarbonate is added the N-hydroxysuccinimide ester of p-nitro-N-tBoc-Phe (4.06 g, 10 mmoles) in 50 ml of dioxane. After 3 hr at room temperature, the solution is concentrated to remove dioxane and is acidified with 10% citric acid and extracted with ethyl acetate. The ethyl acetate solution is washed with water, dried with sodium sulfate, and concentrated to a small volume. On addition of petroleum ether, the product precipitates. Yield, 4 g; m.p. 95°–96°.

p-Nitro-N-tBoc-Phe-(Gly)$_4$ COOH is synthesized similarly, using the same ratio of reagents as for the (Gly)$_2$ compound. The product precipitates upon addition of citric acid. Yield, 3 g; m.p., 168°.

2. (AP)(Gly)$_n$-N-hydroxysuccinimide Ester (n = 2, 4). p-Nitro-N-tBoc-Phe-(Gly)$_2$ COOH (880 mg, 2 mmoles), dissolved in 100 ml of methanol, is hydrogenated in the presence of 100 mg of 10% palladium on charcoal for 2.5 hr. The product, p-amino-N-tBoc-Phe-(Gly)$_2$COOH, after removal of the catalyst by filtration and evaporation of solvent, is dissolved in 2 ml of 2 N hydrochloric acid and 0.4 ml of water at 2°, and sodium nitrite (145 mg, 2.1 mmoles) in 1 ml of water is added. After stirring for 1.5 hr at 2°, the solution is filtered to remove insoluble material and degassed under reduced pressure to expel excess nitrous acid. To this solution, sodium azide (130 mg, 2.0 mmoles) in 1 ml of water is added slowly. After stirring for 45 min at 0°, the aromatic azide compound is extracted into ethyl acetate. The solution is evaporated to dryness to give (AP)(Gly)$_2$COOH. To prepare the N-hydroxysuccinimide ester, the product is dissolved in 5 ml of ethyl acetate, and N-hydroxysuccinimide (172 mg, 1.5 mmoles) is added followed by dicyclohexylcarbodiimide (310 mg, 1.5 mmoles) in 6 ml of ethyl acetate. The mixture is left overnight at 0°. After removal of dicyclohexylurea, the solution is evaporated to dryness, yielding 350 mg of product.

(AP)(Gly)$_4$COOH is prepared similarly to the (Gly)$_2$ compound. The (Gly)$_4$ compound is, however, barely soluble in ethyl acetate. Therefore, at the end of the reaction, sodium hydroxide is added to adjust the pH to about 4.0 and the mixture is lyophilized. The residue is dissolved

[4] Z. Vogel, A. Zamir, and D. Elson, *Proc. Natl. Acad. Sci. U.S.A.* **61,** 701 (1968).

in ethanol and filtered to remove insoluble salts; product is obtained after evaporation of the ethanol. Thin-layer chromatography on silica gel with chloroform:methanol (3:1), or pure methanol, indicates the presence of a single product. In order to prepare the N-hydroxysuccinimide ester, 276 mg (0.5 mmole) of (AP)(Gly)$_4$COOH are dissolved in 12 ml of dimethylformamide, and N-hydroxysuccinimide (56 mg, 0.5 mmole) is added, followed by dicyclohexylcarbodiimide (103 mg, 0.5 mmole) in 1.5 ml of dimethylformamide. The mixture is left overnight at 0°. After removal of dicyclohexylurea, the dimethylformamide solution is used directly for coupling to Phe-tRNA.

3. (AP)(Gly)$_4$-[^{3}H]Phe-tRNA ($n = 2, 4$). These compounds are prepared in a manner similar to that for (AP)[^{3}H]Phe-tRNA as described above. The (Gly)$_2$ derivative is prepared in dimethyl sulfoxide and the (Gly)$_4$ derivative in dimethylformamide. The preparation of (AP)(Gly)$_2$-[^{3}H]Phe-tRNA contains between 15 and 25% of unreacted [^{3}H]Phe-tRNA, whereas the preparation of (AP)(Gly)$_4$-[^{3}H]Phe-tRNA contains between 20 and 30% of unreacted [^{3}H]Phe-tRNA.

Reversible Binding of (AP)(Gly)$_n$-Phe-tRNA to 70 S Ribosomes ($n = 0, 2, 4$)

The reaction mixture, in 0.15 M ammonium chloride, 30 mM magnesium acetate, and 50 mM Tris chloride (pH 7.4), contains the following per ml: 120–240 μg of poly(U), 1.4 mg of 70 S ribosomes, and 640 pmoles of (AP)(Gly)$_n$-[^{3}H]Phe-tRNA ($n = 0, 2, 4$). Reaction mixtures are incubated at 37° for 20 min to allow formation of binding complexes.

Covalent Binding of (AP)(Gly)$_n$-Phe-tRNA ($n = 0, 2, 4$) to 70 S Ribosomes

Binding complexes are irradiated with a high-pressure mercury lamp (Hanovia, 450 watts) at an average distance of 8 cm for 5 min in a water bath at room temperature. Light of wavelengths below 250 nm is eliminated by a Corex filter. Covalent binding is determined as described.[5]

Comments

The series of compounds described here is suitable for screening the ribosomal components involved in peptidyltransferase activity. All compounds described are peptidyl-tRNA analogs, and the shortest member

[5] N. Sonenberg, M. Wilchek, and A. Zamir, *Proc. Natl. Acad. Sci. U.S.A.* **72**, 4332 (1975).

in the series, (AP)Phe-tRNA, has been shown to bind preferentially to the ribosomal donor site.[5] The mode of reversible binding has not been determined for the longer peptidyl tRNAs. The compounds described are of potential value in characterizing the ribosomal environment at increasing distances from the site of interaction of the —CCA terminus of tRNA. With the same idea in mind, peptidyl-tRNAs of varying lengths with a bromoacetyl group blocking the terminal amino group have been previously synthesized and tested as affinity probes.[6]

Information gathered so far with the present series indicates that the irradiation of poly(U)-directed ribosomal complexes of each of the compounds described results in a reaction with 23 S rRNA. Specifically, labeling takes place within the 18 S fragment that includes about 1700 nucleotides extending from the 3′ end. The results point to the functional significance of rRNA in the peptidyltransferase center. Sequencing studies of the different modified rRNAs may clarify the mode of arrangement of the rRNA within the functional site. Labeling of 23 S rRNA by chemical and photoreactive analogs of peptidyl-tRNA has been reported from several other laboratories.[7-11]

Acknowledgment

We are indebted to Mr. I. Jacobson for synthesizing some of the intermediate compounds. This research was supported by a grant from the United States–Israel Binational Science Foundation (BSF), Jerusalem, Israel.

[6] D. Eilat, M. Pellegrini, H. Oen, Y. Lapidot, and C. R. Cantor, *J. Mol. Biol.* **88,** 831 (1974).
[7] L. Bispink and H. Matthaei, *FEBS Lett.* **37,** 291 (1973). See also this volume [74].
[8] A. S. Girshovich, E. S. Bochkareva, U. M. Kramarov, and Y. A. Ovchinnikov, *FEBS Lett.* **45,** 213 (1974). See also this volume [77] and [78].
[9] M. Yukioka, T. Hatayama, and S. Morisawa, *Biochim. Biophys. Acta* **390,** 192 (1975).
[10] A. Barta, E. Kuechler, C. Branlant, J. Sriwidada, A. Krol, and J. P. Ebel, *FEBS Lett.* **56,** 170 (1974). See also this volume [81].
[11] J. B. Breitmeyer and H. F. Noller, *J. Mol. Biol.* **101,** 297 (1976).

[85] Photoaffinity Labeling of Ribosomes with the Unmodified Ligands Puromycin and Initiation Factor 3

By BARRY S. COOPERMAN

There are two general approaches to photoaffinity labeling. In the first, a radioactive photolabile ligand derivative is irradiated in the presence of its receptor. In the second, radioactive native ligand, having no special photolability, is irradiated in the presence of its receptor, with, in some cases, an added photosensitizer as well. There are two good reasons for preferring the first approach. First, using a photolabile derivative it should be possible to choose experimental conditions such that the major photochemical event is photolysis of the derivative, with little or no accompanying photochemical destruction of the receptor. Such discrimination will in general not be possible using the second approach. Second, derivatives may be synthesized that, on photolysis, form highly reactive intermediates having low chemical selectivity, thus increasing the likelihood that covalent bond formation accurately reflects the site of noncovalent bonding.[1] In the second approach the photochemical mechanism leading to incorporation will be difficult to establish, leaving open the question of the accuracy of covalent bond formation. On the other hand, the second approach presents the obvious practical advantage that once radioactive native ligand is available, labeling studies can begin immediately, there being no need, as there is with the first approach, either to synthesize a new photolabile derivative or to show that such a derivative binds noncovalently to receptor in a correct manner.[1] Furthermore, the photolabile groups currently being used in making photolabile derivatives all photolyze to give reactive intermediates having some chemical specificity, making it desirable to conduct affinity labeling studies with more than one derivative. Labeling results obtained from the second approach would therefore be useful for comparison with labeling results using the first approach. Below we present results, some of which are preliminary, using the second approach to study two different ligand interactions with ribosomes of *Escherichia coli*.

The Photoincorporation of Puromycin into 70 S Ribosomes

Puromycin has a structure closely analogous to the 3′ terminus of a tyrosine-tRNA, and it inhibits protein synthesis by accepting, in place of an aminoacyl-tRNA, an incomplete polypeptide chain from ribosome-bound peptidyl-tRNA, thus prematurely terminating protein synthesis.[2] We have found that irradiation at 2537 Å (low-pressure Hg lamp) of solutions containing puromycin and ribosomes from *E. coli* leads to significant covalent incorporation of puromycin into the ribosome.[3] For

[1] B. S. Cooperman, *in* "Aging, Carcinogenesis, and Radiation Biology" (K. C. Smith, ed.), p. 315. Plenum, New York, 1976.

[2] M. B. Yarmolinsky and G. L. de la Haba, *J. Biol. Chem.* **237**, 1190 (1962).

[3] B. S. Cooperman, E. N. Jaynes, D. J. Brunswick, and M. A. Luddy, *Proc. Natl. Acad. Sci. U.S.A.* **72**, 2974 (1975).

example, with our apparatus, irradiation of ribosomes in the presence of 0.3 mM puromycin for 3 min leads to an incorporation of 0.18 puromycin/ribosome. In an earlier paper[3] we showed that a portion (as much as 50% at low puromycin concentrations) of the incorporation proceeds by way of saturable, or site-specific process (K_D = 0.65 mM), the remainder being nonspecific; that incorporation occurs into the RNA and protein portions of both 30 S and 50 S particles; and that a single protein, L23, is labeled to a major extent above background. More recently[4] we have conclusively shown that incorporation into L23 proceeds via affinity labeling. The evidence is, first, that incorporation into L23 (as measured using polyacrylamide gel electrophoresis) shows saturation behavior as a function of puromycin concentration, with a K_D (0.7 mM) identical to that found for the saturable portion of the overall incorporation process; and second; that two structural and functional analogs of puromycin, N-phenylalanylpuromycin aminonucleoside[5] and cytidylyl(3',5')-3'-O-phenylalanineadenosine,[6] specifically block puromycin incorporation into L23. Moreover, the relative (compared to puromycin) K_D values for the analog calculable from the blocking experiments are similar in magnitude to the relative (compared to puromycin) K_m values for the analog in assays of peptidyltransferase activity, supporting the notion that L23 labeling is taking place from a functional puromycin binding site. Other experiments suggest that a region of the 50 S RNA fraction also provides part of the puromycin binding site, although here the evidence is more tentative.

The Photoincorporation of IF-3 into 30 S Subunits

Initiation factor 3 (IF-3) is one of three protein cofactors necessary for polypeptide chain initiation. It is now known that such initiation proceeds via dissociation of 70 S ribosomes into 30 S and 50 S subunits. IF-3, by binding tightly to the 30 S subunit, strongly favors this dissociation, and also appears to have the additional function of directing the binding of 30 S ribosomal subunits to start signals in messenger RNA.[7] Recently, a structurally detailed hypothetical mechanism account-

[4] E. N. Jaynes, Jr., P. G. Grant, G. Giangrande, R. Wieder, and B. S. Cooperman, submitted for publication.

[5] R. J. Harris, J. E. Hanlon, and R. H. Symons, *Biochim. Biophys. Acta* **240**, 244 (1971).

[6] D. Ringer and S. Chladek, *FEBS Lett.* **39**, 75 (1974).

[7] S. Ochoa and R. Mazumder, *in* "The Enzymes" (P. Boyer, ed.), 3rd ed., Vol. 10, p. 1. Academic Press, New York, 1974.

ing for the dual IF-3 functionality has been proposed.[8] Localization of IF-3 binding site on the 30 S subunit would provide an important test for this mechanism. In a recent attempt to do so, the IF-3·30 S complex was treated with periodic acid and sodium borohydride, leading to formation of a covalent link between IF-3 and the 3' terminus of 16 S RNA. The complex was also allowed to react with tartarylazide, leading to IF-3 cross-linking to protein S7.[9] These results represent a promising start toward localizing the IF-3 binding site and offer a point of comparison for the results obtained[10] as described below.

Purified IF-3[11] was made radioactive by reductive methylation with [^{14}C]formaldehyde (44 mCi/mole) and sodium borohydride, a procedure that had previously been shown not to affect the activity of IF-3.[12] Typically, this method was able to achieve specific activities of 5×10^3 cpm per microgram of protein, corresponding to 10^5 cpm/nmole.[13] Incubation mixtures of [^{14}CH$_3$]IF-3 and 30 S ribosomes were irradiated under various conditions and tested for covalent IF-3 incorporation by measurement of the radioactivity comigrating with a 30 S particle in a sucrose gradient containing 0.7 M NaCl; the high salt serves to break up the noncovalent 30 S·IF-3 complex.[14] Direct irradiation at 2537 Å for periods sufficient to yield significant puromycin incorporation leds to only minor IF-3 incorporation, as did direct irradiation at 3650 Å.[15] However, irradiation at 3650 Å in the presence of a photosensitizing system (flavin mononucleotide, 0.1 mM; riboflavin, 0.1 mM; FeCl$_3$, 0.5 μM; NADH, 0.125 mM)[16] led to significant IF-3 incorporation. Furthermore, this incorporation was specific in the sense of proceeding by prior formation of a noncovalent complex. This was demonstrated, first by showing that ionic conditions (0.7 M NaCl) or added ligands (spermine and aurintricarboxylic acid), which block noncovalent bonding[12,14] also block co-

---

[8] J. Van Duin, C. G. Kurland, J. Dondon, M. Grunberg-Manago, C. Branlant, and J. P. Ebel, *FEBS Lett.* **62**, 111 (1976).

[9] J. Van Duin, C. G. Kurland, J. Dondon, and M. Grunberg-Manago, *FEBS Lett.* **59**, 287 (1975).

[10] B. S. Cooperman, J. Dondon, J. Finelli, M. Grunberg-Manago, and A. M. Michelson, submitted for publication.

[11] T. Godefrey-Colburn, A. D. Wolfe, J. Dondon, M. Grunberg-Manago, P. Dessen, and D. Pantaloni, *J. Mol. Biol.* **94**, 461 (1975).

[12] C. L. Pon, M. Friedman, and C. Gualerzi, *Mol. Gen. Genet.* **116**, 192 (1972).

[13] Specific activities as high at 2×10^4 cpm/μg have been reported.[14]

[14] C. L. Pon and C. Gualerzi, *Biochemistry* **15**, 804 (1976).

[15] Blak-Ray B-100A lamp, Ultraviolet Products Co.

[16] This system was suggested by A. M. Michelson as a way to generate several radicals in solution simultaneously, including O$_2^-$· and HO·. Recent experiments show that only the flavins are essential for incorporation.

valent incorporation, and second, by demonstrating that incorporation results upon irradiation of the isolated IF-3·30 S complex.

In a typical experiment, irradiation for 7 min led to incorporation of 0.2 moles of IF-3 per mole of noncovalently bound IF-3; the incorporated radioactivity was distributed approximately 3:1 between the protein and 16 S RNA fractions of the 30 S particle. SDS polyacrylamide gel electrophoresis[17] of the labeled protein fraction shows the presence of two or three new radioactive protein bonds in the region of 40,000–55,000 daltons. This molecular weight range is appropriate for IF-3 cross-linking to 30 S ribosomal proteins.[18] Work to further characterize the cross-linked protein products and the sites of insertion into 16 S RNA is currently underway.

Comments

The two examples discussed above show that, at least in certain cases, direct or photosensitized irradiation of solutions containing a receptor (in this case, *E. coli* ribosomes) and a radioactive but not especially photolabile native ligand can yield true affinity labeling results.[19] Several of the problems that arise in the course of utilizing this approach are discussed below.

As anticipated, ribosomal destruction, as measured by the loss of poly(U)-directed polyphenylalanine synthesis activity, occurs during the irradiation leading to puromycin incorporation. This raises the question of whether L23 labeling resulted from the native or a photodenatured puromycin site. Under standard conditions the half-time for loss of activity was 10 min. Therefore, the labeling patterns obtained on irradiation at 2, 8, 16, and 32 min were inspected and it was found that L23 labeling becomes more specific as irradiation time is reduced.[4] Thus, L23 labeling occurs from a native puromycin site. This procedure of extrapolating a labeling pattern back to zero irradiation time should be a general one for dealing with the problem of receptor destruction, but requires the availability of a ligand of high specific radioactivity. [³H]-Puromycin is an especially fortunate case since it is available commercially at 4 Ci/mmole,[20] i.e., 2×10^3 cpm/pmole. We can routinely obtain a preliminary labeling pattern from one-dimensional polyacrylamide gel

[17] U. K. Laemmli, *Nature (London)* **227**, 680 (1970).

[18] H. G. Wittmann, *in* "Ribosomes" (M. Nomura, A. Tissières, and P. Lengyel, eds.), p. 93. Cold Spring Harbor Laboratory, Cold Spring Harbor, New York, 1974.

[19] Other examples are discussed by Cooperman[1] and in articles in the same text by P. R. Schimmel *et al.* (p. 123) and C. Hélène (p. 149).

[20] Available from New England Nuclear.

electrophoretic analysis of 100 pmoles of 50 S protein (60 μg) containing a total of 1500 cpm. Since approximately one-third of total incorporation into 70 S ribosomes takes place into 50 S protein, this corresponds to a fairly low level of total incorporation, i.e., 0.02 puromycin/ribosome. Ribosomal destruction is also taking place during IF-3 incorporation, and we expect to study the labeling pattern as a function of light fluence here as well.

We have failed in attempts to transfer the N-acetylphenylalanine portion of ribosome-bound N-acetylphenylalanine-tRNA to photoincorporated puromycin.[4] A possible explanation for our failure is that the process, as yet unknown, leading to photoincorporation changes the puromycin structure in such a way as to make the covalently incorporated material inactive as a peptidyl acceptor. This is not unlikely given the known stringent structural requirements for peptidyl acceptors.[5,21,22] Since such requirements are a general property of receptor–ligand interactions, it is to be anticipated that the lack of functionality in ligands photoincorporated by the above procedure might be a common occurrence. This result reflects an inherent limitation of this approach, and we plan to further examine the puromycin–ribosome interaction by using photolabile puromycin derivatives. Derivatives of this kind, i.e., those that successfully mimic puromycin function,[23] should incorporate into ribosomes by photochemical mechanisms which leave the essential puromycin structure intact, thereby improving the chances of retaining puromycin functionality on covalent incorporation. The labeling patterns obtained with these new derivatives should also provide a test of the importance of L23 in the formation of part of the ribosomal puromycin binding site. Testing of the functionality of covalently bound IF-3 has not yet been carried out.

At present little is known of the photochemical mechanisms for either the incorporation of puromycin by direct photolysis or the photosensitized incorporation of IF-3.[24] However, two observations may be relevant. Both incorporation reactions are inhibited by thiols, suggesting that they

[21] Nathans and A. Neidle, *Nature* (*London*) **197**, 1076 (1963).

[22] L. V. Fisher, W. W. Lee, and L. Goodman, *J. Med. Chem.* **13**, 775 (1970).

[23] We have already synthesized N-ethyl-2-diazomalonylpuromycin (see Cooperman *et al.*[3]), but this derivative, although it binds to *E. coli* ribosomes, does so at a site apparently removed from the site of puromycin binding.

[24] We had originally (Cooperman[3]) suggested that puromycin incorporation might proceed via S-alkylation,[25,26] but have recently shown[4] using [8-^{3}H]puromycin that this is not the case.

[25] J. Salomon and D. Elad, *Photochem. Photobiol.* **19**, 21 (1974).

[26] D. Leonov, J. Salomon, S. Sasson, and D. Elad, *Photochem. Photobiol.* **17**, 465 (1973).

might be proceeding by free radical mechanisms. Second, both direct and photosensitized irradiation of a [^{3}H]dihydrostreptomycin·30 S complex led to no significant covalent incorporation of radioactivity into 30 S particles, suggesting, not surprisingly perhaps, that the approach described in this article may be limited to ligands containing conjugated groups.

Acknowledgments

This work was supported by Research Grants AM13212 from the U.S. Public Health Service and NP-176 from the American Cancer Society, during the tenure of an Alfred P. Sloan fellowship. The work with IF-3 was performed in the laboratory of Dr. Grunberg-Manago (Institut de Biologie Physico-Chimique, Paris, France) for whose warm hospitality the author is grateful.

Author Index

Numbers in parentheses are footnote reference numbers and indicate that an author's work is referred to although his name is not cited in the text.

Crestfield, A. M., 135
Cretcher, L. H., 539
Creveling, C. R., 555
Crown, D., 192
Crusberg, T. C., 309
Csányi, V., 532
Csuros, Z., 407
Cuatrecasas, P., 154, 358, 359(1), 360(6), 361(6), 448, 490, 503, 669, 672(1), 674(1)
Cunningham, L. W., 207, 234, 235(10)
Cunnion, S. O., 684
Curphey, T. J., 418
Cushley, R. J., 308
Cysyk, R., 88, 89(124), 110, 111(124)
Czernilofsky, A. P., 193, 627, 628(27, 28, 29), 629(29), 630(29), 631(28), 632(28, 29)
Czombos, J., 200
Czoncha, L., 603

D

Dahl, L. F., 24
Dahlquist, F. W., 70
Dall-Larsen, T., 295
Dalton, J. C., 470(10), 471
Daly, J. W., 555
D'Ambrosio, S. M., 153, 608, 609(12), 611(12), 612(12)
Danchin, A., 317, 318, 319, 320, 321(3, 4)
Danenberg, P. B., 308
Danenberg, P. V., 308, 309(12)
Daniel, D. S., 418
Daring, W., 365
D'Arrigo, J. S., 288
Das Gupta, U., 80(73), 81, 84(73), 99(73)
Datta, A., 276, 277
Dattagupta, J. K., 347
Dauben, W. G., 470(13), 471
Daves, K., 470(10), 471
Davey, W., 545
Davidoff, F., 548, 549(3), 550(3), 552(5), 554
Davidson, E., 420
Davidson, L., 80(113), 82, 87(113)
Davies, D. R., 198, 200(12), 207(12)
Davis, R. H., Jr., 143
David, I. B., 35, 36(12), 416, 420, 421(16), 423(32), 424(16, 32, 39), 427 (39)

Day, R. A., 35, 36(12), 416, 420, 421(16), 424(16), 532, 536
Deal, W., 578(44), 581, 582, 583(5), 585(5), 587(5)
Deana, A. A., 584
De Asua, L. J., 551
de Boer, T. J., 511
Debov, S. S., 420
de Bruyne, C. K., 80(98, 103), 82, 86(98, 103), 363
De Graw, J. I., 604, 605
de Groot, N., 631, 633(39), 638, 672, 708
de la Haba, G. L., 712
De La Mare, S., 141, 142(38), 143(38)
Delaney, P., 143, 695, 701(24)
Delpierre, G. R., 73
De Mairena, J., 162
Dempsey, W. B., 24
Dence, J. B., 448
Dénes, G., 139
de Nobrega Bastos, R., 80(72), 81, 84(72), 99(72)
Denniss, I. S., 198, 199(15), 206(15), 207(15)
de Paulet, A. C., 452, 457(7, 8)
De Renzo, E., 207
de Robichon-Szulmajster, H., 399
Descomps, B., 148, 150, 452, 457(7, 8)
De Sombre, E. R., 453
Dessen, P., 714
Dietrich, L. S., 420
Dimroth, O., 644
Dion, H. W., 24, 415
Dion, R. L., 418
Dixon, G. J., 354
Dixon, M., 476
Döbeln, U. V., 322
Doerr, I. L., 322
Dole, W. P., 602
Dolowy, W. C., 420
Donato, H., 309
Dondon, J., 714
Doolittle, R. F., 487, 493
Doyle, B., 473
Dragoje, B., 206
Dressler, K., 669
Dreyer, W. J., 440
Driscoll, G. A., 149
Drummond, G. I., 293
Dubois, J. M., 80(94), 81, 85(94)

G

Gagliano, R., 52
Gahn, D. T., 226, 228(6)
Galardy, R. E., 80(87), 81, 85(87), 87(87), 89(87), 91(87), 103(87), 104(87), 111(87), 471, 479, 656, 676, 699
Galiazzo, G., 562
Galivan, J. H., 309, 310
Gamble, R. C., 170, 178
Ganguly, M., 452
Garrison, D. R., 603
Garuti, L., 329
Gaskin, F., 571
Gassen, A., 75(32), 77
Gassen, H. G., 622, 623(9, 10, 12), 624(10, 12), 625(10, 12), 626(10)
Gavrilova, L. P., 191
Gehring, H., 44
Geisow, M. J., 441
Gennari, G., 562
Gero, S. D., 370, 372
Gershon, E., 369, 370(13), 377(13)
Gerson, K., 289
Gersten, N. B., 186
Gertler, A., 206
Ghosh, A., 517
Ghosh, S., 420
Giangrande, G., 713, 715(4)
Giartosio, A., 445
Giegé, R. C., 169, 170(10), 180(10)
Giles, K. A., 401
Gilham, P. T., 378
Gillett, J. W., 387
Gilman, A. G., 591
Giovanelli, I., 31
Giovanninetti, G., 329
Girshovich, A. S., 80(76), 81, 84(76), 114(76), 187, 627, 628(30, 32), 630(32), 632(32), 634, 635(58), 636(58), 637, 656, 659(1, 2, 3), 660, 661, 662(1), 711
Gish, D. T., 230
Gitler, C., 80(108), 82, 86(108)
Givol, D., 131, 154, 157, 480, 481, 484, 487, 488, 491(21, 23), 492(14, 21), 493, 496, 497(8), 501, 503, 505, 515
Gladish, Y. C., 161
Glaid, A. J., 20, 24
Glazer, A. N., 413

Gleisner, J. M., 68
Glick, D. M., 136, 156
Glinski, R. P., 671
Glover, G. I., 28, 89(130), 91, 153, 201, 206(34), 229, 230, 233, 478, 608, 609(12), 610(15), 611(12, 17, 18), 612(12, 25)
Glukhova, M. A., 187
Glusker, J. P., 137, 138
Glynn, I. M., 349, 351(12), 529
Gmelin, R., 418
Godefrey-Colburn, T., 714
Godfrey, J. C., 532
Goebel, W. F., 518
Göthe, P. O., 139
Goetzl, E. J., 488, 507
Gold, A. H., 149
Gold, A. M., 128, 256
Goldberg, M. E., 398, 399(3)
Goldstein, A., 80(89), 81, 85(89), 100(89), 602, 603(3), 604, 606(3, 19)
Goldstein, I. J., 362
Goldstein, J. A., 73
Good, A. H., 487
Goodman, D. S., 440
Goodman, L., 308, 572, 716
Gopalakrishnan, P. V., 516, 517, 518(6, 7), 519(6, 12) 521(6, 7) 522(6), 523
Gordon, A. S., 607
Gordon, E., 80(81), 81, 85(81), 187, 632(42), 633(41), 693, 695(21), 698(21), 700(21)
Gorecki, M., 360, 362, 483, 488, 491(23), 493
Gorelic, L. S., 89(135), 91, 169, 174, 640
Goren, H. J., 136, 156
Gorini, L., 622
Gorshkova, I. I., 702
Goto, Y., 420, 422, 423(34)
Gotschlich, E. C., 37
Gottikh, B. P., 260, 261
Govier, M. A., 226, 228(6), 547
Gozes, I., 570
Gracy, R. W., 143
Gradenwitz, A., 117
Gradenwitz, F., 117
Grahn, D. T., 547
Grant, P. G., 713, 715(4)
Graves, D., 648
Gray, C. J., 214, 215(12, 13), 216(12)

Neurath, H., 154, 207, 226, 228(5, 6), 229, 547

Nevinsky, G. A., 80(82), 81, 85(82), 102(82), 103(82), 699, 702(29)

Nickerson, M., 572

Nicklin, H. G., 548

Nicolaides, E. D., 416

Nicolas, J., 148, 150

Nicolas, J. C., 452, 455(7, 8)

Nicolson, G. L., 572

Nielsen, P. E., 80, 84(51), 99(51), 634, 706

Niemann, C., 208, 214(2), 215(2)

Niemczyk, M., 470(10), 471

Nierhaus, K. H., 621

Nikodejevic, B., 555

Nirenberg, M. W., 573, 603

Nishihara, H., 328

Nishimura, J. S., 428

Nishimura, M., 389, 391, 397

Nishimura, S., 701

Nishino, N., 214, 215(11)

Nishizuka, Y., 281, 282(33)

Nisonoff, A., 494

Nixon, J., 346, 352(3), 353, 357(2)

Noll, H., 187, 641, 681

Noll, M., 187, 641, 681

Noller, H. F., 627, 628(23), 630(23), 632(23), 674, 711

Noltmann, E. A., 24, 144, 390

Nomura, A., 301, 302(6)

Nomura, M., 643

Nordström, B., 150

Noroa, W. B., 24

Norris, C., 199

North, A. C. T., 403, 404, 412(3)

Northrop, D., 24

Norton, I. L., 142, 143, 153, 392, 395(42), 396(42), 397(42, 49)

Novick, R. P., 533

Novoa, W. B., 20

Novogrodsky, A., 427, 428

Nowak, H., 445

Nowell, I. W., 79

Nyman, P. O., 139

O

Obrig, T., 336, 337(11), 338(11), 339(11)

Ochoa, S., 388, 550, 713

O'Connell, E. L., 144, 145(50), 381, 382(1), 383(3), 384(2), 387(2), 432, 547

O'Connell, W., 328

Oda, T., 391

Odin, E., 328, 334

Oen, H., 182, 187, 190(19), 193(9, 19), 194(19), 621, 627, 628(20, 21, 22, 25), 630(20), 631(22, 38), 632(22), 633(38, 39), 638, 643, 676, 682, 711

O'Farrel, P. H., 570

Ofengand, J., 80(79, 80, 81), 81, 85(79, 80, 81), 90(140), 91, 102(79), 187, 632(42, 43), 633(40, 41), 637(7), 638, 693, 695(21), 696(25), 697(18), 698(18, 21), 700(18, 21), 701(20, 24)

Offord, R. E., 141, 142(38), 143(38)

Ogawara, H., 531, 532(6, 7), 533

Ogren, W. L., 388, 389, 391(23)

Ogura, K., 470(11, 12), 471

Ohlsson, I., 150

Ohnoki, S., 423, 425

Ojala, D., 274

Oka, T., 198, 199(10), 200(10), 206(10), 207(10)

Okamoto, M., 37, 45, 46(14), 67, 432

Okazaki, R., 310

O'Keefe, K. R., 274

Okina, E. I., 44

Olcott, H. S., 223

Olesker, A., 370

Oleson-Larson, P., 418

O'Malley, B. W., 449

Omar Osama, L. M., 369

Olmsted, J. B., 567

Ondetti, M. A., 222

Ono, Y., 181, 194(2)

Opheim, K. E., 602, 603(3), 606(3)

Oplatka, A., 318, 319, 320, 321(3)

Orekhovick, V. N., 206

Orlacchio, A., 441, 445

Ormand, R. E., 542(9), 543

Ormerad, W. E., 597

Orsi, B. A., 24

Ortanderl, F., 290, 293(2), 294(2), 295

Osawa, T., 80(86), 81, 85(86), 112(86), 363, 368(15), 405

Oshima, T., 701

Osman, A. M., 47, 48(15)

Otieno, S., 362

Otsuka, E., 259

Subject Index

3-Acetylpyridinio-*n*-butyladenosine pyrophosphate, 257

2-Acetylpyridiniopropyladenosine pyrophosphate, 257

3-Acetylpyridiniopropyladenosine pyrophosphate, 257

4-Acetylpyridiniopropyladenosine pyrophosphate, 257

Acid carboxypeptidases, 206

Acid phosphatase, 3

Acridine orange, 563

Acrolein, 33

Acrosin, 206

Actin, 294

polymerization, 293, 294

Activation by light, 478

Acyl azides, 78

Acyl hydrazide, 209

Adenine, 19

Adenine nucleotide
analogs, 259, 302–307
spin-labeled, 260
transport, 83

Adenosine, 19
derivatives, 240–249

Adenosine deaminase, 19, 25, 327–335
difference spectrum with inhibitors, 333
6-halopurine ribosides as substrates, 328
inactivation kinetics, 334
reaction of, 332, 333

Adenosine diphosphate, 64, 65
analogs of, 241
binding proteins, 248, 249
Co(III) complex, 313
spin-labeled, 285, 286

Adenosine monophosphate, 296, 297
aminoalkyl esters of, 26
analogs, 299
chromatography of, 281
Co(III) complex, 313
sites, 241
spin-labeled, 285, 286

Adenosine nucleotide analogs, 289

Adenosine 5'-phosphate aminohydrolase, 305

Adenosine 5'-phosphoromorpholidate, 252

Adenosine triphosphatase, 62, 75, 87, 89, 260, 295, 318

F₁, 83, 277, 278, 287
inhibition by arylazido-β-alanine ATP, 276, 277

Adenosinetriphosphatase (Na⁺ + K⁺), 523–531

Adenosine triphosphate
analogs of, 241, 259, 295, 302
Co(III) complex, 313, 314, 318
Co(III)-phen complex, 314, 315, 318–321
spin-labeled 285, 286

Adenosine triphosphate nitrenes, 264

Adenosine triphosphate triethylammonium salt, 263

S-Adenosylmethionine, 560

Adenylate cyclase, 591, 598
inhibition, 600

Adenylate kinase, 21, 25, 298, 299

Adenylosuccinate AMP-lyase 302, 305, 306

Adipic acid, 506

β-Adrenergic receptor, 131, 578, 579, 591–601
action of propranolol, 600
inhibition, 599, 600

Affinity chromatography, 216

Affinity labeling, 3–14
acetylenic inhibitors, 158–164
controls, 110, 111
covalent bond formation, 10, 11
criteria for, 5, 6, 29, 55, 482, 483, 575
definition, 3–5
diradicals, 82–91
free radicals, 82–91
inactivation, half-time, 6
instability of, 186
kinetics of, 5–7, 28, 203–205, 211–214, 481, 482, 576, 577, 581, 582
macromolecules, 183–186
multicomponent systems, 180–194
multisubstrate analogs, 17
polyfunctional labels, 12
reagents, *see also* specific types, substances
antibody, 486
choice of, 483–486
concentration, 107, 108
k_{cat}, 9, 10
K_s, 8, 9

I